COLLECTED WORKS OF

HERVÉ JACQUET

Hervé Jacquet
Summer 2007, Shahidi's conference
Purdue University

COLLECTED WORKS OF

HERVÉ JACQUET

Dorian Goldfeld
Editor

American Mathematical Society
Providence, Rhode Island

2000 *Mathematics Subject Classification.* Primary 11F67, 11F70, 11F72, 11R39, 22E55.

Library of Congress Cataloging-in-Publication Data

Jacquet, Hervé, 1939–
[Works. 2011]
Collected works of Hervé Jacquet / Dorian Goldfeld, editor.
p. cm.
Includes bibliographical references.
ISBN 978-0-8218-5356-6 (alk. paper)
1. Number theory. I. Goldfeld, D. II. Title.

QA241.J3326 2011
512.7—dc23

2011014420

A complete list of acknowledgments can be found at
the back of this publication.

∞ The paper used in this book is acid-free and falls within the guidelines
established to ensure permanence and durability.
Visit the AMS home page at http://www.ams.org/

10 9 8 7 6 5 4 3 2 1   16 15 14 13 12 11

Contents

INTRODUCTION

DORIAN GOLDFELD

Hervé Jacquet is one of the founders of the modern theory of automorphic representations and their associated L-functions. This volume represents a selection of his most influential papers, although a number of important works are not included because they are already available in book form. Based on several discussions with Hervé Jacquet, I have written the following introduction to try to give a flavor of the significance and prominence of his pioneering mathematical work.

The introduction of the theory of representations of adelic groups in the theory of automorphic forms had been proposed by Gelfand and his collaborators and, separately, by Godement. But it is only with the impact of the Langlands program that the proposal took full effect. Indeed, the book of Jacquet and Langlands [JL70] reexamines the work of Hecke and others, in particular the converse theorem, from the point of view of group representations. The objects of study are the automorphic representations of $GL(2)$. Such a representation may be written as a tensor product of local representations over the places of the ground field. To each local representation is attached an L-factor and an ϵ-factor. The global L-function attached to the representation is the product of the local ones and the global ϵ-factor (or root number) which appears in the functional function is the product of the local ϵ-factors. Thus, the theory of representations reveals a more profound understanding of the functional equation of the automorphic L-functions than was previously seen.

There are several other, deeper motivations for the introduction of group representations in the theory of automorphic forms. For instance, the relation between modular forms for $GL(2)$ and the multiplicative group of a quaternion algebra can be formulated in a very striking way in terms of representations. This is the celebrated Jacquet-Langlands correspondence [JL70]. As a follow up, Jacquet reexamined, from the point of view of representations, the so called Rankin-Selberg convolution of two automorphic L-functions for $GL(2)$ [Jac72]. In the same vein, the book by Godement and Jacquet [GJ72] defined, for the first time, the standard L-functions attached to automorphic representations of $GL(n)$, now called Godement-Jacquet L-functions and proved their basic properties. Of course, in order to carry this through, it was first necessary to obtain essential properties of the irreducible representations of a p-adic group [Jac71].

A series of joint papers with Shalika and Piatetski-Shapiro (in various combinations) pertain to the L-functions of pairs, called, by analogy with the case of $GL(2)$, the Rankin Selberg L-functions. The Rankin Selberg L-functions are attached to representations of $GL(n)$ and $GL(m)$. They are the main ingredients of the converse theorem, which can be used to prove the existence of certain automorphic representations, as predicated by Langlands' principle of functoriality ([JPSS83], [JPSS79a], [JPSS79b]). Primary examples are the construction of automorphic forms attached to cubic extensions of the ground field, the base change over a cubic extension ([JPSS79b], [JPSS81d]), and the construction of automorphic representations of $GL(3)$ obtained as the functorial image of representations of $GL(2)$ ([**GJ78**], joint with Gelbart). Yet another significant application is to the notion of conductor of representations of $GL(n)$ [**JPSS81a**]. A basic ingredient of this effort was an elaboration of the properties of Whittaker models and functions ([JPSS83], [**JS85**], [JS90b]), which Jacquet had made contributions to since his thesis [Jac67].

One of the most important applications of the theory of L-functions of pairs is contained in the two papers with Shalika ([**JS81a**], [**JS81b**]). It is a result of Langlands that an automorphic representation of $GL(n)$ is induced by automorphic cuspidal representations of various $GL(m)$. Applying results on the poles and the non-vanishing of L-functions of pairs, the papers proved that this decomposition is unique, a crucial result in the applications. The required non-vanishing result was first proved in [**JS77**] for the Godement-Jacquet L-functions. The method, which uses the Theory of Eisenstein series and was known to the specialists for the case of $GL(1)$, was subsequently generalized by Shahidi to the case of L-functions of pairs.

In the mid-eighties, Jacquet forayed into new territory and introduced a novel tool to investigate period integrals, that is, integrals of an automorphic form over a subgroup H of the ambient reductive group G, possibly against a one dimensional character of H. In particular, automorphic representations which contain a vector whose period integral is non-zero are said to be distinguished by the subgroup H. One of the goals of this endeavor is to determine which automorphic representations are distinguished by the given subgroup. Another goal is to relate the period integrals to special values of automorphic L-functions. While the usual Selberg trace formula, and its generalizations by Arthur, develop an expression for the integral of a certain kernel over the diagonal, in the new investigation (relative trace formula), one integrates the kernel over appropriate subgroups. In many cases, this appears as a generalization of the Kuznetsov and Petersson formulae from the classical set-up.

In more detail, in order to study the period integral over the subgroup H, it is natural to integrate the kernel over the product $H \times H$. Of course, as in the standard trace formula, the integral needs to be regularized. Examples of such a construction are contained in [**JL85**] and [**Jac86**] where the integral for the pair (G, H) is compared with the integral for a pair (G', H') where G' is an inner form of G. In the case of [**Jac86**], the integral for (G', H') is explicitly related to the value of an L-function and this information is transfered to the period integral over H.

Still another construction is contained in a series of papers: [**Jac01**], [**Jac03a**], [**Jac03b**], [Jac04b], [**Jac05**]. There, the data is a quadratic extension E/F of number fields. The goal is to prove that if a cuspidal automorphic representation of $GL(n, E)$ is a base change of a cuspidal representation of $GL(n, F)$ then it is distinguished by a unitary group (the converse is easy in this case). The relative trace formula compares the (adelic) Kuznetsov-Petersson formula for $GL(n, F)$ with a relative Kuznetsov-Petersson formula for $GL(n, E)$, where one of the variables of the kernel is integrated over a unitary group. Once this characterization is obtained (or known for other reasons in the case of $GL(2)$) one can try to make more comparisons and obtain further results on period integrals. See [Jac87b] and [**JC01**] which gives a more detailed version of [Jac87b]. In [**Jac91**] a conjecture is stated, which describes representations distinguished by an orthonormal group.

Finally, [**JLR93**] and [JLR04] explore another idea. It is natural to try to compare a relative trace formula (with double integral over $H \times H$) and a standard trace formula. However, in order to make this idea works, one needs to introduce suitable weight factors.

References

References in bold appear in this volume.

[AGJ09] Avraham Aizenbud, Dmitry Gourevitch, and Hervé Jacquet. Uniqueness of Shalika functionals: the Archimedean case. *Pacific J. Math.*, 243(2):201–212, 2009.

[BJ79] A. Borel and H. Jacquet. Automorphic forms and automorphic representations. In *Automorphic forms, representations and L-functions (Proc. Sympos. Pure Math., Oregon State Univ., Corvallis, Ore., 1977), Part 1*, Proc. Sympos. Pure Math., XXXIII, pages 189–207. Amer. Math. Soc., Providence, R.I., 1979. With a supplement "On the notion of an automorphic representation" by R. P. Langlands.

[FGJR99] Solomon Friedberg, Stephen Gelbart, Hervé Jacquet, and Jonathan Rogawski. Représentations génériques du groupe unitaire à trois variables. *C. R. Acad. Sci. Paris Sér. I Math.*, 329(4):255–260, 1999.

[FJ93a] Solomon Friedberg and Hervé Jacquet. Le lemme fondamental pour le sous-groupe de Shalika de GL(4). *C. R. Acad. Sci. Paris Sér. I Math.*, 316(5):407–410, 1993.

[FJ93b] Solomon Friedberg and Hervé Jacquet. Linear periods. *J. Reine Angew. Math.*, 443:91–139, 1993.

[FJ96] Solomon Friedberg and Hervé Jacquet. The fundamental lemma for the Shalika subgroup of GL(4). *Mem. Amer. Math. Soc.*, 124(594):x+149, 1996.

[GJ72] Roger Godement and Hervé Jacquet. *Zeta functions of simple algebras*. Lecture Notes in Mathematics, Vol. 260. Springer-Verlag, Berlin, 1972.

[GJ78] Stephen Gelbart and Hervé Jacquet. A relation between automorphic representations of GL(2) and GL(3). *Ann. Sci. École Norm. Sup. (4)*, 11(4):471–542, 1978.

[GJR01] Stephen Gelbart, Hervé Jacquet, and Jonathan Rogawski. Generic representations for the unitary group in three variables. *Israel J. Math.*, 126:173–237, 2001.

[Jac67] Hervé Jacquet. Fonctions de Whittaker associées aux groupes de Chevalley. *Bull. Soc. Math. France*, 95:243–309, 1967.

[Jac71] Hervé Jacquet. Représentations des groupes linéaires p-adiques. In *Theory of group representations and Fourier analysis (Centro Internaz. Mat. Estivo (C.I.M.E.), II Ciclo, Montecatini Terme, 1970)*, pages 119–220. Edizioni Cremonese, Rome, 1971.

[Jac72] Hervé Jacquet. *Automorphic forms on* GL(2). *Part II.* Lecture Notes in Mathematics, Vol. 278. Springer-Verlag, Berlin, 1972.

[Jac77] Hervé Jacquet. Generic representations. In *Non-commutative harmonic analysis (Actes Colloq., Marseille-Luminy, 1976)*, pages 91–101. Lecture Notes in Math., Vol. 587. Springer, Berlin, 1977.

[Jac79] Hervé Jacquet. Principal L-functions of the linear group. In *Automorphic forms, representations and L-functions (Proc. Sympos. Pure Math., Oregon State Univ., Corvallis, Ore., 1977), Part 2*, Proc. Sympos. Pure Math., XXXIII, pages 63–86. Amer. Math. Soc., Providence, R.I., 1979.

[Jac81] Hervé Jacquet. Dirichlet series for the group GL(n). In *Automorphic forms, representation theory and arithmetic (Bombay, 1979)*, volume 10 of *Tata Inst. Fund. Res. Studies in Math.*, pages 155–163. Tata Inst. Fundamental Res., Bombay, 1981.

[Jac84a] Hervé Jacquet. On the residual spectrum of GL(n). In *Lie group representations, II (College Park, Md., 1982/1983)*, volume 1041 of *Lecture Notes in Math.*, pages 185–208. Springer, Berlin, 1984.

[Jac84b] Hervé Jacquet. Sur un résultat de Waldspurger. *C. R. Acad. Sci. Paris Sér. I Math.*, 299(7):199–204, 1984.

[Jac86] Hervé Jacquet. Sur un résultat de Waldspurger. *Ann. Sci. École Norm. Sup. (4)*, 19(2):185–229, 1986.

[Jac87a] Hervé Jacquet. On the nonvanishing of some L-functions. *Proc. Indian Acad. Sci. Math. Sci.*, 97(1-3):117–155 (1988), 1987.

[Jac87b] Hervé Jacquet. Sur un résultat de Waldspurger. II. *Compositio Math.*, 63(3):315–389, 1987.

[Jac91] Hervé Jacquet. Représentations distinguées pour le groupe orthogonal. *C. R. Acad. Sci. Paris Sér. I Math.*, 312(13):957–961, 1991.

[Jac92] Hervé Jacquet. Relative Kloosterman integrals for GL(3). II. *Canad. J. Math.*, 44(6):1220–1240, 1992.

[Jac95] Hervé Jacquet. The continuous spectrum of the relative trace formula for GL(3) over a quadratic extension. *Israel J. Math.*, 89(1-3):1–59, 1995.

[Jac97a] Hervé Jacquet. Automorphic spectrum of symmetric spaces. In *Representation theory and automorphic forms (Edinburgh, 1996)*, volume 61 of *Proc. Sympos. Pure Math.*, pages 443–455. Amer. Math. Soc., Providence, RI, 1997.

[Jac97b] Hervé Jacquet. Note on the analytic continuation of Eisenstein series: An appendix to "Theoretical aspects of the trace formula for GL(2)" [in *representation theory and automorphic forms (edinburgh, 1996)*, 355–405, Proc. Sympos. Pure Math., 61, Amer. Math. Soc., Providence, RI, 1997; MR 98k:11062] by A. W. Knapp. In *Representation theory and automorphic forms (Edinburgh, 1996)*, volume 61 of *Proc. Sympos. Pure Math.*, pages 407–412. Amer. Math. Soc., Providence, RI, 1997.

[Jac97c] Hervé Jacquet. Principal L-functions for $\mathrm{GL}(n)$. In *Representation theory and automorphic forms (Edinburgh, 1996)*, volume 61 of *Proc. Sympos. Pure Math.*, pages 321–329. Amer. Math. Soc., Providence, RI, 1997.

[Jac98] Herve Jacquet. A theorem of density for Kloosterman integrals. *Asian J. Math.*, 2(4):759–778, 1998. Mikio Sato: a great Japanese mathematician of the twentieth century.

[Jac01] Hervé Jacquet. Factorization of period integrals. *J. Number Theory*, 87(1):109–143, 2001.

[Jac03a] Hervé Jacquet. Facteurs de transfert pour les intégrales de Kloosterman. *C. R. Math. Acad. Sci. Paris*, 336(2):121–124, 2003.

[Jac03b] Hervé Jacquet. Smooth transfer of Kloosterman integrals. *Duke Math. J.*, 120(1):121–152, 2003.

[Jac04a] Hervé Jacquet. Integral representation of Whittaker functions. In *Contributions to automorphic forms, geometry, and number theory*, pages 373–419. Johns Hopkins Univ. Press, Baltimore, MD, 2004.

[Jac04b] Hervé Jacquet. Kloosterman identities over a quadratic extension. *Ann. of Math. (2)*, 160(2):755–779, 2004.

[Jac05a] Hervé Jacquet. A guide to the relative trace formula. In *Automorphic representations, L-functions and applications: progress and prospects*, volume 11 of *Ohio State Univ. Math. Res. Inst. Publ.*, pages 257–272. de Gruyter, Berlin, 2005.

[Jac05b] Hervé Jacquet. Kloosterman identities over a quadratic extension. II. *Ann. Sci. École Norm. Sup. (4)*, 38(4):609–669, 2005.

[Jac05c] Hervé Jacquet. Kloosterman integrals for $\mathrm{GL}(2,\mathbb{R})$. *Pure Appl. Math. Q.*, 1(2, part 1):257–289, 2005.

[Jac09] Hervé Jacquet. Archimedean Rankin-Selberg integrals. In *Automorphic forms and L-functions II. Local aspects*, volume 489 of *Contemp. Math.*, pages 57–172. Amer. Math. Soc., Providence, RI, 2009.

[JC01] Hervé Jacquet and Nan Chen. Positivity of quadratic base change L-functions. *Bull. Soc. Math. France*, 129(1):33–90, 2001.

[JL70] H. Jacquet and R. P. Langlands. *Automorphic forms on* GL(2). Lecture Notes in Mathematics, Vol. 114. Springer-Verlag, Berlin, 1970.

[JL83] Hervé Jacquet and King F. Lai. Sur une formule des traces relative. *C. R. Acad. Sci. Paris Sér. I Math.*, 296(23):959–963, 1983.

[JL85] H. Jacquet and K. F. Lai. A relative trace formula. *Compositio Math.*, 54(2):243–310, 1985.

[JLR93] Hervé Jacquet, King F. Lai, and Stephen Rallis. A trace formula for symmetric spaces. *Duke Math. J.*, 70(2):305–372, 1993.

[JLR99] Hervé Jacquet, Erez Lapid, and Jonathan Rogawski. Periods of automorphic forms. *J. Amer. Math. Soc.*, 12(1):173–240, 1999.

[JLR04] Hervé Jacquet, Erez Lapid, and Stephen Rallis. A spectral identity for skew symmetric matrices. In *Contributions to automorphic forms, geometry, and number theory*, pages 421–455. Johns Hopkins Univ. Press, Baltimore, MD, 2004.

[JM07] Hervé Jacquet and Kimball Martin. Shalika periods on $\mathrm{GL}_2(D)$ and GL_4. *Pacific J. Math.*, 233(2):341–370, 2007.

[JPSS79a] Hervé Jacquet, Ilja Iosifovitch Piatetski-Shapiro, and Joseph Shalika. Automorphic forms on GL(3). I. *Ann. of Math. (2)*, 109(1):169–212, 1979.

[JPSS79b] Hervé Jacquet, Ilja Iosifovitch Piatetski-Shapiro, and Joseph Shalika. Automorphic forms on GL(3). II. *Ann. of Math. (2)*, 109(2):213–258, 1979.

[JPSS79c] Hervé Jacquet, Ilja Iosifovitch Piatetski-Shapiro, and Joseph Shalika. Facteurs L et ε du groupe linéaire. *C. R. Acad. Sci. Paris Sér. A-B*, 289(2):A59–A61, 1979.

[JPSS81a] H. Jacquet, I. I. Piatetski-Shapiro, and J. Shalika. Conducteur des représentations du groupe linéaire. *Math. Ann.*, 256(2):199–214, 1981.

[JPSS81b] Hervé Jacquet, Ilja Piatetski-Shapiro, and Joseph Shalika. Conducteur des représentations génériques du groupe linéaire. *C. R. Acad. Sci. Paris Sér. I Math.*, 292(13):611–616, 1981.

[JPSS81c] Hervé Jacquet, Ilja I. Piatetski-Shapiro, and Joseph Shalika. Facteurs L et ε du groupe linéaire: théorie archimédienne. *C. R. Acad. Sci. Paris Sér. I Math.*, 293(1):13–18, 1981.

[JPSS81d] Hervé Jacquet, Ilja I. Piatetski-Shapiro, and Joseph Shalika. Relèvement cubique non normal. *C. R. Acad. Sci. Paris Sér. I Math.*, 292(12):567–571, 1981.

[JPSS83] H. Jacquet, I. I. Piatetskii-Shapiro, and J. A. Shalika. Rankin-Selberg convolutions. *Amer. J. Math.*, 105(2):367–464, 1983.

[JR92a] Hervé Jacquet and Stephen Rallis. Kloosterman integrals for skew symmetric matrices. *Pacific J. Math.*, 154(2):265–283, 1992.

[JR92b] Hervé Jacquet and Stephen Rallis. Symplectic periods. *J. Reine Angew. Math.*, 423:175–197, 1992.

[JR96] Hervé Jacquet and Stephen Rallis. Uniqueness of linear periods. *Compositio Math.*, 102(1):65–123, 1996.

[JS81a] H. Jacquet and J. A. Shalika. On Euler products and the classification of automorphic forms. II. *Amer. J. Math.*, 103(4):777–815, 1981.

[JS81b] H. Jacquet and J. A. Shalika. On Euler products and the classification of automorphic representations. I. *Amer. J. Math.*, 103(3):499–558, 1981.

[JS81c] Hervé Jacquet and Joseph Shalika. Sur le spectre résiduel du groupe linéaire. *C. R. Acad. Sci. Paris Sér. I Math.*, 293(11):541–543, 1981.

[JS83] Hervé Jacquet and Joseph Shalika. The Whittaker models of induced representations. *Pacific J. Math.*, 109(1):107–120, 1983.

[JS85] Hervé Jacquet and Joseph Shalika. A lemma on highly ramified ϵ-factors. *Math. Ann.*, 271(3):319–332, 1985.

[JS90a] Hervé Jacquet and Joseph Shalika. Exterior square L-functions. In *Automorphic forms, Shimura varieties, and L-functions, Vol. II (Ann Arbor, MI, 1988)*, volume 11 of *Perspect. Math.*, pages 143–226. Academic Press, Boston, MA, 1990.

[JS90b] Hervé Jacquet and Joseph Shalika. Rankin-Selberg convolutions: Archimedean theory. In *Festschrift in honor of I. I. Piatetski-Shapiro on the occasion of his sixtieth birthday, Part I (Ramat Aviv, 1989)*, volume 2 of *Israel Math. Conf. Proc.*, pages 125–207. Weizmann, Jerusalem, 1990.

[JS77] Hervé Jacquet and Joseph A. Shalika. A non-vanishing theorem for zeta functions of GL_n. *Invent. Math.*, 38(1):1–16, 1976/77.

[JY90] Hervé Jacquet and Yangbo Ye. Une remarque sur le changement de base quadratique. *C. R. Acad. Sci. Paris Sér. I Math.*, 311(11):671–676, 1990.

[JY92] Hervé Jacquet and Yangbo Ye. Relative Kloosterman integrals for GL(3). *Bull. Soc. Math. France*, 120(3):263–295, 1992.

[JY96] Hervé Jacquet and Yangbo Ye. Distinguished representations and quadratic base change for GL(3). *Trans. Amer. Math. Soc.*, 348(3):913–939, 1996.

[JY99] Hervé Jacquet and Yangbo Ye. Germs of Kloosterman integrals for GL(3). *Trans. Amer. Math. Soc.*, 351(3):1227–1255, 1999.

[JZ87] H. Jacquet and D. Zagier. Eisenstein series and the Selberg trace formula. II. *Trans. Amer. Math. Soc.*, 300(1):1–48, 1987.

Curriculum Vitae

Born: August 4, 1939, France

Education:

École Normale Supérieure , 1959 - 1963

Doctorat d' état under the direction of Roger Godement, 1967

Academic History:

Centre National de la Recherche Scientifique, 1963 - 1969

Member, Institute for Advanced Study, 1967 - 1969

Associate Professor, University of Maryland, 1969 - 1970

Associate Professor, Graduate Center of the City University of New York, 1970 - 1973

Professor, Graduate Center of the City University of New York, 1974

Professor, Columbia University, 1974 - 2007

Professor Emeritus, Columbia University, 2007 –

Honors:

Corresponding member of the Institut de France, Académie des Sciences, 1980

Acknowledgment by Author

Hervé Jacquet

I am very grateful to the American Mathematical Society for publishing this volume of *Collected Works* and to Dorian Goldfeld for editing it. I thank especially Ms Cristin Zannella from the editorial Department of the AMS for all her work on the project. Finally, I thank Jeff Mozzochi for giving permission to use one of the many pictures he took of me.

A number of papers could have been included in this volume, the book with Langlands, the book with Godement and several joint papers with Piatetski-Shapiro and Shalika. They are available in book form from Springer-Verlag or as articles in the *Selected Works of Ilya Piatetski-Shapiro* published by the AMS.

A number of papers which appear in this volume were written in collaboration with other mathematicians. To those alive today, I want to say that the collaboration was enjoyable and fruitful. Alas, two of my most important coauthors, Piatetski-Shapiro and Shalika, passed away. I would like to dedicate this volume to their memories.

Inventiones math. 38, 1 – 16 (1976)

Inventiones mathematicae

A Non-Vanishing Theorem for Zeta Functions of GL_n

To R. Godement

Hervé Jacquet[1] * and Joseph A. Shalika[2] **

[1] Department of Mathematics, Columbia University, New York, New York 10027, USA

[2] Department of Mathematics, The Johns Hopkins University, Baltimore, Maryland 21218, USA

§ 1. The Main Theorem

Let F be an $\mathbb{A}$-field. Denote by $F_{\mathbb{A}}$ and $F_{\mathbb{A}}^{\times}$ or simply $\mathbb{A}$ and $\mathbb{I}$ its ring of adeles and group of ideles. We regard the group $G_n = GL(n)$ as an F-group and, for $n \geqq 2$, denote by Z_n its center. Assume $n \geqq 2$. A function ϕ on the quotient

$$G_n^* = G_n(F) \diagdown G_n(\mathbb{A})$$

is said to be cuspidal if the integral

(1.1) $\quad \phi_U(g) = \int_{U^*} \phi(u\,g)\,du, \qquad U^* = U(F) \diagdown U(\mathbb{A}),$

vanishes for all $g \in G_n(\mathbb{A})$ and all horicycles U of G_n. (A horicycle is the unipotent radical of a proper F-parabolic subgroup of G_n.) If ω is a character (of modulus one) of $\mathbb{I}/F^{\times}$, we denote by $L^2(\omega, n)$ the space of all functions f on $G_n(\mathbb{A})$ such that

$$f(\gamma\, a\, g) = \omega(a)\, f(g) \quad \text{ for } \gamma \in G_n(F), \quad a \in Z_n(\mathbb{A}) \approx \mathbb{I},$$

$$\int_{Z_n(\mathbb{A})\, G_n(F) \diagdown G_n(\mathbb{A})} |f(g)|^2\, dg < +\infty.$$

We denote by $L_0^2(\omega, n)$ the closed subspace of cuspidal elements of $L^2(\omega, n)$. The representation of $G_n(\mathbb{A})$ *on* $L_0^2(\omega, n)$ (by right translations) decomposes into a *discrete* sum of irreducible representations, each occurring with multiplicity one [15, 2, 13]. To an irreducible component π of $L_0^2(\omega, n)$, one can attach as in [5] an L-function $L(s, \pi)$ which satisfies a functional equation

(1.2) $\quad L(s, \pi) = \varepsilon(s, \pi)\, L(1-s, \tilde{\pi}),$

where $\tilde{\pi}$ is the representation contragredient to π. (For other references see [1]). Of course $\tilde{\pi}$ is a component of $L_0^2(\omega^{-1}, n)$. Our main result is as follows:

(1.3) **Theorem.** *With the above notations,*

$$L(s, \pi) \neq 0 \quad \text{ for } \operatorname{Re}(s) = 1.$$

---

* Partially supported by N.S.F. Grant MPS75-08545

** Partially supported by N.S.F. Grant MPS71-03321

In the notation of [5], we have trivially that $L(s+t, \pi)=L(s, \pi\otimes\alpha^t)$. Hence the theorem amounts to saying that $L(1,\pi)\neq 0$. For $n=2$, one can define, as in [7], another L-function $L(s, \pi\times\tilde{\pi})$ whose poles occur exactly at $s=0$ or $s=1$ (more precisely when $\alpha^s=0$ or $\alpha^s=1$). It then follows from standard facts on Dirichlet series with non-negative coefficients that $L(s,\pi)$ converges for $\mathrm{Re}(s)>1$. In principle, a similar argument is valid for $GL(n)$.

We remark that for $n=1$ and any characteristic, the theorem is true and classical: for a number field π is a character of $\mathbb{I}/F^\times$ and $L(s,\pi)$ the product of a Hecke L-function by the appropriate Γ-factor. For a function field an even stronger result is true for $n=1$! The proof of (1.3), with the minor modifications given in § 7, applies also to that case. It is quite likely that, in this case, the idea of the present proof occurred to other authors.

If $F=\mathbb{Q}$ and π corresponds to a holomorphic new-form of weight k which is an eigen-function of the Hecke operators for some group $\Gamma_0(N)$, then $L\left(s+\frac{k-1}{2}, \pi\right)$ is the usual Dirichlet series attached to a form multiplied by the appropriate Γ-factor. The non-vanishing of this Dirichlet series and related series has attracted the attention of several authors [12, 14]. Again in this case the technique of proof was probably known to other authors, at least in special cases.

Of course if K is a quadratic (separable) extension of F and χ a character of $K_{\mathbb{A}}^\times/K^\times$ not of the form $\eta\circ N_{K/F}$, then $L(s,\chi)=L(s,\pi)$ for an appropriate component π of $L_0^2(\omega,2)$. Here ω is the restriction of χ to $F_{\mathbb{A}}^\times$ multiplied by the quadratic character of $F_{\mathbb{A}}^\times$ attached to K by class-field theory. A similar remark is valid for a cubic extension of F (cf. [8]). Thus the result is not unexpected. We can only hope that our result will find application in the theory of uniform distribution of primes. The reader who is familiar with the density theorems of Hecke and Tschebotareff may agree that there is ample historical precedent. Finally the principle of applying Eisenstein series to the study of L-functions (rather than the reverse direction) in a systematic fashion is due to Langlands and appears in his monograph on Euler products [11].

Before embarking on the proof of Theorem (1.3), we observe that $\tilde{\pi}$ is equivalent to the representation $\bar{\pi}$ imaginary conjugate to π. Actually one can write π as an infinite tensor product $\bigotimes_v \pi_v$ over all places v of F. Then for each v, $\tilde{\pi}_v$ is equivalent to $\bar{\pi}_v$. Since the matrix coefficients of $\bar{\pi}_v$ are the imaginary conjugates of the matrix coefficients of π_v, the definition of $L(s,\pi_v)$ (Theorem 3.3 in [5]) shows that

$$\overline{L(s,\pi_v)}=L(\bar{s},\bar{\pi}_v)=L(\bar{s},\tilde{\pi}_v).$$

Thus:

(1.4) **Proposition.** *With the above notations,*

$$\overline{L(s,\pi)}=L(\bar{s},\tilde{\pi}).$$

Another remark on $\tilde{\pi}$ will be useful. Namely let g^ι denote the transpose inverse of a matrix g. Thus if π is a representation of $G_n(\mathbb{A})$ or $G_n(F_v)$, one can define another representation π^ι on the same space by setting $\pi^\iota(g)=\pi(g^\iota)$. Actually because g

and ${}^t g$ are always conjugate in $G_n(F_v)$, it can be seen ([2]) that π_v^ι and $\pi_v^\sim$ are equivalent. It thus follows that, globally, $\tilde{\pi}$ is equivalent to π^ι.

We let K be the standard maximal compact subgroups of $G_n(\mathbb{A})$.

§ 2. Some Eisenstein Series

Recall that $n \geqq 2$. (For $n=1$ see § 7.) We fix an irreducible component π of $L_0^2(\omega, n)$. We denote by H the corresponding closed subspace of $L_0^2(\omega, n)$ and by V the space of C^∞-vectors in H. Elements of V are actually C^∞-functions on G_n^*. If F is a number field, the fact that the elements of V are cuspidal, C^∞, and square-integrable imply they are "rapidly decreasing" [4] on G_n^*. If F is a function field the elements of V are again rapidly decreasing but in a very strong sense: they are compactly supported modulo the closed subgroup $Z_n(\mathbb{A})\, G_n(F)$.

For any function f on $G_n(\mathbb{A})$, let $f^\iota(g) = f(g^\iota)$. Then H^ι is a closed subspace of $L_0^2(\omega^{-1}, n)$ and the representation of $G_n(\mathbb{A})$ on H^ι equivalent to $\pi^\iota = \tilde{\pi}$. Thus the space of C^∞-vectors for $\tilde{\pi}$ is

$$V^\iota = \{\phi^\iota \mid \phi \in V\}.$$

We want to consider some Eisenstein-series for the standard parabolic subgroup Q of type $(1, n)$ in G_{n+1}. Thus Q consists of all matrices of the form

$$q = \begin{pmatrix} a & u \\ 0 & b \end{pmatrix}, \qquad a \in G_1,\ b \in G_n.$$

If we let U_Q be, as for any parabolic subgroup, the unipotent radical of Q and denote by M_Q the subgroup of those q for which $u=0$, we have the Levi-decomposition

$$Q = M_Q U_Q.$$

Note that M_Q is isomorphic to $G_n \times G_1$.

For any element ϕ of V and any function Φ in the space

$$\mathscr{S} = \mathscr{S}(n \times (n+1), \mathbb{A})$$

of Schwartz-Bruhat functions on the space of matrices with n rows and $n+1$ columns, we define a function f on $G_{n+1}(\mathbb{A}) \times G_n(\mathbb{A})$ by:

$$(2.1) \quad f(g, h) = |\det g|^{ns} \int_{G_n(\mathbb{A})} \Phi[(0, x)\, g]\, \phi(h\, x^{-1})\, |\det x|^{(n+1)s}\, dx.$$

Here 0 stands for the zero column of size n so that $(0, x)$ is a matrix with n rows and $n+1$ columns. Of course $(0, x)\, g$ is a matrix-multiplication. Since ϕ is bounded, the integral converges for $\mathrm{Re}(s) > n/n+1$ ([5], § 12), uniformly for h in $G_n(\mathbb{A})$, g in a compact set of $G_{n+1}(\mathbb{A})$, $\mathrm{Re}(s)$ in a compact set, and Φ in a bounded set. For q in $Q(\mathbb{A})$, one easily checks that:

$$(2.2) \quad f(q\, g, h) = f(g, h\, b)\, |a|^{ns}\, |\det b|^{-s}, \quad \text{if } q = \begin{pmatrix} a & u \\ 0 & b \end{pmatrix}.$$

Of course (2.2) expresses the fact that f belongs to the space of an induced representation. In addition, since the module of $Q(\mathbb{A})$ is

$$q \to |a|^n |\det b|^{-1},$$

one sees that, for $\operatorname{Re}(s)=\frac{1}{2}$, this induced representation is unitary. In particular, for $\gamma \in Q(F)$ and $a \in Z_{n+1}(\mathbb{A}) \approx \mathbb{I}$, one sees that $f(g)=f(g,e)$ satisfies

$$f(a\gamma g)=f(g)\,\omega(a),$$

and thus one can set, for $g \in G_{n+1}(\mathbb{A})$,

$$(2.3)\quad E(g;\Phi,\phi,s)=\sum_{Q(F)\setminus G_{n+1}(F)} f(\gamma g).$$

(Compare with [3].)

This is an "Eisenstein series" for the parabolic subgroup Q. Set $h=e$ in (2.1), change x to x^{-1} and break up the integration into a sum over $G_n(F)$ and an integral over G_n^*. Thus

$$f(g)=|\det g|^{ns}\int_{G_n^*}\sum_{\xi\in G_n(F)}\Phi[x^{-1}(0,\xi)\,g]\,\phi(x)\,|\det x|^{-(n+1)s}\,dx.$$

Substituting this in (2.3), we arrive at

$$E(g;\Phi,\phi,s)=|\det g|^{ns}\int_{G_n^*}\sum_{\xi,\gamma}\Phi[x^{-1}(0,\xi)\,\gamma\,g]\,\phi(x)\,|\det x|^{-(n+1)s}\,dx,$$

where γ varies in $Q(F)\setminus G_{n+1}(F)$ and $\xi \in G_n(F)$. But then the matrix

$$\eta=(0,\xi)\,\gamma$$

varies among all matrices of rank n ($rk(\eta)=n$) in $M(n\times(n+1),F)$ (the space of matrices with n rows and $n+1$ columns). Thus we can write:

$$(2.4)\quad E(g;\Phi,\phi,s)=|\det g|^{ns}\int_{G_n^*}\sum_{rk(\eta)=n}\Phi[x^{-1}\eta\,g]\,\phi(x)\,|\det x|^{-(n+1)s}\,dx.$$

To justify all our steps we have only to show that, if s is real, $s>n(n+1)$, Φ is positive or zero, and ϕ is replaced by $|\phi|$, then the total expression is finite. This will also show that, for $\operatorname{Re}(s)>n(n+1)$, (2.3) is absolutely convergent.

We proceed as in §11 of [5]. First we let G^0 be the set of $g \in G_n(\mathbb{A})$ such that $|\det g|=1$. We select in any way a system of representatives $g_1, g_2, \ldots, g_N$ for the cosets of $Z_n(\mathbb{A})\,G^0$ (a normal subgroup) in $G_n(\mathbb{A})$. Of course $N=1$ if F is a number field. We also set

$$G'=\bigcup G^0 g_i, \qquad G''=\bigcup G^0 g_i^{\iota},$$

so that for any function ξ on G_n^* we have

$$\int_{G_n^*}\xi(g)\,dg=\int_{\mathbb{I}/F^\times}da\int_{G_n(F)\setminus G'}\xi(ag)\,dg=\int_{\mathbb{I}/F^\times}da\int_{G_n(F)\setminus G''}\xi(ag)\,dg.$$

It will be also convenient to introduce two C^∞ functions F_0, F_1 on $\mathbb{R}_+^\times$ with the following properties: $F_0 \geqq 0$, $F_1 \geqq 0$, $F_0+F_1=1$, $F_1(t)=F_0(t^{-1})$, and there are t_0, t_1 with $t_0<1<t_1$ such that

$$F_0(t)=0 \quad \text{for } t<t_0, \qquad F_0(t)=1 \quad \text{for } t>t_1.$$

Set also, for $i=0, 1$, any character μ (of modulus one) of $\mathbb{I}/F^\times$, and Φ in $\mathscr{S}$,

(2.5) $$\theta^i(s, \mu, \Phi) = \int_{\mathbb{I}/F^\times} \sum_{\xi \neq 0} \Phi(a\xi)\,|a|^s \mu(a)\, F_i(|a|)\, d^\times a,$$

and, for $0<r\leqq n$,

(2.6) $$\theta_r^i(s, \mu, \Phi) = \int_{\mathbb{I}/F^\times} \sum_{rk(\xi)=r} \Phi(a\xi)\,|a|^s \mu(a)\, F_i(|a|)\, d^\times a.$$

In both cases the sum is over ξ in $M(n\times(n+1), F)$. Then, by [5] §11, $\theta^0(s, \mu, \Phi)$ is defined by a convergent integral for all s while $\theta^1(s, \mu, \Phi)$ is defined only for $\operatorname{Re}(s)>n(n+1)$. Moreover the group $G_{n+1}\times G_n$ operates (by matrices multiplication) on $M(n\times(n+1))$ and $G_{n+1}(\mathbb{A})\times G_n(\mathbb{A})$ on $\mathscr{S}$. We set accordingly

$$g\cdot\Phi\cdot h(x) = \Phi(h\,x\,g), \qquad h\in G_n(\mathbb{A}),\ g\in G_{n+1}(\mathbb{A}).$$

The resulting functions

$$(g, h) \to \theta^i(s, \mu, g\cdot\Phi\cdot h^{-1})$$

are "slowly increasing" on $G_{n+1}^*\times G_n^*$ (loc. cit), uniformly for $\operatorname{Re}(s)$ in a compact (and s in the convergence domain). Of course similar assertions are valid for the integrals θ_r^i.

Coming back to (2.4) and keeping in mind that $\phi(ax)=\omega(a)\,\phi(x)$ for $a\in Z_n(\mathbb{A})$, we get

(2.7) $$\begin{aligned} E(g; \Phi, \phi, s) &= |\det g|^{ns} \int_{G_n(F)\setminus G'} \theta_n^0[(n+1)\,n\,s, \omega, g\cdot\Phi\cdot x^{-1}]\,\phi(x)\,|\det x|^{-(n+1)s}\,dx \\ &+ |\det g|^{ns} \int_{G_n(F)\setminus G'} \theta_n^1[(n+1)\,n\,s, \omega, g\cdot\Phi\cdot x^{-1}]\,\phi(x)\,|\det x|^{-(n+1)s}\,dx. \end{aligned}$$

Replacing ϕ by $|\phi|$ and taking $s>n(n+1)$, $\Phi\geqq 0$, we get a finite result, since ϕ is "rapidly decreasing". Of course the first integral in (2.7) is actually convergent for all s.

§3. Analytic Continuation

If Φ is in $\mathscr{S}$ we define its Fourier transform, a function on the *same* space, by

$$\hat{\Phi}(x) = \int \Phi(y)\,\psi[tr(y\cdot{}^t x)]\,dy,$$

where ψ is a non-trivial additive character of $\mathbb{A}/F$ and dy self-dual. Note that

$$(g\cdot\Phi\cdot h)^\wedge = |\det h|^{-(n+1)}\,|\det g|^{-n}\, g^\iota\cdot\Phi\cdot h^\iota.$$

Moreover the Poisson summation formula gives:

(3.1) $$\begin{aligned} \theta_n^1(s, \mu, \Phi) &= \theta_n^0(n(n+1)-s, \mu^{-1}, \hat{\Phi}) + \sum_{1\leqq r<n} \theta_r^0(n(n+1)-s, \mu^{-1}, \hat{\Phi}) \\ &- \sum_{1\leqq r<n} \theta_r^1(s, \mu, \Phi) + \hat{\Phi}(0)\,\lambda(s-n(n+1), \mu) - \Phi(0)\,\lambda(s, \mu), \end{aligned}$$

where we have set (for $\mathrm{Re}(s)>1$)

$$\lambda(s,\mu)=\int_{\mathrm{I}/F^\times}|a|^s\mu(a)F_1(|a|)\,d^\times a.$$

Using this in (2.7), we obtain that $E(g;\Phi,\phi,s)$, for $\mathrm{Re}(s)>n(n+1)$, is the sum of the following terms:

(3.2.1) $$|\det g|^{ns}\int_{G_n(F)\backslash G'}\theta_n^0[(n+1)ns,\omega,g\cdot\Phi\cdot x^{-1}]\,\phi(x)|\det x|^{-(n+1)s}dx,$$

(3.2.2) $$|\det g|^{n(s-1)}\int_{G_n(F)\backslash G'}\theta_n^0[(n+1)n(1-s),\omega^{-1},g^\iota\cdot\hat{\Phi}\cdot x^{-\iota}]\cdot\phi(x)|\det x|^{-(n+1)(s-1)}dx,$$

(3.2.3) $$\sum_{1\leqq r<n}|\det g|^{n(s-1)}\int_{G_n(F)\backslash G'}\theta_r^0[(n+1)n(1-s),\omega^{-1},g^\iota\cdot\hat{\Phi}\cdot x^{-\iota}]\cdot\phi(x)|\det x|^{-(n+1)(s-1)}dx,$$

(3.2.4) $$-\sum_{1\leqq r<n}|\det g|^{ns}\int_{G_n(F)\backslash G'}\theta_r^1[(n+1)ns,\omega,g\cdot\Phi\cdot x^{-1}]\,\phi(x)|\det x|^{-(n+1)s}dx,$$

(3.2.5) $$\lambda((n+1)n(s-1),\omega)|\det g|^{n(s-1)}\hat{\Phi}(0)\int_{G_n(F)\backslash G'}\phi(x)|\det x|^{-(n+1)(s-1)}dx,$$

(3.2.6) $$-\lambda((n+1)ns,\omega)|\det g|^{ns}\Phi(0)\int_{G_n(F)\backslash G'}\phi(x)|\det x|^{-(n+1)s}dx.$$

If ω is not a principal character, the λ-factors in (3.2.5) and (3.2.6) vanish. If ω is a principal character then the integrals on $G_n(F)\backslash G'$ in (3.2.5) and (3.2.6) vanish, because $L_0^2(\omega,n)$ is orthogonal to the one dimensional invariant subspaces of $L^2(\omega,n)$. (See 12.11 in [5].) Thus (3.2.5) and (3.2.6) vanish in all cases.

The terms (3.2.3) and (3.2.4) vanish too. The proof is similar to the one in 12.13 of [5]. For (3.2.4) say, it suffices to show that

$$\int_{G_n(F)\backslash G^0}\phi(x)\,\theta_r^1[(n+1)ns,\omega,\Phi\cdot x^{-1}]\,dx=0,$$

for all cusp forms ϕ.

Explicitly this integral reads

$$\int_{\mathrm{I}/F^\times}|a|^{(n+1)ns}\omega(a)F_1(|a|)\,d^\times a\int_{G_n(F)\backslash G^0}\phi(x)\,dx\sum_{rk(\xi)=r}\Phi[ax^{-1}\xi].$$

Thus it suffices to prove that

(3.3) $$\int_{G_n(F)\backslash G^0}\phi(x)\,dx\sum_{rk(\xi)=r}\Phi(x^{-1}\xi)=0,$$

for $0<r<n$. Note that this is the integral of a rapidly decreasing function against a slowly increasing function. Every matrix ξ of rank r can be written uniquely as

$$\xi=\gamma_1\begin{pmatrix}1_r&0\\0&0\end{pmatrix}\gamma_2,$$

where γ_1, γ_2 are respectively taken in a set of representatives for the cosets:

$$G_n(F)/S(F), \qquad S=\left\{\begin{pmatrix}1_r & * \\ 0 & *\end{pmatrix}\right\},$$

with row blocks r, $n-r$ and column blocks r, $n-r$;

$$T(F)\backslash G_{n+1}(F), \qquad T=\left\{\begin{pmatrix}* & 0 \\ * & *\end{pmatrix}\right\}.$$

with row blocks r, $n-r+1$ and column blocks r, $n-r+1$.

Thus (3.3) is a sum of terms of the form:

$$\int_{G_n(F)\backslash G^0} \phi(x)\,dx \sum_{\gamma\in G_n(F)/S(F)} \phi\left[x^{-1}\gamma\begin{pmatrix}1_r & 0\\ 0 & 0\end{pmatrix}\gamma_2\right]$$

or

$$\int_{S(F)\backslash G^0} \phi(x)\,\Phi\left[x^{-1}\begin{pmatrix}1_r & 0\\ 0 & 0\end{pmatrix}\gamma_2\right]dx.$$

Since

$$x\to\Phi\left[x^{-1}\begin{pmatrix}1_r & 0\\ 0 & 0\end{pmatrix}\gamma_2\right]$$

is invariant under $S(\mathbb{A})$, this is also

$$\int_{G^0\cap S(\mathbb{A})\backslash G^0} \Phi\left[x^{-1}\begin{pmatrix}1_r & 0\\ 0 & 0\end{pmatrix}\gamma_2\right]dx \int_{S(F)\backslash S(\mathbb{A})\cap G^0} \phi(s x)\,ds.$$

Since S contains a horicycle as a normal subgroup, the inner integral factors through an integral of the form (1.1) and thus vanishes. This shows as announced that (3.2.4) or (3.2.3) vanish.

Finally E is the sum of (3.2.1) and (3.2.2). Changing x into x^ι in (3.2.2), we get at last:

(3.4) $$\begin{aligned} &E(g;\Phi,\phi,s)\\ &=|\det g|^{ns}\int_{G_n(F)\backslash G'} \theta_n^0[n(n+1)s,\omega,g\cdot\Phi\cdot x^{-1}]\,\phi(x)\,|\det x|^{-(n+1)s}\,dx\\ &\quad+|\det g^\iota|^{n(1-s)}\int_{G_n(F)\backslash G''}\theta_n^0[n(n+1)(1-s),\omega^{-1},g^\iota\cdot\hat{\Phi}\cdot x^{-1}]\\ &\quad\cdot\phi^\iota(x)\,|\det x|^{-(n+1)(1-s)}\,dx.\end{aligned}$$

This formula, whose right hand side is defined for all s, gives the analytic continuation of E as an entire function of s, a continuous function of (g, s), slowly increasing in g, uniformly for $\mathrm{Re}(s)$ in a compact. Moreover since G' and G'' play the same role, we get the functional equation:

(3.5) $E(g;\Phi,\phi,s)=E(g^\iota;\hat{\Phi},\phi^\iota,1-s)$.

Note that the right hand side may be thought of as being an Eisenstein series for the group $R=Q^\iota$, opposed to Q.

§4. Constant Terms

For any parabolic subgroup P of G_{n+1}, we define, as usual, the constant term relative to P:

(4.1) $$E_P(g;\Phi,\phi,s)=\int_{U_P^*} E(ug;\Phi,\phi,s)\,du.$$

We sketch a proof of the results we need on E_P ([4, 6, 9]). If in (4.1) we replace E by (2.3), we arrive at

(4.2) $$E_P(g;\Phi,\phi,s)=\int_{U_P^*} du \sum_{Q(F)\backslash G_{n+1}(F)} f(\gamma u g).$$

If we replace ϕ by $|\phi|$, assume $s>n(n+1)$, $\Phi\geqq 0$, then the expressions for f, E and E_P are positive and finite. Thus, for $\mathrm{Re}(s)>n(n+1)$ we may compute formally and obtain for E_P the expression:

(4.3) $$\sum_{\gamma\in Q(F)\backslash G_{n+1}(F)/U_P(F)} \int_{Q(F)^\gamma\cap U_P(F)\backslash U_P(\mathbb{A})} f(\gamma u g)\,du,$$

where we set $Q^\gamma=\gamma^{-1}Q\gamma$.

We may assume that P is standard, i.e. contains the group of upper triangular matrices. Let $P=M_P U_P$, $Q=M_Q U_Q$ be the respective Levi-decompositions. Denoting by W the group of permutation matrices, we may select as a system of representatives of the double cosets in (4.3) the matrices

(4.4) $$\gamma=w\mu,\qquad w\in W\cap M_Q\backslash W/W\cap M_P,\ \mu\in Q^w\cap M_P(F)\backslash M_P(F).$$

After a change of variables, (4.3) reads:

(4.5) $$\sum_{w,\mu}\int_{Q(F)\cap U_P^{w^{-1}}(F)\backslash U_P^{w^{-1}}(\mathbb{A})} f(uw\mu g)\,du.$$

To compute this, consider the integral

$$\int_{Q(F)\cap U_P^{w^{-1}}(F)\backslash U_P^{w^{-1}}(\mathbb{A})} f(ug)\,du.$$

Set $M^1=M_Q\cap M_P^{w^{-1}}$, $U^1=M_Q\cap U_P^{w^{-1}}$. An explicit computation shows that M^1U^1 is a Levi-decomposition of a parabolic subgroup of $M_Q\approx GL_1\times GL_n$. Rewrite the integral above as

(4.6) $$\int_{Q(\mathbb{A})\cap U_P^{w^{-1}}(\mathbb{A})\backslash U_P^{w^{-1}}(\mathbb{A})} du \int_{Q(F)\cap U_P^{w^{-1}}(F)\backslash Q(\mathbb{A})\cap U_P^{w^{-1}}(\mathbb{A})} f(u_1 u g)\,du_1.$$

The image of $Q\cap U_P^{w^{-1}}$ under the natural homomorphism from Q to M_Q is U^1. Thus for an appropriate g, taking into account the fact that f is invariant on the left under $U_Q(\mathbb{A})$, the inner integral in (4.6) has the form

$$\int_{U^{1*}} f(vg)\,dv=\int_{U^{1*}} f(g;v)\,dv$$
$$=|\det g|^{ns}\int_{G_n(\mathbb{A})}\Phi[(0,x)g]\,|\det x|^{(n+1)s}\,dx\int_{U^{1*}}\phi(vx^{-1})\,dv.$$

Here we have identified U^1 with a horicycle in GL_n. The above expression is absolutely convergent for $\mathrm{Re}(s)>n/n+1$. Since ϕ is a cusp form, it vanishes unless $U^1=\{e\}$, or, equivalently $M_Q \subset M_P^{w^{-1}}$. Since Q is maximal, this is equivalent to $M_Q = M_P^{w^{-1}}$. Thus $E_P=0$ unless either $M_P=M_Q$ or $M_P=w_0 M_Q w_0^{-1}$, where w_0 is the order reversing permutation. Thus, up to conjugation, we may assume $P=Q$ or $P=R(=Q^t)$.

In the first case ($P=Q$), we must have $w \in M_P$ and we may take $w=e$, $\mu=e$ in (4.4). Thus

$$E_Q(g; \Phi, \phi, s) = \int_{U_Q^*} f(u g)\, du = f(g)$$

(taking $\mathrm{vol}(U_Q^*)=1$).

In the second case, if we take $P=w_0 Q^t w_0^{-1}$, $M_P = w_0 M_Q w_0^{-1} = w^{-1} M_Q w$. Hence $w w_0 \in M_Q$ and we may take $w=w_0$, $\mu=e$ in (4.4). Moreover $Q \cap U_P^{w^{-1}}=\{e\}$. Hence in this case, if $\mathrm{Re}(s) > n(n+1)$ we have

$$E_P(g; \Phi, \phi, s) = \int_{U_P(\mathbb{A})} f(w_0 u g)\, du.$$

Finally

$$E_R(g; \Phi, \phi, s) = \int_{U_R(\mathbb{A})} f(u g)\, du.$$

Explicitly

(4.6) $$E_Q(g; \Phi, \phi, s) = |\det g|^{ns} \int_{G_n(\mathbb{A})} \Phi[(0, x) g]\, \phi(x^{-1})\, |\det x|^{(n+1)s}\, dx$$

and

$$E_R(g; \Phi, \phi, s) = |\det g|^{ns} \int_{G_n(\mathbb{A}) \times \mathbb{A}^n} \Phi[(x y, x) g]\, \phi(x^{-1})\, |\det x|^{(n+1)s}\, dx\, dy,$$

or changing variables,

(4.7) $$E_R(g; \Phi, \phi, s) = |\det g|^{ns} \int_{G_n(\mathbb{A}) \times \mathbb{A}^n} \Phi[(y, x) g]\, \phi(x^{-1})\, |\det x|^{(n+1)s-1}\, dx\, dy.$$

(In the latter formulas, y is a column of size n.)

§ 5. Vanishing of the Constant Terms

We begin the proof of (1.3) with Proposition (5.1):

(5.1) **Proposition.** *Suppose* $L(1, \pi)=0$. *Then for* ϕ *K-finite,*

$$E_Q(g; \Phi, \phi, \tfrac{1}{2})=0, \qquad E_R(g; \Phi, \phi, \tfrac{1}{2})=0.$$

Proof. By (1.2) and (1.4) we have in fact

$$L(1, \pi) = L(1, \tilde{\pi}) = L(0, \pi) = L(0, \tilde{\pi}) = 0.$$

To proceed further we need to recall certain results of [5]. For ϕ a K-finite element of V (or H) and Ψ arbitrary in $\mathscr{S}(n \times n, \mathbb{A})$, set

$$Z(\Psi, s, \phi) = \int_{G_n(\mathbb{A})} \Psi(x)\, \phi(x)\, |\det x|^s\, dx,$$

$$Z(\Psi, s, \phi^{\vee}) = \int_{G_n(\mathbb{A})} \Psi(x)\, \phi(x^{-1})\, |\det x|^s\, dx.$$

Here, as usual, dx is a Haar measure on $G_n(\mathbb{A})$. These integrals converge for $\mathrm{Re}(s)>n$, have analytic continuations as entire functions of s and, as such, satisfy a functional equation. This is Theorem 12.4 of [5]. It could also be derived form the study of constant terms of §4.

(5.2) **Lemma.** *For any s,*

(1) $L(s,\pi)=0 \Rightarrow Z\left(\Psi, s+\frac{n-1}{2}, \phi\right)=0.$

(2) $L(s,\tilde{\pi})=0 \Rightarrow Z\left(\Psi, s+\frac{n-1}{2}, \phi^{\check{}}\right)=0.$

Taking the lemma for granted at the moment, we prove (5.1). We have for $\mathrm{Re}(s)>n(n+1)$ at first, but then by analytic continuation for all s,

$$E_Q(g;\Phi,\phi,s)=|\det g|^{ns} Z(\Psi_1,(n+1)s,\phi^{\check{}})$$

and

$$E_R(g;\Phi,\phi,s)=|\det g|^{ns} Z(\Psi_2,(n+1)s-1,\phi^{\check{}}),$$

where

$$\Psi_1(u)=\Phi[(0,u)g], \qquad \Psi_2(u)=\int_{\mathbb{A}^n}\Phi[(y,u)g]\,dy.$$

To obtain (5.1), it suffices to apply (5.2.2) for $s=0$ and $s=1$.

We pass to the proof of (5.2). We prove (5.2.1) since (5.2.2) can be obtained similarly. (See the remarks at the beginning of §2.)

The expression (12.14.1) of [5] for $Z(\Psi,s,\phi)$ shows that, for ϕ K-finite, $\Psi\to Z(\Psi,s,\phi)$ is a distribution. Hence it suffices to prove our assertion for Ψ of the form

$$\Psi(x)=\prod \Psi_v(x_v),$$

where, for v real,

$$\Psi_v(x)=e^{-\pi tr\,{}^t x x} P(x),$$

for v complex,

$$\Psi_v(x)=e^{-\pi tr\,{}^t \bar{x} x} P(x,\bar{x}),$$

P being, in both cases, a polynomial. Note that Ψ is bi-K-finite.

Now for $\mathrm{Re}(s)$ large,

$$Z(\Psi,s,\phi)=\int_{G_n(\mathbb{A})}\int_K \Psi(kx)\,\phi(kx)\,|\det x|^s\,dk\,dx.$$

Hence for such s, $Z(\Psi,s,\phi)$ is a finite linear combination of integrals of the form

$$Z(\Psi,s,f)=\int_{G_n(\mathbb{A})}\Psi(x)\,f(x)\,|\det x|^s\,dx.$$

Here Ψ is still of the above form and f has the form

$$f(x)=\int_K m(k)\,\phi(kx)\,dk,$$

with m a coefficient of a finite-dimensional representation of K. Since ϕ and f are bounded all the steps are justified for $\mathrm{Re}(s)>n$. Let

$$T(\phi)=\int_K m(k)\,\phi(k)\,dk \qquad (\phi\in V).$$

Then T is a K-finite linear form on V and $f(x)=T(\pi(x)\,\phi)$. Thus f is a bi-K-finite coefficient of π. Hence by Theorem 13.8 of [5], $Z(\Psi, s, f)$ has a continuation to an entire function of s. Thus $Z(\Psi, s, \phi)$ is, for all s, a finite linear combination of functions of the form $Z(\Psi, s, f)$. The proof of the same theorem shows that

$$L(s,\pi)=0 \Rightarrow Z\left(\Psi, s+\frac{n-1}{2}, f\right)=0$$

for any bi-K-finite coefficient f of π. Thus (5.2) is established.

Part of the proof of the preceeding proposition consists in showing that if one constant term vanishes on the unitary line, so does the other. In the general context, this follows from the fact that the two terms are related by an operator which is unitary, in particular defined, on the unitary line [6], [9].

(5.4) **Corollary.** *If* $L(1,\pi)=0$, *then* $E(g;\Phi,\phi,\tfrac{1}{2})=0$ *for all* $g\in G_n(A)$, $\Phi\in\mathcal{S}$, *and* ϕ *K-finite in V.*

Proof. Since by analytic continuation $E_P=0$ if P is not associate to Q, we see that E is, under the assumptions, cuspidal for $s=1/2$. Since E is slowly increasing, we see that for any $h\in C_c^\infty(G_{n+1}(\mathbb{A}))$, bi-$K$-finite, the function k defined by

(5.5) $\quad k(g)=\int E(g\,x,\tfrac{1}{2})\,h(x)\,dx, \qquad E(g,s)=E(g;\Phi,\phi,s),$

is cuspidal and rapidly decreasing.

On the other hand, if F is any function on $G(\mathbb{A})$ satisfying

$$F(\gamma\, a\, g)=\omega(a)\,F(g), \qquad a\in Z_{n+1}(\mathbb{A}),\ \gamma\in G_{n+1}(F),$$

which is cuspidal and rapidly decreasing, then the integral

(5.6) $$\int_{G_{n+1}(F)Z_{n+1}(\mathbb{A})\backslash G_{n+1}(\mathbb{A})} E(g,s)\,\bar{F}(g)\,dg$$

vanishes identically. Indeed since it is a holomorphic function of s, it suffices to prove this for $\mathrm{Re}(s)>n(n+1)$. But then we may replace E by its expression as a series and get for (5.6)

$$\int_{Q(F)Z_{n+1}(\mathbb{A})\backslash G_{n+1}(\mathbb{A})} f(g)\,\overline{F(g)}\,dg.$$

Let $U=U_Q$. The integral factors through an integral of the form

$$\int_{U^*} f(u\,g)\,\bar{F}(u\,g)\,du=f(g)\int_{U^*}\bar{F}(u\,g)\,du=0.$$

Thus we find that (5.6) vanishes. In particular:

$$\int_{G_{n+1}(F)Z_{n+1}(\mathbb{A})\backslash G_{n+1}(\mathbb{A})} E(g,\tfrac{1}{2})\,\bar{k}(g)\,dg=0.$$

Replacing h by $h^* * h$, where $h^*(x) = \bar{h}(x^{-1})$, we get

$$\int_{G_{n+1}(F) Z_{n+1}(\mathbb{A}) \backslash G_{n+1}(\mathbb{A})} k(g)\, \bar{k}(g)\, dg = 0.$$

Thus $k = 0$ and $E(g, \frac{1}{2}) = 0$. Q.E.D.

§6. Non-Vanishing of the Non-Constant Term

Let N_n be the group of n by n upper triangular matrices with diagonal entries equal to one (maximal horicycle). Let $\bar{N}_n = N_n^t$ be the "opposed" horicycle and define a character θ_n of $\bar{N}_n$ by

$$\theta_n(u) = \prod \psi(u_{i,i+1}), \quad 1 \leqq i \leqq n-1,$$

(i is the column index). We set:

(6.1) $$W(g; \Phi, \phi, s) = \int_{\bar{N}^*_{n+1}} E(ug; \Phi, \phi, s)\, \bar{\theta}_{n+1}(u)\, du.$$

(6.2) **Proposition.** *Given s, there are Φ and a K-finite element ϕ of V such that*

$$W(e; \Phi, \phi, s) \neq 0.$$

This will give a contradiction to (5.4) and Theorem (1.3) will be proved.

Proof of (6.2). Once more we may compute formally for $\mathrm{Re}(s) > n(n+1)$. We obtain for (6.1)

(6.3) $$\int_{\bar{N}^*_{n+1}} \bar{\theta}_{n+1}(u)\, du \sum_{\gamma \in Q(F) \backslash G_{n+1}(F) / \bar{N}_{n+1}(F)} \sum_{\delta \in Q^\gamma \cap \bar{N}_{n+1}(F) \backslash \bar{N}_{n+1}(F)} f(\gamma \delta u g).$$

Exchanging sums and integrals and using $W \cap M_Q \backslash W$ as a set of representatives for the double cosets in (6.3), we get

(6.4) $$\sum_w \int_{Q^w \cap \bar{N}_{n+1}(F) \backslash \bar{N}_{n+1}(\mathbb{A})} \bar{\theta}_{n+1}(u)\, f(wug)\, du.$$

The inner integral factors through an integral of the form

$$\int_{(U_Q^w \cap \bar{N}_{n+1})^*} f(wvg)\, \bar{\theta}_{n+1}(v)\, dv = f(wg) \int_{(U_Q^w \cap \bar{N}_{n+1})^*} \bar{\theta}_{n+1}(v)\, dv.$$

Thus it vanishes if θ_{n+1} is non-trivial on $U_Q^w \cap \bar{N}_{n+1}$ or, equivalently, U_Q^w "contains the opposite of a simple root". Thus all terms in (6.4) vanish but those for which this does not happen, that is, for which w is in M_Q. So (6.4) reduces to one term:

(6.5) $$\int_{Q(F) \cap \bar{N}_{n+1}(F) \backslash \bar{N}_{n+1}(\mathbb{A})} \bar{\theta}_{n+1}(u)\, f(ug)\, du.$$

But $\bar{N}_{n+1}$ is the semi-direct product of U_R and the subgroup

$$Q \cap \bar{N}_{n+1} = M_Q \cap \bar{N}_{n+1} \simeq \bar{N}_n.$$

Thus (6.5) can also be written as an integral over

$$U_R(\mathbb{A}) \times \bar{N}_n^*$$

or, using (2.2), as

$$(6.6) \qquad \int_{U_R(\mathbb{A})} \bar{\theta}_{n+1}(u)\, du \int_{\bar{N}_n^*} \bar{\theta}_n(v) f(u g, v)\, dv.$$

Now let ε be the column of size n defined by

$$\varepsilon = {}^t(1, 0, \ldots, 0),$$

so that on

$$\begin{pmatrix} 1 & 0 \\ u & 1_n \end{pmatrix}$$

θ_{n+1} takes the value $\psi({}^t\varepsilon \cdot u)$. Then replacing f by its definition, we find, for $\mathrm{Re}(s) > n(n+1)$, (6.6) and hence (6.1) is

$$(6.7) \quad |\det g|^{ns} \iiint \Phi[(x u, x) g]\, \psi[-{}^t\varepsilon \cdot u]\, \phi(v x^{-1})\, \bar{\theta}_n(v)\, |\det x|^{(n+1)s}\, dx\, du\, dv,$$

the integration being for

$$x \in G_n(\mathbb{A}), \qquad u \in \mathbb{A}^n, \qquad v \in \bar{N}_n^*.$$

Changing variables we get:

$$W(e; \Phi, \phi, s) = \int_{G_n(\mathbb{A})} |\det x|^{(n+1)s-1}\, dx \int_{\mathbb{A}^n} \Phi(u, x)\, \psi[-{}^t(x^t \varepsilon)\, u]\, du \cdot \int_{\bar{N}_n^*} \phi(v x^{-1})\, \bar{\theta}_n(v)\, dv.$$

For $s > n(n+1)$, in fact even for $s > 1$, $\Phi \geqq 0$, the integral is dominated by

$$\int |\det x|^{(n+1)s-1}\, dx \int \Phi(u, x)\, du \int dv < +\infty,$$

which justifies our step. Now set

$$W(h, \phi) = \int_{\bar{N}_n^*} \phi(v h)\, \bar{\theta}_n(v)\, dv,$$

and also

$$\Phi'(a, b) = \int_{\mathbb{A}^n} \Phi(u, b)\, \psi(-{}^t a \cdot u)\, du.$$

Then, for $\mathrm{Re}(s) > n(n+1)$,

$$(6.8) \quad W(e; \Phi, \phi, s) = \int_{G_n(\mathbb{A})} \Phi'(x^t \varepsilon; x)\, |\det x|^{ns}\, W(x^{-1}, \phi)\, dx.$$

Since ϕ has a "Fourier expansion" in terms of $W(x, \phi)$ ([13], [15]), one can find a $\phi \in V$ so that $W(e, \phi) \neq 0$. We refer to [15] for the definition of the spaces $\mathscr{D}(\pi)$ and $\mathscr{D}_S(\pi)$. We have $\mathscr{D}(\pi) = \bigcup \mathscr{D}_S(\pi)$, the union taken over sufficiently large finite sets S of places of F. Since $\mathcal{D}(\pi)$ is $G_n(\mathbb{A})$ stable, we may also choose $\phi \in \mathcal{D}(\pi)$ so that $W(e, \phi) \neq 0$. Let $K_v = GL(n, R_v)$, where R_v is the ring of integers in F_v. Let $K_S = \prod_{v \notin S} K_v$ and $K'_S = \prod_{v \in S} K_v$. It is clear that the K'_S-finite vectors in $\mathcal{D}_S(\pi)$ are dense in that space. Moreover, Lemma 5.6 of [15] shows that $\phi \to W(e, \phi)$ is a continuous linear form on $\mathcal{D}_S(\pi)$. Thus in fact we may choose ϕ K-finite so that $W(e, \phi) \neq 0$.

Fix such a ϕ and let S be a finite set of places containing all archimedean ones such that for all $v \notin S$, ϕ is invariant under K_v. We also assume that, for $v \in S$, R_v is the largest ideal on which ψ_v is trivial. Then the relations

$$W(ux, \phi) = \theta(u)\, W(x, \phi) \qquad \text{for } u \in \bar{N}_n,$$

$$W(xk, \phi) = W(x, \phi) \qquad \text{for } k \in K_v,\ v \notin S,$$

$$W(x, \phi) \neq 0$$

imply that, for v not in S, $x_v = b_v k_v$, where b_v is lower triangular and its diagonal entries satisfy:

$$|a_{1v}| \geqq |a_{2v}| \geqq \cdots \geqq |a_{nv}|.$$

On the other hand, Φ and thus Φ' is *arbitrary*. So we may take Φ' of the form

$$\Phi'(a, b) = \Psi'(a)\, \Psi(b) \qquad (a \in \mathbb{A}^n,\ b \in M(n \times n, \mathbb{A})),$$

with

$$\Psi'(a) = \Psi'_S(a_S) \prod_{v \notin S} \Psi'_v(a_v),$$

$$\Psi(b) = \Psi_S(b_S) \prod_{v \notin S} \Psi_v(b_v),$$

where Ψ_v (resp. Ψ'_v) is the characteristic function of $M(n \times n, R_v)$ (resp. R_v^n) and Ψ_S is any C^∞-function of compact support on $G_n(S) = \prod_{v \in S} G_n(F_v)$.

It is then clear that

$$\Phi'(x^\iota \varepsilon, x)\, W(x^{-1}, \phi) \neq 0$$

implies that, for $v \notin S$,

$$x_v = k_v b_v,$$

where b_v is lower triangular, its entries being in R_v and its diagonal entries in $R_v^\times$. In other words, in (6.8) the support of the integrand is contained in a compact subset of

$$G_n(S) \times K_S$$

Moreover the integrand is invariant on the left under K_S. It follows first that the right hand side of (6.8) converges for all s (a fact which could be proved for all Φ') and (6.8) is true for all s. Moreover, with a suitable choice of the Haar measures:

$$W(e; \Phi, \phi, s) = \int_{G_n(S)} \Psi'_S(x^\iota \varepsilon)\, \Psi_S(x)\, |\det x|^{ns}\, W(x^{-1}, \phi)\, dx.$$

If for a given s this vanishes for all Ψ_S, we find

$$W(e,\phi)\,\Psi'_S(\varepsilon)=0$$

and thus $W(e,\phi)=0$, a contradiction. Thus we have established (6.1) and Theorem (1.3).

§ 7. The Case $n=1$

If $n=1$, π is just a character of $F_{\mathbb{A}}^{\times}/F^{\times}$. Then $L(s,\pi)$ is holomorphic at $s=1$ except when π is the trivial character. Moreover in computing the constant term, one must remember that Q and R are conjugate so that

$$E_Q(g;\Phi,\phi,s)=f(g)+\int_{U^*} f(w\,u\,g)\,du,$$

where

$$w=\begin{pmatrix}0&1\\1&0\end{pmatrix}.$$

Explicitly this is

$$|\det g|^s\int_{\mathbb{I}}\Phi[(0,x)\,g]\,\pi(x^{-1})\,|x|^{2s}\,d^\times x$$
$$+|\det g|^s\int_{\mathbb{I}}\int_{\mathbb{A}}\Phi[(x,y)\,g]\,\pi(x^{-1})\,|x|^{2s-1}\,d^\times x\,dy.$$

Again this will vanish for $s=\frac{1}{2}$ (and E will be cuspidal) if $L(1,\pi)=0$. The rest of the proof is the same.

§ 8. An Application

We give a simple application which provides a good example of our own motivation. Namely let σ be a component of $L_0^2(\omega,2)$ and μ a character of $F_{\mathbb{A}}^{\times}/F^{\times}$. Together σ and μ define a representation of $M_Q(\mathbb{A})$ or $Q(\mathbb{A})$. The representation π of $G_3(\mathbb{A})$ it induces is irreducible.

(8.1) **Proposition.** *The representation π is not a component of $L_0^2(\omega,3)$.*

Proof. If it were, $L(s,\pi\otimes\chi)$ would be entire for all characters χ of $F_{\mathbb{A}}^{\times}/F^{\times}$. But

$$L(s,\pi\otimes\mu^{-1})=L(s,\sigma\otimes\mu^{-1})\,L(s,1_F).$$

Here 1_F is the trivial character of $F_{\mathbb{A}}^{\times}$ and the corresponding L-function has a pole at $s=1$. It cannot be "cancelled" by a corresponding zero of $L(s,\sigma\otimes\mu^{-1})$ and thus $L(s,\pi\otimes\mu^{-1})$ has a pole at $s=1$, a contradiction. Q.E.D.

References

1. Andrianov, A.N.: On the Zeta-functions of the general linear groups. Congrès International des mathematiciens, pp. 273–276. Paris: Gauthier-Villars 1970
2. Gelfand, I.M., Kajdan, D.A.: Representations of the group $GL(n,K)$ where K is a local field. Lie groups and their representations, Summer School of Group Representations, Budapest, pp. 95–118. New York: Halsted-Press 1971

3. Godement, R.: Analyse spectrale des fonctions modulaires. Séminaire Bourbaki, (1964/65), No. 278, 26 p.
4. Godement, R.: Introduction à la théorie de Langlands. Séminaire Bourbaki, (1966/67), No. 321, 30 p.
5. Godement, R., Jacquet, H.: Zeta functions of simple algebras. Springer Lecture Notes 260 (1972)
6. Harish-Chandra: Automorphic forms on Semi-Simple Lie Groups. Springer Lecture Notes 62 (1968)
7. Jacquet, H.: Automorphic forms on $GL(2)$. II. Springer Lecture Notes 278 (1972)
8. Jacquet. H., Pyatetskii-Shapiro, I. I., Shalika, J.: Construction de formes automorphes pour le groupe $GL(3)$. Comptes Rendus, Acad. Sci., Paris, Vol. 282, (1976), pp. 91–94
9. Langlands, R. P.: On the functional equations satisfied by Eisenstein Series, preprint
10. Langlands, R. P.: Problems in the theory of automorphic forms. Modern Analysis and its applications. Springer Lecture Notes 170 (1970), pp. 18–61
11. Langlands, R. P.: Euler products. Lecture notes, Yale University (1967)
12. Ogg, A. P.: On a convolution of L-series. Inventiones Math., **7**, 297–312 (1969)
13. Pyatetskii-Shapiro, I.: Euler subgroups, Lie groups and their representations. Summer School of Group Representations, Budapest, pp. 597–620. New York: Halsted-Press 1971
14. Rankin, R.: Contributions to the theory of Ramanujan's function $\tau(n)$ and similar arithmetical functions. I. Proc. Camb. Phil. Soc. **35**, 351–356 (1939)
15. Shalika, J.: The multiplicity one theorem for GL_n. Annals of math., Vol. 100, No. 1, 171–193 (1974)

Received May 31, 1976

Ann. scient. Éc. Norm. Sup.,
4e série, t. 11, 1978, p. 471 à 542.

A RELATION BETWEEN AUTOMORPHIC REPRESENTATIONS OF GL (2) AND GL (3)*

BY STEPHEN GELBART** AND HERVÉ JACQUET

Some notations

We will denote by G_r the group GL (r), by B_r the standard Borel subgroup of GL (r) (upper triangular matrices), by N_r the unipotent radical of B_r, and by Z_r the center of G_r. On the other hand, we will denote by G the group SL (2), by B its standard Borel subgroup, by $N = N_2$ the unipotent radical group of B, and by A the subgroup of diagonal matrices in G. Finally P will be the standard parabolic subgroup of type (2, 1) in G_3 and U its unipotent radical.

If F is a local non archimedean field, we denote by $\mathfrak{R}_F$ or $\mathfrak{R}$ the ring of integers in F, by $\mathfrak{P}_F$ or $\mathfrak{P}$ its maximal ideal, and by $\tilde{\omega}$ a generator of $\mathfrak{P}$. Finally we set $K_r = GL(r, \mathfrak{R}_F)$, $K = SL(2, \mathfrak{R}_F)$. If F is archimedean the same notations will be used for the standard maximal compact subgroups. When F is local, we will often write G_r, B_r, ... for G_r (F), B_r (F) ...

Introduction

The purpose of this paper is to establish a relation between automorphic forms on GL (2) and GL (3). To formulate our main result more precisely we need first to recall some basic facts.

Suppose F is an A-field and π is an irreducible unitary representation of GL (r, A). Then π can be written as an infinite tensor product $\pi = \bigotimes_v \pi_v$ where π_v is an irreducible representation of $G_v = GL(r, F_v)$ for each place v. Moreover, for almost all finite v the representation π_v is unramified, i. e., contains the trivial representation of a maximal compact subgroup of G_v. But such representations are well known to be parameterized by semi-simple conjugacy classes in GL (r, **C**). Thus one can associate to π a family of semi-simple conjugacy classes (a_v) in GL (r, **C**), a_v being defined for almost all finite v.

* Both authors have been supported by grants from the N.S.F.
** Alfred P. Sloan Fellow.

Now suppose σ is an automorphic representation of GL (2, **A**), i. e., σ occurs in the space of automorphic forms on GL (2). If (b_v) is the corresponding family of conjugacy classes in GL (2, **C**), define a family of conjugacy classes in GL (3, **C**) by

$$a_v = \begin{pmatrix} mn^{-1} & 0 & 0 \\ 0 & 1 & 0 \\ 0 & 0 & m^{-1}n \end{pmatrix} \quad \text{if} \quad b_v = \begin{pmatrix} m & 0 \\ 0 & n \end{pmatrix}.$$

Our main result then asserts that there is an essentially unique automorphic representation of GL (3, **A**)—the "lift of σ—whose corresponding family is (a_v).

To see how this "lifting" appears as a special case of the more general conjectures of Langlands the reader can consult [Bo]. Roughly speaking, if W'_v denotes the Weil-Deligne group of F_v, and $\sigma = \otimes \sigma_v$, then each σ_v corresponds (sometimes only conjecturally) to a two-dimensional representation φ_v of W'_v. So if p denotes the map

$$\begin{array}{ccc} \mathrm{GL}(2, \mathbf{C}) & \xrightarrow{\;p\;} & \mathrm{GL}(3, \mathbf{C}) \\ & \searrow \quad \nearrow & \\ & \mathrm{PGL}(2, \mathbf{C}) & \end{array}$$

determined by the adjoint action of PGL (2, **C**), then (according to Langland's general philosophy) the resulting three dimensional representation

$$p \circ \varphi_v : \quad W'_v \to \mathrm{GL}(3, \mathbf{C})$$

should correspond to an irreducible admissible representation $\pi_v = p^* (\sigma_v)$ of GL (3, F_v). This is the local "lifting" of σ_v to GL (3). Globally, the product $\pi = \bigotimes_v \pi_v$ should (and does!) define an automorphic representation of GL (3, **A**).

The idea of our proof can be explained as follows. Following [GoJa] one can attach to each irreducible representation π of GL (r, **A**) an infinite Euler product

$$L(s, \pi) = \prod_v L(s, \pi_v),$$

with

$$L(s, \pi_v) = \det(1 - q_v^{-s} a_v)^{-1},$$

for almost all v. This product converges in some right half-plane. Moreover, if π is automorphic cuspidal, i. e., π occurs in the space of cusp forms, then L (s, π) continues to a holomorphic function of s and satisfies a functional equation. For $r = 2$ or 3 there is also a converse; (*cf.* [JPSS]). For each character χ of the idèle class group consider the representation $g \mapsto (g)\, \chi(\det(g))$ (also denoted $\pi \otimes \chi$). Then if L (s, $\pi \otimes \chi$) satisfies the above conditions for each χ, π *must* be automorphic cuspidal.

Now recall that to each *pair* of irreducible representations (π_1, π_2) of GL (2, **A**) one can attach an infinite Euler product

$$L(s, \pi_1 \times \pi_2) = \prod_v L(s, \pi_{1,v} \times \pi_{2,v}),$$

with

$$L(s, \pi_{1,v} \times \pi_{2,v}) = \det(1 - q_v^{-s} a_{1,v} \otimes a_{2,v})^{-1},$$

for almost all v; (*cf.* [Ja]). If π_1 and π_2 are automorphic then this infinite product continues to a meromorphic function of s and satisfies a functional equation In particular, if $\pi_1 = \sigma \otimes \chi$ and $\pi_2 = \tilde{\sigma}$ (the contragredient), then:

$$L(s, (\sigma \otimes \chi) \times \tilde{\sigma}) = L_2(s, \sigma, \chi) L(s, \chi),$$

where L (s, χ) is the usual Hecke L-function attached to χ and $L_2(s, \sigma, \chi)$ is a new Euler product of degree 3. If π is automorphic, then $L_2(s, \sigma, \chi)$ is also meromorphic and satisfies a functional equation. These facts are discussed in Sections 1 and 2.

In Section 3 we introduce a representation π of GL (3, **A**)—the "lift" of σ—by requiring that

$$L(s, \pi \otimes \chi) = L_2(s, \sigma, \chi).$$

At almost all places, σ_v is unramified, and the relation between π_v and σ_v is the one already explained. The situation at the remaining places is discussed in Section 3.

The fact that $L_2(s, \sigma, \chi)$ has no poles at all is proved using the method of [Sh]. Our main theorem then results from the converse theorem for GL (3) already alluded to.

To establish the holomorphy of $L_2(s, \sigma, \chi)$ we generalize [Sh] by considering integrals on the metaplectic group Mp (**A**). This group is an extension of SL (2, **A**) by the torus **T** which splits over SL (2, F). Thus any form on SL (2, **A**) may be regarded as a form on Mp (**A**). The "main integral" we consider is an integral over **T** SL (2, F)$\backslash$(Mp (**A**)) of the product a form on GL (2, **A**) belonging to π, an Eisenstein series on Mp (**A**), and a form on the metaplectic group given by a theta series. In Sections 5 and 6 we show that this integral decomposes as a product of integrals over local metaplectic groups and equals $L_2(s, \sigma, \chi)$ for a suitable choice of functions involved. Thus we need only analytically continue the Eisenstein series. This is carried out in Section 8 after analyzing the "constant term" in Section 7.

Actually, our paper deviates a bit from the simple scheme just outlined—for technical reasons discussed within the text. We also have to complete the results of [JPSS] in Section 4.

Many of the results of this paper were announced earlier by us in [GeJ 2]. We are grateful to Arletta Havlik, Joanne Martin, Margeret Reif and Diana Shye for their expeditious typing of the manuscript. The second-named author would also like to thank R. P. Langlands for pointing out the results of Corollary 1.7, and W. Casselman for bringing to his attention the results of Theorem 2.2.

1. GL (2) × GL (2): local theory

In this Section (and the next) we complete the theory of automorphic forms on GL (2) × GL (2) developed in [Ja]. In particular, we describe the poles of $L(s, \pi_1 \times \pi_2)$ both locally and globally. Although some of our results are not used later in this paper they all seem to be of independent interest.

(1.1) Let F denote a local non-archimedean field, q the cardinality of the residue class field, v_F or v the normalized valuation on F, and $|\ |_F$ or $|\ |$ the corresponding absolue value. Then $|x|_F = q^{-v(x)}$. When convenient, we also let let α_F or α denote the absolute value. By ψ_F or ψ we denote a fixed non-trivial additive character of F.

If (π_1, π_2) is a pair of irreducible admissible infinite-dimensional representations of GL (2, F) then $L(s, \pi_1 \times \pi_2)$ will denote the Euler-factor defined in [Ja].

In this section we shall recall the definition of $L(s, \pi_1 \times \pi_2)$ and complete the results of [Ja] by computing this factor in all cases.

Let $\mathscr{W}(\pi_i; \psi_F)$ denote the *Whittaker model* of π_i (with respect to ψ_F). Recall that this is a space of functions W on GL (2, F) satisfying the following properties:

(1.1.1) $\mathscr{W}(\pi_i; \psi_F)$ is invariant for the right action of $G_2(F)$ and the resulting representation of $G_2(F)$ on $\mathscr{W}(\pi_i; \psi_F)$ is equivalent to π_i;

(1.1.2) for each W in $\mathscr{W}(\pi_i; \psi_F)$:

$$W\left[\begin{pmatrix} 1 & x \\ 0 & 1 \end{pmatrix} g\right] = \psi_F(x) W[g] \quad \text{for each } x \in F \quad \text{and} \quad g \in G_2(F).$$

The existence and uniqueness of such a space is proved in Chapter I of [JL].

For $W_i \in \mathscr{W}(\pi_i; \psi)$, Φ a Schwartz-Bruhat function on F^2, and $s \in \mathbf{C}$, set

$$(1.1.3) \qquad \Psi(s, W_1, W_2, \Phi) = \int_{N_2(F) \backslash G_2(F)} W_1(g) W_2(\eta g) \Phi[(0, 1) g] |\det g|^s dg,$$

where

$$\eta = \begin{pmatrix} 1 & 0 \\ 0 & -1 \end{pmatrix}.$$

Then the factor $L(s, \pi_1 \times \pi_2)$ arises as the "g. c. d." of these integrals. More precisely, these integrals converge for Re (s) sufficiently large and define rational functions of q^{-s}; since the subvector space of $\mathbf{C}(q^{-s})$ spanned by them is a fractional ideal of $\mathbf{C}[q^{-s}, q^s]$ which contains 1 it has a unique generator of the form $P^{-1}(q^{-s})$ with $P \in \mathbf{C}[q^{-s}]$ and $P(0) = 1$: this generator is—by definition—$L(s, \pi_1 \times \pi_2)$.

Recall also that if W is in $\mathscr{W}(\pi_1; \psi_F)$ then the function W defined by

$$(1.1.4) \qquad \tilde{W}(g) = W[w^t g^{-1}], \qquad w = \begin{pmatrix} 0 & 1 \\ -1 & 0 \end{pmatrix},$$

belongs to the Whittaker model $\mathscr{W}(\tilde{\pi}_i; \psi_F)$ of the representation $\tilde{\pi}_i$ contragredient to π_i.

Finally the factor $\varepsilon(s, \pi_1 \times \pi_2; \psi_F)$—which is a monomial in q^{-s}—is defined by the functional equation:

(1.1.5)
$$\Psi(1-s, \tilde{W}_1, \tilde{W}_2, \hat{\Phi})/L(1-s, \tilde{\pi}_1 \times \tilde{\pi}_2) = \varepsilon(s, \pi_1 \times \pi_2; \psi_F)\Psi(s, W_1, W_2, \Phi)/L(s, \pi_1 \times \pi_2).$$

Here

(1.1.6)
$$\Phi^{\wedge}(x, y) = \iint \Phi(u, v)\psi_F(ux+vy)\,du\,dv$$

is the Fourier transform of Φ (with respect to the self-dual Haar measure). Denote by ω_i the central quasi-character of π_i.

PROPOSITION (1.2). — *Suppose π_2 is supercuspidal. Then the poles of $L(s, \pi_1 \times \pi_2)$ are simple. If $\alpha^s \otimes \pi_1$ denotes the representation $g \mapsto |\det g|^s \pi_1(g)$ then s_0 is a pole of $L(s, \pi_1 \times \pi_2)$ if and only if $\alpha^{s_0} \otimes \pi_1$ is equivalent to $\tilde{\pi}_2$.*

Proof. — It is clear that s_0 is a pole of $L(s, \pi_1 \times \pi_2)$ if and only if it is a pole $\Psi(s, W_1, W_2, \Phi)$ for at least one choice of W_1, W_2, and Φ. Moreover, the multiplicity of s_0 as a pole of $L(s, \pi_1 \times \pi_2)$ is the maximum multiplicity of s_0 as a pole of each of these integrals.

By Iwasawa's decomposition,

$$\Psi(s, W_1, W_2, \Phi) = \int_K \int_{F^\times} W_1\left[\begin{pmatrix} a & 0 \\ 0 & 1 \end{pmatrix} k\right] W_2\left[\begin{pmatrix} -a & 0 \\ 0 & 1 \end{pmatrix} k\right] f(k) |a|^{s-1} d^\times a\, dk,$$

with

$$K = GL(2, R_F) \quad \text{and} \quad f(g) = |\det g|^s \int \Phi[(0, t)g] |t|^{2s} \omega_1 \omega_2(t)\, d^\times t.$$

But π_2 is supercuspidal. Therefore ([JL], Prop. (2.16)) the function

$$a \mapsto W_2\left[\begin{pmatrix} a & 0 \\ 0 & 1 \end{pmatrix}\right]$$

has compact support in $F^\times$. Thus every pole of Ψ must come from the integral defining f. In particular, if s_0 is a pole, then s_0 is a simple pole, and $\alpha^{2s_0} = (\omega_1 \omega_2)^{-1}$.

So suppose now that s_0 satisfies $\alpha^{2s_0} = (\omega_1 \omega_2)^{-1}$. Then

$$\lim_{s \to s_0} (s-s_0) \int \Phi[(0, t)] |t|^{2s} (\omega_1 \omega_2)^{-1}(t)\, d^\times t = c\Phi(0, 0),$$

with c a non-zero constant independent of Φ. It follows that

$$\lim_{s \to s_0} (s-s_0)\Psi(s, W_1, W_2, \Phi) = c\Phi(0, 0)\, b(W_1, W_2),$$

with

$$b(W_1, W_2) = \int_{Z_2(F) N_2(F) \backslash G_2(F)} W_1(g) W_2(\eta g) |\det g|^{s_0} dg.$$

Thus s_0 is a pole of $L(s, \pi_1 \times \pi_2)$ if and only if the bilinear form b is not identically zero. Note that integrand defining b is indeed invariant on the left by $N_2(F) Z_2(F)$ and compactly supported mod $N_2(F) Z_2(F)$. But since π_i acts by right translations in $\mathscr{W}(\pi_i; \psi)$,

$$b(\pi_1(g) W_1, \pi_2(g) W_2) = |\det g|^{-s_0} b(W_1, W_2).$$

Thus if s_0 is a pole of $L(s, \sigma_1 \times \sigma_2)$, b is non-zero and $\alpha^{s_0} \otimes \pi_1$ is contragredient to π_2.

Conversely, suppose $\alpha^{s_0} \otimes \pi_1 \simeq \tilde{\pi}_2$. Since we may replace the pair (π_1, π_2) by $(\pi_1 \otimes \alpha^t, \pi_2 \otimes \alpha^{-t})$, we may assume without loss of generality that ω_2 is unitary. Then π_2 is preunitary, and $\alpha^{s_0} \otimes \pi_1$ is imaginary conjugate to π_2. In particular, if W_1 is in $\mathscr{W}(\pi_1; \psi_F)$, then the function $g \mapsto |\det g|^{s_0} \overline{W_1(g)}$ belongs to $\mathscr{W}(\pi_2; \overline{\psi}_F)$. But the function $g \mapsto W_2[\eta g]$ also belongs to $\mathscr{W}(\pi_2; \overline{\psi}_F)$. Therefore b is not identically zero, and s_0 is a pole.

Q.E.D.

COROLLARY (1.3). — *Suppose π is a supercuspidal representation and η is the unramified quadratic character of $F^\times$. If π is not equivalent to $\pi \otimes \eta$, then:*

$$L(s, \pi \times \tilde{\pi}) = (1 - q^{-s})^{-1};$$

otherwise

$$L(s, \pi \times \tilde{\pi}) = (1 - q^{-2s})^{-1}.$$

Proof. — The number s_0 is a pole of $L(s, \pi \times \tilde{\pi})$ if and only if $\pi \otimes \alpha^{s_0} \approx \pi$. But this relation implies $\alpha^{2s_0} = 1$, that is $\alpha^{s_0} = \eta$ or 1. Since the poles of L are simple, the Corollary follows.

For completeness, and future reference, we now give the values of L and ε when π_2 is *not* supercuspidal.

PROPOSITION (1.4). — *Suppose $\pi_2 = \pi(\mu_2, \nu_2)$ (using the notation of* [JL], § 3). *Then:*

(1.4.1)
$$\begin{cases} L(s, \pi_1 \times \pi_2) = L(s, \pi_1 \otimes \mu_2) L(s, \pi_1 \otimes \nu_2), \\ L(s, \tilde{\pi}_1 \times \tilde{\pi}_2) = L(s, \tilde{\pi}_1 \otimes \mu_2^{-1}) L(s, \tilde{\pi}_1 \otimes \nu_2^{-1}) \end{cases}$$

and

$$\varepsilon(s, \pi_1 \times \pi_2; \psi) = \varepsilon(s, \pi_1 \otimes \mu_2; \psi_F) \varepsilon(s, \pi_1 \otimes \nu_2; \psi_F).$$

Suppose π_2 is the special representation $\sigma(\mu_2, \nu_2)$ with $\mu_2 \nu_2^{-1} = \alpha$ but π_1 is not special (loc. cit.). *Then:*

(1.4.2)
$$L(s, \pi_1 \times \pi_2) = L(s, \pi_1 \otimes \mu_2),\ L(s, \tilde{\pi}_1 \otimes \tilde{\pi}_2) = L(s, \tilde{\pi}_1 \otimes \nu_2^{-1})$$

and

$$\varepsilon(s, \pi_1 \times \pi_2; \psi_F) = \varepsilon(s, \pi_1 \otimes \mu_2; \psi_F)\varepsilon(s, \pi_1 \otimes \nu_2; \psi_F)\frac{L(1-s, \tilde{\pi}_1 \otimes \mu_2^{-1})}{L(s, \pi_1 \otimes \nu_2)}.$$

Finally suppose $\pi_i = \sigma(\mu_i, \nu_i)$, $\mu_i.\nu_i^{-1} = \alpha_F$, $i = 1,2$. *Then:*

$$(1.4.3)\qquad \begin{cases} L(s, \pi_1 \times \pi_2) = L(s, \mu_1 \mu_2) L(s, \nu_1 \mu_2), \\ L(s, \tilde{\pi}_1 \times \tilde{\pi}_2) = L(s, \nu_1^{-1} \nu_2^{-1}) L(s, \mu_1^{-1} \nu_2^{-1}), \end{cases}$$

and

$$\varepsilon(s, \pi_1 \times \pi_2; \psi_F) = \varepsilon(s, \pi_1 \otimes \mu_2; \psi_F)\varepsilon(s, \pi_1 \otimes \nu_2; \psi_F).$$

Here the L *and* ε *factors on the right are those of* [JL].

Proof. — For the first two assertions *see* Theorem 15.1 in [Ja]. We prove the third one since—contrary to what is implied in [Ja]— Theorem 15.1 does not apply to this case.

Recall that $\mathscr{W}(\pi_1; \psi_F)$ is a subspace of codimension one of the space denoted $\mathscr{W}(\mu_1, \nu_1; \psi_F)$ in [JL] (§ 3, p. 94). Although the representation of $G_2(F)$ in this larger space is reducible the assertions recalled in (1.1) still apply to this space. Similarly (1.4.1) and (1.4.2) apply with the obvious modifications. In particular the "g. c. d." of the integrals $\Psi(s, W_1, W_2, \Phi)$ with $W_1 \in \mathscr{W}(\mu_1, \nu_1; \psi_F)$ and $W_2 \in \mathscr{W}(\pi_2; \psi_F)$ is

$$L(s, \pi_2 \otimes \mu_1) L(s, \pi_2 \otimes \nu_1) = L(s, \mu_1 \mu_2) L(s, \nu_1 \mu_2).$$

Similarly the "g. c. d." of the integrals $\Psi(s, \tilde{W}_1, \tilde{W}_2, \Phi)$ with W_1 and W_2 in the same spaces is

$$L(s, \tilde{\pi}_2 \otimes \mu_1^{-1}) L(s, \tilde{\pi}_2 \otimes \nu_1^{-1}) = L(s, \mu_1^{-1} \nu_2^{-1}) L(s, \nu_1^{-1} \nu_2^{-1})$$

and there is a similar assertion for the ε-factor.

We conclude that the quotients

$$L(s, \pi_1 \times \pi_2)/L(s, \mu_1 \mu_2) L(s, \nu_1 \mu_2)$$

and

$$L(s, \tilde{\pi}_1 \times \tilde{\pi}_2)/L(s, \nu_1^{-1} \nu_2^{-1}) L(s, \mu_1^{-1} \nu_2^{-1})$$

are polynomials. To see that they are identically one—as we must— we observe that appropriate functional equations imply that the quotients $L(1-s, \tilde{\pi}_1 \times \tilde{\pi}_2)/L(s, \pi_1 \times \pi_2)$ and

$$L(1-s, \nu_1^{-1} \nu_2^{-1}) L(1-s, \mu_1^{-1} \nu_2^{-1})/L(s, \mu_1 \mu_2) L(s, \nu_1 \mu_2)$$

can differ only by a constant times a power of q^{-s}. But the second quotient has coprime numerator and denominator. Thus the required identities follow.

Remark (1.4.4). — If π is the one-dimensional representation $g \mapsto \mu(\det g)$ then it can be denoted $\pi(\mu\alpha^{1/2}, \mu\alpha^{-1/2})$ as in [JL] (p. 104). Thus one can use (1.4) to define the factors L and ε for *all* pairs of irreducible admissible representations of $G_2(F)$.

(1.5) Recall that (conjecturally) there is a bijection between the set of classes of 2 dimensional representations of the Weil-Deligne group of F and the set of classes of admissible irreducible representations of $G_2(F)$. This bijection, denoted

$$\sigma \mapsto \pi(\sigma),$$

should satisfy the following properties:

$$(1.5.1) \qquad \pi(\sigma)^{\sim} = \pi(\tilde{\sigma}), \qquad \pi(\sigma \otimes \chi) = \pi(\sigma) \otimes \chi,$$

$$(1.5.2) \qquad \det(\sigma) = \text{central quasi-character of } \pi(\sigma),$$

$$(1.5.3) \qquad L(s, \pi(\sigma) \otimes \chi) = L(s, \sigma \otimes \chi),$$

$$(1.5.4) \qquad L(s, \tilde{\pi}(\sigma) \otimes \chi) = L(s, \tilde{\sigma} \otimes \chi),$$

and

$$(1.5.5) \qquad \varepsilon(s, \pi(\sigma) \otimes \chi; \psi) = \varepsilon(s, \sigma \otimes \chi; \psi);$$

(*cf.* [De]); the factors on the left are those of [JL] and the ones on the right are those of [De], [La 2].

According to the conjectures of Langlands one should also have

$$(1.5.6) \qquad L(s, \pi(\sigma) \times \pi(\tau)) = L(s, \sigma \otimes \tau),$$

and

$$(1.5.7) \qquad \varepsilon(s, \pi(\sigma) \times \pi(\tau); \psi_F) = \varepsilon(s, \sigma \otimes \tau; \psi_F).$$

The factors on the left are now the ones described in (1.1) and the ones on the right are again those of [De], [La 2].

We note that (1.5.6) and (1.5.7) follow from (1.4) except when *both* σ and τ are irreducible representations of the Weil-group. In that case the existence of $\pi(\sigma)$ and $\pi(\tau)$ is perhaps not yet known. Nevertheless we have the following proposition (which completes Corollary 19.16 of [Ja]):

PROPOSITION (1.6). – *Regard* F *as the completion at some place v of some global field* K. *Let σ and τ be two-dimensional irreducible representations of the Weil-group* W_F *satisfying the following properties:*

(i) *there exist two dimensional representations Σ and* T *of the Weil-group* W_K *whose compositions with the morphism* $W_F \to W_K$ *are σ and τ;*

(ii) *for each place w of* K *let Σ_w and T_w denote the compositions of Σ and* T *with the morphism* $W_w \to W_K$*; then $\pi(\Sigma_w)$ and $\pi(T_w)$ exist;*

(iii) *the representations $\otimes\, \pi(\Sigma_w)$ and $\otimes\, \pi(T_w)$ are automorphic cuspidal.*

Then (1.5.6) *and* (1.5.7) *hold.*

Proof. – By Corollary 19.16 of [Ja],

$$L(1-s, \tilde{\pi}(\sigma)\times\tilde{\pi}(\tau))\varepsilon(s, \pi(\sigma)\times\pi(\tau), \psi_F)/L(s, \pi(\sigma)\times\pi(\tau))$$
$$= L(1-s, \tilde{\sigma}\otimes\tilde{\tau})\varepsilon(s, \sigma\otimes\tau; \psi_F)/L(s, \sigma\otimes\tau).$$

Thus it suffices to prove (1.5.6). Since there is no harm in assuming σ and τ unitary and irreducible, $\pi(\sigma)$ and $\pi(\tau)$ can also be taken to be preunitary and super cuspidal.

If s_0 is a pole of $L(s, \sigma\otimes\tau)$ then the representation $\alpha^{s_0}\otimes\sigma$ is equivalent to the representation $\tilde{\tau}$ ([JL], Lemma 12.4); thus s_0 is purely imaginary, and the rational fraction

$$L(s, \sigma\otimes\tau)/L(1-s, \tilde{\sigma}\otimes\tilde{\tau})$$

is irreducible in the ring $\mathbf{C}[q^{-s}, q^s]$. But by Proposition (1.2) the same is true of

$$L(s, \pi(\sigma)\times\pi(\tau))/L(1-s, \tilde{\pi}(\sigma)\times\tilde{\pi}(\tau)),$$

and since these fraction differ only by a unit $\mathbf{C}[q^{-s}, q^s]$, they have in fact the same numerator.

Q.E.D.

COROLLARY (1.7). – *Suppose σ and τ are as in* (1.6). *If $\pi(\sigma)\simeq\pi(\sigma)$, then $\sigma\simeq\tau$.*

Proof. – Indeed 0 is a pole of $L(s, \pi(\sigma)\times\tilde{\pi}(\tilde{\tau}))$. Since $\tilde{\pi}(\tau)=\pi(\tilde{\tau})$ and the pair $(\sigma, \tilde{\tau})$ satisfies the assumptions of (1.6) we find that 0 is also a pole of $L(s, \sigma\otimes\tilde{\tau})$. Thus our conclusion follows again from [JL] (Lemma 12.4).

(1.8) Suppose σ and τ are "dihedral representations" of the Weil group W_F. This means that there are separable quadratic extensions K and H of F, quasi-characters χ and θ of $K^\times$ and $H^\times$ so that

(1.8.1) $$\sigma=\mathrm{Ind}(W_F, W_K, \chi) \quad \text{and} \quad \tau=\mathrm{Ind}(W_F, W_H, \theta).$$

Then $\pi(\sigma)$ and $\pi(\tau)$ exist and (1.5.6) and (1.5.7) are satisfied. This is clear by (1.4.1) if either σ on τ is reducible. Otherwise it follows from (1.6).

In particular, if the residual characteristic of F is not 2, the bijection $\sigma\mapsto\pi(\sigma)$ exists and the relations [(1.5.6), (1.5.7)] apply to all pairs of representations.

(1.9) Now we complete the results of [Ja] (§ 20) (although this will not be needed later in the paper). Let K be a separable quadratic extension of F and set $\psi_K=\psi_F\circ\mathrm{Tr}_{K/F}$. Then for any quasi-character θ of $F^\times$,

(1.9.1) $$L(s, \theta\circ N_{K/F})=L(s, \theta)L(s, \theta\zeta),$$

and

1.9.2) $$\lambda(K/F, \psi_F)\varepsilon(s, \theta\circ N_{K/F}; \psi_K)=\varepsilon(s, \theta; \psi_F)\varepsilon(s, \theta\zeta; \psi_F).$$

Here ζ is the quadratic character of $F^\times$ attached to K and $\lambda(K/F, \psi_F)$ is a contant whichs depends on the extension and the choice of ψ_F (*cf.* [JL], p. 6).

If χ is any character of $K^{\times}$ let σ_{χ} be the representation defined by (1.8.1). If σ is any admissible irreducible reprensentation of $G_2(F)$ and π is an admissible irreducible representation of $G_2(K)$ we way that π *is a base change lifting of* σ (a strict lifting in the terminology of [Ja]) if the following conditions are satisfied:

(1.9.3) the central quasi-characters ω and ω' of σ and π satisfy the relation

$$\omega' = \omega \circ N_{K/F};$$

(1.9.4) for any quasi-character χ of $K^{\times}$,

$$L(s, \sigma \times \pi(\sigma_{\chi})) = L(s, \pi \otimes \chi)$$
$$L(s, \tilde{\sigma} \times \tilde{\pi}(\sigma_{\chi})) = L(s, \tilde{\pi} \otimes \chi)$$

and

$$\varepsilon(s, \sigma \times \pi(\sigma_{\chi}); \psi_F) = \lambda(K/F, \psi_F)^2 \varepsilon(s, \pi \otimes \chi; \psi_K).$$

We note that if σ exists at all it is unique. Moreover, suppose $\sigma = \pi(\tau)$ where τ is a 2-dimensional representation of the Weil-Deligne group of F. Suppose also that (1.5.6) and (1.5.7) are true for each pair $(\pi(\sigma_{\chi}), \pi(\tau))$. Let τ' be the restriction of τ to the Weil-Deligne group of K. If $\pi(\tau')$ exists it is clearly a lifting of $\pi(\tau)$. Thus the existence of a lifting is established in all cases except the following: the residual characteristic of F is 2 and π is a supercuspidal representation not of the form $\pi(\sigma_{\chi})$, χ a quasi-character of $K^{\times}$. Nevertheless the following Lemma holds:

LEMMA (1.9.5). — *Suppose σ is a supercuspidal representation of* $G_2(F)$ *not of the form* $\pi(\sigma_{\chi})$. *Suppose π is an irreducible admissible representation of* $G_2(K)$ *satisfying* (1.9.3) *and the following condition:*

$$\varepsilon(s, \sigma \times \pi(\sigma_{\chi}); \psi_F) L(1-s, \tilde{\sigma} \times \tilde{\pi}(\sigma_{\chi}))/L(s, \sigma \times \pi(\sigma_{\chi}),$$
$$= \varepsilon(s, \pi \otimes \chi; \psi_K) L(1-s, \tilde{\pi} \otimes \chi^{-1})/L(s, \pi \otimes \chi).$$

Then π is a supercuspidal representation of $G_2(F)$ *and a lifting of* σ.

Proof. — By Proposition (1.2):

$$L(s, \sigma \times \pi(\sigma_{\chi})) = L(s, \tilde{\sigma} \times \tilde{\sigma}(\sigma_{\chi})) = 1.$$

Thus

$$L(1-s, \tilde{\pi} \otimes \chi^{-1})/L(s, \pi \otimes \chi)$$

is monomial for all χ. Since this can happen only if π is supercuspidal ([JPSS], § 7),

$$L(s, \pi \otimes \chi) = L(s, \tilde{\pi} \otimes \chi^{-1}) = 1,$$

and (1.9.4) is satisfied.

Examples (1.9.6). — In the Table below, the representation on the right is a base change lifting of the representation on the left.

$$\sigma = \pi(\mu_1, \mu_2), \qquad \pi = \pi(\mu_1 \circ N_{K/F}, \mu_2 \circ N_{K/F}),$$

$$\sigma = \sigma(\mu_1, \mu_2), \qquad \mu_1 . \mu_2^{-1} = \alpha_F, \qquad \pi = \sigma(\mu_1 \circ N_{K/F}, \mu_2 \circ N_{K/F}),$$

$\sigma = \pi(\sigma_\chi)$, χ quasi-character of $K^\times$, $\pi = \pi(\chi, \chi')$ where χ' is the conjugate of χ by the non-trivial element of Gal (K/F).

(1.10) The discussion above applies to the case of an archimedean field too. In this case the bijection $\sigma \mapsto \pi(\sigma)$ is known and the factors $L(s, \pi_1 \times \pi_2)$ — defined by (1.5.6) and (1.5.7) — are still related to integrals like (1.1.3) (*cf.* [Ja], § 17, § 18). The notion of a base-change lifting is also defined and the first and third examples of (1.9.6) now cover all cases.

Remark (1.10.1). — For all fields F the representation $\sigma(\mu_1, \mu_2)$ where $\mu_1 . \mu_2^{-1} = \alpha_F$ can be defined as the infinite dimensional component of the induced representation:

$$\mathrm{Ind}(G_2, B_2, \mu_1, \mu_2).$$

Of course if $F = \mathbf{R}$ or $\mathbf{C}$ then $\sigma(\mu_1, \mu_2) = \pi(\lambda)$ for a suitable λ.

2. GL (2) × GL (2): global theory

(2.1) In this section F will denote a global field and $\psi = \prod_v \psi_v$ a non-trivial additive character of A/F. By π_i, $i = 1,2$, we denote an irreducible admissible representation of (the Hecke algebra of) $G_2(A)$, with central quasi-character ω_i. We assume ω_i is trivial on $F^\times$. Then $\pi_i = \otimes \pi_{i,v}$ and we set:

$$L(s, \pi_1 \times \pi_2) = \prod_v L(s, \pi_{1,v} \times \pi_{2,v}), \tag{2.1.1}$$

$$L(s, \tilde{\pi}_1 \times \tilde{\pi}_2) = \prod_v L(s, \tilde{\pi}_{1,v} \times \tilde{\pi}_{2,v}), \tag{2.1.2}$$

$$\varepsilon(s, \pi_1 \times \pi_2) = \prod_v \varepsilon(s, \pi_{1,v} \times \pi_{2,v}; \psi_v). \tag{2.1.3}$$

For almost all v the representation π_i, v is unramified and has the form

$$\pi_{i,v} = \pi(\mu_{i,v}, \nu_{i,v}), \ \mu_{i,v}(x) = |x|_v^{s_{i,v}}, \ \nu_{i,v}(x) = |x|^{t_{i,v}}. \tag{2.1.4}$$

We assume also that there is a constant c such that

$$-c \leqq \mathrm{Re}(s_{i,v}) \leqq c, \ -c \leqq \mathrm{Re}(t_{i,v}) \leqq c \tag{2.1.5}$$

for almost all v. Note that this condition is always satisfied (with $c = 1/2$) if π_1 and π_2 are preunitary. In any case (2.1.5) implies (2.1.1) and (2.1.2) converge absolutely in some half-plane $\mathrm{Re}(s) > s_0$. As for (2.1.3), it is actually independent of ψ and almost all its factors are equal to 1; in particular, as a function of s, it is just a constant times an exponential function of s.

THEOREM (2.2). — (1) *Suppose* π_1 *and* π_2 *are automorphic. Then the functions* $L(s, \pi_1 \times \pi_2)$ *and* $L(s, \tilde{\pi}_1 \times \tilde{\pi}_2)$*—originally defined only in some half-plane—continue as meromorphic functions to* $\mathbf{C}$.

(2) *If* F *is a number field these functions have only finitely many poles and are bounded at infinity in vertical strips. If* F *is a function field whose field of constants has* Q *elements then they are rational functions of* Q^{-s}.

(3) *These functions satisfy the functional equation*

$$L(s, \pi_1 \times \pi_2) = \varepsilon(s, \pi_1 \times \pi_2) L(1-s, \tilde{\pi}_1 \times \tilde{\pi}_2).$$

(4) *If* π_1 *is cuspidal but* π_2 *is not then* $L(s, \pi_1 \times \pi_2)$ *is entire. If* π_1 *and* π_2 *are cuspidal then the poles of* $L(s, \pi_1 \times \pi_2)$ *are simple. Moreover* s_0 *is a pole if and only if*

$$\alpha^{s_0} \otimes \pi_1 \simeq \tilde{\pi}_2 \qquad \text{or} \qquad \alpha^{1-s_0} \otimes \tilde{\pi}_1 = \pi_2.$$

Proof. — Suppose π_2 is not cuspidal. Then by Theorem (10.10) of [JL] there are two quasi-characters μ, ν of $F_{\mathbf{A}}^{\times}/F^{\times}$ such that for all v, $\pi_{2,v}$ is a component of the representation $\rho(\mu_v, \nu_v)$ of $G_{2,v}$ induced by the quasi-character of $B_{2,v}$ given by the pair (μ_v, ν_v) (*loc. cit.*). Actually, if $\rho(\mu_v, \nu_v)$ is irreducible then $\pi(\mu_v, \nu_v) = \rho(\mu_v, \nu_v)$; if not, then $\rho(\mu_v, \nu_v)$ has two irreducible components $\pi(\mu_v, \nu_v)$ and $\sigma(\mu_v, \nu_v)$. If μ_v, ν_v are unramified then $\pi(\mu_v, \nu_v)$ is unramified. In any case, *for almost all* v,

$$(2.2.5) \qquad \begin{cases} \pi_{2,v} = \pi(\mu_v, \nu_v), \\ L(s, \pi_{1,v} \times \pi_{2,v}) = L(s, \pi_{1,v} \otimes \mu_v) L(s, \pi_{1,v} \otimes \nu_v), \\ L(s, \tilde{\pi}_{1,v} \times \tilde{\pi}_{2,v}) = L(s, \tilde{\pi}_{1,v} \otimes \mu_v^{-1}) L(s, \tilde{\pi}_{1,v} \otimes \nu_v^{-1}), \end{cases}$$

and

$$\varepsilon(s, \pi_{1,v} \times \pi_{2,v}; \psi_v) = \varepsilon(s, \pi_{1,v} \otimes \mu_v; \psi_v) \varepsilon(s, \pi_{1,v} \otimes \nu_v; \psi_v).$$

On the other hand, for all v, the ratios

$$L(s, \pi_{1,v} \times \pi_{2,v})/L(s, \pi_{1,v} \otimes \mu_v) L(s, \pi_{1,v} \otimes \nu_v),$$
$$L(s, \tilde{\pi}_{1,v} \times \tilde{\pi}_{2,v})/L(s, \tilde{\pi}_{1,v} \otimes \mu_v^{-1}) L(s, \tilde{\pi}_{1,v} \otimes \nu_v^{-1})$$

are entire and

$$\begin{aligned} &\varepsilon(s, \pi_{1,v} \times \pi_{2,v}; \psi_v) L(1-s, \tilde{\pi}_{1,v} \times \tilde{\pi}_{2,v})/L(s, \pi_{1,v} \times \pi_{2,v}) \\ &= \varepsilon(s, \pi_{1,v} \otimes \mu_v; \psi_v) L(1-s, \tilde{\pi}_{1,v} \otimes \mu_v^{-1})/L(s, \pi_{1,v} \otimes \mu_v) \\ &\times \varepsilon(s, \pi_{1,v} \otimes \nu_v; \psi_v) L(1-s, \tilde{\pi}_{1,v} \otimes \nu_v^{-1})/L(s, \pi_{1,v} \otimes \nu_v), \end{aligned}$$

as can be checked directly. Thus our assertion follows from the known analytic properties of $L(s, \pi_1 \otimes \mu)$ and $L(s, \pi_1 \otimes \nu)$. (Corollary 11.2 *in* [JL].)

Suppose now both π_1 and π_2 are cuspidal. By Theorem 19.14 of [Ja] it suffices to prove (4). For $W_i \in \mathscr{W}(\pi_i; \psi)$ (notations of [JL], § 10 and § 11) and Φ a Schwartz-Bruhat function on $\mathbf{A}^2$, define an integral $\Psi(s, W_1, W_2, \Phi)$ by (1.1.3) the integral now being on $N_2(\mathbf{A})\backslash G_2(\mathbf{A})$. These integrals converge in some right half-space and $L(s, \pi_1 \times \pi_2)$ is a linear combinaison of integrals Ψ. Conversely, at least for Φ in a certain dense subset, each integral Ψ is equal to the L-factor times an entire function of s.

Now set

$$\varphi_i(g) = \sum_{\alpha \in F^\times} W_i\left[\begin{pmatrix} \alpha & 0 \\ 0 & 1 \end{pmatrix} g\right].$$

Then φ_i is a cusp form which belongs to a space realizing the representation π_i. Consider also for each quasi-character ω of $F_{\mathbf{A}}^\times/F^\times$ the Eisenstein series $E(g, \Phi, \omega, s)$ defined by

$$E(g, \Phi, \omega, s) = \Sigma f(\gamma g), \qquad \gamma \in B_2(F)\backslash G_2(F),$$

$$f(g) = \int_{F_{\mathbf{A}}^\times} \Phi[(0, t) g] \, |t|^{2s} \omega(t) \, d^\times t.$$

Then, for Res large enough,

$$\Psi(s, W_1, W_2, \Phi) = \int \varphi_1(g) \varphi_2(g) E(g, \Phi, \omega_1 \omega_2, s) \, dg,$$

$$g \in Z_2(\mathbf{A}) G_2(F)\backslash G_2(\mathbf{A}).$$

(*cf.* [Ja], § 19.) This shows the left hand side is meromorphic. Moreover, if s_0 is a pole of the left hand-side, it is a pole of the Eisenstein-series. Thus s_0 is a simple pole and $\alpha^{2-2s_0} = \omega_1 \omega_2$ or $\alpha^{2s_0} = \omega_1^{-1} \omega_2^{-1}$. In the first case

$$\lim_{s \to s_0} (s - s_0) \Psi(s, W_1, W_2, \Phi)$$
$$= c_1 \Phi(0, 0) \int |\det g|^{s_0 - 1} \varphi_1(g) \varphi_2(g) \, dg, \quad g \in G_2(F) Z_2(\mathbf{A})\backslash G_2(\mathbf{A}).$$

In the second case

$$\lim_{s \to s_0} (s - s_0) \Psi(s, W_1, W_2, \Phi)$$
$$= c_2 \Phi(0, 0) \int |\det g|^{s_0} \varphi_1(g) \varphi_2(g) \, dg, \quad g \in G_2(F) Z_2(\mathbf{A})\backslash G_2(\mathbf{A}).$$

Here c_1, c_2 are non zero constants.

Arguing as in the local case we find first that any pole s_0 of $L(s, \pi_1 \times \pi_2)$ is simple and such that $\alpha^{2s_0} = (\omega_1 \omega_2)^{-1}$ or $\alpha^{2-2s_0} = (\omega_1 \omega_2)^{-1}$. If s_0 is any number such that $\alpha^{2s_0} = (\omega_1 \omega_2)^{-1}$ it will be a pole of L if and only if the bilinear form

$$(\varphi_1, \varphi_2) \mapsto \int |\det g|^{s_0} \varphi_1(g) \varphi_2(g) \, dg$$

is not identically zero, that is, $\alpha^{s_0} \otimes \pi_1$ is equivalent to $\tilde{\pi}_2$. Conversely if s_0 is any number such that $\alpha^{s_0} \otimes \pi_1 \simeq \tilde{\pi}_2$ then $\alpha^{s_0} = (\omega_1 \omega_2)^{-1}$ and the above argument applies. The other case is handled similarly.

(2.3) We can now complete the global results of [Ja] (§ 20). Suppose π is an admissible irreducible representation of the (Hecke algebra) of $G_2(F_A)$ and K is a separable quadratic extension of F. Then an admissible irreducible representation Π of $G_2(K_A)$ is said to be a *base-change lifting* of π if the following conditions are satisfied:

(2.3.1) Suppose v is a place of F which splits in K, with w_1, w_2 the corresponding places of K, so that $F_v \simeq K_{w_1} \simeq K_{w_2}$ and $G_{2,v} \simeq G_{2,w_1} \simeq G_{2,w_2}$; then $\pi_v \simeq \Pi_{w_1} \simeq \Pi_{w_2}$.

(2.3.2) Suppose v is a place of F which does not split in K, and w is the place of K above v, so that K_w is a quadratic extension of F_v; then Π_w is a base change lifting of π_v in the sense of (1.9) above.

Suppose χ is a character of $K_A^\times/K^\times$. If v is a place of F which does not split let $\pi_v = \pi(\sigma_{\chi_w})$ [using the notation of (2.3.2)]; otherwise let $\pi_v = \pi(\chi_{v_1}, \chi_{v_2})$ [notation of (2.3.1)]. Then set $\pi(\sigma_\chi) = \otimes \pi_v$. It is clear that $\pi(\sigma_\chi)$ admits a lifting Π; indeed $\Pi_w = \pi(\chi_w, \chi'_w)$ where χ' is the conjugate of χ under the action of the Galois group. One can show that both $\pi(\sigma_\chi)$ and Π are automorphic; ([La 5]). If π is automorphic cuspidal but not of the form $\pi(\sigma_\chi)$ then the following is true:

PROPOSITION (2.3.3). — *Let π be an automorphic cuspidal representation of $G_2(F_A)$ not of the form $\pi(\sigma_\chi)$. Then π admits a base change lifting Π to $G_2(K_A)$ and Π is automorphic cuspidal.*

Proof. — Let S be the set (finite and possibly empty) of places v of F of residual characteristic 2 where π_v is supercuspidal but not of the form $\pi(\sigma_\chi)$ [here χ is a quasi-character of $K_w^\times$ and we use the notations of (2.3.2)]. If v does not split and is not in S then π_v admits a lifting Π_w. If v is in S (and does not split) then by Theorem (20.6) of [Ja] we at least know there exists a representation Π_w satisfying the conditions of Lemma (1.9.5). So by this same Lemma, Π_w is actually a lifting of π_w and we conclude π has a lifting Π. Since we may assume π to be preunitary it is easily checked that Π is preunitary and the Π_w are infinite-dimensional. Moreover (*cf.* (20.6) of [Ja]),

$$L(s, \Pi \otimes \chi) = L(s, \pi \times \pi(\sigma_\chi)),$$

$$L(s, \tilde{\Pi} \otimes \chi^{-1}) = L(s, \tilde{\pi} \times \tilde{\pi}(\sigma_\chi)),$$

and

$$\varepsilon(s, \Pi \otimes \chi) = \varepsilon(s, \pi \times \pi(\sigma_\chi)).$$

It follows now from (2.2) that $L(s, \Pi \otimes \sigma_\chi)$ is entire, bounded in vertical strips if F is a number field, and satisfies

$$L(s, \Pi \otimes \chi) = \varepsilon(s, \Pi \otimes \chi) L(1-s, \tilde{\Pi} \otimes \chi^{-1}).$$

Thus our conclusion follows from [JL] (Th. 11.3).

Remark (2.3.4). — From the "strong multiplicity" one theorem for G_2 it follows that Π is indeed the base change lifting of π as defined in [La 4].

3. The notion of lifting

Henceforth, σ will be used to denote a typical representation of GL (2), and π a representation of GL (3).

In this Section we shall introduce the notion of lifting a representation σ of GL (2) to GL (3). We do this in terms of the L-functions

$$L_2(s, \sigma, \chi) = \frac{L(s, (\sigma \otimes \chi) \times \tilde{\sigma})}{L(s, \chi)}.$$

(3.1) Let F denote a local field and $\psi_F = \psi$ a non-trivial additive character of F. For each irreducible admissible representation σ of G_2 (F), set

$$L_2(s, \sigma, \chi) = L(s, (\sigma \otimes \chi) \times \tilde{\sigma})/L(s, \chi) \tag{3.1.1}$$

and

$$\varepsilon_2(s, \sigma, \chi; \psi) = \varepsilon(s, (\sigma \otimes \chi) \times \tilde{\sigma}; \psi)/\varepsilon(s, \chi; \psi). \tag{3.1.2}$$

We note that the factors on the right side have been defined for all σ and all F [(1.4.4) and (1.10)].

Definition 3.1.3. — *Let* π *be an admissible irreducible representation of*

$$G_3(F) = GL(3, F)).$$

We shall say that π *is a lift of* σ *if the following conditions are satisfied:*

(i) *the central quasi-character of* π *is trivial;*

(ii) $\pi \simeq \tilde{\pi}$;

(iii) *for any quasi-character* χ *of* $F^\times$*;*

$$L(s, \pi \otimes \chi) = L_2(s, \sigma, \chi), \qquad \varepsilon(s, \pi \otimes \chi; \psi) = \varepsilon_2(s, \sigma, \chi; \psi).$$

We note that if the last condition is satisfied for one choice of ψ, then by (i) it is satisfied for all choices of ψ. Moreover if π is a lift of σ, it is also a lift of $\sigma \otimes \chi$ for any χ and, in particular, of $\tilde{\sigma} \simeq \sigma \otimes \omega^{-1}$, ω denoting the central quasi-character of σ.

PROPOSITION 3.2. — *Suppose* F *is archimedean. Then any* σ *admits a lift* π, *unique up to equivalence.*

Proof. — Let us denote by $\tau \mapsto \pi(\tau)$ the "natural" bijection between the semi-simple representations of degree n of the Weil-group W_F and the irreducible admissible representations of $G_n(F)$. Write accordingly

(3.2.1) $$\sigma = \pi(\tau), \qquad \deg(\tau) = 2.$$

Then:

(3.2.2) $$\tau \otimes \tilde{\tau} = \lambda \oplus 1, \qquad \text{where} \quad \deg(\lambda) = 3.$$

Set

(3.2.3) $$\pi = \pi(\lambda).$$

We contend that π is a lift of σ. Indeed

(3.2.4) $$(\tau \otimes \tilde{\tau})^{\sim} = \tilde{\tau} \otimes \tau = \tau \otimes \tilde{\tau}.$$

Moreover $\det(\tau \otimes \tilde{\tau}) = 1$. Therefore

(3.2.5) $$\det(\lambda) = 1,$$

and—since the central quasi-character of $\pi(\lambda)$ is $\det(\lambda)$—we see that condition (i) is satisfied. From (3.2.4) one concludes that

(3.2.6) $$\lambda^{\sim} = \lambda.$$

So since $\pi(\lambda^{\sim}) = \pi(\lambda)^{\sim}$, we also see that condition (ii) is satisfied. Now note

(3.2.6) $$L(s, (\pi \otimes \chi) \times \tilde{\pi}) = L(s, (\tau \otimes \chi) \otimes \tilde{\tau}),$$
$$\varepsilon(s, (\pi \otimes \chi) \times \tilde{\pi}; \psi) = \varepsilon(s, (\tau \otimes \chi) \otimes \tilde{\tau}; \psi).$$

But

(3.2.7) $$(\tau \otimes \chi) \otimes \tilde{\tau} = (\tau \otimes \tilde{\tau}) \otimes \chi = (\lambda \otimes \chi) \oplus \chi.$$

Thus:

(3.2.8) $$L(s, (\tau \otimes \chi) \otimes \tilde{\tau}) = L(s, \lambda \otimes \chi) L(s, \chi),$$
$$\varepsilon(s, (\tau \otimes \chi) \otimes \tilde{\tau}; \psi) = \varepsilon(s, \lambda \otimes \chi; \psi) \varepsilon(s, \chi; \psi).$$

So comparing (3.2.6), (3.2.8), (3.1.1), and (3.1.2) we see that condition (iii) is also satisfied.

The uniqueness follows from the following Lemma, whose tedious but straightforward proof is left to the reader.

LEMMA (3.2.9), — *Suppose that* λ *and* λ' *are semi-simple representations of degree* 3 *of* W_F. *Suppose moreover that* $\det(\lambda) = \det(\lambda')$,

$$L(s, \lambda \otimes \chi) = L(s, \lambda' \otimes \chi), \qquad L(s, \tilde{\lambda} \otimes \chi) = L(s, \tilde{\lambda}' \otimes \chi),$$

and

$$\varepsilon(s, \lambda \otimes \chi; \psi) = \varepsilon(s, \lambda' \otimes \chi; \psi) \quad \textit{for all } \chi.$$

Then $\lambda' \simeq \lambda$.

In the non-archimedean case we will eventually manage to prove a result analogous to Proposition 3.2. For the time being, however, we content ourselves with the following weaker result.

PROPOSITION 3.3. — *Suppose* F *is non-archimedean.*

(1) *If a lift of* σ *exists, it is unique up to equivalence.*

(2) *If* σ *is not an "extraordinary representation", then* σ *admits a lift.*

(3) *If* σ *is extraordinary and admits a lift* π, *then* π *is cuspidal.*

Proof. — The first assertion follows from Definition (3.1.3), and Lemma (7.5.3) of [JPSS].

Recall that an extraordinary representation is a cuspidal representation which is not of the form $\pi(\tau)$ where τ is a two-dimensional induced representation of W_F. If σ is not extraordinary, then $\sigma = \pi(\tau)$ where τ is a two-dimensional representation of the Weil-Deligne group. Then (3.2.4) to (3.2.5), as well as (3.2.7) and (3.2.8) still hold. In principle, the lift π of σ is given by (3.2.3). But since $\pi(\lambda)$ has not yet been defined, we give now an explicit description of the lift of σ.

Suppose first that τ is a representation of the Weil group. Suppose moreover that

$$(3.3.4) \qquad \tau = \mu_1 \oplus \mu_2, \quad \mu_i = \chi_i \alpha^{t_i}, \quad \text{with} \quad t_i \in \mathbf{R}, \quad \chi_i \bar{\chi}_i = 1.$$

In other words, suppose $\sigma = \pi(\mu_1, \mu_2)$. Then:

$$(3.3.5) \qquad \mu_1 . \mu_2^{-1} = \chi_1 \chi_2^{-1} \alpha^{t_1 - t_2}, \qquad \mu_2 . \mu_1^{-1} = \chi_2 . \chi_1^{-1} \alpha^{t_2 - t_1}.$$

Let us form the induced representation

$$(3.3.6) \qquad \xi = \operatorname{Ind}(G_3, B_3; \mu_1 . \mu_2^{-1}, 1, \mu_2 . \mu_1^{-1}).$$

If $t_1 - t_2 = 0$, ξ is a unitary irreducible (principal series) representation. We contend then that $\pi = \xi$ is a lift of σ. Indeed the central quasi-character of ξ is $\mu_1 . \mu_2^{-1} . \mu_2 \mu_2^{-2} = 1$. Now

$$(3.3.7) \qquad \tilde{\xi} = \operatorname{Ind}(G_3, B_3; \mu_2 . \mu_1^{-1}, 1, \mu_1 . \mu_2^{-1}),$$

and this representation has the same character as ξ. Thus $\xi = \tilde{\xi}$ or $\tilde{\pi} = \pi$. Finally we have to check that, with the notations of (3.2.2),

$$(3.3.8) \qquad L(s, \pi \otimes \chi) = L(s, \lambda \otimes \chi) \quad \text{and} \quad \varepsilon(s, \pi \otimes \chi; \psi) = \varepsilon(s, \lambda \otimes \chi; \psi).$$

Since $\lambda = \mu_2 . \mu_1^{-1} \oplus 1 \oplus \mu_1 . \mu_2^{-1}$ this follows from [Go Ja] [Theorem (3.4)].

Now suppose $t_1 - t_2 \neq 0$. We may assume $t_1 > t_2$ and then form the representation (3.3.6). It may fail to be irreducible, but it always admits a maximal subrepresentation ξ'.

The quotient representation $\pi = \xi/\xi'$ is then irreducible (*see* [Si] for instance) and again we contend that π is a lift of σ. As before, (i) is satisfied. Now $\tilde{\pi}$ can also be obtained as the unique irreducible quotient of

$$\rho = \mathrm{Ind}(G_3, {}^tB_3; \mu_2.\mu_1^{-1}, 1, \mu_1.\mu_2^{-1}). \tag{3.3.9}$$

(*Cf.* [J 1] for instance). If

$$w = \begin{pmatrix} 0 & 0 & 1 \\ 0 & -1 & 0 \\ 1 & 0 & 0 \end{pmatrix}$$

then $g \mapsto wgw^{-1}$ takes ${}^tB$ to B and ρ to ξ. We conclude that $\pi \simeq \tilde{\pi}$. Finally, since (3.3.8) has also been established in [J 1] [Prop. (3.4)], (iii) is satisfied.

Now suppose that

$$\tau = \mathrm{Ind}(W_F, W_K, \chi), \tag{3.3.10}$$

where K is a separable quadratic extension of F and χ is a quasi-character of K^x, i. e., $\sigma = \pi(\sigma_\chi)$. Let also χ' be the quasi-character of χ conjugate to χ by the action of the nontrivial element of the Galois group. Then:

$$\tau = \mathrm{Ind}(W_F, W_K, \chi'), \qquad \tilde{\tau} = \mathrm{Ind}(W_F, W_K, \chi^{-1}), \qquad \text{and} \qquad \tilde{\tau} \mid W_K = \chi^{-1} \oplus \chi'^{-1}.$$

Thus:

$$\begin{aligned} \tau \otimes \tilde{\tau} &\simeq \mathrm{Ind}(W_F, W_K, \chi \otimes (\tilde{\tau} \mid W_K)) = \mathrm{Ind}(W_F, W_K, 1 \oplus \chi.\chi'^{-1}) \\ &= \mathrm{Ind}(W_F, W_K, 1) \oplus \mathrm{Ind}(W_F, W_K, \chi.\chi'^{-1}) \\ &= 1 \oplus \eta \oplus \mathrm{Ind}(W_F, W_K, \chi.\chi'^{-1}), \end{aligned} \tag{3.3.11}$$

where η is the quadratic character attached to K, i. e.,

$$\tau \otimes \tilde{\tau} = \lambda \oplus 1, \qquad \text{where } \lambda = \mu \oplus \eta \qquad \text{and} \qquad \mu = \mathrm{Ind}(W_F, W_K, \chi.\chi'^{-1}). \tag{3.3.12}$$

But $\chi.\chi'^{-1}$ is a character. Thus μ is unitary. Moreover $(\chi.\chi'^{-1})' = (\chi.\chi'^{-1})^{-1}$, and

$$\tilde{\mu} = \mathrm{Ind}(W_F, W_K, (\chi.\chi'^{-1})^{-1}) = \mathrm{Ind}(W_F, W_K, (\chi.\chi'^{-1})') = \mu. \tag{3.3.13}$$

The representation $\pi(\mu)$ is therefore unitary and isomorphic to its contragredient. Its central character is η and it is either in the principal series or cuspidal. It follows that the induced representation

$$\pi = \mathrm{Ind}(G_3, P; \pi(\mu), \eta), \tag{3.3.14}$$

where P is the parabolic subgroup of type (2, 1), is unitary irreducible. (*Cf.* for instance [JPSS], § 6.) We contend it is a lift of σ. Indeed its central quasi-character is $\eta^2 = 1$, and since $\tilde{\eta} = \eta$, and $\tilde{\pi}(\mu) = \pi(\mu)$, we get $\pi = \tilde{\pi}$. Finally (*cf.* [J 1]):

$$L(s, \pi \otimes \chi) = L(s, \pi(\mu) \otimes \chi) L(s, \eta\chi) = L(s, \mu \otimes \chi) L(s, \eta\chi) = L(s, \lambda \otimes \chi).$$

Since a similar relation is true for ε, (3.3.8) [and hence condition (iii)] is satisfied.

Of course all our discussion so far applies to the archimedean case as well. Suppose now τ is a 2-dimensional representation of the Weil-Deligne group which is not a representation of the Weil group. Then σ is special, that is of the form

$$\sigma(\chi\alpha^{1/2}, \chi\alpha^{-1/2}) = \chi \otimes \sigma(\alpha^{1/2}, \alpha^{-1/2}).$$

Let π be the Steinberg representation of $G_3(F)$, that is, the square integrable component of

$$\mathrm{Ind}(G_3, B_3; \alpha, 1, \alpha^{-1}).$$

That π satisfies (i) and (ii) is well known. As for condition (iii), it follows at once from Theorem 7.11 of [Go J] and (1.4.3) of this paper.

Next suppose σ is an extraordinary representation. We want to show first that

$$L_2(s, \sigma, \chi) = 1.$$

For this we appeal to the fact that $\sigma \otimes \chi \not\simeq \sigma$ if $\chi \neq 1$. (*Cf.* [La 4], Lemma 5.16.) Thus s_0 is a pole of $L(s, (\sigma \otimes \chi) \times \tilde{\sigma})$ if and only if $\alpha^{s_0}\chi = 1$, that is, s_0 is a pole of $L(s, \chi)$. Our conclusion then follows. This being so, if π is a lift of σ, we get

$$L(s, \pi \otimes \chi) = L(s, \tilde{\pi} \otimes \chi) = 1,$$

and it is easily checked that this can happen only if π is cuspidal. (*Cf.* [JPSS], § 7).

(3.4) Suppose σ is unitary, but not one dimensional, the field F being archimedean or not. For our purposes, it will be important to determine when π is unitary "generic"; this means (*cf.* [JPSS], § 6 and § 10) that the restriction of π to the subgroup

(3.4.1) $$P^1 = \left\{ \begin{pmatrix} a & b & u \\ c & d & v \\ 0 & 0 & 1 \end{pmatrix} \right\}$$

is equivalent to the irreducible representation of P^1 induced by the character θ of N_3 defined by

(3.4.2) $$\theta \begin{pmatrix} 1 & x & z \\ 0 & 1 & y \\ 0 & 0 & 1 \end{pmatrix} = \psi(x+y).$$

If F is non-archimedean, σ is extraordinary, and σ admits a lift π, then π is cuspidal and has a trivial central quasi-character; thus π is unitary generic (*cf.* [JPSS]). Similarly, if F is non-archimedean and σ is special, its lift is special and again unitary generic (*loc. cit.*).

The remaining discussion will apply to the archimedean and non-archimedean case as well. We write $\sigma = \pi(\tau)$ where τ is a two-dimensional representation of W_F. If τ has the form (3.3.10), then π is given by (3.3.14) which is irreducible generic (*loc. cit.*). If τ has the form (3.3.4) and $t_1 = t_2$ then π is given by (3.3.6) which is irreducible generic. So it remains only to examine the case where π is in the complementary series, that is τ has the form (3.3.4) where χ is a character and $t_1 = -t_2 = t$, $0 < t < 1/2$. Then (3.3.6) takes the form

$$\xi = \mathrm{Ind}(G_3, B_3; \alpha^{2t}, 1, \alpha^{-2t}).$$

Now the maximal subrepresentation ξ' of ξ is nothing but the kernel of the intertwining operator

$$f \mapsto \int f(wng)\,dn, \qquad n \in N_3.$$

It is then easy to determine ξ' and π. One finds that ξ is irreducible unless $t = 1/4$ in which case

$$\pi = \mathrm{Ind}(G_3, P; 1_2, 1)$$

where 1_2 is the trivial representation of G_2 and 1 the trivial representation of $F^\times$. The representation π is then unitary but not generic.

If $0 < t < 1/4$, then the representation $\pi(\alpha^{2t}, \alpha^{-2t})$ is unitary in the complementary series. Since ξ is irreducible,

$$\pi = \xi = \mathrm{Ind}(G_3, B_3; \alpha^{2t}, \alpha^{-2t}, 1) = \mathrm{Ind}(G_3, P, \pi(\alpha^{2t}, \alpha^{-2t}), 1)$$

and this shows that π unitary generic (*loc. cit.*) If $1/4 < t < 1/2$ then $\pi = \xi$ is not unitary.

(3.5) Suppose F is non-archimedean and σ "quasi-unramified". This means that $\sigma = \sigma_0 \otimes \chi$ where σ_0 is unramified, that is contains the trivial representation of $K_2 = GL(2, R_F)$. Then $\sigma = \pi(\alpha^{s_1}\chi, \alpha^{s_2}\chi) = \pi(\tau)$, where $\tau = \alpha^{s_1}\chi \oplus \alpha^{s_2}\chi$. As we have seen, σ admits a lift $\pi = \pi(\lambda)$ where $\lambda = \alpha^{s_1-s_2} \oplus 1 \oplus \alpha^{s_2-s_1}$. More precisely, π is the only irreducible quotient of

$$\xi = \mathrm{Ind}(G_3, B_3; \alpha^{s_1-s_2}, 1, \alpha^{s_2-s_1}),$$

and π contains the trivial representation of $K_3 = GL(3, R_F)$, that is, π is unramified. Note that if χ is also unramified then $\chi = \alpha^t$ and

(3.5.1) $$L_2(s, \sigma, \chi) = L(s, \pi \otimes \chi) = [(1-q^{-s-t+s_1-s_2})(1-q^{-s-t})(1-q^{-s-t+s_2-s_1})]^{-1}$$

and if ψ has exponent zero,

(3.5.2) $$\varepsilon(s, \sigma, \chi; \psi) = \varepsilon(s, \pi \otimes \chi; \psi) = 1.$$

(3.6) Let now F be an A-field and σ an admissible irreducible representation of $G_2(\mathbf{A})$ (or rather of its Hecke algebra). Then we say that an irreducible admissible representation π of $G_3(\mathbf{A})$ is a *lift* of σ if, for all places v, π_v is a lift of σ_v. One of the purposes of this paper is to show that if σ is automorphic cuspidal then σ admits a lift π and π is automorphic (although perhaps not cuspidal).

(3.7) There is already one case we can dispose of immediately. Indeed suppose σ is automorphic cuspidal and there is a character χ of $F_{\mathbf{A}}^{\times}/F^{\times}$ so that $\chi \neq 1$ and $\sigma \otimes \chi \simeq \sigma$. Then $\chi^2 = 1$, and χ determines a quadratic extension H of F. According to the main result of [LL] there is a quasi-character Ω of $H_{\mathbf{A}}^{\times}/H^{\times}$ so that σ is the automorphic representation attached to Ω. More precisely, if $\chi_v \neq 1$ then v does not split in H and $\sigma_v = \pi(\tau_v)$ with $\tau_v = \mathrm{Ind}(W_{F_v}, W_{H_w}, \Omega_w)$ and w the unique place of H above v. Then the lift π_v of σ_v is given by

$$\pi_v = \mathrm{Ind}(G_{3v}, P_v; \pi'_v, \chi_v),$$

where

$$\pi'_v = \pi(\tau'_v),\ \tau'_v = \mathrm{Ind}(W_{F_v}, W_{H_w}, \Omega_w.\Omega'^{-1}_w),$$

and Ω'_w is the quasi-character conjugate to Ω_w. On the other hand, if $\chi_v = 1$, then v splits and, if w_1, w_2 are the two places of H above v, we have

$$\sigma_v = \pi(\tau_v) \qquad \text{with} \quad \tau_v = \Omega_{w_1} \oplus \Omega_{w_2}.$$

The lift π_v of σ_v is then given by

$$\pi_v = \mathrm{Ind}(G_{3,v}, P_v; \pi'_v, \chi_v),$$

where

$$\pi'_v = \pi(\tau'_v),\ \tau'_v = \Omega_{w_1}.\Omega_{w_2}^{-1} \oplus \Omega_{w_2}.\Omega_{w_1}^{-1}.$$

Thus we see that $\pi' = \otimes \pi'_v$ is the representation attached to the character $\Omega.\Omega'^{-1}$ of $H_{\mathbf{A}}^{\times}/H^{\times}$, where Ω' is the conjugate of Ω. If $\Omega.\Omega'^{-1} = \eta \circ N_{K/F}$ where η is some character of $F_{\mathbf{A}}^{\times}/F^{\times}$, then

$$\pi'_v = \pi(\eta_v, \eta_v \chi_v),$$

$$\pi_v = \mathrm{Ind}(G_{3,v}, B_{3,v}; \eta_v, \eta_v \chi_v, \chi_v),$$

and $\pi = \otimes \pi_v$ is automorphic by [La 5]. Otherwise $\pi' = \otimes \pi'_v$ is automorphic cuspidal and again $\pi = \otimes \pi_v$ is automorphic (*loc. cit.*).

(3.8) Suppose σ is arbitrary automorphic cuspidal. Set

(3.8.1) $$L_2(s, \sigma, \chi) = \prod_v L_2(s, \sigma_v, \chi_v)$$

and

(3.8.2) $$\varepsilon_2(s, \sigma_\chi) = \prod_v \varepsilon_2(s, \sigma_v, \chi_v, \psi_v).$$

It follows from (3.5.1) that (3.8.1) converges in some right half-plane. On the other hand almost all factors in (3.8.2) are one, and their product is an exponential function of s which does not depend on ψ. This being so, the analytic properties of $L(s, (\sigma \otimes \chi) \times \sigma)$ and $L(s, \chi)$ imply that $L_2(s, \sigma, \chi)$ continues to a meromorphic function of s and satisfies the functional equation

$$L_2(s, \sigma, \chi) = \varepsilon_2(s, \sigma, \chi) L_2(1-s, \tilde{\sigma}, \chi^{-1}).$$

On the other hand, if π is a lift of σ, then

$$L(s, \pi \otimes \chi) = L_2(s, \sigma, \chi) \quad \text{and} \quad \varepsilon(s, \pi \otimes \chi) = \varepsilon_2(s, \sigma, \chi).$$

4. Complements for GL (3, F) (F archimedean)

This Section collects some technical results about Whittaker models for representations of GL (3, F) when F is archimedean. Thus $F = \mathbf{R}$ or $\mathbf{C}$ in this Section.

(4.1) In [JPSS] there is attached to every unitary *generic* representation π of $G_3(F)$ a space $\mathscr{W}(\pi; \psi_F)$. However, in lifting unitary representations from $G_2(F)$ to $G_3(F)$ irreducible representations arise which are not unitary generic. As we have seen, these are precisely the representations of the form

(4.1.1) $$\pi = \pi(\lambda) \quad \text{where} \quad \lambda = \alpha^t \oplus 1 \oplus \alpha^{-t} \quad \text{and} \quad \frac{1}{2} \leqq t < 1.$$

More explicitly, such a representation is the unique irreducible quotient of the induced representation

(4.1.2) $$\xi = \operatorname{Ind}(G_3, B_3; \alpha^t, 1, \alpha^{-t}).$$

What we are going to do now is define a space $\mathscr{W}(\pi; \psi_F)$ for representations of the form (4.1.1).

Let $\mathscr{V}$ be the space of C^∞ functions f on $G_3(F)$ such that

(4.1.3) $$f\left[\begin{pmatrix} a_1 & x & z \\ 0 & a_2 & y \\ 0 & 0 & a_3 \end{pmatrix} g\right] = |a_1|_F^{t+1} |a_3|^{-t-1} f(g).$$

For f in V set

(4.1.4) $$W_f(g) = \int_{N_3(F)} f(wng) \bar{\theta}(n)\, dn,$$

where θ is defined by (3.4.2) and

$$w = \begin{pmatrix} 0 & 0 & 1 \\ 0 & -1 & 0 \\ 1 & 0 & 0 \end{pmatrix};$$

this integral converges absolutely and uniformly for g in a compact set. We let $\mathscr{W}(\pi; \psi)$ be the space spanned by the functions (4.1.4) for $f \in V$.

LEMMA (4.1.5). – *The map $f \mapsto W_f$ is a bijection.*

Proof. – Suppose $W_f = 0$. Then from (4.1.4) we get

$$\int \psi(-\alpha x)\,dx \int_{U(F)} f\left[w\begin{pmatrix}1 & x & 0\\ 0 & 1 & 0\\ 0 & 0 & 1\end{pmatrix}ug\right]\theta(u)\,du = 0$$

for $\alpha = 1$. Changing g to diag $(\alpha, 1, 1)\,g$ we get the same identity for $\alpha \neq 0$. By continuity it is therefore true for $\alpha = 0$. Thus we get from the uniqueness of the Fourier transform,

$$\int_{U(F)} f[wug]\,\bar{\theta}(u)\,du = 0.$$

Changing g into

$$\begin{pmatrix}1 & 0 & 0\\ \alpha & \beta & 0\\ 0 & 0 & 1\end{pmatrix}^{-1} g$$

we find

$$\iint f\left[w\begin{pmatrix}1 & 0 & z\\ 0 & 1 & y\\ 0 & 0 & 1\end{pmatrix}g\right]\psi(\alpha x + \beta y)\,dz\,dy = 0$$

for all $\alpha \in F$, $\beta \in F^{\times}$. As before, we find $f = 0$, which concludes the proof of (4.1.5).

Note that $\pi \simeq \tilde{\pi}$. Moreover, as in [JPSS], if W is in $\mathscr{W}(\pi; \psi)$, then the function $\tilde{W}$ defined by

(4.1.6) $$\tilde{W}(g) = W[w^t g^{-1}]$$

is still in $\mathscr{W}(\pi; \psi) = \mathscr{W}(\tilde{\pi}; \psi)$; this follows from thef act that the automorphism $g \mapsto w^t g^{-1} w^{-1}$ leaves invariant N_3, B_3, θ, and ξ.

We also denote by $\mathscr{W}_0(\pi; \psi)$ the space of K_3-finite elements in $\mathscr{W}(\pi; \psi)$. It is clear that this space is invariant under right convolution by the elements of the Hecke-algebra and the representation of the Hecke algebra on $\mathscr{W}_0(\pi; \psi)$ is equivalent to ξ; moreover, since π is the unique irreducible quotient of ξ, the representation ξ is semi-simple if and only if it is irreducible. Then $\xi = \pi$ and $t \neq 1/2$. If in addition ξ is unitary generic then $0 < t < (1/2)$. As in [JPSS] we set

(4.1.7) $$\|g\| = (\Sigma g_{ij}^2)^{1/2}\,|\det g|^{-3/2} \quad \text{if } F = \mathbf{R},$$
$$= (\Sigma g_{ij}\cdot\bar{g}_{ij})(\det g\bar{g})^{-3/2} \quad \text{if } F = \mathbf{C}.$$

LEMMA (4.1.8). – *There is a number $r > 0$ such that any element W of $\mathscr{W}_0(\pi, \psi)$ is dominated by $g \mapsto \|g\|^r$. Moreover any element of $\mathscr{W}_0(\pi; \psi)$ is dominated by a gauge.*

Proof. – The notion of gauge was introduced in [JPSS] (§ 8) and the second assertion follows from the first (*loc. cit.*). To prove the first we observe that, given $f \in \mathscr{V}$ and a compact set Ω of $G_3(F)$, there is a $c > 0$ such that

$$|f(gh)| \leqq c f_0(g) \qquad \text{for} \quad h \in \Omega;$$

here f_0 is the element of $\mathscr{V}$ such that $f_0(k) = 1$ for $k \in K_3$. Note that $f_0 > 0$. Thus

$$|W_f(g)| \leqq f_1(g), \qquad f_1(g) = \int f_0(wng)\,dn.$$

Now f_1 is given by

$$f_1\left[\begin{pmatrix} a_1 & x & z \\ 0 & a_2 & y \\ 0 & 0 & a_3 \end{pmatrix} k\right] = |a_1|^{-t+1} |a_3|^{t+1} f_1(e).$$

So $|f_1(g)| \leqq f_1(e) \|g\|^r$, if r is large enough, and our assertion follows.

We want also to replace π by a representation $\pi' = \pi \otimes \chi$ where χ is a character. Then $\tilde{\pi}' = \pi \otimes \chi^{-1}$. We define $\mathscr{W}(\pi'; \psi)$ to be the space spanned by the functions

$$W \otimes \chi : \quad g \mapsto W(g)\chi(\det g), \; W \in \mathscr{W}(\pi; \psi).$$

PROPOSITION (4.1.9). – *Suppose π' is any representation of the form $\pi \otimes \chi$ where π is as in* (4.1.1) *and χ is a character. Then the assertions of theorems* (9.2) *and* (11.2) *of* [JPSS] *apply to the space $\mathscr{W}(\pi'; \psi)$.*

Proof. – Although ξ may fail to be irreducible, the assertions of Theorem (8.7) of [Go Ja] apply to ξ with

$$L(s, \xi \otimes \chi) = L(s, \pi \otimes \chi) \qquad \text{and} \qquad \varepsilon(s, \xi \otimes \chi; \psi) = \varepsilon(s, \pi \otimes \chi; \psi).$$

We note that if W is in $\mathscr{W}_0(\pi; \psi)$ and η is a K_3-finite function on K_3 then the function φ defined by

$$\varphi(g) = \int_{K_3} \eta(k) W(kg)\,dk$$

is a (bi-K_3-finite) matrix coefficient of ξ. Indeed for f in $\mathscr{V}$, set

$$\lambda(f) = \int_{K_3} \eta(k)\,dk \int_{N_3} f(wnk)\bar{\theta}(n)\,dn.$$

It will suffice to show that this is a continuous linear form on $\mathscr{V}$. [Note $\varphi(g) = \lambda(\xi(g)f)$ if $W = W_f$.] Of course $\|f\| = \operatorname{Sup}_{K_3} |f|$ defines the topology of $\mathscr{V}$. So the function f_0 being as in the proof of (4.1.8), we have

$$|f| \leqq f_0 \|f\|$$

and

$$|\lambda(f)| \leqq \|f\| \iint |\eta(k)| f_0(wn)\,dn\,dk = \|f\| \|\eta\|_1 f_1(e).$$

Thus λ is continuous.

Combining these remarks with lemmas (4.1.5) and (4.1.8) it becomes clear that the proof of (9.2) and (11.2) in [JPSS] applies to the case at hand.

5. Product decomposition of the main integral

Our purpose in this section is to decompose into local factors an integral on the metaplectic group involving an Eisenstein series, a theta-function and a cusp form on G_2. Accordingly F will denote an A-field of characteristic not equal to 2, and G will denote the group SL (2) regarded as an algebraic group over F.

(5.1) By Mp (A) we denote the (global) metaplectic group introduced in [We]. This is a group of unitary operators in L^2 (A) which fits into the sequence

(5.1.1) $$1 \to \mathbf{T} \to \mathrm{Mp}(\mathbf{A}) \overset{pr}{\to} \mathrm{G}(\mathbf{A}) \to 1.$$

Here $\mathbf{T} = \{ z \in \mathbf{C} \mid z\bar{z} = 1 \}$ is regarded as the group of unitary scalar operators $\Phi \mapsto \lambda\,\Phi$ in L^2 (A); it is central in Mp (A) and G (A) is (topologically) isomorphic to Mp (A)/T. In other words, the sequence is exact.

If Φ is L^2 (A) and g is in Mp (A) we write $g.\Phi$ for the image of Φ by the unitary operator g. According to paragraph 41 of [We] the sequence (5.1.1) splits over the subgroup G (F), the splitting homomorphism

(5.1.2) $$r_{\mathrm{F}}: \quad \mathrm{G}(\mathrm{F}) \to \mathrm{Mp}(\mathbf{A})$$

being determined by the condition

(5.1.3) $$\sum_{\xi \in \mathrm{F}} (r_{\mathrm{F}}(\gamma)\Phi)(\xi) = \sum_{\xi \in \mathrm{F}} \Phi(\xi) \qquad \text{for} \quad \Phi \in \mathscr{S}(\mathbf{A}), \quad \gamma \in \mathrm{G}(\mathrm{F}).$$

The sequence also splits over $\mathscr{B}$ (A) or, what amounts to the same thing, over $\mathscr{N}$ (A) and $\mathscr{A}$ (A). The corresponding splitting homomorphisms

(5.1.4) $$t: \ \mathscr{N}(\mathbf{A}) \to \mathrm{Mp}(\mathbf{A}) \qquad \text{and} \qquad d: \ \mathscr{A}(\mathbf{A}) \to \mathrm{Mp}(\mathbf{A})$$

are given by

(5.1.5) $$t\begin{pmatrix} 1 & x \\ 0 & 1 \end{pmatrix}.\Phi(y) = \psi\left(\frac{1}{2}xy^2\right)\Phi(y),$$

and

$$d\begin{pmatrix} a & 0 \\ 0 & a^{-1} \end{pmatrix}\Phi(y) = |a|^{1/2}\,\Phi(ay).$$

Note that t agrees with r_{F} on N (F) and d agrees with r_{F} on A (F). Finally, if $w_0.\Phi(\mathrm{X}) = \hat{\Phi}(-x)$ where $\hat{\Phi}$ is the Fourier transform of Φ with respect to ψ, then $\mathrm{pr}(w_0) = w = \begin{pmatrix} 0 & -1 \\ 1 & 0 \end{pmatrix}$ and $r_{\mathrm{F}}(w) = w_0$.

As in [Ge] we say that a function φ on $\mathrm{Mp}(\mathbf{A})$ is *genuine* if $\varphi(\lambda g) = \lambda\varphi(g)$ for all $g \in \mathrm{Mp}(\mathbf{A})$ and $\lambda \in \mathbf{T}$. Thus, if φ_1 and φ_2 are genuine functions on $\mathrm{Mp}(\mathbf{A})$, the product $\varphi_1 . \bar{\varphi}_2$ is invariant under $\mathbf{T}$, and there is a functions f on $\mathrm{G}(\mathbf{A})$ such that $\varphi_1 . \bar{\varphi}_2(g) = f(\mathrm{pr}(g))$. Similarly $|\varphi_1|$ may be regarded as a function on $\mathrm{G}(\mathbf{A})$.

To prove ultimately that cuspidal representations of GL (2) lift to GL (3) we first need to prove the following:

THEOREM 8.1. — *Suppose σ is cuspidal representation of* GL (2, **A**) *and χ is a "highly ramified" character of* $\mathrm{F}^\times\backslash\mathrm{F}^\times_{\mathbf{A}}$ *(see* (5.3) below). *Then the Euler product of degree* 3 *defined by*

$$\mathrm{L}_2(\sigma, \chi, s) = \frac{\mathrm{L}(s, (\sigma\otimes\chi)\times\tilde{\sigma})}{\mathrm{L}(s, \chi)}$$

is entire and bounded in vertical strips of finite width.

The proof of this Theorem will be lengthy. Thus it might help to explain it classical significance as well as the ideas behind its proof. Suppose

$$f(z) = \sum_{n=1}^{\infty} a_n e^{2\pi i n z}$$

is a holomorphic cusp form of weight k and character ω. Suppose also that ψ is a primitive Dirichlet character of $\mathbf{Z}$,

$$\Sigma a_n n^{-s} = \Pi[(1-\alpha_p p^{-s})(1-\beta_p p^{-s})]^{-1},$$

and $\psi\omega(-1) = 1$. Then in [Sh] Shimura proves that the Euler product

$$\mathrm{L}_2(s, f, \psi) = \pi^{-3s/2}\Gamma\left(\frac{s}{2}\right)\Gamma\left(\frac{s+1}{2}\right)\Gamma\left(\frac{s-k+2}{2}\right)$$
$$\times\prod_p[(1-\psi(p)\alpha_p^2 p^{-s})(1-\psi(p)\alpha_p\beta_p p^{-s})(1-\psi(p)\beta_p^2 p^{-s})]^{-1}$$

is holomorphic everywhere except possibly at $s = k$ or $k-1$.

Now suppose that $\mathrm{F} = \mathbf{Q}$, σ is the cuspidal representation of GL (2, **A**) generated by f, and χ is the character $\omega\psi$. Then:

$$\mathrm{L}_2(s+k-1, f, \psi) = \mathrm{L}_2(s, \sigma, \chi) = \frac{\mathrm{L}(s, (\sigma\otimes\chi)\times\tilde{\sigma})}{\mathrm{L}(s, \chi)}.$$

But in Section 9 we characterize those σ whose corresponding L-functions have no poles. Thus our results refine as well as generalize Shimura's.

The idea behind the proof of Theorem 8.1 is to generalize the Rankin-Selberg method used in [Sh]. More precisely, Shimura exploits the fact that $\mathrm{L}_2(s, f, \psi)$ has an integral representation of the form

$$(\star)\qquad \mathrm{L}_2(s, f, \psi) = \iint_{\Gamma_0(\mathrm{N})\backslash\mathrm{H}} f(z)\bar{\theta}(z)\mathrm{E}(z, s)\frac{dx\,dy}{y^2}.$$

Here N is an integer depending on the level of f and ψ,

$$\theta(z) = \sum_{n=-\infty}^{\infty} \psi(n) e^{2\pi i n^2 z},$$

and E (z, s) is a certain real analytic Eisenstein series of "half-integral weight". Since both sides of ($\star$) define holomorphic functions of s for Re (s) sufficiently large, the problem of analytically continuing $L_2(s, f, \psi)$ is therefore reduced to the problem of analytically continuing E (z, s).

To generalize Shimura's method we introduce functions on the metaplectic group Mp (A) which generalize the functions $f(z)$, $\theta(z)$, and E (z, s) in (*). We denote these functions by $\varphi(g)$, $\theta_\Psi(g)$, and E (g, s) respectively. The functions φ, θ_Ψ and E are invariant by $r_F(G(F))$ and our generalization of the integral in ($\star$) is

$$I(s, \chi, F, \Psi, \varphi) = \int_{r_F(G(F))\backslash Mp(A)} \varphi(g)\bar{\theta}_\Psi(g) E(g, s)\, dg.$$

As already indicated, our purpose in this Section is to prove that this "main integral" can be expressed as the product of certain local integrals $I_v(s, F_v, \Psi_v, W_v)$.

(5.2) Let MK denote the subgroup of Mp (A) which projects to the standard maximal compact subgroup K of G (A). Let χ be a character of $F_A^\times/F^\times$ and F a continuous function on $\mathbf{C} \times$ Mp (A). We will often write $F_s(g)$ in place of $F(s, g)$. We assume that for each $g \in$ Mp (A), $s \mapsto F_s(g)$ is holomorphic and F_s is MK-finite on the right, uniformly with respect to s; in other words, there is a MK-finite function ξ on MK such that

$$\int F_s(gk)\xi(k)\, dk = F_s(g),$$

for all s and all g. Finally we assume that

$$F_s\left[\lambda d\begin{pmatrix} a & 0 \\ 0 & a^{-1}\end{pmatrix} t\begin{pmatrix} 1 & x \\ 0 & 1\end{pmatrix} g\right] = \lambda\chi(a)|a|^{s+1/2} F_s(g), \tag{5.2.1}$$

for all $g \in \mathrm{Mp}(\mathbf{A})$, $\lambda \in \mathbf{T}$.

The *Eisenstein series* corresponding to F is

$$E(g, s) = \sum_{B(F)\backslash G(F)} F_s[r_F(\gamma) g]. \tag{5.2.2}$$

As explained above, the function $|F_s|$ may be regarded as a function on G (A). As such, it is majorized by a function h on G (A) which is K-invariant and such that

$$h\left[\begin{pmatrix} a & x \\ 0 & a^{-1}\end{pmatrix} g\right] = |a|^{Re(s)+1/2} h(g).$$

Thus (5.2.2) is dominated by the ordinary Eisenstein series

$$\sum_{B(F)\backslash G(F)} h(\gamma g), \tag{5.2.3}$$

and hence (5.2.2) converges absolutely for $\operatorname{Re} s > (3/2)$, uniformly for $(\operatorname{Re}(s), g)$ in a compact set. Consequently this series defines a continuous function on the set of pairs (s, g) with $\operatorname{Re} s > (3/2)$; the resulting function $s \mapsto E(g, s)$ is holomorphic, and

$$E(r_F(\gamma) g, s) = E(g, s) \quad \text{for} \quad \gamma \in G(F).$$

As on p. 329 of [JL] we may introduce the notion of a slowly increasing (resp. rapidly decreasing) function on $G(F) \backslash G(\mathbf{A})$. If f is a genuine function on $\operatorname{Mp}(\mathbf{A})$ which is invariant under $r_F(G(F))$ on the left, we will say that f is slowly increasing (resp. rapidly decreasing) if $|f|$—regarded as a function on $G(\mathbf{A})$—is slowly increasing (resp. rapidly decreasing). For $\operatorname{Re} s > (3/2)$, it is well known that the series (5.2.3) defines a slowly increasing function on $G(F) \backslash G(\mathbf{A})$. Thus (5.2.2) is, for $\operatorname{Re} s > (3/2)$, a slowly increasing function of g on $r_F(G(F)) \backslash \operatorname{Mp}(\mathbf{A})$. In fact this condition is satisfied uniformly for $\operatorname{Re}(s)$ in a compact set.

Now suppose Ψ belongs to the Schwartz-Bruhat space $\mathscr{S}(\mathbf{A})$ and set

(5.2.4) $$\theta_\Psi(g) = \sum_{\xi \in F} (g.\Psi)(\xi).$$

Since $(\lambda g).\Phi = \lambda (g.\Phi)$ for $\lambda \in T$, this "theta-series" is a genuine function on $\operatorname{Mp}(\mathbf{A})$. By the very definition of r_F (5.1.3), it is also invariant on the left by $r_F(G(F))$. Finally it is slowly increasing (*cf.* [We]). Thus for $\operatorname{Re} s > (3/2)$,

$$g \mapsto \bar{\theta}_\Psi(g) E(g, s)$$

may be regarded as a slowly increasing function on $G(F) \backslash G(\mathbf{A})$.

(5.3) Let σ be an automorphic cuspidal representation of $G_2(\mathbf{A})$ whose central character ω has module one. In particular, σ is unitary. If φ is a (K_2-finite) form belonging to the corresponding space of cusp-forms, then φ is left $G_2(F)$-invariant, rapidly decreasing, and

(5.3.1) $$\varphi\left[\begin{pmatrix} z & 0 \\ 0 & z \end{pmatrix} g\right] = \omega(z) \varphi(g) \quad \text{for} \quad z \in F_{\mathbf{A}}^{\times}, \quad g \in G_2(\mathbf{A}).$$

The main integral we wish to consider is

(5.3.2) $$I(s, \chi, F, \Psi, \varphi) = \int_{G(F) \backslash G(\mathbf{A})} \varphi(h) \bar{\theta}_\Psi(g) E(g, s) \, dg, \; pr(g) = h.$$

If $\operatorname{Re} s > (3/2)$, the series (5.2.2) converges absolutely, and the integral (5.3.2) converges absolutely precisely because φ is rapidly decreasing and $\theta_\Psi E$ is slowly increasing. The convergence is even uniform for $\operatorname{Re}(s)$ on a compact set; thus the integral is a holomorphic function of s in the half-plane $\operatorname{Re}(s) > (3/2)$.

Now let S be the finite set of finite places v of F where σ_v is not quasi-unramified in the sense of paragraph (3.5). Let χ be a character of $F_{\mathbf{A}}^{\times}/F^{\times}$ whose ramification at each place v in S is so high that

(5.3.3) $$L(s, \chi_v) = 1 \quad \text{and} \quad L(s, (\sigma_v \otimes \chi_v) \times \tilde{\sigma}_v) = L(s, (\tilde{\sigma}_v \otimes \chi_v^{-1}) \times \sigma_v) = 1.$$

Given σ and χ satisfying the above conditions we shall show in paragraph 6 that

$$\text{(5.3.4)} \qquad I(s, \chi, F, \Psi, \varphi) = L_2(s, \sigma, \chi),$$

for F, Ψ, and φ appropriately chosen, and Re (s) sufficiently large. We shall also show in paragraphs 7 and 8 that E (g, s) extends to an entire function of s if—in addition—χ^2 is ramified at a least one place. Thus it will easily follow that $L_2(s, \sigma, \chi)$ itself is entire when χ satisfies the above conditions.

In order to establish (5.3.4) we shall need to write the main integral as a product of local ones. Although these integrals are taken over the local groups $G_v = SL(2, F_v)$, we still need to introduce the local metaplectic groups before defining them.

(5.4) Let F now be a *local* field. The (local) *metaplectic group* Mp (F_v) is a group of unitary operators on $L^2(F_v)$ which fits into the exact sequence

$$\text{(5.4.1)} \qquad 1 \to \mathbf{T} \to \mathrm{Mp}(F) \xrightarrow{pr_F} G(F) \to 1.$$

Here $\mathbf{T}$ is again the ordinary torus regarded as a group of operators in $L^2(F)$ and the sequence (5.4.1) again splits over B (F), N (F), and A (F). If $g.\Phi$ again denotes the image of Φ in $L^2(F)$ under g in Mp (F), the splitting homomorphisms $t_F : N(F) \to \mathrm{Mp}(F)$ and $d_F : A(F) \to \mathrm{Mp}(F)$, noted also t and d, are defined by

$$\text{(5.4.2)} \qquad \left[t\begin{pmatrix} 1 & x \\ 0 & 1 \end{pmatrix}\right].\Phi(y) = \psi_F\left[\frac{1}{2}xy^2\right]\Phi(y)$$

and

$$\text{(5.4.3)} \qquad d\begin{pmatrix} a & 0 \\ 0 & a^{-1} \end{pmatrix}.\Phi(y) = |a|^{1/2}\Phi(ay).$$

Also we define w_0 in Mp (F) and w in G_2 (F) by

$$w_0.\Phi(x) = \hat{\Phi}(-x) = \int \Phi(y)\psi_F(-yx)\,dy, \qquad w = \begin{pmatrix} 0 & -1 \\ 1 & 0 \end{pmatrix}$$

so that pr $(w_0) = w$.

If F does not have residual characteristic 2, and if the conductor of ψ_F is $\mathfrak{R}_F$, then the sequence (5.4.1) splits over $K = SL(2, R_F)$. The corresponding homomorphism $r_{\mathfrak{R}_F} : K \to \mathrm{Mp}(F)$ is determined by the condition

$$\int_{\mathfrak{R}_F} r_{\mathfrak{R}_F}(g).\Phi(x)\,dx = \int_{\mathfrak{R}_F} \Phi(x)\,dx \qquad \text{for } \Phi \in \mathscr{S}(F);$$

(*cf.* [We], § 19). In fact if Φ^0 denotes the characteristic function of $\mathfrak{R}_F$, then

$$r_{\mathfrak{R}_F}(g)\Phi^0 = \Phi^0 \quad \text{for all } g \in K,$$

and it is easy to check that

$$r_{\mathfrak{R}_F}(w) = w_0.$$

It is also clear that $r_{\mathfrak{R}}$ coincide with t and d on N (F) $\cap$ K and A (F) $\cap$ K respectively.

(5.5) Now let F be an A-field. For each place v of F we write t_v, d_v, pr_v for t_{F_v}, d_{F_v}, pr_{F_v}; similarly if v is finite, does not have characteristic 2, and ψ_v has exponent zero, we write r_v for $r_{\mathfrak{R}_v}$. Then there is a natural injection

$$(5.5.1) \qquad i_v : \quad G_v \to G(\mathbf{A})$$

of G_v as a pseudo-factor. Similarly there is a homomorphism

$$(5.5.2) \qquad \xi_v : \quad \mathrm{Mp}(F_v) \to \mathrm{Mp}(\mathbf{A})$$

determined by the condition

$$(5.5.3) \qquad \xi_v(g_v).\Phi(x) = g_v.\Phi_v(x_v).\prod_{w \neq v} \Phi_w(x_w)$$

for each $g_v \in \mathrm{Mp}(F_v)$ and $\Phi = \prod_w \Phi_w$ in $\mathscr{S}(\mathbf{A})$. The diagram

$$(5.5.4) \qquad \begin{array}{ccc} \mathrm{Mp}(\mathbf{A}) & \xrightarrow{pr} & G(\mathbf{A}) \\ \xi_v \uparrow & & \uparrow i_v \\ \mathrm{Mp}(F_v) & \xrightarrow{pr_v} & G_v \end{array}$$

is commutative.

Let $h = (h_v)$ in $G(\mathbf{A})$. For each v choose g_v in $\mathrm{Mp}(F_v)$ so that $\mathrm{pr}_v(g_v) = h_v$. If h_v is in K_v, and r_v is defined, we may take $g_v = r_v(h_v)$; let us agree to take this this choice for almost all v. Then there is now exactly one element g of $\mathrm{Mp}(\mathbf{A})$ such that

$$(5.5.5) \qquad g.\Phi(x) = \prod_v g_v.\Phi_v(x_v), \qquad \Phi = \prod_v \Phi_v.$$

Moreover $\mathrm{pr}(g) = h$. We write (g_v) for g [even though $\mathrm{Mp}(\mathbf{A})$ is not a restricted product of the groups $\mathrm{Mp}(F_v)$].

(5.6) We are now ready to return to the main integral (5.3.2). Let us take $\mathrm{Re}(s)$ sufficiently large and compute formally. We can write the integral (5.3.2) as an integral over $\mathbf{T}\, r_F(G(F))\backslash \mathrm{Mp}(\mathbf{A})$. Then we can replace E by its expression (5.2.2) to get

$$\begin{aligned} I(s, \chi, F, \Psi, \varphi) &= \int_{\mathbf{T} r_F(G(F))\backslash \mathrm{Mp}(\mathbf{A})} \varphi(\mathrm{pr}(g)) \Sigma F_s(r(\gamma) g) \bar{\theta}_\Psi(g)\, dg \\ &= \int_{\mathbf{T} r_F(B(F))\backslash \mathrm{Mp}(\mathbf{A})} \varphi(\mathrm{pr}(g)) F_s(g) \bar{\theta}_\Psi(g)\, dg \\ &= \int_{B(F)\backslash G(\mathbf{A})} \varphi(h) F_s(g) \bar{\theta}_\Psi(g)\, dh, \end{aligned}$$

where g is any element which projects to h. But by Iwasawa's decomposition,

$$(5.6.1) \qquad \begin{aligned} I(s, \chi, F, \Psi, \varphi) = \int \varphi &\left[\begin{pmatrix} 1 & x \\ 0 & 1 \end{pmatrix}\begin{pmatrix} a & 0 \\ 0 & a^{-1} \end{pmatrix} k\right] \\ &\times F_s\left[t\begin{pmatrix} 1 & x \\ 0 & 1 \end{pmatrix} d\begin{pmatrix} a & 0 \\ 0 & a^{-1} \end{pmatrix} m\right] \\ &\times \bar{\theta}_\Psi\left[t\begin{pmatrix} 1 & x \\ 0 & 1 \end{pmatrix} d\begin{pmatrix} a & 0 \\ 0 & a^{-1} \end{pmatrix} m\right] dx \frac{d^\times a}{|a|^2}\, dk, \end{aligned}$$

where $m \in \mathrm{Mp}(\mathbf{A})$ projects to k, k is integrated over K, x over $F_{\mathbf{A}}$ and a over $F_{\mathbf{A}}^{\times}$. In view of (5.2.1), this reads:

(5.6.2)
$$\int \mathrm{F}_s[m]\chi(a)|a|^{s-3/2}\,d^{\times}a\,dk \times \int \varphi\left[\begin{pmatrix}1 & x\\ 0 & 1\end{pmatrix}\begin{pmatrix}a & 0\\ 0 & a^{-1}\end{pmatrix}k\right]\bar{\theta}_{\Psi}\left[t\begin{pmatrix}1 & x\\ 0 & 1\end{pmatrix}d\begin{pmatrix}a & 0\\ 0 & a^{-1}\end{pmatrix}m\right]dx.$$

Now we recall that the function

$$x \mapsto \varphi\left[\begin{pmatrix}1 & x\\ 0 & 1\end{pmatrix}g\right]$$

has no constant Fourier coefficient, i. e.,

(5.6.3)
$$\int_{\mathbf{A}/\mathrm{F}} \varphi\left[\begin{pmatrix}1 & x\\ 0 & 1\end{pmatrix}h\right]dx = 0.$$

Moreover, if

(5.6.4)
$$\mathrm{W}(h) = \int \varphi\left[\begin{pmatrix}1 & x\\ 0 & 1\end{pmatrix}h\right]\psi(-x)\,dx,$$

then the ξ-th Fourier coefficient of this function is

(5.6.5)
$$\mathrm{W}\left[\begin{pmatrix}\xi & 0\\ 0 & 1\end{pmatrix}h\right] = \int \varphi\left[\begin{pmatrix}1 & x\\ 0 & 1\end{pmatrix}h\right]\psi(-x\xi)\,dx.$$

Taking (5.1.5) and (5.2.4) into account, we get that—for $h = \mathrm{pr}(g)$—

(5.6.6)
$$\begin{aligned}
&\int \varphi\left[\begin{pmatrix}1 & x\\ 0 & 1\end{pmatrix}h\right]\bar{\theta}_{\Psi}\left[\begin{pmatrix}1 & x\\ 0 & 1\end{pmatrix}g\right]dx\\
&= \int \varphi\left[\begin{pmatrix}1 & x\\ 0 & 1\end{pmatrix}h\right]\sum_{\xi \in \mathrm{F}} \bar{\psi}\left(\frac{1}{2}\xi^2 x\right)\overline{g.\Psi(\xi)}\,dx\\
&= \sum_{\xi \in \mathrm{F}} g.\overline{\Psi}(\xi)\int \varphi\left[\begin{pmatrix}1 & x\\ 0 & 1\end{pmatrix}h\right]\bar{\psi}\left(\frac{1}{2}\xi^2 x\right)dx\\
&= \sum_{\xi \in \mathrm{F}^{\times}} g.\overline{\Psi}(\xi)\,\mathrm{W}\left[\begin{pmatrix}\frac{1}{2}\xi^2 & 0\\ 0 & 1\end{pmatrix}h\right]\\
&= \sum_{\xi \in \mathrm{F}^{\times}} g.\overline{\Psi}(\xi)\,\mathrm{W}\left[\begin{pmatrix}\frac{1}{2}\xi & 0\\ 0 & \xi^{-1}\end{pmatrix}h\right].
\end{aligned}$$

Thus (5.6.2) gives

$$(5.6.7)\qquad I(s,\chi,F,\Psi,\varphi)=\int_K F_s(m)\,dk\int_{F_{\mathbb{A}}^\times/F^\times}\chi(a)\,|a|^{s-3/2}\,d^\times a$$

$$\times\sum_{\xi\in F^\times} d\begin{pmatrix} a & 0\\ 0 & a^{-1}\end{pmatrix} m.\overline{\Psi}(\xi)\,W\left[\begin{pmatrix}\frac{1}{2}\xi a & 0\\ 0 & \xi^{-1}a^{-1}\end{pmatrix}k\right]$$

$$=\int_K F_s(m)\,dk\int_{F_{\mathbb{A}}^\times/F^\times}\chi(a)\,|a|^{s-1}\,d^\times a\sum_{\xi\in F^\times}\overline{m.\Psi(a\xi)}\,W\left[\begin{pmatrix}\frac{1}{2}\xi a & 0\\ 0 & \xi^{-1}a^{-1}\end{pmatrix}k\right]$$

$$(5.6.8)\qquad =\int_{K\times F_{\mathbb{A}}^\times} F_s(m)\,\overline{m.\Psi(a)}\,W\left[\begin{pmatrix}\frac{1}{2}a & 0\\ 0 & a^{-1}\end{pmatrix}k\right]\chi(a)\,|a|^{s-1}\,dk\,d^\times a.$$

Finally, let ω be the central character of σ so that W transforms according to ω. Then:

$$(5.6.9)\qquad I=\int_{K\times F_{\mathbb{A}}^\times} F_s(m)\,\overline{m.\Psi(a)}\,W\left[\begin{pmatrix}\frac{1}{2}a^2 & 0\\ 0 & 1\end{pmatrix}k\right]\chi\omega^{-1}(a)\,|a|^{s-1}\,d^\times a\,dk,\ \mathrm{pr}(m)=k.$$

Before justifying our steps, let us specify F_s. For each place v we select a function $F_{v,s}(g_v)$ on $Mp(F_v)\times \mathbf{C}$ such that

$$(5.6.10)\qquad F_{v,s}\left[t_v\begin{pmatrix}1 & x\\ 0 & 1\end{pmatrix}d_v\begin{pmatrix}a & 0\\ 0 & a^{-1}\end{pmatrix}g_v\right]=\chi_v(a)\,|a|_v^{s+1/2}\,F_{v,s}(g_v).$$

For each g_v in $Mp(F_v)$ the ratio

$$F_{v,s}(g_v)/L(2s,\chi^2)$$

will be an entire function of s, in fact a polynomial in q_v^{-s}, q_v^s if v is finite (with module q_v) and a sum of functions $as^i e^{bs}$ if v is infinite. Moreover, let K_v be the standard maximal compact subgroup of G_v and MK_v its inverse image in $M_p(F_v)$. Then for each s,

$$k_v\mapsto F_{v,s}(k_v)/L(2s,\chi^2)$$

belongs to a fixed finite dimensional right-invariant subspace of continuous functions on MK_v.

Finally, for almost all v, we select

$$(5.6.11)\qquad F_{v,s}(r_v(k_v))=L(2s,\chi^2)\quad \text{if}\quad k_v\in K_v.$$

Then if $g=(g_v)$ as in (5.5.5), we set

$$(5.6.12)\qquad F_s(g)=\prod_v F_s(g_v).$$

Since almost all these factors are equal to $L(2s, \chi_v^2)$, the product in (5.6.12) converges absolutely for Res large enough. In fact it continues to an holomorphic function of s since χ^2 (by assumption) is not principal (it is ramified at some place v). We observe also that

$$F_s(k)/L(2s, \chi^2)$$

is for any $k \in MK$ a sum of functions of the form $as^i e^{bs}$ and for a fixed s belongs to a finite dimensional space of right-invariant functions on MK, independent of s.

Let us recall also that W belongs to the space $\mathscr{W}(\sigma; \psi)$ spanned by functions of the form

$$g \mapsto \prod_v W_v(g_v)$$

where W_v is in $\mathscr{W}(\sigma_v; \psi_v)$ for all v, and for almost all v is the unique element of $\mathscr{W}(\sigma_v; \psi_v)$ invariant under (and equal to one on) $K_{2,v} = GL(2, \mathscr{R}_v)$.

This being so, for $F_{v,s}$ as above, Ψ_v in $\mathscr{S}(F_v)$, and W_v in $\mathscr{W}(\sigma_v; \psi_v)$, we introduce the local integral

(5.6.13)
$$\begin{aligned} I_v &= I(s, F_v, \Psi_v, W_v) \\ &= \int_{K_v \times F_v^\times} |a_v|^{s-1} \chi_v \omega_v^{-1}(a_v) F_{v,s}(m_v) \\ &\quad \times \overline{m_v . \Psi_v(a_v)} W_v\left[\begin{pmatrix} \frac{1}{2} a_v^2 & 0 \\ 0 & 1 \end{pmatrix} k_v\right] dk_v \, d^\times a_v, \\ &\qquad pr_v(m_v) = k_v. \end{aligned}$$

Then we have:

PROPOSITION (5.6.14). – *Suppose* $\Psi = \Pi \Psi_v$, $W = \Pi W_v$, *and* $F_s = \Pi F_{v,s}$ *as above. Then for* Re (s) *large enough, each one of the integrals* (5.6.13) *converges absolutely, their infinite product converges absolutely, and*

$$I(s, F, \Psi, \varphi) = \prod_v I(s, F_v, \Psi_v, W_v).$$

Proof. – We can choose the element m in Mp $(\mathbf{A})$ which projects to $k = (k_v)$ as in (5.5). Then $m = (m_v)$ and $pr_v(m_v) = k_v$. It is then formally clear that (5.6.9) is equal to the product of the integrals (5.6.13). We need however to justify all our steps.

Fix s in a compact set with Re s large. For each place v there is a number c_v and a function $\Psi_v^1 > 0$ in $\mathscr{S}(F_v)$ so that

$$|F_{v,s}(m)(m.\overline{\Psi^1})(a)| \leq c_v |L(2s, \chi_v^2)| \Psi_v^1(a).$$

There is also a function $\Psi_v^2 > 0$ and $t > 0$ independent of v such that

$$\left| W_v\left[\begin{pmatrix} a_v b_v & 0 \\ 0 & b_v \end{pmatrix} k_v\right]\right| \leq |a_v|^{-t} \Psi_v^2(a_v).$$

(*Cf.* [JPSS], § (2.3) and (8.3.3).) Furthermore, we may assume $c_v = 1$ s real, $\chi_v^2 = 1$, and $\Psi_2^1 = \Psi_v^2$ = characteristic function of $\mathfrak{R}_v$ for almost all v. Then the integral (5.6.13) is dominated by

$$c_v \, L(2s, \chi_v^2) \int_{K_v \times F_v^\times} |a_v|^{s-1-t} \Psi_v^1(a) \Psi_v^2\left(\frac{1}{2} a_v^2\right) d^\times a_v \, dk_v.$$

This shows that when Re s is large enough, the integrals (5.6.13) converge absolutely. Similarly the integral on the right hand side of (5.6.9) is dominated by

$$\Pi \, c_v L(2s, \chi^2) \int_{K \times F_{\mathbb{A}}^\times} |a_v|^{s-1-t} \Psi_v^1(a) \Psi_v^2\left(\frac{1}{2} a^2\right) d^\times a \, dk$$

with $\Psi^1 = \Pi \Psi_v^1$, $\Psi^2 = \Psi_v^2$. Thus it converges absolutely and is equal to the product of the local integrals for Re s large enough.

Now we have to prove identity (5.6.9), for Re s large enough. Certainly we may assume

$$|m.\Psi|(x) \leqq \Psi^1(x)$$

for all $m \in$ MK. Then:

$$\left|\theta_\Psi\left[t\begin{pmatrix} 1 & x \\ 0 & 1 \end{pmatrix} d \begin{pmatrix} a & 0 \\ 0 & a^{-1} \end{pmatrix} m\right]\right| \leqq |a|^{1/2} \sum_{\xi \in F} \Psi^1(a\xi).$$

Going back over our computations we see that all our steps will be justified once we show that

$$\int_{A/F \times F_{\mathbb{A}}^\times \times K} |\varphi| \left[\begin{pmatrix} 1 & x \\ 0 & 1 \end{pmatrix} \begin{pmatrix} a & 0 \\ 0 & a^{-1} \end{pmatrix} k\right] \sum_{\xi \in F} \Psi^1(a\xi) |a|^{s-1/2} \, dx \, d^\times a \, dk$$

is finite for s real and large enough. But there is a $c > 0$ such that

$$\sum_{\xi \in F} \Psi^1(a\xi) \leqq c|a|^{-1}.$$

Thus it suffices to show that

$$\int |\varphi| \left[\begin{pmatrix} 1 & x \\ 0 & 1 \end{pmatrix} \begin{pmatrix} a & 0 \\ 0 & a^{-1} \end{pmatrix} k\right] |a|^{s-3/2} \, dx \, d^\times a < +\infty$$

for s real and large. Since φ is bounded and for any r

$$\varphi\left[\begin{pmatrix} 1 & x \\ 0 & 1 \end{pmatrix} \begin{pmatrix} a & 0 \\ 0 & a^{-1} \end{pmatrix} k\right] = 0(|a|^{-r})$$

for $|a|$ large, this is clear.

Q.E.D.

6. Local computations for the main integral: identification with $L_2(s, \sigma, \chi)$

In this Section we prove that

$$(\star) \qquad \mathrm{I}(s, \chi, \mathrm{F}, \Psi, \varphi) = \frac{\mathrm{L}(s, (\sigma \otimes \chi) \times \tilde{\sigma})}{\mathrm{L}(s, \chi)} (= \mathrm{L}_2(s, \sigma, \chi))$$

for Re s sufficiently large, χ "highly ramified", and F_s, Ψ and φ appropriately chosen. But both sides of $(\star)$ decompose as products over the places of F (the left side by Propositions (5.6.14), the right side by definition). Therefore $(\star)$ is actually a local assertion.

(6.1) Throughout this section, F will denote a local field of characteristic not two and ψ_{F} or ψ will denote a fixed non-trivial character of F. If F = **R** we take $\psi(x) = \exp(2i\pi x)$ and if F = **C** we take $\psi(z) = \exp[2i\pi(z+\bar{z})]$. By σ we denote a fixed irreducible unitary generic representation of $\mathrm{G}_2(\mathrm{F})$, ω its central character, and χ a character of $\mathrm{F}^\times$.

If F is non-archimedean and σ is not quasi-unramified we assume that the ramification of χ is so high that

$$(6.1.1) \qquad \mathrm{L}(s, \chi) = \mathrm{L}(s, \chi^{-1}) = 1 \quad \text{and} \quad \mathrm{L}_2(s, \sigma, \chi) = \mathrm{L}_2(s, \sigma, \chi^{-1}) = 1.$$

We will consider functions $\mathrm{F}(s, g) = \mathrm{F}_s(g)$ of $s \in \mathbf{C}$ and $g \in \mathrm{Mp}(\mathrm{F})$ satisfying the following conditions:

$$(6.1.2) \qquad \mathrm{F}_s\left[d\begin{pmatrix} a & 0 \\ 0 & a^{-1} \end{pmatrix} t\begin{pmatrix} 1 & x \\ 0 & 1 \end{pmatrix} g\right] = \chi(a)|a|^{s+1/2}\mathrm{F}_s[g]$$

$$\text{for all } a \in \mathrm{F}^\times, \quad x \in \mathrm{F}, \quad g \in \mathrm{Mp}(\mathrm{F});$$

(6.1.3) there is an entire function $\mathrm{H}(s)$ with values in a fixed, finite dimensional right invariant space of continuous functions on MK such that

$$\mathrm{F}_s(k) = \mathrm{L}(2s, \chi^2)\mathrm{H}(s)(k) \quad \text{for all } k \in \mathrm{MK}.$$

If F = **R** or **C**, $\mathrm{H}(s)$ is a polynomial in s times an exponential factor. If F is non-archimedean $\mathrm{H}(s)$ is a polynomial in q^{-s}, q^s. In fact in all cases but the complex case F_s will have the form

$$(6.1.4) \qquad \mathrm{F}_s(g) = f_s(\mathrm{pr}(g))(g.\Phi)(0),$$

where Φ is in $\mathcal{S}(\mathrm{F})$ (and MK finite) and f_s is a function of $s \in \mathbf{C}$ and $g \in \mathrm{G}(\mathrm{F})$ such that

$$(6.1.5) \qquad f_s\left[\begin{pmatrix} a & x \\ 0 & a^{-1} \end{pmatrix} g\right] = |a|^s \chi(a) f_s(g)$$

and a condition analogous to (6.1.3) holds with MK replaced by K.

For F_s as above, W in $\mathcal{W}(\sigma; \psi)$, and Ψ in $\mathcal{S}(\mathrm{F})$, we set

$$(6.1.6) \qquad \mathrm{I} = \mathrm{I}(s, \mathrm{F}, \Psi, \mathrm{W}) = \int_{\mathrm{N(F)}\backslash\mathrm{G(F)}} \mathrm{F}_s(h)\overline{(h.\Psi)(1)}\,\mathrm{W}\left[\begin{pmatrix} \frac{1}{2} & 0 \\ 0 & 1 \end{pmatrix} g\right] dg,$$

where $\mathrm{pr}(h) = g$.

When F has the form (6.1.4) we also write

$$I = I(s, f, \Phi, \Psi, W).$$

By Iwasawa's decomposition, we have then:

(6.1.7)
$$I = \int_{F^\times \times K} |a|^{s-1} \chi\omega^{-1}(a) f_s(k)(m.\Phi)(0) \times m.\overline{\Psi}(a) W\left[\begin{pmatrix} \frac{1}{2}a^2 & 0 \\ 0 & 1 \end{pmatrix} k\right] dk\, d^\times a, \quad \mathrm{pr}(m) = k.$$

Now we have already observed in (5.6) that such an integral converges for Re (s) sufficiently large. Our purpose in this section is to prove:

(6.2) Proposition. — *For* Re (s) *sufficiently large, and appropriately chosen* F_s *(or* f_s, Φ *if* $F \neq C$), Ψ, *and* W, *the integral* (6.1.6) *equals* $L_2(s, \sigma, \chi)$ *times an exponential factor.*

The exact form of F_s (or f_s and Φ) will be needed for later reference. Moreover, a more precise result will be needed in the "unramified situation" (see below). In the meantime we make two useful remarks.

(6.2.1) *Remark.* — Replace σ by $\sigma \otimes \eta$. Then the L_2-factor does not change. The space $\mathscr{W}(\sigma; \psi)$, however, does change in general; it is replaced by the space spanned by the functions

$$W \otimes \eta : \quad g \mapsto W(g)\eta(\det g), \qquad W \in \mathscr{W}(\sigma; \psi).$$

However, the functions W enters the integral only through their restrictions to the set

$$\begin{pmatrix} \frac{1}{2} & 0 \\ 0 & 1 \end{pmatrix} G(F).$$

It follows that in proving (6.2) we may, without loss of generality, replace σ by $\sigma \otimes \eta$.

(6.2.2) *Remark.* — Replace ψ by ψ' where $\psi'(x) = \psi(ax)$ and $a \neq 0$. Then the group of unitary operators Mp (F) does not change. The projection pr, however, is replaced by another map pr′, and there is a commutative diagram

(6.2.3)
$$\begin{array}{ccc} \mathrm{Mp}(F) & \xrightarrow{pr'} & G(F) \\ {\scriptstyle \mathrm{Id}}\downarrow & & \downarrow \\ \mathrm{Mp}(F) & \xrightarrow{pr} & G(F) \end{array}$$

where the second vertical arrow is the restriction to G (F) of the inner automorphism

(6.2.4)
$$g \mapsto \begin{pmatrix} a & 0 \\ 0 & 1 \end{pmatrix} g \begin{pmatrix} a^{-1} & 0 \\ 0 & 1 \end{pmatrix}$$

of $G_2(F)$. Similarly there is a commutative diagram

$$\begin{array}{ccc} N(F) & \xrightarrow{t} & Mp(F) \\ \downarrow & & \downarrow id \\ N(F) & \xrightarrow{t'} & Mp(F) \end{array} \tag{6.2.5}$$

where t (resp. t') is the splitting morphism defined by (5.4.3) and $\psi_F = \psi$ (resp. $\psi_F = \psi'$), and the first vertical arrow is induced by (6.2.4). Note that (6.2.4) induces the identity on A (F) and d does not depend on ψ. Moreover the inner automorphism (6.2.4) leaves σ invariant and takes the function W in $\mathcal{W}(\pi; \psi)$ to the function

$$g \mapsto W\left[\begin{pmatrix} a & 0 \\ 0 & 1 \end{pmatrix} g \begin{pmatrix} a^{-1} & 0 \\ 0 & 1 \end{pmatrix}\right]$$

which is in $\mathcal{W}(\pi; \psi')$. Thus it follows that in proving (6.2) we may replace ψ by ψ'.

Our proof will now proceed case by case.

(6.3) F *is non-archimedean,* σ *is quasi-unramified,* χ *is unramified.*

By remark (6.2.1) we may assume that σ is actually unramified. Suppose for the time being the residual characteristic of F is not two. By remark (6.2.2) we may also assume that the conductor of ψ is $\mathfrak{R}_F$; the situation at hand is then the "unramified one". Let W_0 be the element of $\mathcal{W}(\sigma, \psi)$ which is one on K_2 and invariant under K_2. Take $\Phi = \Psi$ to be the characteristic function Φ_0 of $\mathfrak{R}_F$. Finally define f by

$$f_s\left[\begin{pmatrix} a & x \\ 0 & a^{-1} \end{pmatrix} k\right] = |a|^s \chi(a) L(2s, \chi^2) \tag{6.3.1}$$

LEMMA (6.3.2) ("unramified situation"). — *Assume the Haar measures of* $F^\times$ *and* K *are so chosen that* $\mathrm{Vol}(\mathfrak{R}^\times) = 1$, $\mathrm{Vol}(K) = 1$. *Then* f_s *being given by* (6.3.1):

$$I(s, f, \Phi_0, \Phi_0, W_0) = L_2(s, \sigma, \chi).$$

Proof. — The extension splits over K and Φ_0 is $r_F(K)$-invariant. On the other hand f_s and W_0 are K-invariant. So from (6.1.7) we get

$$I = L(2s, \chi^2) \int |a|^{s-1} \chi\omega^{-1}(a) W_0 \begin{pmatrix} \frac{1}{2}a^2 & 0 \\ 0 & 1 \end{pmatrix} d^\times a.$$

Now

$$W\begin{pmatrix} ab & 0 \\ 0 & 1 \end{pmatrix} = W\begin{pmatrix} b & 0 \\ 0 & 1 \end{pmatrix} \quad \text{for} \quad a \in \mathfrak{R}^\times.$$

So I is the product of $L(2s, \chi^2)$ and the series

$$\sum_{n \geqslant -\infty} X^n q^n \chi\omega^{-1}(\tilde{\omega}^n) W_0 \begin{pmatrix} \tilde{\omega}^{2n} & 0 \\ 0 & 1 \end{pmatrix}, \quad X = q^{-s}, \tag{6.3.3}$$

and what remains to be shown is that the sum of this series is $L_2(s, \sigma, \chi) . L(2s, \chi^2)^{-1}$.

Since σ is unramified, we have $\sigma = \pi(\mu, \nu)$ for some choice of quasi-characters μ, ν of $F^\times$. Then, by Proposition (1.4),

$$L_2(s, \sigma, \chi) = (1-\mu\nu^{-1}\chi(\tilde{\omega})q^{-s})^{-1}(1-\chi(\tilde{\omega})q^{-s})^{-1}(1-\mu^{-1}\nu\chi(\tilde{\omega})q^{-s})^{-1}.$$

On the other hand ([JL]), Prop. (9.5)):

$$\int_{F^\times} W\begin{pmatrix} a & 0 \\ 0 & 1 \end{pmatrix} |a|^{s-1/2} d^\times a = \Sigma X^n q^{n/2} W\begin{pmatrix} \tilde{\omega}^n & 0 \\ 0 & 1 \end{pmatrix} = L(s, \sigma) = L(s, \mu) L(s, \nu),$$

while

$$L(2s, \chi^2) = (1+\chi(\tilde{\omega})q^{-s})^{-1}(1-\chi(\tilde{\omega})q^{-s})^{-1}.$$

Thus our assertion follows from the following Lemma, whose proof is left to the reader:

LEMMA (6.3.4). — *Let* X *denote an indeterminate. If*

$$\sum_{n \geqq 0} c_n X^n = (1-\alpha X)^{-1}(1-\beta X)^{-1}$$

then:

$$\sum_{n \geqq 0} c_{2n} X^n = (1+\alpha\beta X)(1-\alpha^2 X)^{-1}(1-\beta^2 X)^{-1}.$$

Suppose now F has residual characteristic 2. Let ψ_0 be a character whose conductor is $\mathfrak{R}$. We may assume $\psi(x) = \psi_0(2x)$ [Remark (6.2.2)]. Let f_s, Φ_0, W_0 be as before.

LEMMA (6.3.5). — *There is $c > 0$ so that:*

$$\int_K (m.\Phi_0)(0)\overline{m.\Phi_0}(a)\,dk = c\,\Phi_0(a), \qquad (\mathrm{pr}(m) = k).$$

Proof. — In the integral above let us change m into

$$t\begin{pmatrix} 1 & x \\ 0 & 1 \end{pmatrix}.m$$

with $x \in \mathfrak{R}$ and let us integrate over $\mathfrak{R}$. We find that the integral gets multiplied by

$$\int_{\mathfrak{R}} \psi_0(xa^2)\,dx.$$

Thus the integral vanishes unless a is in $\mathfrak{R}$. For $a = 0$, it has some positive value c. So it will suffice to show that the integral does not depend on a, provided a is in $\mathfrak{R}$.

So assume a is in $\mathfrak{R}$. Then the integrand is invariant under $t(N \cap K)$ on the left. On the other hand, Φ_0 is invariant under $t(N \cap K)$ and $d(A \cap K)$, so the integrand too is invariant on the right for these groups. Thus there are two positive constants c_1, c_2 so that the integral is:

$$(6.3.6) \qquad c_1(w_0.\Phi_0)(0)\overline{w_0.\Phi_0}(a) + c_2\int_{\mathfrak{P}} \left(w_0 t\begin{pmatrix} 1 & z \\ 0 & 1 \end{pmatrix} w_0^{-1}\right).\Phi(0)$$

$$\times\overline{\left(w_0 t\begin{pmatrix} 1 & z \\ 0 & 1 \end{pmatrix} w_0^{-1}\right).\Phi(a)}\,dz \qquad \text{where} \quad \mathrm{pr}(w_0) = \begin{pmatrix} 0 & -1 \\ 1 & 0 \end{pmatrix}.$$

We also have

$$w_0 . \Phi(x) = \hat{\Phi}(-x).$$

Thus:

$$w_0 . \Phi_0(x) = \Phi_0(2x) | 2 |^{1/2}$$

and the first term in (6.3.6) is indeed independent of a. We also have

$$\left(t \begin{pmatrix} 1 & z \\ 0 & 1 \end{pmatrix} w_0^{-1}\right) . \Phi_0(x) = \Phi_0(2x) \psi_0(x^2 z) | 2 |^{1/2}.$$

So the second term can be written as

$$c_2 | 4 | \int_{\mathfrak{P}} dz \int \Phi_0(2x_1) \psi_0(x_1^2 z) \, dx_1 \int \Phi_0(2x_2) \psi_0(x_2^2 z - 2x_2 a) \, dx_2.$$

In this integral $x_2 \in 1/2\, \mathfrak{R}$. So $2 x_2 a$ is in $\mathfrak{R}$ and the integral does not depend on a.

The Lemma being proved, set

$$W(g) = W_0 \left[\begin{pmatrix} 2 & 0 \\ 0 & 1 \end{pmatrix} g \right].$$

Then W is in $\mathscr{W}(\sigma; \psi)$ and invariant under K. From (6.1.7) and (6.3.5) we seet hat the integral $I(s, f, \Phi_0, \Phi_0, W)$ is thus equal to (6.3.3) and our assertion is proved in the case at hand just as in the previous case.

(6.4) F *is nonarchimedean,* σ *is not quasi-unramified, or* χ *is ramified.*

If σ is quasi-unramified and χ ramified then

(6.4.1) $$L(s, (\sigma \otimes \chi) \times \tilde{\sigma}) = L_2(s, \sigma, \chi) = L(s, \chi) = 1.$$

On the other hand recall that if σ is not quasi-unramified then we assume that χ is so highly ramified that (6.4.1) holds. Thus Proposition (6.2) amounts to the assertion that

(6.4.2) $$I(s, f, \Phi, \Psi, W) = 1$$

for appropriately chosen f, Φ, Ψ, and W.

To prove this we introduce the function

(6.4.3) $$H(k, s) = (m . \Phi)(0) \int_{F^\times} | a |^{s-1} \chi\omega^{-1}(a) \overline{(m . \Psi)}(a) \times W\left[\begin{pmatrix} \frac{1}{2} a^2 & 0 \\ 0 & 1 \end{pmatrix} k \right] d^\times a \quad \text{where} \quad \operatorname{pr}(m) = k.$$

Then:

(6.4.4) $$H\left[\begin{pmatrix} \varepsilon & \eta \\ 0 & \varepsilon^{-1} \end{pmatrix} k, s \right] = \chi^{-1}(\varepsilon) H(k, s), \qquad \text{for} \quad \varepsilon \in \mathfrak{R}_F^\times \quad \text{and} \quad \eta \in \mathfrak{R}_F$$

and

$$\text{(6.4.5)}\qquad I(s, f, \Phi, \Psi, W) = \int_{B\cap K\backslash K} H(k, s) f_s(k)\, dk.$$

Choose n so large that χ identically 1 on $1+\mathfrak{P}_F^n$ and $H(k, s)$ is right invariant by the congruence subgroup

$$\text{(6.4.6)}\qquad K' = \left\{\begin{pmatrix} a & b \\ c & d \end{pmatrix} \in K \mid a \equiv d \equiv 1,\ b \equiv c \equiv 0 \bmod \mathfrak{P}^n\right\}.$$

Let also

$$K'' = \left\{\begin{pmatrix} a & b \\ c & d \end{pmatrix} \in K \mid c \equiv 0 \bmod \mathfrak{P}^n\right\}.$$

Then $K'' = (B \cap K).K'$. Define

$$f_s(k) = \begin{cases} \chi(a) & \text{if } k = \begin{pmatrix} a & b \\ c & d \end{pmatrix} \in K'', \\ 0 & \text{otherwise.} \end{cases}$$

For this choice of f_s, it follows from (6.4.4) that

$$I(s, f, \Phi, \Psi, W) = \int_{B\cap K\backslash K''} H(k, s) f_s(k)\, dk = c\,H(e, s), \qquad c > 0.$$

Thus it remains to prove we can choose Φ, Ψ, and W so that $H(e, s)$ is a non-zero constant, So choose $\Psi(a)$ equal to $\chi\omega^{-1}(a)$ if $a \in \mathfrak{R}^\times$ and zero otherwise. We assume $\Phi(0) \neq 0$ and we choose W in such a way that

$$a \mapsto W\begin{pmatrix} \frac{1}{2}a & 0 \\ 0 & 1 \end{pmatrix}$$

is the characteristic function of $\mathfrak{R}^\times$ in $F^\times$ (*cf.* [JL], Lemma (2.16.1)). Then:

$$H(e, s) = \Phi(0) \int_{F^\times} |a|^{s-1} \chi\omega^{-1}(a) \overline{\Psi}(a)\, W\begin{pmatrix} \frac{1}{2}a^2 & 0 \\ 0 & 1 \end{pmatrix} d^\times a = \text{Vol}(\mathfrak{R}^\times)\Phi(0),$$

and we are done.

(6.5) *Real case.*

If $F \neq \mathbf{C}$, Mp (F) reduces to a non-trivial extension of $G(F) = SL(2, F)$ by $\{1, -1\}$. We denote the corresponding two sheeted covering of G (F) by $Sp_2(F)$. When $F = \mathbf{R}$, $Sp_2(\mathbf{R})$ is therefore the unique two-sheeted covering of $G(\mathbf{R})$ given by the theory of Lie-groups. Its Lie-algebra is the same as the Lie-algebra of $G(\mathbf{R})$: it consists of all real two by two matrices of trace zero. Recall $\psi(x) = \exp(2 i \pi x)$.

For our purposes it is convenient to single out the one parameter subgroups of $Sp_2(\mathbf{R})$ corresponding to the Lie algebra elements

(6.5.1) $$X_+ = \begin{pmatrix} 0 & 1 \\ 0 & 0 \end{pmatrix} \quad \text{and} \quad H = \begin{pmatrix} 1 & 0 \\ 0 & -1 \end{pmatrix}.$$

These are the subgroups

(6.5.2) $$N_* = \left\{ t\begin{pmatrix} 1 & x \\ 0 & 1 \end{pmatrix} \,\middle|\, x \in \mathbf{R}, \right\} \quad A_* = \left\{ d\begin{pmatrix} a & 0 \\ 0 & a^{-1} \end{pmatrix} \,\middle|\, a > 0 \right\}.$$

(*Cf.* (5.4.2) and (5.4.3).) Recall also that $K = SO(2, \mathbf{R})$ [resp. $MK \cap Sp_2(\mathbf{R})$] is the subgroup of $G(\mathbf{R})$ [resp. $Sp_2(\mathbf{R})$] corresponding to the Lie algebra element

(6.5.3) $$U = \begin{pmatrix} 0 & -1 \\ -1 & 0 \end{pmatrix}.$$

As defined, the group $Sp_2(\mathbf{R})$ is provided with a unitary representation on $L^2(\mathbf{R})$; the elements of $\mathcal{S}(\mathbf{R})$ are C^∞-vectors and the corresponding action of the Lie-algebra on them is given by

(6.5.4) $$X_+ . \Phi(y) = i\pi y^2 \Phi(y),$$

(6.5.5) $$H.\Phi(y) = y\Phi'(y) + \frac{1}{2}\Phi(y),$$

and

(6.5.6) $$U.\Phi(y) = i\pi y^2 \Phi(y) - i\frac{1}{4\pi}\frac{d^2}{dy^2}\Phi(y).$$

Suppose now Φ is an eigenvector for U, that is,

(6.5.7) $$U.\Phi = in.\Phi$$

or, as we shall say, Φ is of weight n. Then Φ satisfies the second order differential equation

$$\frac{d^2\Phi}{dy^2}(y) = (4\pi^2 y^2 - 4n\pi)\Phi(y).$$

It follows that an orthogonal basis of eigenvectors for U is provided by the classical Hermite functions

(6.5.8) $$\Phi_m(x) = e^{-\pi x^2} H_m(x\sqrt{2\pi}),$$

where H_m denotes the classical Hermite polynomial of degree m. The corresponding weight is

(6.5.9) $$\frac{n}{2} = m + 1/2.$$

Thus if $\Phi = \Phi_m$ we find that

$$\exp(t\,\mathrm{U}).\Phi = \exp\left(i\frac{n}{2}t\right)\Phi.$$

Similarly if φ is a C^∞-function on $\mathrm{G}(\mathbf{R})$ or $\mathrm{G}_2(\mathbf{R})$ let us say it is of weight n if

$$\frac{d}{dt}\varphi[g\exp(t\,\mathrm{U})]\Big|_{t=0} = in\,\varphi(g). \tag{6.5.10}$$

In proving (6.2) we may—without loss of generality—assume $\chi = 1$ or $\chi = \mathrm{sgn}$. Indeed χ can be replaced by $\chi\alpha^t$ by changing s to $s-t$. We may also replace σ by $\sigma\otimes\mu$ and assume (whenever convenient) that the central character ω of σ is either 1 or sgn. Thus the cases we have to examine are the following:

TABLE (6.5.11)

Case	σ	χ	$L_2(s, \sigma, \chi)$
a	$\pi(\mu_1, \mu_2)$ $\mu_2 = \alpha^{\sigma_1}$, $\mu_2 = \alpha^{\sigma_2}$	1	$\Gamma\left(\frac{s}{2}\right)\Gamma\left(\frac{s+\sigma_1-\sigma_2}{2}\right)\Gamma\left(\frac{s+\sigma_2-\sigma_1}{2}\right)$
b	$\pi(\mu_1, \mu_2)$ $\mu_1 = \alpha^{\sigma_1}$, $\mu_2 = \alpha^{\sigma_2}$	sgn	$\Gamma\left(\frac{s+1}{2}\right)\Gamma\left(\frac{s+1+\sigma_1-\sigma_2}{2}\right)\Gamma\left(\frac{s+1+\sigma_2-\sigma_1}{2}\right)$
c	$\pi(\mu_1, \mu_2)$ $\mu_1 = \alpha^{\sigma_1}\,\mathrm{sgn}$ $\mu_2 = \alpha^{\sigma_2}$	1	$\Gamma\left(\frac{s}{2}\right)\Gamma\left(\frac{s+1+\sigma_1-\sigma_2}{2}\right)\Gamma\left(\frac{s+1+\sigma_2-\sigma_1}{2}\right)$
d	$\pi(\mu_1, \mu_2)$ $\mu_1 = \alpha^{\sigma_1}\,\mathrm{sgn}$ $\mu_2 = \alpha^{\sigma_2}$	sgn	$\Gamma\left(\frac{s+1}{2}\right)\Gamma\left(\frac{s+\sigma_1-\sigma_2}{2}\right)\Gamma\left(\frac{s+\sigma_2-\sigma_1}{2}\right)$
e	σ_Ω $\Omega(z) = \dfrac{z^m}{(z\bar{z})^{m/2}}$	1	$\Gamma\left(\frac{s+1}{2}\right)\Gamma(s+m)$
f	σ_Ω $\Omega(z) = \dfrac{z^m}{(z\bar{z}^{\,m/2})}$	sgn	$\Gamma\left(\frac{s}{2}\right)\Gamma(s+m)$

In this Table, the appropriate values of $\mathrm{L}(s, \sigma, \chi)$ are given within an exponential factor (*cf.* § 3 and [Ja], § 17). The function Φ (resp. Ψ) will be one of the functions (6.5.8) and so will be determined by its weight. Similarly f_s will be given by

(6.5.12) $$f_s\left[\begin{pmatrix} a & x \\ 0 & a^{-1}\end{pmatrix}\begin{pmatrix} \cos\theta & \sin\theta \\ -\sin\theta & \cos\theta\end{pmatrix}\right] = \chi(a)\,|\,a\,|^s e^{in\theta}\,\mathrm{L}(2s, \chi^2)\,\mathrm{P}(s),$$

where P is a polynomial; so f_s will be determined by its weight and the polynomial P. Finally, since the restriction of σ to $\mathrm{K} = \mathrm{SO}(2, \mathbf{R})$ decomposes with multiplicity one, W will also—up to a scalar factor—be determined by its weight. Thus the appropriate choice of f, Φ, Ψ, W, and P is given by the following Table:

TABLE (6.5.13)

Case	wgt (f_s)	P	wgt (Φ)	wgr (Ψ)	wgt (W)
a	0	1	1/2	1/2	0
b	-1	s	5/2	3/2	0
c	0	1	1/2	3/2	1
d	-1	1	1/2	1/2	1
e	0	1	1/2	$m+3/2$	$m+1$
f	1	1	1/2	$m+1/2$	$m+1$

In each case we have

$$\mathrm{wgt}(f_s) + \mathrm{wgt}(\Phi) - \mathrm{wgt}(\Psi) + \mathrm{wgt}(\mathrm{W}) = 0.$$

Thus formula (6.1.7) reduces to

(6.5.14) $$\mathrm{I}(s, f, \Phi, \Psi, \mathrm{W}) = \mathrm{I} = \Phi(0)\,\mathrm{L}(2s, \chi^2)\,\mathrm{P}(s)$$
$$\times \int_0^\infty |\,a\,|^{s-1}\chi\omega^{-1}(a)\,\overline{\Psi}(a)\,\mathrm{W}\begin{pmatrix} \frac{1}{2}a^2 & 0 \\ 0 & 1\end{pmatrix} d^\times a.$$

Since $\Phi(0) \neq 0$, up to an exponential factor this is

(6.5.15) $$\mathrm{L}(2s, \chi^2)\,\mathrm{P}(s)\int_0^\infty |\,a\,|^{(1/2)(s-1)}\chi\omega^{-1}(\sqrt{a})\,\overline{\Psi}(\sqrt{2a})\,\mathrm{W}\begin{pmatrix} a & 0 \\ 0 & 1\end{pmatrix} d^\times a.$$

We treat this integral case by case in (6.6) and (6.7).

(6.6) *Cases a to d.*

We need to compute the Mellin transform of the functions involved in (6.5.15). For $a \mapsto \overline{\Psi}(\sqrt{2a})$ this is easy enough. Indeed, up to scalar factors,

(6.6.1) $$\overline{\Psi}(\sqrt{2x}) = \begin{cases} \exp(-2\pi x) & \text{cases } a \text{ and } d, \\ x\exp(-2\pi x) & \text{cases } b \text{ and } c. \end{cases}$$

Thus:

$$(6.6.2)\quad \int_0^{+\infty} \Psi(\sqrt{2}a)\,|a|^{s-1/2}\,d^\times a \begin{cases} \pi^{-s}\Gamma\left(\dfrac{s+(1/2)}{2}\right)\Gamma\left(\dfrac{s-(1/2)}{2}\right) & \text{cases } a,\ d, \\ \pi^{-s-1}\Gamma\left(\dfrac{s+(3/2)}{2}\right)\Gamma\left(\dfrac{s+(1/2)}{2}\right) & \text{cases } b \text{ and } c. \end{cases}$$

To compute the other Mellin transform, recall that for any infinite dimensional representation σ of the form $\pi(\mu_1, \mu_2)$, the space $\mathscr{W}(\sigma;\psi)$ can be described in the following way: for each Φ in $\mathscr{S}(\mathrm{F}^2)$ of the form $\exp(-\pi(x^2+y^2))\,\mathrm{P}(x, y)$ — where P is a polynomial — set

$$(6.6.3)\qquad f(g) = f_\Phi(g) = \int \Phi[(0, t)g]\,\mu_1\mu_2^{-1}(t)\,|t|\,d^\times t\,\mu_1(\det g)\,|\det g|^{1/2}.$$

Then the elements of $\mathscr{W}(\sigma;\psi)$ are exactly the functions of the form

$$(6.6.4)\qquad \mathrm{W}(g) = \int f\left[w\begin{pmatrix}1 & x\\ 0 & 1\end{pmatrix}g\right]\psi(-x)\,dx.$$

The integral may be divergent but may be given a meaning as in [JL] (§ 5). In particular

$$(6.6.5)\qquad \mathrm{W}\begin{pmatrix}t & 0\\ 0 & 1\end{pmatrix} = \int \Phi'[tx, x^{-1}]\,\mu_1.\mu_2^{-1}(x)\,\mu_1(t)\,|t|^{1/2}\,d^\times x,$$

where

$$(6.6.6)\qquad \Phi'(x, y) = \int \Phi(x, v)\,\psi[-yv]\,dv$$

(in the notation of [JL], $\mathrm{W} = \mathrm{W}_{\Phi'}$).

Thus:

$$(6.6.7)\qquad \int_0^{+\infty} \mathrm{W}\begin{pmatrix}t & 0\\ 0 & 1\end{pmatrix}|t|^{s-1/2}\,d^\times t$$

$$= \frac{1}{2}\left[\int_{-\infty}^{+\infty} \mathrm{W}\begin{pmatrix}t & 0\\ 0 & 1\end{pmatrix}|t|^{s-1/2}\,d^\times t + \int_{-\infty}^{+\infty} \mathrm{W}\begin{pmatrix}t & 0\\ 0 & 1\end{pmatrix}|t|^{s-1/2}\frac{t}{|t|}\,d^\times t\right]$$

$$= \frac{1}{2}\iint \Phi'(x, y)\,\mu_1(x)\,|x|^s\,\mu_2(y)\,|y|^s\,s^\times x\,d^\times y$$

$$+ \frac{1}{2}\iint \Phi'(x, y)\,\mu_1(x)\,|x|^s\frac{x}{|x|}\times\mu_2(y)\,|y|^s\frac{y}{|y|}\,d^\times x\,d^\times y.$$

To get the right weight we take

$$\Phi(x, y) = \begin{cases} \exp(-\pi(x^2+y^2)) & \text{in cases } a,\ b, \\ \exp(-\pi(x^2+y^2))(x+iy) & \text{in cases } c,\ d. \end{cases}$$

Then:

$$\Phi'(x, y) = \begin{cases} \exp(-\pi(x^2+y^2)) & \text{in cases } a, b, \\ \exp(-\pi(x^2+y^2))(x+y) & \text{in cases } c, d \end{cases}$$

and, up to scalar factors

$$\int_0^{+\infty} W\begin{pmatrix} a & 0 \\ 0 & 1 \end{pmatrix} |a|^{s-1/2} d^\times a = \pi^{-(s+\sigma_1)/2}\Gamma\left(\frac{s+\sigma_1}{2}\right)\pi^{-(s+\sigma_2)/2}\Gamma\left(\frac{s+\sigma_2}{2}\right)$$

in cases a, b

$$= \pi^{-(s+\sigma_1+1)/2}\Gamma\left(\frac{s+\sigma_1+1}{2}\right)\Gamma\left(\frac{s+\sigma_2}{2}\right)$$

$$+\pi^{-(s+\sigma_2+1)/2}\Gamma\left(\frac{s+\sigma_1}{2}\right)\Gamma\left(\frac{s+\sigma_2+1}{\sigma}\right)$$

in cases c, d.

To continue we use a Barnes Mellin Lemma as formulated on page 77 of [Ja]. The result then follows immediately.

(6.7) *Cases e and f.*

This is a new situation, since σ belongs to the "discrete series". More precisely, it is the representation in the discrete series with central character $(\mathrm{sgn})^{m+1}$ and lowest positive weight $m+1$. In particular $\mathscr{W}(\sigma; \psi)$ contains—up to a scalar factor—exactly one element W of weight $m+1$. It is given by

$$W\begin{pmatrix} a & 0 \\ 0 & 1 \end{pmatrix} = \begin{cases} e^{-2\pi a} a^{(m+1)/2} & \text{if } a > 0, \\ 0 & \text{if } a < 0. \end{cases}$$

Then:

$$\int_0^{+\infty} W\begin{pmatrix} a & 0 \\ 0 & 1 \end{pmatrix} |a|^{s-1/2} d^\times a = (2\pi)^{-s}\Gamma\left(s+\frac{m}{2}\right);$$

(*cf.* Section 5 of [JL]).

Now consider case (e) where $\chi = 1$. Then $\omega = (\mathrm{sgn})^{m+1}$,

$$\Phi(x) = e^{-\pi x^2} \quad \text{and} \quad \Psi(x) = e^{-\pi x^2} H_{m+1}(x\sqrt{2\pi}).$$

Our main integral reduces to

$$I = L(2s, 1)\int_{-\infty}^{+\infty} \left(\frac{a}{|a|}\right)^{m+1} |a|^{s-1} W\begin{pmatrix} \frac{1}{2}a^2 & 0 \\ 0 & 1 \end{pmatrix} e^{-\pi a^2} H_{m+1}(a\sqrt{2\pi})\, d^\times a.$$

But

$$H_{m+1}(-a) = (-1)^{m+1} H_{m+1}(a).$$

Therefore:

$$I = 2L(2s, 1)\int_0^{+\infty} a^{s-1} W\left[\begin{pmatrix} \frac{1}{2}a^2 & 0 \\ 0 & 1 \end{pmatrix}\right] \times e^{-\pi a^2} H_{m+1}(a\sqrt{2}\pi)\, d^\times a.$$

So replacing W by its expression gives

$$I = L(2s, 1)\mathscr{I}_{m+1}(s),$$

with

$$\mathscr{I}_k(s) = \int_0^\infty a^{s+k-1} e^{-a^2} H_k(a)\, d^\times a.$$

(Recall that we are ignoring exponential factors).

To continue, we need to compute $\mathscr{I}_k(s)$. This is done by integrating by parts using the Rodriguez formula

$$\frac{d^n}{dx^n}(e^{-x^2}) = (-1)^n e^{-x^2} H_n(x).$$

The end result is that

$$\mathscr{I}_k(s) = (s+k-2)(s+k-3)\ldots s\Gamma\left(\frac{s+1}{2}\right).$$

Therefore:

$$I = \Gamma(s)(s+m-1)(s+m-2)\ldots s\Gamma\left(\frac{s+1}{2}\right)$$

$$= \Gamma(s+m)\Gamma\left(\frac{s+1}{2}\right) = L_2(s, \sigma, \chi).$$

Case (f) is entirely similar and we leave it to the reader.

(6.8) *Complex case.*

In this case Mp (C) is a trivial extension of SL (2, C) by **T**. In other words, there is a splitting homomorphism $r : G(\mathbf{C}) \to Mp(\mathbf{C})$ which provides an (ordinary) representation (also denoted r) of G (C) on $L^2(\mathbf{C})$. In fact if we set

$$L_+ = \{f \in L^2(\mathbf{C}) \mid f(x) = f(-x)\},$$

$$L_- = \{f \in L^2(\mathbf{C}) \mid f(x) = -f(-x)\},$$

then:

$$L^2(\mathbf{C}) = L_+ \oplus L_-$$

and the spaces L_+, L_- are invariant and irreducible under G (C). For our purposes it is convenient to extend the resulting representations r_+ and r_- to G_2 by taking them to be trivial on

$$\left\{\begin{pmatrix} t & 0 \\ 0 & t \end{pmatrix} \middle| \ t>0\right\}.$$

Then:

$$r_+ = \pi(\alpha_{\mathbf{C}}^{1/4}, \alpha_{\mathbf{C}}^{-1/4}), \qquad r_- = \pi(\xi, 1), \qquad \text{where} \quad \xi(z) = z(z\bar{z})^{-1/2};$$

(*cf.* [Ge], p. 95).

Suppose Φ is in L_+ (resp. L_-) and K-finite. Then Φ is actually in $\mathscr{S}(\mathbf{C})$. Set

$$W^*(g) = \left(r\left[\begin{pmatrix} 2 & 0 \\ 0 & 1 \end{pmatrix} g\right].\Phi\right)(1).$$

Then:

$$W^*\left[\begin{pmatrix} 1 & x \\ 0 & 1 \end{pmatrix} g\right] = \psi(x) W^*[g].$$

Moreover

$$W^*\begin{pmatrix} z & 0 \\ 0 & 1 \end{pmatrix} = 0(1) \qquad \text{for} \quad |z| \to +\infty.$$

Finally, the space spanned by the $W^{*\prime}s$ realizes the representation r_+ (resp. r_-) of the Hecke algebra. It follows that W^* is arbitrary in $\mathscr{W}(r_+; \psi_{\mathbf{C}})$, [resp. $\mathscr{W}(r_-; \psi_{\mathbf{C}})$].

Since the extension splits, we may regard F_s as a function on $G(\mathbf{C})$ such that

$$F_s\left[\begin{pmatrix} a & x \\ 0 & a^{-1} \end{pmatrix} g\right] = \chi(a) |a|_{\mathbf{C}}^{s+1/2} F_s(g).$$

Our local integral may be written then as an integral over $N(\mathbf{C})\backslash G(\mathbf{C})$:

$$\begin{aligned} I &= I(s, F, \Psi, W) \\ &= \int_{N(F)\backslash G(F)} F_s(g) \overline{r(g).\Psi(1)}\, W\left[\begin{pmatrix} \frac{1}{2} & 0 \\ 0 & 1 \end{pmatrix} g\right] d^\times g \\ &= \int_{N(F)\backslash G(F)} F_s(g) \overline{W^*}\left[\begin{pmatrix} \frac{1}{2} & 0 \\ 0 & 1 \end{pmatrix} g\right] W\left[\begin{pmatrix} \frac{1}{2} & 0 \\ 0 & 1 \end{pmatrix} g\right] d^\times g. \end{aligned}$$

It remains to select F_s. We take

$$F_s(g) = L(s, \chi) \int_{\mathbf{C}^\times} \Phi[(0, t) g] |t|^{s+1/2} \chi(t)\, d^\times t,$$

where:

$$\Phi(z_1, z_2) = \exp(-2\pi(z_1 \bar{z}_1 + z_2 \bar{z}_2)) P(z_1, \bar{z}_1, z_2, \bar{z}_2),$$

and P is a polynomial.

Then, for $k \in K = SU(2)$,

$$F_s(k) = L(s, \chi) L\left(s + \frac{1}{2}, \chi\right) Q(s)(k)$$

where Q is a polynomial with values in the space of K-finite functions on K. Since

$$L(2s, \chi^2) = L(s, \chi) L\left(s+\frac{1}{2}, \chi\right)$$

up to exponential factors, F_s has the required analytic properties (6.1.3).

Now we want to write I as an integral over $N(\mathbf{C})\backslash GL(2, \mathbf{C})$. To that end, recall that without loss of generality we may replace σ by $\sigma \otimes \mu$. But this replaces the central character ω of σ by $\omega\mu^2$. Therefore we may assume that we are in one of the following two cases:

Case	$(\chi\omega^{-1})$	χ	Ψ is in
g..........	1	$\chi = \omega$	L^2_+
h..........	-1	$\chi = \omega\xi$, $\xi(z) = z(z\bar{z})^{-1/2}$	L^2_-

Then in case (g) [resp. (h)] W^* is in $\mathscr{W}(r_+; \psi)$ [resp. $\mathscr{W}(r_-; \psi)$] and transforms under the central character $\omega_+ = 1$ of r_+ (resp. $\omega_- = \xi$ of r_-). Since $\omega\overline{\omega_+} = \chi$ (resp. $\omega\overline{\omega_-} = \chi$) and $G_2(\mathbf{C}) = \mathbf{C}^x G(\mathbf{C})$ we can write

$$I = L(s, \chi) . \int_{N(\mathbf{C})\backslash GL(2, \mathbf{C})} \overline{W^*}\left[\begin{pmatrix} \frac{1}{2} & 0 \\ 0 & 1 \end{pmatrix} g\right] W\left[\begin{pmatrix} \frac{1}{2} & 0 \\ 0 & 1 \end{pmatrix} g\right] |\det g|^{(s/2)+(1/4)} \Phi[(0, 1) g] \, d^x g.$$

Finally $\overline{W^*}$ has the form

$$\overline{W^*}(g) = W_2[\eta g], \qquad \eta = \operatorname{diag}(-1, 1),$$

with W_2 in $\mathscr{W}(\bar{r}_+; \psi)$ [resp. $\mathscr{W}(\bar{r}_-; \psi)$]. In other words

$$I = L(s, \chi) \Psi\left(\frac{s}{2}+\frac{1}{4}, W_1, W_2, \Phi\right).$$

(Notations of § 1.) By [Ja], Proposition 17.4 (*see* also § 1), for appropriate W_1, W_2 and Φ:

$$\Psi(s, W_1, W_2, \Phi) = L(s, \sigma \times \bar{r}_+) \qquad [\text{resp. } L(s, \sigma \times \bar{r}_+)].$$

Taking into account the explicit expansion for all the L-factors concerned, and the duplication formula, we obtain (6.2). This completes the proof of Proposition (6.2) for $= \mathbf{C}$ (and for that matter for arbitrary F).

7. Local computation of the constant term of the Eisenstein series

Having proved that (for appropriate s, F_s, Ψ and φ):

$$L_2(s, \sigma, \chi) = \int_{G(F)\backslash G(\mathbf{A})} \varphi(g)\bar{\theta}_{\Psi}(g)E(g, s)\,ds$$

it remains to analytically continue $E(g, s)$. This we do by first analytically continuing the constant term of $E(g, s)$ (when F_s is appropriately chosen).

Conveniently, the "intertwining integral" describing the constant term of $E(g, s)$ is a product of local integrals. Since we are therefore dealing with a local problem, we can (and do) keep the notation of Section 6.

(7.1) If F_s is the function on Mp (F) defined in paragraph 6, we want to consider the "intertwining integral"

(7.1.1) $$F_s^*(g) = \int F_s\left[w_0\, t\begin{pmatrix} 1 & x \\ 0 & 1 \end{pmatrix} g\right] dx, \quad \mathrm{pr}(w_0) = w.$$

Replacing F_s by the function $|F_s|$ [which is a function on G (F)] and applying standard results for G (F) we find that this integral converges absolutely for $\mathrm{Re}\, s > 1/2$. Also a formal computation shows that

(7.1.2) $$F_s^*\left[\lambda d\begin{pmatrix} a & 0 \\ 0 & a^{-1} \end{pmatrix} t\begin{pmatrix} 1 & x \\ 0 & 1 \end{pmatrix} g\right] = \lambda\chi^{-1}(a)|a|^{1/2-s}F_s^*[g], \qquad \lambda \in \mathrm{T},$$

whenever F_s^* is defined.

What we want to prove is that F_s^* satisfies a condition analogous to (6.1.3).

(7.2) Proposition. — *Notations being as above, there is an entire function* H^*, *with values in a fixed finite dimensional right-invariant space of continuous functions on* MK, *such that*

$$F_s^*(k) = L(2s-1, \chi^2)H^*(s)(k), \qquad k \in \mathrm{MK}.$$

Moreover, if $F = \mathbf{R}$ *or* $\mathbf{C}$, *then* H^* *is a polynomial in s times an exponential factor; if* F *is non-archimedean of module q, then* H^* *is a polynomial in* q^{-s}, q^s.

The proof will proceed case by case. In the unramified case, additional information will be obtained. Recall also that except in the complex case, F_s is given by (6.1.4).

(7.2) F *is non-archimedean,* χ *is unramified.*

By Remark (6.2.2) we may assume $\mathfrak{R}_F$ is the conductor of ψ_F. Assume also—for the time being—that the residual characteristic of F is not 2. Then we are in the "unramified situation".

Lemme (7.2.1) ("unramified situation"). — *With the notation as above, suppose the Haar measure of* F *is chosen so that* $\mathrm{vol}(\mathfrak{R}_F) = 1$. *Then:*

$$F_s^*(k) = L(2s-1, \chi^2) \quad \text{for} \quad k \in r_{\mathfrak{R}}(\mathrm{K}).$$

Since F_s is $r_{\mathfrak{R}}(K)$-invariant, F_s^* has the same property. Thus it suffices to establish this formula for $k = e$. Recall that

$$F_s(g) = f_s(\mathrm{pr}\, g)(g.\Phi_0)(0), \tag{7.2.2}$$

where Φ_0 is the characteristic function of $\mathfrak{R}_F$. As for the function f_s, it will be convenient to extend it to a function (still denoted by f_s) on $G_2(F)$ defined by

$$f_s\left[\begin{pmatrix} a & x \\ 0 & b \end{pmatrix} k\right] = \left|\frac{a}{b}\right|^{1/2} |a|^{s-1} \chi(a) L(2s, \chi^2). \tag{7.2.3}$$

Thus:

$$\begin{aligned} F_s^*(e) &= \int f_s\left[w\begin{pmatrix} 1 & x \\ 0 & 1 \end{pmatrix}\right]\left(w_0 t\begin{pmatrix} 1 & x \\ 0 & 1 \end{pmatrix}\right).\Phi_0(1)\,dx \\ &= \int f_s\left[w\begin{pmatrix} 1 & x \\ 0 & 1 \end{pmatrix}\right] dx \int \Phi_0(y)\psi\left[\frac{1}{2}y^2 x\right] dy, \end{aligned}$$

since for Φ in $\mathscr{S}(F)$,

$$w_0.\Phi(1) = \int \Phi(y)\,dy.$$

Now

$$\int \left| f_s\left[w\begin{pmatrix} 1 & x \\ 0 & 1 \end{pmatrix}\right]\right| dx < +\infty$$

for $\mathrm{Re}\, s > 1$, and since ψ has module one, we see that for $\mathrm{Re}\, s > 1$ (which is sufficient for our purposes), $F_s^*(e)$ is given by the absolutely convergent double integral

$$F_s^*(e) = \int_F \int_F f_s\left[w\begin{pmatrix} 1 & x \\ 0 & 1 \end{pmatrix}\right] \Phi_0(y)\psi\left[\frac{1}{2}y^2 x\right] dx\,dy. \tag{7.2.4}$$

By a simple change of variables this equals

$$\int_F \int_F f_s\left[w\begin{pmatrix} 1 & y^{-2}x \\ 0 & 1 \end{pmatrix}\right] \Phi_0(y)\psi\left[\frac{1}{2}x\right] dx\,|y|^{-2}\,dy$$

or, taking into account the relation

$$w\begin{pmatrix} 1 & y^{-2}x \\ 0 & 1 \end{pmatrix} = \begin{pmatrix} y & 0 \\ 0 & y^{-1} \end{pmatrix} w \begin{pmatrix} 1 & x \\ 0 & 1 \end{pmatrix}\begin{pmatrix} y & 0 \\ 0 & y^{-1} \end{pmatrix},$$

$$\int_F \chi(y)|y|^s \Phi_0(y)|y|^{-2}\,dy \int_F f_s\left[w\begin{pmatrix} 1 & x \\ 0 & 1 \end{pmatrix}\begin{pmatrix} y & 0 \\ 0 & y^{-1} \end{pmatrix}\right]\psi\left[\frac{1}{2}x\right] dx.$$

Now set $\psi'(x) = \psi(-(1/2)x)$. Then the conductor of ψ' is still $\mathfrak{R}_F$, and for each s,

$$W_s(h) = \int_F f_s\left[w\begin{pmatrix} 1 & x \\ 0 & 1 \end{pmatrix} h\right]\psi\left(\frac{1}{2}x\right) dx \tag{7.2.5}$$

is in the space denoted $\mathcal{W}(\chi\alpha^{s-1}, 1; \psi')$ in paragraph 3 of [JL] (if $\chi\alpha^{s-1} \neq \alpha$, this space is also the space $\mathcal{W}(\pi(\chi\alpha^{s-1}, 1); \psi')$).

A description of that space (valid for all local fields) has been recalled in (6.6) (*cf.* [JL], Prop. 3.2).

Let Φ be the characteristic function of $\mathfrak{R}_F^2$ in F^2. Then let

$$f_{\Phi, s}(g) = \int \Phi[(0, t) g] \chi\alpha^s(t) d^\times t . \chi\alpha^{s-1/2}(\det g) \tag{7.2.6}$$

and

$$W_s^*(g) = \int f_{\Phi, s}\left[w\begin{pmatrix} 1 & x \\ 0 & 1 \end{pmatrix} g\right] \psi'(-x) dx. \tag{7.2.7}$$

This is the unique element of $\mathcal{W}(\chi\alpha^{s-1}, 1; \psi')$ invariant under (and equal to one on) K_2. Note that $f_{\Phi,s}$ is proportional to f_s and

$$f_{\Phi, s}(e) = L(s, \chi); \tag{7.2.8}$$

it follows that

$$\int f_s\left[w\begin{pmatrix} 1 & x \\ 0 & 1 \end{pmatrix} g\right] \psi\left[\frac{1}{2} x\right] dx = \frac{L(2s, \chi^2)}{L(s, \chi)} W_s^*(g). \tag{7.2.9}$$

Thus:

$$F_s^*(e) = \frac{L(2s, \chi^2)}{L(s, \chi)} \int_{F^\times} W_s^*\begin{pmatrix} y & 0 \\ 0 & y^{-1} \end{pmatrix} \Phi_0(y) \chi(y) |y|^{s-2} dy$$

or, since W_s^* transforms under the quasi-character $\omega_s = \chi\alpha^{s-1}$ of the center,

$$\begin{aligned} F_s^*(e) &= \frac{L(2s, \chi^2)}{L(s, \chi)} \int_F W_s^*\left[\begin{pmatrix} y^2 & 0 \\ 0 & 1 \end{pmatrix}\right] \Phi_0(y) \frac{dy}{|y|} \\ &= \left(\frac{L(2s, \chi^2)}{L(s, \chi)} (1-q^{-1})\right) \int_{F^\times} W_s^*\begin{pmatrix} y^2 & 0 \\ 0 & 1 \end{pmatrix} d^\times y. \end{aligned} \tag{7.2.9}$$

Here $d^\times y$ is so chosen that $\text{vol}(\mathfrak{R}^\times) = 1$ or, what amounts to the same

$$\frac{dy}{|y|} = (1-q^{-1}) d^\times y.$$

Also we have used the fact that

$$W_s^*\begin{pmatrix} y & 0 \\ 0 & 1 \end{pmatrix} = 0 \quad \text{if } |y| > 1.$$

But the convergent integral in (7.2.9) is the value at $t = s$ of the integral

$$\int |y|^{t-1} \chi\omega_s^{-1}(y) W_s^*\begin{pmatrix} y^2 & 0 \\ 0 & 1 \end{pmatrix} d^\times y.$$

Since the integrand here vanishes for $|y| > 1$, this integral converges for $\text{Re}(t) > s$.

Moreover, for Re (t) sufficiently large, it is given by (6.3.3) with s replaced by t, ω by ω_s, and σ by $\pi(\chi\alpha^{s-1}, 1)$, i. e., its value is

$$L(t, \chi^2\alpha^{s-1})L(t, \chi)L(t, \alpha^{1-s})L(2t, \chi^2)^{-1}.$$

Setting $t = s$ we get

$$\int W_s^*\begin{pmatrix} y^2 & 0 \\ 0 & 1 \end{pmatrix} d^\times y = \frac{L(2s-1, \chi^2)L(s, \chi)L(1, 1)}{L(2s, \chi^2)}$$
$$= \frac{1+\chi(\tilde{\omega})q^{-s}}{(1-\chi^2(\tilde{\omega})q^{-2s+1})(1-q^{-1})}.$$

Therefore

(7.2.10) $$F_s^*(e) = L(2s-1, \chi^2)$$

as required. This concludes the proof of (7.2.1).

Suppose now the residual characteristic of F is 2. Then we choose ψ as in paragraph 6: $\psi(x) = \psi_0(2x)$, and $\psi'(x) = \psi_0(-x)$, where ψ_0 has conductor $\mathfrak{R}_F$. Again F_s is defined by (7.2.2) and we extend f_s to $G_2(F)$ by (7.2.3). We check that for all k in MK:

(7.2.11) $$F_s^*(k)/L(2s, \chi^2)$$

is a polynomial in q^{-s}, q^s. As before

$$F_s^*(k) = (1+\chi(\tilde{\omega})q^{-s})(1-q^{-1}).\int W_s^*\begin{pmatrix} y^2 & 0 \\ 0 & 1 \end{pmatrix} k.\Phi_0(y)\, d^\times y.$$

For $k = e$, this is (7.2.10). Thus it will suffice to show that if

$$U_s(k) = \int_{|y|=1} W_s^*\begin{pmatrix} y^2 & 0 \\ 0 & 1 \end{pmatrix} k.\Phi_0(y)\, d^\times y$$

then:

$$U_s(k) = U_s(e)\lambda(k),$$

where λ is a character of MK.

Note that any element of MK projects to an element of K of the form

$$\begin{pmatrix} 1 & x_1 \\ 0 & 1 \end{pmatrix} w \begin{pmatrix} 1 & x_2 \\ 0 & 1 \end{pmatrix}\begin{pmatrix} a & 0 \\ 0 & a^{-1} \end{pmatrix}, \qquad x_i \in \mathfrak{R}, \quad a \in \mathfrak{R}^\times$$

or

$$w\begin{pmatrix} 1 & x_1 \\ 0 & 1 \end{pmatrix} w^{-1}\begin{pmatrix} 1 & x_2 \\ 0 & 1 \end{pmatrix}\begin{pmatrix} a & 0 \\ 0 & a^{-1} \end{pmatrix}, \qquad x_1 \in \mathfrak{P}, \quad x_2 \in \mathfrak{R}, \quad a \in \mathfrak{R}^\times.$$

In the first case we may assume

$$k = t\begin{pmatrix} 1 & x_1 \\ 0 & 1 \end{pmatrix} w_0\, t\begin{pmatrix} 1 & x_2 \\ 0 & 1 \end{pmatrix} d\begin{pmatrix} a & 0 \\ 0 & a^{-1} \end{pmatrix}.$$

Then, for $y \in \mathfrak{R}$,

$$k.\Phi_0(y) = |2|^{1/2} \psi_0[x_1^2 y] \Phi_0(2y) = |2|^{1/2}.$$

In the second case we may assume

$$k = w_0\, t\begin{pmatrix} 1 & x_1 \\ 0 & 1 \end{pmatrix} w_0^{-1}\, t\begin{pmatrix} 1 & x_2 \\ 0 & 1 \end{pmatrix} d\begin{pmatrix} a & 0 \\ 0 & a^{-1} \end{pmatrix}.$$

Then for $y \in \mathfrak{R}$,

$$k.\Phi_0(y) = |2|^{1/2} \int \Phi_0(2z) \psi_0[x_1^2 z] \psi_0[-2zy]\, dz.$$

However $\Psi_0(-2zy) = 1$ if $\Phi_0(2z) \neq 0$. Thus $k.\Phi_0(y)$ does not depend in $y \in \mathfrak{R}$, and our conclusion follows.

(7.3) F *is non archimedean*, χ *is ramified.* In this case recall that F_s has the form

(7.3.1) $$F_s(g) = f_s(\mathrm{pr}\, g)(g.\Phi)(0),$$

where f_s satisfies (6.1.5) and $f_s \mid K$ is actually independent of s. Thus it will suffice to prove the following Lemma:

LEMMA (7.3.2). — *Suppose* F_s *is a function of the form* (7.3.1) *where* Φ *is in* $\mathcal{S}$ (F) *and* f_s *is a function on* G_2 (F) *satisfying*

$$f_s\left[\begin{pmatrix} a & x \\ 0 & b \end{pmatrix} k\right] = \left|\frac{a}{b}\right|^{1/2} \chi(a) |a|^{s-1} h(k),$$

where h is a fixed finite function on GL (2, $\mathfrak{R}$). *Then if* F_s^* *is as in* (7.1.1), *for any g*

$$F_s^*(g)/L(2s-1, \chi^2)$$

is a polynomial in q^{-s}.

Proof. — We may assume pr (g) is in K. Replacing F_s by a translate we may even assume $g = e$. As in (7.2) we find that for Re s large enough,

$$F_s^*(e) = \int W_s\left[\begin{pmatrix} y^2 & 0 \\ 0 & 1 \end{pmatrix}\right] \Phi(y)\, d^\times y,$$

where

$$W_s(g) = \int_F f_s\left[w\begin{pmatrix} 1 & y \\ 0 & 1 \end{pmatrix} g\right] \psi\left[\frac{1}{2} x\right] dx.$$

In particular (*cf.* Prop. 3.2 of [JL])

$$W_s\begin{pmatrix} y & 0 \\ 0 & 1 \end{pmatrix} = |y|^{1/2} \int f_s\left[w\begin{pmatrix} 1 & x \\ 0 & 1 \end{pmatrix}\right] \psi\left[\frac{1}{2} xy\right] dx.$$

Since f_s is K-finite, we also see that

$$f_s\left[w\begin{pmatrix}1 & x\\ 0 & 1\end{pmatrix}\right]=|x|^{-s}\chi^{-1}(x)f_s\begin{pmatrix}1 & 0\\ -x^{-1} & 1\end{pmatrix}=|x|^{-s}\chi^{-1}(x)f_s(e)$$

for $|x|$ large enough. Thus

$$f_s\left[w\begin{pmatrix}1 & x\\ 0 & 1\end{pmatrix}\right]=\sum_{i=1}^{N}\theta_i(x)q^{-is}+\theta_0(x, s),$$

where the θ_i are in $\mathscr{S}$ (F) and $\theta_0(x, s)$ is defined by

$$\theta_0(x, s)=\begin{cases}|x|^{-s}\chi^{-1}(x) & \text{for } |x|\geqq 1\\ 0 & \text{for } |x|<1.\end{cases}$$

But the Fourier transform of $\theta_0(x, s)$ for Re (s) large enough is given by

$$\int\theta_0(x, s)\psi\left[\frac{1}{2}xy\right]dx=\int_{|x|\geqq 1}|x|^{-s}\chi^{-1}(x)\psi\left[\frac{1}{2}xy\right]dx,$$

and since χ is ramified it is easily found that this integral has the form $q^{-ms}\varphi(y)\chi(y)|y|^{s-1}$ with φ in $\mathscr{S}$ (F). Thus:

$$W_s\begin{pmatrix}y & 0\\ 0 & 1\end{pmatrix}=\sum\hat{\theta}_i(y)|y|^{1/2}q^{-is}+q^{-ms}\chi(y)|y|^{s-1/2}\hat{\varphi}(y),$$

and

$$F_s^*(e)=\sum_i q^{-is}\int\hat{\theta}_i(y)\,dy+q^{-ms}\int\hat{\varphi}(y)\chi^2(y)|y|^{2s-1}\frac{dy}{|y|}.$$

The sum on i is a polynomial in q^{-s}, q^s; the second term is the product of a polynomial in q^{-s}, q^s by L $(2s-1, \chi^2)$ *cf.* ([Ta], [Go Ja]). This concludes the proof of Lemma (7.3.2).

(7.4) *Real case.* In this case:

(7.4.1) $$F_s(g)=f_s(\text{pr}\, g)(g.\Phi)(0),$$

where Φ is an element of $\mathscr{S}$ (F) of a certain weight and f_s is a function on G_2 (F) such that

(7.4.2) $$f_s\left[\begin{pmatrix}a & x\\ 0 & b\end{pmatrix}\begin{pmatrix}\cos\theta & \sin\theta\\ -\sin\theta & \cos\theta\end{pmatrix}\right]=\chi(a)|a|^{s-1}\left|\frac{a}{b}\right|^{1/2}e^{in\theta}L(2s, \chi^2)P(s).$$

The integer n is the weight of f_s and P is a polynomial. The possibilities for χ, P, f_s, and Φ are as follows (*cf.* Tables (6.5.11), (6.5.13) and the remarks made in establishing those tables):

TABLE (7.4.3)

Case	χ	wgt (f_s)	wgt (Φ)	wgt (F_s)	P (s)
(i).......	1	0	1/2	1/2	1
(ii).......	sgn	-1	1/2	$-1/2$	1
(iii)......	sgn	-1	5/2	3/2	s

Since F_s transforms according to a character of MK, to prove Proposition (7.2) for $F = \mathbf{R}$, it will suffices to show there is a polynomial P^* such that

$$F_s^*(e) = L(2s-1, \chi^2) P^*(s), \tag{7.4.4}$$

up to an exponential factor.

Now, as in (7.2),

$$F_s^*(e) = \int W_s\left[\begin{pmatrix} -\frac{1}{2}y^2 & 0 \\ 0 & 1 \end{pmatrix}\right] \Phi(y)\, d^\times y, \tag{7.4.5}$$

where

$$W_s(g) = \int_{-\infty}^{+\infty} f_s\left[w\begin{pmatrix} 1 & x \\ 0 & 1 \end{pmatrix} g\right] \psi(-x)\, dx. \tag{7.4.6}$$

Again, for each s (with $\operatorname{Re} s > 1$) W_s is in the space denoted $\mathscr{W}(\chi\alpha^{s-1}, 1; \psi_F)$ in [JL]. Therefore we proceed as in (7.2). We let Φ_1 be an element of $\mathscr{S}(F^2)$ such that the function

$$f_{\Phi_1}(g) = \int_{\mathbf{R}^\times} \Phi_1[(0, t) g]\, \chi(t) |t|^s d^\times t\, \chi(\det g) |\det g|^{s-1/2} \tag{7.4.5}$$

has the same weight as f_s. Then the function W_s^* defined by

$$W_s^*(g) = \int f_{\Phi_1}\left[w\begin{pmatrix} 1 & x \\ 0 & 1 \end{pmatrix} g\right] \psi(-x)\, dx \tag{7.4.6}$$

is in $\mathscr{W}(\chi\alpha^{s-1}, 1; \psi_F)$ and proportional to W_s.

More precisely, we take

$$\Phi_1(x, y) = \begin{cases} \exp(-\pi(x^2+y^2)) & \text{case (i)}, \\ \exp(-\pi(x^2+y^2))(x-iy) & \text{cases (i) and (ii)}. \end{cases} \tag{7.4.7}$$

Then Φ_1' [*cf.* (6.6.6)] is given by

$$\Phi_1'(x, y) = \begin{cases} \exp(-\pi(x^2+y^2)) & \text{case (i)}, \\ \exp(-\pi(x^2+y^2))(x+y) & \text{cases (i) and (ii)}. \end{cases} \tag{7.4.8}$$

In all cases

$$f_{\Phi_1}(e) = L(s, \chi)$$

so that

$$F_s^*(e) = \frac{L(2s, \chi^2)}{L(s, \chi)} P(s) \int_{-\infty}^{+\infty} W_s^*\left[\begin{pmatrix} -\frac{1}{2}y^2 & 0 \\ 0 & 1 \end{pmatrix}\right] \Phi(y)\, d^\times y. \tag{7.4.9}$$

Since (in all cases) $\Phi(-y) = \Phi(y)$, up to exponential factors we also have

$$F_s^*(e) = \frac{L(2s, \chi^2)}{L(s, \chi)} P(s) \int_0^{+\infty} W_s^*\left[\begin{pmatrix} -y & 0 \\ 0 & 1 \end{pmatrix}\right] \Phi(\sqrt{2y})\, d^\times y, \tag{7.4.10}$$

$$\text{(7.4.11)}\qquad W_s^*\begin{pmatrix} y & 0 \\ 0 & 1 \end{pmatrix} = \int \Phi_1'(yt, t^{-1})\chi(t)|t|^{s-1}d^\times t\,\chi(y)|y|^{s-1/2}.$$

This time

$$\text{(7.4.12)}\qquad \int_0^{+\infty} W_s^*\left[\begin{pmatrix} -y & 0 \\ 0 & 1 \end{pmatrix}\right]|y|^{t-1/2}d^\times t$$

$$= \frac{1}{2}\int_{-\infty}^{+\infty} W_s^*\left[\begin{pmatrix} y & 0 \\ 0 & 1 \end{pmatrix}\right]|y|^{t-1/2}d^\times t - \frac{1}{2}\int_{-\infty}^{+\infty} W_s^*\left[\begin{pmatrix} y & 0 \\ 0 & 1 \end{pmatrix}\right]|y|^{t-1/2}\frac{y}{|y|}d^\times y$$

$$= \frac{1}{2}\iint \Phi_1'(x, y)\chi(x)|x|^{s+t-1}|y|^t d^\times x\, d^\times y$$

$$- \frac{1}{2}\iint \Phi_1'(x, y)\chi(x)\frac{x}{|x|}|x|^{s+t-1}\frac{y}{|y|}|y|^t d^\times x\, d^\times y.$$

Thus this integral can be easily computed.

Similarly it is easy to compute

$$\int_0^{+\infty} \Phi(\sqrt{2y})|y|^{t-1/2}d^\times t$$

since:

$$\Phi(x) = \begin{cases} \exp(-\pi x^2) & \text{cases (i) and (ii),} \\ \exp(-\pi x^2)(8\pi x^2 - 2) & \text{case (ii).} \end{cases}$$

Using a Barnes-Mellin lemma as before we can (for Ret large enough) compute the integral

$$\text{(7.4.13)}\qquad \int_0^{+\infty} W_s^*\begin{pmatrix} -y & 0 \\ 0 & 1 \end{pmatrix}\Phi(\sqrt{2y})|y|^{t-s}d^\times y.$$

[From (7.4.11) we see this integral converges for Re $(t-s) \geqq 0$.] Thus we may evaluate (7.4.10) by setting $s = t$ in the expression for (7.4.13). This will give the required result [with $P^* \not\equiv 1$ only in case (iii)].

(7.5) *Complex case.* — If $F = \mathbf{C}$ recall that

$$F_s(g) = L(s, \chi)\int \Phi[(0, t)g]|t|^{s+1/2}\chi(t)d^\times t,$$

where Φ is a suitably chosen function of the form

$$\Phi(x, y) = \exp[-2\pi(x\bar{x} + y\bar{y})]P(x, \bar{x}, y, \bar{y}).$$

Now:

$$F_s^*(g) = \int F_s\left[w\begin{pmatrix} 1 & x \\ 0 & 1 \end{pmatrix}g\right]dx = L(s, \chi)\iint \Phi[(t, tx)g]|t|^{s+1/2}\chi(t)d^\times t\,dx$$

$$= L(s, \chi)\iint \Phi[(t, x)g]|t|^{s-1/2}\chi(t)d^\times t\,dx.$$

As in Section 6, if g is in SU (2),

$$F_s^*(g) = L(s, \chi) L\left(s-\frac{1}{2}, \chi\right) Q(s)(g)$$

where Q is a polynomial. So since

$$L(s, \chi) L\left(s-\frac{1}{2}, \chi\right) = L(2s-1, \chi^2)$$

up to exponentional factors, F_s^* has the required analytic properties, and the proof of (7.2) is complete.

8. Analytic continuation of the main integral

In this Section F is a *number field.* As before we let $\sigma = \otimes_v \sigma_v$ be an automorphic cuspidal representation of $G_2(\mathbf{A})$ and S the (finite) set of finite places v where σ_v is not quasi-unramified. We let χ be a character of $F_{\mathbf{A}}^{\times}/F_{\times}$ and we assume χ_v is highly ramified for $v \in S$ [*cf.* (5.3.3)]. We assume in addition that there is at least one finite place w such that χ_w^2 is ramified. Since this is so if χ_w is of high enough ramification, we may take w to be in S if $S \neq \emptyset$.

Our purpose in this section is to prove the following:

THEOREM (8.1). – *Suppose σ and χ satisfy the above conditions. Then $L_2(s, \sigma, \chi)$ is entire and bounded in vertical strips of finite width.*

To prove this theorem we need only consider the "main integral" introduced in Section 5. Indeed for each v let $F_{v,s}$ be the function introduced in paragraph 6. (Recall that the choice of $F_{v,s}$ depends on σ_v and χ_v). Define $F_s(g)$ by

$$F_s(g) = \prod_v F_{v,s}(g_v)$$

if $g = (g_v)$ [*cf.* (5.5.5)]. Then:

$$F_s\left[d\begin{pmatrix} a & 0 \\ 0 & a^{-1}\end{pmatrix} t\begin{pmatrix} 1 & x \\ 0 & 1\end{pmatrix} g\right] = \chi(a) |a|^{s+1/2} F(g).$$

If we set

$$E(g, s) = \sum_{\gamma \in B(F)\backslash G(F)} F_s(r_F(\gamma) g)$$

then this converges for Re s large enough. Moreover, there is a Ψ in $\mathcal{S}(\mathbf{A})$ and a cusp form φ on $G_2(\mathbf{A})$ belonging to σ such that

$$L_2(s, \sigma, \chi) = \int_{G(F)\backslash G(\mathbf{A})} E(g, s) \overline{\theta_\Psi(h)} \varphi(g)\, dg, \qquad (\mathrm{pr}(h) = g)$$

for Re s large enough [*cf.* Prop. (6.2) and (5.6.14)]. Thus Theorem 8.1 reduces to:

PROPOSITION 8.2. — *Suppose* F_s *and* χ *are as above, and* φ *is a rapidly decreasing genuine function on* $\mathrm{Mp}(F)\backslash\mathrm{Mp}(\mathbf{A})$. *Then the function*

$$s \mapsto \int E(h, s)\bar{\varphi}(h)\,dh,$$

$$h \in r_F(G(F))T\backslash \mathrm{Mp}(\mathbf{A}),$$

initially defined for Re (s) *sufficiently large, extends to an entire function of* s *which is bounded in vertical strips of finite width.*

This Proposition is tantamount to the analytic continuation of the Eisenstein series $E(g, s)$. Thus we need to recall a crucial fact from the theory of Eisenstein series.

LEMMA (8.2.1). — *Let* $\tilde{F}_s$ *be a function on* $\mathrm{Mp}(\mathbf{A})$ *such that*

$$\tilde{F}_s\left[\lambda d\begin{pmatrix} a & 0 \\ 0 & a^{-1}\end{pmatrix} t\begin{pmatrix} 1 & x \\ 0 & 1\end{pmatrix} k\right] = \lambda H(k)\,|a|^{s+1/2}\chi(a),$$

where H *is a fixed finite function on* MK. *Then the function* $\tilde{F}_s^*$, *defined for* Re s *large enough by*

$$\tilde{F}_s^*(g) = \int_{\mathbf{A}} \tilde{F}_s\left[r_F(w)\,t\begin{pmatrix} 1 & x \\ 0 & 1\end{pmatrix} g\right] dx,$$

continues to a meromorphic function of s *without poles on the line* Re $(s) = 1/2$. *Moreover, if* Re $s = 1/2$, *then:*

$$\int_{MK} \tilde{F}_s(k)\overline{\tilde{F}_s}(k)\,dk = \int_{MK} \tilde{F}_s^*(k)\overline{\tilde{F}_s^*}(k)\,dk.$$

We take this Lemma for granted; it follows from the more general results of [La 3]. (Recall F is a number field!)

Now our function F_s is clearly of the form

(8.2.2) $$F_s(g) = L(2s, \chi^2)\sum h_i(s)\tilde{F}_s^i(g),$$

where, for each i, $\tilde{F}_s^i$ satisfies the conditions of the Lemma and $h_i(s)$ is an elementary function of s, i. e., a sum of terms like $as^i e^{bs}$. If we set

(8.2.3) $$F_s^*(g) = \int F_s\left[r_F(w)\,t\begin{pmatrix} 1 & x \\ 0 & 1\end{pmatrix} g\right] dx,$$

then we have also

(8.2.4) $$F_s^*(g) = L(2s, \chi^2)\sum h_i(s)\tilde{F}_s^{i*}(g).$$

Thus it follows from Lemma (8.2.2) that

(8.2.5) $$\int F_s(k)\overline{F_s}(k)\,dk = \int F_s^*(k)\overline{F_s^*}(k)\,dk \quad \text{if} \quad \mathrm{Re}\, s = \frac{1}{2}.$$

Actually we have more information about the analytic behavior of F_s^*. Namely (by Proposition 7.2):

(8.2.6) $$F_s^*(k) = L(2s-1, \chi^2)\sum_j H_j'(k)\, h_j'(s)$$

where H_j is some finite function on MK and h_j' is elementary.

To continue, we need to use a familiar "truncation process". (*see* [DL] or [Ge J1] for more details.) For $c > 0$ we let χ_c be the characteristic function of the "Siegel set":

$$\left\{g \mid g = d\begin{pmatrix} a & 0 \\ 0 & a^{-1}\end{pmatrix} t\begin{pmatrix} 1 & x \\ 0 & 1\end{pmatrix} k,\ |a| \geqq c, k \in \mathrm{MK}\right\}.$$

We set

(8.2.7) $$F_s'(g) = E^0(g, s)\,\chi_c(g),$$

where E^0 is the constant term of $E(g, s)$:

(8.2.8) $$E^0(g, s) = \int_{F\backslash A} E\left[t\begin{pmatrix} 1 & x \\ 0 & 1\end{pmatrix} g, s\right] dx.$$

Then by the Bruhat decomposition,

(8.2.9) $$E^0(g, s) = F_s(g) + F_s^*(g).$$

Thus E^0 has a known analytic behavior. But from reduction theory we hnow that if c is large enough then the series

(8.2.10) $$E'(g, s) = \sum_{B(F)\backslash G(F)} F_s'(r_F(\gamma)g)$$

has only finitely non-zero many terms when g is in a siegel set. Let us study then the scalar product

(8.2.11) $$\int E'(h, s)\,\bar{\varphi}(h)\, dh.$$

To evaluate this integral consider what happens when h is in a Siegel set $\mathfrak{S}$. There is a compact subset Ω of $\mathfrak{S}$ such that for $h \in \mathfrak{S} - \Omega$,

$$E'(h, s) = E^0(h, s);$$

cf. [Ge J1]. On the other hand, for h in Ω, the series defining E' has only finitely many non zero-terms, so since φ is rapidly decreasing at infinity in $\mathfrak{S}$, it is clear that (8.2.11) converges for all s and defines an entire function of s bounded at infinity in vertical strips.

Now set

(8.2.12) $$E''(h, s) = E(h, s) - E'(h, s).$$

Let also L^2 denote the space of genuine functions on Mp (**A**) which are $r_F(G(F))$ invariant and such that

$$\int f(h)\bar{f}(h)\,dh < +\infty,\ h \in r_F(G(F))\mathbf{T}\backslash \mathrm{Mp}(\mathbf{A}).$$

Clearly φ is in L^2. Then Proposition (8.2) will follow from:

PROPOSITION 8.3. — *If* Re s *is large enough,* E''_s *is in* L^2, *and the map* $s \mapsto E''_s$ *continues to an entire function of* s *with values in* L^2. *Moreover, the map* $s \mapsto \| E''_s \|$ *is bounded in vertical strips of finite width.*

Proof. — Again, from the known theory of Eisenstein series, one knows that E''_s is square integrable and $s \mapsto E''_s$ holomorphic, for Re s large enough. Moreover for Re $s_1 >$ Re $s_2 \gg 0$,

$$(8.3.1)\qquad \begin{aligned}(E''_{s_1}, E''_{\bar{s}_2}) = & \int_{|a|<c} |a|^{s_1+s_2-1}\,d^\times a\left(\int_{MK} F_{s_1}(k)\overline{F_{\bar{s}_2}}(k)\,dk\right)\\ & -\int_{|a|>c} \chi^2(a)|a|^{s_2-s_1}\,d^\times a\int_{MK} F_{s_1}(k)\overline{F^*_{\bar{s}_2}}(k)\,dk\\ & -\int_{|a|>c} |a|^{1-s_1-s_2}\,d^\times a\int_{MK} F^*_{s_1}(k)\overline{F^*_{\bar{s}_2}}(k)\,dk\\ & +\int_{|a|<c} \bar{\chi}^2(a)|a|^{s_1-s_2}\,d^\times a\int_{MK} F^*_{s_1}(k)\overline{F_{\bar{s}_2}}(k)\,dk.\end{aligned}$$

In each integral above the integration over $F^\times_{\mathbf{A}}$ is taken modulo $F^\times$. But χ^2 is not a principal character of $F^\times\backslash F^\times_{\mathbf{A}}$. Therefore the second and fourth terms drop out, leaving

$$(8.3.2)\qquad (E''_{s_1}, E''_{\bar{s}_2}) = A(s_1, s_2)(s_1+s_2-1)^{-1},$$

with

$$(8.3.3)\qquad A(s_1, s_2) = c^{s_1+s_2-1}\int_{MK} F_{s_1}(k)\overline{F_{\bar{s}_2}}(k)\,dk - c^{1-s_1-s_2}\int_{MK} F^*_{s_1}(k)\overline{F^*_{\bar{s}_2}}(k)\,dk.$$

Now $A(s_1, s_2)$ is holomorphic in $\mathbf{C}^2$. Moreover, if $s_1 = (1/2)+iy$ and $s_2 = (1/2)-iy$ with $y \in \mathbf{R}$, we find that A vanishes [*cf.* (8.2.5)]. Thus $A(s_1, s_2) = 0$ if $s_1+s_2-1 = 0$, and it follows that

$$(8.3.4)\qquad (E''_{s_1}, E''_{\bar{s}_2})$$

extends to a holomorphic function of (s_1, s_2) in $\mathbf{C}^2$. A standard argument (*cf.* [La 3] or [Ge J 2]) then shows that $s \mapsto E''_s$ continues to an entire function (with values in L^2), and by analytic continuation, formula (8.3.2) is still valid when $s_1+s_2-1 \neq 0$.

To complete the proof of Proposition 8.3 (and hence Theorem 8.1) we have to show that (E''_s, E''_s) is bounded in vertical strips. For this we write $s = (1/2)+x+iy$ with

$-a \leqq x \leqq a$ and $y \in \mathbf{R}$. Then:

$$(8.3.5) \qquad (E_s'', E_s'') = A\left(\frac{1}{2}+x+iy, \frac{1}{2}+x-iy\right)(2x)^{-1} = B(x, y)(x)^{-1},$$

where

$$B(x, y) = \frac{1}{2}A\left(\frac{1}{2}+x+iy, \frac{1}{2}+x-iy\right).$$

Now $B(0, y) = 0$. Thus:

$$B(x, y) = x\int_0^1 \frac{\partial B}{\partial x}(tx, y)\,dt$$

and

$$(E_s'', E_s'') = \int_0^1 \frac{\partial B}{\partial x}(tx, y)\,dt.$$

But

$$\frac{\partial B}{\partial x}(x, y) = \frac{1}{2}\frac{\partial A}{\partial s_1}\left(\frac{1}{2}+x+iy, \frac{1}{2}+x-iy\right)+\frac{1}{2}\frac{\partial A}{\partial s_2}\left(\frac{1}{2}+x+iy, \frac{1}{2}+x-iy\right).$$

So since the partial derivatives of A can be expressed in terms of elementary functions and L-functions and their derivatives, it easily follows that $\partial B/\partial x$ is bounded for $-a \leqq x \leqq a$, and the same is true of (E_s'', E_s'').

Q.E.D.

9. The Main Theorem

Again F is a number field.

(9.1) We first review and complete the results of [JPSS]. We let T be a finite set of places (possibly empty). For each $v \notin T$ we let π_v be an admissible irreducible representation of $G_{3,v}$ (or its Hecke algebra). In addition we make the following assumptions (and fix notation as follows):

(9.1.1) For each v, the representation π_v has trivial central quasi-character;

(9.1.2) Suppose v is finite; if π_v is generic, i. e. admits a "Whittaker model" in the sense of [JPSS] (§ 2), then we set $\xi_v = \pi_v$; if π_v is not generic, then it is a quotient of a certain induced representation ξ_v which is said to be attached to π_v (*loc. cit.*) for instance, we may have

$$(9.1.2.1) \qquad \pi_v = \mathrm{Ind}(G_{3,v}, P_v; 1_2, 1),$$

where P_v is the parabolic subgroup of type (2.1), $1_2 = \pi(\alpha_v^{1/2}, \alpha_v^{-1/2})$ the trivial representation of $G_{2,v}$, and

$$(9.1.2.2) \qquad \xi_v = \mathrm{Ind}(G_{3,v}, B_{3,v}; \alpha_v^{1/2}, 1, \alpha_v^{-1/2}).$$

(9.1.3) Suppose v is archimedean; then either π_v is unitary generic, in which case we set $\xi_v = \pi_v$, or π_v is one of the representations considered in paragraph 4, namely:

(9.1.3.1) $$\pi_v = \operatorname{Ind}(G_{3,v}, B_{3,v}; \alpha_v^t, 1, \alpha_v^{-t}), \frac{1}{2} < t < 1,$$

in which case $\pi_v = \xi_v$, or π_v is (9.1.2.1), in which case ξ_v is (9.1.2.2).

In all cases, π_v is a quotient of ξ_v, and the representation ξ_v is semi-simple only if $\xi_v = \pi_v$.

(9.1.4) For almost all finite v, π_v is unramified, thus of the form

$$\pi_v = \pi(\mu_{1,v}, \mu_{2,v}, \mu_{3,v}),$$

where $\mu_{i,v}(x) = |x|^{t_{i,v}}$: this means that π_v is the only unramified component of the induced representation

$$\operatorname{Ind}(G_{3,v}, B_{3,v}; \mu_{i,v}, \mu_{2,v}, \mu_{3,v});$$

we also assume there is a $t > 0$ (independent of v) such that

$$-t \leqq t_{i,v} \leqq t \qquad (i = 1, 2, 3, \text{ almost all } v).$$

Given the notation and conditions above, we can now form the representations

$$\pi^T = \bigotimes_{v \notin T} \pi_v, \qquad \xi^T = \bigotimes_{v \notin T} \xi_v$$

of the Hecke-algebra $\mathscr{H}^T$ of the restricted product group

$$G^T = \prod G_v \qquad (v \notin T)$$

and form for each character χ of $F_A^\times/F^\times$ the infinite products

$$L(s, \pi^T, \chi) = \prod_{v \notin T} L(s, \pi_v, \chi_v),$$

and

$$L(s, \tilde{\pi}^T, \chi) = \prod_{v \notin T} L(s, \tilde{\pi}_v, \chi_v^{-1}).$$

These products converge absolutely for Re (s) sufficiently large.

(9.2) Suppose that the above functions are entire, bounded in vertical strips, and satisfy

(9.2.1) $$L(s, \pi^T, \chi) = \prod_{v \notin T} \varepsilon(s, \pi_v \otimes \chi_v, \psi_v) \prod_{v \in T} \varepsilon(s, \chi_v, \psi_v)^3 L(1-s, \tilde{\pi}^T, \chi^{-1})$$

provided the ramification of χ is sufficiently high at each place v in T. Then we conclude as in [JPSS] (§ 13) that there is a space $\mathscr{V}$ of C^∞ functions on $G_3(F) Z_3(A) \backslash G_3(A)$ which is invariant on the right under convolution by $\mathscr{H}^T$ and realizes the representation ξ^T. Moreover, the elements of $\mathscr{V}$ are "slowly increasing". The various possibilities for T and $\mathscr{V}$ are now discussed case by case.

(i) Suppose first that $T = \emptyset$. Then we set

$$\pi = \pi^T = \otimes \pi_v, \qquad \xi = \xi^T = \otimes \xi_v,$$

so that

$$L(s, \pi^T, \chi) = L(s, \pi \otimes \chi), \qquad L(s, \tilde{\pi}^T, \chi) = L(s, \tilde{\pi} \otimes \chi),$$

and the functional equation reads

$$(9.2.2) \qquad L(s, \pi \otimes \chi) = \varepsilon(s, \pi \otimes \chi) L(1-s, \tilde{\pi} \otimes \chi^{-1}).$$

In this case the elements of $\mathscr{V}$ are cuspidal, thus cusp-forms.

Thus each ξ_v is semi-simple and unitary generic; in particular $\pi_v = \xi_v$ for all v, π_v is unitary generic, and π_v cannot be a representation of the form (9.1.3.1) or (9.1.2.1).

(ii) Now suppose $T \neq \emptyset$ but the elements of $\mathscr{V}$ are cuspidal. Then again they are cusp forms. For $v \notin T$, we again find $\pi_v = \xi_v$ is unitary generic. On the other hand, for $v \in T$, there is a unitary generic representation π_v with trivial central character such that $\pi = \otimes \pi_v$ (all v) is automorphic cuspidal. Moreover (9.2.2) is satisfied.

(iii) Next suppose the elements of $\mathscr{V}$ have zero constant terms along

$$N_3 = \left\{ \begin{pmatrix} 1 & \star & \star \\ 0 & 1 & \star \\ 0 & 0 & 1 \end{pmatrix} \right\}$$

but not along

$$U = \left\{ \begin{pmatrix} 1 & 0 & \star \\ 0 & 1 & \star \\ 0 & 0 & 1 \end{pmatrix} \right\};$$

(*cf.* [JPSS], § 13).

Then (*cf.* the proof of [13.8] in [JPSS]), there is a quasi-character μ of $F_{\mathbb{A}}^\times/F^\times$ and an automorphic cuspidal representation τ of $G_2(\mathbb{A})$ with central quasi-character μ^{-1} satisfying the following conditions: for any v set

$$\eta_v = \text{Ind}(G_{3,v}, P_v; \tau_v, \mu_v);$$

then for all $v \notin T$ the representations ξ_v and η_v have a common irreducible component; moreover, for almost all finite v both ξ_v and η_v contain a (unique) component which is unramified and that component is the same for both representations (and equal to π_v). Note that μ need not be a character and therefore τ need not be unitary. (The precise definition of the induced representation η_v—for v infinite—is irrelevant since allt hat matters to us is the infinitesimal class of η_v and its components.) If v is finite then τ_v is generic and η_v has exactly one generic component. Therefore, for $v \in T$, let π'_v be that component and form the representation

$$\pi' = (\bigotimes_{v \in T} \pi'_v) \otimes \pi^T.$$

We will write also $\pi' = \otimes \pi'_v$ so that $\pi'_v \simeq \pi_v$ for $v \notin T$. Then (again *see* the proof of (13.8) in [JPSS]) $L(s, \pi' \otimes \chi)$ is meromorphic and satisfies (9.2.2) for all χ.

(iv) Suppose now that the elements of $\mathscr{V}$ have zero constant terms along N but not along

$$V = \left\{ \begin{pmatrix} 1 & \star & \star \\ 0 & 1 & 0 \\ 0 & 0 & 1 \end{pmatrix} \right\}.$$

Then a similar conclusion can be reached.

(v) Finally suppose that the elements of $\mathscr{V}$ have non zero-constant term along N. Then there are 3 quasi-characters ν_i of $F_A^\times/F^\times$ such that $\nu_1 \nu_2 \nu_3 = 1$ and, if we set,

$$\eta_v = \operatorname{Ind}(G_{3,v}, B_{3,v}; \nu_{1v}, \nu_{2v}, \nu_{3v}),$$

then for all $v \notin T$ the representations ξ_v and η_v have a common component. For almost all v they contain also the same unramified component which is then π_v. For $v \in T$ let π'_v be the unique generic component of η_v and set

$$\pi' = \bigotimes_{v \in T} \pi'_v \otimes \pi^T.$$

Then $L(s, \pi' \otimes \chi)$ satisfies (9.2.2) for all χ (*loc. cit.*).

We can now formulate the main theorem of this paper.

(9.3) THEOREM. — *Let σ be a unitary irreducible representation of $G_2(A)$ which is automorphic cuspidal. Assume that for any character χ of $F_A^\times/F^\times$, $\chi \neq 1$, the representations σ and $\sigma \otimes \chi$ are inequivalent. Then:*

(1) *for any χ, $L_2(s, \sigma, \chi)$ is entire;*

(2) *for any place v the representation σ_v admits a lift π_v to $G_{3,v}$;*

(3) *set $\pi = \bigotimes_v \pi_v$ (all v). Then π is automorphic cuspidal;*

(4) *no representation of the form $\pi(\lambda\alpha_v^t, \lambda\alpha_v^{-t})$ with $(1/4) \leqq t$ occurs as a component of σ.*

The proof will occupy the rest of paragraph 9. (The case when $\sigma \otimes \chi \approx \sigma$ for some χ has already been discussed in Section 3.7; *see* also Remark 9.9).

(9.4) We prove first (9.3.2). Let S_0 be the set of places where σ_v is extraordinary. There is nothing to prove if $S_0 = \emptyset$. So assume $S_0 \neq \emptyset$, and let $S \supset S_0$ be the set of places v where σ_v is not quasi-unramified. For $v \notin S$, let π_v be the lift of σ_v. By (8.1) the assumptions of (9.2) are satisfied with $T = S$. From (9.2) we conclude that there is, for each $v \in S$, a generic representation π'_v, trivial on the center, such that, if

$$\pi' = \bigotimes_{v \in S} \pi'_v \otimes \bigotimes_{v \notin S} \pi_v,$$

hen $L(s, \pi' \otimes \chi)$ is meromorphic and satisfies the functional equation

$$L(s, \pi' \otimes \chi) = \varepsilon(s, \pi' \otimes \chi) L(1-s, \pi' \otimes \chi^{-1})$$

for all χ. On the other hand, $L_2(s, \sigma, \chi)$ is meromorphic and

$$L_2(s, \sigma, \chi) = \varepsilon_2(s, \sigma, \chi) L_2(1-s, \tilde{\sigma}, \chi^{-1})$$

for all χ. Since the local L and ε factors in both equations are in fact the same for all $v \notin S$ we conclude that

$$\prod_{v \in S} L(1-s, \tilde{\pi}'_v \otimes \chi_v^{-1}) \varepsilon(s, \pi'_v \otimes \chi_v; \psi_v)/L(s, \pi'_v \otimes \chi_v) = \prod_{v \in S} L_2(1-s, \tilde{\sigma}_v, \chi_v^{-1}) \varepsilon_2(s, \sigma_v, \chi_v; \psi_v)/L_2(s, \sigma_v, \chi_v). \tag{9.4.1}$$

Now fix $w \in S_0$ and let η be a character of $F_w^\times$. Choose χ in such a way that $\chi_w = \eta$ and χ_v is highly ramified for $v \in S$, $v \neq w$. Then for $v \in S$, $v \neq w$,

$$\begin{cases} L(s, \pi_v \otimes \chi_v) = L(s, \tilde{\pi}_v \otimes \chi_v^{-1}) = 1, \\ \varepsilon(s, \pi'_v \otimes \chi_v; \psi_v) = \varepsilon(s, \chi_v; \psi_v)^3, \end{cases} \tag{9.4.2}$$

and

$$\begin{cases} L_2(s, \sigma_v, \chi_v) = L_2(s, \tilde{\sigma}_v, \chi_v^{-1}) = 1, \\ \varepsilon_2(s, \sigma_v, \chi_v; \psi_v) = \varepsilon(s, \chi_v; \psi_v)^3; \end{cases} \tag{9.4.3}$$

[*cf.* [JPSS] Theorem (5.6) and [Ja] Theorem (16.1)]. We conclude that

$$\varepsilon(s, \pi'_w \otimes \eta; \psi_w) L(1-s, \tilde{\pi}'_w \otimes \eta^{-1})/L(s, \pi'_w \otimes \eta) = \varepsilon_2(s, \sigma_w, \eta; \psi_w) L_2(1-s, \tilde{\sigma}_w, \eta^{-1})/L_2(s, \sigma_w, \eta). \tag{9.4.4}$$

But now σ_w is extraordinary. Therefore

$$L_2(s, \sigma_w, \eta) = L_2(s, \tilde{\sigma}_w, \eta^{-1}) = 1$$

and the left hand side of (9.4.4) is a nomomial in q_w^{-s}. This, however, can happen only if π'_w is super cuspidal ([JPSS], § 7). Therefore

$$L(s, \pi'_w \otimes \eta) = L_2(s, \sigma_w, \eta) \quad (=1),$$

and

$$\varepsilon(s, \pi'_w \otimes \eta; \psi_w) = \varepsilon_2(s, \sigma_w, \eta; \psi_w).$$

Note that the functional equation for π' also reads

$$L(s, \tilde{\pi}' \otimes \chi) = \varepsilon(s, \tilde{\pi}' \otimes \chi^{-1}) L(1-s, \pi' \otimes \chi^{-1}),$$

so since $\pi'_v \simeq \pi_v \simeq \tilde{\pi}_v \simeq \tilde{\pi}'_v$ for $v \notin S$, we find

$$\prod_{v \in S} L(1-s, \tilde{\pi}'_v \otimes \chi_v^{-1}) \varepsilon(s, \tilde{\pi}'_v \otimes \chi_v; \psi_v)/L(s, \pi'_v \otimes \chi_v) = \prod_{v \in S} L(1-s, \pi'_v \otimes \chi_v^{-1}) \varepsilon(s, \pi'_v \otimes \chi_v; \psi_v)/L(s, \tilde{\pi}'_v \otimes \chi_v).$$

Threfore we conclude as before that

$$L(1-s, \tilde{\pi}'_w \otimes \eta^{-1})\,\varepsilon(s, \tilde{\pi}'_w \otimes \eta; \psi_w)/L(s, \pi'_w \otimes \eta)$$

is equal to the same expression for $\tilde{\pi}'_w$. But as π'_w is trivial on the center, this implies $\pi'_w \simeq \tilde{\pi}'_w$ (*cf.* (7.5.3) of [JPSS]). Thus π'_w is indeed the lift of σ_w.

(9.5) Now that (9.3.2) is proved, for each v we let π_v denote the lift of σ_v. For $v \in S$ [S being as in (9.4)], π_v is unitary generic and we set $\xi_v = \pi_v$. For $v \notin S$ we define ξ_v as in (9.2). Note that the only places where $\xi_v \not\doteq \pi_v$ are those for which

$$\text{(9.5.1)} \qquad \sigma_v = \pi(\lambda_v \alpha^{1/4}, \lambda_v \alpha^{-1/4})$$

in which case π_v is given by (9.1.2.1) and ξ_v by (9.1.2.2). Then ξ_v has another component ξ'_v which is unitary generic, namely

$$\text{(9.5.2)} \qquad \xi'_v = \operatorname{Ind}(G_{3,v}, P_v; \sigma(\alpha_v^{1/2}, \alpha_v^{-1/2}), 1_v),$$

where $\sigma(\mu_1, \mu_2)$ is defined as in (1.10.1).

Now set $\pi = \otimes\, \pi_v$ (all v). Then:

$$\text{(9.5.3)} \qquad \begin{cases} L(s, \pi \otimes \chi) = L_2(s, \sigma, \chi), \\ \varepsilon(s, \pi \otimes \chi) = \varepsilon_2(s, \sigma, \chi), \end{cases}$$

and $L(s, \pi \otimes \chi)$ is meromorphic and satisfies the functional equation

$$\text{(9.5.4)} \qquad L(s, \pi \otimes \chi) = \varepsilon(s, \tilde{\pi} \otimes \chi)\, L(1-s, \tilde{\pi} \otimes \chi^{-1}).$$

Finally, set $T = S$ if $S \neq \emptyset$, and $T = \{ w \}$ (where w is any finite place) if $S = \emptyset$.

If χ is *sufficiently ramified at each place of* T, then we know [*cf.* (8.1)] that $L(s, \pi \otimes \chi)$ is entire and bounded in vertical strips; moreover (*cf.* (5.1) of [JPSS]),

$$\text{(9.5.5)} \quad \varepsilon(s, \pi_v \otimes \chi_v, \psi_v) = [\varepsilon(s, \chi_v, \psi_v)]^3, \quad \text{and} \quad L(s, \pi_v \otimes \chi_v) = 1 \quad \text{for} \quad v \in T.$$

Therefore the discussion in (9.2) applies. In particular, we are in one of the cases 9.2 (ii) to 9.2 (v). What we are going to show is that 9.2 (ii) is the only possibility and this will imply the theorem.

(9.6) Suppose we are in case (iii). With the notations of (iii) we know therefore that the representations ξ_v and η_v have a common irreducible component for all $v \notin T$. Although the representations η_v, ξ_v, $\eta = \otimes\, \eta_v$, $\xi = \otimes\, \xi_v$ may fail to be irreducible, the factors $L(s, \xi_v)$, $L(s, \eta_v)$, $L(s, \xi)$, $L(s, \eta)$, $\varepsilon(s, \xi_v; \psi_v)$, $\varepsilon(s, \eta_v; \psi_v)$, $\varepsilon(s, \xi)$, and $\varepsilon(s, \eta)$ are defined. Moreover, for all character χ of $F_{\mathbb{A}}^\times/F^\times$,

$$L(s, \xi \otimes \chi) = \varepsilon(s, \xi \otimes \chi)\, L(1-s, \tilde{\xi} \otimes \chi^{-1}),$$

with a similar statement for $L(s, \eta \otimes \chi)$.

Arguing as we did in (9.4) we conclude that if v is any place in T and λ is a character of $F_v^\times$, then:

$$\begin{aligned}&\varepsilon(s, \xi_v \otimes \lambda; \psi_v) L(1-s, \tilde{\xi}_v \otimes \lambda^{-1})/L(s, \xi_v \otimes \lambda)\\&= \varepsilon(s, \eta_v \otimes \lambda; \psi_v) L(1-s, \tilde{\eta}_v \otimes \lambda^{-1})/L(s, \eta_v \otimes \lambda).\end{aligned}$$

But ξ_v and η_v have unique irreducible generic components ξ'_v and η'_v and the previous identity is true for these pairs too. So once more we conclude that $\xi'_v \simeq \eta'_v$. In particular, *for all* v, ξ_v and η_v have a common irreducible component. Recall also that for almost all finite v, π_v is unramified and equal to the common unramified component of ξ_v and η_v. To prove that (iii) is impossible we first suppose that μ is a character.

(9.6.1) If μ is a character then τ is unitary, each representation τ_v unitary generic, and it follows that η_v is irreducible (in fact unitary generic, *cf.* [JPSS], § 6). Thus η_v is an irreducible component of ξ_v for all v (at the infinite places in the infinitesimal sense only). If, in addition, ξ_v is irreducible then $\eta_v = \xi_v = \pi_v$ and

(9.6.1.1) $$L(s, \pi_v \otimes \mu_v^{-1}) = L(s, \eta_v \otimes \mu_v^{-1}) = L(s, \tau_v \otimes \mu_v^{-1}) L(s, 1_v).$$

On the other hand, if ξ_v is not irreducible, then π_v is given by (9.1.2.1), ξ_v by (9.1.2.2) (v being finite or not), and ξ_v admits another irreducible component which is unitary generic, namely ξ'_v (9.5.2). Since η_v is unitary generic, we find

$$\eta_v = \xi'_v;$$

in particular, η_v is ramified if v is finite. Thus the set E of places where ξ_v is reducible (and $\eta_v = \xi'_v$) is *finite*, and for $v \in E$ we find

(9.6.1.2) $$L(s, \pi_v \otimes \mu_v^{-1}) = L(s, \tau_v \otimes \mu_v^{-1}) L(s, 1_v) \cdot \frac{L(s, \pi(\alpha_v^{1/2}, \alpha_v^{-1/2}))}{L(s, \sigma(\alpha_v^{1/2}, \alpha_v^{-1/2}))}.$$

The factor on the extreme right is $(1-q_v^{1/2-s})^{-1}$ if v is finite, and $ca^s(s-(1/2))^{-1}$ if v is infinite.

In either case, it *does not vanish* for $s = 1$. Thus we find from (9.6.1.1) and (9.6.1.2) that

$$L(s, \pi \otimes \mu^{-1}) = L(s, \tau \otimes \mu^{-1}) L(s, 1) \times \prod_{v \in E} \frac{L(s, \pi(\alpha_v^{1/2}, \alpha_v^{-1/2}))}{L(s, \sigma(\alpha_v^{1/2}, \alpha_v^{-1/2}))}$$

and

$$L(s, (\sigma \otimes \mu^{-1}) \times \tilde{\sigma}) = L(s, \tau \otimes \mu^{-1}) L(s, 1) L(s, \mu^{-1}) \times \prod_{v \in E} \frac{L(s, \pi(\alpha_v^{1/2}, \alpha_v^{-1/2}))}{L(s, \sigma(\alpha_v^{1/2}, \alpha_v^{-1/2}))}.$$

But if μ is note a principal character then $L(1, \mu^{-1}) \neq 0$. Also $L(1, \tau \otimes \mu^{-1}) \neq 0$ by the main result of [J Sh]. So since the finite product above does not vanish at 1, we find that $L(s, (\sigma \otimes \mu^{-1}) \times \tilde{\sigma})$ has a pole at $s = 1$. This means $\sigma \otimes \mu^{-1} \simeq \sigma$, which is a contradiction. Thus μ cannot be a character.

(9.6.2) Now suppose $\mu = \alpha^{2t}\mu_0$ where μ_0 is a character and t is real and $\neq 0$. For almost all finite v, μ_0 is unramified and the representation τ_v is of the form $\tau_v = \pi(\alpha_v^{a_v}, \alpha_v^{b_v})$ The unramified component of η_v is therefore the representation $\pi(\alpha_v^{a_v}, \alpha_v^{b_v}, \alpha^{2t}\mu_{0,v})$ which must (for almost all v) coincide with the unramified component of ξ_v which is π_v. Now π_v has the form $\pi(\alpha_v^{p_v}, 1, \alpha_v^{-p_v},)$ and so the triples $(\alpha_v^{a_v}, \alpha_v^{b_v}, \alpha_v^{c_v})$ and $(\alpha_v^{p_v}, 1, \alpha_v^{-p_v})$ must be the same (up to order).

From the discussion of paragraph 3.3 and 3.4 it follows that σ_v must be in the complementary series for almost all finite v and, after a change of notations,

$$a_v = -2t, \qquad b_v = 0, \qquad \mu_{0,v} = 1, \qquad 0 < |t| < \frac{1}{2}.$$

Thus $\tau_v = \pi(\alpha_v^{-2t}, 1)$ for almost all v. This situation, however, is impossible. Indeed it implies the existence of a unitary cuspidal automorphic representation ρ of $G_2(\mathbf{A})$ with $\rho_v = \pi(\alpha_v^t, \alpha_v^{-t})$, $0 < t < (1/2)$, for all v not in some finite set of places E. This in turn implies [by (9.7)] that

$$L(s, \rho\times\tilde{\rho}) = L(s+t, 1)\,L(s-t, 1)\,L(s, 1)^2 \times \prod_{v\in E'} \frac{L(s, \rho_v\times\tilde{\rho}_v)}{L(s+t, 1_v)\,L(s-t, 1_v)\,L(s, 1_v)^2}.$$

But at $s = t$, $L(s+t, 1)$ and $L(s, 1)$ do not vanish. Neither does the finite product, as is easily checked. Thus $L(s, \rho\times\tilde{\rho})$ has a pole at $s = t \neq 1$, an obvious contradiction.

In summary, possibility (iii) cannot occur; similar arguments show that (iv) too is impossible.

(9.7) Suppose finally we are on case (v). Using the notations of (v) we see (just as before) that η_v and ξ_v have a common irreducible component for all v.

(9.7.1) Suppose first that the ν_i are characters. Then for each v the representation η_v is irreducible unitary generic ([JPSS], § 6 and 10). Thus it is a component of ξ_v. If ξ_v is irreducible we find $\xi_v = \pi_v = \eta_v$ and

(9.7.1.1) $$L(s, \pi_v\otimes\chi_v) = L(s, \eta_v\otimes\chi_v) = \prod_i L(s, \nu_{i,v}\chi_v).$$

If ξ_v is not irreducible then, since π_v is the lift of σ_v, π_v is given by (9.1.2.1) and ξ_v by (9.1.2.2). Since π_v is not unitary generic, η_v must be the other component of ξ_v [*cf.* (9.5.2)] i. e.:

$$\eta_v = \mathrm{Ind}(G_{3,v}, P_v; \sigma(\alpha_v^{1/2}, \alpha_v^{-1/2}), 1_v).$$

On the other hand,

$$\eta_v = \mathrm{Ind}(G_{3,v}, B_{3,v}; \nu_{1,v}, \nu_{2,v}, \nu_{3,v})$$

and—except in the complex case—these two equalities are incompatible.

Thus we let E denote the set of (complex) places where this happens. For $v \in \mathrm{E}$,

$$(9.7.1.2)\quad \mathrm{L}(s, \pi_v \otimes \chi_v) = \mathrm{L}(s, \xi_v \otimes \chi_v)$$
$$= \prod_{i=1}^{3} \mathrm{L}(s, \nu_{i,v}\chi_v) \cdot \frac{\mathrm{L}(s+1/2, \chi_v)\mathrm{L}(s, \chi_v)\mathrm{L}(s-1/2, \chi_v)}{\prod \mathrm{L}(s, \nu_{i,v}\chi_v)}.$$

Moreover we then have (up to order):

$$\nu_{1,v}(z) = z(z\bar{z})^{-1/2}, \qquad \nu_{2,v} = 1, \qquad \nu_{3,v} = \bar{z}(z\bar{z})^{-1/2}$$

and

$$\sigma(\alpha_v^{1/2}, \alpha_v^{-1/2}) = \pi(\nu_{1,v}, \nu_{3,v}).$$

From (9.7.1.1) and (9.7.1.2) we find that

$$\mathrm{L}(s, (\sigma \otimes \chi) \times \tilde{\sigma}) = \prod_{i=1}^{3} \mathrm{L}(s, \nu_i \chi)\mathrm{L}(s, \chi)$$
$$\times \prod_{v \in \mathrm{E}'} \frac{\mathrm{L}(s+1/2, \chi_v)\mathrm{L}(s, \chi_v)\mathrm{L}(s-1/2, \chi_v)}{\prod \mathrm{L}(s, \nu_{i,v}\chi_v)}.$$

Now set $\chi = \nu_1^{-1}$. Then $\mathrm{L}(s, \nu_1 \chi)$ has a pole at $s = 1$ which cannot be compensated for by a zero of the other L-factors (nor by a zero of the finite product, as is easily checked). Thus we get a pole for $\mathrm{L}(s, (\sigma \otimes \nu_1^{-1}) \times \tilde{\sigma})$ at $s = 1$. But if $\nu_1 \neq 1$ this contradicts our hypothesis (since it implies $\sigma \simeq \sigma \otimes \nu_1^{-1}$). On the other hand, if $\nu_1 = 1$, we get a pole of order at least two, and this is also a contradiction.

(9.7.2) Now suppose the ν_i are not all characters. Using again the fact that ξ_v and η_v have (for almost all v) the same unramified component (namely π_v) we conclude as before that σ_v is in the complementary series for almost all v and that (changing notation if necessary):

$$\nu_1 = \alpha^{2t}\nu'_1, \qquad \nu_3 = \alpha^{-2t}\nu'_3, \qquad \nu_2 = \nu'_2,$$

where ν'_i is a character and $t \neq 0$ is real. In fact we even find $\nu'_{i,v} = 1$ for almost all v so that $\nu_1 = \alpha^{2t}$, $\nu_3 = \alpha^{-2t}$, $\nu_2 = 1$. Moreover $0 < t < (1/2)$.

Suppose first that $t \neq (1/4)$. Then:

$$\eta_v = \mathrm{Ind}(\mathrm{G}_{3,v}, \mathrm{B}_{3,v}, \alpha_v^{2t}, 1, \alpha_v^{-2t})$$

is irreducible for all v and is a component of ξ_v, for all v. If ξ_v is also irreducible then $\xi_v = \pi_v = \eta_v$. On the other hand, if ξ_v is not irreducible then ξ_v has (as seen before) 2 components; one is π_v, the other

$$\xi'_v = \mathrm{Ind}(\mathrm{G}_{3,v}, \mathrm{P}_v; \sigma(\alpha_v^{1/2}, \alpha_v^{-1/2}), t_v).$$

But ξ'_v does not contain the trivial representation of $\mathrm{K}_{3,v}$ and η_v does. Thus $\pi_v = \eta_v$ again.

Summing up, if $t \neq 1/4$, then for all v,

$$L(s, \pi_v) = L(s, \eta_v) = L(s+2t, 1_v)L(s, 1_v)L(s-2t, 1_v)$$

and

$$L(s, \sigma\times\tilde{\sigma}) = L(s, \pi)L(s, 1) = L(s+2t, 1)L(s-2t, 1)L(s, 1)^2.$$

Again the pole of $L(s-2t, 1)$ at $2t$ is not cancelled by a zero and we conclude $L(s, \sigma\times\tilde{\sigma})$ has a pole at $s = 2t \neq 1$, a contradiction.

Now suppose $t = 1/4$. Then η_v has always two irreducible components:

$$\eta'_v = \mathrm{Ind}(G_{3,v}, P_v; 1_{2,v}, 1_v),$$
$$\eta''_v = \mathrm{Ind}(G_{3,v}, P_v; \sigma(\alpha_v^{1/2}, \alpha_v^{-1/2}), 1_v).$$

One of these is always a component of ξ_v. Since η'_v is unramified if v is finite, we find $\eta'_v = \pi_v$ for almost all v. If ξ_v is irreducible then either η'_v or η''_v is a component of ξ_v. If ξ_v is reducible then $\xi_v = \eta_v$ and $\pi_v = \eta'_v$. In conclusion, for almost all v, $\eta'_v = \pi_v$ and

$$L(s, \pi_v) = L(s, \eta'_v) = L\left(s+\frac{1}{2}, 1_v\right)L(s, 1_v)L\left(s-\frac{1}{2}, 1_v\right).$$

However, at a finite set of places (say E) we may have $\eta''_v = \pi_v$ and

$$L(s, \pi_v) = L\left(s+\frac{1}{2}, 1_v\right)L(s, 1_v)L\left(s-\frac{1}{2}, 1_v\right)\frac{L(s, \sigma(\alpha_v^{1/2}, \alpha_v^{-1/2}))}{L(s, 1_{2,v})}.$$

So

$$L(s, \sigma\times\tilde{\sigma}) = L\left(s+\frac{1}{2}, 1\right)L\left(s-\frac{1}{2}, 1\right)L(s, 1)^2 \prod_{v\in E}\frac{L(s, \sigma(\alpha_v^{1/2}, \alpha_v^{-1/2}))}{L(s, 1_{2,v})}.$$

Again the pole of $L(s-(1/2), 1)$ at $s = 3/2$ cannot be cancelled by a zero of the other L-factors or the finite product and we again get a contradiction.

(9.8) We have now shown that the only possibility is (ii). Thus $\pi_v = \xi_v$ is unitary generic for any $v \notin T$. If $S \neq \emptyset$, $S = T$ and the same conclusion is true for $v \in S$ [*cf.* (3.41]). On the other hand, if $s = \emptyset$, then $T = \{ w \}$. Since w is really arbitrary the same conclusion is true for all w. Thus $\pi_v = \xi_v$ is unitary generic *for all v*.

We have seen in [9.2 (ii)] that there is a cuspidal automorphic representation π' with trivial central character and $\pi'_v \simeq \pi_v$ for $v \notin S$. Now for any χ:

(9.8.1) $$L(s, \pi'\otimes\chi) = \varepsilon(s, \pi'\otimes\chi)L(1-s, \tilde{\pi}'\otimes\chi^{-1}).$$

Similarly $L(s, \pi\otimes\chi)$ satisfies (9.5.4). So using an argument used before, these two equations imply that for any $v \in T$ and any character λ of $F_v^\times$,

$$L(1-s, \tilde{\pi}_v\otimes\lambda^{-1})\varepsilon(s, \pi_v\otimes\chi; \psi_v)/L(s, \pi_v\otimes\lambda)$$
$$= L(1-s, \tilde{\pi}'_v\otimes\lambda^{-1})\varepsilon(s, \pi'_v\otimes\lambda; \psi_v)/L(s, \pi'_v\otimes\lambda).$$

However, since both π_v and π'_v are generic, this implies $\pi_v \simeq \pi'_v$. Thus $\pi = \otimes \pi_v$ is automorphic cupsidal and we have proved part (3) of Theorem (9.3). But part 1 follows from the identity $L_2(s, \sigma, \chi) = L_2(s, \pi \otimes \chi)$, and since the lift π_v of σ_v is always unitary generic, the last assertion of the Theorem also follows.

Remark (9.9). — If $\sigma \simeq \sigma \otimes \eta$, with $\eta \neq 1$, then $\eta^2 = 1$. Let H be the corresponding quadratic extension of F. Recall that there is a character Ω of $H_A^\times/H^\times$ such that σ is the automorphic representation $\pi(\sigma_\Omega)$ of $G_2(A)$ attached to Ω. Let Ω' be the conjugate of Ω. Then the lift π of σ is

$$\pi = \mathrm{Ind}(G_3(A), P(A); \pi(\sigma_{\Omega\Omega'^{-1}}), \eta)$$

and

$$L(s, \pi \otimes \chi) = L_2(s, \sigma, \chi) = L(s, \Omega\Omega'^{-1}. \chi \circ N_{H/F}) L(s, \eta\chi).$$

Recall that π is automorphic. On the other hand, $L(s, \pi \otimes \chi)$ has some pole (for $\chi = \eta$ for instance), so π is *not* cuspidal.

Concluding Remark. — We wish to comment finally on the classical significance of part (4) of Theorem 9.3. Suppose

$$f(z) = \prod_{n=1}^{\infty} a_n e^{2\pi i n z}$$

is a cusp form of weight k on $\Gamma_0(N)$ (which is not a theta-series with grossencharakter à la Hecke-Maass). By the Ramanujan-Petersson conjecture recently established by Deligne we know that

$$(\star) \qquad a_n = 0(n^{k/2-1/2+\varepsilon}).$$

On the other hand, if $F = \mathbf{Q}$ and π corresponds to the form f above, then part (4) of Theorem (9.3) implies that

$$(\star\star) \qquad a_n = O(n^{k/2-1/4+\varepsilon}).$$

This estimate is stronger than the one made by Rankin in his well-known work [Ra] but obviously weaker than ($\star$). According to [Se] ($\star\star$) also results from Weil's work on exponential sums (*cf.* [W 2]). For number fields, however, our result seems to be new. Indeed the generalized Ramanujan-Petersson conjecture—though established for function fields now by Drinfeld—has not yet been proven in general.

REFERENCES

[Bo] A. Borel, *Automorphic L-functions* (*Proc. of Symposia in Pure Math.*, Corvallis Conference, Oregon 1977, *American Math. Soc.*, Vol. 33, 1979).

[De] P. Deligne, *Les constantes des équations fonctionnelles des fonctions* L (*Springer Lecture Notes*, Vol. 349, 1973, pp. 501-598).

[Ge] S. Gelbart, *Weil's Representation and the Spectrum of the Metaplectic Group* (*Springer Lecture Notes*, Vol. 530, 1976).

[Ge Ja] S. GELBART and H. JACQUET, *A Relation between Automorphic Forms on* GL (2) *and* GL (3) (*Proc. Nat. Acad. Sc. U.S.A.*, Vol. 73, No. 10, october 1976, pp. 3348-3350).

[Ge J 1] S. GELBART and H. JACQUET, *Forms of* GL (2) *from the Analytic Point of View* (*Proc. of Symposia in Pure Math.*, *American Math. Soc.*, Vol. 33, 1979).

[Go Ja] R. GODEMENT and H. JACQUET, *Zeta-functions of Simple Algebras* (*Springer Lecture Notes*, Vol. 260, 1972).

[Ja] H. JACQUET, *Automorphic Forms on* GL (2): *Part II* (*Springer Lecture Notes*, Vol. 278, 1972).

[Jl] H. JACQUET, *Principal* L*-functions of the linear group* (*Proc. of Symposia in Pure Math.*, *American Math. Soc.*, Vol. 33, 1979).

[JL] H. JACQUET and R. LANGLANDS, *Automorphic Forms on* GL (2) (*Springer Lecture Notes*, Vol. 114, 1970).

[JPSS] H. JACQUET, I. I. PIATETSKI-SHAPIRO and J. SHALIKA, *Automorphic Forms on* GL (3) (*Annals of Math.*, Vol. 109, 1979).

[JSh] H. JACQUET and J. SHALIKA, *A non-vanishing Theorem for zeta-functions of* GL_n (*Inventiones Math.*, 38, 1976, pp. 1-16).

[LL] J.-P. LABESSE and R. P. LANGLANDS, L*-indistinguishability for* SL (2), Institute for Advanced Study Princeton, 1977, (preprint).

[La 1] R. P. LANGLANDS, *Problems in the Theory of Automorphic Forms* (*Springer Lecture Notes in Mathematics*, Vol. 170, 1970).

[La 2] R. P. LANGLANDS, *On the Functional Equation of the Artin* L*-Functions*, mimeographed notes, Yale University, 1968.

[La 3] R. P. LANGLANDS, *On the Functional Equations satisfied by Eisenstein Series*, mimeographed notes, Yale University, 1967.

[La 4] R. P. LANGLANDS, *Base Change for* GL (2) : *The Theory of Saïto-Shintani with Applications*, Institute for Advanced Study, Princeton, 1976, (preprint).

[La 5] R. P. LANGLANDS, *On the Notion of an Automorphic Rrepresentation* [*Proc. of Symposia in Pure Mathematics*, Corvallis Conference, Oregon 1977 (to appear)].

[Ra] R. RANKIN, *Contributions to the Theory of Ramanujan's Function* $\tau(n)$ *and Similar Arithmetical Functions* (*Proc. Camb. Phil. Soc.*, Vol. 35, 1939, pp. 351-356).

[Si] A. SILBERGER, *The Langlands quotient theorem for p-adic groups* (preprint).

[Se] A. SELBERG, *On the Estimation of Fourier Coefficients of Modular Forms* (*Proc. of Symposia in Pure Math.*, Vol. 8, A.M.S., Providence, R.I., 1965).

[Sh] G. SHIMURA, *On the Holomorphy of Certain Dirichlet Series* (*Proc. London Math. Soc.*, (3) 31, 1975, pp. 79-95).

[Ta] J. TATE, *Fourier Analysis in Number Fields and Hecke's zeta-function*, in *Algebraic Number Theory* (*Proc. of the Brighton Conference*, Academic Press, New York, 1968).

[We 1] A. WEIL, *Sur certains groupes d'opérateurs unitaires* (*Acta Math.*, Vol. 111, 1964, pp. 143-211).

[We 2] A. WEIL, *On some Exponential Sums* (*Proc. Nat. Acad. sc., U.S.A.*, Vol. 34, 1948, pp. 204-207).

(Manuscrit reçu le 4 janvier 1978,
révisé le 21 mars 1978.)

S. GELBART,
Cornell University,
Ithaca, New York 14853,
U.S.A.

Math. Ann. 256, 199–214 (1981)

Mathematische Annalen

Conducteur des représentations du groupe linéaire

H. Jacquet[1], I. I. Piatetski-Shapiro[2], and J. Shalika[3]

[1] Department of Mathematics, Columbia University, New York, NY 10027, USA
[2] Department of Mathematics, The Johns Hopkins University, Baltimore, MD 21218, USA
[3] Department of Mathematics, Yale University, New Haven, CT 06520, USA

1. Introduction et notations

(1.1) Soit F un corps local non archimédien, ψ un caractère additif non trivial de F, π une représentation admissible et irréducible du groupe $G_r = \mathrm{GL}(r, F)$ sur un espace complexe V. Dans [G–J], on a défini des facteurs $L(s,\pi)$ et $\varepsilon(s,\pi,\psi)$ qui sont de la forme:

$$\begin{aligned} L(s,\pi) &= P_\pi(q^{-s})^{-1}, \quad P_\pi \in \mathbb{C}[X], \quad P_\pi(0) = 1\,; \\ \varepsilon(s,\pi,\psi) &= \varepsilon_\pi(q^{-s},\psi), \quad \varepsilon_\pi(X,\psi) = cX^m\,; \end{aligned} \tag{1}$$

où l'on désigne par q le module du corps F. Nous ferons toujours l'hypothèse que l'exposant du caractère ψ est 0, c'est à dire que le plus grand idéal de F sur lequel ψ est trivial est l'anneau des entiers $\mathfrak{R}$. Lorsqu'il en est ainsi, l'entier m ne dépend que de π.

Par exemple, si $r=1$, alors π est un caractère de $F^\times$ (non nécessairement de module un) et l'entier m est positif. De plus si $m=0$ le caractère est non ramifié c'est-à-dire de la forme $\pi(x) = |x|^s$. Si au contraire m est strictement positif alors π est ramifié; de plus, si $\mathfrak{P}$ est l'idéal maximal de l'anneau $\mathfrak{R}$, on sait que π est trivial sur le groupe $1+\mathfrak{P}^m$ et non trivial sur les groupes $1+\mathfrak{P}^a$ avec $0<a<m$.

Pour $r>1$, on peut se demander si l'entier m est positif. On peut se demander aussi s'il existe une interprétation de l'entier m analogue à celle rappelée ci-dessus. Nous allons démontrer que tel est bien le cas, pourvu que la représentation π soit «générique» [cf. (1.2)]. Par exemple si $r=2$, alors une représentation générique est simplement une représentation de dimension infinie et le résultat est maintenant classique [C, D, N].

Cette question avait été d'abord soulevée par A. Weil (pour $r>2$) et résolue, tout au moins dans le cas cuspidal, par D. Kazdan et M. Novodvorsky. Peut-être d'ailleurs d'autres mathématiciens ont-ils, avant nous, résolu cette question complètement. Aussi ne prétendons nous à aucune espèce d'originalité. Mais comme beaucoup de mathématiciens se sont posés ou nous ont posé cette question, il ne nous à pas paru tout à fait inutile de publier notre démonstration. Elle est du reste calquée sur celle donnée dans [C] pour le cas $r=2$.

0025-5831/81/0256/0199/$03.20

(1.2) Introduisons maintenant les notations et les définitions dont nous aurons besoin. Nous noterons $\mathfrak{R}$, $\mathfrak{P}$, ω, v, q l'anneau des entiers, l'idéal maximal, une uniformisante, la valuation normalisée et le module du corps F respectivement. Nous poserons $\alpha(x)=|x|=q^{-v(x)}$. En particulier $|\omega|=q^{-1}$.

Soit A_r le sous-groupe des matrices diagonales dans G_r. Si les coefficients diagonaux de la matrice $a\in A_r$ sont les nombres $(a_1, a_2, \ldots, a_r)$ nous écrirons

$$a=\operatorname{diag}(a_1, a_2, \ldots, a_r). \tag{1}$$

Soit N_r le groupe trigonal strict supérieur dans G_r. Un caractère additif non trivial ψ de F étant choisi, nous définirons un caractère $\theta_{\psi,r}$ de N_r par la formule

$$\theta_{\psi,r}(n)=\prod_{i=1}^{r-1} \psi(n_{i,i+1}). \tag{2}$$

Une représentation admissible et irréductible π de G_r est dite *générique* s'il existe un espace $\mathscr{W}$ de fonctions sur G_r se transformant à gauche par $\theta_{\psi,r}$ qui soit invariant par translations à droite, la représentation de G_r sur $\mathscr{W}$ étant équivalente à π. L'espace $\mathscr{W}$ est alors unique et se note $\mathscr{W}(\pi;\psi)$. La condition d'être générique ne dépend pas de ψ. D'ailleurs si ψ_1 est un autre caractère additif non trivial de F, on a $\psi_1(x)=\psi(a_1 x)$ avec a_1 dans $F^\times$. Pour W dans $\mathscr{W}(\pi;\psi)$ la fonction W_1 définie par

$$W_1(g)=W(ag), \qquad a=\operatorname{diag}(a_1^{r-1}, a_1^{r-2}, \ldots, a_1, 1),$$

est dans $\mathscr{W}(\pi;\psi_1)$. Enfin si π est générique la représentation contragrediente $\tilde{\pi}$ l'est aussi. De plus, si W est dans $\mathscr{W}(\pi;\psi)$, alors la fonction $\tilde{W}$ définie par

$$\tilde{W}(g)=W(w_r {}^t g^{-1}), \qquad w_r=\begin{pmatrix} 0 & & 1 \\ & 1 & \\ 1 & & 0 \end{pmatrix}, \tag{3}$$

est dans $\mathscr{W}(\tilde{\pi};\bar{\psi})$, où l'on désigne par $\bar{\psi}$ l'imaginaire conjugué de ψ.

(1.3) Quelques remarques sur l'organisation de cet article. Nous vérifions au Par. 2 des propriétés simples mais essentielles des polynômes $P_\pi(X)$ pour π générique. Au Par. 3 nous étudions les propriétés des fonctions se transformant à gauche par $\theta_{\psi,r}$ qui sont, à droite, fonctions propres de l'algèbre de Hecke. Au Par. 4, imitant la méthode du cas $r=2$, nous définissons le «vecteur essentiel» de l'espace $\mathscr{W}(\pi;\psi)$. Pour cela nous utilisons l'équation fonctionnelle (4.2)(1). Avec les résultats rappelés ou prouvés au Par. 2, c'est la seule propriété des facteurs L et ε dont nous aurons besoin. Les propriétés du vecteur essentiel sont établies au Par. 5: elles découlent aussitôt de l'identité (4.2)(1).

2. Le facteur L des représentations génériques

(2.1) Le but de ce paragraphe est de prouver le résultat suivant:

Proposition. *Soit π une représentation générique irréductible de G_r. Alors les polynômes*

$$P_\pi(X) \quad et \quad P_{\tilde{\pi}}(q^{-1}X^{-1})$$

sont premiers entre eux dans l'anneau $\mathbb{C}[X, X^{-1}]$.

(2.2) Nous aurons à utiliser quelques résultats simples sur les représentations génériques et les représentations induites. Soit Q le sous-groupe parabolique standard de type $(r_1, r_2, \ldots, r_n)$ de G_r. Pour chaque entier $i, 1 \leqq i \leqq n$, soit π_i une représentation admissible et irréductible de G_{r_i}. Nour noterons $I(\pi_1, \pi_2, \ldots, \pi_n)$ la représentation de G_r induite par la représentation suivante de Q:

$$\begin{pmatrix} g_1 & & \\ 0 & g_2 & * \\ \vdots & & \ddots \\ 0 & & & g_n \end{pmatrix} \mapsto \pi_1(g_2) \otimes \pi_2(g_2) \otimes \ldots \otimes \pi_n(g_n).$$

Cette représentation admet une suite de composition finie; pour que l'un des quotients irréductibles de la suite de composition soit générique, il faut et il suffit que chacun des π_i le soit, et il y a alors un seul quotient irreductible générique (cf. [R]). Si tous les π_i sont génériques nous noterons $\sigma(\pi_1, \pi_2, \ldots, \pi_n)$ l'unique composant générique de $I(\pi_1, \pi_2, \ldots, \pi_r)$. La «transitivité» de l'opération d'induction montre aussitôt que si les π_i sont génériques, alors:

$$\sigma(\pi_1, \pi_2, \ldots, \pi_n) = \sigma[\sigma(\pi_1, \pi_2, \ldots, \pi_{j-1}), \sigma(\pi_j, \pi_{j+1}, \ldots, \pi_n)]. \tag{1}$$

Comme la représentation contragrédiente d'une représentation générique est elle même générique, la relation évidente

$$I(\pi_1, \pi_2, \ldots, \pi_n)^{\sim} = I(\tilde{\pi}_1, \tilde{\pi}_2, \ldots, \tilde{\pi}_n). \tag{2}$$

entraîne

$$\sigma(\pi_1, \pi_2, \ldots, \pi_n)^{\sim} = \sigma(\tilde{\pi}_1, \tilde{\pi}_2, \ldots, \tilde{\pi}_n). \tag{3}$$

De même, on a

$$I(\pi_1, \pi_2, \ldots, \pi_n) \otimes \mu = I(\pi_1 \otimes \mu, \pi_2 \otimes \mu, \ldots, \pi_n \otimes \mu) \tag{4}$$

pour tout caractère μ de $F^{\times}$. D'où:

$$\sigma(\pi_1, \pi_2, \ldots, \pi_r) \otimes \mu = \sigma(\pi_1 \otimes \mu, \pi_2 \otimes \mu, \ldots, \pi_r \otimes \mu). \tag{5}$$

Enfin toute représentation admissible et irréductible de G_r est un composant d'une représentation induite $I(\pi_1, \pi_2, \ldots, \pi_2)$ où les π_i sont cuspidales («supercuspidal» dans la terminologie d'Harisch-Chandra); de plus, si $I(\pi_1, \pi_2, \ldots, \pi_n)$ et $I(\zeta_1, \zeta_2, \ldots, \zeta_s)$ ont un composant irréductible commun, alors $n = s$ et $(\pi_1, \pi_2, \ldots, \pi_n) = (\zeta_1, \zeta_2, \ldots, \zeta_n)$, à l'ordre près. Il s'en suit que toute représentation générique est de la forme

$$\pi = \sigma(\pi_1, \pi_2, \ldots, \pi_n),$$

où les π_i sont cuspidales et $(\pi_1, \pi_2, \ldots, \pi_n)$ est bien déterminé à l'ordre près.

Par exemple, si Q est le sous-groupe parabolique de type $(1, 1, \ldots, 1)$ dans G_r et si $\pi_1, \pi_2, \ldots, \pi_r$ sont des caractères de $F^{\times}$ tels que $\pi_i(\pi_{i+1})^{-1} = \alpha$, alors on a

$$\pi_i = \mu \alpha^{\frac{1}{2}(r+1)-i}, \quad 1 \leqq i \leqq r,$$

et

$$\sigma(\pi_1, \pi_2, \dots, \pi_r) = \mu \otimes \sigma_r,$$

où σ_r est la représentation spéciale de Steinberg : cela résulte par exemple de ce que σ_r est de carré intégrable, donc générique (cf. [G–J, Par. 7 ; J2]).

(2.3) Vérifions maintenant la proposition (2.1) pour une représentation de la forme

$$\pi = \mu \otimes \sigma_r. \tag{1}$$

On a alors $\tilde{\pi} = \mu^{-1} \otimes \sigma_r$. Si μ est ramifié la proposition est évidente car alors

$$P_\pi(X) = 1, \qquad P_{\tilde{\pi}}(X) = 1. \tag{2}$$

Si μ n'est pas ramifié, alors $\mu = \alpha^t$ pour un t convenable et [G–J, Par. 7] :

$$P_{\mu \otimes \sigma_r}(X) = P_{\sigma_r}(q^{-t}X) = 1 - q^{-t-\frac{1}{2}(r-1)}X, \tag{3}$$

$$P_{\mu^{-1} \otimes \sigma_r}(X) = 1 - q^{t-\frac{1}{2}(r-1)}X. \tag{4}$$

La proposition est donc encore évidente.

Passons maintenant au cas d'une représentation de la forme

$$\pi = \sigma(\pi_1, \pi_2, \dots, \pi_r) \tag{5}$$

où les π_i sont des caractères non ramifiés.

Nous pouvons supposer que la suite $(\pi_1, \pi_2, \dots, \pi_r)$ est de la forme

$$\left.\begin{array}{l} \mu_1^1, \mu_2^1, \dots, \mu_{r_1}^1, \mu_1^2, \mu_2^2, \dots, \mu_{r_2}^2, \dots, \mu_1^m, \mu_2^m, \dots, \mu_{r_m}^m \\ \mu_k^i (\mu_{k+1}^i)^{-1} = \alpha, \qquad \mu_{r_i}^i (\mu_1^j)^{-1} \neq \alpha, \quad \text{si} \quad i \neq j. \end{array}\right\} \tag{6}$$

En posant $\mu_1^i = \alpha^{\mu_i}$, nous avons donc

$$\sigma(\mu_1^i, \mu_2^i, \dots, \mu_{r_i}^i) = \sigma_{r_i} \otimes \alpha^{u_i - \frac{r_i - 1}{2}},$$

$$\pi = \sigma\left(\sigma_{r_1} \otimes \alpha^{u_1 - \frac{r_1 - 1}{2}}, \dots, \sigma_{r_m} \otimes \alpha^{u_m - \frac{r_m - 1}{2}}\right).$$

Il résulte alors de [J2, (2.7.4)] que $P_\pi[X]$ divise le produit

$$\prod_i P_{\sigma_{r_i}}\left(q^{-u_i + \frac{1}{2}(\pi_i - 1)}X\right).$$

De même $P_{\tilde{\pi}}(X)$ divise le produit

$$\prod_i P_{\sigma_{r_i}}\left(q^{u_i - \frac{1}{2}(r_i - 1)}X\right).$$

Si $P_\pi(X)$ et $P_{\tilde{\pi}}(q^{-1}X^{-1})$ ont un facteur commun (non trivial) dans l'anneau $\mathbb{C}[X, X^{-1}]$, il existe au moins un couple (i, j) tel que les polynômes

$$P_{\sigma_{r_i}}\left(q^{-u_i + \frac{1}{2}(r_i - 1)}X\right) = 1 - q^{-u_i}X,$$

$$P_{\sigma_{r_j}}\left(q^{u_j - \frac{r_j - 1}{2}} q^{-1} X^{-1}\right) = 1 - q^{u_j - r_j} X^{-1},$$

soient identiques, au produit près par une unité. C'est impossible si $i=j$. Si $i\neq j$, cela impose $\alpha^{u_j-r_j}=\alpha^{u_i}$ ou encore $\mu_1^i(\mu_{r_j}^j)^{-1}=\alpha^{-1}$. Or ceci contredit (6). La proposition est donc prouvée dans ce cas.

(2.4) En général, écrivons π sous la forme

$$\pi=\sigma(\pi_1,\pi_2,\ldots,\pi_j,\pi_{j+1},\ldots,\pi_n),$$

où les représentations cuspidales $\pi_1,\pi_2,\ldots,\pi_j$ ne sont pas des caractères non-ramifiés tandis que $\pi_{j+1},\ldots,\pi_n$ sont des caractères non ramifiés de $F^\times$. Alors:

$$\pi=\sigma(\pi_1,\pi_2,\ldots,\pi_j,\zeta) \quad \text{avec} \quad \zeta=\sigma(\pi_{j+1},\ldots,\pi_n).$$

Toujours d'après [J2, (2.7.4)] le polynôme $P_\pi(X)$ divise le produit

$$P_\zeta(X)\prod_{1\leqq i\leqq j}P_{\pi_i}(X)$$

Mais $P_{\pi_i}(X)=1$ [loc. cit. (3.1)]. Donc $P_\pi(X)$ divise en fait $P_\zeta(X)$. De même $P_{\tilde\pi}(q^{-1}X^{-1})$ divise $P_{\tilde\zeta}(q^{-1}X^{-1})$. D'aprés (2.3) ces polynômes sont premiers entre eux. Le résultat s'en suit.

(2.5) Nous aurons aussi besoin du résultat auxiliaire suivant:

Proposition. *Soit π une représentation irréducible générique de G_r. Le degré de P est au plus r. Pour que π contienne la représentation unité du sous groupe compact maximal $K_r=\mathrm{GL}(r,\mathfrak{R})$, if faut et il suffit que le degré de P_π soit r.*

Démonstration. Ecrivons π sous la forme

$$\pi=\sigma(\pi_1,\pi_2,\ldots,\pi_m)$$

où les π_i sont cuspidales et $m\leqq r$. Une fois de plus $P_\pi(X)$ divise le produit $\prod P_{\pi_i}(X)$. Comme chaque P_{π_i} est de degré au plus un, on voit que P_π est de degré au plus r. Supposons maintenant que P_π soit de degré r. Alors $m=r$ et chaque P_{π_i} doit être de degré un. Cela signifie que chaque π_i est un caractère non ramifié de $F^\times$. Supposons la représentation induite $\zeta=I(\pi_1,\pi_2,\ldots,\pi_r)$ nonirréductible. Alors il existe au moins un couple (i,j) tel que $\pi_i\cdot\pi_j^{-1}=\alpha$ [B–Z, Theorem (4-11)]. On peut supposer que c'est le couple $(1,2)$. Alors on a:

$$\pi=\sigma(\zeta,\pi_3,\ldots,\pi_r), \qquad \zeta=\sigma(\pi_1,\pi_2)=\sigma_2\otimes\mu.$$

A nouveau P_π divise le produit

$$P_\zeta\prod_{i=3}^{r}P_{\pi_i}.$$

Mais comme P_ζ est de degré un [G–J, Par. 7] on obtient une contradiction. Ainsi ξ est irréductible et $\pi=\xi$. Donc π contient la représentation unité de K_r.

En sens inverse, si π est une représentation irréductible qui contient la représentation unité de K, ou comme nous dirons encore, une représentation *non-ramifiée*, alors π est le composant non-ramifié d'une représentation induite

$\xi = I(\pi_1, \pi_2, \ldots, \pi_r)$ où les π_i sont des caractères non-ramifiés de $F^\times$. Alors

$$P_\pi(X) = \prod_{1 \leqq i \leqq r} P_{\pi_i}(X)$$

et chaque P_{π_i} est de degré un [G–J, Par. 6]. Donc P_π est de degré r.

(2.6) *Remarque.* On peut montrer que l'énoncé obtenu en supprimant dans la Proposition (2.5) le mot générique est encore vrai.

3. Fonctions de Whittaker formelles

(3.1) Nous aurons à utiliser des résultats élémentaires sur les fonctions se transformant à gauche par le caractère $\theta = \theta_{\psi, r}$. Nous les énonçons une fois pour toutes.

Pour g dans G_r posons

$$\begin{aligned} &\alpha_i(g) = a_i / a_{i+1}, \qquad 1 \leqq i \leqq r-1 \\ &\alpha_r(g) = a_r \end{aligned} \tag{1}$$

si $g = nak$ avec $n \in N_r$, $k \in K_r$ et

$$a = \operatorname{diag}(a_1, a_2, \ldots, a_r).$$

Soit φ une fonction lisse à droite telle que

$$\varphi(ng) = \theta(n)\varphi(g).$$

Alors il existe une constante C telle que

$$\varphi(g) \neq 0 \Rightarrow |\alpha_i(g)| \leqq C, \qquad 1 \leqq i \leqq r-1. \tag{2}$$

En effet, comme φ n'a qu'un nombre fini de translatées à droite par K_r, il suffit de montrer l'existence d'un C tel que

$$\varphi(a) \neq 0 \Rightarrow |\alpha_i(a)| \leqq C, \qquad 1 \leqq i \leqq r-1.$$

Mais en prenant

$$n = \begin{pmatrix} 1 & x_1 & & 0 \\ & 1 & \ddots & \\ & & \ddots & \\ 0 & & 1 & x_{r-1} \end{pmatrix}$$

on doit avoir

$$\varphi(an) = \varphi(ana^{-1}a) = \prod_{i=1}^{r-1} \psi(\alpha_i(a) x_i) \varphi(a).$$

Or $\varphi(a) = \varphi(an)$ si les $|x_i|$ sont assez petits. D'où la conclusion. On voit même que si φ est K_r-invariant alors on peut prendre $C = 1$.

De même on a

$$\varphi\left[\begin{pmatrix} g & 0 \\ 0 & 1 \end{pmatrix}\begin{pmatrix} 1_{r-1} & u \\ 0 & 1 \end{pmatrix}\right] = \varphi\left[\begin{pmatrix} g & 0 \\ 0 & 1 \end{pmatrix}\right] \psi[(0, \ldots, 0, 1) g u]. \tag{3}$$

On en conclut que la relation

$$\varphi\left[\begin{pmatrix} g & 0 \\ 0 & 1 \end{pmatrix}\begin{pmatrix} 1_{r-1} & u \\ 0 & 1 \end{pmatrix}\right]=\varphi\left[\begin{pmatrix} g & 0 \\ 0 & 1 \end{pmatrix}\right], \quad \text{pour tout } \begin{pmatrix} 1_{r-1} & u \\ 0 & 1 \end{pmatrix}\in K_r,$$

a lieu pour tout g, si et seulement si la relation

$$\varphi\begin{pmatrix} g & 0 \\ 0 & 1 \end{pmatrix} \neq 0$$

implique que la dernière ligne de g a tous ses coefficients entiers.

(3.2) Nous aurons à utiliser des résultats connus sur les représentations «non-ramifiées» du groupe G_r. Soit H_r l'algèbre de convolution des fonctions bi-invariantes par $K_r = \mathrm{GL}(r, \mathfrak{R})$ et à support compact. On sait que cette algèbre est isomorphe à l'algèbre S_r des polynômes en $X_1, X_1^{-1}, X_2, X_2^{-1}, \ldots, X_r, X_r^{-1}$ qui sont invariants par permutation des variables. Notons $T_1, T_2, \ldots, T_r$ les fonctions symétriques élémentaires des X_i. Ainsi:

$$T_1 = \sum_i X_i, \quad T_2 = \sum_{i<j} X_i X_j, \ldots, T_r = \prod_i X_i. \tag{1}$$

Alors S_r est encore isomorphe à l'algèbre de polynômes $\mathbb{C}[T_1, T_2, \ldots, T_r, T_r^{-1}]$.

Dans la suite nous adopterons les conventions suivantes pour les mesures de Haar et les mesures invariantes. La mesure de Haar de G_r (et de K_r) est normalisée par la condition $\mathrm{mes}(K_r)=1$, celle de N_r par $\mathrm{mes}(K_r \cap N_r)=1$. La mesure invariante de $N_r \backslash G_r$ est le quotient des mesures précédentes.

Soit maintenant $(x_1, x_2, \ldots, x_r)$ une suite de r nombres *non nuls*. Elle détermine un morphisme $S_r \to \mathbb{C}$, à savoir

$$P \mapsto P(x_1, x_2, \ldots, x_r). \tag{2}$$

L'homomorphisme correspondant $\lambda: H_r \to \mathbb{C}$ s'obtient ainsi. Soit $I(x_1, x_2, \ldots, x_r)$ la représentation de G_r induite par le caractère

$$\begin{pmatrix} a_1 & & \\ & a_2 & * \\ 0 & & \ddots \; a_r \end{pmatrix} \mapsto x_1^{v(a_1)} x_2^{v(a_2)} \ldots x_r^{v(a_r)} \tag{3}$$

du groupe trigonal supérieur. Alors l'espace de cette représentation contient un vecteur et un seul fixé par K_r et égal à un sur K_r. Soit f ce vecteur. Pour tout $\varphi \in H$ on a:

$$\int f(gh)\varphi(h)\,dh = \lambda(\varphi) f(g). \tag{4}$$

(3.3) Soit maintenant ψ un caractère additif non trivial de F d'exposant 0. Pour toute suite de nombres complexes non nuls $(x_1, x_2, \ldots, x_r)$, il existe une fonction W et une seule telle que

$$W(ngk) = \theta(n) W(g), \qquad n \in N, \qquad k \in K; \tag{1}$$

$$W(e) = 1; \tag{2}$$

pour tout $\varphi \in H$

$$\int W(gh)\varphi(h)dh = \lambda(\varphi)W(g), \tag{3}$$

où λ est le morphisme de H dans $\mathbb{C}$ décrit dans (3.2) (cf. [S]). Nous noterons cette fonction

$$W(g, x_1, x_2, \ldots, x_r; \psi). \tag{4}$$

Il résulte de l'unicité que

$$W(w{}^tg^{-1}, x_1, x_2, \ldots, x_r; \psi) = W(g, x_1^{-1}, x_2^{-1}, \ldots, x_r^{-1}; \bar{\psi}). \tag{5}$$

De plus si $q^{-s} = x$, alors

$$W(g, x_1, x_2, \ldots, x_r; \psi)|\det g|^s = W(g, x_1 x, x_2 x, \ldots, x_r x; \psi). \tag{6}$$

La formule explicite donnée en [S] montre qu'il existe pour chaque $g \in G_r$ un élément $W(g, X_1, X_2, \ldots, X_r; \psi)$ de S_r dont $W(g, x_1, x_2, \ldots, x_r; \psi)$ soit la spécialisation. On a ainsi défini une fonction sur G_r à valeurs dans S_r; d'ailleurs ses valeurs sur un ensemble compact modulo N_r sont dans un sous-espace vectoriel de dimension *finie* de S_r. D'après (3.1), le support de $W(g, X_1, X_2, \ldots, X_r; \psi)$ est contenu dans l'ensemble des g tels que

$$|\alpha_i(g)| \leqq 1, \quad \text{pour} \quad 1 \leqq i \leqq r-1. \tag{7}$$

Si de plus $|\alpha_r(g)| \leqq 1$ alors $W(g, X_1, X_2, \ldots, X_r; \psi)$ est en fait dans $\mathbb{C}[X_1, X_2, \ldots, X_r]$.

(3.4) Nous aurons à considérer des «intégrales» de la forme

$$\int_{N_r \backslash G_r} W(g, X_1, X_2, \ldots, X_r; \psi)\varphi(g)|\det g|^s dg, \tag{1}$$

où φ est une fonction se transformant à gauche par $\theta_{\bar{\psi}, r} = \bar{\theta}_{\psi, r}$. La fonction φ sera supposée lisse à droite, c'est-à-dire invariante à droite par un sous-groupe compact ouvert de G_r. De plus le support de φ sera toujours contenu dans l'ensemble des g tels que

$$|\alpha_r(g)| \leqq C, \tag{2}$$

où C est une constante non nulle. L'intégrale est à interpréter comme série formelle

$$\sum a_n X^n, \quad X = q^{-s}, \tag{3}$$

à coefficients dans S_r, les coefficients étant donnés par l'intégrale ordinaire

$$a_n = \int_{N_r \backslash G_r} W(g, X_1, X_2, \ldots, X_r; \psi)\varphi(g)\mu_n(\det g)dg, \tag{4}$$

où μ_n est la fonction caractéristique de l'ensemble

$$\{x \in F^\times | v(x) = n\} \tag{5}$$

dans $F^\times$. Il est clair que les conditions (3.3)(7), (3.4)(2) et $v(\det g) = n$ impliquent que g est dans un ensemble compact mod N_r: la fonction à intégrer dans (4) est donc à support compact mod N_r et l'intégrale a bien un sens. De plus notons que les relations (3.3)(7) et (3.4)(2) entrainent $v(\det g) \geqq m$, où m est un entier qui dépend de C. Il en résulte que dans la série (3) on a $a_n = 0$ pour $n < m$.

(3.5) **Lemme.** *Gardons les hypothèses de* (3.4) *mais supposons de plus que* φ *soit invariante à droite par* K_r. *Alors la relation*

$$\int W(g, X_1, X_2, \ldots, X_r; \psi)\varphi(g)|\det g|^s dg = 0$$

entraine $\varphi = 0$.

Démonstration. If suffit de démontrer le lemme lorsque φ a un support compact modulo N_r. L'hypothèse du lemme s'écrit alors:

$$\int_{N_r \backslash G_r} W(g, x_1, x_2, \ldots, x_r; \psi)\varphi(g)dg = 0, \tag{1}$$

pour tout $(x_1, x_2, \ldots, x_r)$. Montrons que cela entraine $\varphi = 0$. Pour cela considérons la représentation π de G_r induite, au sens de Mackey, par le caractère (de module 1) $\theta_{\bar{\psi}, r}$. Décomposons la en somme continue de représentations irréductibles:

$$\pi = \int \pi_x d\mu(x).$$

Soit V l'ensemble des fonctions lisses à droite et support compact modulo N_r dans l'espace de la représentation π. Comme V a une base linéaire dénombrable, il existe, pour presque tout x, une application linéaire A_x de V dans l'espace de la représentation π_x telle que

$$\langle \varphi_1, \varphi_2 \rangle = \int \langle A_x\varphi_1, A_x\varphi_2 \rangle d\mu(x), \qquad \varphi_1, \varphi_2 \in V. \tag{2}$$

De plus, pour presque tout x, on a

$$A_x \pi(g)\varphi = \pi_x(g) A_x \varphi, \qquad g \in G_r, \qquad \varphi \in V.$$

Soit maintenant φ_0 un élément de V invariant sous K_r. Alors $A_x\varphi_0 = 0$ à moins que π_x ait un vecteur $v_x \neq 0$ invariant sous K_r. Supposons qu'il en soit ainsi; alors il existe une fonction W et une seule se transformant à gauche par $\theta_{\psi, r}$ telle que, pour tout $\varphi \in V$,

$$(A_x\varphi, v_x) = \int_{N_r \backslash G_r} \varphi(g) W(g) dg.$$

Comme v_x est K_r-invariant la fonction W est K_r-invariant à droite. De plus v_x doit être un vecteur propre pour l'algèbre H_r. Il en est donc de même de la fonction W. Par conséquent cette fonction est un multiple d'une fonction $W(x_1, x_2, \ldots, x_r; \psi)$. Si φ_0 satisfait (1) alors $(A_x\varphi_0, v_x) = 0$ ou encore $A_x\varphi_0 = 0$. Ainsi $A_x\varphi_0 = 0$ pour presque tout x. D'après (2) on a $\varphi_0 = 0$. Q.E.D.

4. Le vecteur essentiel

(4.1) Nous sommes maintenant en mesure de définir le vecteur essentiel d'une représentation générique π de G_r. Rappelons une fois de plus que l'exposant de ψ est supposé être 0. Pour tout $W \in \mathscr{W}(\pi; \psi)$, posons

$$\begin{aligned} &\Psi(X, W, X_1, X_2, \ldots, X_{r-1}) \\ &= \int_{N_{r-1} \backslash G_{r-1}} W\begin{pmatrix} g & 0 \\ 0 & 1 \end{pmatrix} W(g, X_1, X_2, \ldots, X_{r-1}; \bar{\psi}) \\ &\qquad |\det g|^{s-1/2} dg, \qquad X = q^{-s}. \end{aligned} \tag{1}$$

La fonction $g \mapsto W\begin{pmatrix} g & 0 \\ 0 & 1 \end{pmatrix}$ satisfait évidemment les conditions de (3.4) et l'intégrale représente une série de Laurent formelle à coefficients dans S_{r-1}. D'après (3.4)(6) on a, si U est une indéterminée,

$$\Psi(UX, W, X_1, X_2, \ldots, X_{r-1}) = \Psi(X, W, UX_1, UX_2, \ldots, UX_{r-1}). \qquad (2)$$

Théorème. *Il existe un élément et un seul W de $\mathscr{W}(\pi;\psi)$ tel que*

(i) $W\left(g\begin{pmatrix} h & 0 \\ 0 & 1 \end{pmatrix}\right) = W(g)$ *pour* $h \in K_{r-1}$,

(ii) $\Psi(X, W, X_1, X_2, \ldots, X_{r-1}) = \prod_{i=1}^{r-1} P_\pi^{-1}(XX_i)$,

le second membre de (ii) *étant la série formelle en X à coefficients dans S_{r-1} obtenue en développant $P_\pi^{-1}(XX_i)$ suivant les puissances croissantes de X.*

Démonstration de l'unicité. D'après le lemme (3.4) la restriction de W au sous-groupe

$$\left\{\begin{pmatrix} g & 0 \\ 0 & 1 \end{pmatrix} \middle| g \in G_{r-1}\right\} \qquad (3)$$

est uniquement déterminée. Or, d'après les résultats de [B–Z, Théorème (4.9)], un élément de $\mathscr{W}(\pi;\psi)$ est toujours déterminé par sa restriction au sous-groupe (3). L'unicité est donc évidente.

(4.2) *Démonstration de l'existence.* Nous nous appuierons sur une identité dans l'espace des séries de Laurent à coefficients dans S_{r-1}. Notons que cet espace peut être regardé comme un module sur l'anneau des polynômes en X, X^{-1} à coefficients dans S_{r-1}. De plus cet espace admet un automorphisme changeant la série $F(X) = \sum a_n X^n$ en la série

$$F(X^{-1}) = \sum a_n X^{-n} = \sum a_{-n} X^n.$$

Cet automorphisme est compatible avec l'automorphisme correspondant de l'algèbre des polynômes en X, X^{-1}. L'identité que nous avons en vue s'écrit, pour tout élément W de $\mathscr{W}(\pi;\psi)$:

$$\Psi(q^{-1}X^{-1}, \tilde{W}, X_1^{-1}, X_2^{-2}, \ldots, X_{r-1}^{-1}) \prod_{\pi=1}^{r-1} P_{\tilde{\pi}}(q^{-1}X^{-1}X_i^{-1})$$
$$= \prod_{i=1}^{r-1} \varepsilon_\pi(XX_i, \psi)\Psi(X, W, X_1, X_2, \ldots, X_{r-1}) \prod_{i=1}^{r-1} P_\pi(XX_i). \qquad (1)$$

Dans cette formule $\tilde{W}$ est défini par (1.2)(3). Rappelons que c'est un élément de l'espace $\mathscr{W}(\tilde{\pi};\bar{\psi})$ attaché à la représentation contragrédiente $\tilde{\pi}$ et au caractère $\bar{\psi}$ imaginaire conjugué de ψ.

L'identité (1) est équivalente à la réunion d'une infinité d'identités dans l'anneau S_{r-1}. Pour démontrer (1) il suffit donc de démontrer l'identité analogue obtenue en remplaçant les X_i par des nombres complexes $x_i \neq 0$. On peut même se restreindre au cas où tous les x_i sont de module 1. Alors la représentation induite $\pi' = I(x_1, x_2, \ldots, x_{r-1})$ de G_{r-1} est irréductible et générique et la fonction

$W' = W(x_1, x_2, \ldots, x_{r-1}; \bar{\psi})$ est un élément de l'espace $\mathcal{W}(\pi'; \bar{\psi})$. De même $W(g, x_1^{-1}, x_2^{-1}, \ldots, x_{r-1}^{-1}; \psi)$ n'est autre que la fonction $(W')\tilde{}$. En posant, pour $\mathcal{W}' \in W(\pi'; \bar{\psi})$,

$$\Psi(X, W, W') = \int W\begin{pmatrix} g & 0 \\ 0 & 1 \end{pmatrix} W'(g) |\det g|^{s - 1/2} dg, \qquad X = q^{-s},$$

les résultats énoncés dans [J–P–S2] montrent que

$$\Psi(q^{-1}X^{-1}, \tilde{W}, \tilde{W}') \prod_i P_{\tilde{\pi}}(q^{-1}X^{-1}x_i^{-1})$$
$$= \prod_i \varepsilon_\pi(Xx_i, \psi) \Psi(X, W, W') \prod_i P_\pi(Xx_i), \tag{3}$$

pout tout W'. En prenant $W' = W(x_1, x_2, \ldots, x_{r-1}; \bar{\psi})$ nous obtenons bien la spécialisation de (1).

D'après les remarques de (3.4) la série formelle $\Psi(X, W, X_1, X_2, \ldots, X_{r-1})$ n'a qu'un nombre fini de termes avec un exposant négatif. Il en est donc de même du membre de droite de (1). Au contraire le membre de gauche est une série formelle avec un nombre fini d'exposants positifs. L'identité (1) montre donc que chaque membre est un polynôme en X, X^{-1} à coefficients dans S_{r-1}. D'après (4.1)(2) il existe même un élément $\Xi(W, X_1, X_2, \ldots, X_{r-1})$ de S_{r-1} tel que

$$\Xi(W, XX_1, XX_2, \ldots, XX_{r-1})$$
$$= \Psi(X, W, X_1, X_2, \ldots, X_{r-1}) \prod_{i=1}^{r-1} P_\pi(XX_i). \tag{4}$$

D'une manière imagée, on peut écrire.

$$\Xi(W, X_1, X_2, \ldots, X_{r-1})$$
$$= \int_{N_{r-1}\backslash G_{r-1}} W\begin{pmatrix} g & 0 \\ 0 & 1 \end{pmatrix} W(g, X_1, X_2, \ldots, X_{r-1}; \bar{\psi}) |\det g|^{-1/2} dg \prod_{i=1}^{r-1} P_\pi(X_i). \tag{5}$$

Soit alors $I(\pi)$ le sous espace vectoriel de S_{r-1} engendré par les $\Xi(W, X_1, X_2, \ldots, X_{r-1})$ avec W dans $\mathcal{W}(\pi; \psi)$. C'est en fait un idéal de l'anneau S_{r-1}. En effet soit P un élément quelconque de S_{r-1} et φ l'élément correspondant de H_{r-1}. Alors

$$\int W(gh, X_1, X_2, \ldots, X_{r-1}) \varphi(h) dh$$
$$= W(g, X_1, X_2, \ldots, X_{r-1}; \bar{\psi}) P(X_1, X_2, \ldots, X_{r-1}).$$

Si W est dans $\mathcal{W}(\pi; \psi)$ la fonction W_1 définie par

$$W_1(g) = \int W\left(g \begin{pmatrix} h & 0 \\ 0 & 1 \end{pmatrix}\right) \varphi(h^{-1}) |\det h|^{-1/2} dh$$

est encore dans $\mathcal{W}(\pi; \psi)$. En partant de la formule (5) un calcul formel facile à justifier montre que

$$\Xi(W_1, X_1, X_2, \ldots, X_{r-1}) = \Xi(W, X_1, X_2, \ldots, X_{r-1}) \cdot P(X_1, X_2, \ldots, X_{r-1}).$$

D'où notre assertion.

L'idéal $I(\pi)$ contient le produit $\prod P_\pi(X_i)$. En effet d'après les résultats de [G–K, Théorème (5.2)], il existe un élément W de $\mathscr{W}(\pi;\psi)$ tel que

$$W\begin{pmatrix} g & 0 \\ 0 & 1 \end{pmatrix} = \theta_{\psi,r-1}(n), \quad \text{si} \quad g=nk, \quad n\in N_{r-1}, \quad k\in K_{r-1}; =0 \text{ sinon}.$$

Pour ce W on a évidemment d'après (5):

$$\Xi(W, X_1, X_2, \ldots, X_{r-1}) = \prod_{i=1}^{r-1} P_\pi(X_i).$$

D'où la conclusion.

Finalement l'identité (1) s'écrit aussi:

$$\begin{aligned} &\Xi(\tilde{W}, q^{-1}X_1^{-1}, q^{-1}X_2^{-1}, \ldots, q^{-1}X_{r-1}^{-1}) \\ &= \prod_{i=1}^{r-1} \varepsilon_\pi(X_i, \psi)\Xi(W, X_1, X_2, \ldots, X_{r-1}). \end{aligned} \tag{7}$$

En échangeant les rôles de π et $\tilde{\pi}$ cela montre que le produit $\prod_{i=1}^{r-1} \mathrm{P}_\pi(\mathrm{q}^{-1}\mathrm{X}_i^{-1})$ appartient aussi à $I(\pi)$. D'après la proposition (2.1) on voit qu'il n'existe pas de suite $(x_1, x_2, \ldots, x_{r-1})$ avec $x_i \neq 0$, sur laquelle tous les éléments de $I(\pi)$ s'annulent. Si J est l'ideal correspondant de l'anneau $\mathbb{C}[T_1, T_2, \ldots, T_{r-1}, T_{r-1}^{-1}]$ [notations de (3.1)], cela signifie que les éléments de J n'ont pas de zero commun $(t_1, t_2, \ldots, t_{r-1})$ avec $t_{r-1} \neq 0$. Il en résulte que

$$J = \mathbb{C}[T_1, T_2, \ldots, T_{r-1}, T_{r-1}^{-1}]$$

et $I(\pi) = S_{r-1}$.

En particulier il existe un élément W de $\mathscr{W}(\pi;\psi)$ tel que

$$\Xi(W, X_1, X_2, \ldots, X_{r-1}) = 1.$$

En revenant aux définitions, cela s'écrit

$$\prod_{i=1}^{r-1} P_\pi(XX_i)\Psi(X, W, X_1, X_2, \ldots, X_{r-1}) = 1.$$

Or $\Psi(X, W, X_1, X_2, \ldots, X_{r-1})$ appartient à l'espace des séries de Laurent n'ayant qu'un nombre fini de termes avec exposant négatif; cet espace est un module *sans torsion* sur l'anneau $S_{r-1}[X, X^{-1}]$. La fonction W satisfait donc bien à la condition (ii) du théorème (4.1). Alors la formule

$$W_1(g) = \int_{K_{r-1}} W\left[g\cdot\begin{pmatrix} k & 0 \\ 0 & 1 \end{pmatrix}\right] dk$$

definit un element de $\mathscr{W}(\pi;\psi)$ qui satisfait à la condition (i). Comme l'intégrale Ψ ne change pas lorsqu'on remplace W par W_1 la fonction W_1 satisfait aussi (ii). Le théorème est donc démontré.

(4.3) *Exemple.* Supposons que π contienne la représentation unité de K_r. Alors π est équivalente à une représentation induite $I(y_1, y_2, \ldots, y_r)$ et

$W = W(y_1, y_2, \ldots, y_r; \psi)$ appartient à $\mathscr{W}(\pi; \psi)$. On a alors

$$P_\pi(X) = \prod_{i=1}^{r} (1 - y_i X).$$

D'après les résultats de [J–P–S2], on a aussi:

$$\Psi(X, W, X_1, X_2, \ldots, X_{r-1}) = \prod_{1 \leqq i \leqq r,\, 1 \leqq j \leqq r-1} (1 - y_i X_j X)^{-1}.$$

Comme W est invariant sous K_r, on voit que W satisfait les conditions du théorème (4.1).

(4.4) *Définition.* Sous les hypothèses du théorème (4.1), on dira que W est le vecteur essentiel de $\mathscr{W}(\pi; \psi)$.

5. Propriétés du vecteur essentiel

(5.1) Soit à nouveau π une représentation irréductible et générique de G_r. Rappelons que ψ désigne un caractère additif d'exposant 0. Soit m l'entier tel que

$$\varepsilon_\pi(X, \psi) = C X^m. \tag{1}$$

Dans ce paragraphe, nous établissons le théorème suivant:

Théorème. (i) *On a* $m \geqq 0$.

(ii) *Pour tout entier* $a > 0$ *soit* $K_r(a)$ *le sous-groupe de* K_r *forme des matrices du type*

$$\begin{pmatrix} h & v \\ u & x \end{pmatrix}, \quad h \in K_{r-1}, \quad x \in \mathfrak{R}^\times, \quad x \equiv 1 \bmod \mathfrak{P}^a, \quad u \equiv 0 \bmod \mathfrak{P}^a.$$

On pose aussi $K_r(0) = K_r$. *Alors le vecteur essentiel est invariant sous* $K_r(m)$ *et la dimension de l'espace des vecteurs fixés par* $K_r(m)$ *est un.*

(iii) *Si* $a \geqq 0$ *et s'il existe dans la représentation* π *un vecteur non nul fixé par* $K_r(a)$, *alors* $a \geqq m$.

(5.2) Pour démontrer le théorème nous introduirons, pour chaque entier $a \in \mathbb{Z}$, le sous-groupe $H(a)$ de G_r formé des matrices du type

$$\begin{pmatrix} h & u \\ 0 & 1 \end{pmatrix},$$

où h est dans K_{r-1} et où les éléments de la matrice colonne u sont dant $\mathfrak{P}^a$. Cela étant, le premier pas est de démontrer que le vecteur essentiel W est invariant sous $H(0)$. Par définition W est invariant sous le sous-groupe

$$\left\{ \begin{pmatrix} h & 0 \\ 0 & 1 \end{pmatrix} \middle| h \in K_{r-1} \right\}.$$

If suffit donc de démontrer que W est invariant sous le sous-groupe de $H(0)$ formé des matrices de $H(0)$ pour lesquelles $h = 1_{r-1}$. Comme W est déterminée par sa restriction au sous-groupe (4.1)(3), il suffit, d'après les remarques de (3.1), de prouver le lemme suivant:

Lemme. *Si W est le vecteur essentiel de $\mathscr{W}(\pi;\psi)$ alors la relation*

$$W\begin{pmatrix} g & 0 \\ 0 & 1 \end{pmatrix} \neq 0, \qquad g \in G_{r-1},$$

entraine que la dernière ligne de g est à coefficients entiers.

Démonstration du lemme. Soit u le degré de P_π. On a donc $u \leqq r$ (2.5). Supposons que $u=r$. Alors π contient la représentation unité de K_r (loc. cit.). D'après (4.3) W est alors invariante sous K_r et notre assertion est alors évidente. Supposons maintenant $u \leqq r-1$ et écrivons

$$P_\pi(X) = \prod_{i=1}^{u} (1 - y_i X).$$

Nous distinguerons deux cas: $u<r-1$ et $u=r-1$.

Dans le cas $u<r-1$, on a l'identité (cf. [J–P–S2]):

$$\int_{N_u \backslash G_u} W\left[\begin{pmatrix} g & 0 \\ 0 & 1 \end{pmatrix}, X_1, X_2, \ldots, X_{r-1}; \bar{\psi}\right] W(g, y_1, y_2, \ldots, y_u; \psi) |\det g|^{s-\frac{1}{2}(r-u-1)} dg$$
$$= \prod (1 - y_j X_i X)^{-1}, \quad 1 \leqq j \leqq u, \quad 1 \leqq i \leqq r-1.$$

Définissons maintenant une fonction φ sur G_{r-1} par les conditions suivantes:

$$\varphi(ngk) = \theta_{\psi, r-1}(n) \varphi(g) \quad \text{si} \quad n \in N_{r-1}, \qquad k \in K_{r-1};$$

si $a = \operatorname{diag}(a_1, a_2, \ldots, a_{r-1})$ alors

$$\varphi(a) = W(b, y_1, y_2, \ldots, y_u; \psi) \quad \text{avec} \quad b = \operatorname{diag}(a_1, a_2, \ldots, a_u)$$

si $|a_{u+1}| = \ldots |a_{r-1}| = 1$; $=0$ sinon. Il est clair que φ satisfait les conditions de (3.3). On a évidemment

$$\int_{N_{r-1} \backslash G_{r-1}} \varphi(g) W(g, X_1, X_2, \ldots, X_{r-1}; \bar{\psi}) |\det g|^{s+\frac{1}{2}(r-u-1)} dg = \prod P_\pi(XX_i)^{-1}.$$

D'après le lemme (3.4), on a donc

$$W\begin{pmatrix} g & 0 \\ 0 & 1 \end{pmatrix} = \varphi(g) |\det g|^{\frac{1}{2}(r-u)} \quad .$$

Le lemme en découle aussitôt.

Si $u=r-1$, on a (cf. [J–P–S2])

$$\int_{N_{r-1} \backslash G_{r-1}} W(g, y_1, y_2, \ldots, y_{r-1}; \psi) W(g, X_1, X_2, \ldots, X_{r-1}; \bar{\psi})$$
$$\Phi[(0, 0, \ldots, 0, 1) g] |\det g|^s dg = \prod (1 - y_i X_j X)^{-1},$$

où Φ est la fonction caractéristique de $\mathfrak{R} \times \mathfrak{R} \times \ldots \times \mathfrak{R}$ dans F^{r-1}. En raisonnant comme plus haut on trouve

$$W\begin{pmatrix} g & 0 \\ 0 & 1 \end{pmatrix} = W(g, y_1, y_2, \ldots, y_{r-1}; \psi) \Phi[(0, 0, \ldots, 0, 1) g] |\det g|^{1/2}$$

et, à nouveau, le lemme découle aussitôt de cette relation.

(5.3) Dénotons toujours par W le vecteur essentiel de $\mathscr{W}(\pi;\psi)$. Alors d'après (4.2)(1):

$$\Psi(q^{-1}X^{-1},\tilde{W},X_1^{-1},X_2^{-1},\ldots,X_{r-1}^{-1})\prod_{i=1}^{r-1}P_{\tilde{\pi}}(q^{-1}X^{-1}X_i^{-1})=\prod_{i=1}^{r-1}\varepsilon_\pi(XX_i,\psi).$$

Comme $\Psi(X,\tilde{W},X_1,X_2,\ldots,X_{r-1})$ n'a que des termes avec exposants positifs cela s'écrit aussi [cf. (5.1)(1)]:

$$\Psi(X,\tilde{W},X_1,X_2,\ldots,X_{r-1})=C^{r-1}q^{-m(r-1)}X^{-m(r-1)}\prod_{i=1}^{r-1}X_i^{-m}\prod_{i=1}^{r-1}P_{\tilde{\pi}}^{-1}(XX_i).$$

Soit W' le vecteur essentiel de $\mathscr{W}(\tilde{\pi};\bar{\psi})$. Alors

$$\Psi(X,W',X_1,X_2,\ldots,X_{r-1})=\prod_{i=1}^{r-1}P_{\tilde{\pi}}^{-1}(XX_i).$$

Définissons maintenant un élément W_1 de $\mathscr{W}(\tilde{\pi};\bar{\psi})$ par la formule

$$W_1(g)=W'(ga),\qquad a=\operatorname{diag}(\omega^m,\omega^m,\ldots,\omega^m,1).$$

Alors, il y a une constante C' telle que

$$\Psi(X,\tilde{W},X_1,X_2,\ldots,X_{r-1})=C'\Psi(X,W_1,X_1,X_2,\ldots,X_{r-1}).$$

D'après le lemme (3.4) cela implique $\tilde{W}=C'W_1$ sur le sous-groupe (4.2)(3). Donc, d'après une remarque déjà faite, $\tilde{W}=C'W_1$. Comme W' est invariante sous $H(0)$, cela implique que $\tilde{W}$ est invariante sous $H(m)$. En revenant à la définition de $\tilde{W}$ on voit que cela implique l'invariante de W sous le groupe transposé ${}^tH(m)$.

Si on avait $m<0$, les groupes $H(0)$ et ${}^tH(m)$ engendreraient G_r. La fonction W serait donc constante ce qui est absurde. On a donc $m\geqq 0$. Les groupes $H(0)$ et ${}^tH(m)$ engendrent alors $K(m)$ de sorte que W est bien invariante sous $K(m)$.

Soit maintenant u un entier $\geqq 0$ et W_1 un vecteur de $\mathscr{W}(\pi;\psi)$ invariant sous $K(u)$. L'invariance de W_1 sous $H(0)$ implique que

$$W_1\begin{pmatrix} g & 0\\ 0 & 1\end{pmatrix}\neq 0\Rightarrow|\alpha_i(g)|\leqq 1,\qquad 1\leqq i\leqq r-1$$

[cf. (3.1)]. En particulier dans l'intégrale (4.1)(1) pour $\Psi(X,W_1,X_1,X_2,\ldots,X_{r-1})$ on peut supposer que $|\det g|\leqq 1$ et $W(g,X_1,X_2,\ldots,X_{r-1};\bar{\psi})$ est dans $\mathbb{C}[X_1,X_2,\ldots,X_{r-1}]$ [cf. (3.3)]. La série n'a donc que des termes à exposants positifs et ses coefficients sont dans $\mathbb{C}[X_1,X_2,\ldots,X_{r-1}]$. Il en résulte que $\Xi(W_1,X_1,X_2,\ldots,X_{r-1})$ est un polynôme P en $X_1,X_2,\ldots,X_{r-1}$. D'autre part W_1 étant invariante sous ${}^tH(u)$, la fonction $\tilde{W}_1$ est invariante sous $H(u)$. La fonction W_2 définiee par

$$W_2(g)=\tilde{W}_1(ga),\qquad a=\operatorname{diag}(\omega^{-u},\omega^{-u},\ldots,\omega^{-u},1),$$

est donc invariante sous $H(0)$. Comme plus haut, il en résulte que $\Xi(W_2,X_1,X_2,\ldots,X_{r-1})$ est un élément de $\mathbb{C}[X_1,X_2,\ldots,X_{r-1}]$. Alors il existe un multiple constant Q de $\Xi(W_2,X_1,X_2,\ldots,X_{r-1})$ tel que:

$$\Xi(\tilde{W}_1,X_1,X_2,\ldots,X_{r-1})=(X_1X_2\ldots X_{r-1})^{-u}Q(X_1,X_2,\ldots,X_{r-1}).$$

D'après (4.2)(7) on a alors:

$$(X_1X_2\dots X_{r-1})^u Q(q^{-1}X_1^{-1},q^{-1}X_2^{-1},q^{-1}X_2^{-1},\dots,q^{-1}X_{r-1}^{-1})$$
$$=C^{r-1}q^{-u(r-1)}(X_1,X_2,\dots,X_{r-1})^m P(X_1,X_2,\dots,X_{r-1}).$$

Si on avait $u<m$, $Q(q^{-1},X_1^{-1},\dots,q^{-1},X_{r-1}^{-1})$ serait un polynôme en $X_1,X_2,\dots,X_{r-1}$ divisible par le produit $X_1X_2\dots X_{r-1}$, ce qui est absurde. On a donc $u\geqq m$. Enfin si $u=m$ la relation ci-dessus montre que P est une constante. D'après l'unicité dans le théorème (4.1) la fonction W_1 est alors un multiple de W. Q.E.D.

(5.4) *Remarque*. Si π est une représentation admissible irréductible de G_r, générique ou non, on peut encore poser $\varepsilon_\pi(X,\psi)=CX^m$. On peut montrer que l'on a encore $m\geqq 0$ et que $m=0$ si et seulement si π contient la représentation unité de K_r. Toutefois, si $m>0$ la dimension de l'espace des vecteurs fixés par $K(m)$ peut être 0, comme le montre l'exemple des représentations de la forme $g\mapsto\chi(\det g)$, où χ est un caractère ramifié de $F^\times$.

Bibliographie

[B–Z] Bernstein, I.N., Zelevensky, A.V.: Induced representations of reductive *p*-adic groups. I. Ann. Sci. École Norm. Sup. **4**, 441–472 (1977)

[C] Casselman, W.: On some results of Atkin and Lehner. Math. Ann. **206**, 311–318 (1973)

[D] Deligne, P.: Formes modulaires et représentations de GL(2). In: Modular functions of one variable. II. Lecture Notes in Mathematics, Vol. 349. Berlin, Heidelberg, New York: Springer 1973

[G–J] Godement, R., Jacquet, H.: Zeta functions of simple algebras. In: Lecture Notes in Mathematics, Vol. 260. Berlin, Heidelberg, New York: Springer 1972

[G–K] Gelfand, I.M., Kazdan, D.A.: Representations of GL(n, K) where K is a local field. In: Lie groups and their representations, pp. 95–118. New York: Wiley 1975

[J1] Jacquet, H.: Generic representations. In: Non commutative harmonic analysis, Marseille-Luminy 1976. In: Lecture Notes in Mathematics, Vol. 587, pp. 91–100. Berlin, Heidelberg, New York: Springer 1976

[J2] Jacquet, H.: Principal *L*-functions of the linear group. In: automorphic forms, representations, and *L*-functions, Part 2. Am. Math. Soc. 63–95 (1979)

[J–P–S1] Jacquet, H., Piatetski-Shapiro, I.I., Shalika, J.: Automorphic forms on GL(3). I. Ann. of Math. **109**, 169–212 (1979)

[J–P–S2] Jacquet, H., Piatetski-Shapiro, I.I., Shalika, J.: Facteurs L et ε du groupe linéaire. C. R. Acad. Sci. Ser. A **289**, 59–61 (1979)

[N] Novodvorsky, M.: On certain stationary groups in infinite dimensional representations of Chevalley groups. Funkcional. Anal. i. Prilozen **5**, 87–88 (1971). Transl. Functional Anal. Appl. **5**, 74–76 (1971)

[R] Rodier, F.: Whittaker models for admissible representations of real algebraic groups. In: Harmonic analysis on homogeneous spaces. Proc. Symp. Pure Math. Am. Math. Soc. 425–430 (1976)

[S] Shintani, T.: On an explicit formula for class-1 "Whittaker functions" on GL_n over *p*-adic fields, Proc. Japan Acad. **52**, 180–182 (1976)

Reçu le 4 août 1980

ON EULER PRODUCTS AND THE CLASSIFICATION OF AUTOMORPHIC REPRESENTATIONS I*

By H. JACQUET AND J. A. SHALIKA

0. Introduction and Notations.

(0.1) In this paper we prove two theorems concerning automorphic forms on GL_r, both having their origin in the existence of poles of L-functions.

Let F be a global field and $\mathbf{A}$ be the ring of adeles of F. To each automorphic cuspidal representation π of $GL_r(\mathbf{A})$ one can attach as in [G-J] an L-function $L(s, \pi)$. For $r = 1$, π is an idele-class character and $L(s, \pi)$ the classical L-function of Hecke. For $r > 1$, $L(s, \pi)$ is entire and satisfies a functional equation

$$L(s, \pi) = \epsilon(s, \pi)L(1 - s, \tilde{\pi}),$$

$\tilde{\pi}$ being the contragradient representation. In an earlier paper [J-S], the authors proved that $L(s, \pi)$ does not vanish on $\mathrm{Re}(s) = 1$. One of our results complements this. We prove in Section 5 that the Euler product for $L(s, \pi)$ is absolutely convergent in the half-plane $\mathrm{Re}(s) > 1$. For $r = 2$ this was first proved by R. Rankin [R.R.]. Our proof is based on his. In the same series of papers, Rankin also proves the non-vanishing of $L(s, \pi)$ on $\mathrm{Re}(s) = 1$, by means entirely distinct from ours. With more work his method should also apply to the general case. As in the recent paper of Moreno [C.J.M.], one should be able to obtain zero free regions as well. We also refer the reader to the classical paper of A. Selberg [A.S.].

Our second result concerns the problem of classification of automorphic forms on $GL_r(\mathbf{A})$. We refer to Langlands' Corvallis lecture [R.P.L.] for the exact statement of this problem as well as a discussion of the interesting concept of an isobaric representation. In this paper we consider a special case. We prove that, if S is any finite set of places and if π and π'

Manuscript received February 4, 1980.

*This research was partially supported by National Science Foundation Grants MCS 79-01712 and MCS 79-13799.

American Journal of Mathematics, Vol. 103, No. 3, pp. 499–558
0002-9327/81/1033-0499 $01.50

are automorphic cuspidal representations which are locally equivalent outside S, then π and π' are themselves equivalent. For $r = 1$ this is another formulation of the approximation theorem. For $r = 2$ it is a well-known result of Miyake [T.M.]. Piatetskii-Shapiro [I.P.] has also given a proof in the general case. Both authors suppose however that S does not contain any infinite places. Moreover the assumption needed in our paper is somewhat weaker than the one just stated.

In a subsequent paper we will use the results of this paper to establish the full classification theorem.

We give an outline of the proof of our second theorem. Suppose first that $r = 1$ and that χ and χ' are two idele-class characters of absolute value one which agree locally outside S. Set

$$A(s) = \prod_{v \in S} L(s, \chi_v' \bar{\chi}_v), \qquad B(s) = \prod_{v \in S} L(s, 1_v).$$

The hypothesis implies that

$$B(s)L(s, \chi'\bar{\chi}) = A(s)L(s, 1).$$

But $L(s, 1)$ has a pole at $s = 1$. On the other hand, a local L-factor cannot be zero. Thus the right side has a pole at $s = 1$. Next, and this is an essential point, the factor $L(s, 1_v)$ does not have a pole at $s = 1$. Thus $L(s, \chi'\bar{\chi})$ has a pole at $s = 1$ and that implies $\chi'\bar{\chi} = 1$ or $\chi = \chi'$.

Before passing to the case $r \geq 2$, note that one may obtain the analytic properties of $L(s, \chi)$, by utilizing an integral representation of the form

$$L(s, \chi) = \int_{F_{\mathbf{A}}^\times/F^\times} \chi(t)|t|^s \sum_{\xi \in F^\times} \Phi(t\xi) d^\times t, \tag{1}$$

where Φ is a suitable Schwartz-Bruhat function on $F_{\mathbf{A}}$. Similarly if χ_v is a character of absolute value one of $F_v^\times$ one has

$$L(s, \chi_v) = \int_{F_v^\times} \chi_v(t)|t|^s \Phi_v(t) d^\times t, \tag{2}$$

where Φ_v is a suitable Schwartz-Bruhat function on F_v. The assertion that $L(s, \chi_v)$ is holomorphic at $s = 1$ may be proved by showing that the integral on the right is convergent for $\operatorname{Re}(s) \geq 1$.

Suppose now $r \geq 2$. The exact analogue of the integral (1) is the integral

$$\int_{G(F)\backslash G(\mathbf{A})} \phi'\bar{\phi}(g)|\det g|^s \sum_{\xi \in F^r - \{0\}} \Phi(\xi g)dg, \tag{3}$$

where $G = GL_r$, ϕ and ϕ' are cusp forms belonging to π and π' respectively, and Φ is a Schwartz-Bruhat function on $F_{\mathbf{A}}{}^r$. The analytic properties of this integral are evident. We express this integral in terms of local integrals which are the analogues of (2); the essential difficulty is to prove that these integrals converge for $\text{Re}(s) \geq 1$ (Proposition (1.5) and Proposition (3.17)). For this one needs certain properties of the *unitary* representations π_v (Proposition (1.3) and (3.7)). With this the proof that $\pi \sim \pi'$ is entirely analogous to the case $r = 1$ (Section 4).

In Section 5, we give an application to the absolute convergence in $\text{Re}(s) > 1$ of certain Euler products $L(s, \pi \times \pi')$ attached to pairs of unitary forms respectively on GL_r and GL_p. For $p = 1$ and π' trivial they reduce to the functions $L(s, \pi)$ considered earlier. These functions enter implicitly in the proof of Theorem (4.7). In fact, for $r = p$ they are essentially given by (3) (see also (2.1.2) below). For $r = 2$ they are the classical Rankin-Selberg convolutions.

We note finally that recently, by methods entirely similar to [J-S], but using the full theory of Eisenstein series for GL_r, F. Shahidi [F.S.] has shown that $L(s, \pi \times \pi')$ is non-zero for $\text{Re}(s) = 1$. This together with our results will be used in proving the classification theorem for forms on GL_r.

(0.2) Upon first reading one should concentrate on Section 1, Sections (2.1), (2.2), the statement of Proposition (2.3), then (3.1), the statement of Proposition (3.5), (3.13), and finally Section 4 and Section 5.

(0.3) We now give a list of our most frequently used notations.

The ground field F is either a local field or an $\mathbf{A}$ field. The group $G_r = GL_r = GL(r)$ is regarded as an F-group. We denote by Z_r the center of G_r; often we identify Z_r with the group GL_1. The group of matrices of the form

$$p = \begin{pmatrix} m & u \\ 0 & 1 \end{pmatrix}, \qquad m \in G_{r-1}, \tag{1}$$

will be denoted P_r. The unipotent radical of P_r, the subgroup of $p \in P_r$ with $m = 1$, is noted U_r while the transpose of U_r is often noted $\overline{U_r}$.

A parabolic subgroup R of G_r will always be an F-parabolic subgroup. The unipotent radical of R will be denoted by U_R. For the most part we will only consider standard parabolic subgroups, i.e. groups of matrices of the form

$$g = \begin{pmatrix} m_1 & & & & * \\ & m_2 & & & \\ & & \cdot & & \\ & & & \cdot & \\ 0 & & & & m_t \end{pmatrix} \tag{2}$$

where $m_i \in G_{r_i}$ and $(r_1, r_2, \ldots, r_t)$ is a t-tuple of integers such that

$$r_1 + r_2 + \cdots + r_t = r.$$

The t-tuple is the *type* of the parabolic subgroup. The transpose of a standard parabolic subgroup is termed a lower standard parabolic subgroup. The standard parabolic of type $(1, 1, \ldots, 1)$ (the standard Borel subgroup) is denoted B_r and its unipotent radical N_r. We denote by A_r the group of diagonal matrices; the set of simple roots Δ is the set of simple roots of A_r with respect to B_r, i.e. the homomorphisms $\alpha_i \in \mathrm{Hom}(A_r, GL(1))$ defined by

$$\alpha_i(a) = a_i/a_{i+1} \qquad \text{if } a = \mathrm{diag}(a_1, a_2, \ldots, a_r), \qquad 1 \leq i \leq r-1.$$

The index r is often dropped when this does not create confusion.

When F is local we use standard notation: $|x|$ also denoted by $\alpha_F(x)$ or even $\alpha(x)$ is the module. If F is non-archimedean, $\mathfrak{R}_F = \mathfrak{R}$ will denote the ring of integers of F, $\tilde{\omega}$ will be a generator of the unique maximal ideal of $\mathfrak{R}$ and we will set $q = |\tilde{\omega}|^{-1}$. For any F we will denote by $\mathcal{S}(F^r)$ the space of Schwartz-Bruhat functions on F^r. We often view F^r as a space of row vectors. If ψ is a non-trivial additive character of F, the Fourier transform of a function $\Phi \in \mathcal{S}(F^r)$ is the function $\hat{\Phi} \in \mathcal{S}(F^r)$ defined by

$$\hat{\Phi}(x) = \int_{F^r} \Phi(y)\psi(y^t x)dy. \tag{3}$$

Of course dy is the self-dual Haar measure on F^r.

Similarly when F is an **A**-field we denote by $|x|$, $\alpha_F(x)$ or $\alpha(x)$ the module of F. If v is a place of F we denote by F_v the corresponding local field. We fix, once and for all, a non-trivial character ψ of $F_{\mathbf{A}}/F$. Then we write $\psi = \Pi_v \psi_v$. Again we denote by $\mathcal{S}(F_{\mathbf{A}}{}^r)$ the space of Schwartz-Bruhat functions on $F_{\mathbf{A}}{}^r$ and by $\hat{\Phi}$ the Fourier transform of a function Φ in $\mathcal{S}(F_{\mathbf{A}}{}^r)$.

We will make extensive use of standard terminology in representation theory. We will deal mainly with unitary representations although admissible representations will occasionally make an appearance. If G is a locally compact group and H a closed subgroup, a representation σ of H on a vector space $\mathcal{V}$ induces a representation π of G; it will be denoted by

$$\mathrm{Ind}(G, H; \sigma). \tag{4}$$

The exact definition depends on the context, but in any case the vectors ϕ in the space of π are functions (or classes of functions) on G with values in $\mathcal{V}$ which satisfy

$$\phi(hg) = \delta_H{}^{1/2}(h)\delta_G{}^{-1/2}(h)\sigma(h)\phi(g);\ h \in H, g \in G. \tag{5}$$

Here δ_G(resp. δ_H) is the module of G(resp. H). More precisely if $d_r(g)$ and $d_\ell(g)$ are respectively right and left Haar measures on G, then $d_r(xg) = \delta_G(x)d_r(g)$ and $d_\ell(gx^{-1}) = \delta_G(x)d_\ell(g)$. Of course π acts by right translations:

$$\pi(x)\phi(g) = \phi(gx);\ g, x \in G.$$

For instance, if $G = G_r(F)$ (F being local), if R is the standard parabolic subgroup of type $(r_1, r_2, \ldots, r_t)$ and for each i, $1 \leq i \leq t$, π_i is a representation of $G_{r_i}(F)$, we can define a representation σ of $R(F)$ by

$$\sigma(g) = \otimes\ \pi_i(m_i) \tag{6}$$

for $g \in R(F)$ of the form (2). The representation of $G_r(F)$ induced by σ will be denoted by

$$\mathrm{Ind}(G_r(F), R_r(F); \pi_1, \pi_2, \ldots, \pi_t). \tag{7}$$

If π is a representation of $G_r(F)$ on a complex vector space $\mathcal{V}$ we will denote by $\pi \otimes \alpha^s$ the representation on the same space defined by

$$(\pi \otimes \alpha^s)(g) = |\det g|^s \pi(g). \tag{8}$$

We will denote by $\overline{\mathcal{V}}$ the complex vector space conjugate to $\mathcal{V}$ and by $\bar{\pi}$ the natural representation of $G_r(F)$ on $\overline{\mathcal{V}}$.

If π is admissible we denote by $\tilde{\pi}$ the representation contragredient to π. In case π is irreducible there is a character ω (perhaps not of absolute value one) of $F^\times$ such that

$$\pi(a) = \omega(a)1, \qquad \text{for } a \in Z_r(F) \simeq F^\times. \tag{9}$$

We call ω the central character of π.

If F is local (resp. global) we define a character θ_r (or simply θ) of $N_r(F)$ (resp. $N_r(F_{\mathbf{A}})$) by

$$\theta(n) = \psi\big(\sum_{1 \leq i \leq r} n_{i,i+1}\big). \tag{10}$$

Finally when F is local we often write G_r for $G_r(F)$, P_r for $P_r(F)$, etc.

We let K be the standard maximal compact subgroup of $G_r(F)$: $K = 0(r, \mathbf{R})$ if $F = \mathbf{R}$, $K = U(r)$ if $F = \mathbf{C}$, $K = GL(r, \mathfrak{R}_F)$ if F is non-archimedean. When F is an $\mathbf{A}$-field we let K be the compact subgroup of $G_r(F_{\mathbf{A}})$ defined by

$$K = \prod_v K_v$$

where K_v is the above subgroup of $G_r(F_v)$.

1. Generic representations: non-archimedean case. In this section F is a non-archimedean local field. We first review the notion of a "generic representation" (Prop. (1.3)) adding appropriate remarks in the case of unitary representations. We then discuss an analogue for $r \geq 2$ of the integral (2) appearing in the introduction (0.1).

(1.1) Let π be an irreducible admissible representation of G on a complex vector space $\mathcal{V}$. We say that π is generic if there exists a linear form $\lambda \neq 0$ on $\mathcal{V}$ such that

$$\lambda(\pi(n)v) = \theta(n)v, \qquad \text{for } n \in N,\ v \in \mathcal{V}, \tag{1}$$

(c.f. (0.10)).

Up to multiplication by a scalar factor this form is unique ([J.S.], [G-K]). We let $\mathcal{W}(\pi; \psi)$ be the space of all functions W of the form

$$W(g) = \lambda(\pi(g)v) \tag{2}$$

where v is in $\mathcal{V}$. The space $\mathcal{W}(\pi;\psi)$ is invariant under right translations and the representation of G on that space is equivalent to π. Clearly for any W in $\mathcal{W}(\pi;\psi)$ we have

$$W(ng) = \theta(n)W(g), n \in N, g \in G. \tag{3}$$

The map $W \mapsto W|P$ is a bijection of $\mathcal{W}(\pi;\psi)$ on a space $\mathcal{K} = \mathcal{K}(\pi;\psi)$ of functions ϕ on P_0. Thus we may identify $\mathcal{V}$ with $\mathcal{K}$. Then

$$\pi(p_0)\phi(p) = \phi(pp_0), \phi \in \mathcal{K}, p_0 \in P. \tag{4}$$

Let $\mathcal{K}_0$ be the space of functions ϕ on P which transform on the left like θ under N, are right invariant under some open compact subgroup of P, and have compact support mod N. Let τ_0 be the representation of P on $\mathcal{K}_0$ by right translations. Then for all generic representations π, $\mathcal{K}_0$ is contained in $\mathcal{K}(\pi;\psi)$. In particular $\pi|P$ contains, as a subrepresentation, the representation τ_0 (c.f. also [F.R.]).

(1.2) Let π now be an irreducible unitary representation of G on a Hilbert space $\mathcal{H}$. Let $\mathcal{H}^\infty$ be the subspace of smooth vectors, that is, the space of vectors fixed by some open subgroup of G. Let π^∞ be the representation of G on $\mathcal{H}^\infty$.

Finally let τ be the unitary representation of P induced—in Mackey's sense—by θ. Let $\mathcal{L}$ be the Hilbert space of τ and $\mathcal{L}^\infty$ the space of smooth vectors in $\mathcal{L}$. Clearly $\mathcal{L}^\infty \supset \mathcal{K}_0$.

(1.3) **Proposition.** *Notations being as above, the following conditions are equivalent:*

(i) *the restriction of π to P contains τ;*

(ii) *the representation π^∞ is generic.*

Assume these conditions satisfied and identify $\mathcal{W}(\pi;\psi)$ with $\mathcal{H}^\infty$. Then there is a constant $c > 0$ such that

$$\int_{N\backslash P} |W|^2(pg)d_r(p) \leq c\,||W||^2 \tag{1}$$

for $W \in \mathcal{W}(\pi;\psi)$ and $g \in G$.

Proof. Let A be a non-zero operator in $\mathrm{Hom}_P(\pi | P, \tau)$. Then $A(\mathcal{H}^\infty)$ is contained in $\mathcal{L}^\infty$. Thus for $v \in \mathcal{H}^\infty$, Av is a smooth function on P transforming on the left like θ under N. We set

$$\lambda(v) = A(v)(e), \qquad v \in \mathcal{H}^\infty.$$

We then have

$$\begin{aligned}\lambda(\pi(n)v) &= A(\pi(n)v)(e) \\ &= (\tau(n)A(v))(e) = A(v)(n) = \theta(n)Av(e) \\ &= \theta(n)\lambda(v).\end{aligned}$$

Since $A \neq 0$ and $\mathcal{H}^\infty$ is dense in $\mathcal{H}$ there is a $v \in \mathcal{H}^\infty$ such that $A(v) \neq 0$. Choose $p \in P$ with $A(v)(p) \neq 0$. Then $\lambda(\pi(p)v) = A(v)(p) \neq 0$. Hence $\lambda \neq 0$ and π^∞ is generic. Thus (i) implies (ii).

Now assume π^∞ is generic. Let $(\cdot\ ,\ \cdot)$ denote the Hermitian scalar product on $\mathcal{H}$. We may identify $\mathcal{H}^\infty$ with the space $\mathcal{K}$ defined in (1.1). Then the restriction of the scalar product to $\mathcal{K}_0$ is a positive Hermitian form, invariant under the representation τ_0. It follows from [B-Z], Proposition 3.7, that there is a positive constant c such that

$$(\phi_1, \phi_2) = c \int_{N\backslash P} \phi_1(p)\bar{\phi}_2(p)d_r(p)$$

for $\phi_i \in \mathcal{K}_0$. Thus the injection $\mathcal{K}_0 \to \mathcal{K}$ is continuous for the norm induced on $\mathcal{K}_0$ by $\mathcal{L}$ and the norm induced on $\mathcal{K}$ by $\mathcal{H}$. Thus A extends to a bounded operator $A : \mathcal{L} \to \mathcal{H}$ commuting with P. Thus (ii) implies (i).

Finally assume the equivalent conditions of Proposition (1.3) are satisfied. Again let $A \neq 0$ be in $\mathrm{Hom}_P(\pi | P, \tau)$. If $W \in \mathcal{W}(\pi; \psi)$, let v be the corresponding vector in $\mathcal{H}^\infty$:

$$W(g) = A(\pi(g)v)(e).$$

Fix $g \in G$ and let ϕ be $A(\pi(g)v)$. We have

$$\begin{aligned}\int_{N\backslash P} |\phi|^2(p)d_r(p) &= ||\phi||^2 = ||A\pi(g)v||^2 \\ &\leq ||A||^2||\pi(g)v||^2 = ||A||^2||v||^2.\end{aligned}$$

Now $W(pg) = \phi(p)$. Thus

$$\int_{N\backslash P} |W|^2(pg)d_r(p) \leq ||A||^2||v||^2. \qquad \square$$

(1.4) *Remark.* In a future paper, we will show that under the assumptions of (1.3), $\pi|P$ is actually equivalent to τ.

(1.5) Suppose now that π and π' are irreducible *unitary* generic representations of G_r. For $\Phi \in \mathcal{S}(F^r)$, $W \in \mathcal{W}(\pi; \psi)$, and $W' \in \mathcal{W}(\pi'; \psi)$, we set

$$\Psi(s, W', W, \Phi) = \int_{N\backslash G} W'(g)\overline{W}(g)\Phi(\epsilon g)|\det g|^s dg \tag{1}$$

where ϵ is the following row vector:

$$\epsilon = (0, 0, \ldots, 1). \tag{2}$$

Proposition. (i) *The integral* (1) *converges absolutely in the half space* Re$(s) \geq 1$, *normally for* Re(s) *in a compact subset of* $[1, \infty[$.

(ii) *Given* $W' \neq 0$ *and* s *with* Re$(s) \geq 1$, *there exist* W *and* Φ *such that*

$$\Psi(s, W', W, \Phi) \neq 0. \tag{3}$$

Proof. The integral can be written explicitly as

$$\int_K dk \int_{N\backslash P} d_r p\,|\det p|^{s-1} W'(pk)\overline{W}(pk) \int_{F^\times} \Phi(\epsilon ak)|a|^{rs}\omega'\omega(a)d^\times a,$$

where ω'(resp. ω) is the central character of π (resp. π'). There is a $\Phi_0 \geq 0$ in $\mathcal{S}(F^r)$ such that

$$|\Phi(xk)| \leq \Phi_0(x)$$

for all $x \in F^r$ and $k \in K$. Then for s real,

$$\int_{F^\times} |\Phi(\epsilon ak)||a|^{rs}|\omega'\overline{\omega}(a)|d^\times a \leq \int_{F^\times} \Phi_0(\epsilon a)|a|^{rs}d^\times a.$$

This is certainly uniformly bounded for s in a compact subset of $[1, \infty[$. Thus it suffices to show that the integral

$$\int_K dk \int_{N\backslash P} d_r(p)|\det p|^{s-1}|W'(pk)||W(pk)|$$

converges uniformly for s in a compact subset of $[1, \infty[$. From (1.3.1) and the Schwartz inequality we already know that

$$\int_K dk \int_{N\backslash P} d_r(p)|W'(pk)||W(pk)| < +\infty. \tag{4}$$

Let ϕ_1(resp. ϕ_2) be the characteristic function of the set of $g \in G_r$ such that $|\det g| \leq 1$ (resp. $|\det g| > 1$). It will be enough to show that the integrals

$$I_i = \int_K dk \int_{N\backslash P} |W'W(pk)|\phi_i(p)|\det p|^{s-1}d_r(p), \qquad i = 1, 2,$$

converge uniformly for s in a compact subset of $[1, \infty[$. In the first case we may take $s = 1$ and our assertion follows from (4). In the second case we may take s large and replace ϕ_2 by the constant function one.

Recall that W and W' are majorized by a "gauge" ([J-P-S] (2.3)): a gauge ξ is a function on G_r invariant on the left under NZ, on the right under K and given on the diagonal matrices by the formula

$$\xi(a) = \phi(\alpha_1(a), \alpha_2(a), \ldots, \alpha_{r-1}(a))|\alpha_1(a)\alpha_2(a) \cdots \alpha_{r-1}(a)|^{-t}$$

where $t \geq 0$ and $\phi \geq 0$ is in $\mathcal{S}(F^{r-1})$. All we have to do then is show that if ξ and ξ' are gauges then

$$\int_{N\backslash P} \xi\xi'(p)|\det p|^{s-1}d_r(p) < +\infty \qquad \text{for } s \text{ large.}$$

This is a straight-forward matter left to the reader.

Proof of (ii). Suppose $W' \neq 0$ and s with $\mathrm{Re}(s) \geq 1$ are given. Then $W'(g_0) \neq 0$ for some $g_0 \in G$. On the other hand

$$\int_{N\backslash G} W'(gg_0)\overline{W}(g)\Phi(\epsilon g)|\det g|^s dg$$

$$= |\det g_0|^{-s} \int_{N\backslash G} W'(g)\overline{W}(gg_0{}^{-1})\Phi(\epsilon gg_0{}^{-1})|\det g|^s dg.$$

Thus at the cost of replacing W' by a right translate we may assume $W'(e) \neq 0$. Since W' is smooth there is a ϕ in the space $\mathcal{K}_0$ such that

$$\int_{N\backslash P} W'(p)\bar{\phi}(p)|\det p|^{s-1}d_r(p) \neq 0.$$

Choosing W in $\mathcal{W}(\pi; \psi)$ such that $W|P = \phi$(c.f. (1.1)) we get the relation

$$\int_{N\backslash P} W'(p)\overline{W}(p)|\det p|^{s-1}d_r(p) \neq 0.$$

For $k \in K$ set

$$F(k) = \int_{N\backslash P} W'(pk)\overline{W}(pk)|\det p|^{s-1}d_r(p).$$

Then $F(e) \neq 0$ and

$$F(pk) = F(k) \qquad \text{for } p \in P \cap K,$$

$$F(ak) = \omega'\bar{\omega}(a)F(k) \qquad \text{for } a \in Z \cap K.$$

It follows that there is a function Φ on F^r such that $\Phi(x) = \overline{F(k)}$ if $x = \epsilon k$ with $k \in K$, $\Phi(x) = 0$ otherwise. Moreover it is clear from the identity $G_r = PZK$ that if $x \in F^r$ is non-zero, then xK_0 is open for any open compact subgroup K_0 of K. It follows readily that $\Phi \in \mathcal{S}(F^r)$. Now, for appropriate normalizations of the measures,

$$\int_{F^\times} \Phi(\epsilon ak)|a|^{rs}\omega'\bar{\omega}(a)d^\times a = \int_{\mathfrak{R}^\times} \Phi(\epsilon ak)\omega'\bar{\omega}(a)d^\times a = \overline{F}(k).$$

Thus

$$\Psi(s, W', W, \Phi) = \int_K F(k)\overline{F}(k)dk.$$

Since this is clearly non-zero, the proof of Proposition (1.5) is now complete.

(1.6) *Remark.* One can show that the integrals (1.5.1) possess a "g.c.d" in the following sense. Given π and π', an arbitrary pair of generic irreducible representations of G_r, there is a polynomial $P(x)$ such that $P(0) = 1$ with the property that for each triple W, W', Φ as above, there is a polynomial $Q(X, Y)$ such that $\Psi(s, W', \overline{W}, \Phi) = P(q^{-s})^{-1}Q(q^s, q^{-s})$; moreover the vector space $\mathbf{C}[X; Y]$ is spanned by the Q's. We write $L(s, \pi \times \pi') = P(q^{-s})^{-1}$. The L-factor $L(s, \pi \times \pi')$ may be computed explicitly in terms of Langlands' classification. The proposition implies that if π and π' are also unitary, then $L(s, \pi \times \pi') \neq 0$, for $\mathrm{Re}(s) \geq 1$.

2. Euler factors in the unramified case. We compute our basic integral (1.5.1) when W', W and Φ are "unramified."

(2.1) Suppose π is an irreducible, admissible representation of G. We will say that π is *unramified* if it contains the unit representation of K; then the multiplicity of this representation is one. Unramified representations are parametrized by semi-simple conjugacy classes in $GL(r, \mathbf{C})$. In more detail if π is unramified, then, as is well known, π is the unique unramified component of an induced representation of the form

$$\mathrm{Ind}(G_r, B_r; \mu_1, \mu_2, \ldots, \mu_r)$$

where the μ_i's are unramified quasi-characters. The r-tuple $(\mu_1, \mu_2, \ldots, \mu_r)$ is uniquely determined up to permutation. The class A attached to π is by definition

$$A = \mathrm{diag}(\mu_1(\tilde{\omega}), \mu_2(\tilde{\omega}), \ldots, \mu_r(\tilde{\omega})).$$

We set

$$L(s, \pi) = \det(1 - q^{-s}A)^{-1}. \tag{1}$$

If π' is another unramified representation corresponding to A' we set

$$L(s, \pi \times \pi') = \det(1 - q^{-s}A \otimes A')^{-1}. \tag{2}$$

(The definition is consistent with the one given in (1.6), as we will prove in a subsequent paper.)

It will be useful to remark that if π corresponds to A then the imaginary conjugate $\bar{\pi}$ corresponds to $\bar{A}$. Moreover if ω is the central character of π, then $\omega(\tilde{\omega}) = \det A$.

(2.2) Now suppose π is generic. Suppose also that the largest ideal of F on which ψ is trivial is $\mathfrak{R}$. Then the space $\mathcal{W}(\pi; \psi)$ contains a unique element W invariant under K. Moreover $W(e) \neq 0$ and we may normalize W by requiring that

$$W(e) - 1. \tag{1}$$

This element will be called the *essential element* in $\mathcal{W}(\pi; \psi)$. There is also an explicit formula for W ([C-S], [T.S.]).

Indeed for each r-tuple J of integers set $\tilde{\omega}^J = \operatorname{diag}(\tilde{\omega}^{j_1}, \tilde{\omega}^{j_2}, \ldots, \tilde{\omega}^{j_r})$, if $J = (j_1, j_2, \ldots, j_r)$. Then every $g \in G_r$ can be written in the form $g = n\tilde{\omega}^J k$ where J is uniquely determined, $n \in N$ and $k \in K$. Thus W is completely determined by its values on the matrices $\tilde{\omega}^J$. Moreover if $T(r)$ is the set of r-tuples $J = (j_1, j_2, \ldots, j_r)$ with $j_1 \geq j_2 \geq \cdots \geq j_r$, then $W(\tilde{\omega}^J) = 0$ unless $J \in T(r)$. This being so, for $J \in T(r)$, let ρ_J be the rational representation of $GL(r, \mathbf{C})$ whose dominant weight is the character Λ_J defined by

$$\Lambda_J(a) = a_1^{j_1} a_2^{j_2} \cdots a_r^{j_r} \qquad \text{if } a = \operatorname{diag}(a_1, a_2, \ldots, a_r).$$

Let δ be the module of the group $A_r(F)N_r(F)$. The formula we have in mind is

$$W(\tilde{\omega}^J) = \delta^{1/2}(\tilde{\omega}^J)\operatorname{Tr}(\rho_J(A)) \qquad \text{if } J \in T(r). \tag{2}$$

(2.3) Proposition. *Let π and π' be admissible irreducible representations of G_r. Suppose they are generic and unramified. Let W and W' be the essential elements of $\mathcal{W}(\pi; \psi)$ and $\mathcal{W}(\pi'; \psi)$ respectively.*

Suppose Φ is the characteristic function of $\mathfrak{R}$. Then, as meromorphic functions in $\mathrm{Re}(s) > 1$,

$$\Psi(s, W', W, \Phi) = L(s, \pi \times \bar{\pi}'),$$

the measures being normalized by the conditions $\mathrm{vol}(N \cap K) = 1$, $\mathrm{vol}(K) = 1$.

Proof. For the proof we may suppose $\mathrm{Re}(s)$ large. Because of the Iwasawa decomposition the integral, which converges for $\mathrm{Re}(s)$ large, can be written as an integral on $A_r/A_r \cap K$, or equivalently as a sum over all r-tuples $J = (j_1, j_2, \ldots, j_r)$:

$$\sum_J W'\overline{W}(\tilde{\omega}^J)\delta^{-1}(\tilde{\omega}^J)\Phi(0, 0, \ldots, \tilde{\omega}_r^{j_r})|\det \tilde{\omega}^J|^s. \tag{1}$$

As we have noted $W'(\tilde{\omega}^J) = 0$ unless $J \in T(r)$. Moreover $\Phi(0, 0, \ldots, \tilde{\omega}_r^{j_r}) = 1$ if $j_r \geq 0$ and zero otherwise. Call $T_+(r)$ the set of r-tuples $J \in T(r)$ whose last entry j_r is ≥ 0. Then (1) can also be written as

$$\sum_{J \in T_+(r)} W'\overline{W}(\tilde{\omega}^J)\delta^{-1}(\tilde{\omega}^J)|\det \tilde{\omega}^J|^s. \tag{2}$$

Set now $\mathrm{tr}(J) = j_1 + \cdots + j_r$ and $X = q^{-s}$. Then $|\det \tilde{\omega}^J|^s = X^{\mathrm{tr}(J)}$ and (2) is also

$$\sum_{J \in T_+(r)} \mathrm{Tr}\,\rho_J(A')\,\overline{\mathrm{Tr}\,\rho_J(A)}\,X^{\mathrm{Tr}(J)}.$$

After replacing A by $\overline{A}$ we can see that what we have to prove is the identity

$$\sum_{J \in T_+(r)} \mathrm{Tr}\,\rho_J(A')\,\mathrm{Tr}\,\rho_J(A)\,X^{\mathrm{Tr}(J)} = \det(1 - A \otimes A'X)^{-1}. \tag{3}$$

Let us denote by $S^m A$ the mth symmetric power of the matrix A. Then the right side is

$$\sum_{m \geq 0} \mathrm{Tr}\,S^m(A \otimes A')X^m.$$

Thus we are reduced to proving the identity

$$\sum_{J \in T_+(r), \mathrm{Tr}(J)=m} \mathrm{Tr}(\rho_J(A') \otimes \rho_J(A)) = \mathrm{Tr}\, S^m(A \otimes A'). \tag{4}$$

(2.4) We will prove a somewhat more general statement which will be used elsewhere to compute similar integrals—but now associated to a product $GL_p \times GL_r$ with r not necessarily equal to p. For that we introduce more notation.

Let T_+ denote the set of all infinite vectors

$$J = (j_1, j_2, \ldots, j_r, 0, 0, \ldots) \tag{4}$$

with only a finite number of non-zero entries and satisfying $j_1 \geq j_2 \geq \cdots \geq j_r \geq 0$. We will identify the element $(j_1, j_2, \ldots, j_r)$ of $T_+(r)$ with the corresponding element (4) of T_+. Then if $p \leq r$, as subsets of T_+, $T_+(p) \subset T_+(r)$. If $J \in T_+(r)$ we write $\rho(r, J)$ for the representation of $GL_r(\mathbf{C})$ with highest weight $\wedge_J$. Let ρ_r denote the standard representation of $GL_r(\mathbf{C})$.

PROPOSITION. *Suppose $p \leq r$. For each $m \geq 0$, the representation $S^m(\rho_p \otimes \rho_r)$ of the group $G_p(\mathbf{C}) \times G_r(\mathbf{C})$ is the direct sum with multiplicity one of the representations $\rho(p, J) \otimes \rho(r, J)$ where $J \in T_+(p)$ and $tr(J) = m$.*

When $p = r$, this assertion is equivalent to (2.3.4).

Proof. We first recall some classical results ([H.W.]).

Let $\mathfrak{S}_m$ denote the symmetric group on m letters. By the classical theory of "Young symmetrizers" *there is a one-one correspondence $J \mapsto \sigma_J$ between the elements J of T_+ of trace m and the irreducible representations of $\mathfrak{S}_m$.*

Let V_r be the standard complex vector space of dimension r. $\mathfrak{S}_m$ operates in a natural way on $\otimes^m V_r$. We denote the corresponding representation by $\sigma_r{}^m$:

$$\sigma_r{}^m(\xi^{-1})(v_1 \otimes v_2 \otimes \cdots \otimes v_m) = v_{\xi(1)} \otimes v_{\xi(2)} \otimes \cdots \otimes v_{\xi(m)}.$$

A given representation σ_J of $\mathfrak{S}_m$ occurs with positive multiplicity $\mu_{r,J}$ if and only if $J \in T_+(r)$. Let $V_r(m, J)$ be the corresponding isotypic component.

Then:

$$\otimes^m V_r = \bigoplus_J V_r(m, J),$$

the sum being over all $J \in T_+(r)$ such that $\mathrm{tr}(J) = m$. Next the representations $\rho_r^{\otimes m}$ of $GL(r, \mathbf{C})$ and $\sigma_r{}^m$ of $\mathfrak{S}_m$ on $\otimes^m V_r$ commute. Moreover the operators $\sigma_r{}^m(\xi)$, $\xi \in \mathfrak{S}_m$, generate the commuting algebra of $\rho_r^{\otimes m}$. Thus $V_r(m, J)$ is also a multiple m_J of an irreducible representation of $GL(r, \mathbf{C})$. That representation is precisely the representation $\rho(r, J)$ of highest weight $\wedge_J$. Note that $m_J = \deg \sigma_J$ and therefore $\mu_{r,J} = \deg \rho(r, J)$.

We now proceed to the proof of Proposition (2.4). Let

$$W = \otimes^m(V_p \otimes V_r) \cong (\otimes^m V_p) \otimes (\otimes^m V_r). \tag{1}$$

Let ρ' be the representation $(\rho_p)^{\otimes m} \otimes (\rho_r)^{\otimes m}$ of $GL(p, \mathbf{C}) \times GL(r, \mathbf{C})$ on W. Its decomposition into isotypic components is:

$$W = \oplus V_p(m, J) \otimes V_r(m, K), \tag{2}$$

the sum extended over all $J \in T_+(p)$ of trace m and all $K \in T_+(r)$ of trace m.

Now $\mathfrak{S}_m$ acts on W via the representation $\sigma_{p+r}{}^m$. The space W_0 of invariants of this action is just $S^m(V_p \otimes V_r)$ and affords the representation $S^m(\rho_p \otimes \rho_r)$ of $GL_p(\mathbf{C}) \times GL_r(\mathbf{C})$. On the other hand, via (1), the representation of $\mathfrak{S}_m$ on $(\otimes^m V_p) \otimes (\otimes^m V_r)$ is $\sigma_p{}^m \otimes \sigma_r{}^m$. Since that representation leaves each term on the right side of (2) invariant, we have

$$W_0 = \oplus\ W_0 \cap V_p(m, J) \otimes V_r(m, K)$$

the sum over J, K as before. Now $W_0 \cap V_p(m, J) \otimes V_r(m, K) \neq 0$ implies that the representation $\sigma_J \otimes \sigma_K$ contains the unit representation of $\mathfrak{S}_m$. Since, as is well known, the irreducible representations of $\mathfrak{S}_m$ are all self-contragradient, this implies that $J = K$ (in T_+). Moreover $\sigma_J \otimes \sigma_J$ then contains the trivial representation of $\mathfrak{S}_m$ exactly once. Thus

$$\mathrm{Dim}(W_0 \cap V_p(m, J) \otimes V_r(m, J)) = \mu_{p,J}\mu_{r,J} = \deg \rho(p, J) \deg \rho(r, J).$$

Since the representation of $GL_p(\mathbf{C}) \times GL_r(\mathbf{C})$ is isotypic of type $\rho(p, J) \otimes \rho(r, J)$ it must be $\rho(p, J) \otimes \rho(r, J)$. Thus we have established that the representation of $GL_p(\mathbf{C}) \times GL_r(\mathbf{C})$ on W_0 is the direct sum, the sum being over all J satisfying $J \in T_+(p)$, $\mathrm{Tr}(J) = m$,

$$\bigoplus_{J \in T_+(p),\ \mathrm{Tr}(J) = m} \rho(p, J) \otimes \rho(r, J),$$

as required.

(2.5) We have the following corollary to Propositions (1.5) and (2.3).

Corollary. Let π be an admissible irreducible representation of G_r. Suppose π is unitary, generic and unramified. Let A be the class of π. Then the eigenvalues of A are in absolute value $< q^{1/2}$.

Proof. We apply Proposition (2.3) with $\pi = \pi'$, $W = W'$ and Φ the characteristic function of $\mathfrak{R}$. Then

$$\det(1 - q^{-s}A \otimes \overline{A})\Psi(s, W, W, \Phi) = 1,$$

at first for $\mathrm{Re}(s) > 1$. But then, by Proposition (1.5), $\Psi(s, W, W, \Phi)$ has a continuous extension to the half-plane $\mathrm{Re}(s) \geq 1$. Thus $\det(1 - q^{-s}A \otimes \overline{A})$ cannot vanish in this closed half-plane. In particular, if λ is an eigenvalue of A, we must have

$$1 - q^{-\sigma}|\lambda|^2 \neq 0$$

for all $\sigma \geq 1$. Our assertion follows at once.

The proof should be compared with a similar argument of Rankin [R.R.].

(2.6) *Remarks.* (1) The proof of Proposition (3.14) below together with (3.15) can be used to show that the representations

$$\mathrm{Ind}(G_r, B_r; \alpha^s, \alpha^{-s}, 1, \ldots, 1)$$

of G_r are irreducible, unitary, and generic provided $0 < s < 1/2$. Thus the conclusion of Corollary (2.5) is the best possible.

(2) We note that, if π is the trivial representation of G_r, then the maximum eigenvalue of A is $q^{(n-1)/2}$.

(3) We remark finally that the corollary applies to local components of cusp forms.

3. Generic representations: archimedean case. In this section $F = \mathbf{R}$ or $\mathbf{C}$. We extend the results of Section 1 to include the archimedean case. After the analogue of (1.3) is established, this will be an easy matter. However, in order to establish this analogue we will need considerable preliminaries.

(3.1) Let G be a (real) Lie group. Denote by $\mathfrak{L}(G)$, $\mathfrak{U}(G)$, $\mathfrak{Z}(G)$, the Lie algebra of G, the enveloping algebra of $\mathfrak{L}(G) \otimes \mathbf{C}$, and the center of $\mathfrak{U}(G)$ respectively. On $\mathfrak{U}(G)$ we have a natural filtration. We let $\mathfrak{U}^n(G)$ be the n-th term of this filtration.

Let π be a unitary representation of G on a Hilbert space $\mathcal{H}$. We will denote by $\mathcal{H}_G{}^\infty$ or, when this does not create confusion, by $\mathcal{H}^\infty$ the space of smooth vectors in $\mathcal{H}$. Thus v is in $\mathcal{H}^\infty$ if and only if the map $g \mapsto \pi(g)v$ from G to $\mathcal{H}$ is C^∞. As is well known this is equivalent to saying that, for all $w \in \mathcal{H}$, the coefficients $(\pi(g)v, w)$ are smooth functions on G. Moreover $\mathcal{H}^\infty$ is dense in $\mathcal{H}$.

As is customary, if $X \in \mathfrak{L}(G)$, $v \in \mathcal{H}^\infty$, we set

$$\pi(X)v = (d/dt)\pi(\exp tX)v\,|_{t=0},$$

the limit taken in the topology of $\mathcal{H}$. Then $\pi(X)v$ is again in $\mathcal{H}^\infty$. The semi-norms

$$v \mapsto \|\pi(D)v\|, \qquad D \in \mathfrak{U}(G),$$

$\|\ \|$ being the given norm on $\mathcal{H}$, define on $\mathcal{H}^\infty$ the topology of a complete locally convex space. The space $\mathcal{H}^\infty$ is invariant under G and the representation π of G on $\mathcal{H}^\infty$ is smooth.

Let $\mathcal{H}'$ be the dual of $\mathcal{H}$ and π' the representation contragradient to π. We may define similarly the dual $\mathcal{H}_G{}^{-\infty}$ of $\mathcal{H}_G{}^\infty$ and the representation $\pi^{-\infty}$ contragradient to π^∞. Since $\mathcal{H}_G{}^\infty$ is dense in $\mathcal{H}$, we may and will regard $\mathcal{H}'$ as a subspace of $\mathcal{H}_G{}^{-\infty}$. Then for $v \in \mathcal{H}'$, we have $\pi'(g)v = \pi^{-\infty}(g)v$. Every vector v in $\mathcal{H}_G{}^{-\infty}$ is a finite sum of the form

$$v = \Sigma\pi^{-\infty}(X_i)v_i, \qquad X_i \in \mathfrak{U}(G), \qquad v_i \in \mathcal{H}'.$$

(3.2) $\mathcal{H} \mapsto \mathcal{H}^\infty$ is a functor: if π_1 and π_2 are unitary representations on the Hilbert spaces $\mathcal{H}_1$ and $\mathcal{H}_2$ respectively and A belongs to $Hom_G(\pi_1, \pi_2)$ then $A(\mathcal{H}_1{}^\infty) \subset \mathcal{H}_2{}^\infty$ and A defines a continuous operator from $\mathcal{H}_1{}^\infty$ to $\mathcal{H}_2{}^\infty$. In particular:

Lemma. *Suppose that π is a unitary representation of G on $\mathcal{H}$. Suppose $\mathcal{H}$ is the direct sum of two orthogonal closed invariant subspaces $\mathcal{H}_1$ and $\mathcal{H}_2$. Then $\mathcal{H}_i^\infty = \mathcal{H}^\infty \cap \mathcal{H}_i$ and $\mathcal{H}^\infty$ is the topological direct sum of $\mathcal{H}_1^\infty$ and $\mathcal{H}_2^\infty$.* □

(3.3) We will repeatedly use the following lemma ([P] Theorem (1.3)):

Lemma. *Suppose $\mathcal{V}$ is a G-invariant subspace of $\mathcal{H}$ dense for the norm topology. Then if $\mathcal{V}$ is contained in $\mathcal{H}_G^\infty$ it is dense in $\mathcal{H}_G^\infty$.*

(3.4) Let R be a closed subgroup of G. Then we may consider the restriction of π to R and the corresponding subspace $\mathcal{H}_R^\infty$ of $\mathcal{H}$. Clearly $\mathcal{H}_G^\infty \subset \mathcal{H}_R^\infty$. Since $\mathcal{H}_G^\infty$ is dense in $\mathcal{H}$, it follows from Lemma (3.3) (applied to R) that $\mathcal{H}_G^\infty$ is dense in $\mathcal{H}_R^\infty$. Thus we may regard $\mathcal{H}_R^{-\infty}$ as a subspace of $\mathcal{H}_G^{-\infty}$.

Now suppose there is a subgroup V of G such that

$$\mathfrak{L}(G) = \mathfrak{L}(R) \oplus \mathfrak{L}(V).$$

Call $\mathcal{H}^{-n}$ the space of vectors $v \in \mathcal{H}_G^{-\infty}$ of the form

$$v = \Sigma \pi(D) w_D; \qquad D \in \mathfrak{U}^n(V), \qquad w_D \in \mathcal{H}_R^{-\infty}.$$

Then:

Lemma. *With the above notations, each $\mathcal{H}^{-n}$ is an $\mathfrak{U}(R)$ submodule of $\mathcal{H}^{-\infty}$, $\mathcal{H}^{-n} \subset \mathcal{H}^{-(n+1)}$, and*

$$\bigcup_{n \geq 0} \mathcal{H}^{-n} = \mathcal{H}_G^{-\infty}.$$ □

(3.5) In the following paragraphs we will apply the above to the study of the representations of P_r which contain the unitary representation

$$\tau_r = \operatorname{Ind}(P_r, N_r, \theta_r) \tag{1}$$

induced by the character θ_r.

We recall first an elementary lemma from direct integral theory. Let G be a separable locally compact group and V a closed normal subgroup:

LEMMA. *Let π be a unitary representation of G. Suppose $\pi | V$ does not contain the trivial representation of V. Then π is a direct integral of irreducible representations*

$$\pi = \int_X \xi_x d\mu(x)$$

where, for almost all $x \in X$, $\xi_x | V$ is not the trivial representation.

(3.6) We describe what Mackey's semi-direct product theory tells us about the representations of P_r. Let η_r be the restriction of θ_r to U_r. Since U_r is normal in P_r, the group P_r operates on U_r as well as its dual group $\hat{U}_r$. There are two orbits for P_r acting on $\hat{U}_r$. One is the orbit of the trivial character and the other is the orbit of η_r. The fixer of η_r is the semi-direct product of U_r and the subgroup $P_{r-1} \subset G_{r-1}$ identified with a subgroup of P_r via the map

$$p \mapsto \begin{pmatrix} p & 0 \\ 0 & 1 \end{pmatrix}.$$

Let ξ be an irreducible unitary representation of P_r. It now follows from Mackey's theory ([G.W.M.], Theorems 14.1 and 14.2) that, either ξ has a U_r fixed vector in which case ξ is really a representation of $P_r/U_r \simeq G_{r-1}$, or else ξ has the form

$$\xi = \operatorname{Ind}(P_r, P_{r-1}U_r; \sigma \otimes \eta_r)$$

where σ is an irreducible unitary representation of P_{r-1}. We may now apply Lemma (3.5) to $G = P_r$, $V = U_r$, noting that inducing commutes with direct integrals, to obtain the following lemma:

LEMMA. *Suppose π is a unitary representation of P_r on a Hilbert space $\mathcal{H}$. Suppose $\mathcal{H}$ does not contain a non-zero vector fixed by $\pi | U_r$. Then*

$$\pi \simeq \operatorname{Ind}(P_r, P_{r-1}U_r; \rho \otimes \eta_r)$$

for some unitary representation ρ of P_{r-1}. □

Note that Mackey's theory implies that τ_r is irreducible.

(3.7) *Proposition. Let π be a unitary representation of P_r on a Hilbert space $\mathcal{H}$. The following conditions are equivalent:*

(i) *the representation π contains τ_r;*

(ii) *there is a $\lambda \neq 0$ in $\mathcal{H}_P^{-\infty}$ such that*

$$\lambda(\pi(n)v) = \theta(n)\lambda(v) \tag{1}$$

for all v in $\mathcal{H}_P^{\infty}$ and all $n \in N_r$.

Proof. We prove first that (ii) implies (i).

Suppose first that $\pi = \tau_r$. Then every $f \in \mathcal{H}_P^{\infty}$ is actually a C^{∞}-function on P transforming on the left like θ under N. Moreover there are $D_j \in \mathfrak{U}(P)$ such that for every $f \in \mathcal{H}_P^{\infty}$

$$|f(e)| \leq \sum_j ||\tau(D_j)f||.$$

Thus

$$\lambda(f) = f(e)$$

defines a continuous linear form on $\mathcal{H}_P^{\infty}$ which clearly satisfies (1). Of course $\lambda \neq 0$ and so (ii) is true.

More generally suppose π is a unitary representation of P_r on $\mathcal{H}$. Suppose $\mathcal{H}_\tau$ is an invariant subspace affording the representation τ_r. Let $A\colon \mathcal{H} \to \mathcal{H}_\tau$ be the orthogonal projection. Then by functoriality

$$A(\mathcal{H}^{\infty}) = \mathcal{H}_\tau^{\infty}.$$

Thus

$$\lambda(v) = A(v)(e), \qquad v \in \mathcal{H}^{\infty},$$

defines a non-zero linear form satisfying (1).

Suppose now π is given and $\lambda \neq 0$ belongs to $\mathcal{H}_P^{-\infty}$ and satisfies (1). Let $\mathcal{H}^U$ be the space of $U(= U_r)$ fixed vectors for π, and let $\mathcal{H}_U$ be the orthogonal complement of $\mathcal{H}^U$ in $\mathcal{H}$. Thus

$$\mathcal{H} = \mathcal{H}^U \oplus \mathcal{H}_U,$$

and by Lemma (3.2)

$$\mathcal{H}^\infty = (\mathcal{H}^U)^\infty \oplus (\mathcal{H}_U)^\infty.$$

Clearly λ vanishes on $(\mathcal{H}^U)^\infty$. Thus we may as well replace $\mathcal{H}$ by $\mathcal{H}_U$. By Lemma (3.6) this amounts to saying that π has the form

$$\pi = \mathrm{Ind}(P_r, P_{r-1}U_r; \sigma \otimes \eta_r) \qquad (2)$$

for some unitary representation σ of P_{r-1}. If $r = 2$, then $P_{r-1} = \{1\}$, $U_r = N_r$, $\eta_r = \theta_r$ and π is actually a multiple of τ_r. Suppose then $r > 2$ and the theorem true for $r - 1$. Let $\mathcal{K}$ be the space of σ. The inclusion of G_{r-1} in G_r given by

$$g \mapsto \begin{pmatrix} g & 0 \\ 0 & 1 \end{pmatrix}$$

induces an isomorphism $P_{r-1}\backslash G_{r-1} \simeq P_{r-1}U_r\backslash P_r$. It follows that $\pi | G_{r-1}$ is the representation

$$\mathrm{Ind}(G_{r-1}, P_{r-1}; \sigma).$$

Its space $\mathcal{H}$ is the space of all $\mathcal{K}$-valued functions F on G_{r-1} satisfying

$$F(pg) = \delta_{P_{r-1}}^{1/2}(p)\sigma(p)F(g)$$

and such that

$$\int_{P_{r-1}\backslash G_{r-1}} ||F(g)||^2 dg < \infty.$$

Moreover, for $u \in U_r$,

$$(\pi(u)F)(g) = \eta_r(gug^{-1})F(g).$$

Next for $f \in C_c^\infty(G_{r-1}; \mathcal{K}^\infty)$, $g \in G_{r-1}$, set

$$F_f(g) = \int_{P_{r-1}} \delta_{P_{r-1}}^{-1/2}(p)\sigma(p)^{-1}f(pg)d_r(p).$$

Then the space of functions F_f is a dense subspace of $\mathcal{K}^\infty$ (Lemma (3.3)). Moreover $f \mapsto F_f$ is continuous. Thus λ is completely determined by the $\mathcal{K}^\infty$-distribution T defined by

$$T(f) = \int_{G_{r-1}} f(g)dT(g) = \lambda(F_f).$$

Let us denote by χ_u the function on G_{r-1} defined by $\chi_u(g) = \eta_r(gug^{-1})$, $u \in U_r$. We have by (1)

$$\lambda(\chi_u F) = \eta_r(u)\lambda(F), \qquad F \in \mathcal{K}^\infty.$$

Moreover $F_{\chi_u f} = \chi_u F_f$. Hence $T(\chi_u f) = T(f)$, $f \in C_c^\infty(G_{r-1}; \mathcal{K}^\infty)$. It follows that T has support in the fixer P_{r-1} of the character η_r.

A formal manipulation shows that

$$dT(pmn^{-1}) = \sigma(p^{-1})\delta_{P_{r-1}}^{1/2}(p)\theta(n)dT(p), \tag{3}$$

for $m \in G_{r-1}$, $p \in P_{r-1}$, $n \in N_{r-1}$. We now apply the results of (5.2.3) of [G.W.] to our present case. Following their notation we set $M = G_{r-1}$, $G = P_{r-1} \times N_{r-1}$ with G acting on M via

$$g \circ m = pmn^{-1}, \qquad g = (p, n).$$

We let Q be the closed orbit P_{r-1}. The isotropy group H at $1 \in Q$ is the diagonal in $N_{r-1} \times N_{r-1}$. In particular the module of H is 1. Let U be the *differentiable* representation of G on $\mathcal{K}^\infty$ defined by

$$U(p, n) = \delta_{P_{r-1}}^{-1/2}(p)\sigma(p)\bar{\theta}(n), \qquad p \in P_{r-1}, \qquad n \in N_{r-1}.$$

By (3), T is quasi-invariant relative to U. Thus (loc. cit.) we obtain a finite-dimensional vector space V, a continuous bilinear form $z \neq 0$ on $\mathcal{K}^\infty \times V$ and an *algebraic* representation μ of N_{r-1} on V such that

$$z(U(n, n)a, \mu(n)b) = z(a, b),$$

$$a \in \mathcal{K}^\infty, b \in V, n \in N_{r-1},$$

or explicitly

$$z(\sigma(n)a, \mu(n)b) = \theta(n)z(a, b).$$

There is a vector $b_0 \neq 0$ in V fixed by μ. At the cost of replacing V by a quotient we may assume that the linear form

$$a \mapsto z(a, b_0)$$

on $\mathcal{K}^\infty$ is non-zero. Clearly it satisfies (1). Thus, by the induction hypothesis, σ contains τ_{r-1}.

Since π has the form (2), we see that π contains

$$\operatorname{Ind}(P_r, P_{r-1}U_r; \tau_{r-1} \otimes \eta_r) \simeq \tau_r.$$

This concludes the proof of Proposition (3.7). □

(3.8) We now prove the analogue of Proposition (1.3) for archimedean fields.

Proposition. *Let π be a unitary representation of G_r on a Hilbert space $\mathcal{H}$. The following conditions are equivalent:*

(i) *the restriction of π to P_r contains τ_r;*

(ii) *there is a linear form $\lambda \neq 0$ on $\mathcal{H}_G{}^\infty$ such that*

$$\lambda(\pi(n)v) = \theta(n)v$$

for all $v \in \mathcal{H}_G{}^\infty$ and $n \in N$.

Proof. Suppose (i) is satisfied. By Proposition (3.7) there is a linear form $\lambda \neq 0$ on $\mathcal{H}_P{}^\infty$ satisfying (3.7.1). Since $\mathcal{H}_G{}^\infty$ is dense in $\mathcal{H}_P{}^\infty$ with continuous injection, the restriction of λ to $\mathcal{H}_G{}^\infty$ has the required properties.

Now suppose that (ii) is satisfied. Then λ is an element of $\mathcal{H}_G{}^{-\infty}$ and for any $X \in \mathfrak{U}(N)$

$$\pi^\infty(X)\lambda = \dot{\theta}(X)\lambda \tag{1}$$

where $\dot{\theta}$ denotes the homomorphism $\mathfrak{U}(N) \to \mathbf{C}$ induced by the character θ.

If we could show that this implies that λ is actually in $\mathcal{H}_P{}^{-\infty}$ then (i) would follow from Proposition (3.7). To see what this entails let us use the notation of (3.4) with $G = G_r$, $R = P_r Z_r$, $V = \overline{U}_r = {}^t U_r$. Then since Z operates by scalars on $\mathcal{H}_G{}^\infty$ the operators $\pi^\infty(X)$ for $X \in \mathfrak{U}(Z)$ are also scalar. Thus

$$\mathcal{H}_R{}^{-\infty} = \mathcal{H}_{P_r Z_r}{}^{-\infty} = \mathcal{H}_{P_r}{}^{-\infty}$$

as (topological) vector spaces.

Now $\lambda \in \mathcal{H}^{-n}$ for some n by Lemma (3.4). Thus it would suffice to show that λ is in $\mathcal{H}^{-n}$ implies λ is in $\mathcal{H}^{-n+1}$. Call $\mathfrak{J}$ the subalgebra of $\mathfrak{U}(G_r)$ generated by $\mathfrak{U}(P_r)$ and $\mathfrak{Z}(G_r)$. Then each $\mathcal{H}^{-n}$ is an $\mathfrak{J}$-module and our assertion would follow from the following statement:

(2) *Let $\mathcal{V}$ be a $\mathfrak{U}(G)$-module and λ an element of $\mathcal{V}$ such that*

$$X\cdot\lambda = \dot\theta(X)\lambda \qquad \text{for } X \in \mathfrak{L}(N).$$

Suppose there is a subspace $\mathcal{W}$ of $\mathcal{V}$ stable under $\mathfrak{J}$ such that λ is in $\mathfrak{L}(\overline{U})\mathcal{W} + \mathcal{W}$. Then λ is in $\mathcal{W}$.

It is easy enough to establish this assertion for $r = 2$. Indeed suppose $F = \mathbf{R}$ and let E^+ and E^- be the following elements of $\mathfrak{L}(G)$:

$$E^+ = \begin{pmatrix} 0 & 1 \\ 0 & 0 \end{pmatrix}, \qquad E^- = \begin{pmatrix} 0 & 0 \\ 1 & 0 \end{pmatrix}.$$

By considering the Casimir element of $\mathfrak{Z}(G)$ we see that

$$E^+E^- \equiv E^-E^+ \equiv 0 \pmod{\mathfrak{J}}.$$

Suppose

$$\lambda \equiv E^-\mu \pmod{\mathcal{W}}$$

where μ is in $\mathcal{W}$. Then, applying E^+ to both sides of this congruence, we get

$$\dot\theta(E^+)\lambda \equiv E^+E^-\mu \equiv 0 \pmod{\mathcal{W}}.$$

Since $\dot\theta(E^+) \neq 0$ we indeed have $\lambda \in \mathcal{W}$. The proof for $r = 2$, $F = \mathbf{C}$ is similar.

Unfortunately we have not been able to prove (2) for all r. What we will do instead is prove (2) for $r = 4$, $F = \mathbf{R}$. At this point we will have proved that (ii) implies (i) in the following cases:

$$r = 2,\ F = \mathbf{R} \text{ or } \mathbf{C} \text{ and } r = 4,\ F = \mathbf{R}.$$

We will then prove that (ii) implies (i) in full generality by using a certain reduction technique (c.f. (3.12)).

(3.9) *Proof of Proposition* (3.8) *when* $r = 4$, $F = \mathbf{R}$.

Let E_{ij} denote the 4×4 matrix with one in the ith row and jth column and zero elsewhere. Let also

$$H_1 = \operatorname{diag}(1, 0, 0, 0), \qquad H_2 = \operatorname{diag}(0, 1, 0, 0),$$

$$H_3 = \operatorname{diag}(0, 0, 1, 0), \qquad H_4 = \operatorname{diag}(0, 0, 0, 1).$$

The E_{ij}, $1 \le i, j \le 4$, H_i, $1 \le i \le 4$, form a basis for $\mathfrak{g}(G)$. In what follows we will normalize $\dot{\theta}$ by requiring $\dot{\theta}(E_{i,i+1}) = 1$ for $1 \le i \le 3$.

Our assumption is that

$$\lambda \equiv E_{41}\mu_1 + E_{42}\mu_2 + E_{43}\mu_3 \,(\operatorname{mod} \mathfrak{W}) \tag{1}$$

with $\mu_i \in \mathfrak{W}$. Apply $E_{34} \in \mathfrak{U}(P)$ to this congruence to get

$$E_{34}\lambda = \lambda \equiv E_{34}E_{41}\mu_1 + E_{34}E_{42}\mu_2 + E_{34}E_{43}\mu_3.$$

But mod $\mathfrak{J}$ we have

$$[E_{34}, E_{41}] = E_{31} \equiv 0, \qquad [E_{34}, E_{42}] = E_{32} \equiv 0,$$

$$[E_{24}, E_{42}] \equiv 0, \qquad [E_{14}, E_{41}] \equiv 0,$$

and

$$E_{34}E_{43} + E_{24}E_{42} + E_{14}E_{41} \equiv 0, \tag{2}$$

the last relation from consideration of the Casimir operator. Thus

$$\lambda \equiv E_{41}\nu_1 + E_{42}\nu_2$$

where

$$\nu_1 = E_{34}\mu_1 - E_{14}\mu_3, \qquad \nu_2 = E_{34}\mu_2 - E_{24}\mu_3.$$

Thus changing notations, we may assume, instead of (1),

$$\lambda \equiv E_{41}\mu_1 + E_{42}\mu_2 \,(\mathrm{mod}\ \mathcal{W}) \tag{3}$$

with μ_1, μ_2 in $\mathcal{W}$.

To proceed we need to consider elements of $\mathfrak{Z}(G)$ other than the Casimir operator. Now $\mathfrak{Z}(G)$ is the image under the symmetrization map of the invariant elements of the symmetric algebra $S(\mathfrak{L}(G))$ of $\mathfrak{L}(G)$ in $\mathfrak{U}(G)$. We may use the invariant bilinear form

$$(X, Y) \mapsto \mathrm{tr}(XY)$$

on $\mathfrak{L}(G)$ to identify $S(\mathfrak{L}(G))$ with the polynomial algebra on $\mathfrak{L}(G)$, the coordinate function g_{ij} corresponding to the monomial E_{ji}. Let ω_r denote the *invariant* polynomial function on $\mathfrak{L}(G)$ defined by

$$\omega_r(g) = \mathrm{tr}(\Lambda^r(g)).$$

The image in $S(\mathfrak{L}(G))$ is obtained by formally replacing g by the matrix of E_{ji}'s. We denote by Ω_r the corresponding element of $\mathfrak{Z}(G)$. Thus for example if $r = 3$ we have to compute the sum of the four principal 3×3 minors of

$$\begin{bmatrix} H_1 & E_{21} & E_{31} & E_{41} \\ E_{12} & H_2 & E_{32} & E_{42} \\ E_{13} & E_{23} & H_3 & E_{43} \\ E_{14} & E_{24} & E_{34} & H_4 \end{bmatrix}$$

and then symmetrize. We then find the following expression for Ω_3 mod $\mathfrak{U}(P)$:

$$\Omega_3 \equiv A_1'E_{41} + A_2'E_{42} + A_3'E_{43} \,(\mathrm{mod}\ \mathfrak{U}(P))$$

where $A_3' = E_{24}E_{32} + 2E_{34} + E_{14}E_{31} - (H_1 + H_2)E_{34}$ and A_1', A_2' are obtained from A_3' by obvious symmetries. Using (2) again, we arrive at the relation

$$A_1E_{41} + A_2E_{42} + A_3E_{43} \equiv 0\,(\mathfrak{J}), \tag{4}$$

where

$$\begin{aligned} A_1 &= E_{34}E_{13} + 2E_{14} + E_{24}E_{12} + H_1E_{14} - H_3E_{14} \\ A_2 &= E_{34}E_{23} + 2E_{24} + E_{14}E_{21} + H_2E_{24} - H_3E_{24} \\ A_3 &= E_{24}E_{32} + 2E_{34} + E_{14}E_{31}. \end{aligned} \tag{5}$$

Using (2), we have the following commutation relations mod $\mathfrak{J}$:

$$\begin{aligned} [A_1, E_{41}] &\equiv 0\,(\mathrm{mod}\,\mathfrak{J}) \\ [A_2, E_{41}] &\equiv 0 \\ [A_2, E_{42}] &\equiv 0 \\ [A_3, E_{43}] &\equiv E_{43}E_{34} \\ [A_3, E_{42}] &\equiv 0 \\ [A_3, E_{41}] &\equiv 0. \end{aligned} \tag{6}$$

From (5), we get immediately that $A_i \in \mathfrak{U}(PZ)$ and in addition, noting that $A_3 = E_{32}E_{24} + E_{31}E_{14}, A_2 = E_{23}E_{34} + E_{21}E_{14} + (H_2 - H_3)E_{24}$,

$$A_2\lambda = \lambda, \qquad A_1\lambda = A_3\lambda = 0. \tag{7}$$

Next multiply (4) by E_{34} on the right to get

$$A_1E_{41}E_{34} + A_2E_{42}E_{34} + A_3E_{43}E_{34} \equiv 0\,(\mathfrak{J}). \tag{8}$$

Using (2) and the fact that $A_i \in \mathfrak{U}(PZ)$, this can be written as

$$(A_1E_{34} - A_3E_{14})E_{41} + (A_2E_{34} - A_3E_{24})E_{42} \equiv 0\,(\mathfrak{J}); \tag{9}$$

or, using (6) for A_1 and A_3, as

$$E_{41}U + VE_{42} \equiv 0\,(\mathrm{mod}\,\mathfrak{J}), \tag{10}$$

where

$$U = A_1 E_{34} - A_3 E_{14}, \qquad V = A_2 E_{34} - A_3 E_{24}. \tag{11}$$

From (7) we get

$$V\lambda = \lambda. \tag{12}$$

Note that U and V belong to $\mathfrak{U}(PZ)$. Thus (3) now implies

$$\lambda \equiv VE_{41}\mu_1 + VE_{42}\mu_3 \pmod{\mathcal{W}},$$

or from (10)

$$\lambda \equiv VE_{41}\mu_1 - E_{41}U\mu_2.$$

Since by (6) V commutes with E_{41} mod $\mathfrak{J}$, this may be written as

$$\lambda \equiv E_{41}(V\mu_1 - U\mu_2).$$

But $V\mu_1 - U\mu_2 \in \mathcal{W}$ and thus we may replace (3) by

$$\lambda \equiv E_{41}\mu \pmod{\mathcal{W}}, \text{ with } \mu \in \mathcal{W}. \tag{13}$$

Before introducing one more element of $\mathfrak{Z}(G)$, using (6), we rewrite (9) in the form

$$(A_1E_{34} - A_3E_{14})E_{41} + E_{42}(A_2E_{34} - A_3E_{24}) \equiv 0 \pmod{\mathfrak{J}},$$

or, as

$$U'E_{41} + E_{42}V' \equiv 0 \pmod{\mathfrak{J}}, \tag{14}$$

where

$$U' = A_1E_{34} - A_3E_{14}, \qquad V' = A_2E_{34} - A_3E_{24}.$$

Once more, by (7),

$$U'\lambda = 0, \qquad V'\lambda = \lambda. \tag{15}$$

Finally we make use of Ω_4. Expanding the determinant and symmetrizing, we get

$$\Omega_4 \equiv B_1E_{41} + B_2E_{42} + B_3E_{43} \,(\mathrm{mod}\ \mathfrak{U}(PZ)) \tag{16}$$

with $B_i \in \mathfrak{U}(PZ)$. Moreover

$$B_1\lambda = c\lambda \text{ with } c \neq 0. \tag{17}$$

It will not be necessary to know the explicit form of the B_i.

Next multiply (16) by E_{34} on the right to get

$$B_1E_{41}E_{34} + B_2E_{42}E_{34} + B_3E_{43}E_{34} \equiv 0\,(\mathfrak{J}), \tag{18}$$

or, using (2) once more,

$$(B_1E_{34} - B_3E_{14})E_{41} + (B_2E_{34} - B_3E_{24})E_{42} \equiv 0\,(\mathfrak{J}).$$

Multiplying on the right by V' (which is in $\mathfrak{J}$) and using (14) we get

$$(B_1E_{34} - B_3E_{14})E_{41}V' - (B_2E_{34} - B_3E_{24})U'E_{41} \equiv 0\,(\mathfrak{J}).$$

It follows from (6) that V' commutes, mod $\mathfrak{J}$, with E_{41}. Thus we get at last

$$AE_{41} \equiv 0\,(\mathrm{mod}\ \mathfrak{J})$$

where A is the following element of $\mathfrak{J}$:

$$A = (B_1E_{34} - B_3E_{14})V' - (B_2E_{34} - B_3E_{24})U'.$$

From (15) and (17) we now have

$$A\lambda = (B_1E_{34} - B_3E_{14})\lambda = B_1\lambda = c\lambda.$$

But then applying A to both sides of (13) we get

$$c\lambda = A\lambda \equiv AE_{41}\mu \equiv 0\,(\mathrm{mod}\ \mathfrak{W}).$$

So λ is in $\mathcal{W}$. This concludes the proof of the statement that (ii) implies (i) in the case $r = 4$, $F = \mathbf{R}$.

(3.10) To complete the proof of Proposition (3.8) in general we will first show that every unitary representation of G_r satisfying (ii) is induced, in the unitary sense, by similar representations of G_4 and G_2. We will then show that (i) is inductive. The conjunction of these two results will show that (ii) implies (i) for all r and F.

We first prove the following proposition:

Proposition. *Suppose R is the standard parabolic of type $(n_1, n_2, \ldots, n_u)$ so that $R/U_R \simeq \Pi_i G_{n_i}$. For each i, let σ_i be a unitary representation of G_{n_i}. Let π be the representation induced—in the unitary sense—by the σ_i:*

$$\pi = \operatorname{Ind}(G, R; \sigma_1, \sigma_2, \ldots, \sigma_u).$$

Then if π satisfies condition (ii) *of* (3.8) *then so does each σ_i.*

Proof. The proof of the corresponding assertion for the p-adic case is due to Rodier ([F.R]). The proof in the archimedean case is essentially the same, except for technical difficulties due to the existence of transversal derivatives.

Let $\mathcal{K}_i$ be the Hilbert space on which σ_i operates and $\mathcal{K} = \hat{\otimes}\, \mathcal{K}_i$ the space of the representation $\sigma = \otimes\, \sigma_i$ of $M \simeq R/U_R \simeq \Pi G_{n_i}$. As in (3.7), given $f \in C_c^\infty(G_r; \mathcal{K}^\infty)$, set for g in G_r,

$$F_f(g) = \int_R \delta_R^{-1/2}(r)\sigma(r^{-1})f(rg)dr$$

where dr is a right-invariant measure on R. Given $\lambda \neq 0$ satisfying (3.8)(ii), define a $\mathcal{K}^\infty$-distribution T on G_r by setting

$$T(f) = \lambda(F_f).$$

Then

$$dT(rgn^{-1}) = \sigma(r^{-1})\delta_R^{1/2}(r)\theta(n)dT(g).$$

We let the group $R \times N$ act on the manifold G by

$$(r, n) \circ g = rgn^{-1}.$$

Then G is a finite union of orbits. For any orbit

$$Q = RgN,$$

let $\Omega(Q)$ be the union of Q with the orbits of strictly larger dimension. Then Q is a closed submanifold of the open set $\Omega(Q)$. Starting with (open) orbits of maximal dimension and proceeding inductively, we find a Q such that $S = T|\Omega(Q)$ is non-zero and has support in Q.

We now proceed exactly as in (3.7). We write $Q = RwN$ for some permutation matrix w. The isotropy group H at w is the subgroup of $R \times N$ of elements (r, n) satisfying $rwn^{-1} = w$. This group is isomorphic to $N \cap w^{-1}Rw$ via the map $n \mapsto (wnw^{-1}, n)$. In particular it is unimodular. We now apply (5.2.3) of [G.W.], this time with $M = \Omega(Q)$ and the *differentiable* representation U of $R \times N$ on $\mathcal{K}^\infty$ being given by

$$U(r, n) = \delta_R{}^{-1/2}(r)\sigma(r)\bar{\theta}(n), \qquad r \in R, n \in N.$$

As before we get a finite-dimensional vector space V, a continuous bilinear form $z \neq 0$ on $\mathcal{K}^\infty \times V$ and an algebraic representation μ of H on V such that

$$z(U(h)a, \mu(h)b) = z(a, b),$$

for $a \in \mathcal{K}^\infty$, $b \in V$, $h \in H$, or explicitly

$$z(\sigma(wnw^{-1})a, \mu(n)b) = \theta(n)z(a, b), \tag{1}$$

for all $n \in N \cap w^{-1}Rw$. Suppose α is a simple root and let N_α be the corresponding subgroup of N. If now $wN_\alpha w^{-1}$ is contained in U_R, then since σ is trivial on U_R, we get

$$z(a, \mu(n)b) = \theta(n)z(a, b), \; n \in N_\alpha.$$

But $\mu(n)$ is unipotent. Thus $\theta(n) = 1$ for all $n \in N_\alpha$, a contradiction. Thus there are no simple roots α for which $wN_\alpha w^{-1} \subset U_R$ and this implies that Q is open. In fact $Q = Rw_0N$ where w_0 is the order reversing permutation matrix. Now write $R = MU_R$. Then $w_0Nw_0{}^{-1} \cap M = N'$ is a maximal unipotent subgroup of M. Exactly as before we may, at the cost of replacing V by a quotient, assume there is b_0 in V fixed by $N \cap w^{-1}Rw$ and such that the linear form

$$\lambda'(a) = z(a, b_0), \qquad a \in \mathcal{H}^\infty,$$

is non-zero. From (1), we get

$$\lambda'(\sigma(n')a) = \theta(w_0{}^{-1}n'w_0)\lambda'(a) \tag{2}$$

for $n' \in N'$ and $a \in \mathcal{H}^\infty$.

We may write $M_R = \Pi_i M_i$ with $M_i \simeq G_{n_i}$, $N = \Pi \overline{N_i}$ where $\overline{N_i}$ is the group of lower unipotent matrices in G_{n_i}. By Lemma (3.3) the space spanned by the vectors of the form $\otimes\ v_i$ with $v_i \in \mathcal{H}_i{}^\infty$ is dense in $\mathcal{H}^\infty$. Thus given an index j there are vectors $v_i \in \mathcal{H}_i{}^\infty$, for $i \neq j$, such that the linear form

$$\lambda_j'(v) = \lambda'(v \otimes \otimes_{i \neq j} v_i)$$

on $\mathcal{H}_j{}^\infty$ is non-zero. It clearly satisfies

$$\lambda_j'(\sigma_j(n_j)v) = \theta(w_0{}^{-1}n_jw_0)\lambda_j(v), \qquad n_j \in N_j.$$

Finally let w_j be the order-reversing permutation matrix in G_{n_j}. Set, for $v \in \mathcal{H}_j{}^\infty$,

$$\lambda_j(v) = \lambda_j'(\sigma_j(w_j)v).$$

Then the (non-zero) linear form λ_j on $\mathcal{H}_j{}^\infty$ clearly satisfies condition (ii) of (3.8) for the representation σ_i of G_{n_i}. □

(3.11) We will need an elementary result on induced representations. If $(\xi, \mathcal{V})$ is an admissible irreducible representation of the pair $(\mathfrak{U}(G), K)$ we will often allow ourselves to speak of ξ as a representation of G.

We say that an admissible representation $(\xi, \mathcal{V})$ is *pre-unitary* if it is the underlying representation of some unitary representation π of G on a Hilbert space $\mathcal{H}$. Since any admissible $(\mathfrak{U}(G), K)$ module has finite length, it is easy to see that $(\pi, \mathcal{H})$ is unique (up to unitary equivalence).

Now let R be a parabolic subgroup of G and η an admissible representation of $R/U_R = M$. Then one can define the representation induced by η

$$\xi = \mathrm{Ind}(G, R; \eta) \tag{1}$$

even though η is not a representation of the group M. If η underlies the unitary representation σ of M, then ξ underlies the unitary representation π of G induced by σ. Conversely:

Lemma. *Notations being as above, suppose $\bar{\eta} \simeq \tilde{\eta}$. Suppose that ξ is irreducible and underlies the unitary representation π. Then η is the underlying admissible representation of a unitary representation σ of M. Moreover π is equivalent to the unitary representation of G induced by σ.*

Proof. Denote by $(\cdot\ ,\ \cdot)$ a non-zero invariant Hermitian form on the space $\mathcal{H}_0$ of η. The assumption on η is precisely that there is such a form. A (K-finite) vector in the space of ξ may be regarded as a K-finite function on K with values in $\mathcal{H}_0$. Thus the formula

$$(f_1, f_2) = \int_K (f_1(k), f_2(k))dk$$

defines a non-zero Hermitian form on the space for ξ. It is invariant and therefore must be proportional to a positive form. Changing the original product on $\mathcal{H}_0$, we may assume it is positive. Thus we see that

$$\int_{M\cap K\backslash K} (f(k), f(k))dk \geq 0 \tag{1}$$

for all (right) K-finite functions on K with values in $\mathcal{H}_0$ satisfying

$$f(k_0k) = \sigma(k_0)f(k)$$

for all $k_0 \in M \cap K$. In (1) we are free to replace f by $f\phi$, where ϕ is any right K-finite scalar function on $K \cap M\backslash K$. Such functions are precisely of the form

$$\phi(k) = \int_{M\cap K} \psi(mk)dm \tag{2}$$

where ψ is right K-finite. If ϕ is a continuous function on $M \cap K\backslash K$, it may also be expressed in the form (2) for some continuous function ψ on K. By Stone-Weierstrass there is a sequence ψ_m of right K-finite functions on K converging uniformly to ψ. If ϕ_n is the corresponding sequence of functions then clearly ϕ_n tends to ϕ uniformly.

Suppose now α is any positive continuous function on $K \cap M\backslash K$. Choose ϕ_n as above converging uniformly to $\alpha^{1/2}$. Replacing f by $f\phi_n$ in (1) and taking limits we get

$$\int_{M\cap K\backslash K} \alpha(k)(f(k), f(k))dk \geq 0.$$

This implies $(f(e), f(e)) \geq 0$. Since $f(e)$ can be any vector in $\mathcal{H}_0$, we see that $(\cdot\ ,\ \cdot)$ is positive (semi-definite) on $\mathcal{H}_0$. Since η is irreducible it is actually positive definite. Thus η is the underlying admissible representation of a unitary representation σ of M and therefore π is infinitesimally equivalent to the unitarily induced representation

$$\mathrm{Ind}(G, R; \sigma).$$

Since π is unitary, it is actually equivalent in the unitary sense to this induced representation. □

(3.12) To continue we recall Langlands' classification of representations (c.f. [N.W.] for example). Let again R be a standard parabolic subgroup of type $(n_1, n_2, \ldots, n_u)$. As usual set $R/U_R = \Pi M_i$, $M_i = G_{n_i}$. For each i let π_i be a (unitary) *tempered* representation and s_i some real number. Suppose $s_1 > s_2 > \cdots > s_u$. Then the induced representation

$$\xi = \mathrm{Ind}(G, R; \pi_1 \otimes \alpha^{s_1}, \ldots, \pi_u \otimes \alpha^{s_u}) \tag{1}$$

contains a unique maximal invariant subspace $\mathcal{V}$. The representation η of G on $\mathcal{H}/\mathcal{V}$ is denoted by

$$\eta = J(G, R; \pi_1 \otimes \alpha^{s_1}, \ldots, \pi_u \otimes \alpha^{s_u}). \tag{2}$$

Every irreducible representation π of G is equivalent to a representation of the form (2) where R, the class of the π_i, and the s_i are uniquely determined. We shall call ξ the *induced representation associated to* η.

We also need to recall certain results from [B.K.] and [D.V.], specialized to $G = GL_n$. We refer to [D.V.], Section 6, for the concept of a "large" representation. As above we shall refer to a representation of R/U_R of the form

$$(\pi_1 \otimes \alpha^{s_1}) \otimes \cdots \otimes (\pi_u \otimes \alpha^{s_u})$$

where the π_i are tempered and the s_i arbitrary real numbers as *quasi-tempered.* Then we have the following theorem which is a simple restatement of Theorem 6.2 of [D.V.]:

THEOREM A. *Suppose η is an irreducible admissible representation of $G = GL_r(F)$, $F = \mathbf{R}$ or $\mathbf{C}$. Then η is large if and only if η has the form*

$$\eta = \mathrm{Ind}(G, R; \sigma)$$

where R is a parabolic in G and σ is a quasi-tempered representation of R/U_R.

We shall also need the following Theorem—a special case of Theorem D of [B.K.]:

THEOREM B. *Suppose $(\eta, \mathcal{V})$ is an irreducible admissible representation of $G = GL_r(F)$. Suppose there is a non-zero linear form λ on $\mathcal{V}$ such that*

$$\lambda(\eta(X)v) = \dot{\theta}(X)\lambda(v)$$

for all $X \in \mathfrak{U}(N)$. Then η is large.

Combining the two theorems in a trivial way we obtain the proposition we need.

PROPOSITION. *Let $(\eta, \mathcal{V})$ be an irreducible admissible representation of $GL_r(F)$. Suppose there is a non-zero linear form λ on $\mathcal{V}$ such that*

$$\lambda(\eta(X)v) = \dot{\theta}(X)\lambda(v)$$

for all $X \in \mathfrak{U}(N)$. Then the induced representation ξ associated to η is irreducible (and thus equivalent to η).

Proof. By the results of Kostant and Vogan we have

$$\eta = \mathrm{Ind}(G, R; \sigma)$$

where R is a parabolic in G and σ a quasi-tempered representation of R/U_R. Say

$$\sigma = (\pi_1 \otimes \alpha^{s_1}) \otimes \cdots \otimes (\pi_u \otimes \alpha^{s_u}).$$

Let R' be a parabolic obtained from R by permuting the blocks. Then $R'/U_{R'} \simeq R/U_R$. Let σ' be the representation of R' obtained by transport of structure. Then the representation

$$\eta' = \mathrm{Ind}(G, R'; \sigma')$$

has the same character as η. Since η is irreducible it must be equivalent to η'. Thus we may assume from the outset that $s_1 \geq s_2 \geq \cdots \geq s_u$.

Suppose R is of type $(n_1, n_2, \ldots, n_u)$. Suppose say $t_1 = s_1 = \cdots = s_p > s_{p+1}$. Let $m_1 = n_1 + n_2 + \cdots + n_p$ and R_1 be the parabolic of type $(n_1, n_2, \ldots, n_p)$ in G_{m_1}. Let σ_1 be the tempered representation of G_{m_1} defined by

$$\sigma_1 = \mathrm{Ind}(G_{m_1}, R_1; \pi_1, \pi_2, \ldots, \pi_p).$$

Repeating this procedure for each string of equal exponents s_i we obtain a new partition $(m_1, \ldots, m_\ell)$ of r, for each j, $1 \leq j \leq \ell$, an exponent t_j, and a tempered representation σ_j of GL_{m_j}. By construction $t_1 > t_2 > \cdots > t_\ell$. Let Q be the (standard) parabolic in G_r associated to this partition. Then, by the lemma on inducing in stages, we have

$$\eta = \mathrm{Ind}(G, Q; \sigma_1 \otimes \alpha^{t_1}, \ldots, \sigma_\ell \otimes \alpha^{t_\ell}),$$

and our proof is complete. □

(3.13) Combining Lemma (3.11) and Proposition (3.12) we will obtain the following result:

Lemma. *Suppose π is an irreducible unitary representation of G_r satisfying condition* (ii) *of Proposition* (3.8). *Then there is a partition $(n_1, n_2, \ldots, n_u)$ of r with $n_i = 1, 2$ or 4 if $F = \mathbf{R}$, $n_i = 1, 2$ if $F = \mathbf{C}$ and for each i an irreducible unitary representation π_i of G_{n_i} so that*

$$\pi = \mathrm{Ind}(G, R; \pi_1, \pi_2, \ldots, \pi_u),$$

where R is the parabolic subgroup of type $(n_1, n_2, \ldots, n_u)$.

Proof. Let π be an irreducible unitary representation of G. By the results stated in (3.12) we can choose R, π_i and the s_i so that π is (infinitesimally equivalent) to the representation η of (3.12.2). Clearly

$$\bar{\eta} = J(G, R; \bar{\pi}_1 \otimes \alpha^{s_1}, \bar{\pi}_2 \otimes \alpha^{s_2}, \ldots, \bar{\pi}_u \otimes \alpha^{s_u}).$$

If $(n_1, n_2, \ldots, n_u)$ is the type of R, call Q the parabolic of type $(n_u, \ldots, n_2, n_1)$. Then

$$\tilde{\eta} = J(G, Q; \overline{\pi}_u \otimes \alpha^{-s_u}, \ldots, \overline{\pi}_2 \otimes \alpha^{-s_2}, \overline{\pi}_1 \otimes \alpha^{-s_1}).$$

Since π is unitary we must have $\overline{\eta} \simeq \tilde{\eta}$. This can happen only if $R = Q$ and

$$(\pi_1 \otimes \alpha^{s_1}, \pi_2 \otimes \alpha^{s_2}, \ldots, \pi_u \otimes \alpha^{s_u})$$
$$= (\pi_u \otimes \alpha^{-s_u}, \ldots, \pi_2 \otimes \alpha^{-s_2}, \pi_1 \otimes \alpha^{-s_1}).$$

Assume furthermore that π satisfies (3.8)(ii). Then it satisfies the conditions of Proposition (3.12) and so π is actually (infinitesimally) equivalent to the representation

$$\xi = \mathrm{Ind}(G, R; \pi_1 \otimes \alpha^{s_1}, \pi_2 \otimes \alpha^{s_2}, \ldots, \pi_u \otimes \alpha^{s_u}).$$

In particular ξ is irreducible.

Let $m = [u/2]$ and, for $1 \leq i \leq m$, let R_i be the parabolic subgroup of G_{2n_i} of type (n_i, n_i). Set also

$$\eta_i = \mathrm{Ind}(G_{2n_i}, R_i; \pi_i \otimes \alpha^{s_i}, \pi_i \otimes \alpha^{-s_i}).$$

If now u is even, $u = 2m$ and ξ has the same character as the induced representation

$$\xi_1 = \mathrm{Ind}(G, S; \eta_1, \eta_2, \ldots, \eta_m),$$

where S is the parabolic subgroup of type $(2n_1, 2n_2, \ldots, 2n_m)$. Since ξ is irreducible it is equivalent to ξ_1 and each η_i must be irreducible also. But

$$\overline{\eta}_i = \mathrm{Ind}(G_{n_i}, R_i; \overline{\pi}_i \otimes \alpha^{s_i}, \overline{\pi}_i \otimes \alpha^{-s_i})$$
$$\tilde{\eta}_i = \mathrm{Ind}(G_{n_i}, R_i; \overline{\pi}_i \otimes \alpha^{-s_i}, \overline{\pi}_i \otimes \alpha^{-s_i}).$$

Thus $\overline{\eta}_i$ and $\tilde{\eta}_i$ have the same character. Since they are irreducible $\overline{\eta}_i = \tilde{\eta}_i$. We can now appeal to Lemma (3.11): each η_i is the underlying admissible representation of an irreducible unitary representation η_i' and π is equivalent to the unitary representation

$$\operatorname{Ind}(G, S; \eta_1', \eta_2', \ldots, \eta_m').$$

If u is odd, $u = 2m + 1$ and we obtain a similar conclusion. Each η_i, $1 \leq i \leq m$, (resp. π_{m+1}) underlies an irreducible unitary representation η_i' (resp. π_{m+1}') and, denoting by S the parabolic subgroup of G_r of type $(2n_1, 2n_2, \ldots, 2n_m, n_{m+1})$, π is equivalent to the unitary representation

$$\operatorname{Ind}(G, S; \eta_1', \eta_2', \ldots, \eta_m', \pi_{m+1}').$$

Now π_{m+1}' is a tempered representation, thus certainly satisfies the conclusion of (3.13) (c.f. [H.J.] for example). If we could prove similarly that each η_i' satisfies the conclusion of (3.13) it would follow from the transitivity of Mackey's construction that π itself satisfies it.

Thus we are reduced to proving the following fact. Let π be a tempered representation of G_n, $s > 0$, R the parabolic subgroup of type (n, n) in G_{2n}. Suppose that the induced representation

$$\xi = \operatorname{Ind}(G_{2n}, R; \pi \otimes \alpha^s, \pi \otimes \alpha^{-s})$$

is irreducible and equivalent to a unitary representation η. Then η satisfies the conclusion of (3.13).

Indeed π has the form

$$\pi = \operatorname{Ind}(G_n, S; \pi_1, \pi_2, \ldots, \pi_u),$$

where S is the parabolic subgroup of type $(n_1, n_2, \ldots, n_u)$ in G_n, $n_i = 1, 2$ if $F = \mathbf{R}$ and $n_i = 1$ if $F = \mathbf{C}$, and π_i is an irreducible square-integrable (unitary) representation of G_{n_i}. Let then T be the parabolic subgroup of type $(2n_1, 2n_2, \ldots, 2n_u)$ in G_{2n}. Since ξ is irreducible it is equivalent to the representation

$$\operatorname{Ind}(G_{2n}, T; \sigma_1, \sigma_2, \ldots, \sigma_u),$$

where

$$\sigma_i = \operatorname{Ind}(G_{2n_i}, R_i; \pi_i \otimes \alpha^s, \pi_i \otimes \alpha^{-s}),$$

R_i again being the parabolic subgroup of type (n_i, n_i) in G_{2n_i}. Since ξ is irreducible so are the σ_i. Once more $\tilde{\sigma}_i = \bar{\sigma}_i$ and again by Lemma (3.11) the representation σ_i underlies a unitary representation σ_i' and η is equivalent to the unitary representation

$$\mathrm{Ind}(G_{2u}, T; \sigma_1', \sigma_2', \ldots, \sigma_u').$$

This completes the proof of Lemma (3.13). □

(3.14) *End of the proof of Proposition* (3.8).

Suppose that π satisfies condition (3.8)(ii). Then Lemma (3.13) applies to π. The representations π_i of that lemma must then satisfy (3.8)(ii) by Proposition (3.10). (Perhaps it would be useful to observe that conditions (i) and (ii) of Proposition (3.8) are empty for $r = 1$.) By what we have seen in (3.8) and (3.9) they must also satisfy condition (i) of Proposition (3.8). Thus Proposition (3.8) will at last be a consequence of the following proposition.

PROPOSITION. *Suppose R is a parabolic subgroup of type $(r_1, r_2, \ldots, r_u)$ and for each i, $1 \leq i \leq u$, π_i is a unitary representation of G_i. Let π be the unitary representation of G induced by the π_i's:*

$$\pi = \mathrm{Ind}(G, R; \pi_1, \pi_2, \ldots, \pi_u).$$

If each π_i satisfies condition (3.8)(i) *so does π.*

Proof. By induction on u and the transitivity of inducing representations it suffices to prove the theorem when $u = 2$. It will be more convenient to take for R the lower standard parabolic of type (r_1, r_2). Suppose first $r_2 = 1$. Then R is the transpose of $P_r Z_r$. In any case the complement of $RU(U = U_r)$ has measure zero in G_r. Thus all functions in the space of π are determined by their restriction to U. More precisely, let us identify U with the group F^{r-1} of column vectors of size $r - 1$. Then we may regard π as acting on

$$L^2(F^{r-1}, \mathcal{W}),$$

where $\mathcal{W}$ is the space of π_1, the action of $\pi | P_r$ being given by:

$$\pi \begin{pmatrix} 1 & u_0 \\ 0 & 1 \end{pmatrix} f(u) = f(u + u_0),$$

$$\pi\begin{pmatrix} m & 0 \\ 0 & 1 \end{pmatrix} f(u) = \pi_1(m)\,|\det m|^{-1/2} f(m^{-1}u).$$

Let us use the Fourier transform on F^{r-1} to obtain a new representation $\hat{\pi}$ of P_r—on the same space—equivalent to $\pi|P_r$. More precisely, $\pi(g)\hat{f} = (\hat{\pi}(g)f)\hat{}$ for $g \in P_r$. We have

$$\hat{\pi}\begin{pmatrix} 1 & u \\ 0 & 1 \end{pmatrix} f(v) = f(v)\psi({}^t uv)$$

and

$$\hat{\pi}\begin{pmatrix} m & 0 \\ 0 & 1 \end{pmatrix} f(v) = \pi_1(m)\,|\det m|^{1/2} f({}^t mv).$$

Let $\eta = {}^t(0, 0, \ldots, 1) = {}^t\epsilon$. We have

$$F^{r-1} = G_{r-1}(F)\eta \cup \{0\},$$

the stabilizer of η being just $\overline{P_{r-1}}(F)$. Hence every f is determined by the function ϕ from $G_{r-1}(F)$ to $\mathcal{W}$ defined by

$$\phi(m) = \pi_1(m)\,|\det m|^{1/2} f({}^t m\eta).$$

Such ϕ satisfy

$$\phi(pm) = \pi_1(p)\,|\det p|^{1/2}\phi(m), \qquad p \in P_{r-1}(F),$$

$$\int ||f(v)||^2 dv = \int_{P_{r-1}(F)\backslash G_{r-1}(F)} ||\phi(m)||^2 dm.$$

Moreover on the ϕ's the representation, say π', corresponding to $\hat{\pi}$ has the form:

$$\pi'\begin{pmatrix} m' & 0 \\ 0 & 1 \end{pmatrix} \phi(m) = \phi(mm'), \qquad \pi'\begin{pmatrix} 1 & u \\ 0 & 1 \end{pmatrix} \phi(m) = \phi(m)\psi(\epsilon mu).$$

This shows that $\pi | P_r$ is equivalent to the induced representation

$$\xi = \mathrm{Ind}(P_r, P_{r-1}U_r; \sigma \otimes \eta_r),$$

where P_{r-1} is identified with a subgroup of M via the map

$$p \mapsto \begin{pmatrix} p & 0 \\ 0 & 1 \end{pmatrix},$$

and $\sigma_1 = \pi_1 | P_{r-1}$. The assumption is that σ contains the representation τ_{r-1} of P_{r-1}. It follows that ξ contains the induced representation

$$\mathrm{Ind}(P_r, P_{r-1}U_r; \tau_{r-1} \otimes \eta_r),$$

a representation equivalent to τ_r. Since $\pi | P_r$ is equivalent to ξ, we are done in this case.

Suppose now $r_2 > 1$. Consider the subgroup $R \cap P_r$ and the representation σ of that group defined by

$$\sigma \begin{pmatrix} m_1 & 0 \\ u & \\ & p \\ 0 & \end{pmatrix} = \pi_1(m_1) \otimes \pi_2(p);$$

here $m_1 \in G_{r_1}$, $p \in P_{r_2}$, and u is a matrix with $r_2 - 1$ rows and r_1 columns. Again functions in the space of π are determined by their restrictions to P_r. Thus $\pi | P_r$ is just the representation

$$\mathrm{Ind}(P_r, R \cap P_r; \sigma).$$

Since the restriction of π_2 to P_{r_2} contains τ_{r_2} we see that $\pi | P_{r_2}$ contains the induced representation

$$\xi' = \mathrm{Ind}(P_r, R \cap P_r; \sigma'),$$

where σ' is now the representation of $R \cap P_r$ defined by

$$\sigma'\begin{pmatrix} m_1 & 0 \\ u & \\ & p \\ 0 & \end{pmatrix} = \pi_1(m_1) \otimes \tau_{r_2}(p).$$

All we have to show is that ξ' contains τ_r. This is an assertion independent of π_2, provided $\pi_2|P_{r_2}$ contains τ_{r_2}. Thus it is enough to prove it for one particular choice of π_2. For instance, we may take π_2 in the unitary principal series; as is very well known $\pi_2|P_{r_2}$ is then equivalent to τ_{r_2} (c.f. for example [H.J.]).

Thus it suffices to prove that $\pi|P_r$ contains τ_r whenever π is a representation of G_r of the form

$$\pi = \mathrm{Ind}(G, R; \pi_1, \pi_2);$$

where $\pi_1|P_{r_1}$ contains τ_{r_1},

$$\pi_2 = \mathrm{Ind}(G_{r_2}, B_{r_2}; \mu_1, \mu_2, \ldots, \mu_{r_2}),$$

B_{r_2} is the parabolic subgroup of type $(1, 1, \ldots, 1)$, and μ_i a character of $F^\times$. We have proved this assertion when $r_2 = 1$. The transitivity of inducing unitary representations allows us to prove it in full generality by induction on r_2. This completes the proof of Proposition (3.8). □

(3.15) *Remarks.* (1) Suppose π is an irreducible unitary representation which satisfies the equivalent conditions of Proposition (3.8). Then in fact:

(iii) *$\pi|P_r$ is equivalent to τ_r.*

Indeed the uniqueness of λ can be used first of all to show that $\pi|P_r$ contains τ_r with multiplicity one. If $r = 2$ we see that $\pi|P_r$ is the sum of τ and a representation trivial on U; but a representation of G_2 cannot contain the trivial representation of U without being one-dimensional. Thus $\pi|P_2 \simeq \tau_2$. If $r = 4$ and $F = \mathbf{R}$, a similar but much more complicated argument can be used to show that $\pi|P_4 \simeq \tau_4$. In general, by Lemma (3.13) we see that π has the form indicated in Proposition (3.14) where the π_i satisfy the stronger condition that $\pi_i|P_{r_i} = \tau_{r_i}$. Then the proof of Proposition (3.14) shows in fact that $\pi|P_r$ is equivalent to τ_r.

(2) In general if π is any irreducible unitary representation of G then, by extending the results of (3.6), one can show that $\pi|P_r$ is a *finite* sum of

irreducible representations. Kirillov has sketched a proof of the fact that $\pi | P_r$ is irreducible but we have not been able to fill in the details.

(3.16) From now on an irreducible unitary representation π of G will be said to be *generic* if it satisfies the equivalent conditions of Proposition (3.8). Let $\mathcal{H}$ be the (Hilbert) space of π and λ any element of $\mathcal{H}^{-\infty}$ satisfying (3.8)(ii). We will denote by $\mathcal{W}(\pi; \psi)$ the space the functions on G of the form

$$W(g) = \lambda(\pi(g)v),$$

where v is in $\mathcal{H}^\infty$; we will also denote by $\mathcal{W}_0(\pi; \psi)$ the subspace of those W for which v is in $\mathcal{H}_0$, the space of K-finite vectors in $\mathcal{H}$. Clearly $v \mapsto W$ defines an isomorphism $\mathcal{H}_0 \xrightarrow{\sim} \mathcal{W}_0(\pi; \psi)$ in the admissible category. We may then identify these two spaces. Then exactly as in Proposition (1.5) we have the following Proposition which we will use repeatedly in the next section.

PROPOSITION. *Notations being as above, there is a positive constant c such that*

$$\int_{N\backslash P} |W|^2(pg)d_r(p) \leq c||W||^2$$

for all $w \in \mathcal{W}_0(\pi; \psi)$ *and all* $g \in G$. □

(3.17) Let now π and π' be two irreducible unitary generic representations of G_r. We again introduce, as in (1.5), the integrals

$$\Psi(s, W', W, \Phi)$$

for $W' \in \mathcal{W}_0(\pi'; \psi)$, $W \in \mathcal{W}_0(\pi; \psi)$, and Φ in $\mathcal{S}(F^r)$.

PROPOSITION. (i) *The integral* $\Psi(s, W', W, \Phi)$ *converges absolutely in the half space* $\mathrm{Re}(s) \geq 1$, *normally for* $\mathrm{Re}(s)$ *in a compact subset of* $[1, \infty[$.

(ii) *Given* $W' \neq 0$ *in* $\mathcal{W}_0(\pi'; \psi)$, *and s with* $\mathrm{Re}(s) \geq 1$, *there exist* W *in* $\mathcal{W}_0(\pi; \psi)$ *and* Φ *in* $\mathcal{S}(F^r)$ *such that*

$$\Psi(s, W', W, \Phi) \neq 0.$$

Proof. The proof of (i) is similar to the proof of (i) in Proposition (1.5). As for (ii) we may assume at the cost of replacing W' by a

K-translate, that $W'|P$ is not identically zero. Then by Proposition (3.16) $W'|P$ is a non-zero element of the space $\mathcal{K}$ of τ_r. Moreover it follows from the same proposition that the function

$$\phi'(p) = W'(p)|\det p|^{s-1}$$

is also in $\mathcal{K}$, whenever $\mathrm{Re}(s) \geq 1$. In more detail suppose $t \geq 0$ and write the integral

$$\int_{N\backslash P} |W(p)|^2 |\det p|^t d_r(p)$$

as a sum of two integrals, the first over $|\det p| \leq 1$ and the second over $|\det p| \geq 1$. In the first integral we may decrease t to 0 in which case the convergence follows from Proposition (3.16). In the second we may replace W by a gauge and take t large, in which case the convergence readily follows.

On the other hand, let $\mathcal{K}$ be the space of π. $\mathcal{K}_0$ is dense in $\mathcal{K}$. Set for $W \in \mathcal{W}_0(\pi; \psi) \simeq \mathcal{K}_0$, $A(W) = W|P$. Then by Proposition (3.16), A is non-zero element of $\mathrm{Hom}_P(\pi|P, \tau_r)$. Thus $A(\mathcal{K}_0)$ is dense in $\mathcal{K}$. Otherwise said, the functions $p \mapsto W(p)$ with W in $\mathcal{W}_0(\pi; \psi)$ are dense in $\mathcal{K}$. Thus there is a W such that

$$\int_{N\backslash P} W'(p)\overline{W}(p)|\det p|^{s-1} d_r(p) \neq 0.$$

Thus the function F on K defined by

$$F(k) = \int_{N\backslash P} W'(pk)\overline{W}(pk)|\det p|^{s-1} d_r(p)$$

is non-zero and K-finite. As in (1.5),

$$\Psi(s, W', W, \Phi) = \int_K dk \int_{N\backslash P} d_r(p)|\det p|^{s-1} W'(pk)\overline{W}(pk)$$

$$\times \int_{F^\times} \Phi(\epsilon a k)|a|^{rs} \omega' \overline{\omega}(a) d^\times a$$

$$= \int_K \int_{F^\times} F(k)\Phi(\epsilon a k)|a|^{rs}\omega'\bar{\omega}(a)d^\times a\,dk$$

$$= \int_K \int_{\mathbf{R}_+{}^\times} F(k)\Phi(\epsilon t k)|t|^{rs}\omega'\bar{\omega}(t)d^\times t\,dk.$$

Now choose $\phi \in C_c{}^\infty(\mathbf{R}_+{}^\times)$ such that $\int_{F^\times} \phi(t)|t|^{rs}\omega'\bar{\omega}(t)d^\times t \neq 0$. Then take Φ to be the following function:

$$\Phi(x) = \phi(t)\overline{F}(k) \text{ if } x = t\epsilon K,\ t > 0,\ k \in K;\ \Phi(x) = 0 \text{ otherwise.}$$

Then $\Phi \in \mathcal{S}(F^r)$ and we get finally

$$\Psi(s, W', W, \Phi) = \int_{F^\times} \phi(t)\omega'\bar{\omega}(t)|t|^{rs}d^\times t \int_K \overline{F}(k)F(k)dk \neq 0. \quad \square$$

4. Global Theory. In this section F is an $\mathbf{A}$-field. Although the main results are stated in full generality, we have found it convenient to give the proofs only in the case of number fields. The case of function fields is easier and left to the reader (see also Remark (4.8)). Accordingly until subsection (4.4) the ground field F is assumed to be a number field.

(4.1) We first recall the properties of certain special Eisenstein series. Let η be a character of $F^\times \backslash F_{\mathbf{A}}{}^\times$ of absolute value one. For $\Phi \in \mathcal{S}(F_{\mathbf{A}}{}^r)$ set

$$f(g, s, \eta) = |\det g|^s \int_{F_{\mathbf{A}}{}^\times} \Phi(a\epsilon g)|a|^{rs}\eta(a)d^\times a, \tag{1}$$

where, as before, ϵ is the row matrix with r entries given by

$$\epsilon = (0, 0, \ldots, 0, 1).$$

This integral converges for $\mathrm{Re}(s) > 1/r$. Set also

$$E(g, \Phi, s, \eta) = \Sigma f(\gamma g, s, \eta) \tag{2}$$

the sum over γ in $Z_r(F)P_r(F)\backslash G_r(F)$. This is the Eisenstein series we have in mind. A useful expression can be found if we write

$$f(g, s, \eta) = |\det g|^s \int_{F_{\mathbf{A}}^\times/F^\times} \sum_{\alpha \in F^\times} \Phi(a\alpha\epsilon g)|a|^{rs}\eta(a)d^\times a$$

and replace f by this expression in (2):

$$E(g, \Phi, s, \eta) = \sum_{\gamma} |\det \gamma g|^s \int_{F_{\mathbf{A}}^\times/F^\times} \sum_{\alpha \in F^\times} \Phi(a\alpha\epsilon\gamma g)|a|^{rs}\eta(a)d^\times a.$$

Now every vector $\xi \neq 0$ in F^r can be written uniquely as $\xi = \alpha\epsilon\gamma$. Thus after exchanging the sum over γ and the integral we get

$$E(g, \Phi, s, \eta) = |\det g|^s \int_{F_{\mathbf{A}}^\times/F^\times} \sum_{\xi} \Phi(a\xi g)|a|^{rs}\eta(a)d^\times a, \tag{3}$$

the sum extended over all ξ in $F^r - \{0\}$.

It now follows from Lemmas (11.5) and (11.6) of [G-J] that the right side of (3) converges absolutely for $\operatorname{Re}(s) > 1$; the same is therefore true for the right side of (2) and they are equal.

It will be convenient to consider a certain set of conditions for functions on $G(F)\backslash G(\mathbf{A})$ depending on a complex parameter. Let I be the open interval $]0, \infty[$, Ω the set of $s \in \mathbf{C}$ such that $\operatorname{Re}(s) \in I$. For $t > 0$ let A_t be the set of $a \in A_\infty$ such that

$$\det a = 1 \text{ and } |\alpha(a)| \geq t \qquad \text{for all } \alpha \in \Delta.$$

If ϕ is a function on $G(F)\backslash G(\mathbf{A}) \times \Omega$ the conditions we have in mind are the following: (4)

(i) ϕ is continuous;

(ii) for each g, the function $s \mapsto \phi(g, s)$ is holomorphic on Ω;

(iii) (if F is a number field), given a compact subset C of $G(\mathbf{A})$, a compact subset J of Ω, and $t > 0$, there exists a constant $c > 0$ and a positive integer m such that

$$|\phi(ax, s)| \leq c \sup_{\alpha \in \Delta} |\alpha(a)|^m$$

for all $a \in A_t$, $x \in C$, and $s \in J$.

(4.2) Lemma. *The functions $E(g, \Phi, s, \eta)$ extend to meromorphic functions on Ω.*

(i) *If* η *is non-trivial on the ideles of* F *of absolute value one, they are holomorphic and the function* $(g, s) \to E(g, \Phi, s, \eta)$ *satisfies condition* (4.1.4).

(ii) *There is a constant* $c \neq 0$ *such that if* $\eta = \alpha^{i\sigma}$ *then*

$$E(g, \Phi, s, \eta) = c \cdot \hat{\Phi}(0) |\det g|^{-i\sigma/r} \cdot (s + i\sigma/r - 1)^{-1} + R(g, s),$$

where R *is a function on* $G(F)\backslash G(\mathbf{A}) \times \Omega$ *satisfying* (4.1.4).

Proof. We break (4.1.3) into the sum of two integrals:

$$|\det g|^s \int_{|a| \leq 1} \sum_{\xi \neq 0} \Phi(a\xi g) |a|^{rs} \eta(a) d^\times a, \tag{1}$$

and

$$|\det g|^s \int_{|a| \geq 1} \sum_{\xi \neq 0} \Phi(a\xi g) |a|^{rs} \eta(a) d^\times a. \tag{2}$$

Now by the Poisson summation formula

$$\sum_{\xi \neq 0} \Phi(a\xi g) = \sum_{\xi \neq 0} \hat{\Phi}(a^{-1} \xi^t g^{-1}) |a|^{-r} |\det g|^{-1}$$

$$+ \hat{\Phi}(0) |a|^{-r} |\det g|^{-1} - \Phi(0),$$

where as usual

$$\hat{\Phi}(x) = \int_{\mathbf{A}} \Phi(y) \psi(y^t x) dy.$$

Thus, after changing a to a^{-1} where needed, we can write the first integral (1) as a sum of three terms:

$$|\det g|^{s-1} \int_{|a| \geq 1} \sum_{\xi \neq 0} \hat{\Phi}(a\xi^t g^{-1}) |a|^{r(1-s)} \bar{\eta}(a) d^\times a \tag{3}$$

$$|\det g|^{s-1} \hat{\Phi}(0) \int_{|a| \leq 1} |a|^{r(s-1)} \eta(a) d^\times a \tag{4}$$

$$- |\det g|^s \Phi(0) \int_{|a| \leq 1} |a|^{rs} \eta(a) d^\times a. \tag{5}$$

This computation is valid for $\mathrm{Re}(s) > 1$.

Now by Lemma (11.5) of [G-J] expressions (2) and (3) are defined for all s and, as functions of (g, s) satisfy the conditions (4.1.4). If now η is non-trivial on the ideles of absolute value one (4) and (5) vanish. This proves the first assertion of the lemma. On the other hand if $\eta = \alpha^{i\sigma}$ where σ is real (4) and (5) are respectively

$$c|\det g|^{s-1} \hat{\Phi}(0)(s + i\sigma/r - 1)^{-1}, \tag{6}$$

$$-c|\det g|^s \Phi(0)(s + i\sigma/r)^{-1}, \tag{7}$$

where c is a certain positive constant. Clearly (7), as a function of (g, s), satisfies (4.1.4). On the other hand we may write

$$a^s = 1 + \phi(a, s)s \qquad (a > 0)$$

where ϕ is a continuous function on $I \times \mathbf{C}$, holomorphic in s. Then (6) is the sum of two terms:

$$c|\det g|^{-i\sigma/r} \hat{\Phi}(0)(s + i\sigma/r - 1)^{-1} \tag{8}$$

and

$$c|\det g|^{-i\sigma/r} \hat{\Phi}(0) \phi(|\det g|, s + i\sigma/r - 1). \tag{9}$$

Again (9) satisfies (4.1.4), and we have proved (ii), R being the sum of (2), (3), (7) and (9).

(4.3) In this paragraph we introduce what is basically an integral representation for the L-functions $L(s, \pi \times \pi')$ attached to a pair of cusp forms on $G_r(\mathbf{A})$. While these functions appear somewhat implicitly here they are behind the scene and will make their full appearance in subsequent publications. As we have already said their treatment in the case $r = 2$ is due to Rankin and Selberg..

Suppose then that π and π' are irreducible unitary representations of $G(\mathbf{A})$. Let ω and ω' be their central characters. We assume that π and π'

are "cuspidal" (c.f. [B-J]). Let $\mathfrak{A}$ (resp. $\mathfrak{A}'$) be the corresponding space of cusp forms. Each ϕ in $\mathfrak{A}$ (resp. ϕ' in $\mathfrak{A}'$) is K-finite on the right, transforms like the character ω (resp. ω') under $Z(\mathbf{A})$, and is invariant on the left under $G(F)$. Moreover each ϕ is "rapidly decreasing." Namely in the notation of (4.1.4) for any compact set C in $G(\mathbf{A})$, any $t > 0$, and any positive integer m there is a positive constant c such that

$$|\phi(ax)| \leq c(\sup_{\alpha\in\Delta}|\alpha(a)|)^{-m}$$

for all $a \in A_t$ and x in C. Finally note that ω and ω' are trivial on $F^\times$.

Now set $\eta = \omega'\bar{\omega}$ and consider the integral

$$I(s, \Phi, \phi, \phi') = \int_{Z(\mathbf{A})G(F)\backslash G(\mathbf{A})} E(g, \Phi, s, \eta)\phi'(g)\bar{\phi}(g)dg. \tag{1}$$

For $a \in Z(\mathbf{A}) \simeq F_{\mathbf{A}}^\times$ we have

$$E(ag, \Phi, s, \eta) = \bar{\eta}(a)E(g, \Phi, s, \eta).$$

Indeed this is clear for Re(s) large, and by analytic continuation it is true for all s. Thus the integrand is indeed invariant on the left under $Z(\mathbf{A})G(F)$.

Suppose $\eta = \omega'\bar{\omega}$ has a non-trivial restriction to the ideles of F of absolute value one. Then by (4.2)(i) the integrand is a continuous function on $G(\mathbf{A}) \times \Omega$, holomorphic in s, and uniformly bounded for s in a compact subset of Ω. Since the volume of $G(F)Z(\mathbf{A})\backslash G(\mathbf{A})$ is finite the integral converges for all s in Ω and defines a holomorphic function of s on Ω.

Suppose now that $\eta = \alpha^{i\sigma}$ with σ real. Then

$$I(s, \Phi, \phi, \phi') = c\cdot\hat{\Phi}(0)(s + i\sigma/r - 1)^{-1}$$

$$\cdot\int_{Z(\mathbf{A})G(F)\backslash G(\mathbf{A})} |\det g|^{-i\sigma/r}\phi'\bar{\phi}(g)dg + \int_{Z(\mathbf{A})G(F)\backslash G(\mathbf{A})} R(g, s)\phi'\bar{\phi}(g)dg. \tag{2}$$

Note that since the function $g \mapsto |\det g|^{-i\sigma/r}$ transforms like $\bar{\eta}$ under $Z(\mathbf{A})$ this is true also for $g \mapsto R(g, s)$. The second integral is again holomorphic on Ω.

(4.4) We will use the above remarks to establish the following lemma.

LEMMA. (i) *Suppose* $\hat{\Phi}(0) \neq 0$, $\pi \simeq \pi'$, *and* $\phi = \phi' \neq 0$. *Then* $I(s, \Phi, \phi, \phi')$ *has a* (simple) *pole at* $s = 1$.

(ii) *Suppose there exist* Φ, ϕ, *and* ϕ' *such that* $I(s, \Phi, \phi, \phi')$ *has a pole at* $s = 1$. *Then* $\pi \simeq \pi'$.

Proof. Under the assumptions of (i) we have $\eta = 1$, $\sigma = 0$, and

$$\int_{Z(\mathbf{A})G(F)\backslash G(\mathbf{A})} |\phi(g)|^2 dg \neq 0,$$

and so by (4.3.2) $I(s, \Phi, \phi, \phi')$ has a (simple) pole at $s = 1$.

Suppose $I(s, \Phi, \phi, \phi')$ has a pole at $s = 1$. Then η must be trivial on the ideles of absolute value one, or $\eta = i\sigma$ for some real σ. Then by (4.3) we must have $\sigma = 0$ and so $\omega = \omega'$. We have then

$$\int_{G(F)Z(\mathbf{A})\backslash G(\mathbf{A})} \phi'\bar{\phi}(g) dg \neq 0$$

and this implies $\pi \simeq \pi'$. □

(4.5) We transform the integrals of (4.2) into the global analogues of the integrals of (1.3). Recall that if

$$\pi = \bigotimes_v \pi_v$$

then each π_v is generic. Fix once for all a non-trivial character ψ of $F_{\mathbf{A}}$ trivial on F and write

$$\psi = \prod_v \psi_v.$$

Then the space $\mathcal{W}(\pi_v; \psi_v)$ is defined as well as the space $\mathcal{W}_0(\pi_v; \psi_v)$ if v is archimedean. Denote by $\mathcal{W}_0(\pi; \psi)$ the space of functions on $G(\mathbf{A})$ spanned by the products

$$W(g) = \prod_v W_v(g_v), \tag{1}$$

where W_v is in $\mathcal{W}(\pi_v; \psi_v)$ if v is non-archimedean, in $\mathcal{W}_0(\pi_v; \psi_v)$ if v is archimedean, and, for almost all places v, W_v is the essential element of $\mathcal{W}(\pi_v; \psi_v)$ ([J-P-S] Sections 11, 12).

For each W in $\mathcal{W}(\pi; \psi)$ the function ϕ defined by

$$\phi(g) = \Sigma\, W(\xi g), \qquad \xi \in N_r(F)\backslash P_r(F), \tag{2}$$

is in $\mathfrak{A}$. The function W can be recovered from ϕ:

$$W(g) = \int_{N(\mathbf{A})} \phi(ng)\bar{\theta}(n)dn. \tag{3}$$

Moreover each W is majorized by a gauge (loc. cit.). Recall that a gauge ξ on $G(\mathbf{A})$ is a function ξ on $G(\mathbf{A})$ invariant under $N(\mathbf{A})Z(\mathbf{A})$ on the left, under K on the right, and given on $A(\mathbf{A})$ by

$$\xi(a) = \Phi(\alpha_1(a), \alpha_2(a), \ldots, \alpha_{r-1}(a))\,|\alpha_1(a) \cdots \alpha_{r-1}(a)|^{-t}$$

where $\Phi > 0$ is in $\mathcal{S}(\mathbf{A}^{r-1})$ and $t \geq 0$.

Consider now the integral

$$\Psi(s, W', W, \Phi) = \int_{N(\mathbf{A})\backslash G(\mathbf{A})} W'(g)\overline{W}(g)\Phi(\epsilon g)\,|\det g|^s dg, \tag{4}$$

where $W \in \mathcal{W}(\pi; \psi)$, $W' \in \mathcal{W}(\pi'; \psi)$ and $\Phi \in \mathcal{S}(\mathbf{A}^r)$. From the estimates we have just recalled it is easy to see that the integral converges for $\mathrm{Re}(s)$ large. Moreover we have for $\mathrm{Re}(s)$ large

$$I(s, \Phi, \phi', \phi) = \Psi(s, W', W, \Phi), \tag{5}$$

where $\phi' \in \mathfrak{A}'$ is defined in terms of W' as ϕ is in terms of W. Indeed if we replace in the left side of (4.3.1), $E(g, \Phi, s, \eta)$ by its expression (4.1.2) as a series, we get

$$\int_{Z(\mathbf{A})P_r(F)\backslash G_r(\mathbf{A})} f(g, s, \eta)\phi'(g)\bar{\phi}(g)\,|\det g|^s dg.$$

Then from the definition (4.1.1) of f we get

$$I(s, \Phi, \phi', \phi) = \int_{P_r(F)\backslash G_r(\mathbf{A})} \Phi(\epsilon g)\phi'(g)\bar{\phi}(g)\,|\det g|^s dg.$$

Now replace ϕ (or rather $\bar{\phi}$) by its expression as the series (2). We get

$$\int_{N(F)\backslash G(\mathbf{A})} \Phi(\epsilon g)\phi'(g)\overline{W}(g)\,|\det g|^s dg.$$

Since the functions $g \mapsto \Phi(\epsilon g)$ and $g \mapsto |\det g|^s$ are invariant under $N(\mathbf{A})$ while W transforms like θ this is also

$$\int_{N(\mathbf{A})\backslash G(\mathbf{A})} \Phi(\epsilon g)\overline{W}(g)\,|\det g|^s dg \int_{N(F)\backslash N(\mathbf{A})} \phi'(ng)\bar{\theta}(n)dn$$

and the inner integral is W'. Thus (5) is established—provided we justify our formal manipulations.

Since ϕ' is bounded and $N(F)\backslash N(\mathbf{A})$ has finite volume it suffices to show that the integral

$$\int_{N(\mathbf{A})\backslash G(\mathbf{A})} \Phi(\epsilon g)\, W(g)\,|\det g|^s dg$$

converges absolutely for $\mathrm{Re}(s)$ large. Since W is majorized by a gauge the proof is essentially the same as the convergence of the integral in (4) (for $\mathrm{Re}(s)$ large). □

(4.6) From (4.5.5) and (4.4), we obtain the following lemma:

LEMMA. *The function* $\Psi(s, W', W, \Phi)$ *extends to a meromorphic function of* s *on* Ω, *holomorphic for* $\mathrm{Re}(s) > 1$.

(i) *If* $\hat{\Phi}(0) \neq 0$, $\pi = \pi'$ *and* $W = W' \neq 0$ *that function has a* (simple) *pole at* $s = 1$.

(ii) *Conversely, if there exist* W, W' *and* Φ *such that* $\Psi(s, W', W, \Phi)$ *has a pole at* $s = 1$ *then* $\pi \simeq \pi'$. □

(4.7) We now write $\Psi(s, W', W, \Phi)$ as a finite sum of products of the local integrals of Section 1 and Section 3. Suppose W and W' are of the form (4.5.1) and that

$$\Phi(x) = \prod_v \Phi_v(x_v), \tag{1}$$

where Φ_v is in $\mathcal{S}(F_v{}^r)$ for all v and is, for almost all v, the characteristic function of $\mathfrak{R}_v{}^r$ in $F_v{}^r$. Now for $\mathrm{Re}(s) \geq 1$ all the local integrals $\Psi(s, W_v', W_v, \Phi_v)$ converge [(1.5) and (3.17)] and so, for $\mathrm{Re}(s)$ large,

$$\Psi(s, W', W, \Phi) = \prod_v \Psi(s, W_v', W_v, \Phi_v), \tag{2}$$

the right side being an absolutely convergent product. Our first theorem is as follows:

(4.8) THEOREM. *Suppose π and π' are irreducible unitary cuspidal representations of $G(\mathbf{A})$. Suppose S is a finite set of places containing the infinite places and all the finite places where either π_v or π_v' is ramified. Suppose that for $v \notin S$* (notation of (2.1.2):

$$L(s, \pi_v' \times \bar{\pi}_v) = L(s, \pi_v' \times \bar{\pi}_v').$$

Then $\pi \simeq \pi'$.

Proof. (for a number field). At the cost of enlarging S we may assume that for $v \notin S$ the largest ideal on which ψ_v is trivial is $\mathfrak{R}_v$. We take W and W' of the form (4.5.1), Φ and Φ' in $\mathcal{S}(\mathbf{A}^r)$ of the form (4.7.1) where

$$\Phi_v'(0) \neq 0, \Psi(1, W_v', W_v, \Phi_v) \neq 0 \qquad \text{for } v \in S$$

(c.f. (1.5) and (3.17));

$$\Phi_v' = \Phi_v = \text{the characteristic function of } \mathfrak{R}_v{}^r \text{ for } v \notin S;$$

W_v' and W_v are the essential elements of $\mathcal{W}(\pi_v'; \psi_v)$ and $\mathcal{W}(\pi_v; \psi_v)$ resp. for $v \notin S$. Then for $v \notin S$, by Proposition (2.3),

$$\Psi(s, W_v', W_v', \Phi_v') = L(s, \pi_v' \times \bar{\pi}_v'),$$

$$\Psi(s, W_v', W_v, \Phi_v) = L(s, \pi_v' \times \bar{\pi}_v),$$

and so by hypothesis

$$\Psi(s, W_v', W_v', \Phi_v') = \Psi(s, W_v', W_v, \Phi_v) \qquad \text{for } v \notin S. \tag{1}$$

Set

$$A(s) = \prod_{v \in S} \Psi(s, W_v', W_v, \Phi_v),$$

$$B(s) = \prod_{v \in S} \Psi(s, W_v', W_v', \Phi_v').$$

Then $A(s)$ and $B(s)$ are holomorphic in $\mathrm{Re}(s) > 0$, continuous in $\mathrm{Re}(s) \geq 1$ (cf. (1.5) and (3.17)) and $A(1) \neq 0$.

From (1) we get the identity

$$B(s)\Psi(s, W', W, \Phi) = A(s)\Psi(s, W', W', \Phi')$$

at first for $\mathrm{Re}(s)$ large but then, by analytic continuation, for all s with $\mathrm{Re}(s) > 1$.

Now by (4.6)(i), $\Psi(s, W', W', \Phi')$ has a pole at $s = 1$, since by choice $W' \neq 0$ and $\hat{\Phi}'(0) = \Pi_v \hat{\Phi}_v'(0) = \Pi_{v \in S} \hat{\Phi}_v'(0) \neq 0$. Since $A(1) \neq 0$ we have

$$\lim_{s \to 1} |A(s)\Psi(s, W', W', \Phi')| = +\infty$$

the limit taken for $s > 1$. Since $B(1)$ is finite we must have similarly

$$\lim_{s \to 1} |\Psi(s, W', W, \Phi)| = +\infty.$$

Thus $\Psi(s, W', W, \Phi)$ has a pole at $s = 1$. By (4.6)(ii), $\pi \simeq \pi'$. □

(4.9) REMARK. For the function field case, the proof is substantially the same, the changes being mostly notational. Since a cusp form has compact support mod $Z(\mathbf{A})G(F)$ no estimate is needed for $E(g, \Phi, s, \eta)$.

(4.10) COROLLARY. *Suppose π and π' are irreducible unitary representations of $G_r(\mathbf{A})$. Suppose S is a finite set of places such that $\pi_v \simeq \pi_v'$ for $v \notin S$. Then $\pi \simeq \pi'$.* □

(4.11) *Remark. Let S* be as in (4.8). Set

$$L_S(s, \pi' \times \bar{\pi}) = \prod_{v \notin S} L(s, \pi_v' \times \bar{\pi}_v).$$

As we noted the product is absolutely convergent for Re(s) large. One can actually prove that $L_S(s, \pi' \times \bar{\pi})$ is meromorphic—by methods quite distinct from those of the present paper. If one takes this for granted the proof of Theorem (4.8) shows that $L_S(s, \pi' \times \bar{\pi})$ has at most a simple pole at $s = 1$ with non-zero residue if and only if $\pi' \simeq \bar{\pi}$. In the next section we show that the Euler product for $L_S(s, \pi' \times \bar{\pi})$ is absolutely convergent for Re(s) > 1 and any pair (π, π').

5. Convergence of Euler Products. The results of this section complement the earlier non-vanishing theorem for L-functions obtained by the present authors ([J-S]). See also [F.S.]

(5.1) Suppose that π(resp. π') is an irreducible unitary cuspidal representation of $G_r(\mathbf{A})$ (resp. $G_p(\mathbf{A})$). Let S be a finite set of places containing the infinite places and all the finite places where either π_v or π_v' is ramified. As in Section 4, we fix once and for all a non-trivial character ψ on $F_{\mathbf{A}}$ trivial on F. We *assume* throughout this section that S is large enough so that for $v \notin S$ the largest ideal on which ψ_v is trivial is $\mathfrak{R}_v$. As in Section 2 let A_v(resp. A_v') be the (semi-simple) conjugacy class in $G_r(\mathbf{C})$ (resp. $G_p(\mathbf{C})$) associated to π_v (resp. π_v'). We set as before

$$L(s, \pi_v \times \pi_v') = \det(1 - q_v^{-s} A_v \otimes A_v')^{-1}. \tag{1}$$

It follows from (2.5) that, if

$$A_v = \operatorname{diag}(\mu_{1v}(\tilde{\omega}_v), \mu_{2v}(\tilde{\omega}_v), \ldots, \mu_{rv}(\tilde{\omega}_v)), \tag{2}$$

then

$$|\mu_{jv}(\tilde{\omega}_v)| \leq q_v^{1/2} \tag{3}$$

for all $v \notin S$, $1 \leq j \leq r$. Thus it is clear that the Euler product

$$L_S(s, \pi \times \pi') = \prod_{v \notin S} L(s, \pi_v \times \pi_v') \tag{4}$$

converges absolutely to an analytic function in some right half-plane. Before turning to our second theorem we prove an intermediate lemma.

(5.2) LEMMA. *Let π be an irreducible unitary cuspidal representation of $G_r(\mathbf{A})$. Then the function $L_S(s, \pi \times \bar{\pi})$ has an analytic continuation to the half plane* Re(s) > 1.

Proof. We proceed as in (4.8). Fix s_0 with $\mathrm{Re}(s_0) \geq 1$. We take W and W' of the form (4.5.1), Φ in $\mathcal{S}(\mathbf{A}^r)$ of the form (4.7.1) where

$$\Psi(s_0, W_v', W_v, \Phi_v) \neq 0 \qquad \text{for } v \in S$$

(c.f. (1.5) and (3.17));

$$\Phi_v = \text{the characteristic function of } \mathfrak{R}_v{}^r \text{ for } v \notin S;$$

$W_v' = W_v$ is the essential element of $\mathcal{W}(\pi_v; \psi_v)$ for $v \notin S$. Set $A(s) = \prod_{v \in S} \Psi(s, W_v', W_v, \Phi_v)$. Then as in (4.8) (c.f. also (2.3)) we have

$$\Psi(s, W', W, \Phi) = A(s) L_S(s, \pi \times \bar{\pi})$$

for $\mathrm{Re}(s)$ large. By (4.6) the left side is holomorphic for $\mathrm{Re}(s) > 1$. Since $A(s_0) \neq 0$ and $A(s)$ is also holomorphic for $\mathrm{Re}(s) > 1$, it follows that $L_S(s, \pi \times \bar{\pi})$ is holomorphic in a neighborhood of s_0. This being true now for any s_0 with $\mathrm{Re}(s_0) > 1$ our assertion follows.

(5.3) THEOREM. *Let* π (resp. π') *be an automorphic cuspidal representation of* $G_r(\mathbf{A})$ (resp. $G_p(\mathbf{A})$). *Suppose in addition that* π *and* π' *are unitary. Then the infinite product*

$$L_S(s, \pi \times \pi') = \prod_{v \notin S} L(s, \pi_v \times \pi_v')$$

is absolutely convergent in the half-plane $\mathrm{Re}(s) > 1$. *In particular the function* $L_S(s, \pi \times \pi')$ *does not vanish for* $\mathrm{Re}(s) > 1$.

Proof. We suppose A_v given by (5.1.2) and

$$A_v' = \mathrm{diag}(\nu_{1v}(\tilde{\omega}), \nu_{2v}(\tilde{\omega}), \ldots, \nu_{pv}(\tilde{\omega})).$$

Then

$$L(s, \pi_v \times \pi_v') = \prod (1 - \mu_i(\tilde{\omega}) \nu_j(\tilde{\omega}) q_v{}^{-s})^{-1},$$

the product over $1 \leq i \leq r$, $1 \leq j \leq p$. The individual terms in this finite product are of the form $(1 - z)^{-1}$ with $|z| < 1$. Thus if we set $\pi(v^n) = \mathrm{tr}(A_v{}^n)$, $\pi'(v^n) = \mathrm{tr}(A_v{}'^n)$, we have

$$L(s, \pi_v \times \pi_v') = \exp \sum_{n \geq 1} \pi(v^n)\pi'(v^n)/nq_v^{ns} \tag{1}$$

the series being absolutely convergent for $\operatorname{Re}(s) > 2$. Taking the product over $v \notin S$, and using (5.1.3), we get for $\operatorname{Re}(s)$ large

$$L_S(s, \pi_v \times \pi_v') = \exp \sum_{v \notin S} \sum_{n \geq 1} \pi(v^n)\pi'(v^n)/nq_v^{ns}, \tag{2}$$

the sum over $v \notin S$ being taken in any order.

To proceed further we assume first that $\pi' = \bar{\pi}$. Then

$$f(s) = \sum_{v \notin S} \sum_{n \geq 1} |\pi(v^n)|^2/nq_v^{ns} \tag{3}$$

is a Dirichlet series with non-negative coefficients, converging for $\operatorname{Re}(s)$ large. Let a be the real point of the line of convergence of this series. By a well-known theorem on Dirichlet series [E.C.T.] a is a singularity of $f(s)$. Suppose $a > 1$. Then from (1) we get by analytic continuation

$$L_S(s, \pi \times \bar{\pi}) = \exp f(s) \tag{4}$$

for $\operatorname{Re}(s) > a$. For s real, $s > a$, $f(s) \geq 0$ is increasing as $s \to a$. On the other hand by Lemma (5.2) the left side of (4) approaches a finite limit as $s \to a$. It follows that $L_S(a, \pi \times \bar{\pi}) \geq 1$ and in particular is non-zero. Thus in the intersection of a small disc about a with the half-plane $\operatorname{Re}(s) > a$, we get

$$\log L_S(s, \pi \times \bar{\pi}) = f(s).$$

This contradicts the fact that a is a singularity of f. Thus $a \leq 1$ and the series (3) converges for $\operatorname{Re}(s) > 1$.

Next an application of Cauchy-Schwartz shows that the double series

$$\sum_{v \notin S} \sum_{n \geq 1} \pi(v^n)\pi'(v^n)/nq_v^{ns} \tag{5}$$

also is absolutely convergent for $\operatorname{Re}(s) > 1$. It follows immediately from (2) that the product for $L_S(s, \pi \times \pi')$ is commutatively and therefore absolutely convergent for Re(s) > 1 ([N.B. Chap. 8]). This completes the proof of Theorem (5.3).

(5.4) *Remark.* The theorem applies in particular to the case $p = 1$, $\pi' = 1$. Then:

$$L_S(s, \pi \times \pi') = \prod_{v \notin S} \det(1 - A_v q_v{}^{-s})^{-1}.$$

Since a local L-factor is never zero we conclude from the results of [J-S] that

$$L(s, \pi) = \prod_v L(s, \pi_v)$$

does not vanish in the half-plane Re $s \geq 1$.

THE JOHNS HOPKINS UNIVERSITY
COLUMBIA UNIVERSITY

REFERENCES

[B-Z] I. N. Bernstein and A. V. Zelevinsky, "Induced representations of reductive 𝔓-adic groups," *Ann. Scient. Éc. Norm.* Sup., 10 (1977) pp. 441–472.

[R.J.B.] R. J. Blattner, "On induced representations," *Amer. J. Math.*, **83** (1961) pp. 79–98.

[B-J] A. Borel and H. Jacquet, "Automorphic forms and automorphic representations," Automorphic forms, representations and L-functions, *Proc. of Symposia in Pure Math.*, Vol. XXXIII, Part 1, A.M.S. (1979) pp. 189-202.

[N.B.] N. Bourbaki, *Éléments de Mathématique*, Vol. 5, Livre III, Topologie Générale, Hermann and Co. (1955).

[F.B.] F. Bruhat, [1] "Sur les représentations induites de groupes de Lie," *Bull. Soc. Math. France*, **84** (1956) pp. 97-205.

[C-S] W. Casselman and J. Shalika, "The unramified principal series of p-adic groups II," *Comp. Math.*, **41** (1980), pp. 207-231.

[G-K] I. M. Gelfand and D. A. Kazdan, "Representations of $GL(n, K)$ where K is a local field," in *Lie Groups and their Representations*, John Wiley and Sons (1975) pp. 95-118.

[H.J.] H. Jacquet, "Generic representations," Non-commutative harmonic analysis, Marseille-Luminy 1976, Springer-Verlag Lecture Notes 587 (1976) pp. 376–378.

[J-P-S] H. Jacquet, I. I. Piatetski-Shapiro, and J. Shalika, "Automorphic forms on $GL(3)$, I and II," *Annals of Math.*, **109** (1979) pp. 169-212.

[J-S] H. Jacquet and J. A. Shalika, "A non-vanishing theorem for zeta functions of GL_n," *Inventiones math.*, **38** (1976) pp. 1-16.

[A.A.K.] A. A. Kirillov, *Infinite dimensional representations of the total matrix group*, Dokladi Academii Nauk CCR, **144** (1962) pp. 37-39 (= Soviet Math. **3** (1962) pp. 652-655).

[B.K.] B. Kostant, "On Whittaker vectors and representation theory," *Inventiones Math.* **48** (1978) pp. 101-184.

[R.P.L.] R. Langlands, "Automorphic representations, Shimura varieties, and motives. Ein märchen," *Proc. of Symposia in Pure Math.*, Vol. XXXIII, Part 2, A.M.S. (1979), pp. 205-246.

[G.W.M.] G. W. Mackey, "Induced representations of locally compact groups I," *Annals of Math.*, **55** (1952), pp. 101-139.

[T.M.] T. Miyake, "On automorphic forms on GL_2 and Hecke operators," *Ann Math.*, **94** (1971), pp. 174-189.

[C.J.M.] C. J. Moreno, "Explicit formulas in the theory of automorphic forms," Number Theory Day, New York 1976, Springer-Verlag Lecture Notes 626 (1976) pp. 73-216.

[I.P.] I. Piatetski-Shapiro, "Multiplicity one theorems, Automorphic forms, representations and L-functions," *Proc. of Symposia in Pure Math.*, Vol. XXXIII, Part 1, A.M.S. (1979) pp. 209-212.

[N.S.P.] N. S. Poulsen, "On C^∞-vectors and intertwining bilinear forms for representations of Lie groups," *J. Func. Analysis*, **9** (1972) pp. 87-120.

[R.R.] R. Rankin, "Contributions to the theory of Ramanujan's function $\tau(n)$ and similar arithmetical functions, I and II," *Proc. Camb. Phil. Soc.* **35** (1939) pp. 351-356 and pp. 357-372.

[F.R.] F. Rodier, "Whittaker models for admissible representations of reductive $\mathfrak{P}$-adic split groups," *Proc. of Symposia in Pure Math.*, Vol. XXVI, A.M.S. (1973) pp. 425-430.

———, Décomposition Spectrale des Répresentations Lisses, Non-commutative harmonic analysis, Marseille-Luminy 1976, *Springer-Verlag Lecture Notes* **587** (1976) pp. 177-195.

[A.S.] A. Selberg, "Bemerkungen über eine Dirichletsche Reihe, die mit der Theorie der Modulformen nahe verbunden ist," *Arch. Math. Naturvid.* **43** (1940), pp. 47-50.

[F.S.] F. Shahidi, "On certain L-functions," *Amer. J. Math.*, **103** (1981), pp. 297-356.

[J.S.] J. A. Shalika, "The multiplicity one theorem for GL_n," *Ann of Math.*, **100** (1974) pp. 171-193.

[T.S.] T. Shintani, "On an explicit formula for class-1 "Whittaker functions" on GL_n over $\mathfrak{P}$-adic fields," *Proc. Japan Acad.*, **52** (1976) pp. 180-182.

[E.C.T.] E. C. Titchmarsh, *The theory of functions*, Oxford University Press (1939).

[D.V.] D. A. Vogan Jr., "Gelfand-Kirillov dimension for Harish-Chandra modules," *Invent. Math.* **48** (1978) pp. 75-98.

[N.W.] N. R. Wallach, "Representations of reductive Lie groups," Automorphic forms, representations and L-functions, *Proc. of Symposia in Pure Math.*, Vol. XXXIII, Part 1, A.M.S. (1979) pp. 71-86.

[G.W.] G. Warner, *Harmonic Analysis on Semi-simple Lie Groups* I, Springer-Verlag, Vol. 188 (1972).

[H.W.] Hermann Weyl, *The Classical Groups*, Princeton University Press (1939).

ON EULER PRODUCTS AND THE CLASSIFICATION OF AUTOMORPHIC FORMS II*

By H. JACQUET and J. A. SHALIKA

0. Introduction.

(0.1) Let f be an integer >1 and a an integer prime to f. A classical theorem of Dirichlet asserts that there are infinitely primes in the residue class of a mod f. This may be reformulated as follows. Let $\chi_1, \chi_2, \ldots, \chi_t$ be distinct characters of $(\mathbf{Z}/f)^\times$ and $c_1, c_2, \ldots, c_t$ be complex numbers. If

$$\sum_{1 \le j \le t} c_j \chi_j(p) = 0 \tag{1}$$

for all but finitely many primes p, prime to f, then $c_1 = c_2 = \cdots = c_t = 0$. A somewhat weaker statement than Dirichlet's theorem is the existence of infinitely many prime powers in the class of a. This may again be formulated in terms of Dirichlet characters, replacing (1) by the condition

$$\sum_{1 \le j \le t} c_j \chi_j(p^n) = 0, \tag{2}$$

for all integers $n \ge 1$, and all but finitely many primes p, prime to f.

We may of course view the χ_j as idele-class characters or, what amounts to the same, as automorphic representations of GL(1), regarded as a $\mathbf{Q}$-group. Then we may replace $\mathbf{Q}$ by any number field. The purpose of this paper is to further extend this result to all the linear groups. In more detail, let π be a automorphic cuspidal representation of $\mathrm{GL}(r, \mathbf{A})$ where $\mathbf{A}$ is the ring of adeles of a global field F. We write $r = \deg \pi$. Then π decomposes into a tensor product $\pi = \otimes_v \pi_v$, over all the places v of F. Let S be a finite set of places containing all the infinite places. Suppose S is so large that π_v is unramified for $v \notin S$. For $v \notin S$ let A_v be the semi-

Manuscript received June 6, 1980.

*This research was partially supported by National Science Foundation Grants MCS 79-01712 and MCS 79-13799.

American Journal of Mathematics, Vol. 103, No. 4, pp. 777–815 777
0002-9327/81/1034-0777 $1.50

simple conjugacy class, the "Langlands class," in GL(r, **C**) attached to π_v (c.f. (1.3) below). For $n \geq 1$ we set

$$\pi(v^n) = \operatorname{tr} A_v{}^n.$$

Suppose now that π_i, $1 \leq i \leq p$, is a family of such automorphic representations, possibly of varying degrees. Our main result is that, if the π_i are unramified outside S and

$$\sum_{1 \leq i \leq p} c_i \pi_i(v^n) = 0,$$

for all $v \notin S$ and all $n \geq 1$, then $c_i = 0$ for $1 \leq i \leq p$.

We use this result to answer a question raised by Langlands (R.P.L. II) on the classification of automorphic representations (Theorem (4.4)), proving, in conjunction with the theory of Eisenstein series, the existence of the category of "isobaric forms."

We also given an application to the strong form of Artin's conjecture. Here however our assertion will not have substance until further progress has been made.

Our proof is based on properties of the L-functions $L(s, \pi \times \pi')$ attached to pairs (π, π') of automorphic, cuspidal representations. More precisely, if π and π' are unramified outside S, we consider the partial L-functions

$$L_S(s, \pi \times \pi') = \Pi_{v \notin S} \det(1 - q_v{}^{-s} A_v \otimes A_v{}')^{-1}.$$

For $r = r' = 1$, $L_S(s, \pi \times \pi')$ is just the L-function attached to the product $\pi\pi'$. For $r = r' = 2$ the L-functions $L(s, \pi \times \pi')$ were first considered by R. Rankin and A. Selberg and to a great extent their work is the inspiration for ours (cf. [R.R.] I, II and [A.S.]). In general, these L-functions were first considered in joint work with Piatetski-Shapiro. We hope to publish a more detailed description of their properties—with applications to the determination of the poles of Eisenstein series and the formulation and proof of "converse theorems" similar to those in [J-L] and [J-P-S].

Just as in the proof of Dirichlet's Theorem our proof depends on the existence and non-existence of poles and zeros of L-functions. Essential use is made of a recent result of F. Shahidi [F.S.] on the non-vanishing of these L-functions on the edge of the critical strip.

(0.2) Upon first reading one should read Section (1), Section (2.1), the statement of Proposition (2.6), Section (3.2), the statements of Propositions (3.3), (3.6), (3.7) and finally Section (4).

(0.3) Notations are the same as in part I. We shall use standard terminology concerning automorphic forms and automorphic representations. We refer the reader to [B-J], [R.P.L. I] and in general to the two Corvallis volumes for the basic definitions and concepts. In addition we write $\mathcal{A}_r$ for the set of equivalence classes of (irreducible) automorphic, cuspidal representations of $G_r(\mathbf{A})$ and $\mathcal{A}$ for the disjoint union

$$\mathcal{A} = \cup\, \mathcal{A}_r. \tag{1}$$

We write $\deg \pi = r$ if $\pi \in \mathcal{A}_r$. Any π in $\mathcal{A}_r$ is uniquely expressible as an infinite (algebraic) tensor product of representations of the G_v:

$$\pi = \otimes_v \pi_v. \tag{2}$$

We call π_v the local component (at v) of π.

Finally, if $\mathfrak{G}$ is a finite group and π a (finite-dimensional) representation of $\mathfrak{G}$, χ_π will denote the character of π. If f and g are two complex-valued functions on $\mathfrak{G}$ we write

$$(f, g) = [\operatorname{Card} \mathfrak{G}]^{-1} \sum_{x \in \mathfrak{G}} f(x)\overline{g(x)} \tag{3}$$

for their inner product.

1. The non-archimedean case.

(1.1) In this section, F is a non-archimedean local field. For the convenience of the reader we recall certain definitions and results.

Let r be an integer ≥ 1 and π an irreducible admissible representation of $G_r(F)$. Let $\mathcal{V}$ be the space of π. We say that π is generic if there exists a linear form $\lambda \neq 0$ on $\mathcal{V}$ such that

$$\lambda(\pi(n)v) = \theta(n)\lambda(v) \tag{1}$$

for $n \in N_r(F)$, $v \in \mathcal{V}$. The form λ is then unique up to a scalar factor. The space generated by the functions of the form

$$W(g) = \lambda(\pi(g)v) \tag{2}$$

depends only on ψ and π. We denote it by $\mathcal{W}(\pi; \psi)$.

If the central character of π is unitary then any element of $\mathcal{W}(\pi; \psi)$ is majorized by a gauge ([J-P-S] Prop. 2.3.6), that is to say a function $\xi > 0$ on G_r, invariant by the center, on the left by N, on the right by K, and given on A by

$$\xi(a) = \phi(\alpha_1(a), \alpha_2(a), \ldots, \alpha_{r-1}(a)) |\alpha_1\alpha_2 \cdots \alpha_{r-1}(a)|^{-t},$$

with $t \geq 0$ and $\phi \geq 0$ in $\mathcal{S}(F^{r-1})$.

(1.2) Let f be a function on P_r, transforming on the left under N_r according to θ, invariant on the right under an open compact subgroup of G_r and compactly supported mod N_r. Then given π, there exists W in $\mathcal{W}(\pi; \psi)$ whose restriction to P_r coincides with f ([G-K]. (5.2)).

(1.3) Suppose that π is unramified, that is it contains the trivial representation of K. Suppose further that the largest ideal on which ψ is trivial is $\mathfrak{R}$. Then up to scalars there exists exactly one element W_0 in $\mathcal{W}(\pi; \psi)$ invariant under K on the right. Moreover $W_0(e) \neq 0$ and we normalize W_0 by requiring $W_0(e) = 1$ ([C-S], [T.S.]). We call W_0 the *essential element* of $\mathcal{W}(\pi; \psi)$.

We recall that the unramified representations of G_r (generic or not) are parameterized by the semi-simple conjugacy classes in GL(r, **C**). More precisely, if π is unramified then π is the unique unramified component of an induced representation of the form

$$\mathrm{Ind}(G_r, B_r; \mu_1, \mu_2, \ldots, \mu_r),$$

where the μ_i are unramified quasi-characters. The class attached to π is by definition the class of the diagonal matrix

$$A = \mathrm{diag}\,(\mu_1(\tilde{\omega}), \mu_2(\tilde{\omega}), \ldots, \mu_r(\tilde{\omega})).$$

(1.4) Let r and r' be two integers, $r > r'$. Let π (resp. π') be an irreducible admissible representation of G_r (resp. $G_{r'}$). Set for $W \in \mathcal{W}(\pi; \psi)$, $W' \in \mathcal{W}(\pi'; \psi)$,

$$\Psi(s, W, W') = \int_{N_{r'}\backslash G_{r'}} W\begin{pmatrix} g & 0 \\ 0 & 1 \end{pmatrix} W'(\epsilon g) |\det g|^{s-(r-r')/2}\, dg, \tag{1}$$

where

$$\epsilon = \operatorname{diag}(1, -1, 1, -1, \ldots) \tag{2}$$

is a row vector of length r, and dg is an invariant measure on $N_{r'} \backslash G_{r'}$. It follows from the majorization of W and W' by gauges that the integral converges for $\operatorname{Re}(s)$ large.

We will also need the following. First if $W'(e) \neq 0$, then by (1.2) one can choose W so that the integrand in (1) is compactly supported. Moreover we can even assume that the resulting integral (which then exists for all s) is then identically one. Second suppose that π and π' are unramified, the largest ideal on which ψ is trivial is $\mathfrak{R}$, and that the measure dg is the quotient of the Haar measures of $G_{r'}$ and $N_{r'}$ normalized so that K and $K \cap N_{r'}$ have volume one. Then if W and W' are the essential elements of $\mathcal{W}(\pi; \psi)$ and $\mathcal{W}(\pi'; \psi)$ respectively, one has

$$\Psi(s, W, W') = \det(1 - A \otimes A' q^{-s})^{-1} \tag{3}$$

where A (resp. A') is the conjugacy class associated to π (resp. π'). In fact referring to Section 2 of [J-S], we have in the notation of that paper

$$\Psi(s, W, W') = \Sigma_J W\begin{pmatrix} \tilde{\omega}^J & 0 \\ 0 & 1 \end{pmatrix} W(\tilde{\omega}^J)\delta_{r'}^{-1}(\tilde{\omega}^J)\,|\det \tilde{\omega}^J|^{s-(r-r')/2}, \tag{4}$$

the sum being over all r'-tuples of integers

$$J = (j_1, j_2, \ldots, j_{r'})$$

satisfying

$$j_1 \geq j_2 \geq \cdots \geq j_{r'} \geq 0.$$

Here $\delta_{r'}$ is the module of the group $B_{r'}$. As there set $\operatorname{tr} J = \Sigma j_i$, $X = q^{-s}$. Using the explicit formulae for W and W' ((2.1) of [J-S]), we find that the right side of (3) is also equal to

$$\sum_{J \in T^+(r')} \operatorname{Tr}(\rho(r, J)(A)) \operatorname{Tr}(\rho(r', J)(A')) X^{\operatorname{Tr}(J)}. \tag{5}$$

The equality (3) then follows directly from Prop. (2.4) of [J-S]. *We will denote the Euler factor in* (3) *by* $L(s, \pi \times \pi')$.

Finally, if $r' = 1$, then π' is a (unitary) character of $F^\times$ and, up to a constant, $W'(x) = \pi'(x)$. Then the integral is simply

$$\int_{F^\times} W\begin{pmatrix} a & 0 \\ 0 & 1_{r-1} \end{pmatrix} |a|^{s-(r-1)/2} \pi'(a) d^\times a.$$

(1.5) PROPOSITION. (i) *With the above notations, given W' there is a W such that the integral* (1.4.1) *converges and is equal to one for all s.*

(ii) *Assume π and π' are unitary. Then the integral* (1.4.1) *converges for* $\mathrm{Re}(s) \geq 1$, *normally for* $\mathrm{Re}(s)$ *in a compact set.*

(iii) *If π and π' are unitary and unramified then $L(s, \pi \times \pi')$ does not have a pole in* $\mathrm{Re}(s) \geq 1$.

Proof. The first assertion follows from (1.2). Assume π is unitary. Then by Proposition (1.3) of [J-S] the restriction of any element of $\mathcal{W}(\pi; \psi)$ to P_r is square-integrable mod N_r. The same is true for π'. The proof is then similar to the proof of (2.6)(i) below, but easier; it is left to the reader. The third assertion follows from the second one. □

2. The archimedean case.

(2.1) Let F be either $\mathbf{R}$ or $\mathbf{C}$. We will regard $G_r(F)$ throughout as a real Lie group.

We begin by recalling the definition of a "generic representation." We then study the convergence properties of the integrals, analogous to those introduced in (1.4), associated to a pair of such representations. The basic result is Prop. (2.6) below.

We denote by $\mathfrak{L}$ and $\mathfrak{U}$ the Lie-algebra of $G_r(F)$ and the enveloping algebra of $\mathfrak{L} \otimes_{\mathbf{R}} \mathbf{C}$ respectively. Let $r \geq 1$ be an integer and π an irreducible unitary representation of $G_r(F)$ on a separable Hilbert space $\mathcal{H}$. Let $\mathcal{H}^\infty$ be the space of smooth vectors in $\mathcal{H}$ with its canonical topology. We say that π is generic if there exist a non-zero continuous linear form λ on $\mathcal{H}^\infty$ such that

$$\lambda(\pi(n)v) = \theta(n)v \tag{1}$$

for $n \in N_r$ and $v \in \mathcal{H}^\infty$. That form is then unique up to scalars ([J.S.]) As in Section 1, the space of functions of the form

$$W(g) = \lambda(\pi(g)v), \tag{2}$$

where v is in $\mathcal{H}^\infty$, is then uniquely determined by π and ψ; we denote that space by $\mathcal{W}(\pi; \psi)$.

(2.2) We will need estimates for the elements of $\mathcal{W}(\pi; \psi)$ similar to those in (1.1). For $\phi \in C_c^\infty(G)$, set as usual $\check{\phi}(x) = \phi(x^{-1})$, $x \in G$. Then all elements v in $\mathcal{H}^\infty$ are of the form

$$v = \pi(\check{\phi})v_1$$

where ϕ is in $C_c^\infty(G)$ and v_1 is again in $\mathcal{H}^\infty$. (c.f. [D-M]). In terms of the corresponding functions W and W_1 that amounts to the relation

$$W(g) = \int_G W_1(gh)\phi(h)dh.$$

By the proof of Lemma (8.3.3) in [J-P-S] we find then that W is majorized by a gauge, the definition of a gauge for F archimedean being precisely the same as in (1.1).

(2.3) On the other hand let τ_r be the irreducible unitary representation of P_r induced by the character θ_r of N_r. By [J-S] Prop. (3.8), there is an $A \neq 0$ in $\operatorname{Hom}_{P_r}(\pi | P_r, \tau_r)$. On the other hand one can identify the space $\mathcal{H}_r$ of τ_r with the Hilbert space of functions on G_{r-1} transforming on the left under N_{r-1} like θ_{r-1} and square-integrable mod N_{r-1}. For ϕ in this space, $h \in G_{r-1}(F)$, we have obviously

$$\left(\tau_r\begin{pmatrix} h & 0 \\ 0 & 1\end{pmatrix}\phi\right)(g) = \phi(gh), \qquad g \in G_{r-1}(F).$$

Thus, if $v \in \mathcal{H}^\infty$, then Av, being a differentiable vector for τ_r is actually a smooth function. Thus $v \mapsto \mathrm{Av}(e)$ is a linear form satisfying (2.1.1). We may then assume $\lambda(v) = \mathrm{Av}(e)$ and we deduce immediately the following consequence:

Lemma. (i) *There exists a positive constant c such that for all $v \in \mathcal{H}^\infty$ one has*

$$\int_{N_{r-1}\backslash G_{r-1}} |W|^2 \begin{pmatrix} g & 0 \\ 0 & 1\end{pmatrix} dg \leq c\|v\|^2,$$

where W is the element of $\mathcal{W}(\pi; \psi)$ defined by (2.1.2).

(ii) *If v runs through a dense subset of $\mathcal{H}$ contained in $\mathcal{H}^\infty$, then the corresponding functions*

$$g \mapsto W\begin{pmatrix} g & 0 \\ 0 & 1 \end{pmatrix}$$

run through a dense subset of $\mathcal{K}_r$.

(2.4) We will also need an elementary result whose proof we give for the convenience of the reader. Let U be a unipotent subgroup of G_r; as an algebraic variety U is isomorphic to a vector space over F. Regarding that vector space as a real vector space, we may then speak of polynomial functions, Schwartz functions, and so forth on U.

PROPOSITION. *Let π be a unitary representation of G on a Hilbert space $\mathcal{H}$. Let ϕ be a Schwartz function on U. Then for all $v \in \mathcal{H}^\infty$ the vector*

$$\pi(\phi)v = \int_U \pi(u)v\phi(u)du$$

is again in $\mathcal{H}^\infty$ and the map $v \mapsto \pi(\phi)v$ of $\mathcal{H}^\infty$ to itself is continuous.

Here the integral is the usual Bochner integral for Hilbert space-valued functions.

Proof. Clearly the integral is absolutely convergent. Let $\langle\ ,\ \rangle$ denote the given inner product on $\mathcal{H}$. To prove the first assertion it suffices to show that for all $w \in \mathcal{H}$, $g \in G$, $x \in \mathfrak{L}(G)$, the function f on $\mathbf{R}$ defined by

$$f(t) = \langle \pi(\exp tXg)\pi(\phi)v, w\rangle \tag{1}$$

is smooth at $t = 0$. We have

$$\pi(g)\pi(\phi)v = \pi(\phi')\pi(g)v$$

where ϕ' is a Schwartz function on the unipotent group gUg^{-1}. Since the unipotent group U is arbitrary we may assume $g = e$. Next we have

$$\frac{d}{dt}\pi(\exp tX)v = \pi(\exp tX)\pi(X)v.$$

Thus if

$$g(t, u) = \langle \pi(\exp tX)\pi(u)v, w\rangle$$

we get immediately

$$\frac{\partial g}{\partial t}(t, u) = \langle \pi(\exp tX)\pi(X)\pi(u)v, w\rangle. \tag{2}$$

Next write $\mathrm{ad}(u^{-1})X = \Sigma\, P_i(u)X_i$ where the P_i are polynomials. Then

$$\pi(X)\pi(u) = \Sigma\, P_i(u)\pi(u)\pi(X_i). \tag{3}$$

Substituting in (2) we find that $u \mapsto (\partial g/\partial t)(t, u)\phi(u)$ is integrable on U uniformly in t. Thus the double integral

$$\int_0^x \int \frac{\partial g}{\partial t}(t; u)\phi(u)du dt$$

is absolutely convergent and represents a differentiable function of x. Interchanging the order of integration we get for this integral $f(x) - f(0)$. Thus f is differentiable; moreover

$$f'(t) = \int \langle \pi(\exp tX)\pi(X)\pi(u)v, w\rangle\, \phi(u)du. \tag{4}$$

Using (3) again, we find that $f'(t)$ is a finite sum of functions of the same type as f. Thus f is C^∞ everywhere and in particular at zero.

Finally evaluating (4) at $t = 0$ we get

$$\pi(D)\pi(\phi)v = \int \pi(D)\pi(u)v\phi(u)du, \tag{5}$$

at first for $D \in \mathfrak{U}$ of degree ≤ 1, but then by an easy induction for all D. The second assertion follows. We leave the details to the reader.

Now suppose λ is any continuous linear form on $\mathcal{H}^\infty$. Then

$$\lambda(v) = \Sigma\, \lambda_D(\pi(D)v),\ v \in \mathcal{H}^\infty,$$

(finite sum) where the λ_D are continuous linear forms on $\mathcal{H}$. It follows from (5) applied to $\mathrm{ad}(g^{-1})D$ that

$$\lambda(\pi(g)\pi(\phi)v) = \int \lambda(\pi(gu)v)\phi(u)du.$$

In particular if π is irreducible and generic, W an element in $\mathcal{W}(\pi; \psi)$, ϕ a Schwartz function on U, then the integral

$$\int_U W(gu)\phi(u)du$$

converges and the function of g it defines is again in $\mathcal{W}(\pi; \psi)$.

(2.5) Suppose that r and r' are two integers with $1 \le r' \le r$, π (resp. π') an irreducible (unitary) generic representation of G_r (resp. $G_{r'}$) on the Hilbert space $\mathcal{H}$ (resp. $\mathcal{H}'$). As in (1.4), for $W \in \mathcal{W}(\pi; \psi)$ and W' in $\mathcal{W}(\pi'; \psi)$, set

$$\Psi(s, W, W') = \int_{N_{r'}\backslash G_{r'}} W\begin{pmatrix} g & 0 \\ 0 & 1 \end{pmatrix} W'(\epsilon g)|\det g|^{s-(r-r')/2}dg \quad (1)$$

with ϵ and dg as before. As there the integral converges for $\mathrm{Re}(s)$ large.

(2.6) Proposition. (i) *With the preceding notation, one can choose W and W' such that the integral* (2.5.1) *converges normally for* $\mathrm{Re}(s) \ge 1$. *Accordingly it defines a continuous function of s, holomorphic for* $\mathrm{Re}(s) > 1$.

(ii) *If s_0 is given with* $\mathrm{Re}(s_0) \ge 1$, *one can choose W and W' with the additional property that* $\Psi(s_0, W, W') \ne 0$.

Proof. Let p and m be two integers with $1 \le p \le m$. We will use the following integration formula:

$$\int_{N_m\backslash G_m} \phi(g)dg = \int_{Q_{m,p}\backslash S_m} dh \int_{N_p\backslash G_p} \phi\left[\begin{pmatrix} x & 0 \\ 0 & 1 \end{pmatrix} h\right] |\det x|^{-(m-p)}dx,$$

(1)

where S_m is the group SL(m), $Q_{m,p}$ is the subgroup of all matrices of the form

$$q = \begin{pmatrix} g & u \\ 0 & n \end{pmatrix}, \qquad g \in S_p, n \in N_{m-p},$$

and the measures are the appropriate invariant measures on the respective quotients.

For example, by Lemma (2.3), one has for all $W \in \mathcal{W}(\pi; \psi)$ the relation

$$\int_{Q_{r-1,r'}\backslash S_{r-1}} dh \int_{N_{r'}\backslash G_{r'}} |W|^2 \left[\begin{pmatrix} g & 0 \\ 0 & 1_{r-r'} \end{pmatrix} \begin{pmatrix} h & 0 \\ 0 & 1 \end{pmatrix} \right] |\det g|^{-(r-r'-1)} dg$$

$$= \int_{N_{r-1}\backslash G_{r-1}} \left| W \begin{pmatrix} g & 0 \\ 0 & 1 \end{pmatrix} \right|^2 dg < +\infty.$$

There is of course at least one element of $\mathcal{W}(\pi; \psi)$ whose restriction to G_{r-1} is $\neq 0$, i.e. the above integral is positive. Thus there exists at least one $h \in S_{r-1}$ for which the integral

$$\int_{N_{r'}\backslash G_{r'}} |W|^2 \left[\begin{pmatrix} g & 0 \\ 0 & 1_{r-r'} \end{pmatrix} h \right] |\det g|^{1-(r-r')} dg$$

is both finite and positive. Then by replacing W by a translate we obtain the existence of an element W_0 of $\mathcal{W}(\pi; \psi)$ for which

$$0 < \int_{N_{r'}\backslash G_{r'}} |W_0|^2 \left[\begin{pmatrix} g & 0 \\ 0 & 1_{r-r'} \end{pmatrix} \right] |\det g|^{1-(r-r')} dg < \infty. \tag{2}$$

For this W_0 we also have, for $s \geq 0$,

$$\int_{N_{r'}\backslash G_{r'}} |W_0|^2 \left[\begin{pmatrix} g & 0 \\ 0 & 1_{r-r'} \end{pmatrix} \right] |\det g|^{s+1-(r-r')} dg < +\infty. \tag{3}$$

To see this break up the integral into an integral over $|\det g| \leq 1$ and one over $|\det g| \geq 1$. In the first integral we may decrease s to zero and use (2). In the second we may take s large, integrate over all of $N_{r'} \backslash G_{r'}$ and replace W by a gauge to obtain the necessary convergence.

Next let Φ be a Schwartz function on $F^{r'}$ given any W in $\mathcal{W}(\pi; \psi)$, set

$$W_1(g) = \int_{F^{r'}} W_0 \left[g \begin{pmatrix} 1_{r'} & u & 0 \\ 0 & 1 & 0 \\ 0 & 0 & 1_{r-r'-1} \end{pmatrix} \right] \Phi(u) du \tag{4}$$

(u being a column of height r'). By Prop. (2.4) and the remark following that Proposition, W_1 is again in $\mathcal{W}(\pi; \psi)$. Moreover, for $g \in G_{r'}$, $g^{-1}N_r g$ contains the group we are integrating over in (4). Changing variables we get

$$W_1 \begin{pmatrix} g & 0 \\ 0 & 1 \end{pmatrix} = W_0 \begin{pmatrix} g & 0 \\ 0 & 1 \end{pmatrix} \phi(\eta g) \tag{5}$$

where ϕ is the Fourier transform of Φ defined by

$$\phi(x) = \int \Phi(u) \psi(x \cdot u) du, \tag{6}$$

and η is the row-vector of length r' given by

$$\eta = (0, 0, \ldots, 0, 1). \tag{7}$$

Since as Φ varies, ϕ represents an arbitrary Schwartz function in r' variables, we may choose Φ so that ϕ is smooth and compactly supported with support contained in the open orbit

$$\eta G_{r'} \simeq Q_{r', r'-1} \backslash S_{r'}.$$

Next with ϕ and W_0 chosen in the above fashion, we will show that for *all* W' in $\mathcal{W}(\pi'; \psi)$ the integral $\Psi(s, W_1, W')$ converges (absolutely) for $\text{Re}(s) \geq 1$, normally for $\text{Re}(s)$ in a compact subset of $\text{Re}(s) \geq 1$. By (5) we are reduced to proving the convergence for $s \geq 1$ of the integral

$$\int_{N_{r'}\backslash G_{r'}} |W_0| \begin{pmatrix} g & 0 \\ 0 & 1 \end{pmatrix} |W'|(\epsilon g)|\phi(\eta g)||\det g|^{s-(r-r')/2} dg, \tag{8}$$

uniformly for $b > s > a \geq 1$. Breaking the integral up as before over $|\det g| \leq 1$ and $|\det g| \geq 1$ we need only prove the convergence for $s \geq 1$. Then, by Cauchy-Schwartz, the integral is majorized by the square-root of the product of the integral

$$\int_{N_{r'}\backslash G_{r'}} |W_0|^2 \begin{pmatrix} g & 0 \\ 0 & 1 \end{pmatrix} |\det g|^{s-(r-r')} dg \tag{9}$$

and the integral

$$\int_{N_{r'}\backslash G_{r'}} |W'|^2(\epsilon g)|\phi|^2(\eta g)|\det g|^s dg. \tag{10}$$

By (3) the integral (9) is finite.

On the other hand there is a Schwartz function ϕ_1 on $F^{r'}$ such that $|\phi(x)| \leq \phi_1$. Then $\phi_2 = \phi_1{}^2$ is again a Schwartz function and (10) is majorized by the integral

$$\int |W'|^2(\epsilon g)\phi_2(g)|\det g|^s dg,$$

whose convergence has been proved in [J-S] Proposition (3.17). We have therefore proved the first part of Proposition (2.6).

We now prove the second part. We *fix* W_0 satisfying (2), and s_0 with $\mathrm{Re}(s_0) \geq 1$. We will prove that there is a ϕ as above with support in $\eta G_{r'}$ and a $W' \in \mathcal{W}(\pi'; \psi)$ such that

$$\Psi(s_0, W_1, W') \neq 0.$$

Suppose in fact that this integral vanishes for all such ϕ and W'. Using once more the formula (1) we get:

$$\int_{Q_{r',r'-1}\backslash S_{r'}} \phi(\eta h) dh \int_{N_{r'-1}\backslash G_{r'-1}} W_0 \left[\begin{pmatrix} g & 0 \\ 0 & 1_{r-r'+1} \end{pmatrix} \begin{pmatrix} h & 0 \\ 0 & 1_{r-r'} \end{pmatrix} \right]$$

$$\times W' \left[\epsilon \begin{pmatrix} g & 0 \\ 0 & 1 \end{pmatrix} h \right] |\det g|^{s_0-1-(r-r')/2} dg = 0. \tag{14}$$

Fix W'. The inner integral represents a locally integrable function f on $Q_{r',r'-1}\backslash S_{r'}$. That function is orthogonal to all smooth functions of compact support on $Q_{r',r'-1}\backslash S_{r'}$. Since the latter space is countable at infinity it follows that f vanishes almost everywhere. Thus given W', *there exists a set $M(W')$ of measure zero in $Q_{r',r'-1}\backslash S_{r'}$ such that for all $h \notin M(W')$, the integral*

$$\int_{N_{r'-1}\backslash G_{r'-1}} W_0\left[\begin{pmatrix} g & 0 \\ 0 & 1_{r-r'+1}\end{pmatrix}\begin{pmatrix} h & 0 \\ 0 & 1_{r-r'}\end{pmatrix}\right]$$

$$\times W'\left[\epsilon\begin{pmatrix} g & 0 \\ 0 & 1\end{pmatrix} h\right] |\det g|^{s_0-1-1/2(r-r')}\, dg \tag{15}$$

is convergent and equal to zero.

Next apply the integration formula (1) this time to (3) (with $s = 2(\mathrm{Re}(s_0) - 1)$). We get

$$\int_{Q_{r',r'-1}\backslash S_{r'}} dh \int_{N_{r'-1}\backslash G_{r'-1}} |W_0|^2\left[\begin{pmatrix} g & 0 \\ 0 & 1_{r-r'+1}\end{pmatrix}\begin{pmatrix} h & 0 \\ 0 & 1_{r-r'}\end{pmatrix}\right]$$

$$\times\ |\det g|^{2(\mathrm{Re}\, s_0-1)-(r-r')}\, dg < +\infty. \tag{16}$$

There exists then a set M_0 of measure zero in $Q_{r',r'-1}\backslash S_{r'}$ such that for $h \notin M_0$ the function

$$g \mapsto W_0\left[\begin{pmatrix} g & 0 \\ 0 & 1_{r-r'+1}\end{pmatrix}\begin{pmatrix} h & 0 \\ 0 & 1_{r-r'}\end{pmatrix}\right] |\det g|^{s_0-1-(r-r')/2} \tag{17}$$

on $N_{r'-1}\backslash G_{r'-1}$ is square-integrable. Thus for h $\notin M_0$, each such function is in the space $\mathcal{K}_{r'}$.

On the other hand, let $\mathcal{H}'$ denote the space of π' and fix X a *countable* dense subset of $\mathcal{H}'$ contained in $(\mathcal{H}')^\infty = \mathcal{W}(\pi'; \psi)$. For any $h \in G_{r'}$, the set $\pi'(h)X$ has similar properties. Let Y denote the image of X in

$\mathcal{W}(\pi'; \psi)$. Applying Lemma (2.3), we find in particular, that, for a given $h \in S_{r'}$, the set of functions

$$g \mapsto \overline{W}'\left[\epsilon\begin{pmatrix} g & 0 \\ 0 & 1\end{pmatrix} h\right], \qquad W' \in Y,$$

is dense in $\mathcal{K}_{r'}$.

Now let $M = \bigcup M(W') \cup M_0$, the countable union being over $W' \in Y$. We conclude that for a given $h \notin M$, the inner product (15) of the function defined by (17) with the elements of a dense subset of $\mathcal{K}_{r'}$ is zero. Let α be the canonical projection

$$\alpha: S_{r'} \to Q_{r',r'-1}\backslash S_{r'}.$$

The same statement is true for $h \notin \alpha^{-1}(M)$. Thus by the continuity of W_0, we have

$$W_0\left[\begin{pmatrix} g & 0 \\ 0 & 1_{r-r'+1}\end{pmatrix}\begin{pmatrix} h & 0 \\ 0 & 1_{r-r'}\end{pmatrix}\right] = 0, \tag{18}$$

at first, for $g \in G_{r'-1}$, $h \notin \alpha^{-1}(M)$, but then again by continuity for all g in $G_{r'-1}$ and $h \in S_{r'}$. Taking then $h = e$ in (18), we conclude that

$$W_0\begin{pmatrix} g & 0 \\ 0 & 1\end{pmatrix} = 0$$

for $g \in G_{r'}$ in obvious contradiction to (2). □

3. Partial L-functions.

(3.1) Let F be a global field and $F_{\mathbf{A}} = \mathbf{A}$ the associated ring of adeles. If π is an irreducible unitary representation of $G_r(\mathbf{A})$ on a Hilbert space $\mathcal{H}$, then π is uniquely representable as a topological tensor product

$$\pi = \hat{\bigotimes_v} \pi_v, \qquad \mathcal{H} = \hat{\bigotimes_v} \mathcal{H}_v,$$

where π_v is an irreducible unitary representation of $G_r(F_v)$ on $\mathcal{H}_v$. As above let $\mathcal{H}_v^\infty$ denote the space of smooth vectors in $\mathcal{H}_v$. (If v is finite, this is the space of K_v-finite vectors in $\mathcal{H}$.) Let then

$$\mathcal{H}^\infty = \bigotimes_v \mathcal{H}_v^\infty$$

be the algebraic tensor product of the $\mathcal{H}_v^\infty$, that is, the linear span of the pure tensors:

$$w = \bigotimes_v w_v, \tag{1}$$

w_v being for almost all v a fixed K_v-invariant vector e_v in $\mathcal{H}_v$ (or $\mathcal{H}_v^\infty$). Fix once for all a non-trivial character

$$\psi = \Pi_v \psi_v$$

of $F_\mathbf{A}$ trivial on F. If each π_v is generic then the local spaces $\mathcal{W}(\pi_v; \psi_v)$ have been defined. Let

$$\mathcal{W}(\pi; \psi) = \bigotimes_v \mathcal{W}(\pi_v; \psi_v)$$

be the algebraic tensor product. We have for each v an isomorphism of $\mathcal{H}_v^\infty$ with $\mathcal{W}(\pi_v; \psi_v)$, given by (1.1.2) if v is finite and similarly (2.1.2) if v is infinite. If π_v is unramified we normalize this isomorphism so that e_v corresponds to the essential element of $\mathcal{W}(\pi_v; \psi_v)$. Taking tensor products we get then an isomorphism

$$\mathcal{H}^\infty \simeq \mathcal{W}(\pi; \psi). \tag{2}$$

It is clear from (2.2) that any pure tensor $w \in \mathcal{H}^\infty$ is of the form

$$w = \pi(\phi)w_1 \tag{3}$$

where $\phi \in C_c^\infty(G_r(\mathbf{A}))$ and w_1 is again in $\mathcal{H}^\infty$. A similar statement applies to $\mathcal{W}(\pi; \psi)$.

Now suppose ω is a character (unitary) of $F^\times \backslash F_\mathbf{A}^\times$. Let ${}^0L_2(\omega)$ denote the space of cusp forms ([B-J]) on $G_r(\mathbf{A})$ transforming under $Z_r(\mathbf{A})$

according to ω. Suppose that π is a component of ${}^0L_2(\omega)$ and let $\mathfrak{A}(\pi)$ denote the corresponding irreducible subspace. Let also $\mathfrak{A}^\infty(\pi)$ denote the image of $\mathcal{H}^\infty$ in $\mathfrak{A}(\pi)$. By (3) any function ϕ in $\mathfrak{A}^\infty(\pi)$ is a smooth function of $G_r(\mathbf{A})$. Moreover by the general theory, if F is a number field, ϕ is "rapidly-decreasing."

We will need to consider the "Fourier series" of an element of $\mathfrak{A}^\infty(\pi)$. For $0 \le m \le r-1$, let $N_{r,m}$ be the unipotent radical of the parabolic of type $(m+1, 1, \ldots, 1)$, that is the subgroup of N_r consisting of all matrices (n_{ij}) for which $n_{ij} = 0$ for $1 \le i \le m+1$, $1 \le j \le m+1$, $i \ne j$. We set

$$V_{m,\phi}(g) = \int_{N_{r,m}(F)\backslash N_{r,m}(\mathbf{A})} \phi(ug)\bar{\theta}_r(u)du \tag{3}$$

so that $V_{r-1,\phi} = \phi$. Note that

$$V_{m,\phi}(\gamma g) = V_{m,\phi}(g), \qquad \gamma \in G_m(F). \tag{4}$$

Moreover $N_{r,0} = N_r$ and $V_{0,\phi} \in \mathcal{W}(\pi; \psi)$ (c.f. [S], Section 4). Thus in particular $V_{0,\phi}$ is majorized by a gauge ([J-P-S], Props. (2.3.6), (2.4.1) and Section 12).

Now we have at once

$$\int_{\mathbf{A}^m/F^m} V_{m,\phi}\left[\begin{pmatrix} 1_m & u & 0 \\ 0 & 1 & 0 \\ 0 & 0 & 1_{r-m-1} \end{pmatrix} g\right] \bar{\theta}_{m+1}\left(\begin{bmatrix} 1_m & u \\ 0 & 1 \end{bmatrix}\right) du = V_{m-1,\phi}(g). \tag{5}$$

On the other hand, if U is the unipotent radical of a standard parabolic of type $(m, r-m)$, the constant term in the Fourier expansion of the function

$$u \mapsto V_{m,\phi}\left[\begin{pmatrix} 1_m & u & 0 \\ 0 & 1 & 0 \\ 0 & 0 & 1_{r-m-1} \end{pmatrix} g\right] \tag{6}$$

on the group $\mathbf{A}^m/F^m$ factors through an integral of the form

$$\int_{U_F\backslash U_{\mathbf{A}}} \phi(ug)du = 0$$

and is therefore zero. Next any non-trivial character of $\mathbf{A}^m/F^m$ has the form

$$u \mapsto \psi(\eta\gamma u),$$

where as before

$$\eta = (0, 0, \ldots, 0, 1)$$

is a row vector of length m, $\gamma \in P_m(F)\backslash G_m(F)$, and u is regarded as a column of height m. Thus if in (5) we replace u by γu and use (4) we get

$$V_{m,\phi}(g) = \Sigma_{P_m(F)\backslash G_m(F)} V_{m-1,\phi}\left[\begin{pmatrix}\gamma & 0\\ 0 & 1_{r-m}\end{pmatrix} g\right]. \tag{6}$$

Then by recursion starting with m we get, at first formally,

$$V_{m,\phi}(g) = \Sigma_{N_m(F)\backslash G_m(F)} V_{0,\phi}\left[\begin{pmatrix}\gamma & 0\\ 0 & 1_{r-m}\end{pmatrix} g\right]. \tag{7}$$

Since, as we have noted, $V_{0,\phi}$ is majorized by a gauge, by Proposition (1.2.2) of [J-P-S], the series in (7) is absolutely convergent. Thus the passage from (6) to (7) by combining the various Fourier series is justified.

(3.2) Let again r and r' be integers. Let π and π' be admissible irreducible representations of $G_r(\mathbf{A})$. Then, if S is a sufficiently large finite set of places containing the places at infinity, the local components π_v and π_v' are unramified. Let A_v and A_v' be the associated conjugacy classes, respectively in $GL_r(\mathbf{C})$ and $GL_{r'}(\mathbf{C})$ (c.f. (1.3)). We set

$$L(s, \pi_v \times \pi_v') = \det(1 - q_v^{-s} A_v \otimes A_v')^{-1} \tag{1}$$

as before and then set

$$L_S(s, \pi \times \pi') = \prod_{v \notin S} L(s, \pi_v \times \pi_v'). \tag{2}$$

If now π and π' are unitary, then as is well known the eigenvalues λ of A_v (resp. A_v') are in absolute value $\leq q_v^{(r-1)/2}$ (resp. $q_v^{(r'-1)/2}$) (c.f. [R.G.-J] (6.12)). In fact we even have $\lambda < q_v^{1/2}$ (c.f. [J-S] (2.5)). Thus in that case the Euler product in (2) is absolutely convergent for $\mathrm{Re}(s)$ large.

Now suppose in addition that π and π' are automorphic cuspidal representations. Then by Theorem (5.3) of [J-S] the product (2) is actually absolutely convergent for $\mathrm{Re}(s) > 1$. Thus $L_S(s, \pi \times \pi')$ is defined in the half-plane $\mathrm{Re}(s) > 1$. It follows from (5.3.2) and the uniform convergence of the series (5.3.5) of [J-S] in compact subsets of $\mathrm{Re}(s) > 1$, that $L_S(s, \pi \times \pi')$ is also holomorphic and $\neq 0$ in $\mathrm{Re}(s) > 1$.

(3.3) PROPOSITION. *Suppose $r \neq r'$ and the unitary representations π and π' are automorphic and cuspidal. Then the function $L_S(s, \pi \times \pi')$ has a continuous extension to the closed half plane* $\mathrm{Re}(s) \geq 1$.

Proof. 1st part: We note that by (1.5)(iii) we may enlarge S at will. We may assume then that for $v \notin S$, $\mathfrak{R}_v$ is the largest ideal on which ψ_v is trivial.

That being so, let for $v \notin S$, W_v (resp. W_v') be the essential element of $\mathcal{W}(\pi_v; \psi_v)$. (resp. $\mathcal{W}(\pi_v'; \psi_v)$). For v finite in S, choose W_v and W_v' so that each (local) integral $\Psi(s, W_v, W_v')$ converges for all s and is equal to one (c.f. (1.5)). For v infinite choose W_v and W_v' so that the integral $\Psi(s, W_v, W_v')$ converges for $\mathrm{Re}(s) \geq 1$ (Prop. (2.6)). Then $\otimes_v W_v$ (resp. $\otimes_{v'} W_v'$) belongs to $\mathcal{W}(\pi; \psi)$ (resp. $\mathcal{W}(\pi'; \psi)$). Let ϕ and ϕ' be the corresponding elements of $\mathcal{A}^\infty(\pi)$ and $\mathcal{A}^\infty(\pi')$ respectively. Now for an appropriate normalization of the measure of $N_r(F)\backslash N_r(\mathbf{A})$, by Theorem (4.5) of [S], the isomorphism of (3.1.2) is actually given by $\phi \mapsto V_{0,\phi}$. Thus in fact

$$V_{0,\phi}(g) = \prod_v W_v(g_v), \qquad V_{0,\phi'}(g) = \prod_v W_v'(g_v). \tag{1}$$

Next consider the integral

$$\int_{G_{r'}(F)\backslash G_{r'}(\mathbf{A})} V_{r',\phi}\left[\begin{pmatrix} g & 0 \\ 0 & 1 \end{pmatrix}\right] \phi'(g) |\det g|^{s-(r-r')/2} \, dg. \tag{2}$$

Calculating formally at first, we get by (3.1.7) the equivalent expression

$$\int_{G_{r'}(F)\backslash G_{r'}(\mathbf{A})} \sum_{N_{r'}(F)\backslash G_{r'}(F)} V_{0,\phi}\left[\begin{pmatrix} \gamma g & 0 \\ 0 & 1 \end{pmatrix}\right] \phi'(g)|\det g|^{s-(r-r')/2}\, dg. \tag{3}$$

Since $\phi'(g)|\det g|^{s-(r-r')/2}$ is invariant on the left under $G_{r'}(F)$ we may combine the integral and the sum to get

$$\int_{N_{r'}(F)\backslash G_{r'}(\mathbf{A})} V_{0,\phi}\left[\begin{pmatrix} g & 0 \\ 0 & 1 \end{pmatrix}\right] \phi'(g)|\det g|^{s-(r-r')/2}\, dg. \tag{4}$$

Next $V_{0,\phi}(ng) = \theta_r(n)V_{0,\phi}(g)$ for $n \in N_r(\mathbf{A})$, $g \in G_r(\mathbf{A})$. Note that for $n \in N_{r'}(F)$,

$$\theta_r\left[\begin{pmatrix} n & 0 \\ 0 & 1_{r-r'} \end{pmatrix}\right] = \theta_{r'}(n).$$

Thus integrating in stages, we get

$$\int_{N_{r'}(\mathbf{A})\backslash G_{r'}(\mathbf{A})} |\det g|^{s-(r-r')/2} V_{0,\phi}\left[\begin{pmatrix} g & 0 \\ 0 & 1 \end{pmatrix}\right] dg \times \int_{N_{r'}(F)\backslash N_{r'}(\mathbf{A})} \phi'(ng)\theta_{r'}(n)dn. \tag{5}$$

But one has

$$\int_{N_{r'}(F)\backslash N_{r'}(\mathbf{A})} \phi'(ng)\theta_{r'}(n)dn = V_{0,\phi'}(\epsilon g) \tag{6}$$

and therefore we conclude that

$$\begin{aligned} &\int_{G_{r'}(F)\backslash G_{r'}(\mathbf{A})} V_{r',\phi}\left[\begin{pmatrix} g & 0 \\ 0 & 1 \end{pmatrix}\right] \phi'(g)|\det g|^{s-(r-r')/2}\, dg \\ &= \int_{N_{r'}(\mathbf{A})\backslash G_{r'}(\mathbf{A})} V_{0,\phi}\left[\begin{pmatrix} g & 0 \\ 0 & 1 \end{pmatrix}\right] V_{0,\phi}(\epsilon g)|\det g|^{s-(r-r')/2}\, dg. \end{aligned} \tag{7}$$

Here as before $\epsilon = \operatorname{diag}(1, -1, 1, -1, \ldots)$.

The calculations are justified when Re(s) is sufficiently large. In effect it suffices to prove that the integral (4) is absolutely convergent for Re(s) large. But ϕ' is bounded and as we have noted $V_{0,\phi}$ is majorized by a gauge, say ξ. Thus our integral is majorized by

$$\int \xi\left[\begin{pmatrix} g & 0 \\ 0 & 1 \end{pmatrix}\right] |\det g|^{\mathrm{Re}(s)-(r-r')/2}\, dg,$$

the integration being over $N_{r'}(F)\backslash G_{r'}(\mathbf{A})$, or what amounts to the same over $N_{r'}(\mathbf{A})\backslash G_{r'}(\mathbf{A})$. Since ξ is invariant on the right under K, we may calculate this as an integral over $A_r(\mathbf{A})$ in which case it is easily seen to be finite.

(3.4) *Proof of Proposition* (3.3). 2nd *part*. We show first that the integral on the left in (3.3.7) converges for all s, uniformly for Re(s) in a compact set. This will be a consequence of the following lemma.

LEMMA. (i) *Suppose F is a number field. For all $N \geq 1$ and all compact subsets Ω of $G_r(\mathbf{A})$ there exists a constant C_N such that*

$$\left| V_{r',\phi}\left[\begin{pmatrix} g & 0 \\ 0 & 1 \end{pmatrix}\omega\right] \right| \leq C_N \inf(|\det g|^N, |\det g|^{-N})$$

for all $g \in G_{r'}(\mathbf{A})$ and $\omega \in \Omega$.

(ii) *Suppose F is a function field. For all compact subsets Ω of $G_r(\mathbf{A})$ there exists a constant $C > 0$ such that, for $\omega \in \Omega$, the support of the function*

$$g \mapsto V_{r',\phi}\left[\begin{pmatrix} g & 0 \\ 0 & 1 \end{pmatrix}\omega\right]$$

is contained in the set of $g \in G_{r'}(\mathbf{A})$ satisfying

$$C^{-1} \leq |\det g| \leq C.$$

Proof of the lemma (i).

The unipotent group $N_{r,r'}$ may be described as the set of all matrices of the form

$$\begin{pmatrix} 1_{r'} & u \\ 0 & v \end{pmatrix} \tag{1}$$

where the first column of $u \in M(r' \times r - r')$ is zero and v belongs to $N_{r-r'}$. Thus this group is the semi-direct product of the groups

$$N' = \left\{\begin{pmatrix} 1_{r'} & u \\ 0 & 1_{r-r'} \end{pmatrix}\right\}, \quad N'' = \left\{\begin{pmatrix} 1_{r'} & 0 \\ 0 & v \end{pmatrix}\right\}, \tag{2}$$

u and v being as before, N' being normal in $N_{r,r'}$ and N'' commuting with all matrices of the form

$$h = \begin{pmatrix} g & 0 \\ 0 & 1_{r-r'} \end{pmatrix}, \qquad g \in G_{r'}. \tag{3}$$

Denote by $N^*_{r,r'}$ the quotient $N_{r,r'}(F)\backslash N_{r,r'}(\mathbf{A})$ and use a similar notation for N', N''. We have, immediately from the definition,

$$|V_{r',\phi}(h\omega)| \leq \int_{N^*_{r,r'}} |\phi(nh\omega)|\,dn = \int_{N''*} dn'' \int_{N'*} |\phi|(n'n''h\omega)dn'. \tag{4}$$

Now $n''h = hn''$. Let Ω' be a compact subset of N'' projecting onto $N''*$. Replacing Ω by $\Omega'\Omega$ it will be sufficient to estimate the integral

$$\int_{N'*} |\phi|(n'h\omega)dn', \tag{5}$$

where $\omega \in \Omega$ and h is of the form (3). Let τ be the permutation matrix defined by

$$\tau = \begin{pmatrix} 1_{r'} & 0 \\ 0 & \mu \end{pmatrix}, \qquad \mu = \begin{bmatrix} 0 & & & 1 \\ & & 1 & \\ & \cdot & & \\ 1 & & & 0 \end{bmatrix}. \tag{6}$$

Clearly τ commutes with any matrix of the form (3). Since in addition ϕ is automorphic, we have

$$\int_{N'^*} |\phi|(n'h\omega)dn' = \int_{N'^*} |\phi|(\tau n'\tau^{-1}h\tau\omega)dn'.$$

Now $\tau n'\tau^{-1}$ is contained in the group of matrices of the form

$$\begin{pmatrix} g & 0 \\ 0 & 1 \end{pmatrix}, \qquad g \in G_{r-1}(\mathbf{A}). \tag{7}$$

Finally, since h also has this form and N'^* is compact, it suffices to show that, for all $N > 0$, there is a C_N such that

$$|\phi(h\omega)| \leq C_N \operatorname{Inf}(|\det h|^{-N}, |\det h|^N) \tag{8}$$

for all h of the form (7) and all $\omega \in \Omega$. By reduction theory we have

$$G_{r-1}(\mathbf{A}) = G_{r-1}(F)A_{r-1}(\mathbf{A})\Omega_0$$

where Ω_0 is a compact subset of $G_{r-1}(\mathbf{A})$. Thus replacing Ω by a sufficiently large compact subset of $G_r(\mathbf{A})$, it will suffice to prove (8) when h is a diagonal matrix of the form

$$a = \operatorname{diag}(a_1, a_2, \ldots, a_{r-1}, 1). \tag{9}$$

Next let Λ be the function defined on all of A_r by

$$\Lambda = \prod_{1 \leq 1 \leq r-1} \alpha_i^i. \tag{10}$$

For a of the special form (9), note that

$$\Lambda(a) = \det a. \tag{11}$$

For $a \in A_r(\mathbf{A})$, there exists a permutation matrix w such that $a = wbw^{-1}$ where b is in the positive chamber, that is to say $|\alpha_i(b)| \geq 1$ for $1 \leq i \leq r - 1$. On the other hand by the general theory, if we set

$$\kappa(b) = \sup_i |\alpha_i(b)|,$$

then, for all $N \geq 1$, there is a constant C_N such that

$$|\phi(bw^{-1}\omega)| \leq C_N\kappa(b)^{-N}$$

for all $\omega \in \Omega$, b in the positive chamber. Thus

$$|\phi(a\omega)| = |\phi(wbw^{-1}\omega)| = |\phi(bw^{-1}\omega)| \leq C_N\kappa(b)^{-N}.$$

On the other hand, it is easily verified that there is a constant $m > 0$ such that

$$\kappa(b)^{-m} \leq |\Lambda(wbw^{-1})| \leq \kappa(b)^m \tag{12}$$

for all b in the positive chamber and all w. Thus one has for all $a \in A(\mathbf{A})$, $\omega \in \Omega$,

$$|\phi(a\omega)| \leq C_{mN} \inf(|\Lambda(a)|^{-N}, |\Lambda(a)|^N)$$

and combining this with (11), the assertion (8) follows at once. This completes the proof of the first part of the lemma.

Proof of the lemma (ii). We will use the following fact due to G. Harder ([G.H.] (1.2)). Suppose F is a function field. Let ϕ be a cusp form on $G_r(\mathbf{A})$ and Ω a fixed compact subset of $G_r(\mathbf{A})$. *Then there exists a constant $d > 0$, such that for all b in the positive chamber, $\omega \in \Omega$, and w a permutation matrix, the relation $\phi(bw^{-1}\omega) \neq 0$ implies*

$$|\alpha_i(b)| \leq d, \qquad 1 \leq 1 \leq r - 1.$$

The proof of (ii) proceeds as the proof of the first part. One is reduced to proving that, for a given compact subset Ω of $G_r(\mathbf{A})$, there exists a constant $C > 0$ such that $\phi(a\omega) \neq 0$, $\omega \in \Omega$, $a = \operatorname{diag}(a_1, a_2, \ldots, a_{r-1}, 1)$ implies $C^{-1} \leq |\det a| \leq C$. Again there exists a matrix w such that $a = wbw^{-1}$ where b is in the positive chamber. One then has

$$\phi(a\omega) = \phi(bw^{-1}\omega).$$

Thus by the result we quoted $\kappa(b) \leq d$, and finally by (12)

$$d^{-m} \leq |\Lambda(wbw^{-1})| \leq d^m.$$

Since $|\det a| = |\Lambda(a)| = |\Lambda(wbw^{-1})|$ our assertion follows.

(3.5) *Completion of the proof of Prop.* (3.3). We show first that the integral on the left in (3.3.7) converges absolutely for all s, uniformly for s in a vertical strip of finite width. Suppose first that F is a number field. Since ϕ' is bounded, by the lemma the integral is, for each N, dominated by an integral of the form

$$\int_{G_{r'}(F)\backslash G_{r'}(\mathbf{A})} \inf(|\det g|^N, |\det g|^{-N}) |\det g|^{s-(r-r')/2}\, dg,$$

with s now real. Let G^0 be group of those $g \in G_{r'}(\mathbf{A})$ satisfying $|\det g| = 1$. Integrating in stages, we get for the previous integral

$$\operatorname{vol}(G_r(F)\backslash G^0) \int_{\mathbf{R}_+^\times} \inf(t^N, t^{-N}) t^{s-(r-r')/2}\, d^\times t$$

which is convergent for s in a compact if N is sufficiently large. If finally F is a function field ϕ' is again bounded. On the other hand since ϕ is also bounded say by a number d, we have

$$|V_{r',\phi}(g)| \leq \int_{N^*_{r,r'}} |\phi(ug)|\, du \leq d \operatorname{vol}(N^*_{r,r'}).$$

Our integral is then dominated by an integral of the form

$$\int_{c^{-1} \leq |\det g| \leq c} |\det g|^{\operatorname{Re}(s)-(r-r')/2}\, dg.$$

Since the integrand is bounded on the domain of integration and the latter has finite volume, we have the same conclusion.

That being the case the left side of (3.3.7) represents an entire function of s. Thus the right side extends to an entire function of s. Now, using (3.3.1), we have for Re(s) sufficiently large,

$$\int_{N_{r'}(\mathbf{A})\backslash G_{r'}(\mathbf{A})} V_{0,\phi}\left[\begin{pmatrix} g & 0 \\ 0 & 1 \end{pmatrix}\right] V_{0,\phi'}(g) |\det g|^{s-(r-r')/2}\, dg$$

$$= \prod_v \Psi(s, W_v, W_v'), \qquad (1)$$

each integral being convergent and the infinite product being absolutely convergent. However, with the particular choice of W_v and W_v' we have made, this product is nothing other than

$$\prod_{v\mid\infty} \Psi(s, W_v, W_v')L_S(s, \pi \times \pi'). \tag{2}$$

The conclusion is therefore immediate if F is a function field and results from Proposition (2.6) if F is a number field.

(3.6) **Proposition.** *Suppose that $r = r'$ and the representations π and π' are automorphic, cuspidal, and unitary. Let X be the set of s on the line* $\mathrm{Re}(s) = 1$ *such that $\alpha^{s-1} \otimes \pi$ is equivalent to the contragredient representation $\tilde{\pi}'$ of π'. Then:*

(i) *the function $L_S(s, \pi \times \pi')$ extends to a continuous function on the complement of X in the closed half-space* $\mathrm{Re}(s) \geq 1$;

(ii) *if s_0 is in X, then*

$$\lim_{s\to s_0} (s - s_0)L_S(s, \pi \times \pi'),$$

the limit taken in $\mathrm{Re}(s) \geq 1$, $s \neq s_0$, *exists, is finite, and is non-zero.*

Proof. This follows readily from the results of [J-S, Section 4]. In more detail, for $W \in \mathcal{W}(\pi; \psi)$, $W' \in \mathcal{W}(\pi'; \psi)$ and $\Phi \in \mathcal{S}(\mathbf{A}^r)$, let

$$\Psi(s, W, W', \Phi) = \int_{N_r\backslash G_r} W(g)W'(\epsilon g)\Phi(\eta g)|\det g|^s\, dg. \tag{1}$$

The integral converges for $\mathrm{Re}(s)$ sufficiently large and extends to a meromorphic function of s in the half plane $\mathrm{Re}(s) > 0$, holomorphic in $\mathrm{Re}(s) > 1$. Moreover, except for notation, Lemma (4.6) (loc. cit.) implies that if $\Psi(s, W, W', \Phi)$ has a pole at $s = 1$, then that pole is simple and $\tilde{\pi}$ is equivalent to π'. An application of the identity

$$\Psi(s + i\sigma, W, W', \Phi) = \Psi(s, W \otimes \alpha^{i\sigma}, W', \Phi)$$

(σ real) then shows that the poles of $\Psi(s, W, W', \Phi)$ are simple and contained in X.

Now choose again a finite set S of places (containing the places at infinity) such that π_v and π_v' are unramified outside S. Assume also that, for $v \notin S$, the largest ideal on which ψ_v is trivial is $\mathfrak{R}_v$. Suppose also that $W = \Pi_v W_v$, $W' = \Pi_v W_v'$, and $\Phi = \Pi_v \Phi_v$, where for $v \notin S$, W_v (resp. W_v') is the essential vector of $\mathcal{W}(\pi_v; \psi_v)$ (resp. $\mathcal{W}(\pi_v'; \psi_v)$) and Φ_v the characteristic function of $\mathfrak{R}_v^r$. Then we have, as in (4.7) (loc. cit.), for Re(s) large,

$$\Psi(s, W, W', \Phi) = \prod_v \Psi(s, W_v, W_v', \Phi_v),$$

the local factor being defined by analogy with (3.2.1). Moreover, for $v \notin S$,

$$\Psi(s, W_v, W_v', \Phi_v) = L(s, \pi_v \times \pi_v')$$

(Section 2 loc. cit.). Thus for Re(s) sufficiently large,

$$\Psi(s, W, W', \Phi) = \prod_{v \in S} \Psi(s, W_v, W_v', \Phi_v) L_S(s, \pi \times \pi').$$

By Proposition (1.5) and (3.17) of [J-S] each of the functions $\Psi(s, W_v, W_v', \Phi_v)$ is continuous in the closed half-plane Re(s) ≥ 1. Moreover for such an s, we may choose W_v, W_v', and Φ_v such that $\Psi(s, W_v, W_v', \Phi_v) \neq 0$. The first assertion follows.

To prove the second assertion, let $s_0 \in X$. Replacing π by $\alpha^{1-s_0} \otimes \pi$ we may assume $s_0 = 1$. Let ϕ (resp. ϕ') be the element of $\mathcal{A}^\infty(\pi)$ (resp. $\mathcal{A}^\infty(\pi')$) corresponding to W (resp. W'). Then combining (4.5.5) and (4.3.2) of [J-S] we find

$$\lim_{s \to 1} (s-1)\Psi(s, W, W', \Phi) = c\hat{\Phi}(0) \int_{Z_{\mathbf{A}} G_F \backslash G_{\mathbf{A}}} \phi(g)\phi'(g)dg,$$

where c is a non-zero constant. Choosing W_v, W_v', and Φ_v once more so that $\Psi(1, W_v, W_v', \Phi_v) \neq 0$ for $v \in S$ we see that

$$\lim_{s \to 1} (s-1) L_S(s, \pi \times \pi')$$

exists (and is finite). Call κ this limit. Then we have, for any choice of W_v, W_v', and Φ_v ($v \in S$)

$$\kappa \prod_{v \in S} \Psi(1, W_v, W_v', \Phi_v)$$

$$= c\hat{\Phi}(0) \int \phi(g)\phi'(g)dg$$

$$= c \prod_{v \in S} \hat{\Phi}_v(0) \int \phi(g)\phi'(g)dg.$$

Since π' is the representation contragredient to π, we may take $W_v'(g) = \overline{W_v(\epsilon g)}$. Then by (3.3.1), $\phi'(g) = \overline{\phi(\epsilon g)} = \overline{\phi(g)}$. We may also assume ϕ is non-zero. Then for any choice of the Φ_v we have

$$\kappa \prod_{v \in S} \Psi(1, W_v, W_v', \Phi_v) = c \prod_{v \in S} \hat{\Phi}_v(0) \int_{Z_{\mathbf{A}} G_F \backslash G_{\mathbf{A}}} |\phi(g)|^2 dg.$$

Since we may choose Φ_v so that, for all $v \in S$, $\hat{\Phi}_v(0) \neq 0$, we see that $\kappa \neq 0$. □

(3.7) A THEOREM OF SHAHIDI. Let S be any finite set of places containing the places at infinity and those places for which π_v and π_v' are ramified. By (3.2), $L_S(s, \pi \times \pi')$ has no zeros in $\text{Re}(s) > 1$.

By abuse of language we shall call the points s_0 in $\text{Re}(s) \geq 1$ for which

$$\lim_{s \to s_0, \text{Re}(s) \geq 1} (s - s_0) L_S(s, \pi \times \pi') \neq 0$$

the poles of $L_S(s, \pi \times \pi')$ in $\text{Re}(s) \geq 1$. These poles of course lie on $\text{Re}(s) = 1$. Thus the function $L_S(s, \pi \times \pi')$ is defined and continuous outside the set of poles in $\text{Re}(s) \geq 1$.

THEOREM. *With the hypotheses of* (3.3) *or* (3.6), *the function* $L_S(s, \pi \times \pi')$ *is not zero in* $\text{Re}(s) \geq 1$.

We have already observed the validity of our assertion for $\text{Re}(s) > 1$. The non-vanishing on $\text{Re}(s) = 1$ is due to F. Shahidi [F.S.] and is based on the general theory of Eisenstein series. Shahidi's proof is similar in spirit to that of [J-S].

Remark (3.8). One can actually prove, either by our methods or those of Shahidi, that $L_S(s, \pi \times \pi')$ is *globally meromorphic*. However, we shall not need this in what follows.

4. Classification Theorems.

(4.1) As before F is an A-field and S is a large finite set of places containing those at infinity. Our results will not change if we replace S by a larger set; in particular we may assume that any representation (or finite set of representations) is unramified outside of S.

Let $\pi = \otimes_v \pi_v$ (resp. $\pi' = \otimes_v \pi_v'$) be an automorphic, cuspidal, unitary representation of GL_r (resp. $GL_{r'}$). As in [J-S], we set

$$\pi(v^n) = \operatorname{tr}(A_v{}^n) \tag{1}$$

if A_v is, as before, the class associated to π_v. Similarly for π'. Next for σ real, $\sigma \geq 1$, consider the double series

$$l(\sigma, \pi \times \pi') = \sum_{v \notin S} \sum_{n \geq 1} \frac{\bar{\pi}(v^n)\pi'(v^n)}{nq_v{}^{n\sigma}}. \tag{2}$$

In [J-S, Section 4] we proved that this series is absolutely convergent for $\sigma > 1$. Clearly it is then uniformly convergent for $\sigma > 1 + \epsilon$ and hence represents a continuous function in the half-line $\sigma > 1$. Moreover,

$$L_S(\sigma, \pi \times \pi') = \exp l(\sigma, \pi \times \pi'), \tag{3}$$

also for $\sigma > 1$.

Finally let $\delta_{\pi,\pi'} = 1$ if $\pi \simeq \pi'$ and $= 0$ otherwise.

Theorem. *As σ tends to 1 through values ≥ 1, we have*

$$\lim_{\sigma \to 1} \left(\log \frac{1}{\sigma - 1}\right)^{-1} l(\sigma, \pi \times \pi') = \delta_{\pi,\pi'}.$$

Proof. Suppose first that $\tilde{\pi} \not\simeq \pi'$. Then by Prop. (3.3), $L_S(\sigma, \pi \times \pi')$ has a continuous extension to the half line $\sigma \geq 1$. Moreover by Theorem (3.7), $L_S(1, \pi \times \pi') \neq 0$. Thus there is a determination of the logarithm (principal or otherwise) for which $\log L_{S'}(\sigma, \pi \times \pi')$ is continuous in an interval $1 \leq \sigma < a$. Next from (3) we get

$$\log L_S(\sigma, \pi \times \pi') = l(\sigma, \pi \times \pi') + 2\pi k(\sigma) \tag{4}$$

where k takes integral values and now $1 < \sigma < a$. But then $k(\sigma)$ being the difference of two continuous functions must be constant in $1 < \sigma < a$.

Thus

$$\lim_{\sigma \to 1, \sigma > 1} l(\sigma, \pi, \times \pi') \text{ exists,} \tag{5}$$

which is more than is required for the proof of the theorem.

The proof for $\tilde{\pi} \simeq \pi'$ is similar. In fact by Prop. (3.6)

$$\lim_{\sigma \to 1} (\sigma - 1) L_S(\sigma, \pi \times \bar{\pi}) = \kappa \neq 0. \tag{6}$$

In this case $L_S(\sigma, \pi \times \bar{\pi}) > 0$ for $\sigma > 1$. Thus from (3) we get (with real logarithms)

$$\log(\sigma - 1) + \log L_S(\sigma, \pi \times \bar{\pi}) = \log(\sigma - 1) + l(\sigma, \pi \times \bar{\pi}) \tag{7}$$

which tends to $\log \kappa$ as $\sigma \to 1$. Our assertion follows at once. □

(4.2) We shall prove as a corollary to Theorem (4.1) that *any finite set of distinct automorphic cuspidal representations is linearly independent.*

Suppose ρ is an automorphic cuspidal representation of $\mathrm{GL}_r(\mathbf{A})$. Then ρ may be written uniquely in the form

$$\rho = \pi \otimes \beta, \tag{1}$$

where π is unitary and β is a positive quasi-character ($\beta = \alpha^t$, t real). We have locally $\rho_v = \pi_v \otimes \beta_v$; thus if A_v is the class of π_v, then $B_v = A_v q_v{}^{-t}$ is the class of ρ. As before we set

$$\rho(v^n) = \mathrm{tr}(B_v{}^n). \tag{2}$$

Next let $\mathfrak{A}_r$ denote the set of all equivalence classes of automorphic cuspidal representations of $G_r(\mathbf{A})$. Let $\mathfrak{A} = \bigcup_{r \geq 1} \mathfrak{A}_r$ (disjoint union). The precise form of our result is as follows:

THEOREM. *Let $\rho_1, \rho_2, \ldots, \rho_p \in \mathfrak{A}$. Let S be a finite set of places containing the infinite places. Suppose that the representations $\rho_1, \rho_2, \ldots, \rho_p$ are unramified outside of S. Let $c_1, c_2, \ldots, c_p$ be complex numbers. Suppose that we have a relation*

$$\sum_{1 \leq j \leq p} c_j \rho_j(v^n) = 0 \tag{3}$$

which holds for all $v \notin S$ and all integers $n \geq 1$. Then

$$c_1 = c_2 = \cdots = c_p = 0.$$

Proof. We may assume that all of the c_j's are non-zero. Write $\rho_j = \pi_j \otimes \alpha^{t_j}$ where π_j is unitary and t_j is real. We may assume that $t_1 \leq t_2 \leq \cdots \leq t_p$. From (2) we have

$$\rho_j(v^n) = \pi_j(v^n) q_v^{-nt_j}. \tag{4}$$

Thus multiplying (3) through by $q_v^{nt_1}$ one may also assume $t_1 = 0$. Then from (3), (4) and (4.1.2) we get immediately

$$\sum_{1 \leq j \leq p} c_j l(\sigma + t_j, \pi_1 \times \pi_j) = 0. \tag{5}$$

If a given t_j is positive, then the function $l(\sigma + t_j, \pi_1 \times \pi_j)$ is continuous in σ near $\sigma = 1$ (c.f. (4.1)). Suppose then that $t_1 = t_2 = \cdots = t_l = 0$ and that $t_{l+1} > 0$. We get then from (5) and Theorem (4.1)

$$\sum_{1 \leq j \leq l} c_j \lim_{\sigma \to 1} \left(\log \frac{1}{\sigma - 1} \right)^{-1} l(\sigma, \pi_1 \times \pi_j) = 0. \tag{6}$$

But by the same theorem this sum is also

$$\sum_{1 \leq j \leq l} c_j \delta_{\pi_1, \pi_j}. \tag{7}$$

Here $\pi_j = \rho_j$ is inequivalent to $\pi_1 = \rho_1$ unless $j = 1$. Thus combining (6) and (7) we get $c_1 = 0$. This contradiction completes the proof of the theorem. □

(4.3) We shall obtain a classification theorem for automorphic forms on GL_r which is a precise analogue for this group of the known results for local groups.

Accordingly let P be a standard parabolic subgroup of G_r of type $(r_1, r_2, \ldots, r_u)$. The quotient of P by its unipotent radical $U = U_P$ is isomorphic to the group

$$M = G_{r_1} \times G_{r_2} \times \cdots \times G_{r_u}. \tag{1}$$

For each j, $1 \leq j \leq u$, let σ_j be an automorphic cuspidal representation of $\mathrm{GL}(m_j, \mathbf{A})$. For each place v the representation $\sigma_v = \otimes_j \sigma_{jv}$ of the group M_v can be regarded as a representation of P_v trivial on U_v; it induces an admissible representation of G_v which we will denote by

$$\xi_v = \mathrm{Ind}(G_v, P_v; \sigma_v). \tag{2}$$

One obtains then a family of irreducible admissible representations of $G_r(\mathbf{A})$ by taking for each v an irreducible component π_v of the representation ξ_v and forming the "tensor product" $\pi = \otimes_v \pi_v$. On the other hand, with $\sigma = \sigma_1 \otimes \sigma_2 \otimes \cdots \otimes \sigma_r$, one can define globally an induced representation

$$\xi = \mathrm{Ind}(G_r(\mathbf{A}), P(\mathbf{A}); \sigma). \tag{3}$$

Of course $\xi = \otimes_v \xi_v$. The irreducible representations π which we have described are exactly the components of ξ. In fact the restriction of ξ to the (local) Hecke algebra at v is a discrete multiple of ξ_v. In this way one obtains all of the automorphic representations of $G_r(\mathbf{A})$ [RPL I].

Let Q be another standard parabolic say of type $(s_1, s_2, \ldots, s_w)$ and τ_j an automorphic cuspidal representation of $\mathrm{GL}(s_j, \mathbf{A})$. As before let $\tau_v = \otimes_j \tau_{jv}$,

$$\eta_v = \mathrm{Ind}(G_v, Q_v; \tau_v), \tag{4}$$

and

$$\eta = \mathrm{Ind}(G_r(\mathbf{A}), Q(\mathbf{A}); \tau), \tag{5}$$

where $\tau = \tau_1 \otimes \tau_2 \otimes \cdots \otimes \tau_w$. We may ask whether ξ and η have a common constituent. Suppose P and Q are associate (in which case $u = w$) and that there is a permutation ϕ of $\{1, 2, \ldots, u\}$ such that $s_j = r_{\phi(j)}$. Suppose moreover that $\tau_j \simeq \sigma_{\phi(j)}$. We will say in this situation that *the pairs* (σ, P) *and* (τ, Q) *are associate.* When this is so the representations ξ_v and η_v have the same character and therefore the same components; in particular if both ξ_v and η_v are unramified then their unique unramified components are the same. In other words the irreducible components of ξ and η are the same.

The following theorem will imply that the converse is also true.

(4.4) **Theorem.** *Let P, Q, σ_j and τ_k be as above. Let S be a finite set of places containing all the places at infinity. Suppose that for $v \notin S$ the representations σ_{jv} and τ_{kv} are unramified and that the representations ξ_v and η_v of G_v they induce have the same unramified component. Then the pairs (σ, P) and (τ, Q) are associate.*

Proof. The proof follows directly from Theorem (4.2). In fact, for $v \notin S$, let A_{jv} (resp. B_{kv}) be the semi-simple conjugacy class in $\mathrm{GL}(r_j, \mathbf{C})$ (resp. $\mathrm{GL}(s_k, \mathbf{C})$) associated to σ_{jv} (resp. τ_{kv}). The class associated to the unramified component of ξ_v (resp. η_v) is nothing other than $\bigoplus_j A_{jv}$ (resp. $\bigoplus_k B_{kv}$). The hypothesis is equivalent to the assertion

$$\bigoplus_j A_{jv} = \bigoplus_k B_{kv} \tag{1}$$

(equality as conjugacy classes in $\mathrm{GL}_r(\mathbf{C})$). From (1) we deduce immediately the equality

$$\sum_{1 \le j \le u} \sigma_j(v^n) = \sum_{1 \le k \le w} \tau_k(v^n) \tag{2}$$

which holds for all $v \notin S$ and all integers $n \geq 1$. We regard the σ_j and τ_k as elements of $\mathcal{A}$. Let then $\mathcal{B} = \{\rho_1, \rho_2, \ldots, \rho_a\}$ (resp. $\mathcal{B}' = \{\rho_1', \rho_2', \ldots, \rho_b'\}$) be the set of distinct elements of $\mathcal{A}$ among the σ_j (resp. τ_k). Then (2) may be rewritten as

$$\sum_{1 \le \alpha \le a} m_\alpha \rho_\alpha(v^n) = \sum_{1 \le \alpha \le b} m_\alpha' \rho_\alpha'(v^n) \tag{3}$$

where m_α (resp. m_α') is the number of times ρ_α (resp. ρ_α') occurs in the left (resp. right) side of (2).

But by Theorem (4.2) $\mathcal{B}$, regarded as a set of functions on the set of v^n, $v \notin S$, $n \geq 1$, is linearly independent. Similarly $\mathcal{B}'$ is linearly independent. Thus we get at once $\mathcal{B} = \mathcal{B}'$, $a = b$ and $m_\alpha = m_\alpha'$ for $1 \leq \alpha \leq a$. In particular

$$u = \sum m_\alpha = \sum m_\alpha' = w, \tag{4}$$

and we also see that the maps

$$j \mapsto \sigma_j, \; j \mapsto \tau_j$$

of $\{1, 2, \ldots, u\}$ into $\mathcal{A}$ differ only by a permutation. □

(4.5) As we have seen in (4.4) any automorphic representation π of $G_r(\mathbf{A})$ is a subquotient of an essentially unique induced representation

$$\mathrm{Ind}(G_r(\mathbf{A}), P(\mathbf{A}); \sigma) \tag{1}$$

where $\sigma = \sigma_1 \otimes \cdots \otimes \sigma_u$ is an automorphic cuspidal representation of $M(\mathbf{A})$. Thus it is natural to associate to π the (formal) sum

$$\sum_{1 \le j \le u} \sigma_j, \tag{2}$$

or, if m_ρ is the multiplicity of a given element ρ of $\mathfrak{A}$ in (2), the formal sum

$$\chi_\pi = \sum_{\rho \in \mathfrak{A}} m_\rho \rho. \tag{3}$$

If we define $\deg \pi = r$, we clearly have

$$\deg \pi = \sum m_\rho \deg \rho. \tag{4}$$

Conversely the theory of Eisenstein series allows us, at least when the m_ρ are positive and integral, to attach to a formal sum (3) an automorphic representation of some group $G_r(\mathbf{A})$. More generally we shall refer to a sum (3) where the coefficients m_ρ are allowed to be positive or negative integers or zero as a *virtual form.* Of course we require $m_\rho = 0$ for all but finitely many ρ. If χ is a virtual form we call the ρ which actually appear ($m_\rho \neq 0$) in χ the *components* of χ, and define

$$\deg \chi = \sum m_\rho \deg \rho. \tag{5}$$

The *contragredient* $\tilde{\chi}$ of χ is the virtual form

$$\tilde{\chi} = \sum m_\rho \tilde{\rho}. \tag{6}$$

If $\chi' = \sum m_\rho' \rho$ is another virtual form we define the inner product (χ, χ') to be

$$(\chi, \chi') = \sum_{\rho \in \mathfrak{A}} m_\rho m_\rho'. \tag{7}$$

If S is a finite set of places (containing those at infinity) outside of which the components of χ and χ' are unramified, we may also define the L-function

$$L_S(s, \chi \times \chi') = \prod_{\rho,\tau \in \mathfrak{A}} L_S(s, \rho \times \tau)^{m_\rho m_\tau'}. \tag{8}$$

If then we set, for $v \notin S$, $n \geq 1$,

$$\chi(v^n) = \Sigma\, m_\rho \rho(v^n), \qquad \chi'(v^n) = \Sigma\, m_\rho' \rho(v^n), \tag{9}$$

we get at once

$$L_S(s, \chi \times \chi') = \exp \sum_{v \notin S} \sum_{n \geq 1} \frac{\chi(v^n)\chi'(v^n)}{n q_n^{ns}} \tag{10}$$

at least if Re(s) is large.

Suppose now that the components of χ and χ' are *unitary*; then (8) is defined for Re(s) $\geq$ 1. Then by Prop. (3.3), Prop. (3.6), and Theorem (3.7), $L_S(s, \tilde{\chi} \times \chi')$ has a pole at $s = 1$ exactly of order (χ, χ'). More precisely the limit

$$\lim_{s \to 1, s \geq 1} (s - 1)^{(\chi, \chi')} L_S(s, \tilde{\chi} \times \chi') \tag{11}$$

exists and is non-zero. In what follows we shall apply this to give a criterion for a Galois representation to be "automorphic." We can only hope that an eventual application will be found.

(4.6) We recall the "strong form" of Artin's conjecture due to Langlands and Weil.

Suppose K is a finite Galois extension of the global field F with Galois group $\mathfrak{G}$. If v, a place of F, is unramified in K, let Fr_v denote the corresponding (Frobenius) conjugacy class in $\mathfrak{G}$. If σ is a representation, irreducible or not, of $\mathfrak{G}$ of degree d, then $\sigma(\mathrm{Fr}_v)$ is a well-defined conjugacy class in $\mathrm{GL}_d(\mathbf{C})$. We shall say that σ is automorphic (or satisfies the strong form of Artin's conjecture) if there is an automorphic representation $\hat{\sigma}$ of $\mathrm{GL}_d(\mathbf{A})$, $\hat{\sigma} = \otimes_v \hat{\sigma}_v$, such that, at each place v of F which is unramified in K, $\hat{\sigma}_v$ is unramified and

$$\sigma(\mathrm{Fr}_v) = A_v, \tag{1}$$

where A_v is the (Langlands) class associated to $\hat{\sigma}_v$. Let χ_σ denote the character (trace) of σ. An equivalent form of (1) is

$$\chi_\sigma(\mathrm{Fr}_v{}^n) = \hat{\sigma}(v^n), \tag{2}$$

with v as above and $n \geq 1$. Note then that by Theorem (4.2), the virtual form

$$\chi_{\hat{\sigma}} = \Sigma\, m_\rho \rho \tag{3}$$

associated to $\hat{\sigma}$ is then uniquely determined. It is easy to see that any component ρ of $\chi_{\hat{\sigma}}$ is unitary. In fact let $\omega = \otimes_v \omega_v$ be the central quasi-character of ρ, and for v outside of a sufficiently large finite set S of places of F, let B_v be the class of ρ_v. Then the eigenvalues of B_v are among those of A_v and thus by (1) have absolute value one. Then

$$|\omega_v(\tilde{\omega}_v)| = 1, \qquad v \notin S, \tag{4}$$

and by Dirichlet's theorem (or (4.2)) ω itself is unitary. Since ρ is a cusp form it must be unitary. We remark that even though a representation σ of $\mathfrak{G}$ may be automorphic it is not clear that the same is true for the irreducible components of σ. We can however establish the following theorem.

(4.7) THEOREM. *Let F be a global field, and K be a finite Galois extension of F. Suppose π is an irreducible representation of* Gal(K/F) *and that as a virtual representation*

$$\pi = \sigma - \tau, \tag{1}$$

where σ and τ are both automorphic. Then π is automorphic and cuspidal.

Proof. Let deg $\sigma = a$, deg $\tau = b$. By hypothesis there are automorphic representations $\hat{\sigma}$ and $\hat{\tau}$ of $\mathrm{GL}_a(\mathbf{A})$ and $\mathrm{GL}_b(\mathbf{A})$ respectively, and a finite set of places S of F such that

$$\chi_\sigma(\mathrm{Fr}_v{}^n) = \chi_{\hat{\sigma}}(v^n), \qquad \chi_\tau(\mathrm{Fr}_v{}^n) = \chi_{\hat{\tau}}(v^n), \tag{2}$$

for all $v \notin S$ and all $n \geq 1$. Let χ be the virtual form

$$\chi = \chi_{\hat{\sigma}} - \chi_{\hat{\tau}}. \tag{3}$$

Then

$$\chi_\pi(\mathrm{Fr}_v{}^n) = \chi_\sigma(\mathrm{Fr}_v{}^n) - \chi_\tau(\mathrm{Fr}_v{}^n) = \chi_{\hat{\sigma}}(\nu^n) - \chi_{\hat{\tau}}(\nu^n) = \chi(\nu^n), \tag{4}$$

and consequently

$$\sum_{v \notin S} \sum_{n \geq 1} |\chi_\pi(\mathrm{Fr}_v{}^n)|^2/nq_v{}^{ns} = \sum_{v \notin S} \sum_{n \geq 1} |\chi(\nu^n)|^2/nq_v{}^{ns}. \tag{5}$$

Exponentiating both sides and using (4.5.8), we get at once

$$L_S(s, \pi \times \tilde{\pi}) = L_S(s, \chi \times \tilde{\chi}), \tag{6}$$

the left side being the product of the local Artin L-functions $L_v(s, \pi \times \tilde{\pi})$ over the places v of F not in S. Similarly

$$L_S(s, \sigma \times \tilde{\sigma}) = L_S(s, \chi_{\hat{\sigma}} \times \tilde{\chi}_{\hat{\sigma}}), \qquad L_S(s, \tau \times \tilde{\tau}) = L_S(s, \chi_{\hat{\tau}} \times \tilde{\chi}_{\hat{\tau}}). \tag{7}$$

There is a well-known analogue of (4.5.11) for Artin L-functions: if ξ is a (unitary) representation of $\mathfrak{G}$ then $L_S(s, \xi)$ has a pole at $s = 1$ of order exactly equal to the multiplicity of the trivial representation in ξ. In particular, $L_S(s, \xi \otimes \tilde{\xi})$ has a pole of order (χ_ξ, χ_ξ) at $s = 1$, the inner product normalized to be 1 if ξ is irreducible. Thus from (6), (7) and (4.5.11) we have at once

$$(\chi_{\hat{\sigma}}, \chi_{\hat{\sigma}}) = (\chi_\sigma, \chi_\sigma), \qquad (\chi_{\hat{\tau}}, \chi_{\hat{\tau}}) = (\chi_\tau, \chi_\tau), \qquad (\chi, \chi) = 1. \tag{8}$$

In particular $\chi = \Sigma_{\rho \in \mathfrak{A}}\, m_\rho \rho$, with $\Sigma\, m_\rho{}^2 = 1$. The m_ρ being integers all but one of them, say m_{ρ_0}, must be zero and $\chi = \pm\rho_0$ is, up to sign, a cuspidal representation. If $\chi = \rho_0$ we are done, noting that, by (3), $\deg \rho_0 = \deg \hat{\sigma} - \deg \hat{\tau} = \deg \sigma - \deg \tau = \deg \pi$. Suppose then that $\chi = -\rho_0$ with ρ_0 an automorphic cuspidal representation. Then from (3) we get

$$\chi_{\hat{\tau}} = \chi_{\hat{\sigma}} + \rho_0. \tag{9}$$

Since $\hat{\sigma}$ and ρ_0 are actually automorphic representations and not merely virtual, we get at once from (9)

$$(\chi_{\hat{\tau}}, \chi_{\hat{\tau}}) \geq (\chi_{\hat{\sigma}}, \chi_{\hat{\sigma}}) + (\rho_0, \rho_0) > (\chi_{\hat{\sigma}}, \chi_{\hat{\sigma}}). \tag{10}$$

But in an analogous way we get from (1),

$$(\chi_\sigma, \chi_\sigma) > (\chi_\tau, \chi_\tau). \tag{11}$$

The obvious contradiction which results from combining (8), (10) and (11) shows in fact that χ is automorphic cuspidal and we are done. □

(4.8) *Remark.* With the above notation, if π is a representation of $\mathrm{Gal}(K/F)$, we may always use Brauer's Theorem to write $\pi = \sigma - \tau$ where the representations σ and τ are sums of *monomial representations.* Thus the construction of forms on $\mathrm{GL}(n)$ attached to arbitrary extensions of the ground field combined with the theory of Eisenstein series [R.P.L. I] would lead to a proof of (the strong form of) Artin's conjecture. Unfortunately except in the case of normal cyclic extensions there does not seem to be much hope in proving the existence of these forms in the near future.

COLUMBIA UNIVERSITY
THE JOHNS HOPKINS UNIVERSITY

REFERENCES

[B-J] A. Borel and H. Jacquet, *Automorphic forms and automorphic representations*, Automorphic forms, representations and L-functions, Proc. of Symposia in Pure Math., Vol. XXXIII, Part 1, A.M.S. (1979) pp. 189–202.

[C-S] W. Casselman and J. Shalika, *The unramified principal series of* $\mathfrak{p}$*-adic groups II, the Whittaker function*, Comp. Math., **41** (1980) pp. 207–231.

[D-M] J. Dixmier and P. Malliavin, *Factorisations de fonctions et de vecteurs indéfiniment différentiable*, Bull. Sci. Math., II. Sér. 102 (1978) pp. 305–330.

[S.G.-J] S. Gelbart and H. Jacquet, *A relations between automorphic forms on* GL_2 *and* GL_3, Ann. Scient. Ec. Norm. Sup., 4ᵉ série, t. 11 (1978) pp. 471–542.

[G-K] I. M. Gelfand and D. A. Kazdan, *Representations of* $\mathrm{GL}(n, K)$ *where* K *is a local field*, in Lie Groups and their Representations, John Wiley and Sons (1975) pp. 95–118.

[R.G.-J] R. Godement and H. Jacquet, *Zeta functions of simple algebras*, Springer-Verlag Lecture Notes 260 (1972).

[G.H] G. Harder, *Chevalley groups over function fields and automorphic forms*, Annals of Math., **100** (1974) pp. 249–306.

[J-L] H. Jacquet and R. P. Langlands, *Automorphic Forms on* GL(2), Springer-Verlag Lecture Notes 114 (1970).

[J-P-S] H. Jacquet, I. I. Piatetski-Shapiro, and J. Shalika, *Automorphic forms on* GL(3), I and II, Annals of Math., **109** (1979) pp. 169–212.

[J-S] H. Jacquet and J. Shalika, *On Euler products and the classification of automorphic forms I*, Amer. J. of Math., **103** (1981) pp. 499–558.

[R.P.L. I] R. P. Langlands, *On the notion of an automorphic representation. A supplement to the preceding paper*, Automorphic forms, representations and L-functions, Proc. of Symposia in Pure Math., Vol. XXXIII, Part 1, A.M.S. (1979) pp. 203–207.

[R.P.L. II] ______, *Automorphic representations, Shimura varieties, and motives. Ein märchen*, Automorphic forms, representations and L-functions, Proc. of Symposia in Pure Math. Vol. XXXIII, part 2, A.M.S. (1979) pp. 205–246.

[R.P.L. III] ______, *Euler products*, Yale Mathematical Monographs 1, Yale University Press (1971).

[T.M.] T. Miyake, *On automorphic forms on* GL_2 *and Hecke operators*, Annals of Math., **94** (1971) pp. 174–189.

[I.P.] I. Piatetski-Shapiro, *Multiplicity one theorems*, Automorphic forms, representations and L-functions, Proc. of Symposia in Pure Math. Vol. XXXIII, Part 1, A.M.S. (1979) pp. 209–212.

[R.R.] R. Rankin, *Contributions to the theory of Ramanujan's function* $\tau(n)$ *and similar arithmetical functions* I *and* II, Proc. Camb. Phil. Soc. **35** (1939) pp. 351–356 and 357–372.

[A.S.] A. Selberg, *Bemerkungen über eine Dirichletsche Reihe, die mit der Theorie der Modulformen nahe verbunden ist*, Arch. Math. Naturvid. **43** (1940) pp. 47–50.

[F.S.] F. Shahidi, *On certain L-functions*, Amer. J. Math., **103** (1981) pp. 297–356.

[J.S.] J. A. Shalika, *The multiplicity one theorem for* GL_n, Annals of Math., **100** (1974) pp. 171–193.

[T.S.] T. Shintani, *On an explicit formula for class-*1 *"Whittaker functions" on* GL_n *over* $\mathfrak{p}$*-adic fields*, Proc. Japan Acad., **52** (1976) pp. 180–182.

[N.W.] N. R. Wallach, *Representations of reductive Lie groups*, Automorphic forms, representations and L-functions, Proc. of Symposia in Pure Math., Vol. XXXIII, Part 1, A.M.S. (1979) pp. 71–86.

[1] Compositio Mathematica 54 (1985) 243–310.

A RELATIVE TRACE FORMULA

H. Jacquet and K.F. Lai

§1. Introduction and notations

(1.1) Let F be a number field, E a quadratic extension of F, M' a division algebra of center F and rank 4. We will make the following assumptions:

(1) the algebra $M'(E) = M' \otimes E$ is a division algebra;

(2) let D be the set of places v of F such that M'_v is a division algebra; then every place v in D splits in E.

We regard the multiplicative group of M' as an algebraic group G' defined over F. Thus:

$$G'(F) = M'^{\times}, \quad G'(E) = M'(E)^{\times},$$

$$G'_v = G'(F_v) \quad \text{if } v \text{ is a place of } F,$$

$$G'_v = G'(E_v) \quad \text{if } v \text{ is a place of } E.$$

Moreover $G'(F_{\mathbf{A}})$ is a closed subset of $G'(E_{\mathbf{A}})$. We will denote by Z' the center of G'. If φ is a continuous function on the quotient

$$Z'(E_{\mathbf{A}})G'(E)\backslash G'(E_{\mathbf{A}})$$

we will set

$$B'(\varphi) = \int \varphi(x)\mathrm{d}x, \quad x \in Z'(F_{\mathbf{A}})G'(F)\backslash G'(F_{\mathbf{A}})$$

and we will say that an automorphic irreducible representation π' of $Z'(E_{\mathbf{A}})\backslash G'(E_{\mathbf{A}})$ is distinguished if there is a smooth function φ in the space of π' such that $B'(\varphi) \neq 0$.

(1.2) Instead of M', we may also consider the algebra $M = M(2, F)$ of two by two matrices and regard its multiplicative group $G = \mathrm{GL}(2)$ as an algebraic group defined over F. It φ is a continuous bounded function on the quotient

$$Z(E_{\mathbf{A}})G(E)\backslash G(E_{\mathbf{A}})$$

we will set

$$B(\varphi)=\int\varphi(x)\mathrm{d}x, \quad x\in Z(F_{\mathbf{A}})G(F)\backslash G(F_{\mathbf{A}}).$$

Let π be an irreducible automorphic cuspidal representation of $Z(E_{\mathbf{A}})\backslash G(E_{\mathbf{A}})$; we will say that π is distinguished if there is a smooth function φ in the space of π such that $B(\varphi)\neq 0$.

(1.3) Our main result is the following theorem:

THEOREM: *Suppose π' is an irreducible automorphic representation of $Z'(E_{\mathbf{A}})\backslash G'(E_{\mathbf{A}})$; suppose π' is infinite dimensional and let π be the corresponding irreducible automorphic cuspidal representation of $Z(E_{\mathbf{A}})\backslash G(E_{\mathbf{A}})$. Then π' is distinguished if and only if π is.*

The motivations for this result can be found in the work of Harder, Langlands and Rapoport on algebraic cycles of certain Shimura varieties. This is not the place for a discussion of their work. Suffices to say that it concerns poles of certain Hasse-Weil zeta functions attached to a Shimura surface. These zeta functions can be computed in terms of automorphic L-functions which have been studied directly by Asai (Cf. [S.A.]) The question arises then of deciding when the L-function attached to an automorphic cuspidal representation π has a pole at $s=1$. Now the L-function has an integral representation. Indeed consider an integral

$$\int\varphi(x)E(x,s)\mathrm{d}x, \quad x\in Z(F_{\mathbf{A}})G(F)\backslash G(F_{\mathbf{A}})$$

where φ belongs to the space of π and $E(x,s)$ is an Eisenstein series. This integral has a pole at $s=1$ with residue $B(\varphi)$. On the other hand it is equal to the L-function times an elementary factor; in particular the L-function has a pole at $s=1$ if and only if, for at least one φ, the integral has a pole at $s=1$, that is if and only if B is non zero on the space of π, in other words, π is distinguished. In trying to extend the result of Harder, Langlands, Rapoport to the case of a compact Shimura surface one has to replace the group G by the group G' and use the above theorem.

Although this will not play a role in the present paper, we recall the characterization of the distinguished representations: a representation is distinguished if and only if it is the "Base Change" of an automorphic cuspidal representation of $G(F_{\mathbf{A}})$ whose central character is the quadratic idele class character of F attached to E. The proof is relatively complicated since it involves the "Base Change" and the theory of the Asai L-function. It would be interesting to see if this result could be established by using an appropriate relative trace formula, similar to the one we are using here.

(1.4). In order to prove our theorem we need a "relative trace formula": it is stated in section 2. In section 3 we derive our theorem from this trace formula. The remaining sections are devoted to the proof of our trace formula.

(1.5) Finally we thank R. Langlands for suggesting this problem to us, for his encouragement and his advice. We thank A. Selberg for his advice on classical results on L-functions. We also thank L. Clozel for making his results available to us; although the results are not used in the present paper, they were quite suggestive.

(1.6) We summarize our principal notations.

(1) The group G is the group GL(2), P is the group of upper triangular matrices, A the subgroup of diagonal matrices, N the group of unipotent matrices in P. We also denote by R the algebra of upper triangular matrices.

(2) We write $\mathrm{Mat}[p, q, r, s]$ for the matrix with rows (p, q) and (r, s). We also set:

$$w = \mathrm{Mat}[0, 1, 1, 0], \qquad n(x) = \mathrm{Mat}[1, x, 1, 0],$$

$$\mathrm{diag}(a, b) = \mathrm{Mat}[a, 0, 0, b].$$

(3) So far as the number field F is concerned we follow standard notations. In particular $F_{\mathbf{A}}$ is the ring of adeles and $F_{\mathbf{A}}^{x}$ the group of ideles. We denote by ψ a non trivial character of the quotient $F_{\mathbf{A}}/F$. We denote by $\mathrm{d}x$ the self dual Haar measure on $F_{\mathbf{A}}$ so that $\mathrm{vol}(F_{\mathbf{A}}/F) = 1$ and $\mathrm{d}(ax) = |a|_F \mathrm{d}x$ where $|a|_F$ is the module of the idele a; we drop the subscript F when this does not create confusion. We write $\mathrm{d}x = \Pi \mathrm{d}x_v$, $\psi = \Pi \psi_v$ where $\mathrm{d}x_v$ is the Haar measure on F_v, self dual with respect to ψ_v. We denote by $\mathrm{d}^\times x_v$, the Tamagawa measure on $F_v^\times$:

$$\mathrm{d}^\times x_v = L(1, 1_v)\mathrm{d}x_v/|x_v|.$$

Then we denote by $\mathrm{d}^\times x$ the Tamagawa measure on $F_{\mathbf{A}}^\times$:

$$\mathrm{d}^\times x = \Pi \mathrm{d}^\times x_v/[(s-1)L(s, 1)]\,|_{s=1}.$$

We denote by F^1 the group of ideles of norm one and by F_∞^+ the group of ideles whose finite components are one and whose infinite components are all equal to some positive number, the same for all infinite places. Then $F_{\mathbf{A}}^{x}$ is the product of F^1 and F_∞^+ and the Tamagawa measure is the product of a measure on F^1 and the measure $\mathrm{d}t/t$ where $t = |x|$. Recall that $\mathrm{vol}(F^1/F^\times)=1$. In general algebraic groups over F are provided with the Tamagawa measure.

(4) Let v be a place of F. We denote by K_v the standard maximal compact subgroup of G. Thus $K_v=\mathrm{GL}(2, R_v)$ where R_v is the ring of integers if v is finite. If $F_v=\mathbb{R}$, then $K_v=O(2,\mathbb{R})$. If $F_v=\mathbb{C}$, then $K_v=U(2,C)$. We denote by K_F the product of the K_v. The group $G(F_{\mathbb{A}})$ and $Z(F_{\mathbb{A}})$ being provided with the Tamagawa measures and the quotient $Z(F_{\mathbb{A}})\backslash B(F_{\mathbb{A}})$ with the corresponding measure, we have, for a function f on the quotient:

$$\int f(g)\mathrm{d}g=\int f[n(x)\ \mathrm{diag}(a,1)k]\mathrm{d}x|a|_F^{-1}\mathrm{d}^\times a\mathrm{d}k$$

where $\mathrm{d}k$ is a certain Haar measure on K_F. Note that $\mathrm{vol}(K_F)\neq 1$. Having chosen an invariant differential form of maximum degree ω on G we have for each place v, the measure $|\omega_v|$ and the Tamagawa measure $L(1,1_v)|\omega_v|$ on G_v. If we give tọ the quotient $Z_v\backslash G_v$ the measure quotient of the Tamagawa measures then we have for a function f on the quotient:

$$\int f(g_v)\mathrm{d}g_v=\int f[n(x_v)\ \mathrm{diag}(a_v,1)k]\mathrm{d}x_v|a_v|^{-1}\mathrm{d}^\times a_v\mathrm{d}k_v$$

where $\mathrm{d}k_v$ is a certain Haar measure on K_v. Again $\mathrm{vol}(K_v)\neq 1$.

(5) We have also the quadratic extension E of F, with Galois group $\{1,\sigma\}$. Whenever convenient we write $E=F[\sqrt{\tau}]$. We denote by ψ_E non trivial character of $E_{\mathbb{A}}/E$. Usually we assume that is trivial on $F_{\mathbb{A}}$. If u is a place of E we denote by K_u the standard maximal compact subgroup of G_u, by K_E or simply K the product of the K_u.

§2. A relative trace formula

(2.1) We will denote by S a finite set of places of F containing D and all infinite places; the set S will be enlarged as need dictates. For each set T of places of F we will denote by $T^\sim$ the set of places of E which are above a place of T. For each place v of F not in D we choose an isomorphism $M'_v\cong M(2,F_v)$; let α be a basis of M' over F and α_v the R_v-module generated by α in M'_v; we may assume that, for all places v of F not in S, our isomorphism takes α_v to $M(2,R_v)$. Extending the scalars, we have, for each place v of E not in $D^\sim$, an isomorphism $M'(E_v)\cong M(E_v)$. From these, we get for each place v of E (resp. F) not in D (resp. $D^\sim$) an isomorphism $G'_v\cong G_v$; we use it to identify the two groups. For each place v of E (resp. F) we denote by K_v the standard maximal compact subgroup of G_v. We choose, in the usual way, two invariant F differential forms of maximal degree ω and ω' on G and G'; over an algebraic closure of F they are the same. Thus, for each place v

of F the groups G_v and G'_v come equipped with forms ω_v and ω'_v; if v is not in D we may assume the isomorphism of G'_v onto G_v takes $|\omega'_v|$. Similar remarks apply to E.

We denote by f a smooth function of compact support on $Z(E_{\mathbf{A}})\backslash G(E_{\mathbf{A}})$; we assume that f is a product of local components f_v. For all v the local component f_v is smooth, of compact support, bi-K_v-finite; for almost all finite v, it is the characteristic function of $Z_v K_v$. Similarly, we consider a smooth function of compact support f' on $Z'(E_{\mathbf{A}})\backslash G'(E_{\mathbf{A}})$. It is also assume to be a product of local factors f'_v. We make the following assumptions on f and f':

(1) for each v not in $D^{\sim}$, $f_v = f'_v$;

(2) let v be a place in D and $v1$, $v2$ the two places of E above v; we have isomorphisms

$$G'_{v1} \cong G'_{v2} \cong G'_v, \quad G_{v1} \cong G_{v2} \cong G_v;$$

we demand that the convolution products

$$h'_v = f'^{v}_{v1} * f'_{v2}, \quad h_v = f^{v}_{v1} * f_{v2}$$

on G'_v and G_v respectively have the same regular orbital integrals.

In other words, we demand that the hyperbolic orbital integrals of h_v vanish; on the other hand, if T'_v and T_v are isomorphic F_v-subalgebras of rank 2 of M'_v and M_v respectively and if $t' \in T'_v - F_v$ corresponds to $t \in T_v - F_v$, then we demand that

$$\int h(g^{-1}xg)\,dg = \int h'(g'^{-1}x'g')\,dg', \qquad g \in T_v^{\times}\backslash G_v, \quad g' \in T'^{\times}_v \backslash G'_v.$$

In this formula the Haar measures on G_v and G'_v are the Tamagawa measures attached to ω_v and ω'_v respectively; on the other hand the measures on $T_v^{\times}$ and $T'^{\times}_v$ are the Tamagawa measures attached to differential forms η_v and η'_v which correspond to one another under the isomorphism $T_v \cong T'_v$. Finally the quotients are given the quotient measures.

(2.2) The operator $\rho(f)$ defined by f on the space

$$L^2(Z(E_{\mathbf{A}})G(E)\backslash G(E_{\mathbf{A}}))$$

is defined by a kernel K. We call P the orthogonal projection on the space of cuspidal elements. Then $P\rho(f)P$ is represented by a kernel K_{cusp}; it is a smooth bounded function. Similarly the operator $\rho(f')$ defined by f' on the space

$$L^2(Z'(E_{\mathbf{A}})G'(E)\backslash G'(E_{\mathbf{A}}))$$

is represented by a kernel K'; we call P' the orthogonal projection on the orthogonal complement of the space spanned by the functions of the form $\chi \circ \det$, where χ is a quadratic character of $E^\times \backslash E^\times_{\mathbf{A}}$. Then $P'\rho(f')P'$ is represented by a kernel K'_{cusp}. It is also smooth and bounded.

(2.3) PROPOSITION: *Suppose that the above assumptions, in particular* (2.1.1) *and* (2.1.2), *are satisfied. Then*:

$$\iint K_{\text{cusp}}(x, y)\mathrm{d}x\mathrm{d}y = \iint K'_{\text{cusp}}(x', y')\mathrm{d}x'\mathrm{d}y',$$

$$x, y \in G(F)Z(F_{\mathbf{A}})\backslash G(F_{\mathbf{A}}), \quad x', y' \in G'(F)Z'(F_{\mathbf{A}})\backslash G'(F_{\mathbf{A}}).$$

In this formula $G(F_{\mathbf{A}})$, $G'(F_{\mathbf{A}})$, $Z'(F_{\mathbf{A}})$ *and* $Z(F_{\mathbf{A}})$ *are given the Tamagawa measures.*

The proof will occupy section 4 and the following sections.

§3. Demonstration of the theorem

(3.1) We will assume Proposition (2.2) and derive Theorem (1.3). Let therefore π' and π be as in (1.3). We will prove that if π is distinguished then π' is distinguished. The proof of the converse assertion is similar and left to the reader. So from now on we assume π is distinguished. We first recall a remark of Langlands. Suppose v is a place of F which splits into $v1$ and $v2$ in the extension E; then we have isomorphisms

$$G_{v1} \cong G_{v2} \cong G_v.$$

The restriction of B to the space of π is a non zero $G(F_{\mathbf{A}})$-invariant form; imbedding the space of $\pi_{v1} \otimes \pi_{v2}$ into the space of π appropriately, we find a non-zero linear form which is invariant under the group $G_{v1} \times G_{v2}$. Therefore π_{v1} and π_{v2} are contragredient to one another. Since they are trivial on the center, we see that the following condition is satisfied:

(1) if v splits into $v1$ and $v2$, then $\pi_{v1} \cong \pi_{v2}$.

Of course the representation π' satisfies the analoguous condition. We set $\pi_v = \pi_{v1} = \pi_{v2}$ and $\pi'_v = \pi'_{v1} = \pi'_{v2}$.

(3.2) Recall that S is a finite set of places of F containing D and the infinite places. Let us set

$$G^{S^\sim} = \Pi G_v (\text{restricted product}),$$

$$K^{S^\sim} = \Pi K_v, \quad v \notin S^\sim;$$

$$G_{S^\sim} = \Pi G_v, \quad K_{S^\sim} = \Pi K_v, \quad v \in S^\sim;$$

$$f^{S^\sim} = \Pi f_v, \quad v \notin S^\sim, \qquad f_{S^\sim} = \Pi f_v, \quad v \in S^\sim.$$

We may choose S so large that Π contains the unit representation of $K^{S^{\sim}}$. We than take f_v for $v \notin S^{\sim}$ to be bi-invariant under K_v. In general if σ is an irreducible representation of $Z(E_{\mathbf{A}})\backslash G(E_{\mathbf{A}})$ containing the unit representation of $K^{S^{\sim}}$, we will denote by e_σ the corresponding character of the Hecke algebra of the group $G^{S^{\sim}}$. In particular if σ is an automorphic cuspidal representation we will denote by $V(\sigma)$ the space of forms in the space of σ which are invariant under $K^{S^{\sim}}$. The orthogonal projection onto the space $V(\sigma)$ will be noted P_σ. The space $V(\sigma)$ is invariant under $G_{S^{\sim}}$ and the corresponding representation will be noted $\pi_{S^{\sim}}$. Thus for φ in $V(\sigma)$ we have:

$$\sigma(f)\varphi = e_\sigma(f^{S^{\sim}}) \cdot \pi_{S^{\sim}}(f_{S^{\sim}})\varphi. \tag{1}$$

Accordingly we may write K_{cusp} as the sum of the following uniformly convergent series:

$$K_{\text{cusp}}(x, y) = \sum_\sigma e_\sigma(f^{S^{\sim}}) \cdot K_\sigma(x, y) \tag{2}$$

where K_σ is the kernel attached to the operator $P_\sigma \pi(f_{S^{\sim}}) P_\sigma$ and the sum is over all cuspidal representation σ containing the unit representation of $K^{S^{\sim}}$. In turn, if φ_j is an orthonormal basis of $V(\sigma)$, then K_σ is equal to the finite sum

$$K_\sigma(x, y) = \sum_j \pi_{S^{\sim}}(f_{S^{\sim}})\varphi_j(x) \cdot \bar{\varphi}_j(y).$$

Therefore

$$\iint K_{\text{cusp}}(x, y)\mathrm{d}x\mathrm{d}y = \sum_\sigma e_\sigma(f^{S^{\sim}}) a_\sigma, \text{ with } a_\sigma = \iint K_\sigma(x, y)\mathrm{d}x\mathrm{d}y. \tag{3}$$

We will choose now $f_{S^{\sim}}$ in such a way that a_π is non zero and there exists a function f' satisfying the conditions (2.1.1) and (2.1.2). To that end let us observe that $\pi_{S^{\sim}}$ is equivalent to the tensor product representation

$$\otimes \pi_v, \quad v \in S^{\sim}.$$

We choose therefore an isomorphism

$$\pi_{S^{\sim}} \cong \pi_{T^{\sim}} \otimes \pi_{D^{\sim}}, \quad V(\pi) = V_{T^{\sim}} \otimes V_{D^{\sim}}, \qquad \text{where} \quad S = T \cup D;$$

we have set

$$\pi_{T^{\sim}} = \otimes \pi_v, \quad V_{T^{\sim}} = \otimes V_v, \quad v \in T^{\sim};$$

$$\pi_{D^{\sim}} = \otimes \pi_v, \quad v_{D^{\sim}} = \otimes V_v, \quad v \in D^{\sim},$$

and V_v is the space of π_v. At the cost of enlarging S we may assume there is a unitary vector u in the space V_{T^-} such that the restriction of B to $u \otimes V_{T^-}$ is non zero. We choose f_{T^-} in such a way that $\pi_{T^-}(f_{T^-})$ is the orthogonal projection on u. Then if v_j is any basis of V_{D^-} we have:

$$\iint K_\pi(x, y)\mathrm{d}x\mathrm{d}y = B\left[u \otimes \pi_{D^-}(f_{D^-})v_j\right] \cdot \overline{B}\left[u \otimes v_j\right]. \tag{4}$$

For each v in D let $v1$ and $v2$ be the two places of E above v. Let C_v be an invariant linear form on the tensor product $V_{v1} \otimes V_{v2}$. Then V_{D^-} is the tensor product of the spaces $V_{v1} \otimes V_{v2}$ and the restriction of B to $u \otimes V_{D^-}$ is, up to a constant factor, the tensor product of the C_v. For each v in D let us choose bases of V_{v1} and V_{v2} dual to one another, a_i and b_j say. Then the above expression is, up to a constant factor, the product over all v in D of the sums:

$$\sum_{i,j} C_v\left[\pi_{v1}(f_{v1})a_i \otimes \pi_{v2}(f_{v2})b_j\right] \cdot \overline{C}_v\left[a_i \otimes b_j\right]. \tag{5}$$

Since $C_v[a_i \otimes b_j] = \delta_{ij}$ this sum reduces to

$$\begin{aligned}
&\sum_i C_v\left[\pi_{v1}(f_{v1})a_i \otimes \pi_{v2}(f_{v2})b_i\right] \\
&= \sum C_v\left[\pi_{v1}(f_{v2}^v)\pi_{v1}(f_{v1})a_i \otimes b_i\right] \\
&= \sum C_v\left[\pi_{v1}(h_v)a_i \otimes b_i\right] \\
&= \operatorname{tr} \pi_v(h_v),
\end{aligned} \tag{6}$$

where

$$h_v = f_{v2}^v * f_{v1}.$$

(3.3) At this point we need a lemma:

LEMMA: *We can choose* f_{v1} *and* f_{v2} *in such a way that* $\operatorname{tr} \pi_v(h_v) = 1$ *and* $\operatorname{tr} \sigma(h_v) = 0$ *for every infinite dimensional irreducible representation* σ *of* G_v *which is not equivalent to* π_v.

PROOF OF LEMMA: In any case π_v is in the discrete series. Thus if v is finite there is a function h_v with the required traces and it is trivial that it is a convolution product. So we are done in this case. If v is infinite then v is real. So we may identify E to $\mathbb{R}$ and G_v to $\mathrm{GL}(2, \mathbb{R})$. Let K_0 be the group $\mathrm{SO}(2, R)$ of K_v and, for each $n \in Z$, χ_n the character

$$k(\theta) = \begin{bmatrix} \cos(\theta) & \sin(\theta) \\ -\sin(\theta) & \cos(\theta) \end{bmatrix} \mapsto \exp(in\theta).$$

Say that π_v is the representation of the discrete series of highest weight $n > 0$. Since π_v is trivial on the center n is necessarily even. Choose smooth functions of compact support a and b on $Z_v \backslash G_v$. We will take then with support in the set of matrices with positive determinant so that we may think of them as being functions on SL(2, R). The function b will be taken bi-invariant under K_0. As for a it will be assumed to transform under $\bar{\chi}_n$ on both sides:

$$a[k(\theta_1)gk(\theta_2)] = \exp[-in(\theta_1+\theta_2)]a(g).$$

We will take h_v to be $(a-b)*(a+b)$. Then if σ is any irreducible representation of GL(2, $\mathbb{R}$) we have:

$$\operatorname{tr}\sigma(h_v) = \operatorname{tr}\sigma(a*a) - \operatorname{tr}\sigma(b*b) = \operatorname{tr}\sigma(a)^2 - \operatorname{tr}\sigma(b)^2.$$

If furthemore σ is in the discrete series then the second term is zero; the first term is zero as well unless the highest weight m of σ is less than or equal to n, that is unless σ belongs to a certain finite set. The linear independence of characters implies then that we may choose a such that $\operatorname{tr}\pi_v(a) = 1$ and $\operatorname{tr}\sigma(a) = 0$ if σ is another representation of the discrete series. Thus all we have to show is that we can choose b in such a way that

$$\operatorname{tr}\sigma(b) = \operatorname{tr}\sigma(a)$$

if σ is a representation of the principal series trivial on the center. Unraveling this we see that this relation is equivalent to $F_a = F_b$, where we have set, for $x > 0$,

$$F_a(x) = \iint a(k\,\mathrm{Mat}[x, u, 0, x^{-1}]k^{-1})\mathrm{d}k\mathrm{d}u$$

and F_b is defined similarly. Since F_a is smooth of compact support on $\mathbb{R}_+^+$ and invariant under the substitution $x \mapsto x^{-1}$, there is a function b with the required properly ("Paley Wiener Theorem for SL(2, R)", cf. for instance [S.L.], Th. 3, V, \$2, p. 71). This concludes the proof of the lemma.

(3.4). Coming back to the situation of (3.2) and choosing h_v as in the lemma we have

$$a_\pi = C\Pi \operatorname{tr}\pi_v(h_v), \quad v \in D, \qquad \text{where } C \text{ is a constant.} \tag{1}$$

Hence $a_\pi = C \neq 0$ as desired.

Now we choose a function f' on $G'(E_{\mathbf{A}})$ so that (2.1.1) and (2.1.2) are satisfied: so for a place v of E not in $D^{\sim}$ we take $f'_v = f_v$. For v in D, the Schur orthogonality relations show there is a convolution product h'_v on G'_v so that tr $\pi'_v(h'_v) = -1$ and tr $\sigma^2(h) = 0$ if σ' is not equivalent to π'_v. Since tr $\pi_v(h_v) = 1$ and tr $\sigma(h_v) = 0$ if σ is not equivalent to π_v we see that h_v and h'_v have the same orbital integrals. We therefore take f'_{v1} and f'_{v2} so that

$$h' = f'^{v}_{v1} * f'_{v2} \tag{2}$$

and we can apply Prop. (2.2). The kernel K'_{cusp} has a decomposition analoguous to (3.2.2). In particular we find

$$\iint K'_{\text{cusp}}(x', y')\mathrm{d}x'\mathrm{d}y' = \sum e_{\sigma'}(f^{S^{\sim}})a_{\sigma'}, \tag{3}$$

where

$$a_{\sigma'} = \iint K'_{\sigma'}(x', y')\mathrm{d}x'\mathrm{d}y'.$$

By Prop. (2.2) expressions (3.2.3) and (3) are equal. We may regard them as infinite linear combinations of characters of the Hecke algebra of $G^{S^{\sim}}$. Since the principle of linear independence of characters apply to these infinite sums ([R.L.] $11) and $e_{\pi} = e_{\pi'}$ we see that $a_{\pi'}$ is non zero. Since $K'_{\pi'}(x, y)$ is a sum of terms of the form $\varphi(x)\bar{\varphi}'(y)$ where φ and φ' belong to the space of π' we see that the restriction of B' to the space of π' is non zero. This concludes the proof of Theorem (1.3).

§4. Double cosets

(4.1) In order to prove our trace formula we need to study the double cosets of $G'(F)Z'(E)$ in $G'(E)$ and the double cosets of $G(F)Z(E)$ in $G(E)$. An element of $G'(F)Z'(E)$ or $G(F)Z(E)$ will be termed singular. An element not in this set will be termed regular.

We start with G. We let Q be a set of representatives for the $G(F)$ conjugacy classes in the set of subfields of $M(2, F)$ of rank 2 over F. We let R be the subalgebra of $M(2, F)$ formed of the upper triangular matrices and P the subgroup of $G(F)$ formed of the upper triangular matrices.

(1) Lemma: *If T is in Q then any singular element of $(T \otimes E)^{\times}$ is in $T^{x} \cdot E^{\times}$. Similarly any singular element of $P(E)$ is in $P(F)Z(E)$.*

Proof of (1): Suppose x is singular in $(T \otimes E)^{\times}$. Then there is c in $E^{\times} = Z(E)$ such that $x = cy$ with y in $G(F)$. Then y is in the intersection of the algebras $T \otimes E$ and $M(2, F)$, that is in T. Since it is invertible, it is in $T^\times$ and we are done. The proof of the second assertion is similar, using the algebra R instead of the algebra T.

Recall that the Galois group of E over F is $\{1, \sigma\}$.

(2) LEMMA: *Suppose x is in $G(E)$. Then x is regular (in our sense) if and only if $x^\sigma x^{-1}$ is regular in the usual sense.*

PROOF OF (2): Suppose x is singular in our sense. Then $x = gc$ with g in $G(F)$ and c in E^x. Then $x^\sigma x^{-1} = c^{\sigma-1}$ is singular. Conversely, suppose $x^\sigma x^{-1} = c$ where c is in $E^\times$. Then $c^\sigma c = 1$ so $c = z^{\sigma-1}$ with z in E^x. This gives $(xz^{-1})^\sigma = xz^{-1}$. Hence xz^{-1} is in $G(F)$ and x is singular in our sense.

(3) LEMMA: *If x is in $G(E)$ then either x is singular, or x is in a double coset $G(F)pG(F)$ with p in $P(E) - P(F)$, or x is in a double coset $G(F)tG(F)$, where t is in $(T \otimes E)^\times - T^\times \cdot E^\times$ and T is in Q. Moreover the three possibilities are exclusive of one another.*

PROOF OF (3): Let us prove the first assertion. Let E be the extension $F[\sqrt{\tau}]$. Then any x in $M(2, E)$ can be written uniquely in the form:

$$x = a + b\sqrt{\tau}, \quad \text{with } a \text{ and } b \text{ in } M(2, F).$$

Suppose x is invertible and not singular; recall that in the present context this means that x is not in $Z(E)G(F)$. Then a is not zero, otherwise x would be in $G(F)E^\times$, hence would be singular. Suppose a is non zero but non invertible. We claim there is an element c of $E^\times$ such that $x' = xc$ has the form

$$x' = a' + b'\sqrt{2} \quad \text{with } a' \text{ in } G(F) \text{ and } b' \text{ in } M(2, F).$$

Indeed a has the following form:

$$a = g\,\mathrm{diag}(u, 0)g^{-1}, \quad \text{with } g \text{ in } G(F) \text{ and } u \text{ in } F^\times.$$

Let us write also:

$$b = g\,\mathrm{Mat}[p, q, r, s]g^{-1}.$$

We claim that for some appropriate z in F the matrix $za + b$ has a non zero determinant. Suppose not. Then

$$(zu + p)s - rq = 0$$

for all z in F. This implies $s=0$ and then $rq=0$. Say $r=0$. Then the matrix x has the form

$$g\,\mathrm{Mat}\left[1+p\sqrt{\tau}\,,\,q\sqrt{\tau}\,,0,0\right]g^{-1}$$

and is thus not invertible, a contradiction. Let therefore z be an element of F such that $za+b$ is invertible. Set $c=z+1/\sqrt{\tau}$. Then c is in $E^{\times}$ and xc has the form $a'+b'\sqrt{\tau}$ where a' is invertible. Thus we may assume that a is invertible. Replacing x by xa^{-1} we may even assume that a is 1. If b is not invertible or if it is invertible but not elliptic (in the usual sense) in $G(F)$, there is a g in $G(F)$ such that $b=gpg^{-1}$ where p is in $R(F)$. Then:

$$x=gqg^{-1} \quad \text{with } q=1+p\sqrt{\tau}\,.$$

Since q is in $R(E)$ we are done in this case. If b is an elliptic element of $G(F)$ then there is a T in Q and a t in $T^{\times}$ such that $b=gtg^{-1}$. Then:

$$x=gyg^{-1} \quad \text{with } y=1+t\sqrt{\tau}\,.$$

Since y is in $T\otimes E$ we are again done.

It remains to prove the second assertion of the lemma (3). Because of lemma (1), all we have to prove is that for T in Q an element x of $(T\otimes E)^{\times}-T^{\times}\cdot E^{\times}$ cannot be of the form g_1pg_2, with p in $P(E)$ and g_1 and g_2 in $G(F)$. Suppose it is. After a simple computation we find:

$$x^{\sigma}x^{-1}=g_1qg_1^{-1}, \quad \text{with } q=p^{\sigma}p^{-1}.$$

Suppose the F algebra T is not isomorphic to E. Then $T\otimes E$ is a field, in fact a quadratic extension of E and $x^{\sigma}x^{-1}$ is an element of that field which is not an elliptic element of $G(E)$ (in the usual sense). Thus $x^{\sigma}x^{-1}$ is singular in the usual sense and by lemma (4.1.2) x is singular in our sense, a contradiction. Suppose now that T is isomorphic to E; in passing note that there is exactly one element of Q with that property. Then $T\otimes E$ is not a field; it is a sum of two copies of E. In particular the minimal polynomial of x over F has degree one or two and the same is true of q. This is possible only if the two eigenvalues of g are in F. Since they are elements of E whose F norm is 1, they are $+1$ or -1. In any case:

$$(a-1)(a+1)=0, \quad \text{where } a=x^{\sigma}x^{-1}.$$

This gives $a^2=1$ and $a^{\sigma}a=1$. SInce x is regular $a\neq\pm1$ (lemma (4.1.2)). Let us identify $T\otimes E$ with the sum of two copies of E; then

$$(x,\,y)^{\sigma}=(y,\,x).$$

From $a^2 = 1$ and $a \neq \pm 1$ we get $a = (1, -1)$ or $(-1, 1)$. Then $a^\sigma a = -1$ a contradiction.

We will say that an element of $G(E)$ is elliptic if it is not in the set $G(F)P(E)G(F)$.

(4.2) We will now classify all elliptic elements.

(1) LEMMA: *Suppose T and T' are in Q. Suppose that x and x' are regular elements of $T \otimes E$ and $T' \otimes E$ respectively but have the same double class modulo $Z(E)G(F)$. Then $T = T'$.*

PROOF OF (1): Since E is contained in $(T \otimes E)^\times$ and $(T \otimes E)^\times$ we may as well assume that x and x' have the same double class modulo $G(F)$:

$$x = cx'c' \quad \text{with } c \text{ and } c' \text{ in } G(F).$$

Then $x^{\sigma-1} = cx'^{\sigma-1}c^{-1}$. Morever $x^{\sigma-1}$ is a regular element of $(T \otimes E)^\times$ in the usual sense (lemma (4.1.2)). Hence $c(T \otimes E)c^{-1} = T' \otimes E$. Taking the intersections with $M(2, F)$ we obtain $cTc^{-1} = T'$, hence $T = T'$.

(2) LEMMA: *Suppose that c and c' are in $G(F)$, T is in Q, x and x' are regular elements of $(T \otimes E)^\times$ and z is in $E^\times$. Suppose that*

$$x = cx'c'z.$$

Then c' and c normalize T and cc' is in T.

PROOF OF (2): Again we have $x^\sigma x^{-1} = cx'^\sigma x'^{-1} z^{\sigma-1} c^{-1}$ and $x^{\sigma-1}$ is a regular element in the usual sense. It follows that c normalizes $T \otimes E$, hence also T. Finally $cc' = cx'^{-1}c^{-1}z^{-1}x$ is in $T \otimes E$, hence in T.

(4.3) We now deal with the remaining elements of $G(E)$.

(1) LEMMA: *Every element x of $P(E)$ is in the same double class modulo $G(F)Z(E)$ as one of the following element:*

$$\text{(i) } e, \quad \text{(ii) } n(\sqrt{\tau}), \quad \text{(iii) } \operatorname{diag}(a, 1), \quad a \in E^\times - F^\times.$$

PROOF OF (1): Let us write x in the form

$$x = \operatorname{Mat}[a, b, 0, c].$$

If a/c is in F then after multiplying by an element of $Z(E)A(F)$ we may assume that $a = c = 1$. If b is in F then x is in $N(F)$ and we are in case (i). If b is not in F then after multiplying by a matrix in $N(F)$ we may assume that $b = u\sqrt{\tau}$ with u in F. Then

$$\mathrm{diag}(u, 1) \times \mathrm{diag}(u, 1)^{-1}$$

is the matrix (ii). If u/c is not in F then, after multiplying by an element of $Z(E)$, we may assume $c = 1$. Since a is not in F we may write b in the form

$$b = u + av \quad \text{with } u \text{ and } v \text{ in } F.$$

Then the matrix

$$n(-u) \times n(-v)$$

has the form (iii). This concludes the proof of the lemma.

(2) Lemma: *In lemma* (1) *cases* (i), (ii), (iii) *are exclusive of one another.*

Proof of Lemma (2): In view of lemma (4.1.1), it suffices to show that case (ii) and (iii) are not compatible. Suppose they were. Then there would be g and g in $G(F)$ and a matrix d in $A(E)$, where A is the group of diagonal matrices, such that:

$$n(\sqrt{\tau}) = g_1 d g_2$$

Then we would find:

$$n(2\sqrt{\tau}) = n(\sqrt{\tau})^{\sigma - 1} = g_1 d^{\sigma - 1} g_2.$$

Since $n(2\sqrt{\tau})$ is unipotent and $d^{\sigma - J}$ is a diagonal matrix this is a contradiction.

(3) Lemma: *Suppose* c *and* c' *are in* $G(F)$, x *and* x' *are regular elements of* $A(E)$ *and* z *is in* E. *Suppose that*

$$x = cx'c'z.$$

Then c *and* c' *normalize* A *and* cc' *is in* $A(F)$.

The proof is the same as the proof of lemma (4.2.2).

(4) Lemma: *Let* $n = n(\sqrt{\tau})$. *Suppose* g *and* g' *are in* $G(F)$, z *in* E. *Suppose that*

$$gng' = zn.$$

Then z *is in* F, $gg' = z$ *and* g *and* g' *are in* $N(F)Z(E)$, *where* N *denotes the group of triangular matrices with unit diagonal.*

Proof of (4): Once more we find $gn^{\sigma-1}g^{-1} = z^{\sigma-1}n^{\sigma-1}$. Since $n^{\sigma-1}$ is a regular unipotent element, this implies that $z^{\sigma-1} = 1$ and g centralizes $n^{\sigma-1}$; in turn this implies that g is in $Z(E)N(E)$ hence in fact in $Z(F)N(F)$ and z is in F. Similarly g' is in $Z(F)N(F)$. Since $Z(E)N(E)$ is a commutative group $gg' = z$ and we are done.

(4.4) Altough not necessary for the description of the double classes the two following lemmas will be useful latter on.

(1) Lemma: *Suppose ξ is in $G(E)$ and $\xi^{\sigma}\xi^{-1}$ is in $P(E)$. Then ξ is in $G(F)P(E)$.*

Proof of (1): Set $p = \xi^{\sigma}\xi^{-1}$. By assumption p is in $P(E)$. On the other hand $p^{\sigma}p = 1$. We claim there is a q in $P(E)$ such that $p = q^{\sigma}q^{1}$. This will prove the lemma because then ξq^{-1} is invariant under σ hence in $G(F)$. To prove our claim we write

$$p = \mathrm{Mat}[a, x, b, 0], \; q = \mathrm{Mat}[u, y, v, 0].$$

The condition that $p^{\sigma}p = 1$ gives:

$$\text{(i) } a^{\sigma}a = 1, \quad \text{(ii) } b^{\sigma}b = 1, \quad \text{(iii) } zx^{\sigma} + x = 0 \quad \text{where} \quad z = a/b. \tag{2}$$

The condition $q^{\sigma}q^{-1} = p$ gives:

$$\text{(i) } a = u^{\sigma}u^{-1}, \quad \text{(ii) } b = v^{\sigma}v^{-1}, \quad \text{(iii) } x = -y^{\sigma}z + y. \tag{3}$$

Because of (2.i) and (2.ii) it is possible to solve (3.i) and (3.ii). Next regard (3.iii) as a system of 2 linear equations for r and s where $z = r + s\sqrt{\tau}$. The condition $z^{\sigma}z = 1$ means that the determinant of the system is zero and the condition (3.iii) is then the compatibility condition for the two equations. Our claim follows.

Say that an element of $G(E)$ is semi-simple regular if it is elliptic or in the double class modulo $G(F)Z(E)$ of an element diag$(a, 1)$ with $a \notin F$. Say it is unipotent regular if it is in the double class of an element of the form $n(y)$ with y not in F.

(4) Lemma: *An element x of $G(E)$ is singular (resp. semi-simple regular, resp. unipotent regular) if and only if $x^{\sigma-1}$ is singular (resp. semisimple, resp. unipotent regular) in the usual sense.*

Remark: In this lemma an element of the form $zn(y)$ with z in Z is regarded as unipotent.

PROOF OF (4): We may already assume x regular (lemma ((4.1.2)). Then we know that $x = gyg'$ where g and g' are in $G(F)$ and y is either in $(T \otimes E)^\times - T^\times E^\times$ or of the form diag(a, b) with $a/b \notin F$, or of the form $zn(u)$ with $u \notin F$ and z in $Z(E)$. Then $x^\sigma x^{-1} = gy^\sigma y^{-1} g^{-1}$. In the first case $y^\sigma y^{-1}$ is an element of $T \otimes E$ hence a semi-simple one. In the second case $y^\sigma y^{-1}$ is hyperbolic and in the last case it is unipotent. In any case it is regular (loc. cit.).

(4.5) The previous results apply mutatis mutandis to G'. Every element is either singular or elliptic. The set Q is replaced by a set Q' of conjugacy classes for the subalgebras of M' of rank 2. Of course, any element of Q' is isomorphic to exactly one element of Q.

§5. Terms attached to the elliptic and singular elements

(5.1) To prove our trace formula we now write down K_{cusp} more explicitly:

$$K_{\text{cusp}} = K - K_{\text{eis}} - K_{\text{sp}} \tag{1}$$

where:

$$K(x, y) = \sum f(x^{-1}\xi y),\ \xi \in Z(E)\backslash G(E). \tag{2}$$

A formula for the two other kernels will be recalled below. In turn we write (2) as the sum of three more kernels:

$$K = K_{\text{s}} + K_{\text{e}} + K_{\text{r}} \quad \text{where:}$$

$$K_{\text{s}}(x, y) = \sum f(x^{-1}\xi y), \quad \xi \text{ singular;} \tag{4}$$

$$K_{\text{e}}(x, y) = \sum f(x^{-1}\xi y), \quad \xi \text{ elliptic;} \tag{5}$$

$$K_{\text{r}}(x, y) = \sum f(x^{-1}\xi y), \quad \xi \text{ regular, but not elliptic.} \tag{6}$$

Similarly, we write K'_{cusp} as

$$K'_{\text{cusp}} = K' - K'_{\text{sp}} \tag{7}$$

where

$$K'(x, y) = \sum f'(x^{-1}\xi y), \quad \xi \in Z(E)\backslash G(E). \tag{8}$$

We write (8) as the sum of two other kernels:

$$K' = K'_{\text{s}} + K'_{\text{e}}. \tag{9}$$

where

$$K'_{\mathrm{s}}(x, y) = \sum f'(x^{-1}\xi y), \quad \xi \text{ singular}; \tag{9}$$

$$K'_{\mathrm{e}}(x, y) = \sum f'(x^{-1}\xi y), \quad \xi \text{ elliptic}; \tag{10}$$

In this section we will prove that, under the assumptions of Proposition (2.2), the kernels K_{e} and K'_{e}, K_{s} and K'_{s}, K_{sp} and K'_{sp} respectively, have the same integrals.

(5.2) We first deal with K_{e} and prove that K_{e} is indeed integrable over the product of the quotient $Z(F_{\mathbf{A}})G(F)\backslash G(F_{\mathbf{A}})$ by itself. To that end we first prove a lemma. We let K_E be the standard maximal compact subgroup of $g(E_{\mathbf{A}})$ and for g in $G(E_{\mathbf{A}})$ define the height $H(g)$ of g to be:

$$H(g) = |a/b|_E,$$

$$\text{if} \quad g = ndk,\ n \in N(E_{\mathbf{A}}),\ d = \operatorname{diag}(a, b),\ k \in K_E. \tag{1}$$

We first recall without proof a standard lemma (Cf. [J.A.]):

(2) Lemma: *Let Ω be a compact subset of $G(E_{\mathbf{A}})$. Then there is a number $d > 0$ such that the relations*:

$$h'^{-1}\xi h \in \Omega Z(E_{\mathbf{A}}), \quad \xi \in G(E), \quad H(h) > d, \quad H(h') > d$$

imply ξ is in $P(E)$.

We derive from this lemma the following result:

(3) Lemma: *Suppose Ω is a compact subset of $G(E)$. Then there is a number $d > 0$ such that the relations*

$$x^{-1}\xi h \in \Omega Z(E_{\mathbf{A}}), \quad \xi \in G(E), \quad x \in G(F_{\mathbf{A}}), \quad H(h) > d$$

imply that ξ is in $G(F)P(E)$.

Proof of (3): We have also $h^{-\sigma}\xi^{-\sigma} \in \Omega^{-\sigma}$. After multiplication we get $h^{-\sigma}\xi^{-\sigma}\xi h \in \Omega^{-\sigma}\Omega$. Applying the previous lemma we see that if d is sufficient large and $H(h) > d$ then $\xi^{-\sigma}\xi \in P(E)$. The conclusion follows from lemma (4.4.1).

Recall that

$$K_{\mathrm{e}}(x, y) = \sum f(x^{-1}\xi y), \quad \xi \text{ elliptic}.$$

If x and y are in $G(F_{\mathbf{A}})$ then the previous lemma shows that the relation $f(x^{-1}\xi y)=0$ implies first that x and y are in sets compact modulo $G(F)Z(E_{\mathbf{A}})$; in turn this implies that ξ stays in a set finite modulo $E^{\times}$. Hence K_e is indeed integrable. Furthemore, by taking f positive in what follows we can justify our formal manipulations.

(5.3) The results of the sub-section (4.2) allow us to write:

$$K_e(x, y)=\sum f(x^{-1}\gamma\xi\delta y), \tag{1}$$

$$T\in Q, \quad \gamma\in G(F)/T^{\times}, \quad \delta\in N(T)\backslash G(F),$$

$$\xi \text{ regular in } (T\otimes E)^{\times}/E^{\times};$$

we have noted $N(T)$ the normalizer of T in $G(F)$. Since T has index 2 in $N(T)$ this can also be written as:

$$K(x, y)=1/2\sum f(x^{-1}\gamma\xi\delta y), \tag{2}$$

$$T\in Q, \gamma\in G(F)/Z(F), \quad \delta\in T^{\times}\backslash G(F),$$

$$\xi \text{ regular in } T^{\times}\backslash(T\otimes E)^{\times}/E^{\times}.$$

If we integrate formally over the product of the quotient $G(F)Z(E_{\mathbf{A}})\backslash G(F_{\mathbf{A}})$ by itself we find:

$$\iint K_e(x, y)\mathrm{d}x\mathrm{d}y=1/2\sum v_T\iint f(x\xi y)\mathrm{d}x\mathrm{d}y; \tag{3}$$

the sum is over all T in Q and then ξ regular in $T^{\times}\backslash(T\otimes E)^{\times}/E^{\times}$; the integral in x is over $G(F_{\mathbf{A}})/Z(F_{\mathbf{A}})$ and the integral in y over the quotient $T_{\mathbf{A}}^{\times}\backslash G(F_{\mathbf{A}})$; finally v_T is the volume of the quotient $F_{\mathbf{A}}^{\times}T^{\times}\backslash T_{\mathbf{A}}^{\times}$. In turn, each integral in (3), apart from a constant factor, can be written as a product over all places v of F of local ones.

If v is a place of F which does not splits in E and u is the corresponding place of E then E is a quadratic extension of F and G is a subgroup of G. The corresponding local factor is nothing but:

$$\iint f_u(x\xi y)\mathrm{d}x\mathrm{d}y, \quad x\in G_v/Z_v, \quad y\in T_v^{x}\backslash G_v. \tag{4}$$

The convergence of this integral can be established as follows: if the integrand is not zero then $x\xi y$ belongs to a set $Z_u\Omega$ where Ω is compact. Then $y^{-1}\xi^{1-\sigma}y$ is in a compact set. Since $\xi^{1-\sigma}$ is semi-simple regular (lemma (4.4.4)) y must be in a set compact $\mathrm{mod}(T_v\otimes E_v)^{\times}$. Now

$(T_v \otimes E_v)^\times$ is compact mod $T_v^\times$. So y is in a set $T_v^\times \Omega'$ where Ω' is a compact set of G_v. For y in Ω' the functions $x \mapsto f(x\xi y)$ have support in a fixed compact set of $G(F_v)$ and are uniformly bounded. Thus the double integral (4) converges.

For almost all such v, the integral (4) can be described as follows: f_u is the characteristic function of $Z_u K_u$, E_u is the unramified quadratic extension of F_v and ξ is in $K_u Z_u$. If $f(x\xi y) = 0$ then $x\xi y$ is in $K_u Z_u$ hence $y^{-1}\xi^{1-\sigma}y$ is in K_u. Since $\xi^{1-\sigma}$ is in K_u and is semi-simple regular, this implies that y is in $(T_v \otimes E_u)^\times \cdot K_u \cap G_v$. Because E_u is unramified this is $T_v^\times K_v$. Taking then y in K_v we find that x must be in $Z_v K_v$. So the factor in this case is equal to

$$\mathrm{vol}(Z_v \backslash Z_v K_v) \cdot \mathrm{vol}(T_v \backslash T_v K_v). \tag{5}$$

If on the contrary v is a place of E which splits into $v1$ and $v2$ in E then we have isomorphisms $G_{v1} \cong G_v$ and $G_{v2} \cong G_v$; let μ_1 and μ_2 be the images of $\mu \in G(E)$ under these isomorphisms. Then G_v may be identified with the diagonal of the product $G_{v1} \times G_{v2}$ and our local factor is

$$\iint f_{v1}(x\xi_1 y) \cdot f_{v2}(x\xi_2 y) \mathrm{d}x \mathrm{d}y, \quad x \in G_v/Z_v, \quad y \in T_v^\times \backslash G_v. \tag{6}$$

After a change of variables, (6) can be written also as:

$$\iint f_{v1}(x_{v1}) \cdot f_{v2}(xy^{-1}\xi_1^{-1}\xi_2 y) \mathrm{d}x \mathrm{d}y \tag{7}$$

or

$$\int h_v(y^{-1}\eta_v y) \mathrm{d}y, \quad \text{with } \eta_v = \xi_1^{-1}\xi_2 \text{ and } h_v = f_{v1}^v * f_{v2}. \tag{8}$$

Now $\eta_v = (\xi_2)^{1-\sigma}$. Therefore this is a regular semi-simple element of $G(F_v)$ (lemma (4.4.4)) and this is an orbital integral for the function h_v; in particular it converges.

Again let us see what happens to the integral at almost all v. Then f_{vi} is the characteristic function of $Z_v K_v$ and ξ_1 and ξ_2 are in $Z_v K_v$. If the integrand is non zero we find that $x\xi_1 y$ and $x\xi_2 y$ are in $Z_v K_v$. Hence $y^{-1}\eta_v^{-1}y$ is in the same set. Since η_v is in K and is semisimple regular, this implies that y is in $T_v K_v$. Taking y in K_v we find that x is in $Z_v K_v$ and the integral is again equal to (5). Since the product of the factors (5) is finite our assertion is justified: each one of the local integral converges and their product converges too.

(5.4) Very much the same considerations apply to G' and the kernel K'. It is now time to prove the equality of the integral of K_e and K'_e under the assumptions of Prop. (2.2). To that end, consider a term in (5.3.3). Suppose that T is an algebra which does not imbed into H. Then there is at least one v in D such that T does not imbed into H, that is such that T is not a field. Then with the previous notations η_v is an hyperbolic regular element and the factor corresponding to v is an hyperbolic orbital integral of h_v hence zero by our choice of h_v. Suppose that now that T does imbed into H. It is therefore isomorphic to exactly one element T' of Q'. Then $v_T = v_{T'}$ and all we have to check is the equality of the integrals:

$$\iint f(x\xi y)\mathrm{d}x\mathrm{d}y = \iint f'(x'\xi'y')\mathrm{d}x'\mathrm{d}y'$$

if ξ corresponds to ξ'. Once more we decompose each global integral into a product of local ones and a constant factor, the same on both sides. Suppose v is not in D. Then $G_v = G'_v$ and $f_v = f'_v$. So the factor attached to v is trivially the same in both sides of the equality. Suppose v is in D. Then the factor attached to v in the left hand side is the integral of h_v on the orbit of η_v. Similarly the factor attached to v in the right hand side is the integral of h'_v on the orbit of η'_v, where η'_v corresponds to η_v in the isomorphism of T_v with T'_v. Again, by assumption, these factors must be the same and we have proved the required equality.

(5.5) We quickly dispose of the singular terms. Recall that

$$K_s(x, y) = \sum f(x^{-1}\xi y), \quad \xi \in G(F)/Z(F).$$

Then:

$$\iint K_s(x, y)\mathrm{d}x\mathrm{d}y = \iint f(xy)\mathrm{d}x\mathrm{d}y,$$

$$\text{with } x \in G(F_{\mathbf{A}})/Z(F_{\mathbf{A}}), \quad y \in G(F)Z(F_{\mathbf{A}})\backslash G(F_{\mathbf{A}}).$$

If we integrate with respect to x first the resulting integral does not depend on y so we find also:

$$\iint K_s(x, y)\mathrm{d}x\mathrm{d}y = \mathrm{vol}\int f(x)\mathrm{d}x, \quad x \in Z(F_{\mathbf{A}})\backslash G(F_{\mathbf{A}}), \tag{1}$$

where vol stands for the volume of $Z(F_{\mathbf{A}})G(F)\backslash G(F_{\mathbf{A}})$.

Similarly we find:

$$\iint K'_s(x, y)\mathrm{d}x\mathrm{d}y = \mathrm{vol}'\int f'(x)\mathrm{d}x, \quad x \in Z'(F_{\mathbf{A}})\backslash G'(F_{\mathbf{A}}), \tag{2}$$

where vol′ stands now for the volume of $Z'(F_{\mathbf{A}})G'(F)\backslash G'(F_{\mathbf{A}})$. We want to show that (1) and (2) are equal. To begin with vol = vol′. Next we write the integrals on the right hand sides of (1) and (2) as products of local integrals (and a constant factor). Again the factor attached to a place v of F which is not in D is trivially the same in both integrals. To see that the factors attached to a place v in D are the same we note that in the right hand side of (1) for instance this factor is nothing but the following integral:

$$\int f_{v1}(x)f_{v2}(x)\mathrm{d}x, \quad x\in G_v/Z_v.$$

This is in fact $h_v(e)$. Since h_v and h'_v have the same orbital integrals they have also the same value at e and we are done.

(5.6) We also dispose of the kernels K_{sp} and K'_{sp}. For K_{sp} we have the following expression:

$$K_{\mathrm{sp}}(x, y) = 1/\mathrm{VOL}\sum \int f(g)\chi(\det g)\mathrm{d}g\cdot\chi(\det x)\cdot\bar{\chi}(\det y),$$

where VOL denotes the volume of the quotient $G(E)Z(E_{\mathbf{A}})\backslash G(E_{\mathbf{A}})$, the sum is over all quadratic characters χ of the idele class group of E and the integral is over $Z(E_{\mathbf{A}})\backslash G(E_{\mathbf{A}})$. The integral of the term attached to χ is zero unless the restriction of χ to $F_{\mathbf{A}}^{\times}$ is trivial. Thus we find:

$$\iint K_{\mathrm{sp}}(x, y)\mathrm{d}x\mathrm{d}y = (\mathrm{vol})^2/\mathrm{VOL}\sum \int f(g)\chi(\det g)\mathrm{d}g, \qquad (1)$$

the sum being over all such χ. Similarly, we have:

$$\iint K'_{\mathrm{sp}}(x, y)\mathrm{d}x\mathrm{d}y = (\mathrm{vol}')^2/\mathrm{VOL}'\sum \int f'(g')\chi(\det g')\mathrm{d}g' \qquad (2)$$

where VOL′ denotes the volume of $G'(E)Z'(E)\backslash G'(E)$ and we have written det for the reduced norm. To prove the equality of the integrals (1) and (2), it suffices therefore to check that

$$\int f(g)\chi(\det g)\mathrm{d}g = \int f'(g')\chi(\det g')\mathrm{d}g'. \qquad (3)$$

We decompose each integral into a product over all places of E of analoguous local integrals (and a constant). It is again clear that the factor attached to a place not in $D^{\sim}$ is the same on both sides. If v is a place of D and $v1$ and $v2$ the two corresponding places of E then in the left hand side we have the factor:

$$\int f_{v1}(g_{v1})\chi_{v1}(\det g_{v1})\mathrm{d}g_{v1}\int f_{v2}(g_{v2})\chi_{v2}(\det g_{v2})\mathrm{d}g_{v2}. \qquad (4)$$

Since the restriction of χ to $F_{\mathbf{A}}^{\times}$ is trivial we have $\chi_{v1} = \bar{\chi}_{v2} = \chi_v$ say and this can be written also as

$$\int h_v \chi_v(\det g_v) \mathrm{d} g_v. \tag{5}$$

So what we have to prove is that this is also

$$\int h'_v \chi_v(\det g'_v) \mathrm{d} g'_v. \tag{6}$$

Call σ the special representation of G. Then (6) is minus tr $\sigma \otimes \chi_v(h_v)$. But because h_v has zero trace in any principal series this is also equal to (5) and we are done.

§6. Truncation of *K*

(6.1) In order to finish the proof of Proposition (2.2) we need only show that the integral of the difference $K_r - K_{\text{eis}}$ over the product of $z(F_{\mathbf{A}})G(F)\backslash G(F_{\mathbf{A}})$ by itself is zero. This will be proved under the following assumption:

(1) There is a set X of places of F with at least two elements and the following property: every v in X splits into $v1$ and $v2$ in E and the hyperbolic integrals of $h_v = f_{v1}^{v} * f_{v2}$ vanish.

Since K_{cusp}, K_e, K_s and K_{sp} are integrable over the product, it is clear that the difference $K_r - K_{\text{eis}}$ is indeed integrable. However, just as in the classical case, each term alone is not integrable and therefore we must use what one could call a "restriction truncation operator". It is defined as follows: let h be a continuous function on $Z(E_{\mathbf{A}})G(E)\backslash G(E_{\mathbf{A}})$, let c be a positive number and let χ_c be the characteristic function of $]c, +\infty[$; then we will denote by $T^c h$ the function on $Z(F_{\mathbf{A}})G(F)\backslash G(F_{\mathbf{A}})$ defined by:

$$T^c h(g) = h(g) - \sum h_N(\gamma g)\chi_c(H(\gamma g)), \quad \gamma \in P(F)\backslash G(F); \tag{1}$$

we have denoted by h_N the constant term of f along N. Recall that it is the function on $G(E_{\mathbf{A}})$ defined by:

$$h_N(g) = \int h(ng)\mathrm{d}n, \quad n \in N(E)\backslash N(E_{\mathbf{A}}). \tag{2}$$

Recall also that $H(g)$ is the height: $H(g) = |a/b|_E$ if $g = n(y)$ diag$(a/b)k$. The following lemma is standard (Cf. [J.A]):

(4) LEMMA: *If γ is in $G(E)$ and there is a g in $G(E_{\mathbf{A}})$ such that $H(g)>1$ and $H(\gamma g)>1$ then γ belongs to $P(E)$.*

This lemma implies that the sum in (1) has at most one term and also that if $H(g)>c>1$ then $T^c h(g)=h(g)-h_N(g)$. Of course if h is a cuspidal function on $Z(E_{\mathbf{A}})G(E)\backslash G(E_{\mathbf{A}})$, then $T^c h=h$ for any c. Now $K_{\text{cusp}}(x, y)$ is a cuspidal function of x and a cuspidal function of y. We will denote by T_1^c the truncation operator with respect to the first variable and by T_2^c the truncation operator with respect to the second variable. We will take two numbers c_1 and c_2 with $c_1>c_2>0$. Then $T_1^{c_1}T_2^{c_2}K_{\text{cusp}}=K_{\text{cusp}}$. Thus:

$$K_{\text{cusp}}(x,y)=T_1^{c_1}T_2^{c_2}K(x,y)-T_1^{c_1}T_2^{c_2}K_{\text{eis}}(x,y)-T_1^{c_1}T_2^{c_2}K_{\text{sp}}(x,y), \tag{5}$$

for x and y in $Z(F_{\mathbf{A}})G(F)\backslash G(F_{\mathbf{A}})$. The function f being fixed we will take c_1 and c_2 as large as need dictates. The number c_2 being fixed, we will see that if c_1 is large enough then every term in (5) is integrable over the product of $Z(F_{\mathbf{A}})G(F)\backslash G(F_{\mathbf{A}})$ by itself; furthemore the integrals have limits as c_1 tends to infinity so that we will be able to write:

$$\iint K_{\text{cusp}}(x, y)\mathrm{d}x\mathrm{d}y=\lim\iint T_1^{c_1}T_2^{c_2}K(x, y)\mathrm{d}x\mathrm{d}y$$
$$-\lim\iint T_1^{c_1}T_2^{c_2}K_{\text{eis}}(x, y)\mathrm{d}x\mathrm{d}y$$
$$-\lim\iint T_1^{c_1}T_2^{c_2}K_{\text{sp}}(x, y)\mathrm{d}x\mathrm{d}y.$$

At this point, each term will be a function of c_2. We will perform an asymptotic evaluation of each term for c_2 large: in a precise way we will write each term as the sum of a linear function of log c_2 plus a function tending to 0 as c_2 tends to infinity. The integral of K_{cusp} will then be evaluated by evaluating each linear function at zero. This will give the expression we need: see section (10).

(6.2) In section (6) we study the truncation of K. By definition we have:

$$T_1^{c_1}T_2^{c_2}K(x, y)= \tag{1}$$

$$K(x, y) \tag{1.i}$$

$$-\sum_{\gamma_2}\int\sum_{\xi} f(x^{-1}\xi n\gamma_2 y)\,\mathrm{d}n\,\chi_{c2}(H(\gamma_2 y)) \tag{1.ii}$$

$$-\sum_{\gamma_1}\int\sum_{\xi} f(x^{-1}\gamma_1^{-1}n\xi y)\,\mathrm{d}n\,\chi_{c_1}(H(\gamma_1 x)) \tag{1.iii}$$

$$+\sum_{\gamma_1,\gamma_2}\iint\sum_{\xi} f(x^{-1}\gamma_1^{-1}n_1\xi n_2\gamma_2 y)\,\mathrm{d}n_1\mathrm{d}n_2$$

$$\chi_{c_1}(H(\gamma_1 x))\chi_{c_2}(H(\gamma_2 y)); \tag{1.iv}$$

in this expression the sums for γ_1 and γ_2 are over $P(F)\backslash G(F)$; the sum for ξ is over $Z(E)\backslash G(E)$; the integrals for n, n_1 and n_2 are over $N(E)\backslash N(E_{\mathbf{A}})$. We first prove a lemma:

(2) LEMMA: *Given f there is a number d with the following property: suppose $c > d$; then if c is sufficiently large and x, y are in $G(F_{\mathbf{A}})$ the terms* (1.iii) *and* (1.iv) *cancell while in* (1.ii) *ξ is in $G(F)P(E)$.*

PROOF OF (2): By lemma (5.2.2) there is $d > 0$ such that for $c_1 > d$, $c_2 > d$ in (1.iv) ξ is actually in $P(E)$. Then the integration for n_2 is superfluous. By lemma (5.2.3) we may assume c_1 so large that in (1.iii) ξ is actually in the set $P(E)G(F)$. Then (1.iii) and (1.iv) do cancel. Furthermore we may assume d so large that, for $c_2 > d$, ξ in (ii) is actually in $G(F)P(E)$. This concludes the proof of the lemma.

We may further write the expression in (1.ii) as the sum of two terms written below as (3.i) and (3.ii):

$$\sum_{\gamma_2,\gamma_2}\int\sum_{\xi} f'(x^{-1}\gamma_1^{-1}\xi n\gamma_2 y)\,\mathrm{d}n\,\chi_{c_2}(H(\gamma_2 y)),$$

$$\xi\in N(E),\quad n\in N(E)\backslash N(E_{\mathbf{A}}),\quad \gamma_1\in Z(F)N(F)\backslash G(F),$$

$$\gamma_2\in P(F)\backslash G(F); \tag{3.i}$$

$$\sum_{\gamma_1,\gamma_2}\int\sum_{\xi} f(x^{-1}\gamma_1^{-1}\xi n\gamma_2 y)\,\mathrm{d}n\,\chi_{c_2}(H(\gamma_2 y)),$$

$$\xi\in Z(E)\backslash[P(E)-P(F)N(E)],\quad n\in N(E)\backslash N(E_{\mathbf{A}}),$$

$$\gamma_2\in P(F)\backslash N(F),\quad \gamma_1\in P(F)\backslash G(F). \tag{3.ii}$$

In each term we may recombine an integration on $N(E)\backslash N(E_{\mathbf{A}})$ with a sum on $N(E)$ to get:

$$\sum_{\gamma_1,\gamma_2}\int f(x^{-1}\gamma_1^{-1}n\gamma_2 y)\mathrm{d}n\chi_{c_2}(H(\gamma_2 y)),$$

$$n\in N(E_{\mathbf{A}}),\quad \gamma_1\in Z(F)N(F)\backslash G(F),\quad \gamma_2\in P(F)\backslash G(F);\tag{4.i}$$

$$\sum_{\gamma_1,\gamma_2}\int\sum_{\alpha} f(x^{-1}\gamma_1^{-1}\operatorname{diag}(\alpha,1)n\gamma_2 y)\mathrm{d}n\chi_{c_2}(H(\gamma_2 y)),$$

$$n\in N(E_{\mathbf{A}}),\quad \alpha\in E^\times-F^\times,\quad \gamma_2\in P(F)\backslash G(F),$$

$$\gamma_1\in P(F)\backslash G(F).\tag{4.ii}$$

Similarly we write K as the sum

$$K=K_{\mathrm{s}}+K_{\mathrm{e}}+K_{\mathrm{r}}\tag{5}$$

and use the results of section 4 to break up K_{r} into a sum of two terms written below as (5.i) and (5.ii):

$$\sum f(x^{-1}\gamma_1^{-1}n(v)\gamma_2 y),$$

$$v\in E-F,\ \gamma_1\in Z(F)N(F)\backslash G(F),\quad \gamma_2\in P(F)\backslash G(F);\tag{5.i}$$

$$\tfrac{1}{2}\sum f(x^{-1}\gamma_1^{-1}\operatorname{diag}(d,1)\gamma_2 y),$$

$$\alpha\in E^\times-F^\times,\quad \gamma_1\in A(F)\backslash G(F),\quad \gamma_2\in A(F)\backslash G(F).\tag{5.ii}$$

Note that in (5.ii) we have used the fact that $A(F)$ has index two in its normalizer. We have thus written $T_1^{c_1}T_2^{c_2}K$ as the sum of K_{s}, K_{e}, the difference (6.2.5.i)–(6.2.4.i) and the difference (6.2.5.ii)-(6.2.4.ii).

(6.3) We now deal with the difference (6.2.5.i)–(6.2.4.i).

(1) Lemma: *The difference* (6.2.5.i)–(6.2.4.i) *is integrable over the product of* $Z(F_{\mathbf{A}})G(F)\backslash G(F_{\mathbf{A}})$ *by itself.*

Proof of (1): This difference can be written as

$$\sum F(\gamma_1 x,\gamma_2 y),\quad \gamma_1\in Z(F)N(F)\backslash G(F),\quad \gamma_2\in P(F)\backslash G(F)$$

where we have set (here $c_2=c$):

$$F(x,y)=\sum f[x^{-1}n(\nu)y]-\int f(x^{-1}ny)\mathrm{d}n\chi_c(H(y)),$$

$$\nu\in E-F,\quad n\in N(E_{\mathbf{A}}).\tag{2}$$

Thus it suffices to show that F is integrable over the product

$$N(F)Z(F_{\mathbf{A}})G(F_{\mathbf{A}}) \times P(F)Z(F_{\mathbf{A}})\backslash G(F_{\mathbf{A}}).$$

To see that we will use the Iwasawa decomposition. Treating the variables in the maximal compact as parameters and then ignoring these parameters we see that what we have to show is the following: let H be the function on $E_{\mathbf{A}}^{\times} \times E_{\mathbf{A}}$ defined by

$$H(a, b) = f[\mathrm{diag}(a, 1)n(b)]; \tag{3}$$

then the following expression is finite:

$$\int \left| \sum H[a^{-1}b, b^{-1}(v + n_1 + n_2)] - \chi_c(|b|_E) \int H(a^{-1}b, b^{-1}x)\mathrm{d}x \right|$$

$$|ba|_F^{-1} \mathrm{d}n_1 \mathrm{d}n_2 \mathrm{d}^{\times}a \mathrm{d}^{\times}b. \tag{4}$$

In this expression v is summed over $E - F$, n_1 and n_2 are integrated over the quotient $N(F)\backslash N(F_{\mathbf{A}})$, x is integrated over $E_{\mathbf{A}}$, a is integrated over the idele group of F and b over the idele class group of F. Clearly one of the integrations over $N(F)\backslash N(F_{\mathbf{A}})$ is superfluous. Moreover we may change a into ab. Then a varies in a compact set so we can treat it as a parameter that we ignore. Finally what we have to prove is the following: let Φ be a smooth function of compact support on $E_{\mathbf{A}}$. Then the following expression is finite:

$$\int \left| \sum \Phi[b^{-1}(v + n)] - \chi_c(|b|) \int \Phi(b^{-1}x)\mathrm{d}x \right| |b|_F^{-2} \mathrm{d}n \mathrm{d}^{\times}b; \tag{3}$$

the summation is for v in $E - F$ and the integration for n is over $F\backslash F_{\mathbf{A}}$ and the integration for b is over the idele class group of F. In (3) we first integrate for $|b|_F < c^{1/2}$. We identify E to the sum of 2 copies of F. Then we get:

$$\iint \left| \sum_{\alpha, \beta} \Phi(b^{-1}\alpha, b^{-1}\beta + b^{-1}n) \right| \mathrm{d}n |b|_F^{-2} \mathrm{d}^{\times}b, \quad \beta \in F, \quad \alpha \in F^{\times}. \tag{4}$$

This is less than the following expression:

$$\iint \sum_{\alpha, \beta} |\Phi(b^{-1}\alpha, b^{-1}\beta + b^{-1}n)| |b|_F^{-2} \mathrm{d}^{\times}b \mathrm{d}n.$$

If we combine the sum on β with the integration on n and then change n to bn we get:

$$\int \sum_{\alpha} |\Phi(b^{-1}\alpha, n)| \mathrm{d}n |b|_F^{-1} \mathrm{d}^{\times}b, \quad \alpha \in F^{\times}, \quad n \in F_{\mathbf{A}}, \quad |b|_F < c^{1/2}$$

which is clearly finite. Now we integrate over $|b|_F > c^{1/2}$ in (3). We get:

$$\int \left| \sum_v \Phi[b^{-1}(v+n)] - \int \Phi(b^{-1}x)\mathrm{d}x \right| \mathrm{d}n |b|_F^{-2} \mathrm{d}^\times b. \tag{5}$$

But this is majorized by the sum of the following 2 terms:

$$\int \left| \sum_v \Phi[b^{-1}(\nu+n)] - \int \Phi(b^{-1}x)\mathrm{d}x \right| \mathrm{d}n |b|_F^{-2} \mathrm{d}^\times b, \quad v \in E; \tag{5.i}$$

$$\iint \left| \sum_v \Phi[b^{-1}(\nu+n)] \right| \mathrm{d}n |b|_F^{-2} \mathrm{d}^\times b, \quad v \in F. \tag{5.ii}$$

The term (5.ii) is itself majorized by

$$\int \sum_v |\Phi|[b^{-1}(\nu+n)] \mathrm{d}n |b|_F^{-2} \mathrm{d}^\times b = \int |b|_F^{-2} \mathrm{d}^\times b \int |\Phi|(b^{-1}n)\mathrm{d}n;$$

in the right hand side n is integrated over $F_{\mathbf{A}}$. After changing n to bn we finally get

$$\int |b|_F^{-1} \mathrm{d}^\times b \int |\Phi|(n)\mathrm{d}n, \quad |b|_F > c^{1/2},$$

which is clearly finite. As for (5.i) by Poisson summation formula it is equal to

$$\int \left| \sum \Phi\hat{}(b\gamma)\psi_E(\nu n) \right| \mathrm{d}n \mathrm{d}^\times b, \quad \nu \in E^\times.$$

This in turn is less than

$$\int \sum |\Phi\hat{}|(bv)\mathrm{d}^x b, \quad v \in E^x, \quad |b|_E > c,$$

which is clearly finite. This concludes the proof of lemma (1).

We now compute the integral of the difference (6.2.5.i)–(6.2.4.i). We proceed formally. We let H be the function on $E_{\mathbf{A}}^\times \times E_{\mathbf{A}}$ defined by:

$$H[a, b] = \iint f[k_1 \operatorname{diag}(a, 1) n(b) k_2] \mathrm{d}k_1 \mathrm{d}k_2. \tag{6}$$

Then the integral of the difference (6.2.5.i)–(6.2.4.i) is equal to:

$$\iint\Big\{\sum_{v} H[a^{-1}b,\ b^{-1}(v+n_1+n_2)] -$$

$$\chi_{c_2}(|b|)\int H(a^{-1}b,\ b^{-1}x)\mathrm{d}x\Big\}\mathrm{d}n_1\mathrm{d}n_2|ab|_F^{-1}\mathrm{d}^{\times}a\mathrm{d}^{\times}b. \tag{7}$$

The set in which the variables are is the same as before; in particular ν is in $E-F$. We change a into ab and then, for convenience, b into b^{-1}. We get:

$$\int\Big\{\int\sum_{\nu}\Phi[b(\nu+n)]\mathrm{d}n - \int\Phi(bx)\mathrm{d}x\chi_{c_2}\big(|b|_E^{-1}\big)\Big\}|b|_F^2\mathrm{d}^{\times}b, \tag{8}$$

where we have set:

$$\Phi(b)=\int H(a^{-1},\ b)|a|_F^{-1}\mathrm{d}^{\times}a. \tag{9}$$

In (8) we break up the integral over b into two pieces: $|b|>1$ and $|b|<1$. For the piece $|b|<1$ we remark that $|b|_F^{-1}>C_2^{-1/2}>1$ implies $|b|_F<1$. We also use a simple consequence of the Poisson summation formula:

$$\int\sum_{\nu}\Phi[b(\nu+n)]\mathrm{d}n-\int\Phi(bx)\mathrm{d}x$$

$$=\sum\Phi\hat{\ }(b^{-1}\nu)|b|_F^{-2}-\int\Phi(bn)\mathrm{d}n, \tag{10}$$

$\nu\in E-F$, $x\in E_{\mathbf{A}}$, $n\in F\backslash F_{\mathbf{A}}$ on the left; $\nu\in F-\emptyset$, $n\in F_{\mathbf{A}}$ on the right. This supposes that ψ_E is so chosen that its restriction to $F_{\mathbf{A}}$ is trivial. Thus (8) can be expressed as a sum:

$$\int_{|b|\geqslant 1}\Big\{\sum_{\nu\in E-F}\Phi[b(\nu+n)]\mathrm{d}n\Big\}|b|_F^2\mathrm{d}^{\times}b$$

$$+\Big\{\int_{|b|\leqslant 1}\big(1-\chi_{c_2}\big(|b|_E^{-1}\big)\mathrm{d}^{\times}b\Big\}\Phi\hat{\ }(0).$$

$$+\int_{|b|\leqslant 1}\sum_{v\in F^{\times}}\Phi\hat{\ }(b^{-1}v)\mathrm{d}^{\times}b-\iint\Phi(n)\mathrm{d}n|b|_F\mathrm{d}^{\times}b. \tag{11}$$

Now we write $E=F[\sqrt{\tau}\,]$ and take $\psi_E(z)=\psi_F((z-z)/2\sqrt{\tau})$. We set

$$\Psi(v)=\int\Phi(u+v\sqrt{\tau})\mathrm{d}u \tag{12}$$

and let $\Psi^{\wedge}$ denote the Fourier transform of Ψ as a Schwartz-Bruhat function on $F_{\mathbf{A}}$. We see at last that the integral of the difference (6.2.5.i)–(6.2.4.i) is given by the following expression:

$$\int_{|b|\geqslant 1}\Big\{\sum_{v\in F^{\times}}\Psi(bv)\Big\}|b|_F\mathrm{d}^{\times}b+\int_{|b|\geqslant 1}\Big\{\sum_{v\in F^{\times}}\Psi^{\wedge}(bv)\Big\}\mathrm{d}^{\times}b$$

$$+\Psi^{\wedge}(0)\int_{|b|\leqslant 1}\Big\{1-\chi_{c_1}\big(|b|_E^{-1}\big)\Big\}\mathrm{d}^{\times}b-\Psi(0)\int_{|b|\leqslant 1}|b|_F\mathrm{d}^{\times}b. \quad (13)$$

Recall that the Tate integral, or rather its analytic continuation, has a pole at $s=1$:

$$\int\Psi(b)|b|_F^s\mathrm{d}^{\times}b=c_{-1}(\Psi)/s-1+c_0(\Psi)+\ \ldots .$$

Furthemore recall that $c_0(\Psi)$ vanishes if Ψ is a product of local components Ψ_v and there are at least 2 places v of F such that

$$\int\Psi_v(b_v)\mathrm{d}b_v=0.$$

The expression (13) is nothing but

$$c_0(\Psi)+C\log(c_2),\quad C \text{ a constant.}$$

We are now within reach of our goal in this subsection:

(14) LEMMA: *The integral of the difference* (6.2.5.i)–(6.2.4.i) *is a linear function of* $\log c_2$. *If condition* (6.1.1)) *is satisfied the constant term vanishes.*

PROOF OF (14): The function Ψ introduced in (12) is a product of local functions over all places of F and we need only verify that for v in X the integral of the local component vanishes. Call again $v1$ and $v2$ the two places of E above v. Then, with obvious notations:

$$\int\Psi_v(b_v)\mathrm{d}b_v$$

$$=\int H_{v1}\big[(a^{-1},\ n+b\sqrt{\tau})\big]\,H\big[(a^{-1},\ n-b\sqrt{\tau})\big]\mathrm{d}n\mathrm{d}b\mathrm{d}^{\times}a,$$

$$n\in F_v,\quad b\in F_v,\quad a\in F_v^{\times}.$$

Recalling the definition of H (cf. (6) above) we see this is the integral over k_1, k_2 in K_v, a in $F_v^\times$, x and b in F_v of:

$$f_{v1}\left[k_1^{-1}\operatorname{diag}(a,1)n(x+b\sqrt{\tau})k_2\right]f_{v2}\left[k_1^{-1}\operatorname{diag}(a,1)n(x-b\sqrt{\tau})k_2\right].$$

This is also the following integral, where g is integrated over G_v:

$$\int f_{v1}\left[gn(b\sqrt{\tau})k\right]f_{v2}\left[gn(-b\sqrt{\tau})k\right]\mathrm{d}g\mathrm{d}b\mathrm{d}k.$$

Recall that h_v is the convolution product of f_{v1}^v and f_{v2}. This integral is nothing but:

$$\iint h_v\left[k^{-1}n(-2b\sqrt{\tau})k\right]\mathrm{d}b\mathrm{d}k.$$

This is a unipotent orbital integral of h_v. Since the hyperbolic integrals of h_v vanish the above integral vanish too. This concludes the proof of lemma (13).

(6.4) We now study the difference (6.2.5.ii)–(6.2.4.ii). We start with a remark. Consider the following function of $y\in G(F_{\mathbf{A}})$:

$$\sum f\left[x^{-1}\gamma_1^{-1}\operatorname{diag}(\alpha,1)y\right],\quad \gamma_1\in A(F)\backslash G(F),\quad \alpha\in E^\times-F^\times.$$

It is invariant on the left under the normalizer of A. In particular call w the following element of this normalizer:

$$w=\begin{vmatrix}0&1\\1&0\end{vmatrix}.$$

Then we have:

$$\sum f\left[x^{-1}\gamma_1^{-1}\operatorname{diag}(\alpha,1)\gamma_2 y\right]\chi_{c_2}(H(\gamma_2 y))$$
$$=\sum f\left[x^{-1}\gamma_1^{-1}\operatorname{diag}(\alpha,1)\gamma_2 y\right]\chi_{c_2}(H(\omega\gamma_2 y)),$$
$$\alpha\in E^\times-F^\times,\quad \gamma_1,\gamma_2 A(F)\backslash G(F).$$

In particular the difference (6.2.5.ii)–(6.2.4.ii) can be written as the sum of two terms written below as (1.i) and (1.ii):

$$\frac{1}{2}\sum_{\gamma_1,\gamma_2}\left\{\sum_{\alpha} f\left[x^{-1}\gamma_1^{-1}\operatorname{diag}(\alpha,1)\gamma_2 y\right]\right.$$
$$\left.\left[1-\chi_{c_2}(H(\gamma_2 y)-\chi_{c_2}(H(\omega\gamma_1 y))\right]\right\},$$
$$\gamma_1,\gamma_2\in A(F)\backslash G(F),\quad \alpha\in E^\times-F^\times; \tag{1.i}$$

$$\sum_{\gamma_1 \gamma_2} \left\{ f\left[x^{-1}\gamma_1^{-1} \operatorname{diag}(\alpha, 1)\gamma_2 y\right] \gamma_{c_2}(H(\gamma_2 y)),\right.$$

$$- \sum_{\gamma_1, \gamma_2} \int \sum_{\alpha} f\left[x^{-1}\gamma_1^{-1} \operatorname{diag}(\alpha, 1) n \gamma_2 y\right] \mathrm{d}n \chi_{c_2}(H(\gamma_2 y)), \qquad (1.\text{ii})$$

in the first term $\gamma_1, \gamma_2 \in A(F)\backslash G(F)$, in the second term $\gamma_1, \gamma_2 \in P(F)\backslash G(F)$, $n \in N(E_{\mathbf{A}})$; in both terms $\alpha \in E^{x} - F^{x}$.

We discuss the integral of each term separately.

(2) Lemma: *The term* (1.i) *is integrable. Its integral is a linear function of* $\log c_2$. *Under the assumptions of* (6.1.1) *the constant term is zero.*

Proof of (2): We forego the verification of integrability. It does not differ substantially from the formal computation of the integral that we now present. When we integrate (1.i) on the product of the quotient $Z(F_{\mathbf{A}})G(F)\backslash G(F_{\mathbf{A}})$ by itself we can combine the summations on $A(F)\backslash G(F)$ with the integration to obtain an integral on the product of $Z(F_{\mathbf{A}})A(F)\backslash G(F_{\mathbf{A}})$ by itself:

$$\iint \sum_{\alpha} f\left[x^{-1} \operatorname{diag}(\alpha, 1) y\right]\left(1 - \chi_{c_2}(H(y)) - \chi_{c_2}(H(wy))\right) \mathrm{d}x \mathrm{d}y,$$

$$\alpha \in E^{\times} - F^{\times}, \quad x, y \in Z(F_{\mathbf{A}})A(F)\backslash G(F_{\mathbf{A}}).$$

We then use the Iwasawa decomposition to compute the integrals in x and y. We get:

$$\int \sum_{\alpha} f\left[k_1^{-1} \operatorname{diag}(a^{-1}\alpha b, 1) n(\alpha^{-1} b^{-1} x + y) k_2\right]$$

$$\left[1 - \chi_{c_2}(|b|) - \chi_{c_2}\left(|b|_E^{-1} H(wn)\right)\right] |a|_F^{-1} \mathrm{d}k_1 \mathrm{d}k_2 \mathrm{d}x \mathrm{d}y \mathrm{d}^{\times} a \mathrm{d}^{\times} b;$$

here α is summed over $E^{\times} - F^{\times}$, a and b are integrated over the idele class group of F, x and y over the adeles of F; furthermore

$$n = n(y).$$

In the above integral we can change a to ab and y to yb. The integral

$$\int \sum_{a} f\left[k_1^{-1} \operatorname{diag}(a^{-1}\alpha, 1) n(\alpha^{-1} x + y) k\right] \mathrm{d}x \mathrm{d}y \mathrm{d}k_1 \mathrm{d}k_2 |a|_F^{-1} \mathrm{d}^{\times} a$$

$$\left\{1 - \chi_{a_2}(|b|) - \chi_{c_2}\left(|b|^{-1} H(wn)\right)\right\} \mathrm{d}^{\times} b.$$

In the inner integral we have $H(wn) < 1$. We may assume $c_2 > 1$ and then the range of b is defined by the inequalities $c_2^{-1} < H(wN) < |b|_E < c_2$. Thus after evaluation the inner integral gives us a factor $C' \log_2 + C'' \log H(wn)$ where C' and C'' are constants. Thus the integral of (1.1) can be written as

$$C \log c_2 + C'' \int \sum_{\alpha} f\left[k_1^{-1} \operatorname{diag}(a^{-1}\alpha, 1) n(\alpha^{-1}x + y) k\right]$$

$$\log H(wn) |a|_F^{-1} d^{\times} a dx dy dk_1 dk_2.$$

We can break up the sum on α into a sum on β in $F^{\times}$ followed by a sum on α in $(E^{\times} - F^{\times})/F^{\times}$; furthemore we can change x into $x\beta$ and then combine the integration on a with the summation on β to get an integral on the idele group of F. After a simple formal manipulation we get:

$$C \log c + C'' \sum_{\alpha} \int f\left[k_1^{-1} \operatorname{diag}(a^{-1}, 1) n(x) \operatorname{diag}(\alpha, 1) n(y) k_2\right]$$

$$\log H(wn) dx dy dk_1 dk_2 |a|_F^{-1} d^{\times} a.$$

Using again the Iwasawa decomposition we recombine the integrals on a, x and k into an integral on $Z(F_{\mathbf{A}}) \backslash G(F_{\mathbf{A}})$. So we see that the integral of (1.i) can be written also as:

$$C \log c_2 + C'' \sum_{\alpha} \int f[g \operatorname{diag}(\alpha, 1) nk] \log H(wn) dg dn dk,$$

$$\alpha \in F^{\times} \backslash (E^{\times} - F^{\times}), \quad n \in N(F_{\mathbf{A}}), \quad g \in G(F_{\mathbf{A}}), \quad k \in K_F.$$

We claim that under the assumption of (6.1.1) the second term vanishes. Indeed, it can alos be written as a double (actually finite) sum over α and all places u of F of the following integral:

$$\int f[g \operatorname{diag}(\alpha, 1) nk] \log H(wn) dg dn dk. \tag{4}$$

This integral is a product of local integrals and a constant. Since X has two elements, there is at least one v in X different of u. The corresponding local integral is a factor of (4). We claim that this local integral is actually zero; our claim will then prove lemma (2). Let $v1$ and $v2$ be the two places of E above x; let also α_1 and α_2 be the images of α under the isomorphisms $E_{v1} \cong F_v$ and $E_{v2} \cong F_v$. Then our local integral is:

$$\int f_{v1}[g \operatorname{diag}(\alpha_1, 1) nk] f_{v2}[g \operatorname{diag}(\alpha_2, 1) nk] dg dn dk.$$

After a change of variables, this can be written as:

$$\int f_{v1}[g] f_{v2}[gk^{-1}n^{-1} \operatorname{diag}(\alpha_1^{-1}\alpha_2, 1) nk] \mathrm{d}g \mathrm{d}n \mathrm{d}k.$$

Recall the convolution product h_v of f_{v1}^{v} and f_{v2}. This integral is nothing but

$$\int h_v[k^{-1}n^{-1} \operatorname{diag}(\alpha_1^{-1}\alpha_2, 1) nk] \mathrm{d}n \mathrm{d}k;$$

since α is not in F, $\alpha_1^{-1}\alpha_2 \neq 1$ and this is an hyperbolic orbital integral of h_v. Thus it is zero by assumption.

We now pass to the term (1.ii).

(5) LEMMA: *The term* (1.ii) *is integrable over the product of the quotient* $Z(F_{\mathbf{A}})G(F)\backslash G(F_{\mathbf{A}})$ *by itself. Its integral is zero.*

PROOF OF (5): Consider the first term in the difference (1.ii). It can also be written as follows:

$$\sum f[x^{-1}\gamma_1^{-1} \operatorname{diag}(\alpha, 1) n(\alpha^{-1}\nu_1 + \nu_2)\gamma_2 y] \chi_{c_2}(H(\gamma_2 y)),$$

$$\alpha \in E^\times - F^\times, \quad \gamma_1, \gamma_2 \in P(F)\backslash G(F), \quad \nu_1, \nu_2 \in F.$$

Since α is not in F every element of E can be written as a sum $\alpha^{-1}\nu_1 + \nu_2$, where v_1 and v_2 are in F. Thus the first term in the difference (1.ii) can also be written in the form:

$$\sum f[x^{-1}\gamma_1^{-1} \operatorname{diag}(\alpha, 1) n(\mu)\gamma_2 y] \chi_{c_2}(H(\gamma_2 y)),$$

$$\alpha \in E^\times - F^\times, \quad \gamma_1, \gamma_2 \in P(F)\backslash G(F), \quad \mu \in E.$$

As a consequence the difference (1.ii) can now be written as a double sum

$$\sum F(\gamma_1 x, \gamma_2 y), \gamma_1, \gamma_2 \in P(F)\backslash G(F),$$

where we have set:

$$F(x, y) = \sum_{\alpha, \mu} f[x^{-1} \operatorname{diag}(\alpha, 1) n(\mu) y] \chi_{c_2}(H(y))$$

$$- \sum_{\alpha} f[x^{-1} \operatorname{diag}(\alpha, 1) n(z) y] \mathrm{d}x \chi_{c_2}(H(y)) \mathrm{d}z. \tag{6}$$

To show that (1.ii) is integrable it is therefore enough to show that F is integrable over the product of $Z(F_{\mathbf{A}})P(F)\backslash G(F_{\mathbf{A}})$ by itself. We use the Iwasawa decomposition. To show integrability we treat the variables in the maximal compact as parameters that we ignore. Setting

$$H(a, x) = f[\operatorname{diag}(a, 1)n(x)] \tag{7}$$

we see that we have to prove the following expression is finite:

$$\int \Bigg| \sum_{\alpha,\mu} H\left[a^{-1}\alpha b,\, b^{-1}\left(\alpha^{-1}x_1 + \mu + x_2\right)\right] - \int H[a^{-1}\alpha b,\, b^{-1}z]\mathrm{d}z \Bigg| |a|_F^{-1}|b|_F^{-1}\chi_{c_2}(|b|)\mathrm{d}^\times b\mathrm{d}^\times a\mathrm{d}x_1\mathrm{d}x_2. \tag{8}$$

We may change a to ab to get

$$\int \Bigg| \sum_{\alpha,\mu} H\left[a^{-1}\alpha,\, b\left(\alpha^{-1}x_1 + \mu + x_2\right)\right] - \int H[a^{-1}\alpha,\, b^{-1}z]\mathrm{d}z \Bigg| |a|_F^{-1}|b|_F^{-1}\chi_{c_2}(|b|)\mathrm{d}^\times b\mathrm{d}^\times a\mathrm{d}x_1\mathrm{d}x_2. \tag{9}$$

Now $a^{-1}\alpha$ stays in a fixed compact set. Hence the module of a stays in a compact set of $\mathbb{R}^\times$. This implies that a stays in a relatively compact set; therefore α stays in a compact set, hence in fact in a finite set. Thus our assertion on convergence will be proved if we show that the following expression is finite:

$$\int \sum \Bigg| \Phi\left[b^{-1}\left(\alpha^{-1}x_1 + x_2 + \mu\right)\right] - \int \Phi[b^{-1}z]\mathrm{d}z \Bigg| \mathrm{d}x_1\mathrm{d}x_2|b|_F^{-2}\chi_{c_2}(|b|)\mathrm{d}^\times b; \tag{10}$$

In this expression α is some element of $E - F$, μ is summed over E, z is integrated over E and x_1, x_2 over $F\backslash F_{\mathbf{A}}$. By Poisson summation formula this is also:

$$\int \Bigg| \sum_{\mu} \Phi\hat{\ }(b)\psi_E\left(\alpha^{-1}x_1 + x_2\right)\mu\big) \Bigg| \chi_{c_2}(|b|_E)\mathrm{d}^\times b\mathrm{d}x_1\mathrm{d}x_2,$$

where μ is now in $E^\times$. This is majorized by

$$\int \sum |\Phi\hat{\ }(b\mu)|\mathrm{d}^\times b, \quad \mu \in E^\times, \quad |b|_E > c^{1/2}.$$

This is clearly finite.

It remains to prove that the integral of (1.ii) over the product of $Z(F_{\mathbf{A}})G(F)\backslash G(F_{\mathbf{A}})$ by itself is zero. We compute it as the integral of (6) over the product of $Z(F_{\mathbf{A}})P(F)\backslash G(F_{\mathbf{A}})$ by itself. Setting now instead of (7)

$$H(a, x)=\iint f[k_1 \operatorname{diag}(a, 1)n(x)k_2]\mathrm{d}k_1\mathrm{d}k_2 \tag{11}$$

we get for our integral an analogue of (8) and then (9):

$$\begin{aligned}\int\Big\{&\sum_{\alpha,\mu} H\left[a^{-1}\alpha,\, b^{-1}\left(\alpha^{-1}x_1+\psi+x_2\right)\right]\\ &-\sum\int H[a^{-1}\alpha,\, b^{-1}z]\mathrm{d}z\Big\}|a|_F^{-1}|b|_F^{-2}\chi_{c_2}(|b|)\mathrm{d}^\times b\mathrm{d}^\times a\mathrm{d}x_1\mathrm{d}x_2.\end{aligned} \tag{12}$$

However this is also:

$$\begin{aligned}\int\Big\{&\sum_{\alpha,\mu} H\left[a^{-1}\alpha,\, b^{-1}\left(\alpha^{-1}x_1+\mu+x_2\right)\right]\mathrm{d}x_1\mathrm{d}x_2\\ &-\int H[a^{-1}\alpha,\, b^{-1}z]\mathrm{d}z\Big\}|a|_F^{-1}|b|_F^{-2}\chi_{c_2}(|b|)\mathrm{d}^\times b\mathrm{d}^\times a.\end{aligned} \tag{13}$$

Since α is not in F the map $(x_1, x_2)\mapsto\alpha^{-1}x_1+x_2$ is a bijection of F^2 onto E. So the sum on μ in E and the integrations for x_1 and x_2 can be combined to give an integration on $E_{\mathbf{A}}$ and this expression is zero.

(6.5) We may summarize the results of this section as follows:

Proposition: *If c_1 is sufficiently large then $T_1^{c_1}T_2^{c_2}K$ is independent of c_1; it is integrable over the product of $Z(F_{\mathbf{A}})G(F)\backslash G(F_{\mathbf{A}})$ by itself. Furthermore its integral is a linear function in* $\log c_2$ *whose constant term vanishes if* (6.1.1) *is satisfied, plus the integral of K_e plus the integral of K_A.*

Indeed this integral is the sum of the integral of K_e, K_s, the difference (6.2.5.i)–(6.2.4.i), the term (6.4.1.i) and the term (6.4.1.ii). Our assertion follows then from lemmas (6.3.1), (6.3.14), (6.4.2), (6.4.5) and (6.2.6).

7. Local preliminaries

(7.1) In order to study the truncation of the kernel K_{eis} we need some information on local "intertwining operators" and certain local integrals, not all of which is available. This information is reviewed or presented in this section. Accordingly, in this section the field F is local and G is the group GL(2) regarded as an F-group.

(7.2) Let χ be a character of (module one) of $F^{\times}$. We denote by $\mathbb{H}(s, \chi)$ the space of functions h on $G(F)$ such that:

$$h[n(x)\,\mathrm{diag}(a, b)g] = |a/b|^{s+1/2}\chi(a/b)h(g) \tag{1}$$

which are square integrable on K the standard maximal compact subgroup of $G(F)$. We also denote by $\rho(s, \chi)$ the representation of $G(F)$ on $\mathbb{H}(s, \chi)$ by right shifts. We may regard the collection of the spaces $\mathbb{H}(s, \chi)$ as a fiber bundle of base $\mathbb{C}$. We set $\mathbb{H}(\chi) = \mathbb{H}(-1/2, \chi)$. If h is in that space we define a section of our fiber bundle by:

$$h(g, s) = h(g)H(g)^{s+1/2},$$

$$\text{where } H(g) = |a/b| \text{ if } g = n\,\mathrm{diag}(a, b)k. \tag{2}$$

The fiber bundle is trivial and sections obtained in this way will be called constant. We may identify $\mathbb{H}(s, \chi)$ with $\mathbb{H}(\chi)$ and regard $\rho(s, \chi)$ as operating on $\mathbb{H}(\chi)$. On the product $\mathbb{H}(\chi) \times \mathbb{H}(\chi)$ we define a bilinear form:

$$\langle h, h'\rangle = \int h(k)h'(k)\mathrm{d}k; \tag{3}$$

then:

$$\langle \rho(s, \chi)(g)h, \rho(-s, \bar{\chi})(g)h'\rangle = \langle h, h'\rangle.$$

Similarly on the product $\mathbb{H}(\chi) \times \mathbb{H}(\chi)$ we define a sesquilinear form:

$$(h, h') = \int h(k)\bar{h}'(k)\mathrm{d}k; \tag{4}$$

then:

$$(\rho(s, \chi)(g)h, \rho(-s, \chi)(g)h') = (h, h').$$

In particular when s is purely imaginary the representation $\rho(s, \chi)$ is unitary. As usual we define the intertwining operator $M(s, \chi)$ as an operator from $\mathbb{H}(\chi)$ to $\mathbb{H}(\chi)$ by

$$[M(s, \chi)h](g)H(g)^{-s+1/2} = \int h(wng)H(wng)^{s+1/2}\mathrm{d}n, \quad n \in N(F); \tag{5}$$

the integral converges for Re $s > 0$ but extends meromorphically to the whole complex plane. We may also regard $M(s, \chi)$ as an operator from $\mathbb{H}(s, \chi)$ to $\mathbb{H}(-s, \bar{\chi})$. We define also a "normalized intertwining operator" $R(s, \chi)$ by

$$M(s, \chi) = [L(2s, \chi^2)/L(2s+1, \chi^2)\varepsilon(2s, \chi^2, \psi)] \cdot R(s, \chi). \quad (6)$$

For our purposes it is convenient to use a description of these operators in terms of the Whittaker modele (Cf. [F.S.]): accordingly let $W(s, \chi)$ be the map from $\mathbb{H}(\chi)$ to the Whittaker modele of the representation $\rho(s, \chi)$; it is defined by

$$W(s, \chi)h(g) = \int h[wn(x)g, s] \cdot \psi_F(-x)\mathrm{d}x, \ x \in F. \quad (7)$$

The integral converges for Re $s > 0$ but extends holomorphically to the whole complex plane. In particular we have:

$$[W(s, \chi)h](\mathrm{diag}(a, 1)) = \int h[wn(x)g, s] \cdot \psi_F(-ax)\mathrm{d}x\bar{\chi}(a)|a|^{-s+1/2}. \quad (8)$$

If Re $s = 0$ the integral on the right hand side can be interpreted as the Fourier transform of the square integrable function $x \mapsto f[wn(x)]$. In particular the scalar product (3) can be computed as

$$\langle h, h' \rangle = c\int W(\mathrm{diag}(a, 1))W'(\mathrm{diag}(-a, 1))\mathrm{d}^\times a, \quad (9)$$

where W and W' are the images of h and h' under $W(s, \chi)$ and $W(-s, \bar{\chi})$ respectively and c is a constant. Similarly, the scalar product (4) can be computed as

$$(h, h') = c\int W(\mathrm{diag}(a, 1))\overline{W'}(\mathrm{diag}(a, 1))\mathrm{d}a, \quad (10)$$

where W and W' are the images of h and h' under $W(s, \chi)$. This being so we have the following result:

(11) Lemma: *With the previous notatons for h in $\mathbb{H}(s, \chi)$ we have*:

$$W(-s, \chi)M(s, \chi)h = [L(2s, \chi^2)/L(1-2s, \bar{\chi}^2)\varepsilon(2s, \chi^2, \psi_F)] \times W(s, \chi)h.$$

PROOF OF (11): It is easy to see that a constant section can be represented by an integral:

$$h(g, s) = \int \Phi[(0, t)g]|t|^{2s+1}\chi^2(t)\mathrm{d}^\times t\chi(\det g)|\det g|^{s+1/2},$$

where Φ is a Schwartz-Bruhat function with compact support contained in the orbit $(0, 1)G(F)$. Then

$$W(s, \chi)h(e) = \iint \Phi[(t, tx)]\chi^2(t)|t|^{2s+1}\mathrm{d}^\times t\psi(-x)\mathrm{d}x\chi(-1).$$

After a change of variables this can be written as

$$W(s, \chi)h(e) = \int \Phi_1(t, t^{-1})\chi^2(t)|t|^{2s}\mathrm{d}^\times t\chi(-1)$$

where

$$\Phi_1(x, y) = \int \Phi(x, v)\psi(-vy)\mathrm{d}v.$$

On the other hand if we apply the intertwining operator to $h(g, s)$ we obtain:

$$\int \Phi[(t, tx)g]|t|^{2s+1}\chi^2(t)\mathrm{d}^\times t\chi(\det g)|\det g|^{s+1/2}.$$

We can change x to xt^{-1}; if we introduce the Fourier transform of Φ defined by:

$$\Phi^\wedge(x, y) = \iint \Phi(u, v)\psi(xv - yu)\mathrm{d}u\mathrm{d}v$$

and use the local functional equation of the Tate integral we get:

$$C\int \Phi^\wedge[(0, t)g]|t|^{-2s+1}\chi^{-2}(t)\mathrm{d}^\times t\tilde{\chi}(\det g)|\det g|^{-s+1/2},$$

where C is the factor in the right hand side of the lemma's formula. Next we find:

$$W(-s, \tilde{\chi})M(s, \chi)h(e) = C\chi(-1)\int \Phi_1^\wedge(t, t)|t|^{-2s}\chi^2(t)\mathrm{d}^\times t$$

where

$$\Phi_1^\wedge(x, y) = \int \Phi^\wedge(x, v)\psi(-vy)\mathrm{d}v.$$

If we change t to t^{-1} and apply Fourier inversion formula our result follows.

(7.3) If now we substitute R to M in lemma (7.2.11) we find:

$$W(-s,\bar{\chi})R(s,\chi)h=\left[L(1+2s,\chi^2)/L(1-2s,\bar{\chi}^2)\right]W(s,\chi)h \tag{1}$$

The properties of the normalized intertwining operator are now easily established:

$$R(-s,\bar{\chi})R(s,\chi)=\mathrm{Id}; \tag{2}$$

$$\text{If } \chi^2=1 \text{ then } R(0,\chi)=\mathrm{Id}; \tag{3}$$

$$\text{For } h\in \mathbb{H}(\gamma),\, h'\in \mathbb{H}(\chi),\, \langle R(s,\chi)h,\, R(-s,\chi)h'\rangle=\langle h,h'\rangle; \tag{4}$$

$$\text{For } \operatorname{Re} s=0 \text{ and } h,h'\in \mathbb{H}(\chi),\, (R(s,\chi)h,\, R(s,\chi)h')=(h,h'). \tag{5}$$

We will need also the dependence of $R(s,\chi)h$ on s, for h in $\mathbb{H}(\chi)$.

(6) LEMMA: *Suppose h is a K-finite element of $\mathbb{H}(\chi)$. Then for k in K, $R(s,\chi)h(k)$ is an elementary function of s, without singularity on the line* $\operatorname{Re} s=0$.

SKETCH OF PROOF OF (6): By an elementary function of s we mean a sum of rational functions of s times an exponential if F is Archimedean and a rational fraction in q^{-s} if F is non Archimedean with a residual field of q elements. Now a constant K-finite section can also be represented by an integral:

$$h(g,s)=Q(s)/L(1+2s,\chi^2)$$

$$\times\int\Phi[(0,t)g]\chi^2(t)|t|^{2\lambda+1}\chi(\det g)|\det g|^{s+1/2} \tag{7}$$

where Q is an elementary function of s. Here, if F is Archimedean the function is a standard Schwartz function: that is, if $F=\mathbb{R}$, it has the form $\Phi(x,y)=P(x,y)\exp(-\pi(x^2+y^2))$, and if $F=\mathbb{C}$, it has the form $\Phi(x,y)=P(x,\bar{x},y,\bar{y})\exp(-2\pi(x\bar{x}+y\bar{y}))$, where P is a polynomial.

Applying the normalized intertwining operator we find

$$R(s, \chi)h(g, -s) = Q(s)/L(1-2s, \chi^2)$$
$$\times \int \Phi\hat{}[(0, t)g]\bar{\chi}^2(t)|t|^{-2s+1}\mathrm{d}^\times t$$
$$\times \bar{\chi}(\det g)|\det g|^{-s+1/2}. \tag{8}$$

If F is Archimedean Φ is still a standard Schwartz function and our assertion follows from the properties of the Tate integral.

We also recall the following improvement on Lemma (6) for functions invariants under K:

(9) Lemma: *Suppose that $h(k)=1$ for all k in K. Suppose also that F is non Archimedean, the order ψ of is zero and χ is unramified. Then $R(s, \chi)h(k)=1$ for all k.*

The proof is similar to the proof of lemma (6).

(7.4) We now change our notations: E is a quadratic extension of F and χ a character of $E^\times$, accordingly $\rho(s, \chi)$ is now a representation of $G(E)$. We will be interested in linear forms on the space of smooth vectors in $\mathbb{H}(\chi)$ (or $\mathbb{H}(s, \chi)$) which are invariant under $G(F)$ operating via the representation $\rho(s, \chi)$. Suppose first the restriction of χ to $F^\times$ is trivial and $s=0$. Then the restriction of a smooth function in $\mathbb{H}(s, \chi)$ to $G(F)$ transforms on the left according to the module of the group $P(F)$; therefore the integral of *a* such function on the group K defines a linear form with this invariant property. What we want to establish in this section is the following result:

(1) Lemma: *With the above notations, for any smooth vector h in $\mathbb{H}(\chi)$ we have*:

$$\int R(0, \chi)h(k)\mathrm{d}k = \int h(k)\mathrm{d}k, \quad k \in K_F(=K \cap G(F)).$$

Proof of (1): Once more it is convenient to use the Whittaker modele. As usual ψ_F is a non trivial additive character of F; we choose a non trivial additive character of E whose restriction to F is trivial. For instance we write $E=F[\sqrt{\tau}]$ and then we take $\psi_E(x)=\psi_F[(x^\sigma - x)/2\sqrt{2}]$. We first establish the existence of a constant c such that, for any h,

$$\int h(k)\mathrm{d}k = c\int W[\mathrm{diag}(a, 1)]\mathrm{d}^\times a, \quad k \in K, \quad a \in F^\times. \tag{2}$$

where W is the image of h under $W(0,)$. There is a constant c such that:

$$\int h(k)\mathrm{d}k = c\int h[wn(x), 0]\mathrm{d}x, \quad x \in F. \tag{3}$$

On thc othcr hand we havc:

$$W[(\operatorname{diag}(a, 1)] = \int h[wn(z), 0]\psi_F(-az)\mathrm{d}z|a|_E^{1/2}\bar{\chi}(a);$$

integrating over F we find:

$$\int W[(\operatorname{diag}(a, 1)]\mathrm{d}^\times a = \int |a|_F d^\times a$$

$$\times \iint h[wn(x + \sqrt{\tau}y)]\psi(-ay)\mathrm{d}y\mathrm{d}x.$$

Our assertion follows now from Fourier inversion formula. In view of formula (6.3.1) we have now only to establish the following lemma:

(4) LEMMA: *Suppose χ is a character of $E^\times$ trivial on $F^\times$. We have*:

$$L(s, \chi^2)/L(s, \bar{\chi}^2) = 1.$$

PROOF OF (4): If $F = \mathbb{R}$, $E = \mathbb{C}$, the character χ has the form

$$\chi(z) = z^n(z\bar{z})^{-n/2}$$

and the formula is checked at once. Suppose F is non Archimedean. If χ^2 is ramified both L-factors are equal to one and our assertion is trivial. Suppose χ^2 is unramified. If x_F is a uniformizer for F we have $\chi(x_F) = 1$. Now either x_F is a uniformizer for E or the square of a uniformizer. Thus if χ_E is a uniformizer for E we have $\chi^2(x_E) = 1$. In any case χ^2 is trivial and our assertion follows.

(7.5) We continue with the notations of (7.4). This time we assume that χ is a character of $E^\times$ invariant under σ where $\mathrm{Gal}(E/F) = \{1, \sigma\}$. We first need a description of the double coset space $P(E)\backslash G(E)/G(F)$.

(1) LEMMA: *We have a disjoint union*

$$G(E) = P(E)G(F) \cup P(E)\lambda G(F).$$

Furthemore the algebra $L = \gamma^{-1}R(E)\gamma \cap M(2, F)$ is an algebra of degree 2 on F which is F-isomorphic to E. Finally if $e(l)$ and $e'(l)$ are the two *eigenvalues of* $\gamma l \gamma^{-1}$ *then* $l \mapsto e(l)$ *and* $l \mapsto e'(l)$ *are the two isomorphisms of* L *onto* E.

PROOF OF (1): Recall that R is the algebra of triangular matrices. We will prove the corresponding assertion for the cosets $G(F)\backslash G(E)/P(E)$. We let G operates on the right on the vector space of column vectors of dimension 2. We denote by $\{e, e'\}$ the canonical basis. The group $P(E)$ is the fixator of the line Ee. We let μ be any element of $E - F$ and then set

$$\lambda = \mathrm{Mat}[1, 0, \mu\, 1].$$

Let h be an element of $G(E)$ and D the line hEe, c its "slope". If c is finite this means that D contains the vector $e + ce'$. Suppose c is in F (resp. infinite). Then there is g in $G(F)$ such that $e + ce' = ge$ (resp. $e' = ge$) and then $D = gEe$. Suppose c is not in F. Then $c = p + q\mu$ for some p and q in F and

$$e + ce' = g(e + \mu e'), \qquad \text{where} \quad g = \mathrm{Mat}[1, 0, p, q]$$

Hence $D = g\lambda Ee$. This already proves the first assertion of the lemma. To prove the second assertion we remark that if g is in L then $g\lambda e = m(g)\lambda e$, with $m(g)$ in E. Furthemore $ge = (p + q\mu)e + (r + s\mu)e'$ if $g = \mathrm{Mat}[p, q, r, s]$ so $m(g)$ cannot be zero unless g is. Hence $g \mapsto m(g)$ is an injective morphism from L to E. Since L is not reduced to F it is indeed isomorphic to E. Finally it is clear that e and e' are F-morphisms of L into E. Since L has some non scalar elements e and e' are distinct and we are done.

(7.6) Coming back to our goal we let χ be a character of $E^{\times}$ invariant under σ. If h is a smooth function of $\mathbb{H}(\chi)$ then the function $g \mapsto h(\lambda g, s)$ on $G(F)$ is invariant on the left under $L^{\times}$. Thus we are led to set:

$$I(s, \chi)h = \int h(\lambda g, s)\mathrm{d}g, \quad g \in L^{\times}\backslash G(F). \tag{1}$$

The integral converges only if $\mathrm{Re}\, s > 0$ and we need to show that it has analytic continuation. One way is to use once more the Whittaker modele; we sketch a proof. We are going to show there is a constant $c(s)$ such that

$$I(s, \chi)h = c(s)\int W[\mathrm{diag}(a, 1)]\mathrm{d}^{\times}a, \quad a \in F^{\times}, \tag{2}$$

if W corresponds to h. Recall that W is related to h by the following formula:

$$W[\operatorname{diag}(a,1)] = \iint h\left[wn\left(x + y\sqrt{\tau}\right)\right]$$

$$\times \psi_F(-ay)\mathrm{d}y\mathrm{d}x|a|_E^{-s+1/2}\bar{\chi}(a).$$

Now χ has the form $\mu o N(E/F)$ where μ is a character of $F^\times$. Hence the restriction of χ to $F^\times$ is μ^2. Integrating over $F^\times$ we find:

$$\int W[\operatorname{diag}(a,1)]\mathrm{d}a = \int h\left[wn\left(x + y\sqrt{\tau}\right), s\right]$$

$$\times \psi_F(-ay)\mathrm{d}y\mathrm{d}x|a|_F^{-2s+1}\bar{\mu}^2(a)\mathrm{d}^\times a$$

Using the Tate functional equation we find this is

$$\left[\epsilon\left(2s, \mu^2, \psi_F\right)L\left(1-2s, \bar{\mu}^2\right)/L\left(2s, \mu^2\right)\right]$$

$$\times \int h\left[\left(w\left(x + a\sqrt{\tau}\right), s\right]|a|_F^{2s}\mu^2(a)\mathrm{d}^\times a\mathrm{d}x.$$

Of course we should check that the functions are sufficiently smooth to allow this. But this is not hard. Now the integral on the right can be recognized as being

$$\iint h[\lambda \operatorname{diag}(a,1)n(x), s]|a|\mathrm{d}^\times a\mathrm{d}x$$

Since $G(F) = L^\times P(F)$ with $P(F) \cap L^\times = F^\times$, this integral is proportional to the integral of $h(\lambda g, s)$ over $L^\times \backslash G(F)$ and we are done with:

$$c(s) = cL\left(2s, \mu^2\right)/L\left(1-2s, \bar{\mu}^2\right)\varepsilon\left(2s, \mu^2, \psi_F\right). \tag{3}$$

This gives the analytic properties of $I(s, \chi)$. There is another way to obtain these properties. We take:

$$\mu = wn\left(\sqrt{\tau}\right); \qquad \text{then} \quad L = \{\mathrm{Mat}[a, b, b, a]\}. \tag{4}$$

Next we use once more the device of representing $h(g, s)$ in the form

$$h(g,s) = \left[Q(s)/L\left(2s+1, \chi^2\right)\right]$$

$$\times \int \Phi[(0,t)g]\chi^2(t)|t|^{2s+1}\chi(\det g)|\det g|_E^{s+1/2}, \tag{5}$$

where Φ is a Schwartz-Bruhat function in two variables and Q an elementary function. If $t = a + b\sqrt{\tau}$ then

$$(0, t)\lambda = (1, \sqrt{\tau})1 \qquad \text{with} \quad 1 = \mathrm{Mat}[a, b\tau, b, a] \tag{6}$$

$$|t|_E^{2s+1} = |\det(1)|_E^{2s+1}, \quad \chi(t) = \mu o N(E/F)(t) = \mu(\det(1)),$$

$$\chi(\det g) = \mu(\det g \det g^\sigma) = \mu^2(\det g), \qquad \text{for} \quad g \in G(F)$$

and, for g in $G(F)$, the above integral for $h(\lambda g, s)$ can be written as an integral on L:

$$h(\lambda g, s) = \int \Phi[(1, \sqrt{\tau})lg]\mu^2(\det lg)|\det lg|_F^{2s+1} \mathrm{d}^\times l \tag{7}$$

Therefore the integral on $L^\times \backslash G(F)$ can combined with this integral to give an integral on $G(F)$:

$$\int h(\lambda g, s)\mathrm{d}g = [Q(s)/L(2s+1, \chi^2)]$$

$$\times \int \Phi[(1, \sqrt{\tau})g]\mu^2(\det g)|\det g|_F^{2s+1}\mathrm{d}g. \tag{8}$$

We can use the Iwasawa decomposition on $G(F)$ to compute this integral. We find:

$$[Q(s)/L(2s+1, \chi^2)]$$

$$\times \int \Phi[a, b\sqrt{\tau} + x)k]\mathrm{d}x\mu^2(a)|a|_E^{2s}\mu^2(b)|b|_F^{2s+1}\mathrm{d}^\times a\mathrm{d}^\times b\mathrm{d}k. \tag{9}$$

After integrating over x and k the resulting function of a and b is Schwartz Bruhat and the integral is a double Tate integral. In particular we see that for $\mathrm{Re}s > 0$ all our integrals converge. Furthemore, as a function of s, this integral has the form:

$$Q'(s) \cdot L(2s, \mu^2)L(2s+1, \mu^2)/L(2s+1, \chi^2)$$

$$= Q'(s)L(2s, \mu^2)/L(2s+1, \mu^2\eta), \tag{11}$$

where Q' is another elementary function of s and η the quadratic character attached to the extension E of F. We are led to define a "normalized version" of the linear form $I(s, \chi)$. It will be noted $J(s, \chi)$. It is defined by:

$$I(s, \chi) = L(2s, \mu^2)/L(2s+1, \mu^2\eta)J(s, \chi). \tag{12}$$

The advantage of J is that it is defined for all s with $\operatorname{Re} s = 0$. In contrast if the restriction of χ to $F^\times$ is trivial then $\mu^2 = 1$ and $I(s, \chi)$ has a pole at $s = 0$. Furthemore the analytic dependence of $J(s, \chi)$ on s is simple:

(13) LEMMA: *Suppose h is a constant section. Then, if h is K-finite, $J(s, \chi)h$ is an elementary function of s. In particular if $h = 1$ on K, F is non Archimedean, E is the quadratic unramified extension of F, χ is unramified and ψ_F has order 0 then $J(s, \chi)h = \operatorname{vol}(K_F)$ for all s.*

This follows at once from the above computations.

Finally we remark that if the restriction of χ to $F^\times$ is trivial then there exists a constant c such that:

$$J(0, \chi)h = c\int h(k)\mathrm{d}k.$$

This follows at once from formulas (7.4.2) and (7.6.2).

§8. Truncation of an Eisenstein series

(8.1) We will need to study the truncation of the kernel K_{eis} and the integral of its truncation over the product of $Z(F_{\mathbb{A}})G(F)\backslash G(F_{\mathbb{A}})$ by itself. To that end we first recall a few facts on Eisenstein series; we study their truncation and the integral of their truncation over $Z(F_{\mathbb{A}})G(F)\backslash G(F_{\mathbb{A}})$.

As usual we shall consider functions on $G(E_{\mathbb{A}})$ invariant under the centre $Z(E_{\mathbb{A}})$. For a character χ of the idele class group of E and a complex number s we let $\mathbb{H}(s, \chi)$ be the space of functions h on $G(E_{\mathbb{A}})$ such that

$$h[n(x)\operatorname{diag}(a, b)g] = |a/b|^{s+1/2}\chi(a/b)h(g) \tag{1}$$

and whose restriction to K is square integrable. We also denote by $\rho(s, \chi)$ the representation of $G(E_{\mathbb{A}})$ on $\mathbb{H}(s, \gamma)$. The union of the sets $\mathbb{H}(s, \gamma)$ over s is thus a holomorphic fiber bundle of base $\mathbb{C}$. This fiber bundle is trivial; just as in the local case we define $\mathbb{H}(\gamma)$ and for h in $H(\chi)$ the section $h(g, s)$ whose restriction to K is independent of s. If h is any section of this bundle we form an Eisenstein series:

$$E(g, h, s) = \sum h(\gamma g, s), \quad \gamma \in P(E)\backslash G(E). \tag{2}$$

This series converges for $\operatorname{Re} s > 1/2$ and has analytic continuation to the whole complew plane. If h is in $H(\chi)$ the constant term of E along $N(E_{\mathbb{A}})$ has the form

$$E_N(g, h, s) = h(g, s) + [M(s, \chi)h](g, -s) \tag{3}$$

where $M(s, \chi)$ is the operator from $\mathbb{H}(\chi)$ to $\mathbb{H}(\bar{\chi})$ defined by

$$[M(s, \chi)h](g, -s) = \int h(wng, s)\mathrm{d}n, \qquad \text{if} \quad h \text{ is in } \mathbb{H}(\chi).$$

The integral converges also for $\mathrm{Re}(s) > 1/2$ but $M(s, \chi)$ extends to the whole complex plane. Moreover we have

$$(M(s, \chi)h, h') = (h, M(-s, \bar{\chi})h') \quad \text{if} \quad h \in \mathbb{H}(\chi), \ h' \in \mathbb{H}(\chi), \tag{4}$$

$$M(-s, \bar{\chi})M(s, \chi) = \mathrm{Id}. \tag{5}$$

In particular $M(s, \chi)$ is a unitary operator on the line Re $s = 0$. We also introduce a "normalized intertwining operator":

$$R(s, \chi) = M(s, \chi)L(1 + 2s, \chi)\varepsilon(2s, \chi)/L(2s, \chi)$$
$$= M(s, \chi)L(1 + 2s, \chi)/L(1 - 2s, \chi).$$

We need some information on $M(0, \chi)$. Suppose $\chi^2 = 1$. Then $M(0, \chi)$ is an operator commuting with the representation $\rho(0, \chi)$ on $\mathbb{H}(\chi)$. Since this representation is irreducible the operator is a scalar. By (5) this scalar is $+1$ or -1. In fact:

(6) Lemma: *If* $\chi^2 = 1$ *then* $M(0, \chi) = -\mathrm{Id}$.

This is standard (Cf. [R.L.] for instance). Indeed, since $L(s, 1)$ has a pole at $s = 1$, the ratio $L(1 + 2s, \chi^2)/L(1 - 2s, \chi^2)$ takes the value -1 at $s = 0$. Thus it suffices to prove that $R(0, \chi) = 1$. This follows from the fact that $R(0, \chi_v) = 1$ for all v (Cf. (7.3.3)). If $\chi^2 \neq 1$ then $M(0, \chi)$ is no longer an operator intertwinning a representation with itself. We have however the following result:

(7) Lemma: *If* $\chi^2 \neq 1$ *but the restriction of* χ *to* $F_{\mathbf{A}}^{\times}$ *is trivial we have for any element h of* $\mathbb{H}(\chi)$:

$$\int [M(0, \chi)h](k)\mathrm{d}k = \int h(k)\mathrm{d}k, \tag{8}$$

both integral are over $K_F (= K \cap G(F_{\mathbf{A}}))$.

Proof of (7) Since χ^2 is not trivial $L(1 + 2s, \chi^2)$ is holomorphic at $s = 0$. We claim that $L(1, \chi^2) = L(1, \bar{\chi}^2)$. As a matter of fact we claim that

$$L(s, \chi^2) = L(s, \bar{\chi}^2) \qquad \text{for all} \quad s.$$

Indeed it suffices to prove this for Re $s \gg 0$. Then both sides are product over all places v of F of local factors. If v is a place of F which does not split and u the corresponding place of E the local factors on both sides are $L(s, \chi_u^2)$ and $L(s, \bar{\chi}_u^2)$ respectively. They are equal because the restriction of χ to $F_v^\times$ is trivial (Lemma (7.4.4)). On the other hand if v splits into $v1$ and $v2$ the corresponding factors are

$$L(s, \chi_{v1}^2)L(s, \chi_{v2}^2) \quad \text{and} \quad L(s, \bar{\chi}_{v1}^2)L(s, \bar{\chi}_{v2}^2)$$

and they are trivially equal because $\chi_{v1} = \bar{\chi}_{v2}$. Hence $M(0, \chi) = R(0, \chi)$. Now $R(s, \chi)$ is a product over all places u of E of local "normalized intertwining operators" described in section 7. If we assume that h is the product of local components h_u, the integrals on both sides of (8) are product over all places v of F of local analogous integrals and a constant factor the same on both sides. So it suffices to prove the local analogue of (8) for all places v of F. If v is a place of F which splits into $v1$ and $v2$ in E then the local analogue is:

$$\int [R(0, \chi_{v1})h_{v1}](k_v) \cdot [R(0, \chi_{v2})h_{v2}](k_v) \mathrm{d}k_v = \int h_{v1}(k_v)h_{v2}(k_v)\mathrm{d}k_v.$$

Since the restriction of χ to $F_{\mathbf{A}}^\times$ is trivial the character χ_{v1} is actually the inverse of the character χ_{v2} and this follows from (7.3.4). If v does not split and u is the unique place of E above v then the local equality to be proved reads:

$$\int [R(0, \chi_u)h_u](k_v)\mathrm{d}k_v = \int h_u(k_v)\mathrm{d}k_v, \quad k \in K_v.$$

Now the restriction of χ_v to $F_v^\times$ is trivial and this has been proved in (7.4.1). This concludes the proof of lemma (7).

(8.2) We need some estimates on the Eisenstein series. Denote by $\|g\|$ a "norm function" on the group $G(E_{\mathbf{A}})$: it is the product of the norms of the local components g_v divided by $|\det g|$. It is known that, if Ω is a compact set of $\mathbf{C}$ which does not contain any pole of $E(g, s, h)$ and X an element of the envelopping algebra at infinity, then there are constants C and N such that $|\rho(X)E(g, s, h)| < C\|g\|^N$, for s in Ω. We have denoted by $\rho(X)$ the left invariant operator associated to X. It follows from this that the difference of E and its constant term is bounded (in fact rapidly decreasing) on any Siegel set. Consider now our truncated Eisenstein series $T^cE(g, s, h)$. There is a Siegel set S of $G(F_{\mathbf{A}})$ on which it is equal to the difference between the Eisenstein series and its constant term (cf. Lemma (6.1.4)). Moreover the quotient $Z(F_{\mathbf{A}})G(F)\backslash G(F_{\mathbf{A}})$ is the union of a compact set and the image of S. We conclude that for a given h and s in a given compact set of $\mathbb{C}$ we have a uniform estimate $|T^c E(g, s, h)| < C$. In fact we could obtain much better estimates but this will not be needed. We will need however majorizations uniform with respect to s when s is purely imaginary:

(1) LEMMA: *Given h there is a polynomial $P(t)$ such that*

$$|T^c E(g, it, h)| < P(t), \qquad \text{for all real } t.$$

PROOF OF LEMMA (1): Just as before it suffices to establish majorizations of the form

$$|\rho(X)E(g, it, h)| < P(t)\|g\|^N$$

where P and N may depend on X. To see that let us introduce for any Schwartz-Bruhat function Φ in two variables the action of our fibre bundle defined by:

$$F_\Phi(g, s) = \int \Phi[(0, t)g]\chi^2(t)|t|^{2s+1} d^\times t \chi(\det g)|\det g|^{s+1/2}.$$

On the right we have a Tate integral or rather its analytic continuation. There is in fact a finite set T of places of E containing all places at infinity and a function such that

$$h(g, s) = F_\Phi(g, s)/L^T(1 + 2s, \chi^2).$$

where $L^T(s, \chi^2)$ stands for the product of the $L(s, \chi_u)$ over all u not in T. It is classical fact that $1/L^T(1 + 2s, \chi^2)$ is at most of polynomial growth on the line $\mathrm{Re}(s) = 0$. Thus it suffices to find majorizations for the Eisenstein series $E(g, it, F_\Phi)$ and its left invariant derivatives on the group $G(E_\mathbb{A})$. However we may also write this Eisenstein series as the following expression:

$$\int \sum \Phi(t\xi)|t|^{2s+1}\chi^2(t) dt \chi(\det g)|\det g|^{s+1/2}, \quad \xi \in E - 0, \quad t \in E_\mathbb{A}^\times/E^\times.$$

Poisson summation formula gives the analytic continuation of this expression and, at the same time, the required majorizations.

(8.3) We now study the integral of a truncated Eisenstein series over the quotient $Z(F_\mathbb{A})G(F)\backslash G(F_\mathbb{A})$. The description of the double coset space $P(E)\backslash G(E)/G(F)$ given in Lemma (7.5.1) applies: we have a disjoint union

$$G(E) = P(E)G(F) \cup P(E)\lambda G(F).$$

Furthemore the algebra $L = \lambda^{-1} R(E) \lambda \cap M(2, F)$ is an algebra of degree 2 on F which is F-isomorphic to E.

Consider now the integral

$$\int T^c E(g, s, h) \mathrm{d}g, \quad g \in Z(F_{\mathbf{A}}) G(F) \backslash G(F_{\mathbf{A}}).$$

From the majorizations obtained in (2) we see that the integral converges and represents a meromorphic function of s, without pole on the line $\operatorname{Re} s = 0$. We are interested in this integral for s purely imaginary; however in order to compute it we assume $\operatorname{Re}(s) > 1/2$ and then analytically continue our answer. We have

$$T^c E(g, s, h) = \sum h(\xi g, s) - \sum E_N(\gamma g, s, h) \chi_c(H(\gamma g)). \tag{1}$$

The first sum is over $P(E) \backslash G(E)$, the second sum over $P(F) \backslash G(F)$. Using the description of $P(E) \backslash G(E) / G(F)$ we may rewrite the first sum as a sum on $L^\times \backslash G(F)$ and a sum on $P(F) \backslash G(F)$. In a precise way we get:

$$T^c E(g, s, h) = \sum h(\lambda \xi g, s)$$
$$+ \sum \{ h(\gamma g, s) - E_N(\gamma g, s, h) \chi_c(H(\gamma g)) \}. \tag{2}$$

In the first sum, ξ varies in $L^\times \backslash G(F)$; in the second sum γ varies over $P(F) \backslash G(F)$. Taking in account the formula for E_N we may rewrite this as

$$T^c E(g, s, h) = \tag{3}$$

(i) $\sum h(\lambda \xi g, s)$

(ii) $+ \sum h(\gamma g, s)[1 - \chi_c(H(\gamma g))]$

(iii) $- \sum [M(s, \chi) h](\gamma g, -s) \chi_c(H(\gamma g),$

Combining summation and integration in the usual way we get:

$$\int T^c E(g, s, h) \mathrm{d}g = \tag{4}$$

(i) $\int h(\lambda x, s) \mathrm{d}x$

(ii) $+\int h(y, s)[1-\chi_c(H(y))]\mathrm{d}y$

(iii) $-\int [M(s, \chi)h](y, -s)\chi_c(H(y))\mathrm{d}y,$

the integrals over x in $L^\times Z(F_{\mathbf{A}})\backslash G(F_{\mathbf{A}})$ and y in $Z(F_{\mathbf{A}})P(F)\backslash G(F_{\mathbf{A}})$. Using the Iwasawa decomposition we compute integral (4.ii) as

$$\int \chi(a)(1-\chi_c(|a|_E))|a|_E^{s+1/2}|a|_F^{-1}\mathrm{d}^\times a \int h(k)\mathrm{d}k.$$

As usual we denote by F^1 the group of ideles of norm 1 and by F_∞^+ the group of ideles whose finite components are 1 and whose infinite components are equal to some fixed positive number. Then $F_{\mathbf{A}}^\times$ is the direct product of F^1 and F_∞^+. Similarly $F_{\mathbf{A}}^x$ is the product of E^1 and E_∞^+ and when we imbed $F_{\mathbf{A}}^\times$ into $E_{\mathbf{A}}^\times F_\infty^+$ goes to E_∞^+. We choose the character χ to be trivial on E_∞^+ so that χ is trivial on F^2 if and only if it is trivial on $F_{\mathbf{A}}^\times$. Our Haar measure on $F_{\mathbf{A}}^\times$ is the product of a Haar measure on F^1 and the measure $\mathrm{d}t/t$ where $t=|a|$. Then the above integral for a is zero unless the restriction of χ to $F_{\mathbf{A}}^\times$ is trivial in which case its value is:

$$\mathrm{vol}(F^1/F^\times)\int_0^{c^{1/2}} t^{2s}\mathrm{d}t = c^s/2s\ \mathrm{vol}(F^1/F^\times).$$

If we set $\delta(\chi)=\mathrm{vol}(F^1/F^\times)$ if the restriction of χ to $F_{\mathbf{A}}^\times$ is trivial and zero otherwise we see that (8.3.4.ii) is equal to

$$\delta(\chi)\frac{c^s}{2s}\int h(k)\mathrm{d}k. \tag{4.i}$$

Similarly, using the fact that $M(s, \chi)h(g, -s)$ lies in $\mathbb{H}(-s, \bar{\chi})$ we calculate that (8.3.4.iii) is equal to:

$$-\delta(\chi)\frac{c^{-s}}{2s}\int [M(s, \chi)h](k)\mathrm{d}k. \tag{4.ii}$$

Actually $\delta(\chi)=1$ but we will not have to use this fact. We write (8.3.4.i) as

$$\int h(\lambda lg, s)d^x l\mathrm{d}g, \quad l\in F_{\mathbf{A}}^\times L^\times\backslash L_{\mathbf{A}}^\times, \quad g\in L_{\mathbf{A}}^\times\backslash G(F_{\mathbf{A}}).$$

Next we recall that $\lambda^{-1}l\lambda$ is a triangular matrix whose eigenvalues are $e(l)$ and $e'(l)$ where $l\mapsto e(l)$ and $l\mapsto e'(l)$ are the two isomorphisms of L onto E. Then in the above integral the inner integral takes the following form:

$$\int h(\lambda g, s)|t^{\tau-1}|^{s+1/2}\chi(t^{\sigma-1})\mathrm{d}^{\times}t, \quad t\in F^{\times}E_{\mathbf{A}}^{\times}\backslash E_{\mathbf{A}}^{\times}.$$

This is zero unless $\chi^{\sigma-1}=1$, that is unless χ has the form $\mu o N(E/F)$ for some character μ of $F_{\mathbf{A}}^{\times}$. If we let $\varepsilon(\chi)=\mathrm{vol}(F_{\mathbf{A}}^{\times}E^{\times}\backslash E_{\mathbf{A}}^{\times})$ if $\chi^{\sigma-1}=1$ and zero otherwise then (8.3.4.i) is

$$\varepsilon(\chi)\int h(\lambda g, s)\mathrm{d}g, \quad g\in L_{\mathbf{A}}^{\times}\backslash G(F_{\mathbf{A}}).$$

In fact $\varepsilon(\chi)=2$. We need to study the analytic properties of this function of s. We proceed as in the local case (Cf. (7.6)). We take:

$$\lambda = wn(\sqrt{\tau}); \quad \text{then} \quad L=\{\mathrm{Mat}[a, b\tau, b, a]\}.$$

Next we use the standard device of representing $h(g, s)$ in the form

$$h(g, s)=[Q(s)/L(2s+1, \chi^2)]$$
$$\times\int\Phi[(0, t)g])\chi^2(t)|t|_E^{2s+1}\chi(\det g)|\det g|^{s+1/2},$$

where Φ is a Schwartz-Bruhat function in two variables and Q an elementary function, that is a linear combinations of products of rational and exponential functions of s; then

$$(0, t)\lambda=(1, \sqrt{\tau})l \qquad \text{with} \quad 1=\mathrm{Mat}[a, b\tau, b, a],$$

and the above integral for g in $G(F)$ can be written as an integral on $L_{\mathbf{A}}^{x}$:

$$\int\Phi[(1, \sqrt{\tau})1g]\mu^2(\det 1g)|\det 1g|_F^{2s+1}\mathrm{d}^x1.$$

Therefore the integral on $L_{\mathbf{A}}^{\times}\backslash G(F_{\mathbf{A}})$ can be combined with this integral to give an integral on $G(F_{\mathbf{A}})$:

$$\int h(\lambda g, s)\mathrm{d}g=[Q(s)/L(2s+1, \chi^2)]$$
$$\times\int\Phi[(1, \sqrt{\tau})g]\mu^2(\det g)|\det g|_F^{2s+1}\mathrm{d}g.$$

We can use the Iwasawa decomposition on $G(F_{\mathbf{A}})$ to compute this integral. We find:

$$[Q(s)/L(2s+1, \chi^2)]$$

$$\times \int \Phi[(a, b\sqrt{\tau}+x)k]\mathrm{d}x\mu^2(a)|a|_F^{2s}\mu^2(b)|b|_F^{2s+1}\mathrm{d}^\times a\mathrm{d}^\times b\mathrm{d}k.$$

After integrating over x and k the resulting function of a and b is Schwartz-Bruhat and the integral is a double Tate integral. In particular we see that as a function of s this has the form:

$$Q'(s)\cdot L(2s, \mu^2)L(2s+1, \mu^2)/L(2s+1, \chi^2)$$

$$= Q'(s)L(2s, \mu^2)/L(2s+1, \mu^2\eta)$$

where Q' is another elementary function of s and η the quadratic character attached to the extension E of F. This also shows that the original integral converges for Re $s > 1/2$. Furthemore, just as in the local case, this suggests introducing a normalized version of the integral. In a precise way if h is in $H(\chi)$ and $\chi^{s-1}=1$ we will set

$$I(s, \chi)h = \int h(\lambda g, s)\mathrm{d}g, \quad g \in L_{\mathbf{A}}^\times \backslash G(F_{\mathbf{A}}),$$

$$I(s, \chi)h = L(2s, \mu^2)/L(2s+1, \mu^2\eta)\cdot J(s, \chi),$$

where $\chi = \mu o N(E/F)$. It is clear that $J(s, \chi)h$, for h K-finite is an elementary function of s without singularity on the line Re $s=0$.
(8.4) We summarize the results obtained so far:

PROPOSITION: *The integral of $E(g, h, s)$ over $Z(F)G(F)\backslash G(F)$ is the sum of the following*:

$$\delta(\chi)\frac{c^s}{2s}\int h(k)\mathrm{d}k, \tag{1}$$

$$-\delta(\chi)\frac{c^{-s}}{2s}\int [M(s, \chi)h](k)\mathrm{d}k, \tag{2}$$

$$\varepsilon(\chi)I(s, \chi)h. \tag{3}$$

Furthemore

$$I(s, \chi) = L(2s, \mu^2)/L(2s+1, \mu^2\eta)J(s, \chi)h$$

and $J(s, \chi)h$ is an elementary function of s if h is K-finite.

Because the Eisenstein series has no pole on the line Re $s = 0$ and its truncation is integrable the sum of (1), (2) and (3) is actually holomorphic at $s = 0$. It will be useful to make this statement more precise. To begin with we make the following elementary remark about the character χ:

(4) LEMMA: *Consider the following conditions*:

(i) $\chi^{\sigma-1} = 1$;

(ii) *the restriction of* χ *to* $F_{\mathbb{A}}^{\times}$ *is trivial*;

(iii) *the restriction of* χ *to* $F_{\mathbb{A}}^{\times}$ *is the quadratic character* η *attached to the extension* E;

(iv) $\chi^2 = 1$.

Then: (iv) $\Leftrightarrow$ (i) *and* [(ii) *or* (iii)]

PROOF OF (4): Suppose $\chi^2 = 1$. If χ is actually the trivial character then (i) and (ii) are satisfied. Suppose not and let D be the quadratic extension of E corresponding to χ. Then D is a Galois extension of F of degree 4 hence Abelian. By class field theory we may view χ as a character of Gal(D/E). Assertion (i) amounts to say that if σ' is an element of Gal(D/F) whose image in Gal(E/F) is σ then

$$\chi(\sigma' g \sigma'^{-1}) = \chi(g).$$

Since Gal(D/F) is Abelian this is clear. This being so let us write χ in the form $\mu o N(E/F)$. By (iv) we see that μ^2 is trivial on the image of $N(E/F)$. Thus μ^2 is either the character trivial or the character η. Since the restriction of χ to $F_{\mathbb{A}}^{\times}$ is μ^2, we see that either (ii) or (iii) holds. Now suppose (i) is satisfied. Write therefore χ in the form $\mu o N(E/F)$. Again the restriction of χ to $F_{\mathbb{A}}^{\times}$ is μ^2. If (ii) is satisfied we find $\mu^2 = 1$. A fortiori $\chi^2 = \mu^2 o N(E/F) = 1$. If (iii) is satisfied we find $\mu^2 = \eta$. A fortiori $\chi^2 = \mu o N(E/F) = 1$. So in any case $\chi^2 = 1$.

Let us go back now to the sum of the three terms in Proposition (8.4). Suppose first that the restriction of χ to $F_{\mathbb{A}}^{\times}$ is trivial but $\chi^{\sigma-1} \neq 1$. Then the term (3) vanishes. Furthemore $\chi^2 \neq 1$ otherwise by the lemma $\chi^{\sigma-1} = 1$. Accordingly by (8.1.7):

$$\int M(0, \chi) h(k) \mathrm{d}k = \int h(k) \mathrm{d}k \tag{5}$$

and the residues of the terms (1) and (2) at $s = 0$ do cancell. Suppose now that $\chi^{\sigma-1} = 1$ but the restriction of χ to $F_{\mathbb{A}}^{\times}$ is not trivial. Then the terms (1) and (2) vanish. Let us write χ in the form $\chi = \mu o N(E/F)$ so that the restriction of χ to $F_{\mathbb{A}}^{\times}$ is μ^2. Now μ^2 cannot be trivial otherwise the restriction of χ to $F_{\mathbb{A}}^{\times}$ would be trivial. Thus $L(2s, \mu^2)$ is holomorphic at

$s=0$ and (3) is too. Note that if $\chi^2=1$ then, by the lemma, $\mu^2\eta=1$ and $1/L(2s+1, \mu^2\eta)$ has a zero at $s=0$; hence (3) vanishes at $s=0$. Finally suppose that the restriction of χ to $F_{\mathbb{A}}^{\times}$ is trivial and $\chi^{\sigma-1}=1$. Then $\chi^2=1$ and writing $\chi=\mu o N(E/F)$ we have $\mu^2=1$. Then $M(0,\chi)=-1$ and (1) and (2) have the same residue at $s=0$. Doubling this residue we find then

$$\delta(\chi)\int h(k)\mathrm{d}k+\varepsilon(\chi)\lim_{s\to 0} sI(s,\chi)h=0 \tag{6}$$

or

$$\delta(\chi)\int h(k)\mathrm{d}k+\varepsilon(\chi)2c_0/L(1,\mu^2\eta)J(0,\chi)h=0 \tag{7}$$

where c is the residue of $L(s, 1_F)$ at 0. Of course this relation could be proved by local means as well (See below).

(8.5) For Re $s>1/2$ the integral $\int h(\lambda g, s)\mathrm{d}g$ converges absolutely. Assume that h is a product over all places u of E of local components h_u belonging to $\mathbb{H}(\chi_u)$; of course for almost all place u, $h_u(k_u)=1$ for all k_u. Then, for Re $s>1/2$, the integral can be written as a product of local integrals and a constant factor; the product is over all places of F. We now describe the factor attached to a place v of F. Suppose first that v does not split in the extension and let u be the unique place of E above v. Then the local component is just the local integral

$$\int h_v(\lambda g_v, s)\mathrm{d}g_v=L(2s,\mu_v^2)/L(2s+1,\mu_v^2\eta_v)J(s,\chi_u)h_u.$$

Suppose now that v splits into $v1$ and $v2$ in the quadratic extension. Thus we have isomorphisms $G_{v1}\cong G_v$ and $G_{v2}\cong G_v$. Let λ_1 and λ_2 be the images of λ under these two isomorphisms. The local component is the following integral over $L_v^{\times}\backslash G_v$:

$$\int h_{v1}(\lambda_1 g_v)\cdot h_{v2}(\lambda_2 g_v)\mathrm{d}g_v.$$

It will be convenient to take

$$\lambda_i=wn(\pm\sqrt{\tau}),\quad \text{with } + \text{ for } i=1 \text{ and } - \text{ for } i=2.$$

Note that h_{v1} and h_{v2} are invariant under $N(F_v)$. Thus the local integral may also be written as:

$$\int h_{v1}(\mu_1 g)\cdot h_{v2}(\mu_2 g_v)\mathrm{d}g_v.$$

where

$$\mu_i = n(\mp 1/2\sqrt{\tau})wn(\pm\sqrt{\tau}).$$

Observe that

for $1 = \mathrm{Mat}[a, b\tau, b, a]$ in L we have

$$\mu_1^{-1} 1 \mu_2 = \mathrm{diag}(a + b\sqrt{\tau}, a - b\sqrt{\tau}).$$

Hence $\mu_1^{-1} L \mu_2 = A$. After changing f to $\mu_2^{-1} g$ we get therefore an integral over $A_v \backslash G_v$:

$$\int h_{v1}(\mu_1 \mu_2^{-1} g) h_{v2}(g) \mathrm{d}g$$

$$= c_v |1/4\tau|_v^{2s+1} \chi_v(-1/4\tau) \int h_{v1}(wg) h_{v2}(g) \mathrm{d}g$$

because $\mu_1\mu_2^{-1} = \mathrm{diag}(-1/2\sqrt{\tau}, +2\sqrt{\tau})w$. Here c_v is a constant. There is a question of which measures we are using. We regard $L^{\times}$ and A as algebraic groups over F and choose invariant differential forms on them. Then for any place v of F we inherit the Tamagawa measure on L and A. They are so defined that that for almost all v the measure of the maximal compact subgroup is one. Thus, for almost all v, the isomorphism $1 \mapsto \mu_2^{-1} 1 \mu_2$ is compatible with our choice of measures, that is, $c_v = 1$.

Now because $\chi^{\sigma - 1} = 1$ we have $\chi_{v1} = \chi_{v2} = \mu_v$ and

$$\int h_{v1}(wg) h_{v2}(g) \mathrm{d}g = \int h_{v1}(wn(x_v)k_v) h_{v2}(n(x_v)k_v) \mathrm{d}x_v \mathrm{d}k_v$$

$$= \langle M(s, \mu_v) h_{v1}, h_{v2} \rangle.$$

In turn, apart from an ε factor, this is:

$$L(2s, \mu_v^2)/L(2s+1, \mu_v^2) \langle R(s, \mu_v) h_{v1}, h_{v2} \rangle.$$

We conclude that apart from a factor CA^s, $J(s, \chi)h$ is the product over all places v of F of a factor which is either $J(s, \chi_u)h_u$ or $\langle R(s, \mu_v) h_{v1}, h_{v2} \rangle$.

§9. Truncation of the Eisenstein kernel

(9.1) In this section we deal with the integral of the truncation of $K_{\mathrm{eis}}(x, y)$. We shall prove that the following limit exists:

$$\lim_{c_1 \to \infty} \iint T_1^{c_1} T_2^{c_2} K_{\mathrm{eis}}(x, y) \mathrm{d}x \mathrm{d}y, \quad x, y \in Z(F_{\mathbf{A}}) G(F) \backslash G(F_{\mathbf{A}}).$$

The resulting function of c_2 has an asymptotic expansion of the form $A \log c_2 + B + o(1)$ as $c_2 \to +\infty$, where A, B are constants. We shall prove that, under the assumptions of (6.1.1), $B = 0$.

We recall the formula for K_{eis} (cf. [J.A.] for instance). For each character χ of $E_{\mathbf{A}}^{\times}/E^{\times}$ trivial on E_{∞}^{+}, we choose an orthonormal basis h_α of $\mathbb{H}\chi$). Then:

$$K_{\text{eis}}(x, y) = \sum_{\chi} K(x, y, \chi), \text{ where}$$

$$K(x, y, \chi) = 1/4\pi \sum_{\beta,\alpha} \int_{-i\infty}^{i\infty} (\rho(s, \chi)(f) h_\beta, h_\alpha) \times E(x, h_\alpha, s) \bar{E}(y, h_\beta, s) |ds|. \quad (1)$$

The first sum is over all χ; however for a given f almost all terms are zero. Similarly, the second sum is for all pairs (α, β); however for a given f almost all terms are zero. The integral converges uniformly over compacts set of $G(E_{\mathbf{A}})$. In particular to compute the truncation of our kernel, we may "bring the truncation operator under the integral sign":

$$T_1^{c_1} T_2^{c_2} K_{\text{eis}}(x, y) = \sum T_1^{c_1} T_2^{c_2} K(x, y, \chi),$$

$$T_1^{c_1} T_2^{c_2} K(x, y, \chi) = 1/4\pi \sum \int (\rho(s, \chi) h_\beta, h_\alpha) T_1^{c_1} E(x, h_\alpha, s) \cdot T_2^{c_2} E(y, h_\beta, s) |ds|. \quad (2)$$

Since $K_{\text{cusp}}(x, y)$ is integrable and $T_1^{c_1} T_2^{c_2} K$, $T_1^{c_1} T_2^{c_2} K_{\text{sp}}$ are integrable, it is clear that $T_1^{c_1} T_2^{c_2} K_{\text{eis}}$ is also integrable. Moreover $|T^c E(x, h, s)|$ is bounded independently of x by a polynomial in $\text{Im}(s)$ on the line $\text{Re}(s) = 0$ (Cf. (8.2)). In particular we may interchange the integral in x and y with the summations and the integration in s:

$$\iint T_1^{c_1} T_2^{c_2} K_{\text{eis}}(x, y) dx dy = \sum \iint T_1^{c_1} T_2^{c_2} K(x, y, \chi) dx dy,$$

$$\iint T_1^{c_1} T_2^{c_2} K(x, y, \chi) dx dy = \sum 1/4\pi \int (\rho(s, \chi) h_\beta, h_\alpha) \int T_1^{c_1} E(x, h_\alpha, s) dx \cdot \overline{\int T_2^{c_2} E(y, h_\beta, s) dy} |ds|. \quad (3)$$

(9.2) To evaluate the limit of this as c_1 tends to infinity we will use the following elementary lemmas which are recalled without proof:

(1) LEMMA: *Syppose F is a Schwartz function on* $\mathbb{R}$. *Then*:

$$\lim_{x \to \infty} \int F(y) \exp(\pm 2i\pi xy) \mathrm{d}y/y = \pm i\pi F(0)$$

We remark that the integral is actually an improper integral. Furthemore this function of x tends rapidly toward its limit. Thus:

$$x \int F(y) \exp(\pm 2i\pi xy) \mathrm{d}y/y = \pm i\pi x F(0) + o(1). \tag{2}$$

(3) LEMMA: *If F is a Schwartz function on R and* $F(0) = 0$ *then*

$$\int_{-\infty}^{\infty} (F(y)/y^2) \mathrm{d}y = \int_{-\infty}^{\infty} F'(y) \mathrm{d}y/y.$$

Again this is an equality of improper integrals.

We will need to show that the functions we are dealing with are indeed Schwartz functions. We proceed to that end via a series of lemmas.

(4) LEMMA: *For a fixed and* $|t| \mapsto +\infty$ *we have*:

$$\Gamma^{(n)}(a + it) = \Gamma(a + it) O(\log|t|^n).$$

PROOF OF (4): For $n = 1$ this is a classical estimate on the function $\psi(z) = \Gamma'(z)/\Gamma(z)$. Suppose $n > 1$; then the n-th derivative of ψ can be computed as the sum of $\Gamma^{(m)}/\Gamma$ and products of the form $\pm \Pi \Gamma^{(i)}/\Gamma$ where the orders i are stricly less than n and their sum is n. Our assertion follows then by induction on n from the classical estimate $\psi^{(n)}(a + it) = O(1)$.

Next we recall without proof two estimates on L-functions.

(5) LEMMA: *Let* χ *be an idele class character of a number field. Denote by* $L_\infty(s, \chi)$ *the product of the* $L(s, \chi_v)$ *for v infinite and by* $L^\infty(s, \chi)$ *the product of the* $L(s, \chi_v)$ *for v finite. Then, for* $|t|$ *large, we have*

$$\frac{\mathrm{d}^n}{\mathrm{d}s^n} L^\infty(1 + it, \chi) = O\left(\log|t|^{n+1}\right).$$

On the other hand, there is some $N > 0$ such that for $|t|$ large:

$$1/L^\infty(1 + it) = O\left(|t|^N\right)$$

We are now in position to state the results that we need:

(6) LEMMA: *Let χ and η be idele class characters of a number field F. Assume η is trivial or quadratic. Then any derivative of the ratio*

$$L(2s, \chi)/L(1+2s, \chi\eta)$$

is at most of polynomial growth on the line Re $s = 0$.

PROOF OF (6): Our ratio is the product of an exponential factor,

$$L^{\infty}(1-2s, \bar{\chi})/L^{\infty}(1+2s, \chi\eta) \quad \text{and}$$

$$L_{\infty}(1-2s, \bar{\chi})/L_{\infty}(1+2s, \chi\eta).$$

It follows from lemma (5) that any derivative of the first ratio is at most of polynomial growth on the line Re $s = 0$. So it suffices to prove the second ratio has the same property. In turn it is a product of factors of the form:

$$\Gamma_1(s)/\Gamma_2(s) \quad \text{with} \quad \Gamma_1(s) = \Gamma(a(1-2s)+b),$$
$$\Gamma_2(s) = \Gamma(a(1+2s)+b+c)$$

where a and c are real. All we have to show therefore is that any derivative of a function Γ_1/Γ_2 is at most of polynomial growth on the line Re $s = 0$. By Stirling formula this is true of the function itself. On the other hand any derivative of Γ_1/Γ_2 is a sum of products: in each product we have a factor ± 1, the factor Γ_1/Γ_2 and factors of the form $\Gamma_i^{(j)}/\Gamma_i$. Our assertion follows then from Lemma (4).

(9.3) We set $c_i = \exp(2\pi u_i)$, $i = 1, 2$. We also set:

$$\Phi(\alpha, u, t) = 2it \int T^c E(x, h_\alpha, it)\mathrm{d}x, \qquad \text{if} \quad c = \exp(2\pi u).$$

Then, by Proposition (8.4):

$$\Phi(\alpha, u, t) = \delta(\chi)\exp(2i\pi ut)\int h_\alpha(k)\mathrm{d}k$$
$$-\delta(\chi)\exp(-2i\pi ut)\int [M(it, \chi)h_\alpha](k)\mathrm{d}k$$
$$+\varepsilon(\chi)2itI(it, \chi)h_\alpha. \tag{2}$$

The integral is for k in $K_F = K \cap G(F_{\mathbb{A}})$. We remind ourselves that when $\delta(\chi)$ and $\varepsilon(\chi)$ are not zero $I(it, \chi)$ has a pole at $t = 0$ and

$$\lim_{t\to 0} itI(it, \chi)h_\alpha = -\int h_\alpha(k)\mathrm{d}k \tag{3}$$

(Cf. section (8.5)). Furthemore in any case $\Phi(\alpha, u, 0) = 0$ (loc. cit.). With this notation we have:

$$\iint T_1^{c_1} T_2^{c_2} K_{\text{eis}}(x, y) \mathrm{d}x \mathrm{d}y = \sum_{\chi, \alpha, \beta} G(u_1, u_2, \chi, \alpha, \beta) \tag{4.i}$$

where

$$G(u_1, u_2, \chi, \alpha, \beta) = \int_{-\infty}^{\infty} F_{\beta a}(t) \Phi(\alpha, u_1, t) \overline{\Phi}(\beta, u_2, t) \mathrm{d}t/t^2. \tag{4.ii}$$

$$F_{\beta\alpha}(t) = 1/16\pi\big(\rho(s, \chi)(f) h_\beta, h_\alpha\big). \tag{4.iii}$$

To continue in the expression for G we replace $\Phi(\alpha, u_1, t)$ by its expression (2):

$$G(u_1, u_2, \chi, \alpha, \beta) = \tag{5}$$

(i) $$\int F_{\beta\alpha}(t) \frac{\overline{\Phi}(\beta, u_2, t)}{t} \delta(\chi) \int h_\alpha(k) \mathrm{d}k \exp(2\mathrm{i}\pi u_1 t) \frac{\mathrm{d}t}{t}$$

(ii) $$\int F_{\beta\alpha}(t) \frac{\overline{\Phi}(\beta, u_2, t)}{t} \delta(\chi)$$

$$\int (-M(\mathrm{i}t, \chi) h_\alpha)(k) \mathrm{d}k \exp(-2\mathrm{i}\pi u_1 t) \frac{\mathrm{d}t}{t}$$

(iii) $$\int F_{\beta\alpha}(t) \overline{\Phi}(\beta, u_2, t) \cdot 2\mathrm{i}t\varepsilon(\chi) I(\mathrm{i}t, \chi) h_\alpha \frac{\mathrm{d}t}{t^2}.$$

We may apply lemma (9.2.1) to (i) and (ii). Indeed $F_{\beta\alpha}(t)$ is a linear combination of products of the form $\exp(ait)\hat{F}(t)$ where F is a smooth function of compact support on $\mathbb{R}$. On the other hand

$$\int [M(\mathrm{i}t, \chi)h](k) \mathrm{d}k = L(2\mathrm{i}t, \chi^2)/L(1 + 2\mathrm{i}t, \chi^2)$$

$$\times \varepsilon(2\mathrm{i}t, \chi^2) \int R(\mathrm{i}t, \chi) h(k) \mathrm{d}k$$

$$= L(2\mathrm{i}t, \chi^2)/L(1 + 2\mathrm{i}t, \chi^2) Q(\mathrm{i}t)$$

where Q is an elementary function of t (Lemma (7.3.6)). Similarly:

$$I(\mathrm{i}t, \chi)h = Q'(\mathrm{i}t) L(2\mathrm{i}t, \mu^2)/L(1 + 2\mathrm{i}t, \mu^2\eta)$$

where Q' is an elementary functions of s, μ an idele class character of F and η is the quadratic character attached to E (Cf. Prop (8.4)). It follows from lemma (6) that the derivatives of these functions of t are at most of polynomial growth. Taking (2) in account we see that the same is true for the derivatives of $\Phi(\alpha, u_2, t)$. We conclude that indeed the functions of t under the integral sign in (5.i) and (5.ii) are Schwartz functions. Applying then (9.2.1) we find that the limit of (i) + (ii) as u tends to plus infinity is

$$i\pi F_{\beta\alpha}(0) \cdot \left. \frac{\overline{\Phi}(\beta, u_2, t)}{t} \right|_{t=0} \cdot \delta(\chi)$$

$$\cdot \left[\int h_\alpha(k) \mathrm{d}k + \int M(O, \chi) h_\alpha(k) \mathrm{d}k \right].$$

Of course this is zero unless the restriction of χ to $F_{\mathbf{A}}^\times$ is trivial. Assuming this to be the case, if $\chi^2 = 1$ then $M(0, \chi) = -\mathrm{Id}$ and this vanishes. If, on the contrary, $\chi^2 \neq 1$ then both terms are equal (Cf. lemma (8.1.6) and (8.1.7)). Set $\theta(\chi) = 1$ if $\chi^2 \neq 1$ and zero otherwise. Then we see that the limit of (i) + (ii) as u_1 tends to infinity is:

$$2i\pi F_{\beta\alpha}(0) \cdot \left. \frac{\overline{\Phi}(\beta, u_2, t)}{t} \right|_{t=0} \int h_\alpha(k) \mathrm{d}k \delta(\chi) \theta(\chi). \tag{7}$$

Hence we see that the limit of $G(u_1, u_2, \chi, \alpha, \beta)$ as u_1 tends to infinity is the sum of (7) and (5.iii). Recall that $\Phi(\beta, u_2, 0) = 0$. So $\Phi(\beta, u_1, t)/t|_{t=0}$ is the derivative $\mathrm{d}\Phi(\beta, u_2, t)/\mathrm{d}t$ at $t = 0$. To compute it we use (2). However if (7) is non zero the restriction of χ to $F_{\mathbf{A}}^\times$ is trivial but $\chi^2 \neq 1$. By lemma (8.4.4) $\varepsilon(\chi) = 0$ and the last term in (2) vanishes. Thus our derivative is actually:

$$\delta(\chi) 2\pi i u_2 \left[\int h_\beta(k) \mathrm{d}k + \int (M(0, \chi) h_\beta)(k) \mathrm{d}k \right]$$

$$- \delta(\chi) \frac{\mathrm{d}}{\mathrm{d}t} \left[\int [M(it, \chi) h_\beta](k) \mathrm{d}k \right] \Bigg|_{t=0}.$$

On the other hand, since $\chi^2 \neq 1$, we have by lemma (8.1.7)

$$\int M(0, \chi) h_\beta(k) \mathrm{d}k = \int h_\beta(k) \mathrm{d}k.$$

So (7) under the conditions $\delta(\chi) \neq 0$ and $\theta(\chi) = 1$ is the sum of

$$-\theta\delta^2(\chi) 8\pi^2 F_{\beta\alpha}(0) \cdot u_\alpha \int h_\alpha(k) \mathrm{d}k \cdot \overline{\int h_\beta(k) \mathrm{d}k}. \tag{8}$$

$$-2\pi i\theta\delta^2(\chi)F_{\beta\alpha}(0)\cdot\int h_\alpha(k)\mathrm{d}k\cdot\frac{\mathrm{d}}{\mathrm{d}t}\left[\int M(it,\chi)h_\beta(k)\mathrm{d}k\right]\Bigg|_{t=0}. \tag{9}$$

At this point we have evaluated our limit as c_1 tends to infinity. We state this as a lemma:

(10) LEMMA: *The limit of the integral of $T_1^{c_1}T_2^{c_2}K_{\mathrm{eis}}$ as c_1 tends to infinity exists. It is the sum over all χ, α, β of:*

(i) $$-2\pi i\delta^2\theta(\chi)F_{\beta\alpha}(0)\cdot\int h_\alpha(k)\mathrm{d}k\cdot\frac{\mathrm{d}}{\mathrm{d}t}\left[\overline{\int M(it,\chi)h_\beta(k)\mathrm{d}k}\right]\Bigg|_{t=0}$$

(ii) $$+\int_{-\infty}^{\infty}F_{\beta\alpha}(t)\overline{\Phi}(\beta,u_2,t)\cdot 2it\varepsilon(\chi)I(it,\chi)h_\alpha\frac{\mathrm{d}t}{t^2},$$

and a term $A\log c_2$.

(9.4) We now obtain an asymptotic expansion for c_2 large. We remark that in (9.3.10.ii) the factor $\overline{\Phi}(\beta, u_2, t)$ has a zero at $t=0$ while the factor $tI(it,\chi)h_\alpha$ has no pole at $t=0$. Furthemore the product of these functions by $F_{\beta\alpha}(t)$ is a Schwartz function. Thus we may use the integration by parts formula given in Lemma (9.2.2). We replace $\Phi(\beta, u_2, t)$ by its expression from (9.3.2). We obtain then for (9.3.10.ii):

$$-\varepsilon(\chi)4\int\frac{\mathrm{d}}{\mathrm{d}t}\left[F_{\beta\alpha}(t)\cdot tI(it,\chi)h_\alpha\cdot\overline{tI(it,\chi)h_\beta}\right]\frac{\mathrm{d}t}{t}, \tag{1}$$

$$\delta\varepsilon(\chi)\int\frac{\mathrm{d}}{\mathrm{d}t}\left[F_{\beta\alpha}(t)\cdot 2itI(it,\chi)h_\alpha\right]\cdot\int\overline{h}_\beta(k)\mathrm{d}ke^{-2i\pi u_2t}\frac{\mathrm{d}t}{t}, \tag{2}$$

$$-2i\pi u_2\delta\varepsilon(\chi)\int F_{\beta\alpha}(t)\cdot 2itI(it,\chi)h_\alpha\cdot\int\overline{h}_\beta(k)\mathrm{d}ke^{2i\pi u_2t}\frac{\mathrm{d}t}{t}, \tag{3}$$

$$-\delta\varepsilon(\chi)\int\frac{\mathrm{d}}{\mathrm{d}t}\left[F_{\beta\alpha}(t)\cdot 2itI(it,\chi)h_\alpha\right]$$
$$\overline{\int M(it,\chi)h_\beta(k)\mathrm{d}k}e^{2i\pi u_2t}\frac{\mathrm{d}t}{t}, \tag{4}$$

$$-\delta\varepsilon(\chi)\int F_{\beta\alpha}(T)\cdot 2itI(it,\chi)h_\alpha$$
$$\cdot\frac{\mathrm{d}}{\mathrm{d}t}\left[\overline{\int M(it,\chi)h_\beta(k)\mathrm{d}k}\right]e^{2i\pi u_2t}\frac{\mathrm{d}t}{t}, \tag{5}$$

$$-2\pi i u_2 \delta \varepsilon(\chi) \int F_{\beta\alpha}(t) \cdot 2itI(it, \chi) h_\alpha$$

$$\overline{\int M(it, \chi) h_\beta(k) \mathrm{d}k} e^{2i\pi u_2 t} \frac{\mathrm{d}t}{t}. \tag{6}$$

We can apply lemma (9.2.1) to (3) and (6). We get that each term has the form $A \log c_2 + o(1)$. So we can ignore these terms. On the other hand we can use the same lemma to find the limit of (2), (4) and (5). We get:

$$-\delta\varepsilon(\chi)\pi i \frac{\mathrm{d}}{\mathrm{d}t}\left[F_{\beta\alpha}(t) \cdot 2itI(it, \chi)h_\alpha\right]\Big|_{t=0} \cdot \int \bar{h}_\beta(k)\mathrm{d}k, \tag{7}$$

$$-\delta\varepsilon(\chi)\pi i \frac{\mathrm{d}}{\mathrm{d}t}\left[F_{\beta\alpha}(t) \cdot 2itI(it, \chi)h_\alpha\right]\Big|_{t=0} \cdot \overline{\int M(0, \chi)h_\beta(k)\mathrm{d}k}, \tag{8}$$

$$-\delta\varepsilon(\chi)\pi i \left\{\left[F_{\beta\alpha}(t) \cdot 2itI(it, \chi)h_\alpha \cdot \frac{\mathrm{d}}{\mathrm{d}t}\left[\overline{\int M(it, \chi)h_\beta(k)\mathrm{d}k}\right]\right\}\right|_{t=0}. \tag{9}$$

Again (7) and (8) are zero unless $\chi^{\sigma-1} = 1$ and the restriction of χ to $F_{\mathbf{A}}^\times$ is trivial. Then $\chi^2 = 1$ by lemma (8.4.4) and their sum is zero because $M(0, \chi) = -\mathrm{Id}$ (lemma (8.1.6)). Thus (7) + (8) is zero in any case. Let us look at (9). Again it is zero unless $\chi^{\sigma-1} = 1$ and the restriction of χ to $F_{\mathbf{A}}^\times$ is trivial. Then we find by (8.4.6):

$$\lim_{t\to 0} \varepsilon(\chi) itI(it, \chi)h_\alpha = -\delta(\chi)\int h_\alpha(k)\mathrm{d}k$$

and (9) becomes

$$\pi i \delta^2(\chi) F_{\beta\alpha}(0) \cdot \int h_\alpha(k)\mathrm{d}k \cdot \frac{\mathrm{d}}{\mathrm{d}t}\Big|_{t=0} \overline{\int M(it, \chi)h_\beta(k)\mathrm{d}k}. \tag{10}$$

(9.4) We summarize our results in a proposition:

Proposition: *The integral of* $T_1^{c_1}T_2^{c_2}K_{\mathrm{eis}}$ *over the product of* $Z(F_{\mathbf{A}})G(F)\backslash G(F_{\mathbf{A}})$ *by itself has a limit as* c_1 *tends to infinity. The resulting function of* c_2 *is the sum of a term* $A \log c_2$, *a term* $o(1)$ *and the sum over* χ, α, β *of the following terms*:

$$\varepsilon(\chi)\int_{-\infty}^{\infty} \frac{\mathrm{d}}{\mathrm{d}t}\left[F_{\beta\alpha}(T) \cdot tI(it, \chi)h_\alpha \cdot \overline{tI(it, \chi)h_\beta}\right]\frac{\mathrm{d}t}{t} \tag{1}$$

$$-2\pi i\delta^2(\chi)\chi(\chi)F_{\beta\alpha}(0)\int h_\alpha(k)\mathrm{d}k\,\frac{\mathrm{d}}{\mathrm{d}t}\bigg|_{t=0}\left[\int M(\mathrm{i}t,\chi)h_\beta(k)\mathrm{d}k\right] \tag{2}$$

$$\pi i\delta^2\varepsilon(\chi)F_{\beta\alpha}(0)\int h_\alpha(k)\mathrm{d}k\,\frac{\mathrm{d}}{\mathrm{d}t}\bigg|_{t=0}\left[\int M(\mathrm{i}t,\chi)h_\beta(k)\mathrm{d}k\right]. \tag{3}$$

We remark that in (2) and (3) the notation is somewhat misleading: both terms vanish unless $\delta(\chi)=0$. If $\delta(\chi)\neq 0$ then by lemma (8.4.4) either $O(\chi)=0$ or $\varepsilon(\chi)=0$ (exclusive or). Thus either (2) or (3) or both vanish.

(9.5) We now prove that, under the assumptions of (6.6.1), given χ, the sum over α, β of (1) (resp. (2), (3)) in Prop. (9.4) is zero. We start with (2) and (3). In view of the above remarks what we have to prove is this: given χ whose restriction to $F_{\mathbb{A}}^{\times}$ is trivial the following function of t has a zero of order 2 at $t=0$:

$$\sum_{\alpha,\beta}\left(\rho(\mathrm{i}t,\chi)(f)h_\beta,h_\alpha\right)\int h_\alpha(k)\mathrm{d}k\cdot\overline{\int M(\mathrm{i}t,\chi)h_\beta(k)\mathrm{d}k}. \tag{1}$$

Since $M(\mathrm{i}t,\gamma)=m(t)R(\mathrm{i}t,\chi)$ where m is a scalar function which is regular at $t=0$, it suffices to prove this assertion for the following function:

$$\sum\left(\rho(\mathrm{i}t,\chi)(f)h_\beta,h_\alpha\right)\int h_\alpha(k)\mathrm{d}k\cdot\overline{\int R(\mathrm{i}t,\chi)h_\beta(k)\mathrm{d}k}. \tag{2}$$

Of course we have a finite expansion:

$$\rho(\mathrm{i}t,\chi)(f)h_\beta=\sum_\alpha\left(\rho(\mathrm{i}t,\chi)(f)h_\beta,h_\alpha\right)h_\alpha.$$

Apllying the linear form defined by integration over K_F to this identity we see that our function can be written as:

$$\sum_\beta\int\rho(\mathrm{i}t,\chi)(f)h_\beta(k)\mathrm{d}k\overline{\int R(\mathrm{i}t,\chi)h_\beta(k)\mathrm{d}k}. \tag{3}$$

The space $\mathbb{H}(\chi)$ may be regarded as the tensor product over all places u of E of the local analogue spaces $\mathbb{H}(\chi_u)$. Similarly the operator $R(s,\chi)$ may be identified with the tensor product of the local normalized intertwining operators $R(s,\chi_u)$. Furthemore we may pick up an orthonormal basis of $\mathbb{H}(\chi_u)$ for each u and take for basis of $\mathbb{H}(\chi)$ the tensor products of the elements of the local bases. Accordingly the expression

(3) will be, apart from a constant factor, a product over all places v of F of local analogue expressions. What we have to see is that the factor attached to a place v of X is zero at $t=0$. The expression for this factor is the following:

$$\int [\rho(\mathrm{i}t, \chi_{v1})(f_{v1})] h_{\beta 1}(k_v)[\rho(\mathrm{i}t, \chi_{v2})(f_{v2})] h_{\beta 2}(k_v)\mathrm{d}k_v$$

$$\int R(\mathrm{i}t, \chi_{v1}) h_{\beta 1}(k_v) R(\mathrm{i}t, \chi_{v2}) h_{\beta 2}(k_v)\mathrm{d}k_v.$$

Since the restriction of χ to $F_{\mathbf{A}}^{\times}$ is trivial we have $\chi_{v1}=\bar{\chi}_{v2}$. By (7.3.4) for $t=0$ the above expression can be written as:

$$\sum \langle \rho(0, \chi_{v1})(f_{v1})h_{\beta 1}, \rho(0, \chi_{v2})(f_{v2})h_{\beta 2}\rangle\langle h_{\beta 1}, h_{\beta 2}\rangle.$$

Here $h_{\beta 1}$ denotes an orthonormal basis of $\mathbb{H}(\chi_{v1})$ and h an orthonormal basis of $\mathbb{H}(\chi_{v2})$. We remark that $h \mapsto \bar{h}$ is an antilinear isometric map of $\mathbb{H}(\chi_{v1})$ onto $\mathbb{H}(\chi_{v2})$. Thus we may assume that the bases are indexed by the same set: $h_{1\alpha}$ and $h_{2\alpha}$ will be the bases. More to the point we may assume that $\bar{h}_{1\alpha}=h_{2\alpha}$. Since $\langle h, h'\rangle=(h, \bar{h}')$ we have then:

$$\langle h_{1\beta}, h_{2\alpha}\rangle=\delta_{\beta\alpha}.$$

Thus our sum reduces to

$$\sum_{\beta} \langle \rho(0, \chi_{v1})(f_{v1})h_{1\beta}, \rho(0, \chi_{v2})(f_{v2})h_{2\beta}\rangle.$$

Since the map $h \mapsto \bar{h}$ transforms the representation $\rho(0, \chi_{v1})$ into $\rho(0, \chi_{v2})$ we get:

$$\sum_{\beta} \left(\rho(0, \chi_2)(f_{v1})h_{1\beta}, \rho(0, \chi_{v1})(\bar{f}_{v2})h_{1\beta}\right)$$

$$=\sum \left(\rho(0, \chi_{v1})(f_{v2}^{v})\cdot\rho(0, \chi_{v1})(f_{v1})h_{1\beta}, h_{1\beta}\right)$$

$$=\operatorname{tr}\rho(0, \chi_{v1})(h_v)$$

where as usual h_v is the convolution of f_{v2}^{v} and f_{v1}. Since the hyperbolic integrals of h_v vanish this vanishes too.

(9.6) We now examine the term (1) in Proposition (9.4). We want to prove that given χ such tht $\varepsilon(\chi)\neq 0$ the following function of t is zero, provided the conditions of (6.1.1) are satisfied:

$$\sum_{\alpha,\beta} \left(\rho(\mathrm{i}t, \chi)(f)h_{\beta}, h_{\alpha}\right)\cdot tI(\mathrm{i}t, \chi)h_{\alpha}\cdot\overline{tI(\mathrm{i}t, \chi)h_{\beta}}.$$

We may assume $t=0$ and then ignore the factor t and replace $I(\mathrm{i}t,\chi)$ by its normalized version $J(\mathrm{i}t,\chi)$. Furthemore we have the following finite expansion:

$$\big(\rho(\mathrm{i}t,\chi)(f)h_\beta,h_\alpha\big)h_\alpha=\rho(\mathrm{i}t,\chi)(f)h_\beta.$$

If we apply the linear form $J(\mathrm{i}t,\chi)$ to this we find that the function we have to deal with can be written as:

$$\sum_\beta J(\mathrm{i}t,\chi)\big[\rho(\mathrm{i}t,\chi)(f)h_\beta\big]\cdot\overline{J(\mathrm{i}t,\chi)h_\beta}.$$

We can choose the basis h_β as before and then this function will be the product over all places v of F of factors themselves functions of t (Cf. (8.5)). It will be enough to show that the factor attached to a place v of X is zero. Recall that $\chi=\mu oN(E/F)$ and $\chi_{v1}=\chi_{v2}=\mu_v$. We may therefore assume that the bases for $\mathbb{H}(\chi_{v1})$ and $\mathbb{H}(\chi_{v2})$ are the same. By (8.5) this factor is a multiple of the following function:

$$\sum_{\alpha,\beta}\langle R(\mathrm{i}t,\mu_v)\rho(\mathrm{i}t,\mu_v)(f'_{v1})h_\beta,\rho(\mathrm{i}t,\mu_v)(f_{v2})h_\alpha\rangle$$
$$\overline{\langle R(\mathrm{i}t,\mu_v)h_\beta,h_\alpha\rangle}.$$

This is also:

$$\sum_{\alpha,\beta}\langle\rho(-\mathrm{i}t,\bar\mu_v)(f^v_{v2})R(\mathrm{i}t,\mu_v)\rho(\mathrm{i}t,\mu_v)(f_{v1})h_\beta,h_\alpha\rangle$$
$$\overline{\langle R(\mathrm{i}t,\mu_v)h_\beta,h_\alpha\rangle}.$$

Now we appeal to the following summation formula where a and b are elements of $\mathbb{H}(\bar\mu_v)$:

$$(a,b)=\sum_\alpha\langle a,h_\alpha\rangle\overline{\langle b,h_\alpha\rangle}.$$

We find for our factor:

$$\sum\big(\rho(-\mathrm{i}t,\bar\mu_v)(f^v_{v2})R(\mathrm{i}t,\mu_v)\rho(\mathrm{i}t,\mu_v)(f_{v1})h_\beta,R(\mathrm{i}t,\mu_v)h_\beta\big).$$

Because $R(\mathrm{i}t,\mu_v)$ is an intertwining unitary operator this is also:

$$\sum\big(\rho(\mathrm{i}t,\mu_v)(f_{v2}),\rho(\mathrm{i}t,\mu_v)(f_{v1})h_\beta,h_\alpha\big)$$
$$=\sum_\beta\big(p(\mathrm{i}t,\mu_v)(h_v)h_\beta,h_\beta\big)=\operatorname{tr}\rho(\mathrm{i}t,\mu_v)(h_v)$$

and this vanishes because the hyperbolic orbital integrals of h_v are zero.

(9.5) We summarize our results:

PROPOSITION: *The limit of the integral* $T_1^{c_1}T_2^{c_2}K_{\mathrm{eis}}$ *as* c_1 *tends to infinity exists. The resulting function of* c_2 *has the form*

$$A \log c_2 + B + o(1).$$

Furthermore $B = 0$ *under the assumptions of* (6.1.1).

§10. Summing up

(10.1) We deal first with the truncation of K_{sp} then we summarize our results. Recall that

$$K = V^{-1}\sum \chi(\det x)\bar{\chi}(\det y)\int f(g)\chi(\det g)\mathrm{d}g,$$

the sum over all quadratic characters χ of the idele class group of E. The truncation of K_{sp} is therefore:

$$\begin{aligned} T_1^{c_1}T_2^{c_2}K_{\mathrm{sp}}(x, y) &= K_{\mathrm{sp}}(x, y)\left[1 - \sum \chi c_1(H(\gamma x))\right] \\ &\quad \times \left[1 - \sum \chi_{c2}(H(\gamma y))\right] \\ &= K_{\mathrm{sp}}(x, y) - \sum K_{\mathrm{sp}}(x, \gamma_1 y)\chi_{c_1}(H(\gamma_1 x)) \\ &\quad - \sum K_{\mathrm{sp}}(x, \gamma_2 y)\chi_{c_2}(H(\gamma_2 y)) \\ &\quad + \sum K_{\mathrm{sp}}(\gamma, x, \gamma_2 y)\chi_{c_1}(H(\gamma_1 x))\chi_{c_2}(H(\gamma_2 y)), \end{aligned}$$

where all summations are over $P(F)\backslash G(F)$. When we integrate over the product of $Z(F_{\mathbf{A}})G(F)\backslash G(F_{\mathbf{A}})$ by itself we find a sum of 4 terms:

$$\iint K_{\mathrm{sp}}(x, y)\mathrm{d}x\mathrm{d}y, \quad x, y \in Z(F_{\mathbf{A}})G(F)\backslash G(F_{\mathbf{A}}); \tag{1}$$

$$-\iint K_{\mathrm{sp}}(x, y)\chi_{c_1}(H(x))\mathrm{d}x\mathrm{d}y, \quad x \in Z(F_{\mathbf{A}})P(F)\backslash G(F_{\mathbf{A}}),$$
$$y \in Z(F_{\mathbf{A}})G(F)\backslash G(F_{\mathbf{A}}); \tag{2}$$

$$-\iint K_{\mathrm{sp}}(x, y)\chi_{c_2}(H(y))\mathrm{d}x\mathrm{d}y, \quad y \in Z(F_{\mathbf{A}})P(F)\backslash G(F_{\mathbf{A}}),$$
$$x \in Z(F_{\mathbf{A}})G(F)\backslash G(F_{\mathbf{A}}); \tag{3}$$

$$+\iint K_{\mathrm{sp}}(x, y)\chi_{c_1}(H(x))\chi_{c_2}(H(y))\mathrm{d}x\mathrm{d}y,$$

$$x, y \in Z(F_{\mathbf{A}})P(F)\backslash G(F_{\mathbf{A}}). \tag{4}$$

We claim that c_2 being fixed (2) and (3) tend to 0 as c_1 tends to infinity. Let us prove it for (2): (2) is a finite linear combination of integrals of the forms:

$$\int \bar{\chi}(\det y)\mathrm{d}y \int \chi(\det x)\chi_{c_1}(H(x))\mathrm{d}x.$$

This is zero unless the restriction of χ to $F_{\mathbf{A}}^{\times}$ is trivial. Then it can be computed by using the Iwasava decomposition. We find it is a constant times

$$\int_0^{c_2^{1/2}} t^{-1}\mathrm{d}t = c_1^{-1/2}.$$

Hence the limit of the integral of $T_1^{c_1}T_2^{c_2}K$ as c_1 tends to infinity exists: it is equal to (1) + (2). Similarly as c_2 tends to infinity the term (2) tends to 0; that is, for c_2 large we have the following asymptotic expansion:

$$\lim_{c_1 \to 0} \iint T_1^{c_1}T_2^{c_2}K_{\mathrm{sp}}\mathrm{d}x\mathrm{d}y = \iint K_{\mathrm{sp}}\mathrm{d}x\mathrm{d}y + o(1).$$

(10.2) It is now time to sum up. We have:

$$\iint K_{\mathrm{cusp}}(x, y)\mathrm{d}x\mathrm{d}y =$$

$$\iint K_{\mathrm{e}}(x, y)\mathrm{d}x\mathrm{d}y \tag{1}$$

$$+\iint K_{\mathrm{s}}(x, y)\mathrm{d}x\mathrm{d}y \tag{2}$$

$$+\iint (K_{\mathrm{r}}(x, y) - K_{\mathrm{eis}}(x, y))\mathrm{d}x\mathrm{d}y \tag{3}$$

$$-\iint K_{\mathrm{sp}}(x, y)\mathrm{d}x\mathrm{d}y. \tag{4}$$

The third term is integrable because all others are. On the other hand since K_{cusp} is equal to its truncation we have:

$$\iint K_{\text{cusp}}(x, y)\mathrm{d}x\mathrm{d}y =$$

$$+\iint T_1^{c_1}T_2^{c_2}K(x, y)\mathrm{d}x\mathrm{d}y \tag{5}$$

$$-\iint T_1^{c_1}T_2^{c_2}K_{\text{eis}}(x, y)\mathrm{d}x\mathrm{d}y \tag{6}$$

$$-\iint T_1^{c_1}T_2^{c_2}K_{\text{sp}}(x, y)\mathrm{d}x\mathrm{d}y. \tag{7}$$

We now let c_1 tends to infinity. Under the assumptions of (6.6.1) the functions of c_2 obtained from the limit of (5) and (6) have the form $(1)+(2)+A\log c_2$ and $A'\log c_2+o(1)$ respectively. Similarly the limit of (7) has the form $\iint K_{\text{sp}}\mathrm{d}x\mathrm{d}y+o(1)$. Comparing the two expressions for the integral of K_{cusp} we conclude that (3) vanishes. So we have established the result which was our goal:

PROPOSITION: *Let X be a set of places of F which split in E. Suppose that X has at least two elements. For v in X let $v1$ and $v2$ be the two corresponding places of E and h_v the convolution product $f_{v1}^{v} * f_{v2}$. Suppose that for each v in X the hyperbolic orbital integral of h_v vanish. Then the integral of the difference $K_r - K_{\text{eis}}$ over the product of $Z(F_{\mathbf{A}})G(F)\backslash G(F_{\mathbf{A}})$ by itself vanishes.*

References

[J.A.] J. ARTHUR: The Selberg trace formula for groups of F-rank one, *Annals of Math.* 100(2) pp 326–385.

[T.A.] T. ASAI: On certain Dirichlet series associated with Hilbert modular forms and Rankin's method, *Math. Ann.* 226 (1977) 81–94.

[W.C.] W. CASSELMAN: The Hasse-Weil zeta function of some moduli varieties of dimension greater than one. In *Proceeding of Symposia in Pure Mathematics*, Vol. 33 (1979), part 2, pp 141–163.

[H.L.R.] HARDER and LANGLANDS: Rappoport Algebraische Zyklen auf Hilbert-Blumenthal Flacken, to appear.

[J.L.] H. JACQUET and R. LANGLANDS: Automorphic forms on GL(2), Springer Verlag, *Lecture Notes in Mathematics* No. 114 (1970).

[R.L.] R. LANGLANDS: Base Change for GL(2), *Annals of Math. Studies*, No 96, Princeton University Press.

[F.S] F. Shahidi: Local coefficients and normalization of intertwinning operators for GL(n), *Comp. Math.* 48 (1983) 271–295.

(Oblatum 5-X-1983)

Department of Mathematics
Columbia University
New York, NY 10027
USA

Math. Ann. 271, 319–332 (1985)

Mathematische Annalen

A Lemma on Highly Ramified ε-Factors*

Hervé Jacquet and Joseph Shalika

Department of Mathematics, Columbia University, New York, NY 10027, USA

1. Introduction

(1.1) Let F be a non-archimedean local field and G_r the group $\mathrm{GL}(r, F)$. Fix a non trivial additive character ψ of F. For convenience, we will assume that the largest ideal on which ψ is trivial is the ring of integers $\mathfrak{R}$; we will denote by $\mathfrak{P}$ the maximal ideal in $\mathfrak{R}$ and by q the cardinality of the residual field $\mathfrak{R}/\mathfrak{P}$. Let π be an irreducible admissible representation of G_r on a complex vector space V. To π we can attach functions $L(s, \pi)$ and $\varepsilon(s, \pi, \psi)$ [G–J]. They have the form

$$\begin{aligned} L(s,\pi) &= P(q^{-s})^{-1}, \qquad P \in \mathbb{C}[x]; \\ \varepsilon(s,\pi,\psi) &= cq^{-fs}. \end{aligned} \tag{1}$$

A simple but useful property of these functions is the following one: suppose π_1 and π_2 are two such representations with the *same* central character ω; then, if χ is a multiplicative character of conductor $\mathfrak{P}^a$ we have

$$\begin{aligned} L(s,\pi_1\otimes\chi) &= L(s,\pi_2\otimes\chi) = 1, \\ \varepsilon(s,\pi_1\otimes\chi,\psi) &= \varepsilon(s,\pi_2\otimes\chi,\psi), \end{aligned} \tag{2}$$

provided a is large enough. As a matter of fact, we have used this property several times. For the sake of completeness, we give the (standard) proof in Sect. 2.

More generally, if π is an irreducible admissible representation of G_r and σ an irreducible admissible representation of G_t, then one can define factors $L(s, \pi \times \sigma)$ and $\varepsilon(s, \pi \times \sigma, \psi)$ and they have the property analogous to (2). Again we used this fact before. The purpose of this paper is to give a proof of it (Sect. 4). In Sect. 3 we give a property of the *conductor* of a representation π, that is of the ideal $\mathfrak{P}^f$, with f as in (1); it is used in an essential way in Sect. 4.

* This research was partially supported by N.S.F. Grants MCS79-01712 and MCS79-13799

2. The Case of one Representation

(2.1) Let again π be an irreducible admissible representation of G_r. If f is a matrix coefficient of π and Φ a Bruhat-function on the vector space M_r of r by r matrices we set

$$Z(\Phi, s, f)=\int_{G_r} \Phi(x)f(x)\,|\det x|^s d^\times x\,, \tag{1}$$

where $d^\times x$ is a Haar measure on G_r. The factor $L(s,\pi)$ is the "g.c.d." of the integrals $Z(\Phi, s+\frac{1}{2}(r-1), f)$. As for the ε-factor it is defined by the functional equation of the integrals (1); namely if we set

$$\gamma(s,\pi,\psi)=\varepsilon(s,\pi,\psi)L(1-s,\tilde{\pi})L(s,\pi)^{-1} \tag{2}$$

then

$$\begin{aligned} &Z(\Phi^\wedge, 1-s+\tfrac{1}{2}(r-1), f^\vee)\\ &\quad=\gamma(s,\pi,\psi)Z(\Phi, s+\tfrac{1}{2}(r-1), f)\,, \end{aligned} \tag{3}$$

where $f^\vee(g)=f(g^{-1})$ and $\Phi^\wedge$ is the Fourier-transform of Φ:

$$\Phi^\wedge(x)=\int \Phi(y)\psi(\mathrm{Tr}(yx))dy\,, \tag{4}$$

dy being the self-dual Haar measure on M_r.

(2.2) **Proposition.** *Suppose π is as above; let ω be its central character and $\chi_1, \chi_2, \ldots, \chi_r$ characters of $F^\times$ whose product is ω. There is integer $A>0$ with the following property: if $a \geqq A$ and χ is a character of $F^\times$ with the conductor P^a, then:*

$$L(s,\pi\otimes\chi)=1\,, \tag{1}$$

$$\varepsilon(s,\pi\otimes\chi,\psi)=\prod_{i=1}^{r} \varepsilon(s,\chi_i\chi,\psi)\,. \tag{2}$$

(2.3) *Proof of* (2.2.1). Because of the functorial property of the L-factor with respect to induction, it suffices to prove our assertion for an "atom" of the theory, that is, for a supercuspidal π. If $r=1$ any character π of $F^\times$ is a supercuspidal representation and $L(s,\pi)=1$, as soon as π is ramified at all. If $r>1$ and π is supercuspidal then $L(s,\pi)=1$. □

(2.4) *Proof of* (2.2.2). Assume first $r>1$ and π supercuspidal. The definition of the ε-factor can then be reformulated as follows: let f be a matrix coefficient of π and φ an element of $C_c^\infty(F^\times)$ (locally constant functions of compact support). Then the function Φ on M_r defined by

$$\Phi(g)=f(g^{-1})\varphi(\det g)\,, \quad \text{if} \quad \det g \neq 0;\ =0, \text{ otherwise,} \tag{1}$$

is a Bruhat-function; its Fourier transform is similarly given by

$$\Phi^\wedge(g)=f(g)H_\pi\varphi(\det g) \quad \text{if} \quad \det g \neq 0;\ =0, \text{ otherwise}, \tag{2}$$

where

$$H_\pi : C_c^\infty(F^\times)\to C_c^\infty(F^\times) \tag{3}$$

is a linear map, depending on π. Then:

$$\begin{aligned}\gamma(s,\pi,\psi)\int\varphi(a)|a|^{s+\frac{1}{2}(r-1)}d^{\times}a\\ =\int H_{\pi}\varphi(a)\,|a|^{1-s+\frac{1}{2}(r-1)}d^{\times}a\,,\end{aligned}\tag{4}$$

for any $\varphi\in C_c^{\infty}(F^{\times})$. The proof is an easy exercise which uses Schur-orthogonality relations and Lemma 5.3 p. 59 in [G-J]. Formal properties of the map H_π are:

$$H_{\pi}(\chi^{-1}\varphi)=\chi H_{\pi\otimes\chi}(\varphi)\,,\tag{5}$$

for any character χ of $F^{\times}$, and

$$(H_{\tilde{\pi}}H_{\pi}\varphi)(x)=\omega(-1)\varphi[(-1)^{r}x]\,,\tag{6}$$

where ω is the central character of π and $\tilde{\pi}$ the representation contragredient to π. We shall use (4) to prove our result.

Before we embark on the proof we remark the following: suppose we have proved equality (2.2.2) up to a positive factor, depending only on r. Then the equation

$$\varepsilon(s,\pi,\psi)\varepsilon(1-s,\tilde{\pi},\psi)=\omega(-1)$$

will tell us that the factor is actually one. Thus we may, and will, ignore such positive factors. In particular, we do not bother normalizing Haar measures.

(2.5) We let φ be the characteristic function of $\mathfrak{R}^{\times}$ in $F^{\times}$ and compute $H_{\pi\otimes\chi}\varphi$. We denote by $\tilde{\omega}$ a prime element of $\mathfrak{R}$ and by $\mathfrak{R}_r$ the subring of M_r of matrices with integral entries. For $i>0$ we denote by K_i the congruence subgroup

$$K_i=1+\tilde{\omega}^{i}R_r\tag{1}$$

in G_r. There is an $i>0$ and a matrix coefficient f of π such that $f(e)\neq 0$ and f is invariant on both sides under K_i. On the other hand, we let k be an integer – to be taken sufficiently large. We set

$$j=k/2\quad\text{if}\quad k\text{ is even}\,,\qquad j=(k+1)/2\quad\text{if}\quad k\text{ is odd}\,.\tag{2}$$

We let χ be a character of conductor $\mathfrak{P}^k$. Since in any case $2j\geqq k$, the map

$$a\mapsto h=1+a$$

defines a group-isomorphism:

$$\mathfrak{P}^{j}/\mathfrak{P}^{k}\simeq(1+\mathfrak{P}^{j})/(1+\mathfrak{P}^{k})\,.$$

In particular, there is a c in $\tilde{\omega}^{-k}\mathfrak{R}^{\times}$ such that

$$\chi(h)=\psi(ca)\,,\quad\text{for}\quad a\in\mathfrak{P}^{j}\,.\tag{3}$$

On the other hand, if h is in

$$K_j=1+\tilde{\omega}^{j}\mathfrak{R}_r$$

then it has the form

$$h=1+a\,,\qquad a\in\tilde{\omega}^{j}\mathfrak{R}_r$$

and

$$\begin{aligned}\det(h)=1+\operatorname{Tr}(a)\mod\tilde{\omega}^{2j}\mathfrak{R}\,,\\ \operatorname{Tr}(a)=0\mod\tilde{\omega}^{j}\mathfrak{R}\,.\end{aligned}$$

It follows that

$$\chi(\det h)=\psi[\mathrm{Tr}(ca)]. \tag{4}$$

The functions f and φ being as above we set

$$\Phi(g)=f(g^{-1})\chi(\det g)^{-1}\varphi(\det g), \quad \text{if} \quad \det g \neq 0, \tag{5}$$
$$=0 \qquad\qquad \text{otherwise}.$$

Then, for k large enough,

$$\hat{\Phi}(g)=\int_{G_0/K_k} f(x^{-1})\chi(\det x)^{-1}dx \int_{K_k} \psi(\mathrm{Tr}(gxh))dh, \tag{6}$$

where G_0 is the group of $g\in G_r$ such that $|\det g|=1$. The inner integral can also be written as

$$\psi(\mathrm{Tr}(gx)) \int_{\tilde{\omega}^k R_r} \psi(\mathrm{Tr}(gxa))da;$$

it vanishes unless x belongs to the set

$$X_g=g^{-1}\tilde{\omega}^{-k}\mathfrak{R}_r. \tag{7}$$

Thus

$$\hat{\Phi}(g)=\int_{X_g/K_k} \varphi(\det x)f(x^{-1})\chi(\det x)^{-1}\psi(\mathrm{Tr}(gx))dx$$
$$=\int_{X_g/K_j} \varphi(\det x)f(x^{-1})\chi(\det x)^{-1}dx$$
$$\cdot\int_{K_j/K_k} \chi^{-1}(\det(h))\psi(\mathrm{Tr}(gxh)dh. \tag{8}$$

Again the map $a\mapsto h=1+a$ defines a group isomorphism

$$\tilde{\omega}^j\mathfrak{R}_r/\tilde{\omega}^k\mathfrak{R}_r \simeq K_j/K_k$$

and the inner integral in (8) can be written as

$$\psi(\mathrm{Tr}(gx))\int \chi^{-1}(\det(1+a))\psi(\mathrm{Tr}(gxa))da$$
$$=\psi(\mathrm{Tr}(gx))\int \psi(\mathrm{Tr}(gx-ca))da.$$

The last integral is over $\omega^j\mathfrak{R}_r$ and vanishes unless $gx-c$ is in $\tilde{\omega}^{-j}\mathfrak{R}_r$ or, what amounts to the same,

$$gx\in cK_{k-j}.$$

Thus we find that $\Phi^\wedge$ vanishes outside the set cG_0 and for g in that set is given by

$$\Phi^\wedge(g)=\int f(x^{-1})\chi(\det x)^{-1}\psi(\mathrm{Tr}(gx))dx, \tag{9}$$

the integration being over $g^{-1}cK_{k-j}/K_j$.

Now if k is so large that $k-j\geqq i$ this is simply

$$\Phi^\wedge(g)=f(g)\chi(\det g)\,|\det g|^{-r}\tau(\chi,\omega,r) \tag{10}$$

where we have set

$$\tau(\chi,\omega,r)=\omega^{-1}(c)\chi^{-1}(c^r)\eta(\chi,r), \tag{11}$$
$$\eta(\chi,r)=\int_{K_{k-j}/K_j} \chi^{-1}(\det h)\psi(\mathrm{Tr}(ch))dh. \tag{12}$$

Thus, we have proved that

$$(H_{\pi\otimes\chi}\varphi)(a)=\varphi(c^{-r}a)\tau(\chi,\omega,r)\,|a|^{-r}, \tag{13}$$

or, by (2.4.1):

$$\varepsilon(s,\pi\otimes\chi,\psi)=q^{\ \ krs}\tau(\chi,\omega,r). \tag{14}$$

The right hand side is now independent of π. It remains to compute it.

(2.6) **Lemma.** *Given r, if k is large enough and χ has conductor $\mathfrak{P}^k$, then:*

$$\eta(\chi,r)=\eta(\chi,1)^r.$$

Proof of Lemma (2.6). We may write an element h of K_{k-j} in the form:

$$h=(1+u)(1+\delta)(1+v),$$

where u and v are strictly upper and lower triangular matrices and δ is diagonal – each belonging to $\tilde{\omega}^{k-j}\mathfrak{R}_r$. We have $\det h=\det(1+\delta)$ and

$$\operatorname{Tr}h=\operatorname{Tr}(1+\delta)+\operatorname{Tr}(uv)+\operatorname{Tr}(u\delta v).$$

Hence, if $3(k-j)\geqq k$, then

$$\begin{aligned}&\chi^{-1}(\det(h))\psi(\operatorname{Tr}(ch))\\&\quad=\chi^{-1}(\det(1+\delta))\psi(\operatorname{Tr}(c(1+\delta)))\psi(\operatorname{Tr}(cuv)).\end{aligned}$$

Thus, if $2(k-j)\geqq k$, we obtain

$$\begin{aligned}\eta(\chi,r)=&\int\chi^{-1}(\det(1+\delta))\psi[\operatorname{Tr}(c(1+\delta))]d\delta\\&\cdot\int\psi[\operatorname{Tr}(cuv)]du\,dv.\end{aligned}$$

The second integral is positive and may be ignored and the first is $\eta(\chi,1)^r$.

This concludes the proof of the lemma.

(2.7) Using the lemma we get

$$\varepsilon(s,\pi\otimes\chi,\psi)=[q^{-ks}\chi^{-1}(c)\eta(\chi,1)]^r\omega^{-1}(c).$$

This formula applies to the case $r=1$ as well (the proof we have just given being then the classical one). Hence, for $1\leqq i\leqq r$, and k large enough:

$$\varepsilon(s,\chi_i\chi,\psi)=q^{-ks}\chi^{-1}(c)\eta(\chi,1)\chi_i^{-1}(c), \tag{1}$$

if χ has conductor $\mathfrak{P}^k$. Then, as claimed:

$$\varepsilon(s,\pi\otimes\chi,\psi)=\prod_i\varepsilon(s,\chi_i\chi,\psi). \tag{2}$$

Thus (2.2) is proved when π is supercuspidal.

The previous proof shows that the right hand side of (2) depends only on r and ω, provided k is large enough (as is well known.) Again the functoriality of the ε-factor shows then that the formula is true in general. □

(2.8) *Remark.* Mutatis mutandis, the proof applies to the L and ε factors attached to an irreducible representation of the multiplicative group of a simple algebra.

3. A Review of the Conductor

(3.1) Let π be an admissible irreducible representation of G_r on a complex vector space V. Assume π is generic. If N_r is the group of upper triangular matrices with unit diagonal we define a character θ or θ_r of N_r by

$$\theta(n)=\psi(n_{12}+n_{23}+\ldots+n_{r-1r});\tag{1}$$

to say that π is generic means there is a non zero linear form λ on V such that

$$\lambda[\pi(n)v]=\theta(n)v\,,\quad \text{for}\quad n\in N_r\,.\tag{2}$$

This linear form is then unique, up to a scalar factor, and we denote by $\mathscr{W}(\pi;\psi)$ the space of functions W on G_r of the form

$$W(g)=\lambda(\pi(g)v)\,,\qquad v\in V.\tag{3}$$

We will need a few auxiliary notations. We let P_r be the group of matrices p in G_r of the form

$$p=\begin{pmatrix} g & x\\ 0 & 1\end{pmatrix},\qquad g\in G_{r-1}$$

and $Z_r\simeq F^\times$ the center of G_r. We also introduce the r by r matrix

$$w_r=\begin{pmatrix} 0 & & \cdot 1\\ 0 & \cdot^{\cdot^{\cdot}} & \\ \vdots\ 1\cdot & & \\ 1 & & 0\end{pmatrix}\tag{4}$$

and the row-matrix of length r

$$\eta_r=(0,0,\ldots,0,1)\,.\tag{5}$$

We set

$$K_r=\mathrm{GL}(r,\mathfrak{R})$$

and denote, for $j>0$, by $K_r(j)$ the subgroup of matrices $k\in K_r$ of the form:

$$k=\begin{pmatrix} a & b\\ c & d\end{pmatrix},\qquad a\in K_{r-1}\,,\qquad d\in\mathfrak{R}^\times\,,\qquad c\equiv 0 \bmod \mathfrak{P}^j\,.\tag{6}$$

If $j=0$ we set $K_r(0)=K_r$. Then if ω is a character of $F^\times$ (or $\mathfrak{R}^\times$), trivial on $1+\mathfrak{P}^j$, we define a one-dimensional character ω_j of $K_r(j)$ by setting

$$\begin{aligned}&\omega_j(k)=\omega(d)\,,\quad k \text{ as in (6)}\,,\quad \text{if}\quad j>0;\\ &\omega_j(k)=1\,,\quad \text{for all } k\,,\qquad\ \ \text{if}\quad j=0\,.\end{aligned}\tag{7}$$

If f is the integer in formula (1.1.1) then the ideal $\mathfrak{P}^f$ is called the *conductor* of π ([*J-P-S*] II). If there exists a non-zero vector in V transforming under $K_r(j)$ according to ω_j then ω coincides with the central character of π on $\mathfrak{R}^\times$ and necessarily $j\geqq f$. Moreover, if ω is the central character of π then the conductor of ω is contained in $\mathfrak{P}^f$ and the dimension of the space of vectors in V transforming under $K_r(f)$ according to ω_f is one (loc. cit.; note, however, the notations are slightly different). This property characterizes f.

(3.2) In what follows we will need another characterization of the conductor. It will be based on the following simple lemma. We let H be a smooth complex-valued function on G_r satisfying

$$H(ng)=\bar{\theta}(n)H(g)\,, \qquad n\in N_r\,, \qquad g\in G_r\,, \tag{1}$$

and compactly supported modulo N_r.

Lemma. *With H as above, suppose that the integral*

$$\int_{N_r\backslash G_r} H(g)W(g)dg$$

vanishes, for all admissible irreducible generic representations π of G_r and all $W\in\mathscr{W}(\pi;\psi)$. Then $H=0$.

Proof. With trivial modifications the proof is word for word the same as that of Lemma (3.5) in [J-P-S] II. □

(3.3) Next, for $g\in G_r$ of the form $g=nak$ with $n\in N_r$, $k\in K_r$ and

$$a=diag(a_1, a_2, \ldots, a_r)\,,$$

set

$$\varrho(g)=|a_1|\,. \tag{1}$$

We then have the following corollary:

Corollary. *Fix a character ω of $F^\times$ and an integer $j\geqq 0$. Suppose that H satisfies the conditions of the previous lemma and transforms on the right under $K_r(j)$ according to ω_j^{-1}. Suppose further that H has support in the set of $g\in G_r$ satisfying*

$$\varrho(g)\geqq C\,, \tag{2}$$

where C is a positive constant. Then the following conditions (A) *and* (B) *are equivalent:*

(A)
$$\begin{aligned} H(g)&=\bar{\theta}(n)\omega_j^{-1}(k)\,, \quad \text{if} \quad g=nk\,, \quad n\in N_r\,, \quad k\in K_r(j)\,,\\ &=0 \qquad\qquad =0 \quad \text{if} \quad g\notin N_rK_r(j)\,. \end{aligned}$$

(B) *For any generic irreducible representation π of G_r with central character ω and conductor $\mathfrak{P}^f$, $f\leqq j$, one has*

$$\int_{N_r\backslash G_r} H(g)W(g)\,|\det g|^s dg = \int_{K_r(j)} \omega_j^{-1}(k)W(k)dk \tag{3}$$

for all $W\in\mathscr{W}(\pi;\psi)$.

The integral on the left is to be interpreted as a formal Laurent series

$$\sum_m X^m \int_{N_r\backslash G_r} H(g)W(g)\mu_m(g)dg \tag{4}$$

where $X=q^{-s}$, G_r^m is the set of $g\in G_r$ such that $|\det g|=q^{-m}$ and μ_m the characteristic function of G_r^m. Because of the assumption on the support of H the products $H\mu_m$ have compact support modulo N_r, so that the integrals in (4) are well defined (cf. [J-P-S] I and II).

Proof. Clearly (A) implies (B). Assume (B). If π is any irreducible generic representation of G_r with conductor $\mathfrak{P}^f$, $j<f$, then the integral

$$\int_{K_r(f)} \omega_f^{-1}(k)W(gk)dk$$

vanishes for all $W \in \mathscr{W}(\pi;\psi)$ and all $g \in G_r$. Thus both sides of equality (3) are then zero. Similarly, if π is any irreducible generic representation of G_r, both sides of (3) vanish, unless the central character ω_π of π agrees with ω on $\mathfrak{R}^\times$. Thus condition (B) is equivalent to another condition where equality (3) stands for *all* irreducible generic representations π of G_r and all $W \in \mathscr{W}(\pi;\psi)$. In view of the interpretation of the left hand side of (3) this means that

$$\int H\mu_m dg = 0 \quad \text{for} \quad m \neq 0,$$

$$\int H\mu_0 dg = \int_{K_r(j)} \omega_j^{-1}(k)W(k)dk.$$

Applying the previous lemma to each function $H\mu_m$ we get assertion A. □

4. The Main Result

(4.1) Let π be an irreducible generic representation of G_r and σ an irreducible generic representation of G_t. In [J-P-S] I we have defined functions $L(s,\pi\times\sigma)$ and $\varepsilon(s,\pi\times\sigma,\psi)$; we have also set

$$\gamma(s,\pi\times\sigma,\psi) = \varepsilon(s,\pi\times\sigma,\psi)L(1-s,\tilde{\pi}\times\tilde{\sigma})/L(s,\pi\times\sigma). \tag{1}$$

We let ω_π (resp. ω_σ) be the central character of π (resp. σ).

Proposition. *Suppose π_i, $i=1,2$, (resp. σ) is an irreducible generic representation of G_r (resp. G_t) and $\omega_{\pi_1}=\omega_{\pi_2}$. There is an integer A with the following property: If χ is a character of $F^\times$ with conductor $\mathfrak{P}^a$, $a \geqq A$, then*

$$\gamma(s,(\pi_1\otimes\chi)\times\sigma,\psi) = \gamma(s,(\pi_2\otimes\chi)\times\sigma,\psi). \tag{2}$$

(4.2) We first remark that, by definition, the factors attached to the pair $(\pi_i\otimes\chi,\sigma)$ where σ is a character of $F^\times$ are the same as the factors attached to $\pi_i\otimes\chi\sigma$ (Sect. 1). Thus we already know the proposition in case $t=1$.

We also remark that we may apply this result to an arbitrary irreducible generic representation π_1 of G_r and the irreducible generic component π_2 of a "principal series representation". More precisely, let $\chi_1,\chi_2,\dots,\chi_r$ be characters of $F^\times$ whose product is ω_{π_1}; let B be the group of upper triangular matrices in G_r and π_2 the generic component of the induced representation

$$\xi = \operatorname{Ind}(G_r, B; \chi_1,\chi_2,\dots,\chi_r).$$

By Theorem (3.1) of [J-P-S] I, the right hand side of (4.1.2) is nothing but

$$\prod_i \gamma(s,\sigma\otimes\chi_i\chi,\psi).$$

At this point we may apply the result for $t=1$ and we see that if $\chi_1\chi_2\dots\chi_r=\omega_\pi$ and $\eta_1\eta_2\dots\eta_t=\omega_\sigma$ then

$$\gamma(s,(\pi\otimes\chi)\times\sigma,\psi) = \prod_{i,j}\gamma(s,\chi_i\eta_j\chi,\psi), \tag{1}$$

provided a is large enough and χ has conductor $\mathfrak{P}^a$. In particular, we see that

$$\gamma(s, (\pi \otimes \chi) \times \sigma, \psi) = C_0 X^{art}, \qquad X = q^{-s},$$

where χ has conductor $\mathfrak{P}^a$, a is large enough, and C_0 depends only on χ, ω_{π_1}, and ω_σ.

(4.3) *Reduction to the Case* $t=r-1$. It is easy to see that we can reduce ourselves to the case when $t=r-1$. If $t<r-1$, we can consider an induced representation

$$\eta = \mathrm{Ind}(G_{r-1}, Q; \sigma, \eta_1, \eta_2, \ldots, \eta_{r-1-t})$$

where the η_i are characters of $F^\times$ and Q is a parabolic subgroup of type $(t, 1, \ldots, 1)$ in G_{r-1}. Then, if σ' is the unique irreducible generic component of η, we have (Theorem (13.1) of [J-S-P] I):

$$\begin{aligned} &\gamma(s, (\pi_i \otimes \chi) \times \sigma, \psi) \\ &\quad = \gamma(s, (\pi_i \otimes \chi) \times \sigma', \psi) \Big/ \prod_j \gamma(s, \pi_i \chi \eta_j, \psi). \end{aligned} \tag{1}$$

Suppose Prop. (4.1) true for $t=r-1$. Recall it is true for $t=1$. Then the right hand side of (2) has the same value for $i=1$ and 2, provided the conductor of χ is deep enough and we get our assertion in the case (r, t), $t<r-1$. Similarly, in the case $t \geqq r$ we can consider an induced representation

$$\xi = \mathrm{Ind}(G_r, Q; \pi, \mu_1, \mu_2, \ldots, \mu_{t+1-r}),$$

to reduce ourselves to the case $t=r-1$.

(4.4) Next choose a character ξ of conductor $\mathfrak{P}^B$ such that, for $i=1,2$,

$$L(s, \pi_i \otimes \xi) = L(s, \tilde{\pi}_i \otimes \xi^{-1}) = 1,$$

$$\varepsilon(s, \pi_i \otimes \xi, \psi) = C_i q^{-rBs};$$

this is possible by the case $t=1$. At the cost of restricting a to be larger than B in the proposition, we may replace π_i by $\pi_i \otimes \xi$ and, therefore assume, for $i=1,2$: $L(s, \pi_i) = L(s, \tilde{\pi}_i) = 1$, $\varepsilon(s, \pi_i, \psi) = C_i q^{-fs}$. In particular π_1 and π_2 have the same conductor $\mathfrak{P}^f$. Moreover, the space $\mathscr{W}(\pi_i, \psi)$ contains exactly one vector $W_{i,0}$ transforming under ω_f and taking the value 1 on e: this is the "essential vector" of [J-P-S] II. It is clear from their construction (loc. cit.) that the functions $W_{i,0}$ agree on $P_r Z_r K_r(f)$.

We now prove the identity (4.1.2), provided the conductor of χ is $\mathfrak{P}^a$ where a is large enough. We set $\omega = \omega_{\pi_i}$. A simple formal manipulation shows that the functional equation which defines γ ([J-P-S] I, Theorem (2.7)) can be written in the form:

$$\omega_{\sigma \otimes \chi}(-1)^{r-1} \gamma(s, \pi_i \times (\sigma \otimes \chi)) I_i = J_i, \tag{1}$$

with

$$I_i = \int W_i \left[\begin{pmatrix} g & 0 \\ 0 & 1 \end{pmatrix} w_r \right] W'(g) \, |\det g|^{s-1/2} dg, \tag{2}$$

$$J_i = \int W_i \left[\begin{pmatrix} 1 & 0 \\ 0 & g \end{pmatrix} \right] W'(g w_{r-1}) \, |\det g|^{s-1/2} dg, \tag{3}$$

and where W_i is in $\mathscr{W}(\pi_i;\psi)$ and W' in $\mathscr{W}(\sigma\otimes\chi;\bar{\psi})$. We will take W_i to be of the form

$$W_i(g)=\int W_{i,0}\left[g\begin{pmatrix}1 & u & v\\ 0 & 1 & 0\\ 0 & 0 & 1_{r-2}\end{pmatrix}\right]\varphi(u)\varphi_1(v)du dv, \tag{4}$$

where φ and φ_1 are Bruhat functions on F and F^{r-2} to be chosen below. We are going to show that φ, φ_1 and W' may be so chosen that $J_1=J_2 \neq 0$ and then check that $I_1=I_2$. This will establish our assertion.

To proceed, in the expression for J_i we set

$$g=t\begin{pmatrix}1 & 0\\ 0 & h\end{pmatrix}\begin{pmatrix}1 & 0\\ x & 1_{r-2}\end{pmatrix}$$

with $t\in F^\times$, h in $N_{r-2}\backslash G_{r-2}$ and then we get

$$J_i=\int W_{i,0}\left[\begin{pmatrix}t^{-1} & 0 & 0\\ 0 & 1 & 0\\ 0 & 0 & h\end{pmatrix}\begin{pmatrix}1 & 0 & 0\\ 0 & 1 & 0\\ 0 & x & 1_{r-2}\end{pmatrix}\right]$$
$$W'\left[\begin{pmatrix}1 & 0\\ 0 & h\end{pmatrix}\begin{pmatrix}1 & 0\\ x & 1_{r-2}\end{pmatrix}w_{r-1}\right]$$
$$\varphi^\wedge(t^{-1})\varphi_1^\wedge[-t^{-1}({}^tx)]\omega\chi^{r-1}\omega_\sigma(t)$$
$$|t|^{(r-1)(s-1/2)}|\det h|^{s+1/2}dx d^\times t dh. \tag{5}$$

Of course $\hat{\varphi}$ is the Fourier transform of φ.

Next we choose φ and φ_1 in such a way that

$$\hat{\varphi}(t)=\omega_\sigma\chi^{r-1}(t), \quad \text{if} \quad |t|=1; \quad =0 \text{ otherwise}. \tag{6}$$

$$\hat{\varphi}_1(tx)=\omega(t)\hat{\varphi}_1(x), \quad \text{if} \quad |t|=1. \tag{7}$$

$$\hat{\varphi}_1(x)\neq 0 \quad \text{implies that} \tag{8}$$

$$\begin{pmatrix}1 & 0\\ {}^tx & 1_{r-2}\end{pmatrix}\in K_{r-1}(f).$$

Then (5) becomes

$$J_i=\int_{N_{r-2}\backslash G_{r-2}} W_{i,0}\left[\begin{pmatrix}1 & 0 & 0\\ 0 & 1 & 0\\ 0 & 0 & h\end{pmatrix}\right]F(h)\,|\det h|^{s+1/2}dh, \tag{9}$$

where we have set, for $h\in G_{r-2}$,

$$F(h)=\int W'\left[\begin{pmatrix}1 & 0\\ 0 & h\end{pmatrix}\begin{pmatrix}1 & 0\\ x & 1_{r-2}\end{pmatrix}w_{r-1}\right]\hat{\varphi}_1[-{}^tx]dx. \tag{10}$$

Furthermore the value of the right hand side of (9) does not change if we replace F by the function F_0 defined by

$$F_0(g)=\int_{K_{r-2}(f)} F(gk)\omega_f(k)dk. \tag{11}$$

The next step is to choose W' and φ_1 in such a way that

$$F_0(h)=\bar{\theta}_{r-2}(n)\omega_f^{-1}(k), \quad \text{if} \quad g=nk \tag{12}$$

with $n\in N_{r-2}$ and $k\in K_{r-2}(f)$;
$$=0 \quad \text{otherwise}.$$

Of course φ_1 is still constricted to satisfy (7) and (8). Then we will get from (9) the simple relation

$$J_i=cW_{i,0}(e), \qquad i=1,2,$$

where c is a positive constant independent of i. Since $W_{i,0}(e)=1$ (loc. cit.) we will get $J_1=J_2\neq 0$, as required.

To proceed we remark that F_0 *satisfies the hypotheses of Corollary* (3.3), for it is a sum of functions of the form

$$h\mapsto W_j\begin{pmatrix}1 & 0\\ 0 & h\end{pmatrix}, \quad \text{with} \quad W_j\in\mathscr{W}(\sigma;\bar{\psi}).$$

By this corollary, in order to obtain condition (1.2), we need only choose φ_1 and W' in such a way that for any irreducible generic representation τ of G_{r-2}, with central character ω and conductor $\mathfrak{P}^j$, $j\leqq f$, the identity

$$\begin{aligned}&\int_{N_{r-2}\backslash G_{r-2}} F_0(g)W''(g)\,|\det g|^s dg\\ &= \int_{K_{r-2}(f)} \omega_f^{-1}(k)W''(k)dk,\end{aligned} \tag{13}$$

stands for any $W''\in\mathscr{W}(\tau;\psi)$. Without loss of generality, we may even consider only those W'' which, under right-shifts, transforms according to the character ω_f of $K_{r-2}(f)$. Then we may replace back F_0 by F without changing the left-hand side of (13).

To check that this choice of W' and φ_1 is possible, we start with the functional equation which defines the factor γ for the pair $(\sigma\otimes\chi,\tau)$. We take it in the form [cf. (1)]

$$\begin{aligned}&\omega(-1)^{r-2}\gamma(s,(\sigma\otimes\chi)\times\tau,\psi)\\ &\cdot\int W'\left[\begin{pmatrix}g & 0\\ 0 & 1\end{pmatrix}\begin{pmatrix}w_{r-2} & 0\\ 0 & 1\end{pmatrix}w_{r-1}\right]W''(g)\,|\det g|^{s-1/2}dg\\ &=\int W'\left[\begin{pmatrix}1 & 0\\ 0 & g\end{pmatrix}\right]W''(g)\,|\det g|^{s-1/2}dg.\end{aligned}$$

Taking into account the definition of F ((10)), we easily deduce the following identity:

$$\begin{aligned}&\int F(g)W''(g)\,|\det g|^s dg=\omega(-1)^{r-2}\gamma(s+\tfrac{1}{2},(\sigma\otimes\chi)\times\tau,\psi)\\ &\int W'\left[\begin{pmatrix}g & 0\\ 0 & 1\end{pmatrix}\begin{pmatrix}w_{r-2} & 0\\ 0 & 1\end{pmatrix}\right]\varphi_1[\eta_{r-2}g]W''(g)\,|\det g|^s dg;\end{aligned} \tag{14}$$

and we have to prove this is equal to the right-hand side of (13), for an appropriate choice of W' and φ_1. Now by the induction hypothesis [cf. (4.2.2)]:

$$\gamma(s,(\sigma\otimes\chi)\times\tau,\psi)=C_0X^{a(r-1)(r-2)}, \qquad X=q^{-s}, \tag{15}$$

where C_0 depends solely on χ, ω_σ, and $\omega_\tau=\omega$.

To proceed, we choose W' in such a way that the function

$$H'(g)=W'\left[\begin{pmatrix} g & 0 \\ 0 & 1 \end{pmatrix}\begin{pmatrix} w_{r-2} & 0 \\ 0 & 1 \end{pmatrix}\right] \tag{16}$$

has support in the set $\tilde{\omega}^{-a(r-1)}N_{r-2}K_{r-2}(f)$ and is such that

$$H'(\tilde{\omega}^{-a(r-1)}k)=C_1 C_0^{-1} q^{-a(r-1)(r-2)/2} \tag{17}$$

with

$$C_1=\omega(-1)^{r-2}\omega(\tilde{\omega}^{a(r-1)})$$

and C_0 as in (15). Finally, we specify φ_1 by requiring that, for $x\in F^{r-3}$, $y\in F$,

$$\varphi_1(x,y)=\omega^{-1}(d), \quad \text{if}$$

$$x\equiv 0 \bmod \mathfrak{P}^{-a(r-1)+f}, \quad y=\tilde{\omega}^{-a(r-1)}d, \quad |d|=1, \tag{18}$$

and that φ_1 be zero otherwise. Then (7) is satisfied in any case. Moreover $\hat{\varphi}_1$ has support in the set of $y\equiv 0 \bmod \mathfrak{P}^{a(r-1)-f}$, so that condition (8) is also satisfied, provided a is large compared to f.

With these choices then we see finally that the right-hand side of (14) reduces to the right-hand side of (13). Hence we have established that $J_1=J_2\neq 0$.

It remains to see that $I_1=I_2$ [cf. (2)]. It suffices to show that the functions

$$g\mapsto W_i\left[\begin{pmatrix} g & 0 \\ 0 & 1 \end{pmatrix} w_r\right]$$

agree. Since $W_{1,0}$ and $W_{2,0}$ agree on the subset $P_rZ_rK_r(f)$ and W_i is related to $W_{i,0}$ by (4), it will be enough to show that

$$\varphi(u)\varphi_1(x,y)\neq 0, \quad u\in F, \quad x\in F^{r-3}, \quad y\in F, \tag{19}$$

implies that the matrix

$$w_r\begin{pmatrix} 1 & u & x & y \\ 0 & 1 & 0 & 0 \\ 0 & 0 & 1_{r-3} & 0 \\ 0 & 0 & 0 & 1 \end{pmatrix} \tag{20}$$

belongs to $P_rZ_rK_r(f)$. But this matrix can also be written as the following product:

$$\begin{pmatrix} 1 & 0 & 0 \\ 0 & w_{r-2} & 0 \\ 0 & 0 & 1 \end{pmatrix}\begin{pmatrix} 1 & 0 & y^{-1} \\ 0 & 1_{r-2} & 0 \\ 0 & 0 & 1 \end{pmatrix}$$

$$\begin{pmatrix} -y^{-1} & 0 & 0 \\ 0 & 1_{r-2} & 0 \\ 0 & 0 & y \end{pmatrix}\begin{pmatrix} 1 & 0 & 0 \\ 0 & 1_{r-2} & 0 \\ y^{-1} & 0 & 1 \end{pmatrix}$$

$$\begin{pmatrix} 1 & u & x & 0 \\ 0 & 1 & 0 & 0 \\ 0 & 0 & 1_{r-3} & 0 \\ 0 & 0 & 0 & 1 \end{pmatrix}.$$

The first three matrices are in $P_r Z_r$. The product of the last two can also be written as

$$\begin{pmatrix} 1 & u & x & 0 \\ 0 & 1 & 0 & 0 \\ 0 & 0 & 1_{r-3} & 0 \\ 0 & 0 & 0 & 1 \end{pmatrix} \begin{pmatrix} 1 & 0 & 0 & 0 \\ 0 & 1 & 0 & 0 \\ 0 & 0 & 1_{r-3} & 0 \\ y^{-1} & y^{-1}u & y^{-1}x & 1 \end{pmatrix}. \tag{21}$$

The first of these two matrices is in P_r. Finally we may assume a is so large that $\mathfrak{P}^a$ is contained in the conductor of ω_σ. Then [formula (6)] $\hat{\varphi}$ is constant on the cosets of $\mathfrak{P}^a$ in F. Thus φ is supported on $\mathfrak{P}^{-a}$ and in the second matrix $u \in \mathfrak{P}^{-a}$. By (18) $y^{-1} \in \tilde{\omega}^{a(r-1)}\mathfrak{R}^\times$. Thus $y^{-1}u \in \mathfrak{P}^{a(r-2)} \subset \mathfrak{P}^a$ since $r>2$. Assuming $a \geqq f$ we have $y^{-1}u \in \mathfrak{P}^f$ and $y^{-1} \in \mathfrak{P}^f$. Finally by (18) again, x is in $\mathfrak{P}^{-a(r-1)+f}$ so $y^{-1}x$ is in $\mathfrak{P}^f$. Thus the second matrix in (21) is in $K_r(f)$ and with that the proof is complete. □

5. Complements

(5.1) **Proposition.** *Let π and σ be two irreducible generic representations of G_r and G_t respectively. If a is large enough and χ has conductor $\mathfrak{P}^a$ then*

$$L(s, (\pi \otimes \chi) \times \sigma) = 1.$$

Proof. Suppose first that π and σ are supercuspidal. One may as well assume they are preunitary and χ is a character of module one. Then, by [J-P-S] II Proposition (8.3), the factors

$$L(s, (\pi \otimes \chi) \times \sigma), \qquad L(s, (\tilde{\pi} \otimes \chi^{-1}) \times \tilde{\sigma})$$

have no pole in the region $\mathrm{Re}(s) > 0$. In particular the fraction

$$L(1-s, (\tilde{\pi} \otimes \chi^{-1}) \times \tilde{\sigma}) / L(s, (\pi \otimes \chi) \times \sigma)$$

is in irreducible form. Since it is equal to the γ-factor, up to a monomial factor, we see that it is itself a monomial, if the conductor of χ is deep enough. Then

$$L(s, (\pi \otimes \chi) \times \sigma) = 1.$$

In general π and σ are components of induced representations

$$\xi = \mathrm{Ind}(G_r, Q; \pi_1, \pi_2, \ldots, \pi_m),$$
$$\eta = \mathrm{Ind}(G_t, S; \sigma_1, \sigma_2, \ldots, \sigma_n),$$

where the π_i and the σ_j are supercuspidal. Then:

$$L(s, (\pi \otimes \chi) \times \sigma) = P_\chi(q^{-s}) \prod_{i,j} L(s, (\pi_i \otimes \chi) \times \sigma_j)$$

where P_χ is a polynomial ([J-P-S] I, Theorem (3.1)) and our assertion follows.

(5.2) *Remark.* Even if π and σ are not generic it is possible to define the factors L and ε ([J-P-S] I). Propositions (5.1) and (4.1) are then true for all pairs.

References

[G-J] Godement, R., Jaquet, H.: Zeta functions of simple algebras. Lect. Notes Math., Vol. 260. Berlin, Heidelberg, New York: Springer 1972

[J-P-S] I Jacquet, H., Piatetski-Shapiro, I., Shalika, J.: Rankin-Selberg convolutions. Am. J. Math. (to appear)

[J-P-S] II Jacquet, H., Piatetski-Shapiro, I., Shalika, J.: Conducteur des représentations du groupe linéaire. Math. Ann. **256**, 199–214 (1981)

[J-P-S] III Jacquet, H., Piatetski-Shapiro, I., Shalika, J.: Relèvement cubique non normal. C. R. Acad. Sci. Paris **292**, 567–571 (1981)

[J-L] Jacquet, H., Langlands, R.: Automorphic forms on GL(2). Lect. Notes Math., Vol. 114. Berlin, Heidelberg, New York: Springer 1970

Received July 30, 1982

Ann. scient. Éc. Norm. Sup.,
4e série, t. 19, 1986, p. 185 à 229.

SUR UN RÉSULTAT DE WALDSPURGER ([1])

PAR HERVÉ JACQUET

0. Introduction

(0.1) Nous allons donner une nouvelle démonstration d'un résultat remarquable de Waldspurger ([W3], Théorème 2). La démonstration de Waldspurger repose sur les propriétés de la représentation de Weil. La nôtre repose sur une variante de la formule des traces. Nous espérons qu'elle ne sera pas sans intérêt.

Rappelons d'abord le résultat. Soient F un corps de nombres, E une extension quadratique de F, η le caractère du groupe des classes d'idèles de F attaché à E. Regardons le groupe GL(2) comme un groupe algébrique G défini sur F ; soit Z son centre. Soit A un tore déployé maximal dans G, disons, pour fixer les idées, le groupe des matrices diagonales. Soit π une représentation cuspidale automorphique du groupe $G(F_A)$, triviale sur le centre $Z(F_A)$. Nous dirons que π satisfait à la première condition de Waldspurger (en abrégé W1) s'il existe des formes automorphes ϕ_1 et ϕ_2 dans l'espace de π telles que les intégrales suivantes soient non nulles :

$$(1) \qquad \int \phi_1(a)da, \qquad \int \phi_2(a)\eta\,(\det a)da, \qquad a \in A(F_A)/Z(F_A).$$

Introduisons d'autre part l'ensemble X(E : F), ou simplement X, des classes d'isomorphismes de couples (G′, T′), où G′ est une forme intérieure de G et T′ un tore maximal de G′, isomorphe sur F au groupe multiplicatif de E. Rappelons qu'un tel couple s'obtient au moyen d'un couple (H, L), formé d'une algèbre simple H de rang quatre sur F et d'un sous-corps L de H F-isomorphe à E, en prenant pour G′ le groupe multiplicatif de H et pour T′ celui de L. Le centre de G′ sera noté Z′.

Nous identifierons l'ensemble X à l'un de ses systèmes de représentants. Notons aussi $X(\pi)$ l'ensemble des triplets (G', T', π'), où le couple (G′, T′) est dans X et π' est une représentation automorphe cuspidale de $G'(F_A)$ reliée à π par la condition de [J-L], Th. (15.1) ; celle-ci, rappelons-le, s'énonce encore ainsi : il existe un ensemble fini S de places de F tel que, pour v non dans S, les groupes G_v et G'_v soient isomorphes et les représentations π_v et π'_v équivalentes, après identification des deux groupes. Nous dirons que π satisfait à la

([1]) Partially supported by N.S.F. Grant MCS 82-00551.

ANNALES SCIENTIFIQUES DE L'ÉCOLE NORMALE SUPÉRIEURE. — 0012-9593/86/02 185 45/$ 6.50/

deuxième condition de Waldspurger (W2) s'il existe un triplet (G', T', π') dans $X(\pi)$ et une forme automorphe ϕ dans l'espace de π' telle que l'intégrale suivante soit non nulle :

$$\int \phi(t)dt, \qquad t \in T'(F_A)/Z'(F_A). \tag{2}$$

Alors :

Théorème (Waldspurger). — *Les conditions* W1 *et* W2 *sont équivalentes.*

(0.2) Indiquons les grandes lignes de notre démonstration. Tout d'abord nous pouvons identifier l'ensemble des doubles classes $A\backslash G/A$ avec la réunion disjointe des doubles classes $T'\backslash G'/T'$(§ 1). A vrai dire pour avoir cette identification il nous faut nous limiter aux doubles classes « régulières ». Cela nous conduit à considérer une fonction lisse à support compact f sur $G(F_A)/Z(F_A)$ et, pour chaque (G', T'), une fonction lisse à support compact f' sur $G'(F_A)/Z'(F_A)$. En fait f' sera nulle pour presque tous les (G', T'). A la fonction f est associé le noyau cuspidal K_c et de même à chaque fonction f' est associé un noyau cuspidal K'_c. Les conditions imposées sur ces fonctions sont telles que (§ 7 à § 10) :

$$\left\{\begin{array}{l} \displaystyle\iint K_c(a, b)da\eta(\det b)db = \sum_{(G',T')} \iint K'_c(s, t)dsdt, \\ a, b \in A(F_A)/A(F)Z(F_A), \qquad s, t \in T'(F_A)/T'(F)Z'(F_A). \end{array}\right. \tag{1}$$

La relation entre f et les f' est comme suit. Bien entendu ces fonctions sont des produits de fonctions locales. Si v est une place de F qui se décompose dans E alors pour tout (G', T') les groupes G'_v et G_v sont les « mêmes » et nous prenons pour f_v et f'_v la « même » fonction. Supposons au contraire que v se décompose. Alors l'ensemble $X(E_v : F_v)$ est encore défini mais il est réduit à deux éléments (G_{vi}, T_{vi}), $i=1, 2$, avec G_{v1} déployé. Nous pouvons encore identifier les doubles classes régulières de A_v avec la réunion disjointe des doubles classes régulières de T_1 et T_2. Nous montrons que pour une fonction f donnée il existe des fonctions f_i sur G_i telles que

$$\iint f(agb)da\eta_v(\det b)db = \iint f_i(sg't)dsdt,$$
$$a, b \in A_v/Z_v, \qquad s, t \in T_{iv}/Z_{iv},$$

si g correspond à g' (§ 2 à 4) (L'énoncé exact est un peu différent puisque le membre de gauche n'est pas tout à fait une fonction sur l'ensemble des doubles classes). Si de plus la situation est non ramifiée et f_v est une fonction de Hecke alors nous pouvons prendre et prenons $f_1 = f_v$ et $f_2 = 0$ (§ 5). La condition est maintenant que $f'_v = f_i$ si $G'_v = G_i$. Le résultat de Waldspurger découle facilement de l'identité (1) (*cf.* § 11). Le paragraphe 6 contient des résultats auxiliaires.

La méthode de démonstration de la formule (1) est basée sur une généralisation de la formule des traces qui peut s'énoncer comme suit. Soit G un groupe semi-simple défini sur F et A, B des sous-groupes de G définis sur F, λ et μ des caractères de $A(F_A)/A(F)$ et

$B(F_A)/B(F)$ respectivement. Donnons-nous une fonction f sur $G(F_A)/G(F)$ lisse à support compact et calculons l'intégrale suivante :

$$\iint K_c(a, b)\lambda(a)\mu(b)da\,db$$

où K_c est le noyau cuspidal attaché à f. Ce noyau a une expression compliquée qui contient en tout cas la somme :

$$\sum_\zeta f(x^{-1}\zeta y), \qquad \zeta\in G(F).$$

Choisissons un système de représentants pour les doubles classes des groupes A(F) et B(F). D'autre part si η est un élément de G(F) notons H_η le sous-groupe de $A \times B$ formé des paires (α, β) telles que $\alpha^{-1}\eta\beta = \eta$. Alors tout élément de G(F) peut s'écrire uniquement sous la forme :

$$\zeta = \alpha^{-1}\eta\beta, \qquad \eta\in A(F)\backslash G(F)/B(F), \quad (\alpha, \beta)\in H_\eta(F)\backslash A(F)\times B(F).$$

En imitant le calcul formel usuel nous arrivons aussitôt à l'expression suivante pour l'intégrale ci-dessus :

$$\sum_\eta \mathrm{Vol}\,(H_\eta(F)\backslash H(F_A)\iint f(a^{-1}\eta b)\lambda(a)\mu(b)da\,db,$$
$$a\in A(F)\backslash A(F_A), \qquad b\in B(F)\backslash B(F_A),$$

la somme portant sur tous les η tels que $\lambda(a)\mu(b)=1$ si $a^{-1}b=1$. Bien entendu nous avons ignoré les problèmes de convergence et l'existence des autres termes dans l'expression pour K_c.

(0.3) Il me reste à remercier l'Institute for Adavanced Study et ses membres permanents pour leur hospitalité, la majeure partie de ce travail ayant été écrite durant mon séjour à l'Institute, à l'occasion de l'année spéciale 1983-1984 sur les fonctions L. En particulier je remercie Langlands pour l'intérêt qu'il a pris à ce travail. Enfin je dois beaucoup de reconnaissance à Piatetski-Shapiro qui était aussi à l'Institute la même année. Sa connaissance approfondie de l'œuvre de Waldspurger m'a été très utile ; de plus une conversation avec Piateski-Shapiro a été le point de départ de ce travail.

1. Doubles classes

(1.1) Dans ce paragraphe F sera un corps quelconque, disons de caractéristique zéro, et E une extension quadratique de F. Nous noterons N(E : F) ou simplement N le sous-groupe des normes de E dans le groupe multiplicatif de F. L'ensemble X(E : F) ou simplement X introduit au paragraphe est encore défini. Considérons l'un de ses éléments (G, T). Il existe donc une algèbre simple H de rang 4 sur F et un sous-corps L de H isomorphe à E tels que G soit le groupe multiplicatif de H et T celui de L. Nous nous proposons de donner une paramétrisation des doubles classes T\G/T. A cet effet choisissons un élément ε du normalisateur N(T) de T qui ne soit pas dans T. Alors tout h dans H s'écrit de manière unique sous la forme :

(1) $$h=h_1+\varepsilon h_2, \qquad \text{avec} \quad h_i \in \mathrm{L}.$$

D'autre part, si $z \rightarrow z^-$ désigne l'unique F-automorphisme non trivial de L alors :

(2) $$\varepsilon z \varepsilon^{-1}=z^-.$$

Le carré $c=\varepsilon^2$ est dans Z, ou, autrement dit, dans F. De plus la classe de c modulo N est déterminée par la classe d'isomorphisme du couple (G, T) et, réciproquement, la détermine.

Définissons deux involutions j^+ et j^- de H par les formules :

(3) $$j^{\pm}(h)=h_1^- \pm \varepsilon h_2 \quad \text{avec } h \text{ comme dans (1).}$$

Il est facile de vérifier que ce sont les seules involutions de H qui induisent sur L l'unique F-automorphisme non trivial de L. Pour h dans G nous poserons :

(4) $$\mathrm{X}(h)=\frac{(1/2\mathrm{T}r(hj^+(h))}{(1/2\mathrm{T}r(hj^-(h))}.$$

Comme le dénominateur de cette fraction n'est autre que la norme réduite de h, X(h) est un élément bien défini de F ne dépendant que de la double classe de h modulo T. Nous introduirons aussi la fonction P(h : T) ou simplement P(h) définie par

(5) $$\mathrm{X}(h)=\frac{1+\mathrm{P}(h)}{1-\mathrm{P}(h)},$$

ou encore

(6) $$\mathrm{P}(h)=ch_2h_2^-(h_1h_1^-)^{-1}, \qquad c=\varepsilon^2.$$

Ainsi P est une fonction à valeurs dans la droite projective qui est constante sur les doubles classes de T dans G. Remarquons toutefois que d'après la formule précédente, si P(h) n'est ni zéro ni infini, alors c'est un élément de la classe cN déterminée par le couple (G, T). De plus P(h) ne peut être égal à un, car cela donnerait X(h) infini. Nous dirons que h (ou sa double classe) est T-singulier si P(h) est zéro ou l'infini, T-régulier dans le cas contraire.

PROPOSITION. — *Deux éléments h et h' de* G *ont la même double classe modulo* T *si et seulement si* P(h)=P(h'). *De plus si x est dans c*N *et différent de* 1 *alors il existe un h dans* G *tel que* P(h)=x.

La démonstration est laissée au lecteur.

(1.2) La proposition suivante justifie l'emploi de l'adjectif T-régulier :

PROPOSITION. — *Supposons h* T-*régulier. Alors les relations*

$$sht=hz, \qquad s\in\mathrm{T}, \quad t\in\mathrm{T}, \quad z\in\mathrm{Z}$$

entraînent

$$s\in\mathrm{Z}, \quad t\in\mathrm{Z}, \qquad st=z.$$

La démonstration est laissée au lecteur.

(1.3) Ce qui précède s'applique « *mutatis mutandis* » à un couple de la forme (G, A) où G est le groupe GL(2) et A un tore déployé maximal, disons le groupe des matrices diagonales dans G. Alors H est l'algèbre des matrices 2 par 2, L la sous-algèbre des matrices diagonales et nous pouvons prendre

$$\varepsilon = \begin{vmatrix} 0 & 1 \\ 1 & 0 \end{vmatrix}, \qquad c = 1.$$

Les fonctions X et P(. : A) (ou simplement P) sont définies comme plus haut. En particulier :

$$P(h) = bc(ad)^{-1} \qquad \text{si} \quad h = \begin{vmatrix} a & b \\ c & d \end{vmatrix}.$$

Elles sont constantes sur les doubles classes de A dans G. A nouveau P ne peut prendre la valeur 1. Nous dirons encore qu'un élément h de G est A-singulier si P(h) est zéro ou l'infini, A-régulier dans le cas contraire. Il y a maintenant 6 doubles classes A-singulières : les classes sur lesquelles P prend la valeur zéro :

$$(1) \qquad T, \quad Tn_+T, \quad Tn_-T, \qquad \text{où} \quad n_+ = \begin{vmatrix} 1 & 1 \\ 0 & 1 \end{vmatrix}, \qquad n_- = \begin{vmatrix} 1 & 0 \\ 1 & 1 \end{vmatrix},$$

et les classes sur lesquelles P prend la valeur infinie :

$$(2) \qquad \varepsilon T, \qquad T\varepsilon n_+T, \qquad T\varepsilon n_-T.$$

Ainsi P ne permet pas de distinguer ces classes l'une de l'autre.

Toutefois P sépare les classes A-régulières :

PROPOSITION. — *Soient h et h' des éléments* A-*réguliers de* G. *Alors h et h' sont dans la même classe si et seulement si* $P(h) = P(h')$. *Si x est dans* F *différent de* 1 *et* 0, *alors il existe un élément* A-*régulier h tel que* $P(h) = x$.

Nous laisserons la démonstration au lecteur.

(1.4) Nous avons aussi l'analogue de la proposition (1.2) :

PROPOSITION. — *Supposons h* A-*régulier. Alors les relations*

$$ahb = hz, \qquad a \in A, \quad b \in A, \quad z \in Z$$

entraînent

$$a \in Z, \quad b \in Z, \qquad ab = z.$$

Nous laisserons la démonstration au lecteur.

2. Intégrales orbitales : cas d'un tore compact

(2.1) Gardons les notations du paragraphe 1 mais supposons maintenant que F soit un corps local. Alors T/Z est compact. Choisissons une fois pour toutes un caractère additif non trivial ψ de F. Munissons le groupe additif F de la mesure dx auto-duale pour le caractère ψ, le groupe multiplicatif $F^\times$ de la mesure $L(1, 1_F) \mid x \mid^{-1} dx$ (mesure de Tamagawa relative à ψ). Munissons de même le groupe multiplicatif $E^\times$ de la mesure de Tamagawa relative au caractère $\psi \circ Tr$. Par transport de structure nous obtenons des mesures sur T et Z. Munissons T/Z de la mesure quotient. Soit f une fonction lisse à support compact sur G/Z. Posons :

$$(1) \qquad H(g : f : T) = \iint f(sgt) ds dt, \qquad s, t \in T/Z.$$

Il est clair que $H(g : f : T)$ ne dépend que de la double classe de g modulo T. Soit x un élément de $F^\times$. S'il existe g dans G tel que $P(g : T) = x$, nous poserons $H(x : f : T) = H(g : f : T)$. Sinon nous poserons $H(x : f : T) = 0$. Nous obtenons ainsi une fonction $H(f : T)$ sur $F^\times$ et nous nous proposons de caractériser les fonctions H sur $F^\times$ qui sont de la forme $H = H(f : T)$ pour une fonction f appropriée.

(2.2) Considérons donc une fonction $H = H(f : T)$. Par définition H s'annule, donc est lisse, sur le complémentaire de cN. Considérons un point x de la forme $P(h : T)$. Comme la norme est une application submersive de $E^\times$ dans $F^\times$, l'application $g \to P(g : T)$ est *a fortiori* submersive au point h. Il en résulte que H est lisse au point x. Finalement supposons que 1 soit dans cN (c'est-à-dire que le groupe G soit déployé); nous pouvons alors supposer que $c = 1$. Nous allons montrer que H est nulle au voisinage de 1.

Puisque f est à support compact modulo Z, il existe un sous-ensemble compact C de G tel que $H(g : f : T) \neq 0$ entraîne $g \in TCT$. Il suffira donc de montrer l'existence d'un nombre K tel que la relation $g \in TCT$ entraîne $| P(g : T) - 1 | > K$. Supposons qu'il n'existe pas de tel nombre. Alors il existerait une suite g_i d'éléments de TCT telle que $P(g_i : T)$ tende vers 1. Quitte à agrandir C et à multiplier les éléments de la suite par des éléments de T nous pouvons supposer que

$$g_i = 1 + \varepsilon t_i = c_i z_i$$

avec t_i dans T, c_i dans C et z_i dans Z. Alors

$$P(g_i : T) = t_i t_i^- = 1 + a_i$$

et a_i tend vers zéro. D'autre part nous avons :

$$\det g_i = -a_i \qquad \text{et} \qquad \det g_i = (z_i)^2 \det c_i.$$

Donc z_i tend vers zéro. Il en est donc de même de g_i. Comme la projection de g_i sur L est 1 cela nous donne une contradiction.

(2.3) Examinons le comportement de la fonction H au voisinage de zéro et au voisinage de l'infini. Nous allons montrer qu'il existe un voisinage U de 0 dans F et une fonction lisse A sur U telle que :

$$(1) \qquad H(x) = A(x)(1 + \eta(cx)), \qquad \text{pour} \quad x \in U,$$

$$(2) \qquad 2A(0) = \operatorname{vol}(T/Z) \int f(t) dt.$$

De même nous montrerons qu'il existe un voisinage U de 0 dans F et une fonction lisse B sur U telle que :

(3) $$\mathrm{H}(x)=\mathrm{B}(x^{-1})(1+\eta(cx)), \qquad \text{pour} \quad x^{-1}\in \mathrm{U},$$

(4) $$2\mathrm{B}(0)-\mathrm{vol}\,(\mathrm{T}/\mathrm{Z})\int f(\varepsilon t)dt.$$

En effet comme $\mathrm{P}(\varepsilon g : \mathrm{T})=\mathrm{P}(g : \mathrm{T})^{-1}$ nous avons :

$$\iint f(s\varepsilon gt)dsdt=\mathrm{H}(x^{-1} : f : \mathrm{T})$$

ou encore :

$$\mathrm{H}(x^{-1} : f : \mathrm{T})=\mathrm{H}(x : f' : \mathrm{T}) \qquad \text{avec} \quad f'(g)=f(\varepsilon g).$$

Il suffira donc de démontrer les assertions relatives au point zéro. Il sera commode de traiter d'abord le cas non archimédien. Prenons un x dans $c\mathrm{N}$. Alors $x=cll^-$ d'où $x=\mathrm{P}(h)$ avec $h=1+\varepsilon l$. Nous pouvons donc écrire :

$$\mathrm{H}(x : f : \mathrm{T})=\iint f\,[t_1(1+\varepsilon l)t_2\,]dt_1dt_2$$

ou encore après un changement de variables :

(5) $$\mathrm{H}(x : f : \mathrm{T})=\iint f\left[\left(1+\varepsilon l\,\frac{\bar{t}_1}{t_1}\right)t_2\right]dt_1dt_2.$$

Comme f est lisse il existe un idéal V de E tel que pour l dans V nous ayons :

$$f(g)=f\,[(1+\varepsilon l)g\,] \quad \text{pour tout} \quad g.$$

Il existe alors un idéal U dans F tel que $ll^-\in\mathrm{U}$ soit équivalent à $l\in\mathrm{V}$. Pour x dans $c\mathrm{U}$ nous avons donc $\mathrm{H}(x)=0$ si x n'est pas dans $c\mathrm{N}$; si au contraire x est dans $c\mathrm{N}$ alors $x=cll^-$ avec l dans V et nous avons d'après la formule (5) :

$$\mathrm{H}(x)=\mathrm{vol}\,(\mathrm{T}/\mathrm{Z})\int f(t)dt.$$

Notre assertion est alors immédiate.

Passons au cas archimédien. Alors F est le corps des nombres réels et L le corps des nombres complexes. Posons $\mathrm{K}(x)=\mathrm{H}(cx)$. Soit V un disque $\{\,z\,|\,zz^- < a\,\}$ dans L tel que $1+\varepsilon\mathrm{V}$ soit contenu dans G. Alors le second membre de (5) définit une fonction lisse sur V, disons $\mathrm{C}(l)$, ne dépendant que de la norme de l. Nous avons

$$\mathrm{K}(x)=0 \quad \text{si} \quad x<0,$$
$$\mathrm{K}(x)=\mathrm{C}(l) \quad \text{si} \quad x>0 \quad \text{et} \quad x=ll^- \quad \text{avec} \quad l \text{ dans V}.$$

En particulier la restriction de C à l'axe réel est lisse et paire et nous avons

$$\mathrm{K}(x)=0 \quad \text{si} \quad x<0,$$
$$\mathrm{K}(x)=\mathrm{C}(y) \quad \text{si} \quad 0<x<a \quad \text{et} \quad x=y^2 \quad \text{avec} \quad y \text{ réel}.$$

Le contenu de notre assertion est l'existence d'une fonction lisse D sur F telle que $D(x) = K(x)$ pour $a > x > 0$. Elle est donc conséquence d'un théorème de Whitney.

(2.4) Les propriétés précédentes caractérisent les fonctions $H(f:T)$:

PROPOSITION. — *Soit* H *une fonction sur* $F^\times$. *Pour qu'il existe une fonction lisse à support compact* f *sur* G/Z *telle que* $H = H(f:T)$ *il faut et il suffit que les conditions suivantes soient satisfaites :*

(1) H *est nulle dans le complément de* cN ;

(2) H *est nulle dans un voisinage du point* 1 ;

(3) *il existe une fonction lisse* A *sur un voisinage de* 0 *dans* F *telle que, pour* x *près de* 0, *on ait :*

$$H(x) = A(x)(1 + \eta(cx)) ;$$

(4) *il existe une fonction lisse* B *sur un voisinage de* 0 *de* F *telle que pour* $|x|$ *suffisamment grand on ait :*

$$H(x) = B(x^{-1})(1 + \eta(cx)) .$$

Enfin si f, A *et* B *satisfont ces conditions alors :*

$$2A(0) = \mathrm{vol}\,(T/Z) \int f(t)dt, \qquad 2B(0) = \mathrm{vol}\,(T/Z) \int f(\varepsilon t)dt .$$

Nous venons de montrer que les conditions (1) à (4) sont nécessaires. Nous laisserons au lecteur le soin de montrer qu'elles sont aussi suffisantes. La dernière assertion de la proposition a été prouvée au numéro (2.3).

3. Intégrales orbitales : cas d'un tore déployé

(3.1) Dans ce paragraphe F est un corps local, E une extension quadratique, η le caractère quadratique de $F^\times$ attaché à E, G le groupe GL(2) et A le sous-groupe des matrices diagonales. Munissons encore $F^\times$ de la mesure de Tamagawa, $(F^\times)^2$ du produit tensoriel des mesures de Tamagawa des facteurs. Par transport de structure nous obtenons une mesure sur A. Munissons A/Z de la mesure quotient. Si f est une fonction lisse à support compact sur G/Z et g est A-régulier dans G nous poserons :

$$H(g:f:A) = H(g:f:1) = \iint f(agb)da\,db, \qquad a,\ b \in A/Z, \tag{1}$$

$$H(g:f:\eta) = \iint f(agb)da\,\eta\,(\det b)da\,db, \qquad a,\ b \in A/Z. \tag{2}$$

La première intégrale ne dépend que de $P(g:A)$ et nous noterons $H(x:f:A)$ ou $H(x:f:1)$ sa valeur en un point g telle que $P(g:A) = x$. Nous poserons aussi $H(1:f:A) = H(1:f:1) = 0$. Pour x dans F différent de 0 et 1 nous définirons une matrice $g(x)$ par :

$$g(x) = \begin{vmatrix} 1 & x \\ 1 & 1 \end{vmatrix}. \tag{3}$$

Alors $P(g(x))=x$ de sorte que nous avons défini une section de l'espace des doubles classes de A dans G. Nous poserons $H(x : f : \eta)=H(g(x) : f : \eta)$ si x est différent de 1 et 0; $H(x : f : \eta)=0$ si $x=1$. Posons :

(4) $$w=\begin{vmatrix} 0 & -1 \\ 1 & 0 \end{vmatrix}.$$

Notons N le groupe des matrices trigonales strictes supérieures et N′ le groupe des matrices trigonales strictes inférieures. Alors nous avons un recouvrement de G par deux ouverts :

(5) $$G = ANN' \cup ANwN.$$

Nous pouvons donc écrire f comme une somme f_1+f_2 où f_1 a son support dans le premier ouvert et f_2 dans le deuxième. Posons

(6) $$\phi(g)=\int f(ag)da, \qquad a\in A/Z$$

et définissons de même ϕ_1 et ϕ_2. Alors ϕ est invariante à gauche sous A et à support compact modulo A. Il en est de même de ϕ_1 et ϕ_2 et $\phi=\phi_1+\phi_2$. De plus les fonctions Φ_1 et Φ_2 définies par

(7) $$\Phi_1(u, v)=\phi_1\left[\begin{vmatrix} 1 & u \\ 0 & 1 \end{vmatrix}\times\begin{vmatrix} 1 & 0 \\ v & 1 \end{vmatrix}\right],$$

(8) $$\Phi_2(u, v)=\phi_2\left[\begin{vmatrix} 1 & u \\ 0 & 1 \end{vmatrix} w \begin{vmatrix} 1 & v \\ 0 & 1 \end{vmatrix}\right]$$

sont à support compact sur $F \times F$. Comme

(9) $$g(x)\begin{vmatrix} a & 0 \\ 0 & 1 \end{vmatrix}=\begin{vmatrix} a(1-x) & 0 \\ 0 & 1 \end{vmatrix}\times\begin{vmatrix} 1 & a^{-1}(1-x)^{-1}x \\ 0 & 1 \end{vmatrix}\times\begin{vmatrix} 1 & 0 \\ a & 1 \end{vmatrix}$$
$$=\begin{vmatrix} 1-x & 0 \\ 0 & a \end{vmatrix}\times\begin{vmatrix} 1 & a(1-x)^{-1} \\ 0 & 1 \end{vmatrix} w \begin{vmatrix} 1 & a^{-1} \\ 0 & 1 \end{vmatrix},$$

nous avons pour x différent de 0 et 1 :

(10) $$H(x : f : A)=\int \Phi_1 [a^{-1}(1-x)^{-1}x, a]d^{\times}a+\int \Phi_2 [a(1-x)^{-1}, a^{-1}]d^{\times}a.$$

Pour prouver la convergence de notre intégrale, nous pouvons supposer f positive. Alors le calcul précédent est justifié. Dans le membre de droite de (10) les intégrandes sont à support compact dans $F^{\times}$ et donnent donc des intégrales convergentes. Ainsi H est convergente et égale à (10). De la même façon, nous avons, pour x différent de 0 et 1 :

(11) $$H(x : f : \eta)=\int \Phi_1 [a^{-1}(1-x)^{-1}x, a]\eta(a)d^{\times}a + \int \Phi_2 [a(1-x)^{-1}, a^{-1}]\eta(a)d^{\times}a.$$

(3.2) Nous allons étudier les propriétés des fonctions $H(f : \eta)$. La formule (3.1.11) montre déjà que $H(x : F : \eta)$ est une fonction lisse en tout point x différent de 0 et 1. D'autre part si Φ_1 et Φ_2 ont leur support dans l'ensemble des (x, y) tels que $|x|<C$, $|y|<C$ alors dans la deuxième intégrale nous avons, sur le support de Φ_2, $|a(1-x)^{-1}|<C$ et $|a^{-1}|<C$

ce qui donne : $C^{-2} < |1-x|$ si la deuxième intégrale n'est pas nulle. De même si la première intégrale n'est pas nulle nous trouvons $|(1-x)^{-1}x| < C^2$, ce qui implique aussi $D < |1-x|$ pour une constante D convenable. Il en résulte que $H(x : f : \eta)$ est nulle au voisinage de 1. On peut donc considérer la formule (3.1.11) comme vraie pour tout x non nul.

Étudions maintenant $H(f : \eta)$ au voisinage de 0. Dans (3.1.11) la deuxième intégrale est évidemment une fonction lisse de x au point 0. Pour étudier la première intégrale nous utiliserons le lemme suivant, dont la démonstration est laissée au lecteur :

LEMME. — *Soit Φ une fonction de Schwartz-Bruhat à deux variables. Alors il existe deux fonctions de Schwartz-Bruhat à une variable A_1 et A_2 telles que pour tout x différent de 0 on ait :*

$$\int \Phi(a^{-1}x, a)\eta(a)d^{\times}a = A_1(x) + A_2(x)\eta(x).$$

Si F est réel et Φ est donnée avec un support compact, on peut prendre A_1 et A_2 à support compact.

Revenons à la première intégrale de (3.1.11). Avec les notations de la démonstration du lemme, l'intégrale est égale à

$$A_1(x(1-x)^{-1}) + A_2(x(1-x)^{-1})\eta(x(1-x)^{-1}). \tag{3}$$

Si x est suffisamment près de 0 alors $1-x$ est une norme et $\eta(x(1-x)^{-1}) = \eta(x)$. D'autre part $A_i(x(1-x)^{-1})$ est une fonction lisse de x dans un voisinage de 0. Comme la deuxième intégrale de (3.1.11) est évidemment lisse au point 0 nous en concluons que, dans un voisinage de 0, $H(x : f : \eta)$ a la forme suivante :

$$H(x : f : \eta) = A_1(x) + A_2(x)\eta(x) \tag{4}$$

où A_i, $i = 1, 2$, est lisse.

Pour étudier $H(x : f : \eta)$ pour $|x|$ grand remarquons que

$$\varepsilon g(x) = g(x^{-1}) \begin{vmatrix} 1 & 0 \\ 0 & x \end{vmatrix} \quad \text{si} \quad \varepsilon = \begin{vmatrix} 0 & 1 \\ 1 & 0 \end{vmatrix}.$$

Il en résulte que

$$H(x^{-1} : f : \eta) = H(x : f : \eta)\eta(x) \qquad \text{avec} \quad f'(g) = f(\varepsilon g). \tag{3}$$

Il existe donc deux fonctions B_i, $i = 1, 2$, définies au voisinage de 0 et lisses, telles que :

$$H(x : f : \eta) = B_1(x^{-1}) + B_2(x^{-1})\eta(x) \tag{4}$$

pour $|x|$ suffisamment grand.

(3.3) En résumé :

PROPOSITION. — *Soit H une fonction sur $F^{\times}$ telle qu'il existe une fonction lisse à support compact f sur G/Z avec $H(x : f : \eta) = H(x)$. Alors :*

(1) H *est lisse sur* $F^{\times}$;

(2) H *s'annule sur un voisinage de* 1 ;

(3) *il existe un voisinage* U *de* 0 *et deux fonctions lisses* A_i, $i = 1, 2$, *dans* U *tels que, pour* x *près de* 0, *on ait :*

$$H(x) = A_1(x) + A_2(x)\eta(x);$$

(4) *il existe un voisinage* U *de* 0 *et deux fonctions lisses* B_i, $i=1,2$, *dans* U *tels que, pour* $|x|$ *assez grand, on ait :*

$$H(x)=B_1(x^{-1})+B_2(x^{-1})\eta(x).$$

(3.4) Nous allons discuter la signification des valeurs en zéro des fonctions A_i et B_i de la proposition (3.3). A cet effet rappelons tout d'abord que, si ϕ est une fonction de Schwartz-Bruhat sur F, alors l'intégrale :

$$\int \phi(x)\,|x|^s d^\times x,$$

ou plutôt son prolongement analytique, a un pôle au point $s=0$; le résidu en ce point a la forme $C\phi(0)$, où la constante C dépend du choix de la mesure de Haar sur le groupe $F^\times$. D'autre part l'intégrale :

$$\int \phi(x)\,|x|^s \eta(x) d^\times x$$

a un prolongement holomorphe au point zéro et sa valeur en ce point sera encore notée comme une intégrale :

$$\int \phi(x)\eta(x) d^\times x.$$

Nous introduirons les quantités suivantes :

(1) $$H(n_+ : f : \eta) = \iint f\left[\begin{vmatrix} a & 0 \\ 0 & 1 \end{vmatrix} \times \begin{vmatrix} 1 & b \\ 0 & 1 \end{vmatrix}\right] d^\times a \eta(b) d^\times b,$$

(2) $$H(n_- : f : \eta) = \iint f\left[\begin{vmatrix} a & 0 \\ 0 & 1 \end{vmatrix} \times \begin{vmatrix} 1 & 0 \\ b & 1 \end{vmatrix}\right] d^\times a \eta(b) d^\times b,$$

(3) $$H(\varepsilon n_+ : f : \eta) = \iint f\left[\begin{vmatrix} a & 0 \\ 0 & 1 \end{vmatrix} \varepsilon \begin{vmatrix} 1 & b \\ 0 & 1 \end{vmatrix}\right] d^\times a \eta(b) d^\times b,$$

(4) $$H(\varepsilon n_- : f : \eta) = \iint f\left[\begin{vmatrix} a & 0 \\ 0 & 1 \end{vmatrix} \varepsilon \begin{vmatrix} 1 & 0 \\ b & 1 \end{vmatrix}\right] d^\times a \eta(b) d^\times b.$$

En général ces intégrales sont divergentes, mais on peut les interpréter comme il a été rappelé ci-dessus. Par exemple la première intégrale est la valeur au point 0 de la fonction méromorphe qui, pour Re$s > 0$, est donnée par l'intégrale convergente :

(5) $$\iint f\left[\begin{vmatrix} a & 0 \\ 0 & 1 \end{vmatrix} \begin{vmatrix} 1 & b \\ 0 & 1 \end{vmatrix}\right] d^\times a \eta(b)\,|b|^s d^\times b.$$

Remarquons toutefois que si f a son support dans l'ensemble ouvert ANwN alors les intégrales (1) et (2) sont convergentes. En fait l'intégrale (1) est nulle puisque AN ne rencontre pas ANwN. D'autre part l'intersection de AN′ avec un ensemble compact contenu dans ANwN est un compact disjoint de A. Il en résulte que dans (2) l'intégrande a support compact dans $F^\times \times F^\times$ et l'intégrale converge donc trivialement. De même les intégrales (3) et (4) convergent si f a son support dans ANN′.

PROPOSITION. — *Avec les notations ci-dessus et celles de la proposition (3.3) nous avons :*

(6) $$H(n_+ : f : \eta) = A_2(0)$$

(7) $$H(n_- : f : \eta) = A_1(0)$$

(8) $$H(\varepsilon n_+ : f : \eta) = B_1(0)$$

(9) $$H(\varepsilon n : f : \eta) = B_2(0).$$

Démonstration. — Démontrons les assertions (6) et (7). D'après (3.1), il suffit de faire la démonstration quand f a son support dans ANN′ ou dans ANwN. Supposons d'abord que f ait son support dans ANwN. Alors (*cf.* (3.1.9) et (3.1.11)) :

(10) $$H(x : f : \eta) = \int \Phi[a(1-x)^{-1}, a^{-1}]\eta(a)d^\times a$$

où

(11) $$\Phi(u, v) = \phi\left[\begin{vmatrix} 1 & u \\ 0 & 1 \end{vmatrix} w \begin{vmatrix} 1 & v \\ 0 & 1 \end{vmatrix}\right],$$

(12) $$\phi(g) = \int f(ag)da, \qquad a \in A/Z.$$

Comme H est lisse en zéro nous avons $A_2 = 0$ et $A_1 = H(f : \eta)$. Alors :

$$A_1(0) = \int \phi\left[\begin{vmatrix} 1 & a \\ 0 & 1 \end{vmatrix} w \begin{vmatrix} 1 & a^{-1} \\ 0 & 1 \end{vmatrix}\right]\eta(a)d^\times a.$$

En tenant compte du fait que f est invariante sous le centre nous pouvons écrire ceci sous la forme :

$$A_1(0) = \iint f\left[b\begin{vmatrix} a^2 & 0 \\ a & 1 \end{vmatrix}\right]db\,\eta(a)d^\times a.$$

Un changement de variables montre que ceci n'est autre que $H(n_- : f : \eta)$. Comme A_2 et $H(n_+ : f : \eta)$ sont nuls la relation (6) est aussi vérifiée.

Supposons maintenant que f ait son support dans ANN′. Alors (*cf.* (3.1.7) et (3.1.11)) :

(13) $$H(x : f : \eta) = \int \Phi(a^{-1}(1-x)^{-1}x, a)\eta(a)d^\times a$$

où

(14) $$\Phi(u, v) = \phi\left[\begin{vmatrix} 1 & u \\ 0 & 1 \end{vmatrix} \times \begin{vmatrix} 1 & 0 \\ v & 1 \end{vmatrix}\right],$$

(15) $$\phi(g) = \int f(ag)da, \qquad a \in A/Z.$$

Rappelons la définition de A_1 et A_2 :

(16) $$H(x) = A_1(x) + A_2(x)\eta(x);$$

d'autre part d'après le lemme (3.2) :

(17) $$\int \Phi(a^{-1}x, a)\eta(a)d^{\times}a = C_1(x) + C_2(x)\eta(x).$$

En comparant avec (4) nous voyons que $C_i((1-x)^{-1}x) = \Lambda_i(x)$. Donc C_i et A_i ont la même valeur en zéro. En prenant la transformée de Mellin de l'équation précédente nous obtenons:

$$\iint \Phi(x, a)|x|^s\eta(a)|a|^s d^{\times}a = \int C_1(x)|x|^s d^{\times}x + \int C_2(x)|x|^s\eta(x)d^{\times}x.$$

En calculant le résidu des deux membres à $s=0$ nous trouvons :

$$\int \Phi(0, a)\eta(a)d^{\times}a = C_1(0).$$

D'après (14) et (15) le membre de gauche n'est autre que $H(n_- : f : \eta)$. D'autre part le membre de droite n'est autre que $A_1(0)$. La relation (7) est donc établie. La relation (6) peut s'établir de la même façon.

Les relations (8) et (9) découlent des relations (6) et (7) appliquées à la fonction f' définie par $f'(g) = f(\varepsilon g)$.

4. Fonctions appariées

(4.1) Dans ce paragraphe E est un corps local, E une extension quadratique de F, η le caractère quadratique attaché à E. Nous considérerons encore le couple (G, A) formé du groupe GL(2) et du sous-groupe des matrices diagonales et l'ensemble $X = X(E : F)$. Il est réduit à deux éléments (G_i, T_i), $i=1, 2$, avec disons G_1 déployé. Soit f une fonction lisse à support compact sur G/Z et f_i, $i=1, 2$, une fonction lisse à support compact sur G_i/Z_i. Nous dirons que f et le couple (f_1, f_2) sont appariés si la condition suivante est satisfaite. Pour tout x dans F différent de 1 et 0 soient i et g dans G_i tels que $x = P(g : T_i)$ ($i=1$ si x est une norme de E, $i=2$ sinon) ; alors :

$$H(x : f : \eta) = H(g : f_i : T_i).$$

Proposition. — *Une fonction f étant donnée il existe un couple (f_1, f_2) apparié à f. De plus si f et le couple (f_1, f_2) sont appariés nous avons :*

$$\operatorname{Vol}(T_i/Z_i)\int f_i(t_i)dt_i = H(n_+ : f : \eta) \pm H(n_- : f : \eta),$$

$$\operatorname{Vol}(T_i/Z_i)\int f_i(\varepsilon t_i)dt_i = H(\varepsilon n_- : f : \eta) \pm H(\varepsilon n_+ : f : \eta)$$

avec le signe + si $i=1$, le signe − si $i=2$.

Démonstration. — Cela résulte aussitôt des propositions (2.4), (3.3) et (3.4).

(4.2) Si F est réel nous désignerons par K le sous-groupe orthogonal dans G. Nous noterons U l'ensemble des couples (f_1, f_2) qui sont appariés à une fonction lisse à support compact f sur G/Z, K-finie si F est réel.

Notons U_1 (resp. U_2) la première (resp. seconde) projection de U. Alors les ensembles U_i ont une propriété de densité.

PROPOSITION. — *Soit ϕ une fonction continue sur G_i/Z_i biinvariante sous T_i. Supposons que l'intégrale de ϕ contre toute fonction de U_i soit nulle. Alors ϕ est nulle.*

La démonstration occupera le reste de ce paragraphe.

(4.3) Soit f une fonction continue à support compact sur G/Z ; alors :

$$\int f(g)dg = c \int H(x : f : A) \mid 1-x \mid^{-2} dx, \qquad x \neq 0. \tag{1}$$

où c est une constante indépendante de f.

Nous aurons aussi besoin d'une estimée sur les fonctions $H(x : f : A)$ où f est continue à support compact sur G/Z :

(2) LEMME. — *Soit f une fonction à support compact sur G/Z et $H(x) = H(x : f : A)$. Alors H s'annule dans un voisinage du point 1 et est $0(\log \mid x \mid)$ pour $\mid x \mid$ petit ou grand.*

Démonstration du lemme. — Elle est analogue à celle donnée dans (3.2) excepté que le lemme de (3.2) est remplacé par l'assertion suivante : si Φ est une fonction de Schwartz-Bruhat à deux variables, alors il existe deux fonctions de Schwartz-Bruhat à une variable A_i, $i=1, 2$, telles que:

$$\int \Phi(a^{-1}x, a) d^{\times}a = A_1(x) + A_2(x) \log \mid x \mid.$$

Enfin nous aurons besoin de formules d'intégration pour les groupes G_i analogues à (1) :

$$\int f_1(g)dg = c_1 \int H(x : f_1 : T_1) \mid 1-x \mid^{-2} dx, \qquad x > 0, \tag{3}$$

$$\int f_2(g)dg = c_2 \int H(x : f_2 : T_2) \mid 1-x \mid^{-2} dx, \qquad x < 0, \tag{4}$$

où c_i est une constante et f_i une fonction continue à support compact sur G_i/Z_i.

(4.4) *Démonstration de la proposition (4.2).* — Pour fixer les idées supposons $i=1$. Posons $H(x) = H(x : \phi : T_1)$. En particulier $H(x) = 0$ si x n'est pas une norme. Nous calculerons à des constantes multiplicatives près. Supposons maintenant f et (f_1, f_2) appariés avec f K-finie si F est réel. Alors d'après (4.3.3) :

$$\int \phi(g_1) f_1(g_1) dg_1 = \int H(x) H(x : f_1 : T_1) \mid 1-x \mid^{-2} dx.$$

Soit d'autre part ϕ_0 la fonction sur G définie par

$$\phi_0(g) = H(x)\eta(\det b) \qquad \text{si} \quad g = a g(x) b.$$

Vues les propriétés de H et la formule d'intégration (4.3.1) ϕ_0 est localement intégrable et :

$$\int \phi_0(g) f(g) dg = \int H(x) H(x : f : \eta) \mid 1-x \mid^{-2} dx.$$

Comme $H(x:f:\eta)=H(x:f_1:T_1)$ si x est une norme il vient

$$\int \phi_0(g)f(g)dg=\int \phi(g_1)f_1(g_1)dg_1.$$

Le second membre est nul par hypothèse. Donc ϕ_0 est orthogonale à toute fonction lisse (resp. toute fonction lisse K-finie si F est réel). Il en résulte que ϕ_0 est nulle. Il en est donc de même de H ; comme ϕ est T_1-biinvariante ϕ est déterminée complètement par H et nous obtenons $\phi=0$.

5. Intégrales orbitales : la situation non ramifiée

(5.1) Dans ce paragraphe F est un corps local non archimédien, E une extension quadratique non ramifiée de F. Nous supposerons que la caractéristique résiduelle de F n'est pas 2 et que l'ordre du caractère ψ est 0. Nous considérerons le couple (G, A) formé du groupe GL(2) et du sous-groupe diagonal. Nous noterons R l'anneau des entiers de F, P l'idéal maximal de R, ω une uniformisante et K le groupe GL(2, R). L'ensemble $X=X(E:F)$ est réduit à deux éléments (G_1, T_1) et (G_2, T_2). Nous supposerons $G_1=G$, T_1 contenu dans le sous-groupe ZK. Nous écrirons simplement T pour T_1. Les mesures de $A\cap K/Z\cap K$ et $T\cap K/Z\cap K$ sont donc égales à 1. Le but de ce paragraphe est de démontrer la proposition suivante :

PROPOSITION. — *Soit f une fonction bi-K-invariante et à support compact sur le groupe G/Z. Alors f et le couple $(f, 0)$ sont appariés. De plus :*

$$H(n_+:f:\eta)=H(n_-:f:\eta)=1/2\,\mathrm{Vol}\,(T/Z)\int f(t)dt.$$

Il sera commode de considérer des fonctions à support compact sur G plutôt que des fonctions à support compact sur G/Z. Bien entendu les mesures des ensembles $A\cap K$, $T\cap K$, $Z\cap K$ sont donc égales à 1. Si f est une fonction bi-K-invariante à support compact sur G, alors nous poserons :

$$\text{(1)}\qquad H(g:f:T)=\iint f(sgt)ds dt,\qquad s\in T,\quad t\in T/Z.$$

Puisque T est contenu dans ZK ceci se réduit à :

$$\text{(2)}\qquad H(g:f:T)=\int f(zg)dz,\qquad z\in Z.$$

De même, nous poserons :

$$\text{(3)}\qquad H(g:f:\eta)=\iint f(agb)da\,\eta\,(\det b)db,\qquad a\in A,\quad b\in A/Z.$$

Nous écrirons encore $H(x:f:\eta)$ pour $H(g(x):f:\eta)$. Les relations à prouver sont donc :

(4) $$H(x:f:\eta)=\int f(zg)dz \qquad \text{si} \quad v(x) \text{ est pair et } P(g:T)=x,$$

(5) $$H(x:f:\eta)=0 \qquad \text{si} \quad v(x) \text{ est impair}.$$

Par linéarité nous pouvons supposer que f est, soit la fonction caractéristique f_0 de K, soit la fonction caractéristique f_m de l'ensemble

(6) $$K\begin{vmatrix} \omega^m & 0 \\ 0 & 1 \end{vmatrix}K, \qquad \text{où} \quad m>0.$$

Remarquons que $f_m(g)\neq 0$ si et seulement si les conditions suivantes sont vérifiées :

(7) les coefficients de g sont des entiers ;

(8) $v(\det g)=m$;

(9) l'un au moins des coefficients de g est une unité.

Bien entendu si $m=0$ la condition (9) est superflue.

(5.2) Nous calculerons d'abord $H(x:f_m:\eta)$ que, pour simplicité, nous noterons $H(x:m)$. Nous commencerons par le cas $m>0$.

PROPOSITION. — *Supposons $m>0$. Alors $H(x:m)$ est donné par les formules suivantes :*

(1) *si $v(x)$ est impair* $H(x:m)=0$;

(2) *si $v(x)$ est pair* $H(x:m)=0$, *à moins que $v(x)=0$ et $v(1-x)=m$ auquel cas* $H(x:m)=1$.

Démonstration. — Nous utiliserons le lemme suivant :

LEMME. — *Posons*

$$S=(-1)^{i+j}$$

où la somme porte sur tout les couples d'entiers (i, j) appartenant au bord du rectangle défini par les inégalités

$$0\leq i\leq P, \qquad 0\leq j\leq Q.$$

Alors S est donné par les formules suivantes :

(3) si $PQ>0$, alors $S=0$;

(4) si $P=0$ et $Q>0$, alors $S=1$ si Q est pair et $S=0$ si Q est impair ;

(5) si $Q=0$ et $P>0$, alors $S=1$ si P est pair et $S=0$ si P est impair ;

(6) si $P=0$ et $Q=0$, alors $S=1$.

Démontrons maintenant la proposition. Il sera commode d'écrire Mat $[a, b, c, d]$ pour la matrice dont les coefficients sont les nombres a, b, c, d. Avec cette notation nous avons :

(7) $$H(x:m)=\sum f_m(\text{Mat}\,[\omega^{i+k}, x\omega^{j+k}, \omega^i, \omega^j])(-1)^{i+j},$$

où la somme porte sur tous les triplets d'entiers (i, j, k). Comme le déterminant de la matrice (7) a une valuation égale à $i+j+k+v(1-x)$ la condition (5.1.8) montre que dans la somme ci-dessus nous pouvons nous restreindre aux triplets tels que :

$$i+j+k+v(1-x)=m.$$

Cela nous permet d'éliminer k et, en tenant compte de (5.1.7) et (5.1.9), d'écrire :

(8) $$\mathrm{H}(x:m)=\sum(-1)^{i+j}$$

où la somme porte sur tous les couples d'entiers (i, j) tels que :

(9) $$0 \leq i \leq m-v(1-x)+v(x)$$

(10) $$0 \leq j \leq m-v(1-x)$$

(11) $$ij[m-v(1-x)+v(x)-i][m-v(1-x)-j]=0.$$

La somme est vide et $\mathrm{H}(x:m)$ nul à moins que :

(12) $$m-v(1-x) \geq 0 \quad \text{et} \quad m-v(1-x)+v(x) \geq 0.$$

Supposons les conditions (12) satisfaites. Alors nous pouvons appliquer le lemme. Nous avons donc $\mathrm{H}(x:m)=0$ à moins que

(13) $$[m-v(1-x)][m-v(1-x)+v(x)]=0.$$

La vérification de la proposition est alors élémentaire.

(5.3) Calculons maintenant $\mathrm{H}(x:0)$.

PROPOSITION. — $\mathrm{H}(x:0)$ *est donné par les formules suivantes :*

(1) *si* $v(x)$ *est impair alors* $\mathrm{H}(x:0)=0$;

(2) *si* $v(x)$ *est pair alors* $\mathrm{H}(x:0)=1$, *à moins que* $v(x)=0$ *et* $v(1-x)>0$ *auquel cas* $\mathrm{H}(x:0)=0$.

Démonstration. — Nous avons encore :

(3) $$\mathrm{H}(x:0)=\sum f_0(\mathrm{Mat}\,[\omega^{i+k}, x\omega^{j+k}, \omega^i, \omega^j])(-1)^{i+j},$$

où la somme porte sur tous les triplets d'entiers (i, j, k). Comme plus haut, en tenant compte des conditions (5.1.7) et (5.1.8), nous pouvons éliminer k et écrire :

(4) $$\mathrm{H}(x:0)=\sum(-1)^{i+j}$$

où la somme porte sur tous les couples d'entiers (i, j) tels que :

(5) $$0 \leq i \leq v(x)-v(1-x)$$

(6) $$0 \leq j \leq -v(1-x).$$

La vérification de la proposition est alors élémentaire.

(5.4) Calculons maintenant l'intégrale $\int f_m(zg)dz$. Elle ne dépend que de $x=\mathrm{P}(g:\mathrm{T})$ et nous noterons $\mathrm{H}(x:m:\mathrm{T})$ sa valeur. Rappelons que par définition x est une norme, autrement dit la valuation de x est paire. Nous commencerons par le cas $m>0$.

PROPOSITION. — *Supposons* $m>0$. *Alors* $\mathrm{H}(x:m:\mathrm{T})=0$, *à moins que* $v(x)=0$ *et* $v(1-x)=m$ *auquel cas* $\mathrm{H}(x:m:\mathrm{T})=1$.

Démonstration. — Nous pouvons supposer que E est l'extension engendrée par la racine carré de τ, où τ est une unité. Alors nous pouvons prendre pour T le groupe multiplicatif de l'algèbre :

(1) $$L = \left\{ \begin{vmatrix} a & b \\ b\tau & a \end{vmatrix} \right\}$$

et pour ε la matrice :

(2) $$\varepsilon = \begin{vmatrix} 1 & 0 \\ 0 & -1 \end{vmatrix}.$$

Calculons $H(x : m : T)$ où $x = P(g : T)$. Nous pouvons supposer que :

(3) $$g = 1_2 + \varepsilon \begin{vmatrix} u & v \\ v\tau & u \end{vmatrix}.$$

Alors det $g = 1 - x$ et $x = u^2 - v^2\tau$. Nous avons :

(4) $$H(b : m : T) = \sum_k f_m(\omega^k g),$$

où

(5) $$\omega^k g = \begin{vmatrix} \omega^k(1+u) & \omega^k v \\ -\omega^k v\tau & \omega^k(1-u) \end{vmatrix}.$$

Vue la condition (5.1.8) cette somme a donc au plus un terme dont l'indice k est déterminé par l'équation

(6) $$k = 1/2\,[m - v(1-x)].$$

En particulier $H(b : m : T) = 0$ ou 1. Vues les conditions (5.1.7) à (5.1.9) $H(x : m : T) = 1$ si et seulement si les conditions suivantes sont vérifiées :

(7) $$m \equiv v(1-x) \bmod 2\,;$$

(8) les coefficients de la matrice (5) où k est donné par (6) sont entiers ;

(9) l'un au moins des coefficients de cette matrice est une unité.

Supposons d'abord $v(x) < 0$. Alors $v(1-x) = v(x)$. Puisque $x = u^2 - v^2\tau$ et τ n'est pas un carré $v(x)$ est pair. D'après (7) $H(x : m : T) = 0$ à moins que m ne soit aussi pair. Supposons donc qu'il en soit ainsi. Nous pouvons écrire :

$$u = u_0\omega^{1/2v(x)}, \qquad v = v_0\omega^{1/2v(x)},$$

où u_0 et v_0 sont des entiers, l'un au moins étant une unité. Alors les coefficients de la matrice (5) sont les nombres :

$$\omega^{1/2(m-v(x))}(1 + u_0\omega^{1/2v(x)}), \qquad \omega^{1/2m}v_0,$$
$$-\omega^{1/2m}v_0\tau, \qquad \omega^{1/2(m-v(x))}(1 - u_0\omega^{1/2v(x)}).$$

Tous ces nombres sont dans l'idéal P donc $H(x : m : T) = 0$.

Supposons $v(x) > 0$. Alors $v(1-x) = 0$. D'après (7) $H(x : m) = 0$ à moins que m ne soit pair.

Supposons qu'il en soit ainsi. Alors $k=1/2m$, u et v sont des entiers. Les coefficients de la matrice (5) sont maintenant les nombres

$$\begin{matrix} \omega^{1/2m}(1+u), & \omega^{1/2m}v, \\ -\omega^{1/2m}v\tau, & \omega^{1/2m}(1-u). \end{matrix}$$

Tous ces nombres sont dans l'idéal P donc $H(x:m:T)=0$.

Supposons enfin $v(x)=0$. Alors $v(1-x)\geq 0$. Si $m-v(1-x)$ est impair alors $H(x:m:T)=0$. Supposons donc $m-v(1-x)$ pair. Alors les coefficients de la matrice (5) sont les nombres :

$$\begin{matrix} \omega^{1/2[m-v(1-x)]}(1+u), & \omega^{1/2[m-v(1-x)]}v \\ -\omega^{1/2[m-v(1-x)]}v\tau, & \omega^{1/2[m-v(1-x)]}(1-u). \end{matrix}$$

Comme x est une unité u et v sont des entiers et l'un au moins est une unité. Si $1+u$ et $1-u$ étaient tous les deux dans P nous aurions $2\in P$, une contradiction. Donc au moins l'un des nombres $1+u$ et $1-u$ est une unité. Si donc $H(x:m:T)$ n'est pas nulle alors les conditions (8) et (9) entraînent que $m=v(1-x)$. Les coefficients de la matrice (5) se réduisent donc à

$$1+u, \qquad v, \qquad -v\tau, \qquad 1-u.$$

Ce sont des entiers et l'un au moins est une unité. Donc $H(x:m:T)=1$.

Nous avons donc complètement calculé H et notre résultat est en accord avec la proposition.

(5.5) Calculons enfin $H(x:0:T)$. Rappelons une fois de plus que $v(x)$ est pair.

PROPOSITION. — $H(x:0:T)=1$, *à moins que* $v(x)=0$ *et* $v(1-x)>0$ *auquel cas* $H(x:0:T)=0$.

Démonstration. — Comme plus haut nous avons

(1) $$H(x:m:T)=\sum_k f_0(\omega^k g).$$

La somme a en fait au plus un terme dont l'indice k est donné par :

(2) $$k=-1/2v(1-x).$$

En particulier $H(x:0:T)=0$ ou 1. De plus $H(x:0:T)=1$ si et seulement si $v(1-x)$ est pair et la matrice

(3) $$\omega^k g=\begin{vmatrix} \omega^k(1+u) & \omega^k v \\ -\omega^k v\tau & \omega^k(1-u) \end{vmatrix}.$$

avec k donné par (2) est dans GL(2, R).

Supposons $v(x)<0$ et $v(1-x)$ pair. Alors $v(1-x)=v(x)$, $v(x)$ est pair et

$$u=u_0\omega^{1/2v(x)}, \qquad v=v_0\omega^{1/2v(x)}$$

où u_0 et v_0 sont des entiers, l'un au moins étant une unité. Alors les coefficients de la matrice (3) sont les nombres :

$$\omega^{-1/2v(x)}+u_0, \qquad v_0, \qquad -v_0\tau, \qquad \omega^{-1/2v(x)}-u_0.$$

Ce sont des entiers. Comme le déterminant de la matrice (3) est une unité d'après le choix de k la matrice (3) est dans GL(2, R) et $H(x:0:T)=1$.

Supposons $v(x)\geq 0$ et $v(1-x)=0$ (bien entendu $v(x)>0$ entraîne $v(1-x)=0$. Alors $k=0$ et les coefficients de la matrice (3) se réduisent aux nombres :

$$1+u, \qquad v, \qquad -v\tau, \qquad 1-u.$$

Comme u et v sont des entiers ces nombres sont aussi des entiers et la matrice (3) est dans GL(2, R). Donc $H(x:0:T)=1$.

Supposons enfin $v(x)=0$ et $v(1-x)>0$. Alors H est nul à moins que $v(1-x)$ ne soit pair. Supposons qu'il en soit ainsi. Alors les coefficients de la matrice (3) sont les nombres :

$$\omega^{-1/2v(1-x)}(1+u), \qquad \omega^{-1/2v(1-x)}v$$
$$\omega^{-1/2v(1-x)}v\tau, \qquad \omega^{-1/2v(1-x)}(1-u).$$

Comme $1+u$ ou $1-u$ est une unité l'un au moins de ces nombres n'est pas un entier donc (3) n'est pas dans GL(2, R) et $H(x:0:T)=0$.

Nous avons donc complètement calculé H et notre résultat est en accord avec la proposition.

(5.6) En rapprochant les propositions (5.2), (5.3), (5.4) et (5.5) nous voyons que nous avons prouvé les identités (5.1.4) et (5.1.5). Ceci termine donc la démonstration de la première assertion de la proposition (5.1). La deuxième découle alors de la proposition (4.1).

(5.7) Pour établir la convergence des intégrales orbitales globales nous aurons besoin d'un résultat supplémentaire, dont nous laisserons la démonstration au lecteur :

(1) LEMME. — *Supposons h dans* KZ *et posons* $x=P(h:A)$. *Supposons* $v(x)=0$ *et* $v(1-x)=0$. *Alors les relations*

$$ahb\in KZ, \qquad a\in A, \qquad b\in A$$

entraînent

$$a\in Z(K\cap A), \qquad b\in Z(K\cap A).$$

Le lemme implique évidemment la proposition suivante :

PROPOSITION. — *Soit f la fonction caractéristique de* KZ. *Supposons l'extension quadratique* E *non ramifiée et* T *contenu dans* KZ. *Soit h un élément de* KZ *et* $x=P(h:A)$. *Si x et $1-x$ sont des unités alors :*

$$H(h:f:A)=1, \qquad H(h:f:\eta)=1.$$

6. Rappels sur les représentations locales de GL(2)

(6.1) Soit F un corps local et E une extension quadratique de E. Nous considérerons encore l'ensemble X qui est réduit à deux éléments (G_1, T_1) et (G_2, T_2), avec disons G_1 déployé. Il sera commode d'utiliser le résultat suivant :

PROPOSITION. — *Soit π une représentation unitaire irréductible de* G_i/Z_i. *Alors la dimension de l'espace des formes linéaires continues et* T_i*-invariantes sur l'espace des vecteurs* *lisses de π est au plus un. De plus une telle forme est donnée par le produit scalaire avec un vecteur lisse* T_i*-invariant.*

Si F n'est pas archimédien alors l'assertion sur la dimension est prouvée dans [W2], Propositions 9. Si F est réel, elle est classique. Le reste de la proposition est évident.

(6.2) De même :

PROPOSITION. — *Soit π une représentation unitaire irréductible de* G_1/Z_1 *de dimension infinie. Alors la dimension de l'espace des formes linéaires continues et A-invariantes (respectivement invariantes relativement au caractère* $\eta \circ \det$ *de A) sur l'espace des vecteurs lisses de π est un.*

Ce sont les propositions 9 et 10 de [W1].

(6.3) Considérons, pour $i=1, 2$, une représentation unitaire irréductible π_i de G_i/Z_i. Nous supposerons que le couple (π_1, π_2) satisfait aux conditions du théorème (15.1) de [J. L.] ; en particulier π_1 est dans la série discrète.

PROPOSITION. — *Les représentations* π_i *ne peuvent toutes les deux avoir un vecteur non nul invariant sous le groupe* T_i.

Si F est non archimédien alors notre assertion se trouve dans le théorème 2 de [W2]. Si F est réel, elle est classique.

7. Intégrales orbitales globales : cas d'un tore déployé

(7.1) Dans le reste de ce travail F sera un corps de nombres et E une extension quadratique de F, η le caractère quadratique du groupe des classes d'idèles de F attaché à E. Dans ce paragraphe et le suivant nous considérerons le couple (G, A) et une fonction lisse à support compact f sur $G(F_A)/Z(F_A)$. A la fonction f est attaché le noyau cuspidal K_c. Soit ϕ_j une base orthonormale de l'espace des formes cuspidales pour le groupe G/Z.

Alors, par définition :

$$K_c(x, y)=\sum \rho(f)\phi_j(x)\phi_j^-(y) \tag{1}$$

où :

$$\rho(f)\phi(x)=\int f(g)\phi(xg)dg\,. \tag{2}$$

Nous nous proposons dans ce paragraphe et le suivant de donner une expression utile pour l'intégrale :

$$\iint K_c(a, b)\eta\,(\det b)\,da\,db, \qquad a, b\in A(F_A)/A(F)Z(F_A). \tag{3}$$

Nous avons choisi un caractère non trivial ψ de F_A/F. Nous avons donc à chaque place v la mesure de Tamagawa attachée à ψ_v et par transport de structure sur A_v et Z_v. Nous avons donc la mesure produit sur $A(F_A/F)$ et la mesure quotient sur $A(F_A)/Z(F_A)$. Nous désignerons par S un ensemble fini de places contenant les places infinies, les places ramifiées dans E et les places de caractéristique résiduelle 2. Pour toute place v de F nous noterons K_v

le sous-groupe compact maximal usuel. En particulier $K_v = GL(2, R_v)$ si v est fini. Nous prendrons la fonction f produit de fonctions locales f_v qui sont K_v-finies à toutes les places. Nous supposerons que f_v est bi-K_v-invariante pour tous les v non dans S. Bien entendu f_v est en fait la fonction caractéristique de $K_v Z_v$ pour presque tous les v non dans S. Nous avons une décomposition de K_c en une somme :

(4) $$K_c(x, y) = \sum f(x^{-1}\gamma y) - K_{sp}(x, y) - K_{ei}(x, y),$$

où la somme est sur tous les γ dans G(F)/Z(F), K_{sp} dénote le noyau spécial et K_{ei} le noyau d'Eisenstein (la définition est rappelée plus loin). Nous pouvons écrire le premier terme de cette somme comme la somme de deux autres termes K_r et K_s où :

(5) $$K_r(x, y) = \sum f(x^{-1}\gamma, y), \qquad \gamma \text{ A-régulier};$$

(6) $$K_s(x, y) = \sum f(x^{-1}\gamma, y), \qquad \gamma \text{ A-singulier}.$$

Alors K_c peut s'écrire

(7) $$K_c = K_r + K_s - K_{sp} - K_{ei}.$$

(7.2) Nous considérons d'abord l'intégrale de K_r. Tout élément γ de G(F)/Z(F) peut s'écrire uniquement sous la forme :

(1) $$\gamma = \alpha g(\xi)\beta, \quad \text{avec } \alpha \text{ et } \beta \in A(F)/Z(F) \text{ et } \xi \text{ différent de 0 et 1}.$$

(*cf.* (3.1.3) pour la notation et § 1). Il vient donc aussitôt :

(2) $$\iint K_r(a, b)\eta(\det b)\,da\,db = \sum H(\xi : f : \eta), \qquad \xi \neq 0 \text{ et } 1$$

où nous avons posé :

(3) $$H(\xi : f : \eta) = \iint f(ag(\xi)b)\eta(\det b)\,da\,db, \qquad a, b \in A(F_A)/Z(F_A).$$

Justifions nos calculs formels. Tout d'abord le support de f ne rencontre qu'un nombre fini de classes régulières. En effet la fonction X introduite au paragraphe 1 (*cf.* (1.1.4)) définit une fonction continue du groupe $G(F_A)/Z(F_A)$ dans F_A. Elle ne prend donc qu'un nombre fini de valeurs sur l'intersection du support de f avec l'ensemble des points rationnels ; il en est donc de même de la fonction P(. : A), ce qui nous donne notre assertion. D'autre part chacune des intégrales (3) converge absolument : il suffit en effet de le prouver pour l'intégrale

(4) $$H(\xi : f : A) = \iint f(ag(\xi)b)\,da\,db, \qquad a, b \in A(A)/Z(A).$$

Chacune des intégrales locales $H(\xi : f_v : A_v)$ converge ; presque toutes sont égales à 1 (*cf.* (5.7)). Donc (4) converge. Il en est donc de même de (3) et (3) est le produit des intégrales locales correspondantes :

(5) $$H(\xi : f : \eta) = \prod H(\xi : f_v : \eta_v).$$

Dans ce produit presque toutes les intégrales sont égales à 1 (*cf.* (5.7)).

(7.3) Passons à l'intégrale de K_s. Elle n'est pas absolument convergente, mais elle est « faiblement » convergente au sens suivant. Soit c un nombre plus grand que 1. Convenons de désigner par

$$(1) \qquad \int_{c^{-1}}^{c}\int_{c^{-1}}^{c} K_s(a, b)\eta(\det b)\,da\,db, \qquad a, b \in A(F_A)/A(F)Z(F_A)$$

l'intégrale de $K_s(a, b)\eta(\det b)$ sur l'ensemble des couples (a, b) satisfaisant $c^{-1} < |a_1/a_2| < c$, $c^{-1} < |b_1/b_2| < c$; bien entendu a_1 et a_2 par exemple désignent les coefficients diagonaux de a. Comme l'intégrale est prise sur un ensemble compact elle existe. Nous allons voir que l'intégrale (1) tend vers une limite lorsque c tend vers l'infini. Cette limite sera par définition l'intégrale faible de $K_s(a, b)\eta(\det b)$. Nous avons vu au numéro (1.3) qu'il y avait 6 doubles classes singulières pour A, à savoir les doubles classes des éléments suivants : $e, n_+, n_-, \varepsilon, \varepsilon n_+, \varepsilon_{n-}$. Numérotons les de 1 à 6. Alors nous avons une décomposition de K_s en 6 termes K_i, $1 \le i \le 6$, où K_i est la somme des $f(x^{-1}\gamma y)$ pour tous les γ dans la ième double classe. Étudions l'intégrale de K_1 par exemple. Nous avons

$$K_1(x, y) = \sum f(x^{-1}\alpha y), \qquad \alpha \in A(F)/Z(F).$$

D'où :

$$\int_{c^{-1}}^{c}\int_{c^{-1}}^{c} K_1(a, b)\eta(\det b)\,da\,db = \int_{c^{-1}}^{c}\int_{c^{-1}}^{c} f(ab)\eta(\det b)\,da\,db\,;$$

dans l'intégrale de gauche a et b varient dans le sous-ensemble compact de $A(F_A)/A(F)Z(F_A)$ défini ci-dessus ; dans l'intégrale de droite b varie encore dans le sous-ensemble compact de $A(F_A)/A(F)Z(F_A)$ défini par $c^{-1} < |b_1/b_2| < c$, mais a varie dans le sous-ensemble de $A(F_A)/Z(F_A)$ défini par $c^{-1} < |a_1/a_2| < c$. Changeons a en ab^{-1} dans l'intégrale de droite. Nous obtenons une intégrale double, l'intégrale intérieure ne dépendant que de $|b_1/b_2|$. Cette intégrale intérieure s'écrit :

$$\int_{c^{-1}}^{c} \eta(\det b)\,db, \qquad b \in A(F_A)/A(F)Z(F_A).$$

Elle est nulle parce que la restriction de η au groupe des idèles de valeur absolue 1 n'est pas triviale. L'intégrale de K_1 est donc faiblement convergente et sa valeur est 0. Il en va de même pour l'intégrale de K_4.

Examinons les intégrales des autres termes, K_2 par exemple. Nous avons :

$$(2) \qquad K_2(x, y) = \sum f(x^{-1}\alpha n_+ \beta y), \qquad \alpha, \beta \in A(F)/Z(F).$$

Il en résulte que :

$$\int_{c^{-1}}^{c}\int_{c^{-1}}^{c} K_2(a, b)\eta(\det b)\,da\,db = \int_{c^{-1}}^{c}\int_{c^{-1}}^{c} \sum f(an_+ \beta b)\eta(\det b)\,da\,db\,;$$

dans l'intégrale de droite b varie encore dans le sous-ensemble compact de $A(F_A)/A(F)Z(F_A)$ défini par $c^{-1} < |b_1/b_2| < c$, mais a varie dans le sous-ensemble de $A(F_A)/Z(F_A)$ défini par

$c^{-1}<|a_1/a_2|<c$. Introduisons maintenant la fonction ϕ sur $(F^\times)_A \times F_A$ définie par :

$$\text{(3)} \qquad \phi(x, y)=f\left[\begin{vmatrix} x & 0 \\ 0 & 1 \end{vmatrix} \times \begin{vmatrix} 1 & y \\ 0 & 1 \end{vmatrix}\right].$$

Elle est à support compact. Notre intégrale s'écrit :

$$\int \sum_{\zeta} \int \phi(ab^{-1}, b\zeta)\eta(b)da\,db,$$

$$\zeta\in F^\times, \qquad a\in(F^\times)_A, \qquad c^{-1}<|a|<c, \qquad b\in(F^\times)_A/F^\times, \qquad c^{-1}<|b|<c.$$

En utilisant la formule de Poisson par rapport à la deuxième variable et en prenant la transformée de Fourier par rapport à la deuxième variable nous obtenons pour cette intégrale l'expression :

$$\int \sum_{\zeta} \int \phi(ab^{-1}, b\zeta)\eta(b)da\,db + \int \sum_{\zeta} \int \phi^{\wedge}(ab, b\zeta)\,|b|\,\eta(b)da\,db, \qquad c^{-1}<|a|<c, \quad 1<|b|<c.$$

Il est évident que les deux mêmes intégrales étendues au domaine :

$$a\in F_A^\times, \qquad b\in F_A^\times/F_A^\times, \qquad 1<|b|,$$

convergent absolument. De plus dans les intégrales étendues au domaine précédent nous pouvons changer a en $ab^{\pm 1}$. Nous concluons que l'intégrale de K_2 est faiblement convergente et que sa valeur est la somme :

$$\iint \sum_{\zeta} \phi(a, b\zeta)\eta(b)da\,db + \iint_{\zeta} \phi^{\wedge}(a, b\zeta)\,|b|\,\eta(b)da\,db,$$

$$a\in F_A^\times, \qquad b\in F_A^\times/F_A^\times, \qquad 1<|b|.$$

Ceci n'est autre que la valeur en $s=0$ du prolongement analytique de l'intégrale suivante :

$$\text{(4)} \qquad \iint \phi(a, b)\,|b|^{s}\eta(b)da\,db, \qquad a\in F_A^\times, \qquad b\in F_A^\times.$$

Cette valeur sera aussi notée comme une intégrale :

$$\text{(5)} \qquad \iint \phi(a, b)\eta(b)da\,db, \qquad a\in F_A^\times, \qquad b\in F_A^\times.$$

Avec cette convention nous pouvons écrire que l'intégrale faible de K_2 est égale à :

$$\text{(6)} \qquad H(n_+ : f : \eta)=\iint f\left[\begin{vmatrix} a & 0 \\ 0 & 1 \end{vmatrix} \times \begin{vmatrix} 1 & b \\ 0 & 1 \end{vmatrix}\right]\eta(b)da\,db.$$

Un résultat analogue est valable pour les intégrales des autres K_i. Finalement nous voyons que l'intégrale faible de K_s existe et est égale à la somme :

$$\text{(7)} \qquad H(n_+ : f : \eta)+H(n_- : f : \eta)+H(n\varepsilon_+ : f : \eta)+H(\varepsilon n_- : f : \eta);$$

le premier terme est défini par (6) et les autres termes sont définis de manière analogue :

$$\text{(8)}\qquad H(n_- : f : \eta) = \iint f\left[\begin{vmatrix} a & 0 \\ 0 & 1 \end{vmatrix} \times \begin{vmatrix} 1 & 0 \\ b & 1 \end{vmatrix}\right]\eta(b)da\,db\,;$$

$$\text{(9)}\qquad H(\varepsilon n_+ : f : \eta) = \iint f\left[\begin{vmatrix} a & 0 \\ 0 & 1 \end{vmatrix} \varepsilon \begin{vmatrix} 1 & b \\ 0 & 1 \end{vmatrix}\right]\eta(b)da\,db\,;$$

$$\text{(10)}\qquad H(\varepsilon n_- : f : \eta) = \iint f\left[\begin{vmatrix} a & 0 \\ 0 & 1 \end{vmatrix} \varepsilon \begin{vmatrix} 1 & 0 \\ b & 1 \end{vmatrix}\right]\eta(b)da\,db\,.$$

(7.4) Passons maintenant à l'intégrale de K_{sp}. Rappelons la définition de K_{sp} :

$$K_{sp}(x, y) = \mathrm{Vol}^{-1}\sum \int f(g)\chi(\det g)dg\chi(\det x)\chi^{-1}(\det y)\,,$$

où la somme porte sur tous les caractères quadratiques χ du groupe des classes d'idèles de F et Vol est le volume du quotient $G(F_A)/G(F)Z(F_A)$. Si χ est un tel caractère alors, soit χ, soit $\chi\eta$ a une restriction non triviale aux groupes des classes d'idèles de norme 1. En raisonnant comme pour K_1 nous voyons aussitôt que K_{sp} est faiblement intégrable et son intégrale nulle.

8. Le noyau d'Eisenstein

(8.1) Nous continuons avec les notations du paragraphe 7. Nous allons voir que l'intégrale

$$\text{(1)}\qquad \iint K_{ei}(a, b)\eta(\det b)da\,db, \qquad a, b \in A(F_A)/A(F)Z(F_A)$$

est faiblement convergente. En ce qui concerne la valeur de l'intégrale, tout comme dans les applications maintenant classiques de la formule des traces, nous n'aurons besoin que d'un résultat assez faible. Choisissons en effet une place u en dehors de S qui se décompose dans E. Fixons les composants de f aux autres places et regardons l'intégrale (1) comme une fonction de f_u. Notons $f_u^\wedge$ la transformée de Satake de f_u. Nous prouverons le résultat suivant :

PROPOSITION. — *Il existe une fonction intégrable sur la droite réelle ϕ et une constante c telles que :*

$$\text{(2)}\qquad \iint K_{ei}(a, b)\eta(\det b)da\,db = \int \phi(t)f_u^\wedge(q_u^{-it})dt + cf_u^\wedge(q^{-1}).$$

(8.2) Nous aurons besoin de résultats standards sur la transformée de Mellin d'une série d'Eisenstein. Nous fixerons une fois pour toutes un sous-groupe C de F_A isomorphe au groupe des nombres réels >0 tel que F_A soit le produit de C et F^1, le groupe des idéles de module 1. Le groupe C est muni de la mesure image réciproque de la mesure $t^{-1}\,dt$ par l'application $c \to |c|$ et F^1 de la mesure quotient. Sauf mention expresse du contraire tous les caractères du groupe des classes d'idèles seront supposés triviaux sur C. Soit χ un tel caractère et V(X) l'espace des fonctions ϕ sur K (le produit des K_v) telles que :

$$\text{(1)}\qquad \phi\left[\begin{vmatrix} a & x \\ 0 & b \end{vmatrix} k\right] = \chi(ab^{-1})\phi(k)$$

si $\begin{vmatrix} a & x \\ 0 & b \end{vmatrix}$ est dans K.

Considérons maintenant une fonction ϕ sur $K \times \mathbb{C}$, telle que pour chaque nombre complexe u la fonction $\phi(.,u)$ soit dans $V(\chi)$. La fonction sera supposée être holomorphe, ou même, méromorphe par rapport à u ; par exemple elle peut être indépendante de u. Nous étendrons ϕ en une fonction $\phi(g, u, \chi)$ sur $G(F_A)$ telle que :

$$\phi\left[\begin{vmatrix} a & x \\ 0 & b \end{vmatrix} g, u, \chi\right] = \chi(ab^{-1}) \mid ab^{-1} \mid^{u+1/2} \phi(g, u, \chi). \tag{2}$$

La série d'Eisenstein est alors le prolongement analytique de la série :

$$E(g, \phi, u, \chi) = \sum \phi(\gamma g, u, \chi), \qquad \gamma \in G(F)/A(F)N(F). \tag{3}$$

La série converge absolument si Re $u > 1/2$. Le terme constant de E le long de N, le groupe des matrices trigonales strictes supérieures, est par définition l'intégrale

$$E_N(g, \phi, u, \chi) = \int E(ng, \phi, u, \chi) dn, \qquad n \in N(F_A)/N(F). \tag{4}$$

Il a la forme :

$$E_N(g, \phi, u, \chi) = \phi(g, u, \chi) + M(u, \chi)\phi(g, -u, \chi^{-1}) \tag{5}$$

où $M(u, \chi)$ est l'opérateur d'entrelacement qui va de $V(X)$ à $V(\chi^{-1})$. Nous aurons aussi besoin d'un autre coefficient de Fourier de E, à savoir :

$$W(g, \phi, u, \chi) = \int E\left[\begin{vmatrix} 1 & x \\ 0 & 1 \end{vmatrix} g, \phi, u, \chi\right] \psi(-x) dx, \qquad x \in F_A/F, \tag{6}$$

où ψ est le caractère fixé du groupe F_A/F. La série de Fourier de E s'écrit donc :

$$E(g, \phi, u, \chi) = \phi(g, u, \chi) + M(u, \chi)\phi(g, -u, \chi^{-1}) + \sum W(\alpha g, \phi, u, \chi) \tag{7}$$

où la somme est pour α dans $A(F)/Z(F)$. Nous pouvons aussi considérer une série de Fourier pour le groupe N′ des matrices trigonales strictes inférieures. Comme

$$N' = wNw^{-1}, \qquad \text{avec} \quad w = \begin{vmatrix} 0 & -1 \\ 1 & 0 \end{vmatrix}, \tag{8}$$

elle s'écrit en fait :

$$E(g, \phi, u, \chi) = \phi(wg, u, \chi) + M(u, \chi)\phi(wg, -u, \chi^{-1}) + \sum W(\alpha wg, \phi, u, \chi). \tag{9}$$

La transformée de Mellin $L(s, \lambda : \phi : u, \chi)$ de E est définie par l'intégrale suivante (ou son prolongement analytique) :

$$L(s, \lambda : \phi : u, \chi) = \int \{ E\,[\mathrm{diag}\,(a, 1)] - E_N\,[\mathrm{diag}\,(a, 1)] \} \mid a \mid^{s-1/2} \lambda(a) da, \qquad a \in F_A^{\times}/F^{\times}. \tag{10}$$

Nous avons supprimé de la notation la dépendance de E des variables autres que la première. En remplaçant E par sa série de Fourier nous obtenons aussitôt pour la transformée de Mellin l'expression :

$$\int W\,[\mathrm{diag}\,(a, 1)]\,|\,a\,|^{s-1/2}\lambda(a)da. \tag{11}$$

Nous pouvons aussi écrire la transformée de Mellin de E comme suit :

$$L(s\ldots)=\int_{1}^{+\infty}(E-E_N)+\int_{0}^{1}(E-E_{N'})+\int_{1}^{+\infty}E_N+\int_{0}^{1}E_{N'}. \tag{12}$$

Dans chacune de ces intégrales, la fonction est évaluée au point diag $(a, 1)$ et intégrée contre $|\,a\,|^{s-1/2}\lambda(a)$ sur un sous-ensemble du groupe des classes d'idèles. Pour la première intégrale, par exemple, le sous-ensemble est défini par l'inégalité $1<|\,a\,|$. En utilisant la série de Fourier de E nous obtenons sans peine une autre expression pour la transformée de Mellin de E :

$$\begin{aligned}(13)\quad &\int_{1}^{+\infty} W\,[\mathrm{diag}\,(a, 1)]\,|\,a\,|^{s-1/2}\lambda(a)da\\ &+\int_{1}^{+\infty} W\,[\mathrm{diag}\,(a, 1)w]\,|\,a\,|^{s-1/2}\lambda(a)da\\ &+\int_{1}^{+\infty}[\,|\,a\,|^{s+u}\lambda\chi(a)\phi(e)+|\,a\,|^{s-u}\lambda\chi^{-1}(a)M(u,\chi)\phi(e)\,]da\\ &+\int_{0}^{1}[\,|\,a\,|^{s-u-1}\lambda\chi^{-1}(a)\phi(w)+|\,a\,|^{s+u-1}\lambda\chi(a)M(u,\chi)\phi(w)\,]da.\end{aligned}$$

Les deux premières intégrales convergent pour tout s et les deux dernières pour Re$s > 1/2$. Les deux dernières intégrales se calculent facilement. En particulier, pour $s=1/2$ et u purement imaginaire, nous obtenons l'expression suivante pour la transformée de Mellin de E au point $s=1/2$:

$$\begin{aligned}(14)\quad L(1/2, \lambda:\phi:u,\chi)=&\int_{1}^{+\infty} W\,[\mathrm{diag}\,(a, 1)]\lambda(a)da\\ &+\int_{1}^{+\infty} W\,[\mathrm{diag}\,(a, 1)w]\lambda(a)da\\ &-\frac{1}{u+1/2}[\phi(w)\delta(\lambda\chi^{-1})+\phi(e)\delta(\lambda\chi)]\\ &+\frac{1}{u-1/2}[M(u,\chi)\phi(w)\delta(\lambda\chi)+M(u,\chi)\phi(e)\delta(\lambda\chi^{-1})].\end{aligned}$$

où nous avons posé pour tout caractère χ du groupe des classes d'idèles

$$\delta(\chi)=\int\chi(a)da,\qquad a\in F^1{}_A/F^\times.$$

Nous aurons à calculer la différence entre la transformée de Mellin et l'intégrale suivante :

$$(15)\qquad \int_{c^{-1}}^{c} \mathrm{E}\,[\operatorname{diag}(a,1)]\lambda(a)da\,.$$

Rappelons que cette notation signifie que l'intégrale est prise sur le sous-ensemble compact des classes d'idèles a telles que $c^{-1}<|a|<c$. Au lieu de (12) nous avons pour l'intégrale (15) l'expression :

$$(16)\qquad \int_{1}^{c}(\mathrm{E}-\mathrm{E}_{\mathrm{N}})+\int_{c^{-1}}^{1}(\mathrm{E}-\mathrm{E}_{\mathrm{N}'})+\int_{1}^{c}\mathrm{E}_{\mathrm{N}}+\int_{c^{-1}}^{1}\mathrm{E}_{\mathrm{N}'}\,.$$

En remplaçant à nouveau E par sa série de Fourier nous obtenons pour (15) l'expression :

$$(17)\qquad \int_{1}^{c}\mathrm{W}\,[\operatorname{diag}(a,1)]\lambda(a)da+\int_{1}^{c}\mathrm{W}\,[\operatorname{diag}(a,1)w]\lambda(a)da$$
$$+\int_{1}^{c}[|a|^{1/2+u}\lambda\chi(a)\phi(e)+|a|^{1/2-u}\lambda\chi^{-1}(a)\mathrm{M}(u,\chi)\phi(e)]da$$
$$+\int_{c^{-1}}^{1}[|a|^{-u-1/2}\lambda\chi^{-1}(a)\phi(w)+|a|^{u-1/2}\lambda\chi(a)\mathrm{M}(u,\chi)\phi(w)]da\,.$$

En calculant les deux dernières intégrales et en comparant à (14) nous obtenons finalement l'expression que nous avions en vue :

$$(18)\qquad \int_{c^{-1}}^{c}\mathrm{E}\,[\operatorname{diag}(a,1)]\lambda(a)da=\mathrm{L}(1/2,\lambda:\phi:u,\chi)$$
$$+\frac{c^{u+1/2}}{u+1/2}\,\delta(\chi\lambda)\phi(e)+\frac{c^{-u+1/2}}{-u+1/2}\,\delta(\chi^{-1}\lambda)\mathrm{M}(u,\chi)\phi(e)$$
$$+\frac{c^{u+1/2}}{u+1/2}\,\delta(\chi^{-1}\lambda)\phi(w)+\frac{c^{-u+1/2}}{-u+1/2}\,\delta(\chi\lambda)\mathrm{M}(u,\chi)\phi(w)+\mathrm{R}(c)$$

où $\mathrm{R}(c)$ est donné par :

$$(19)\qquad -\mathrm{R}(c)=\int_{c}^{+\infty}\mathrm{W}\,[\operatorname{diag}(a,1)]\lambda(a)da+\int_{c}^{+\infty}\mathrm{W}\,[\operatorname{diag}(a,1)w]\lambda(a)da\,.$$

Il est clair que $\mathrm{R}(c)$ tend vers zéro lorsque c tend vers l'infini.

(8.3) Nous aurons besoin d'estimées précises sur $\mathrm{R}(c)$. Rappelons que R dépend, non seulement de c, mais aussi de u, λ et ϕ. Nos estimées seront conséquence du lemme suivant :

LEMME (1). — *Supposons la fonction ϕ indépendante de u. Alors il existe une fonction de Schwartz-Bruhat Φ telle que, pour tout u imaginaire, nous ayons :*

$$|\mathrm{W}\,[\operatorname{diag}(a,1),\phi,u,\chi]|\leq\Phi(a)\,|a|^{-1/2}\,|\mathrm{L}(2u+1,\chi^{2\mathrm{S}})|^{-1}\,.$$

La notation $\mathrm{L}(s,\chi^{\mathrm{S}})$ désigne le produit des facteurs locaux $\mathrm{L}(s,\chi_v)$ pour tous les v non dans S. De plus nous supposons ϕ invariante sous K_v pour tous les v non dans S.

Démonstration. — Il existe une fonction de Schwartz-Bruhat à deux variables Φ telle que :

$$\phi(g,u,\chi)=\int\Phi[(0,t)g]\chi^2(t)\,|t|^{2u+1}dt\times\chi(\det g)\,|\det g|^{u+1/2}\times\mathrm{L}(2u+1,\chi^{2\mathrm{S}})^{-1}\,.$$

Un calcul formel (fait en détail dans [J. L.], § 3) donne :

$$W\,[\mathrm{diag}\,(a, 1)\ldots] = L(2u+1, \chi^{2S})^{-1} \times \int \Phi^{\wedge}(ta, t^{-1})\chi^2(t)\,|\,t\,|^{2u+1}dt\chi(a)\,|\,a\,|^{u+1/2},$$

où $\Phi^{\wedge}$ est la transformée de Fourier par rapport à la seconde variable. Il suffira donc de démontrer l'assertion suivante : étant donnée une fonction de Schwartz-Bruhat $\Phi \geq 0$ à deux variables, il existe une fonction de Schwartz-Bruhat à une variable $\phi \geq 0$ telle que pour tout idèle a nous ayons :

$$\int \Phi(at, t^{-1})dt \leq \phi(a)\,|\,a\,|^{-1}.$$

Considérons le problème local analogue. De manière précise, considérons d'abord le cas où le corps local F est non archimédien et la fonction Φ est la fonction caractéristique des entiers. Alors l'intégrale n'est autre que le volume de l'ensemble défini par les inégalités $|\,a\,| \leq |\,t\,| \leq 1$. L'intégrale est donc 0 à moins que a ne soit un entier. En supposant que ce soit le cas l'intégrale vaut $1+v(a)$. Puisque $q \geq a$ ceci est plus petit que $q^{v(a)}$. Donc notre intégrale est au plus $\phi(a)\,|\,a\,|^{-1}$, où ϕ est la fonction caractéristique des entiers. Si F et Φ sont quelconques l'intégrale, regardée comme une fonction de a, a la forme :

$$\int \Phi(at, t^{-1})dt = \phi_1(a) + \phi_2(a)\log|\,a\,|$$

où les ϕ_i sont des fonctions de Schwartz-Bruhat (*cf.* (4.3)). Il est clair que le membre de droite est majoré par $\phi(a)\,|\,a\,|^{-1}$, où ϕ est une fonction de Schwartz-Bruhat convenable. En multipliant ces majorations locales nous obtenons facilement la majoration globale requise.

Il est classique que la fonction $L(2u+1, \chi^{2S})^{-1}$ est à croissance au plus polynomiale sur la droite $\mathrm{Re}\,(u) = 0$. D'autre part, si ϕ est une fonction de Schwartz-Bruhat, il existe pour tout $N > 0$ une constante $C(N)$ telle que

$$\int_c^{+\infty} \phi(a)\,|\,a\,|^{-1}da \leq C(N)c^{-N}.$$

En comparant avec la définition (19) de R nous obtenons aussitôt :

LEMME (2). — *Pour tout* N *il existe des constantes* C(N) *et* M *telles que pour tout u imaginaire nous ayons :*

$$|\,R(c, u)\,| \leq C(N)\,|\,c\,|^{-N}\,|\,u\,|^{M}.$$

De la même façon en utilisant l'expression (8.3.14) pour la transformée de Mellin et le fait que l'opérateur $M(u, \chi)$ est unitaire sur l'axe imaginaire nous obtenons l'estimée suivante :

LEMME (3). — *Sur l'axe imaginaire* $M(u, \chi)\phi(k)$ *et* $L(1/2, \lambda : \phi : u, \chi)$ *sont à croissance au plus polynomiale.*

(8.4) Étudions maintenant l'intégrale du noyau K_{ei}. Rappelons sa définition. Pour tout caractère χ choisissons une base orthonormée ϕ_i de l'espace de Hilbert $V(\chi)$; désignons par $\rho(u, \chi)$ la représentation de $G(F_A)$ par translations à droite dans l'espace des fonctions ϕ telles que

$$\phi\left[\begin{vmatrix} a & x \\ 0 & b \end{vmatrix} g\right] = \chi(ab^{-1}) \mid ab^{-1} \mid^{u+1/2} \phi(g). \tag{1}$$

Nous pouvons identifier l'espace de $\rho(u, \chi)$ avec $V(\chi)$ et poser :

$$F(u, \chi : i, j) = (\rho(u, \chi)\phi_i, \phi_j). \tag{2}$$

Nous écrirons $E_{ei}(x, i, \ldots)$ pour $E_{ei}(x, \phi_i, \ldots)$. Avec ces notations :

$$K_{ei}(x, y) = \sum_{\chi} K_\chi(x, y) \tag{3}$$

où, pour chaque caractère χ du groupe des classes d'idèles,

$$K_\chi(x, y) = (2i\pi)^{-1} \sum_{i,j} \int_{-i\infty}^{+i\infty} F(u, \chi : i, j) E(x, j, u, \chi) E(y, i, u, \chi)^{-} du. \tag{4}$$

Pour un f donné les sommes (3) et (4) sont finies. Posons

$$I(c, \chi) = \int_{c^{-1}}^{c} \int_{c^{-1}}^{c} K_\chi(a, b) \eta (\det b) da\, db. \tag{5}$$

Nous pouvons évidemment échanger l'ordre des intégrations pour u et le couple (a, b). En tenant compte de (8.2.18) nous obtenons pour $I(c, \chi)$ l'expression suivante :

$$\begin{aligned}
(6)\quad & (i\pi)^{-1} \sum_{i,j} \int_{-i\infty}^{+i\infty} F(u, \chi : i, j) \\
& \times \Bigg[L(1/2, 1 : j, u, \chi) + R(c, u) + \frac{c^{u+1/2}}{u+1/2} \delta(\chi)\phi_j(e) + \frac{c^{-u+1/2}}{-u+1/2} \delta(\chi^{-1}) M(u, \chi)\phi_j(e) \\
& + \frac{c^{u+1/2}}{u+1/2} \delta(\chi^{-1})\phi_j(w) + \frac{c^{-u+1/2}}{-u+1/2} \delta(\chi) M(u, \chi)\phi_j(w) \Bigg] \\
& \times \Bigg[L(1/2, \eta, i, u, \chi)^{-} + R'(c, u) + \frac{c^{-u+1/2}}{-u+1/2} \delta(\eta\chi)\phi_i^{-}(e) + \frac{c^{u+1/2}}{u+1/2} \delta(\chi^{-1}\eta) M(u, \chi)\phi_i^{-}(e) \\
& + \frac{c^{-u+1/2}}{-u+1/2} \delta(\chi^{-1}\eta)\phi_i^{-}(w) + \frac{c^{u+1/2}}{u+1/2} \delta(\chi\eta) M(u, \chi)\phi_i^{-}(w) \Bigg] \times du.
\end{aligned}$$

Pour chaque (i, j), les termes $R(c, u)$ et $R'(c, u)$ satisfont aux conclusions du lemme (8.3.2). Pour un f donné $F(u, \chi : i, j)$ est nul sauf pour un nombre fini de couples (i, j). En particulier, $F(u, \chi : i, j)$ est nul à moins que ϕ_i et ϕ_j ne soient tous les deux invariants par tous les K_v avec v non dans S. De plus, sur l'axe imaginaire, $F(u, \chi : i, j)$ décroît rapidement (plus vite que l'inverse d'un polynôme en u). Au contraire, d'après (8.3), les termes $L(\ldots)$ et les termes contenant les puissances de c sont à croissance lente. Il en résulte que lorsque nous développons l'expression (6) nous trouvons un certain nombre de termes qui tendent vers zéro lorsque c tend vers l'infini. Nous pouvons ignorer ces termes. Parmi les termes qui restent il y a une intégrale indépendante de c :

$$(7)\qquad \sum_{i,j}\int_{-i\infty}^{+i\infty} F(u,\chi,i,j)\times L(1/2,1:j:u,\chi)\times[L(1/2,\eta,i,u,\chi)]^{-}du.$$

Les autres termes ne sont présents que si $\chi=1$ ou $\chi=\eta$. Chacun de ces termes est de l'un des types suivants :

$$(8)\qquad \int F(u,1:i,j)L(1/2,\eta:i:u,1)^{-}\frac{c^{1/2+u}}{1/2+u}(\phi_j(e)+\phi_j(w))du,$$

$$(9)\qquad \int F(u,\eta:i,j)L(1/2,1:j:u,\eta)\frac{c^{1/2-u}}{1/2-u}(\phi_i^{-}(e)+\phi_i^{-}(w))du,$$

$$(10)\qquad \int F(u,1:i,j)L(1/2,\eta:i:u,1)^{-}\frac{c^{1/2-u}}{1/2-u}(M(u,1)\phi_j(e)+M(u,1)\phi_j(w))du,$$

$$(11)\qquad \int F(u,\eta:i,j)L(1/2,1:j:u,\eta)\frac{c^{1/2+u}}{1/2+u}(M(u,\eta)\phi_i^{-}(e)+M(u,\eta)\phi_i^{-}(w))du.$$

L'intégrale (7) a visiblement les propriétés requises par la proposition (8.1). Pour démontrer cette proposition, il nous suffira donc de montrer que chacune des expressions (8) à (11) a une limite lorsque c tend vers l'infini et que, de plus, cette limite est nulle si la transformée de Satake de la fonction f_u est nulle au point q^{-1}. Cette dernière condition signifie que l'intégrale de f_u sur G_u/Z_u est nulle et implique que $F(u, 1:i,j)$ et $F(u,\eta:i,j)$ s'annulent aux points $u=1/2$ et $u=-1/2$.

(8.5) Examinons le terme (8) du numéro (8.4). Nous déplacerons le contour d'intégration de la ligne $\operatorname{Re} u=0$ à la ligne $\operatorname{Re} u=-1/2$; toutefois dans cette dernière ligne, nous remplacerons le segment joignant le point $-1/2-i\varepsilon$ au point $-1/2+i\varepsilon$ par le demi-cercle de centre $-1/2$ et de rayon ε qui passe par les points $-1/2-\varepsilon i$, $\varepsilon-1/2$ et $-1/2+i\varepsilon$. Vérifions que ce déplacement du contour est légitime. Le facteur

$$F(u)=F(u,1:i,j)(\phi_j(e)+\phi_j(w))$$

ainsi que ses dérivées, est holomorphe et à décroissance rapide dans la bande verticale $-1/2\leq \operatorname{Re} u\leq 0$. La fonction exponentielle reste bornée. Le facteur $(1/2+u)^{-1}$ reste aussi borné à l'infini dans cette bande verticale. Examinons la transformée de Mellin. Rappelons que nous pouvons trouver une représentation intégrale de $\phi_i(g,u,1)$:

$$\phi_i(g,u,1)=\int \Phi[(0,t)g]\,|t|^{2u+1}dt\times|\det g|^{u+1/2}L(2u+1,1^S)^{-1}.$$

Un calcul formel simple donne alors pour la transformée de Mellin (notée $L(u)$ en abrégé) :

$$(1)\qquad L(u)=L(2u+1,1^S)^{-1}\int \Phi^{\wedge}(a,b)\,|a|^{1/2+u}\eta(a)\,|b|^{1/2-u}\eta(b)\,da\,db,$$

où $\Phi^{\wedge}$ est la transformée de Fourier de Φ par rapport à la deuxième variable. En prenant l'imaginaire conjugué de deux membres nous obtenons donc :

(2) $$L(-u^{-})^{-} = L(-2u+1, 1^{S})^{-1}T(u),$$

où

(3) $$T(u) = \int \Phi_1(a, b) \mid a \mid^{1/2-u} \eta(a) \mid b \mid^{1/2+u} \eta(b) da\, db.$$

Dans cette expression Φ_1 est une fonction de Schwartz-Bruhat; l'intégrale de « Tate » double $T(u)$, ainsi que tous ses dérivées, est bornée dans la bande verticale $-1/2 \leq \operatorname{Re}(u) \leq 0$. Enfin dans la bande en question nous avons $1 \leq \operatorname{Re}(-2u+1) \leq 2$ et la fonction $L(-2u+1, 1_S)^{-1}$ est holomorphe et bornée par un polynôme en $\operatorname{Im}(u)$. Comme notre intégrale s'écrit :

$$\int F(u) L(1-2u, 1^{S})^{-1} T(u) c^{1/2+u} (1/2+u)^{-1} du.$$

Notre déplacement du contour d'intégration est en effet légitime. En remplaçant u par $u-1/2$, nous obtenons pour le terme (8.3.8) l'expression suivante :

(4) $$\int F(u-1/2) L(2-2u, 1^{S})^{-1} T(u-1/2) c^{u} u^{-1} du.$$

Dans (4) le contour d'intégration est maintenant la droite $\operatorname{Re}(u)=0$, excepté que le segment joignant le point $-i\varepsilon$ au point $i\varepsilon$ est remplacé par le demi-cercle de centre 0 qui passe par les points $-i\varepsilon, \varepsilon, i\varepsilon$. Faisons maintenant tendre ε vers 0. Alors l'intégrale sur le demi-cercle tend vers

$$i\pi F(-1/2) L(2, 1^{S})^{-1} T(-1/2),$$

tandis que l'intégrale sur la partie rectiligne du contour tend vers une « valeur principale de Cauchy ». En utilisant une variable d'intégration réelle t nous obtenons donc que (4) est aussi égale à :

(5) $$\int_{-\infty}^{+\infty} F(it-1/2) L(2-2it, 1^{S-1}) T(it-1/2) c^{it} t^{-1} dt + i\pi F(-1/2) L(2, 1^{S})^{-1} T(-1/2).$$

Pour t réel la fonction $L(-2it+2, 1^{S})$ est donnée par un produit infini (ou une série de Dirichlet) absolument et uniformément convergent. Ses dérivées sont donc bornées et son inverse aussi est bornée. Les dérivées du facteur $L(-2it+2, 1^{S})^{-1}$ sont donc bornées. Dans (5) le produit des trois premiers termes est donc une fonction de Schwartz de t. Lorsque c tend vers l'infini l'intégrale de Cauchy tend vers $i\pi$ fois la valeur de la fonction de Schwartz au point 0. Au total nous voyons que (5), i. e. le terme (8.4.8), tend vers une limite finie lorsque c tend vers l'infini, à savoir :

$$2i\pi F(-1/2) L(2, 1^{S})^{-1} T(-1/2);$$

cette limite s'annule en même temps que $F(-1/2, 1 : i, j)$. C'est bien là ce qu'il nous fallait prouver. Une conclusion analogue s'applique au terme (8.4.9).

(8.6) Examinons maintenant le terme (8.3.10). Nous poserons pour simplifier :

$$F(u)=F(u,1:i,j).$$

Nous allons utiliser une expression un peu différente de celle que nous avons utilisé jusqu'à présent pour la transformée de Mellin.

Écrivons ϕ pour ϕ_j et supposons, comme il est loisible, que ϕ soit un produit de fonctions locales ϕ_v. Nous pouvons aussi supposer que, pour chaque place v, ϕ_v est soit K_v invariante, soit au contraire d'intégrale nulle sur K_v. Notons T l'ensemble des places où cette dernière condition est satisfaite. Alors T est fini et contient S. Nous pouvons trouver une représentation intégrale pour $\phi(g, u, 1)$ de la forme :

(1) $$\phi(g,u,1)=\int\Phi[(0,t)g]\,|t|^{2u+1}dt \times |\det g|^{u+1/2}L(2u+1,1^T)^{-1}.$$

Nous en concluons, comme plus haut, que la transformée de Mellin qui apparaît dans (8.3.10) peut s'écrire :

(2) $$L(-2u+1,1^T)^{-1}T(u).$$

où $T(u)$ est défini par une intégrale de Tate double, holomorphe pour tout u. D'autre part nous pouvons écrire l'opérateur d'entrelacement $M(u, 1)$ comme un produit :

(3) $$M(u,1)=L(2u,1)L(2u+1,1)^{-1}N(u,1)$$

où N est l'opérateur d'entrelacement normalisé. Maintenant le quotient de $L(2u, 1)$ par $L(-2u+1, 1)$ est une fonction exponentielle ab^u. Il en résulte que le produit des facteurs (2) et (3) se réduit à :

(4) $$ab^uL(-2u+1,1_T)L(2u+1,1_T)^{-1}L(2u+1,1^T)^{-1}T(u)N(u,1).$$

Ainsi le terme (8.3.10) est donné par l'intégrale suivante :

(5) $$\int F(u)L(2u+1,1^T)^{-1}T(u)c^{1/2-u}(1/2-u)^{-1}A(u)du,$$

avec

$$A(u)=ab^uL(-2u+1,1_T)L(2u+1,1_T)^{-1}[N(u,1)\phi(e)+N(u,1)\phi(w)].$$

Nous allons déplacer le contour d'intégration. Le contour présent est la droite $\mathrm{Re}(u)=0$. Le nouveau contour sera la droite $\mathrm{Re}(u)=1/2$, excepté que le segment joignant les points $1/2-i\varepsilon$ et $1/2+i\varepsilon$ sera remplacé par le demi-cercle passant par les points $1/2-i\varepsilon$, $1/2-\varepsilon$, $1/2+i\varepsilon$. La fin de la démonstration sera alors la même que dans le cas précédent, excepté qu'il nous faut montrer que le facteur $A(u)$ est holomorphe et à croissance lente dans la bande $0 \le \mathrm{Re}(u) \le 1/2$. Le rapport des facteurs L qui apparaît dans A est le produit des rapports

$$L(-2u+1,1_v)L(2u+1,1_v)^{-1}$$

pour tous les v dans T. Si v est une place finie, alors ce rapport est une fraction rationnelle en q_v^{-u} donc est à croissance lente. Si v est infinie, la formule de Stirling montre que ce rapport est à croissance lente. Rappelons que ϕ_v est égale à un sur tout K_v pour tous les v non dans T. Pour un tel v, nous avons $N(u, 1_v)\phi_v(k_v)=1$ pour tout u. Donc $N(u, 1)\phi(e)$ par exemple est en fait le produit sur tous les v dans T de

$$N(u, 1_v)\phi_v(e).$$

Si v est finie ceci est encore à croissance lente. Si v est infinie, ceci est un polynôme en u. Donc A est bien à croissance lente. Prouvons enfin l'holomorphie de A aux pôles du facteur $L(-2u+1, 1_T)$ dans la bande. Prouvons par exemple l'holomorphie en 1/2 de :

$$L(-2u+1, 1_T)L(2u+1, 1_T)^{-1}N(u, 1)\phi(e).$$

Le produit précédent s'écrit en fait :

$$\prod_{v\in T} L(-2u+1, 1_v)L(2u+1, 1_v)^{-1}N(u, 1_v)\phi_v(e).$$

Prenons un v dans T. Comme l'intégrale de φ_v sur K_v est nulle, $N(u, 1_v)\phi_v(e)$ s'annule au point $u=1/2$ et ce zéro compense le pôle du facteur $L(-2u+1, 1_v)$ au même point. Le produit est donc bien holomorphe au point 1/2 et ceci termine notre discussion pour le terme (8.3.10). Une discussion analogue s'applique au terme (8.3.11). Les assertions du numéro (8.1) sont donc complètement prouvées.

9. Intégrales orbitales globales : le cas d'un tore compact

(9.1) Dans ce paragraphe F est encore un corps de nombres et E une extension quadratique de F. Nous fixerons un élément (G, T) de l'ensemble X(E : F) et un élément ε de $N(T)-T$. Alors le carré c de ε est un élément de $F^\times$ et la classe cN du groupe des normes N de E détermine la classe d'isomorphisme de (G, T). Soit f une fonction, lisse et à support compact, sur le groupe $G(F_A)/Z(F_A)$. A la fonction f est attaché le noyau cuspidal K_c. Soit ϕ_i une base orthonormale de l'espace des formes automorphes qui sont cuspidales et orthogonales aux fonctions $g \to \chi(\det g)$, où χ est un caractère de carré trivial du groupe des classes d'idèles de F. Alors, par définition :

(1) $$K_c(x, y)=\sum \rho(f)\phi_j(x)\bar{\phi_j}(y).$$

Nous nous proposons de donner une expression utile pour l'intégrale

(2) $$\iint K_c(s, t)dsdt, \qquad s, t\in T(F_A)/T(F)Z(F_A).$$

Bien entendu $\psi\circ \mathrm{Tr}$ est un caractère de E_A/E et nous avons donc pour chaque place v de E la mesure de Tamagawa sur le groupe $E_v^\times$ et, par transport de structure sur le groupe T_v. Nous avons aussi la mesure produit sur le groupe $T(E_A)$ et la mesure quotient sur $T(F_A)/Z(F_A)$. Nous noterons S un ensemble fini de places de F contenant les places à l'infini, les places ramifiées dans E, les places où G n'est pas déployé, les places où ψ_v n'est pas d'ordre 0 et les places de caractéristique résiduelle 2. Nous choisirons pour tout v un sous-groupe compact maximal K_v de G_v de telle manière que T_v soit contenu dans K_vZ_v si v ne se décompose pas dans E et $G(F_A)$ soit le produit restreint des G_v par rapport aux K_v. Nous supposerons que f est le produit de fonctions locales f_v, lisses et à support compact, sur G_v/Z_v. Nous supposerons f_v bi-K_v-invariante pour chaque v non dans S. Nous ne changeons pas l'intégrale (1) si pour v non décomposée dans E nous remplaçons f_v par la fonction f_v' définie par :

$$f_v'(g) = \mathrm{vol}\,(T_v)^{-1} \int f(s_v g t_v) ds_v dt_v .$$

Nous pouvons donc supposer que pour chaque v qui ne se décompose pas dans E la fonction f_v est bi-T_v-invariante, en particulier bi-K_v-finie. Enfin nous supposerons f_v bi-K_v-finie aux places v qui se décomposent dans E. Nous avons alors :

$$\text{(3)} \qquad K_c(x, y) = \sum f(x^{-1}\gamma y) - K_{sp}(x, y) - K_{ei}(x, y),$$

où la somme porte sur tous les γ dans G(F)/Z(F), K_{sp} dénote le noyau spécial et K_{ei} le noyau d'Eisenstein. Le noyau d'Eisenstein est bien entendu nul si G n'est pas déployé. Le noyau K_{sp} est défini par la somme suivante :

$$\text{(4)} \qquad K_{sp}(x, y) = \sum \mathrm{Vol}^{-1} \int f\,(\det g) dg \chi\,(\det x) \chi^{-1}\,(\det y);$$

la somme porte sur tous les caractères χ de carré trivial du groupe des classes d'idèles de F et Vol est le volume du quotient $G(F_A)/G(F)Z(F_A)$. Nous pouvons définir deux autres noyaux :

$$\text{(5)} \qquad K_r(x, y) = \sum f(x^{-1}\gamma y), \qquad \gamma \text{ T-régulier},$$

$$\text{(6)} \qquad K_s(x, y) = \sum f(x^{-1}\gamma y), \qquad \gamma \text{ T-singulier},$$

Alors K_c est la somme suivante :

$$\text{(7)} \qquad K_c = K_r + K_s - K_{sp} - K_{ei}.$$

Comme le quotient $T(F_A)/T(F)Z(F_A)$ est compact, l'intégrale (2) est simplement la somme des intégrales de chaque terme dans (7).

(9.2) Étudions l'intégrale de K_r. Chaque élément γ de G(F)/Z(F) qui est T-régulier peut s'écrire de manière unique sous la forme :

$$\text{(1)} \qquad \gamma = \sigma^{-1}\mu\tau,$$

où σ et τ parcourent T(F)/Z(F) et μ un ensemble de représentants pour les doubles classes T-régulières de T(F) dans G(F) (Prop. (1.2)). Nous obtenons donc immédiatement :

$$\text{(2)} \qquad \iint K_r(s, t) ds dt = \sum \iint f(s^{-1}\mu t) ds dt,$$

les intégrales du membre de droite portant maintenant toutes les deux sur $T(F_A)/Z(F_A)$. L'intégrale double du membre de droite ne dépend que de $\zeta = P(\mu : T)$ et nous noterons $H(\zeta : f : T)$ sa valeur. Nous pouvons donc écrire :

$$\text{(2)} \qquad \iint K_r(s, t) ds dt = \sum H(\zeta : f : T), \qquad \zeta \in c\mathrm{N} - 1,$$

puisque la fonction P paramétrise les doubles classes régulières et ses valeurs, sur les éléments réguliers, sont tous les points de la classe cN associée au couple (G, T) moins le point 1 (Prop. (1.1)). Bien entendu l'intégrale orbitale $H(\zeta : f : T)$ est le produit des intégrales orbitales locales :

$$(3) \qquad H(\zeta : f : T) = \prod H(\zeta : f_v : T_v).$$

Presque tous les facteurs sont égaux à 1. En effet soit v une place de F qui n'est pas dans S ; supposons que f_v soit la fonction caractéristique de $Z_v K_v$. Si v ne se décompose pas dans E, alors T_v est contenu dans $Z_v K_v$ et l'intégrale vaut 1. Si v se décompose dans E alors l'intégrale locale vaut encore 1 d'après la proposition (5.7).

(9.3) Passons à l'intégrale du terme K_r. Il n'y a que deux doubles classes singulières, T(F) et εT(F). Il vient donc :

$$(1) \qquad \iint K_s(s, t)\,ds\,dt = \mathrm{vol} \int f(t)\,dt + \mathrm{vol} \int f(\varepsilon t)\,dt,$$

où vol dénote le volume du quotient $T(F_A)/T(F)Z(F_A)$ et chacune des intégrales porte sur le quotient $T(F_A)/Z(F_A)$.

(9.4) Passons à l'intégrale du terme K_{sp}. D'après (4), nous avons :

$$(1) \qquad \iint K_{sp}(s, t)\,ds\,dt = \sum \mathrm{Vol}^{-1} \int f(g)\chi(\det g)\,dg \int \chi(\det s)\,ds \int \chi^{-1}(\det t)\,dt\,;$$

chacune des intégrales sur le quotient $T(F_A)/T(F)Z(F_A)$ est 0 à moins que $s \to \chi(\det s)$ ne soit trivial sur $T(F_A)$; ceci est le cas si et seulement si $\chi = 1$ ou $\chi = \eta$. L'intégrale de K_{sp} se réduit donc à deux termes :

$$(2) \qquad \iint K_{sp}(s, t)\,ds\,dt = \mathrm{Vol}^{-1}\,\mathrm{vol}^2 \left[\int f(g)\chi(\det g)\,dg + \int f(g)\,dg\right].$$

En particulier, choisissons comme dans (8.1) une place z de F non dans S, fixons les composants de f aux places autres que z et regardons l'intégrale comme une fonction de $f_z^{\wedge}$. Alors l'intégrale (5) est de la forme $cf_z^{\wedge}(q_z^{-1})$, où c est une constante.

(9.5) Passons au terme K_{ei}. Il est nul si G n'est pas déployé sur F. Supposons G déployé et revenons aux notations de (8.4). Nous avons :

$$(1) \qquad K_{ei}(x, y) = (i\pi)^{-1} \int_{-i\infty}^{+i\infty} A(x, y, u)\,du,$$

où

$$(2) \qquad A(x, y, u) = \sum_{\chi, j} [\rho(f)E](x, j, u, \chi)E(y, j, u, \chi)^{-}.$$

Notons que le sous-groupe compact maximal implicite dans la définition des séries d'Eisenstein est maintenant le produit des groupes K_v, avec $K_v Z_v = T_v$ si v est infinie. En particulier la série (2) est finie. Comme nous intégrons sur un ensemble compact nous obtenons :

(3) $$\iint K_{ei}(s,t)dsdt=(i\pi)^{-1}\int_{-i\infty}^{+i\infty} A(u:f)du,$$

où

(4) $$A(u:f)=\sum_{\chi,j}\int [\rho(f)E](s,j,u,\chi)ds\int E(t,j,u,\chi)^{-}dt.$$

Maintenant $[\rho(f)E](x,j,u,\chi)$ est nul à moins que ϕ_j ne soit K_v-invariante pour toutes les places v non dans S. En particulier, choisissons comme plus haut une place z de F qui n'est pas dans S et se décompose dans E. Alors $f=f^z.f_z$ où f^z est le produit des f_v pour $v\neq z$ et

(5) $$[\rho(f)E](x,j,u,\chi)=f_z^{\wedge}(q_z^{-2iu})[\rho(f^z)E](x,j,u,\chi).$$

Il vient donc :

(6) $$\iint K_{ei}(s,t)dsdt=(2i\pi)^{-1}\int_{-i\infty}^{+i\infty} f_z^{\wedge}(q_z^{-2iu})A(u:f^z)du,$$

où $A(u:f^z)$ est intégrable, résultat qui sera suffisant pour notre objet.

10. L'identité fondamentale

(10.1) Dans ce paragraphe F est toujours un corps de nombres, E une extension quadratique de F, η le caractère quadratique attaché à E. Nous considérerons encore le couple (G, A) formé du groupe GL(2) et du sous-groupe des matrices diagonales A ; nous fixerons un élément ε du normalisateur de A qui ne soit pas dans A. Nous noterons K_v le sous-groupe compact maximal usuel de G_v et nous supposerons ε contenu dans K_v pour tout v. Nous nous donnerons un ensemble fini S de places de F, contenant les places infinies, les places qui se ramifient dans E, les places où ψ n'est pas d'ordre 0 et les places de caractéristique résiduelle 2. Il sera commode de supposer que S a un nombre pair d'éléments. Soit X(S) l'ensemble des couples (G′, T′) dans X(E : F) tels que G′ se déploie en dehors de S. Pour chaque (G′, T′) dans X(S) et chaque place v, nous choisirons un sous-groupe compact maximal K'_v de G'_v de manière que $G'(F_A)$ soit le produit restreint des G'_v par rapport aux K'_v. Nous supposerons que si v ne se décompose pas dans E alors T'_v est contenu dans K'_vZ_v. Pour tout v non dans S les mesures de $T'_v\cap K'_v/K'_v\cap Z'_v$ et $A_v\cap K_v/K_v\cap Z_v$ sont 1. Nous fixerons un élément ε' du normalisateur de T′ qui ne soit pas dans T′ et nous supposerons que ε' est dans K'_v pour tous les v non dans S. Nous nous donnerons une fonction lisse à support compact f sur $G(F_A)/Z(F_A)$ et, pour chaque (G′, T′) dans X(S), une fonction lisse à support compact f' sur $G'(F_A)/Z'(F_A)$. Bien entendu ces fonctions seront supposées être des produits de fonctions locales. Nous ferons de plus les hypothèses suivantes :

(1) Soit v une place de S qui ne se décompose pas dans E. Alors f'_v est T'_v-biinvariante. De plus si x est un élément de F_v différent de 1 et 0, (G′, T′) un élément de X(S) et g' un élément de G'_v tels que $x=P(g':T_v)$ alors (cf. § 4) :

$$H(x:f_v:\eta_v)=H(g':f'_v:T'_v).$$

(2) Soit v une place de S qui se décompose dans E. Alors f_v est K_v-finie et f_v' K_v'-finie. Soit g un élément A_v-régulier de G_v. Si $(G', T') \in X(S)$ et $g' \in G_v$ sont tels que

$$P(g : A_v) = P(g' : T_v')$$

alors :

$$(i) \qquad H(g : f_v : A_v) = H(g' : f_v' : T_v') ;$$

$$(ii) \qquad \int f_v(a_v)\,da_v = \int f_v'(t_v')\,dt_v' ;$$

$$(iii) \qquad \int f_v(\varepsilon\, a_v)\,da_v = \int f_v'(\varepsilon' t_v')\,dt_v' .$$

(3) Si v n'est pas dans S alors f_v est K_v-biinvariante, f_v' K_v'-biinvariante et tout isomorphisme du couple (G_v, K_v) sur le couple (G_v', K_v') transforme f_v en f_v'.

(4) *Remarque.* — Dans la situation de la condition (2) il existe un isomorphisme du couple (G_v, A_v) sur le couple (G_v', T_v'). La condition (2) est satisfaite si nous prenons pour f_v' l'image de f_v par cet isomorphisme. En effet cela est clair pour (2.*i*) et (2.*ii*). Pour (2.*iii*), l'intégrale du membre de droite ne change pas si nous remplaçons ε' par l'image de ε par l'isomorphisme en question et alors notre assertion est évidente.

A la fonction f est associé le noyau cuspidal K_c pour le groupe G. De même, pour chaque (G', T'), à la fonction f' est associé le noyau cuspidal K_c' pour le groupe G'. Nous allons dans ce paragraphe démontrer le résultat suivant :

THÉORÈME. — *Avec les hypothèses et notations précédentes :*

$$(4) \qquad \iint K_c(a, b)\eta\,(\det b)\,da\,db = \sum_{(G', T')} \iint K_c'(s, t)\,ds\,dt, \qquad (G', T') \in X(S).$$

(10.2) Pour prouver notre identité nous écrirons, comme dans les paragraphes 7 et 9 :

$$(1) \qquad K_c = K_r + K_s - K_{sp} - K_{ei} ,$$

$$(2) \qquad K_c' = K_r' + K_s' - K_{sp}' - K_{ei}' .$$

Nous prouverons d'abord les identités suivantes :

$$(3) \qquad \iint K_r(a, b)\eta\,(\det b)\,da\,db = \sum_{(G', T')} \iint K_r'(s, t)\,ds\,dt ,$$

$$(4) \qquad \iint K_s(a, b)\eta\,(\det b)\,da\,db = \sum_{(G', T')} \iint K_s'(s, t)\,ds\,dt .$$

Supposons ces identités démontrées et montrons comment le théorème en découle. Considérons la différence :

$$(5) \qquad \iint K_c(a, b)\eta\,(\det b)\,da\,db - \sum_{(G', T')} \iint K_c'(s, t)\,ds\,dt .$$

Vues (3) et (4) elle s'écrit :

$$-\iint K_{sp}(a,b)\eta(\det b)da\,db+\sum_{(G',T')}\iint K'_{sp}(s,t)ds\,dt$$
$$-\iint K_{ei}(a,b)\eta(\det b)da\,db+\sum_{(G',T')}\iint K'_{ei}(s,t)ds\,dt\,.$$

Rappelons que pour le groupe G il s'agit d'intégrales faibles.

Choisissons maintenant une place z de E qui ne soit pas dans S et qui se décompose dans E. Fixons les composantes de f et des f' aux autres places. A la place z les transformées de Satake de f_z et des f'_z sont les mêmes. Nous pouvons donc regarder nos intégrales comme fonctions de $f_z^\wedge$. Alors d'après (8.1), (9.3) et (9.4) la somme ci-dessus a la forme

(6) $$\int_{-\infty}^{+\infty}\phi(t)f_z^\wedge(q_z^{-2it})dt+cf_z^\wedge(q_z^{-1}),$$

où ϕ est intégrable. On finit la démonstration comme dans [L] en utilisant le fait que les intégrales de K_c et K'_c ont aussi la forme :

(7) $$\sum_t a_t f_z^\wedge(t),$$

où les nombres complexes t sont, soit sur le cercle unité, soit sur l'axe réel entre q_z^{-1} et q_z et la série des a_t est absolument convergente. L'unicité de la décomposition d'une mesure en une mesure atomique et une mesure continue entraîne que la différence (5) est en fait nulle.

(10.3) Prouvons l'égalité (10.2.3). Le membre de gauche s'écrit :

$$\sum_\zeta H(\zeta:f:\eta),\qquad \zeta\neq 0,1\,,$$

tandis que le membre de droite s'écrit comme une somme double :

$$\sum_{(G',S')}\sum_\zeta H(\zeta:f':T')$$

la somme intérieure portant sur tous les ζ dans la classe cN, privée du point 1, déterminée par le couple (G′, T′). Nous pouvons recombiner les deux sommes et écrire le membre de droite comme une somme

$$\sum_\zeta H(\zeta:f':T),\qquad \zeta\in N(S)-1\,,$$

en désignant par N(S) la réunion des classes cN déterminées par les éléments de X(S). D'après la théorie du corps de classe les éléments de $F^\times - N(S)$ sont exactement les ζ dans $F^\times$ qui satisfont à la condition suivante : il existe une place v de F, qui n'est pas dans S, qui ne se décompose pas dans F et qui est telle que ζ ne soit pas une norme de l'extension quadratique E_v de F_v. D'après la Proposition (5.1) nous avons, pour un tel ζ, $H(\zeta:f_v:\eta_v)=0$ si v est la place en question. Il en résulte que $H(\zeta:f:\eta)=0$. Il suffira donc de démontrer l'égalité des intégrales orbitales $H(\zeta:f:\eta)$ et $H(\zeta:f':T')$ si ζ est dans N(S). Décomposons ces intégrales en produits d'intégrales locales $H(\zeta:f_v:\eta_v)$ et $H(\zeta:f'_v:T'_v)$ respectivement.

Pour v dans S l'égalité de ces intégrales résulte des hypothèses (1) et (2). Pour v non dans S l'égalité résulte de l'hypothèse (3) et de la proposition (5.1). L'égalité des intégrales orbitales globales, et la formule (3), sont donc établies.

(10.4) Prouvons maintenant l'égalité (10.2.4). Nous pouvons utiliser la formule (7.3.7) pour calculer le membre de gauche et la formule (9.3.1) pour calculer le membre de droite. L'égalité (10.2.4) sera alors conséquence des deux égalités suivantes :

$$\text{(1)} \qquad H(n_+ : f : \eta) + H(n_- : f : \eta) = \sum_{(G', T')} \operatorname{vol}(T'(F_A)/T'(F)Z'(F_A)) \int f'(t)\,dt,$$

$$\text{(2)} \qquad H(\varepsilon n_+ : f : \eta) + H(\varepsilon n_- : f : \eta) = \sum_{(G', T')} \operatorname{vol}(T'(F_A)/T'(F)Z'(F_A)) \int f'(\varepsilon' t)\,dt.$$

La deuxième identité résulte de la première identité appliquée à la fonction f_1 définie par $f_1(g) = f(\varepsilon g)$ et aux fonctions f'_1 définies par $f'_1(g) = f'(\varepsilon' g)$. Il est en effet facile de vérifier que les conditions (10.1.1) à (10.1.3) sont satisfaites par f_1 et les f'_1. Prouvons donc la première identité.

Calculons le membre de droite de (1). Introduisons un idèle différentiel a de E et b de F. Le prolongement analytique de l'intégrale de Tate

$$\text{(3)} \qquad \int \phi(t)\,|\,t\,|^s \eta(t)\,dt,$$

où ϕ est une fonction de Schwartz-Bruhat, prend au point $s=0$ la valeur :

$$L(0, \eta) \prod_{v \in T} \int \phi_v(t_v)\eta(t_v)\,dt_v L(0, \eta_v)^{-1} \prod_{v \in V} \phi_v(0)\,|\,a_v\,|^{1/2},$$

où T est l'ensemble des places de F qui ne se décomposent dans E et V l'ensemble de celles qui se décomposent.

Appliquons cette formule aux fonctions ϕ_+ et ϕ_- définies par :

$$\phi_+(x) = \int f\left[a \begin{vmatrix} 1 & x \\ 0 & 1 \end{vmatrix}\right] da, \qquad a \in A(F_A)/Z(F_A),$$

$$\phi_-(x) = \int f\left[a \begin{vmatrix} 1 & 0 \\ x & 1 \end{vmatrix}\right] da, \qquad a \in A(F_A)/Z(F_A).$$

Les composants locaux de ϕ_+ et ϕ_- sont définis de manière analogue en termes des composants locaux de f. Alors le membre de droite de (1) n'est autre que la somme des valeurs des intégrales de Tate (3) de ϕ_+ et ϕ_- au point $s=0$. De plus nous avons évidemment pour chaque v dans V :

$$\phi_{+v}(0) = \phi_{-v}(0) = \int f_v(a_v)\,da_v, \qquad a_v \in A_v/Z_v.$$

D'autre part pour chaque v dans T les valeurs au point 0 des intégrales de Tate de ϕ_{+v}

et ϕ_{-v} ne sont autres que les intégrales orbitales singulières des points n_+ et n_-. Posons donc pour simplifier les notations :

$$M_v = \int f_v(a_v)da_v \qquad \text{pour} \quad v \text{ dans V},$$

$$M_{v\pm} = 2H(n_\pm : f_v : \eta_v) \qquad \text{pour} \quad v \text{ dans T}.$$

Alors le membre de gauche de l'égalité (1) s'écrit :

$$(4) \qquad L(0, \eta)\prod_T 1/2L(0, \eta_v)^{-1}\prod_V |a_v|^{1/2} \times \prod_V M_v[\prod_T M_{v+} + \prod_T M_{v-}].$$

Remarquons que dans chacun des produits infinis presque tous les facteurs sont égaux à 1.

Passons au second membre de (1). Le volume qui y apparaît n'est autre que $2L(1, \eta)$, comme il est bien connu. L'intégrale est évidemment le produit des intégrales locales analogues :

$$\int f'(t)dt - \prod_v \int f'_v(t_v)dt_v.$$

Si v est dans V, l'intégrale locale n'est autre que M_v d'après l'hypothèse (10.1.2.*ii*). Si v est dans T alors l'intégrale est égale à

$$\operatorname{vol}(T_v/Z_v)^{-1}1/2[M_{v-} + \eta_v(c)M_{v+}]$$

d'après la proposition (4.1) et la proposition (5.1). Le volume qui apparaît dans cette formule n'est autre que $|b_w|^{1/2}|a_v|^{-1/2}$, où w est l'unique place de E au-dessus de v. Au total le membre de droite de (1) est égal au produit suivant :

$$(5) \quad 2L(1, \eta) \sum_{c\in N(S)/N} \prod_{v\in T} 1/2L(0, \eta_v)^{-1} \prod_{v\in T} |b_w|^{-1/2}|a_v|^{1/2} \prod_{v\in V} M_v \prod_{v\in T} 1/2[M_{v-} + \eta_v(c)M_{v-}].$$

En comparant avec (4) nous voyons qu'il suffit de prouver les égalités suivantes :

$$(6) \qquad L(0, \eta)\prod_{v\in V} |a_v|^{1/2} = L(1, \eta)\prod_{v\in T} |b_w|^{-1/2}|a_v|^{1/2},$$

$$(7) \qquad \prod_{v\in T} M_{v+} + \prod_{v\in T} M_{v-} = 2\sum_{c\in N(s)/N} \prod_{v\in T} 1/2[M_{v-} + \eta_v(c)M_{v+}].$$

L'égalité (6) résulte aussitôt des équations fonctionnelles des fonctions $L(s, 1_E)$ et $L(s, 1_F)$ et leur relation avec $L(s, \eta)$.

Passons à l'égalité (7). Pour $v\in T-S$ nous avons $\eta_v(c)=1$ d'après la définition de N(S) et $M_{v+} = M_{v-}$ (Prop. (5.1)); de plus, pour presque tous les $v \in T-V$, $M_{v+} = M_{v-} = 1$. En posant $U = T\cap S$, nous voyons que l'identité (7) se réduit à :

$$\prod_{v\in U} M_{v+} + \prod_{v\in U} M_{v-} = 2\sum_{c\in N(S)/N} \prod_{v\in U} 1/2[M_{v-} + \eta_v(c)M_{v+}].$$

Soit H le groupe $\{1, -1\}^U$. Pour chaque v dans U définissons un caractère χ_v de H par la formule $\chi_v(h) = h_v$. Soit H' le sous-groupe de H défini par l'équation $\prod_{v\in U} \chi_v(h) = 1$. Alors l'application $c \to (\eta_v(c))$ définit une bijection de N(S)/N sur H'. Avec cette notation la formule à prouver s'écrit

$$\prod_{v\in U} M_{v+} + \prod_{v\in U} M_{v-} = 2\sum_{h\in H}\prod_{v\in U} 1/2[M_{v-} + \chi_v(h)M_{v+}].$$

Comme $[H:e]=2[H':e]$, le membre de droite s'écrit :

$$[H':e]^{-1}\sum_{Y}\sum_{h\in H'}\prod_{v\in Y}\chi_v(h)\prod_{v\in Y}M_{v+}\prod_{v\in U-Y}M_{v-},$$

où la somme extérieure porte sur tous les sous-ensembles Y de U. Le caractère $\prod_{v\in Y}\chi_v$ a une restriction non triviale à H', à moins que Y ne soit vide ou égal à U. Dans la somme précédente, il n'y a donc que les termes correspondant à l'ensemble vide et U ; ceci nous donne notre égalité.

11. Le résultat de Waldspurger

(11.1) Nous allons finalement démontrer le résultat de Waldspurger en utilisant l'identité du paragraphe 10. Nous désignerons par S un ensemble fini de places de F satisfaisant les conditions du paragraphe 10. Comme nous pouvons prendre S arbitrairement grand, il n'y aura aucun inconvénient à ne considérer que des représentations cuspidales de G non ramifiées en dehors de S, des couples (G', T') appartenant à X(S) et, pour un tel couple, des représentations cuspidales de G' non ramifiées en dehors de S. Nous noterons K (resp. K^S) le produit des sous-groupes compacts K_v pour tous les v (resp. tous les v non dans S) et G^S le produit restreint des G_v pour v non dans S. Pour un couple (G', T') dans X(S) les notations K', K'^S et G'^S ont une signification analogue.

Pour préciser la première condition de Waldspurger remarquons que si les intégrales

$$\int\phi(a)da \qquad \text{et} \qquad \int\phi(bc)\eta(\det b)db$$

ne sont pas nulles pour un couple de vecteurs lisses, elles ne sont pas nulles pour un couple de vecteurs K-finis (ϕ, ϕ') ; de plus si S est assez grand, alors ϕ et ϕ' sont K^S-invariants. De même s'il existe un couple (G', T') dans X, une représentation cuspidale π' et un vecteur lisse ϕ dans l'espace de π' tels que l'intégrale $\int\phi(t)dt$ ne soit pas nulle, alors nous pouvons prendre ϕ K'-finie ; de plus, si S est assez grand, (G', T') est dans X(S) et ϕ invariante sous K'^S.

(11.2) Considérons donc un ensemble S et des fonctions f et f' satisfaisant les conditions du paragraphe 10. En particulier f_v (resp. f'_v) est biinvariante sous K_v (resp. K'_v). Considérons le noyau K_c. Nous pouvons l'écrire :

(1) $$K_c = \sum_{\pi} K_{\pi},$$

où, pour chaque représentation automorphe cuspidale (non ramifiée en dehors de S) π, nous avons posé :

(2) $$K_{\pi}(x, y) = \sum_j \rho(f)\phi_j(x)\phi_j^{-}(y),$$

ϕ_j dénotant une base orthonormale du sous-espace des vecteurs K^S-invariants dans l'espace de π. Nous supposerons les ϕ_j K-finies. La série (1) converge non seulement dans l'espace de Hilbert des fonctions de carré intégrable sur le quotient $G(F_A)/G(F)Z(F_A)$, mais aussi dans l'espace des fonctions à décroissance rapide sur le quotient $G(F_A)/G(F)Z(F_A)$. De plus, puisque f_v est K_v-finie pour v infinie, la séric (2), pour un f donné, n'a qu'un nombre fini de termes non nuls. Désignons par H(S) l'algèbre de Hecke du groupe G^S relative au sous-groupe K^S. Écrivons $f = f_S f^S$, où f_S (resp. f^S) est le produit des f_v pour v dans S (resp. non dans S). Soit Λ_π le caractère de H(S) attaché à une représentation π. Alors nous avons :

(3) $$\iint K_c(a, b)\eta\,(\det b)da\,db = \sum_\pi a(\pi, f_S)\Lambda_\pi(f^S),$$

où nous avons posé :

(4) $$a(\pi, f_S) = \sum_j \int \rho(f_S)\phi_j(a)da \int \phi_j^-(b)\eta\,(\det b)db.$$

(11.3) Considérons de même un couple (G′, S′) dans X(S). Nous avons encore une décomposition

(1) $$K'_c = \sum_{\pi'} K_{\pi'},$$

(2) $$K_{\pi'}(x, y) = \sum_j \rho(f')\phi_j(x)\phi_j^-(y),$$

où ϕ_j est une base orthonormale de l'espace des vecteurs K'^S-invariants de π. La série (1) converge encore dans l'espace des fonctions à décroissance rapide et la séric (2) est finie. En intégrant terme à terme nous trouvons :

(5) $$\iint K_{\pi'}(s, t)ds\,dt = a(\pi', f'_S)\Lambda_{\pi'}(f'^S),$$

où nous avons posé

(6) $$a(\pi', f'_S) = \sum_j \int \rho(f)\phi_j(s)ds \int \phi_j^-(t)dt.$$

L'intégrale totale est donc :

$$\iint K'_c(s, t)dsdt = \sum_{\pi'} a(\pi', f'_S)\Lambda_{\pi'}(f'^S).$$

(11.4) Utilisons maintenant notre identité fondamentale. Remarquons que si π' est une représentation cuspidale de G′ et π la représentation cuspidale de G qui lui correspond, alors $\Lambda_\pi(f_S) = \Lambda_{\pi'}(f'_S)$. Puisque π' détermine π nous pouvons écrire $a(\pi, f'_S)$ pour $a(\pi', f'_S)$. D'autre part, pour une représentation π donnée pour le groupe G, il sera commode de poser $a(\pi, f'_S) = 0$ s'il n'existe pas de représentation π' pour G′ correspondant à π. Alors notre identité fondamentale s'écrit :

(1) $$\sum_\pi a(\pi, f_S)\Lambda(\pi, f^S) = \sum_\pi \Big[\sum_{(G', T')} \sum a(\pi, f_S')\Big]\Lambda(\pi, f^S).$$

Dans cette formule f^S est un élément arbitraire de H(S). Soit v une place dans S. Alors la fonction f_v est arbitraire K_v-finie. La fonction f_v' est reliée à f_v par les conditions (10.1.1) ou (10.1.2). Si v se décompose f_v' est en fait arbitraire K_v'-finie (*cf.* (10.1.4)). Si au contraire v ne se décompose pas, alors f_v' n'est plus arbitraire mais satisfait à une condition de densité : si une fonction h continue sur G_v/Z_v est bi-T_v-invariante et orthogonale à toutes les f_v' possibles alors h est nulle (Prop. (4.2)).

Supposons que π satisfasse à la première condition de Waldspurger ; alors il existe des vecteurs K-finis ϕ et ϕ' dans l'espace de π tels que :

$$\int \phi(a)da \neq 0 \qquad \text{et} \qquad \int \phi'(b)\eta\,(\det b)db \neq 0,$$

et nous pouvons supposer ϕ et ϕ' K^S-invariants. Choisissons la base ϕ_j de telle sorte que $\phi_1 = \phi' \,\|\, \phi' \,\|^{-1}$. Il existe f_S telle que $\rho(f_S)\phi_1 = \phi$ et $\rho(f_S)\phi_j = 0$ si $j \neq 1$. Alors nous avons

$$a(\pi, f_S) = \int \phi(a)da \left[\int \phi_1(b)\eta\,(\det b)db \right]^- \neq 0.$$

D'après le principe de l'indépendance linéaire « infinie » des caractères de H(S) ([L], p. 211), il existe au moins un (G', T') tel que $a(\pi, f_S')$ ne soit pas nul. Il en résulte évidemment qu'il existe au moins un ϕ dans l'espace de π' tel que $\int \phi(t)dt$ ne soit pas nulle. Ainsi π satisfait à la seconde condition de Waldspurger.

Supposons maintenant qu'il existe un couple (G', T'), une représentation π' et un vecteur K'-fini ϕ dans l'espace de π' tels que l'intégrale $\int \phi(t)dt$ ne soit pas nulle ; nous pouvons supposer que (G', T') est dans X(S) et ϕ est invariante sous K'^S. Nous allons voir que nous pouvons choisir f_S' de manière que $a(\pi', f_S')$ ne soit pas nul. L'intégrale sur T définit une forme linéaire continue sur l'espace des vecteurs lisses de π' fixés par K'^S. Écrivons-la comme le produit scalaire avec un vecteur « généralisé » e_T :

$$\int \phi(t)dt = (\phi, e_T).$$

Si h est une fonction lisse à support compact sur G_S/Z_S, alors $\pi'(h)(e_T)$ est défini : c'est un vecteur lisse tel que $(\phi, \pi'(h)e_T) = (\pi'(h^*)\phi, e_T)$ pour tout vecteur ϕ, lisse ou non. Avec cette notation nous avons :

$$a(\pi, f_S') = (\pi'(f_S')e_T, e_T).$$

Le sous-espace de π' formé des vecteurs K'^S-invariants est isomorphe au produit tensoriel des espaces des π_v' avec v dans S. Pour chaque v dans S, il existe une forme linéaire continue non nulle e_v sur l'espace des vecteurs lisses de π_v' qui est invariante sous T_v. Cette forme est unique, à un facteur scalaire près (*cf.* (6.1) et (6.2)), et nous pouvons donc écrire :

$$a(\pi', f_S') = (\pi'(f_S')e_T, e_T) = C \prod_{v \in S} (\pi_v(f_v')e_v, e_v),$$

où C est une constante non nulle. Il s'agit de voir que nous pouvons choisir f_v' de telle sorte que $(\pi_v(f_v')e_v, e_v)$ ne soit pas nul. C'est évident si v se décompose puisque f_v'

est alors arbitraire K'_v-finie. Si v se décompose e_v est en fait un vecteur ordinaire puisque T_v est compact. Alors $(\pi_v(f_v')e_v, e_v)$ est le produit scalaire de la fonction f_v' avec le coefficient matriciel continu $(\pi'(g)e_v, e_v)$; il ne peut donc être nul pour tout choix de f_v' d'après la propriété de densité des f_v'. D'autre part si (G″, T″) est un autre élément de X(S), alors $a(\pi, f_S'') = 0$; autrement il existerait au moins une place v dans S où les groupes G'_v et G''_v ne sont pas isomorphes et les représentations π'_v et π''_v admettent des vecteurs non nuls invariants sous T'_v et T''_v respectivement. Or ceci est impossible (Proposition (6.3)). Le coefficient de Λ_π dans le second membre de (1) est donc non nul, pour un choix convenable de f_S' (i. e. f_S). Il en résulte comme plus haut que $a(\pi, f_S)$ n'est pas nul. Cela implique évidemment que π satisfait à la première condition de Waldspurger. Ceci termine donc la démonstration de l'équivalence des deux conditions de Waldspurger.

BIBLIOGRAPHIE

[D.-M.] J. DIXMIER et P. MALLIAVIN, *Factorisations de fonctions et vecteurs indéfiniment différentiable* (*Bull. Sc. Math.*, II, Sér. 102, p. 305-330).

[J.-L.] H. JACQUET et R. LANGLANDS, *Automorphic Forms on* GL(2) (*Lecture Notes in Math.*, n° 114, Springer Verlag, 1970).

[L] R. LANGLANDS, *Base Change for* GL(2) (*Annals of Mathematics Studies*, 96, Princeton University Press).

[W1] J.-L. WALDSPURGER, *Correspondance de Shimura* (*J. Math. pures et appl.*, vol. 59, 1980, p. 1-133).

[W2] J.-L. WALDSPURGER, *Correspondance de Shimura et Quaternions*, prerint.

[W3] J.-L. WALDSPURGER, *Sur les valeurs de certaines fonctions* L *automorphes en leur centre de symétrie*, preprint ⟨ 19 p, 1 p 3 ⟩.

(Manuscrit reçu le 15 mai 1984,
révisé le 30 mai 1985).

Hervé JACQUET,
Columbia University,
Department of Mathematics,
New York, N. Y. 10027,
U. S. A.

C. R. Acad. Sci. Paris, t. 312, Série I, p. 957-961, 1991

Théorie des groupes/*Group Theory*

Représentations distinguées pour le groupe orthogonal

Hervé JACQUET

Résumé – Nous reprenons des travaux antérieurs (eux-mêmes repris de Waldspurger) pour prouver que les représentations automorphes de GL(2) qui correspondent aux représentations du revêtement à deux feuillets de GL(2) sont caractérisées par la propriété d'être distinguées, en un sens convenable, par rapport à un groupe orthogonal que, de plus, l'on peut prendre déployé. L'intérêt de cette Note est de mettre en valeur le rôle joué par les diverses formes du groupe orthogonal et de suggérer une généralisation à GL(m).

Distinguished representations for the orthogonal group

Abstract – *We reeaxim earlier work (which is itself a reexamination of Waldspurger's work) to show that automorphic representations of* GL(2) *which correspond to automorphic representations of the two fold cover of* GL(2) *are distinguished, in a suitable sense, with respect to an orthogonal group, which, furthermore, can be taken to be split. The motivation for this Note is to emphasize the role of the various forms of the orthogonal group and to suggest a generalization to* GL(m).

1. RÉSULTATS LOCAUX. – Soient F un corps de caractéristique 0, S_F ou S l'espace des matrices symétriques inversibles 2×2, regardé comme variété algébrique sur F. Le groupe $G=GL(2, F)$ opère sur S par $s\mapsto {}^t gsg$. Soient N le groupe des matrices trigonales strictes supérieures, A celui des matrices diagonales, Z le centre de G. Posons :

$$w=\begin{pmatrix}0 & 1\\ 1 & 0\end{pmatrix}.$$

Alors les orbites de N dans S rencontrent A ou wZ en un unique point. Fixons un caractère non trivial ψ de F et définissons un caractère θ de N par la formule

$$\theta\begin{pmatrix}1 & x\\ 0 & 1\end{pmatrix}=\psi\left(\frac{x}{2}\right). \tag{1}$$

Une orbite est dite pertinente si θ est trivial sur le fixateur d'un point de l'orbite. Toutes les orbites sont pertinentes. Soient Φ une fonction lisse à support compact et χ un caractère de $F^\times$, de carré trivial. Nous définissons des intégrales orbitales :

$$I(a, \Phi, \chi)=\int \Phi\left[{}^t n\begin{pmatrix}ab & 0\\ 0 & b\end{pmatrix}n\right]\theta(n)\,dn\,\chi(b)\,d^\times b, \tag{2}$$

$$S(\Phi, \chi)=\int \Phi\left[{}^t n\begin{pmatrix}0 & b\\ b & 0\end{pmatrix}n\right]\theta(n)\,dn\,\chi(b)\,d^\times b. \tag{3}$$

Considérons d'autre part le groupe $\tilde{G}$, revêtement à deux feuillets de G. C'est donc l'ensemble des paires (g, z) avec $g\in G$, $z\in\mathbb{T}$, le produit étant défini par un cocycle convenable. Nous identifierons g à l'élément $(g, 1)$ et noterons $g.g'$ le produit $(g, 1)(g', 1)$. Soit $\gamma(a, \psi)$ la constante de Weil; elle est définie par l'équation

$$\int \hat{\Phi}(x)\psi\left(\frac{1}{2}ax^2\right)dx=|a|^{-1/2}\gamma(a, \psi)\int \Phi(x)\psi\left(-\frac{1}{2}a^{-1}x^2\right)dx.$$

Soit (a, b) le symbole de Hilbert. Nous poserons

$$\mu(a, \psi)=\frac{\gamma(a, \psi)}{\gamma(1, \psi)}(-1, a).$$

Note présentée par Hervé JACQUET.

0764-4442/91/03120957 $ 2.00

Rappelons que

$$(\operatorname{diag}(a, a), z) \mapsto \mu(a, \psi) z$$

est un caractère du groupe $\tilde{Z}$ image réciproque de Z. Soit f' une fonction lisse à support compact sur $\tilde{G}$ telle que $f'(g, z) = f'(g, 1) z$. Nous ne considérerons que des fonctions satisfaisant cette condition. Nous définissons des intégrales orbitales :

$$\text{(4)} \quad I(a, f', \chi) = \int f'\left[{}^t n_1 \cdot \begin{pmatrix} a & 0 \\ 0 & 1 \end{pmatrix} \cdot \begin{pmatrix} b & 0 \\ 0 & b \end{pmatrix} \cdot n_2\right] \theta(n_1) \theta(n_2) \chi(b) \mu(b, \psi) d^\times b \, dn_1 \, dn_2,$$

$$\text{(5)} \quad S(f', \chi) = \int f'\left[\begin{pmatrix} b & 0 \\ 0 & b \end{pmatrix} \cdot n\right] \theta(n) \chi(b) \mu(b, \psi) \, dn \, d^\times b.$$

PROPOSITION 1. — *Il existe une constante* c_ψ *telle que, étant donnée* Φ (resp. f'), *il existe* f' (resp. Φ) *avec*

$$c_\psi |a|^{-1/4} I(a, \Phi, \chi) = I(a, f', \chi).$$

Nous avons alors :

$$S(\Phi, \chi) = S(f', \chi).$$

Sous les conditions de la proposition 1, nous écrirons $\Phi \leftrightarrow f'$. Comme dans [1], pour prouver la proposition dans le cas non archimédien, on montre que $I(a, \Phi, \chi) = 0$ pour $|a|$ assez grand, et que, de plus, pour $|a|$ assez petit, $I(a, \Phi) = 0$ à moins que $a = -t^2$ pour un t convenable auquel cas :

$$\text{(6)} \quad I(a, \Phi, \chi) = S(\Phi, \chi)\left[\psi\left(\frac{1}{t}\right)\chi(-t) + \psi\left(-\frac{1}{t}\right)\chi(t)\right].$$

Puis on montre, par la méthode de la phase stationnaire, que l'intégrale de f' a un comportement analogue, au facteur $c_\psi |t|^{-1/2}$ près.

Supposons maintenant F non archimédien. Soient R l'anneau des entiers et π_F une uniformisante. Supposons ψ de conducteur R et $|2| = 1$ (situation non ramifiée). Soient $K = GL(2, R)$ et Φ_0 la fonction caractéristique de $K \cap S$.

Soient $m \geqq 0$ un entier, et f_m la fonction caractéristique de

$$K \begin{pmatrix} \pi_F^m & 0 \\ 0 & 1 \end{pmatrix} K.$$

Définissons une nouvelle fonction Φ_m sur S par la formule :

$$\Phi_m(s) = \int_G \Phi_0({}^t g s g) f_m(g^{-1}) \, dg.$$

Soit d'autre part $\tilde{K}$ un sous-groupe de $\tilde{G}$ isomorphe à K qui se projette sur K et f'_m la fonction bi-invariante sous $\tilde{K}$ qui correspond à f_m [2]. Alors :

PROPOSITION 2. — *On a* $c_\psi = 1$ *et, si* χ *n'est pas ramifié,* $\Phi_m \leftrightarrow f'_m$.

Par exemple, pour $m = 0$, l'intégrale (2) pour Φ_0 est 0 si $|a| > 1$, est 1 si $|a| = 1$; enfin, si $|a| < 1$, l'intégrale est zéro à moins que $a = -t^2$ et dans ce cas est donnée par (6). L'intégrale pour f'_0 a la valeur correspondante. Les démonstrations sont analogues à celles de [1].

Soit maintenant $\{\varepsilon\}$ un système de représentants pour les orbites de $G(F)$ sur S_F. Pour chaque ε, notons H_ε le groupe orthogonal correspondant, T_ε le groupe des similitudes, H_ε^0 et T_ε^0 les composantes connexes de ces groupes. Ainsi T_ε^0 est un sous-tore maximal de G et T_ε son normalisateur. Notons aussi $\lambda(t)$ le rapport de similitude. Alors $\lambda(t) = \det t$

pour t dans la composante connexe et $\lambda(t) = -\det t$ sinon. Soient Φ une fonction lisse à support compact sur S et f une fonction lisse à support compact sur G; nous écrirons

$$\Phi \overset{\varepsilon}{\to} f \tag{7}$$

si

$$\Phi({}^t g \varepsilon g) = \int_{H_\varepsilon} f(hg)\, dh.$$

Alors, pour un choix convenable des mesures :

$$\int_{T_\varepsilon} f(tg)\chi(\lambda(t))\, dt = \int \Phi({}^t g a \varepsilon g)\chi(a)\, d^\times a,$$

où l'intégrale de droite porte sur l'image de T_ε sous λ. Nous dirons que f est paire (pour ε) si

$$\int_{H^0_\varepsilon} f(hh_1 g)\, dh = \int_{H^0_\varepsilon} f(hg)\, dh, \tag{8}$$

pour $h_1 \in H_\varepsilon$. Alors :

$$\int_{T^0_\varepsilon} f(tt_1 g)\chi(\lambda(t))\, dt = \chi(\lambda(t_1)) \int_{T^0_\varepsilon} f(tg)\chi(\lambda(t))\, dt,$$

pour $t_1 \in T_\varepsilon$. Plaçons nous dans la situation non ramifiée. Choisissons un système de représentants contenant les matrices w et $\varepsilon_1 = \mathrm{diag}(1, -\tau)$, où τ est une unité qui n'est pas un carré. Alors on a

$$\Phi_m \overset{\varepsilon}{\to} f$$

avec $f = f_m$ si $\varepsilon = w$ ou si $\varepsilon = \varepsilon_1$ et $f = 0$ dans tous les autres cas. Ces fonctions sont paires.

2. Une formule des traces relatives. — Soient maintenant F un corps de nombres et χ un caractère automorphe de $F^\times$, de carré trivial. Nous choisissons un caractère non trivial ψ de $F_{\mathbb{A}}/F$ et définissons encore θ par 1. Pour chaque place v de F soit Φ_v une fonction lisse à support compact sur S_v; nous supposons que, pour presque tout v, Φ_v est la fonction caractéristique de $K_v \cap S_v$. Soit d'autre part f'_v une fonction sur $\tilde{G}_v$ telle que $\Phi_v \leftrightarrow f'_v$. Nous supposons que f'_v est la fonction « caractéristique » de $\tilde{K}_v$ pour presque tout v. Soient Φ le produit des Φ_v et f' le « produit » des f'_v; formons maintenant un noyau géométrique K' sur $\tilde{G}(F_{\mathbb{A}})$. Alors :

$$\iint K'({}^t n_1, \mathrm{diag}(a, a).n_2)\theta(n_1)^{-1}\theta(n_2)\, dn_1\, dn_2\, \mu(a, \psi)\chi(a)\, d^\times a \tag{9}$$

$$= \sum_\sigma \int \Phi({}^t n a \sigma n)\chi(a)\theta(n)\, dn\, d^\times a, \tag{10}$$

où

$$n, n_1, n_2 \in N(F)\backslash N(F_{\mathbb{A}}), \qquad a \in F^\times_{\mathbb{A}};$$

la somme porte sur un système de représentants des orbites (pertinentes) de N(F) sur S(F), à multiplication près par $F^\times$. Choisissons comme plus haut un système $\{\varepsilon\}$. Pour chaque ε et chaque v choisissons une fonction *paire* (pour ε) f_v sur G_v telle que $\Phi_v \overset{\varepsilon}{\to} f_v$. Pour presque tout v, nous choisissons la fonction caractéristique de K_v. Soit f_ε le produit.

Formons un noyau géométrique K_ε. Considérons la somme :

$$\sum_{\varepsilon} \int K_\varepsilon(t, n)\, \chi(\lambda(t))\, \theta(n)\, dn\, dt$$

où

$$t \in T^0_\varepsilon(F)\backslash T^0_\varepsilon(F_{\mathbb{A}}), \qquad n \in N(F)\backslash N(F_{\mathbb{A}}).$$

Elle est égale à (10). D'où l'égalité :

$$(11)\quad \sum_{\varepsilon} \int K_\varepsilon(t, n)\, \chi(\lambda(t))\, \theta(n)\, dn\, dt$$

$$= \iint K'({}^t n_1, \operatorname{diag}(a, a).n_2)\, \theta(n_1)^{-1}\, \theta(n_2)\, dn_1\, dn_2\, \mu(a, \psi)\, \chi(a)\, d^\times a.$$

3. Application. — Soit π une représentation automorphe cuspidale de G, de caractère central trivial. Nous poserons :

$$c_v = \varepsilon\left(\frac{1}{2}, \pi_v \otimes \chi_v, \psi_v\right) \chi_v(-1).$$

Supposons qu'il existe ε et φ dans l'espace de π telle que l'intégrale

$$(12)\qquad \int_{Z(F_{\mathbb{A}})\, T^0_\varepsilon(F)\backslash T^0_\varepsilon(F_{\mathbb{A}})} \varphi(t)\, \chi(\lambda(t))\, dt$$

soit non nulle. Écrivons T pour T_ε. Alors, pour g_v dans T_v mais non dans T^0_v, nous avons :

$$\int_{Z(F_{\mathbb{A}})\, T^0(F)\backslash T^0(F_{\mathbb{A}})} \varphi(tg_v)\, \chi(\lambda(t))\, dt = c_v\, \chi(\lambda(g_v)) \int_{Z(F_{\mathbb{A}})\, T^0(F)\backslash T^0(F_{\mathbb{A}})} \varphi(tg_v)\, \chi(\lambda(t))\, dt.$$

Il existe donc un caractère μ du quotient

$$T(F)\, T^0(F_{\mathbb{A}})\backslash T(F_{\mathbb{A}})$$

tel que :

$$\int_{Z(F_{\mathbb{A}})\, T^0(F)\backslash T^0(F_{\mathbb{A}})} \varphi(tg)\, \chi(\lambda(t))\, dt = \mu(g)\, \chi(\lambda(g)) \int_{Z(F_{\mathbb{A}})\, T^0(F)\backslash T^0(F_{\mathbb{A}})} \varphi(tg)\, \chi(\lambda(t))\, dt,$$

pour $g \in T(F_{\mathbb{A}})$. De plus μ est trivial si et seulement si $c_v = 1$ pour tout v.

Théorème 1. — *La représentation π correspond à une représentation automorphe de $\tilde{G}$ si, et seulement si, il existe ε et φ telle que l'intégrale* (12) *soit non nulle et* $\mu = 1$. *De plus, on peut alors prendre $\varepsilon = w$.*

La dernière assertion résulte d'une autre forme de la formule des traces [4] (ou des travaux de Waldspurger qui l'ont inspirée). Supposons que la condition du théorème soit satisfaite. Alors $c_v = 1$ pour tout v. Mais si π_v est dans la série principale, c_v ne dépend pas en fait de χ et π satisfait à la condition de Flicker [2] et la conclusion s'en suit.

Supposons donc que π satisfasse à la condition de Flicker et soit $\tilde{\pi}$ la représentation de $\tilde{G}$ correspondant à π. De l'identité (11) pour les intégrales des K_ε et de K' on déduit une identité analogue pour les noyaux cuspidaux correspondants. Il existe un caractère automorphe χ de carré trivial et un élément $\tilde{\varphi}$ de $\tilde{\pi}$ tels que l'intégrale

$$\int_{F^\times_{\mathbb{A}}/F^\times \times N(F)\backslash N(F_{\mathbb{A}})} \tilde{\varphi}(\operatorname{diag}(a, a).n)\, \mu^{-1}(a, \psi)\, \chi(a)\, d^\times a\, \theta(n)\, dn$$

ne soit pas nulle. D'après un argument standard, notre identité implique l'existence d'un ε, d'un f_ε et d'un φ dans l'espace de π tels que l'intégrale

$$\int_{T_\varepsilon^0(F)\backslash T_\varepsilon^0(F_{\mathbb{A}})\times G(F_{\mathbb{A}})} \varphi(tx)\,\chi(\lambda(t))\,dt\,f_\varepsilon(x)\,dx$$

ne soit pas nulle. En tenant compte du fait que les composantes de f_ε sont paires, on en déduit que $c_v=1$ pour tout v. D'où le théorème.

Pour un ε donné, l'existence d'un χ satisfaisant les conditions du théorème exprime le fait que π est distinguée par rapport au groupe orthogonal H_ε.

Si les composantes des fonctions f_ε ne sont pas paires, alors l'identité (11) n'est plus vraie. Il faut introduire dans le membre de droite des termes supplémentaires correspondant aux divers caractères de $T(F_{\mathbb{A}})/T^0(F_{\mathbb{A}})$. On n'obtient toutefois pas de résultat nouveau.

Enfin, pour $GL(m)$, on peut définir la notion d'orbite pertinente pour l'action de N sur S et pour l'action de $N\times N$ sur G (*cf.* [4]). Les deux types d'orbites admettent un système commun de représentants. Il est donc naturel de conjecturer que la formule (11) et le théorème se généralisent de manière convenable à $GL(m)$ (*cf.* [5] et ses références).

Note remise le 17 décembre 1990, acceptée le 7 janvier 1991.

RÉFÉRENCES BIBLIOGRAPHIQUES

[1] H. JACQUET, On the non vanishing of some L-functions, *Proc. Indian Acad. Sci.* (*Math. Sci.*), 97, 1987, p. 117-155.

[2] Y. FLICKER, Automorphic forms on Covering Groups of Gl(2), *Inventiones Math.*, 57, 1980, p. 119-182.

[3] H. JACQUET, Sur un résultat de Waldspurger, *Ann. Scient. Éc. Norm. Sup.*, 19, 1986, p. 185-229.

[4] H. JACQUET et Y. YE, Une remarque sur le changement de base quadratique, *C.R. Acad. Sci. Paris*, 311, série I, 1990, p. 671-676.

[5] Y. FLICKER et A. KAZHDAN, Metaplectic Correspondence, *Publications Mathématiques de l'I.H.E.S.*, n° 64, p. 53-110.

Columbia University, Department of Mathematics, New York, New York 10027, *U.S.A.*

Vol. 70, No. 2 DUKE MATHEMATICAL JOURNAL © May 1993

A TRACE FORMULA FOR SYMMETRIC SPACES

HERVÉ JACQUET, KING F. LAI, AND STEPHEN RALLIS

1. Introduction.

1.1. Let G be reductive group defined over an algebraically closed field F and let H be the fixator of an involution θ of G. Roughly speaking, the space of double cosets $H\backslash G/H$ is parametrized by the conjugacy classes in another group G' (see [R], [KR]).

In more detail, assume G semisimple and simply connected. Let τ: $G \to G$ be the map defined by $\tau(g) = g\theta(g)^{-1}$ and S the image of τ. Clearly, τ is constant on the left cosets of G modulo H and induces an isomorphism σ from G/H to S. Then $\sigma(xgH) = x\sigma(gH)x^{-1}$ for $x \in H$. Thus the double cosets of $H\backslash G/H$ correspond via σ to the adjoint orbits of H on S. The closed H-orbits are those of the semisimple elements contained in S. They can be described as follows. Let A be a maximal θ-stable torus contained in S. All such tori are conjugate under the action of H. Then every closed orbit of H in S intersects A. Furthermore, there is a Chevalley restriction theorem for the situation at hand. Indeed, let $F[S]$ and $F[A]$ be the rings of regular functions on S and A respectively. Let H^0 be the neutral connected component of H and let W_H be the quotient of the normalizer of A in H^0 by the centralizer of A in H^0. Then the restriction map

$$F[S] \to F[A]$$

restricts to an isomorphism

$$F[S]^H \simeq F[A]^{W_H}.$$

Since $F[S]^H$ separates the closed orbits, we can identify the set of closed orbits of H in S to the (maximal ideal) spectrum of the algebra $F[A]^{W_H}$, that is, to the orbits

Received 27 April 1992. Revision received 23 September 1992.
Jacquet partially supported by NSF grant DMS-91-01637.
Rallis partially supported by NSF grant DMS-91-03263.

of W_H in A. Let R be the set of roots of A in G; it is a root system. Let R_r be the corresponding reduced root system.

Now there is a semisimple group G' and a maximal torus A' with Weyl group W' and root system R', and an isomorphism of A' onto A which takes W' to W and R' to R_r. The usual Chevalley restriction theorem implies that the set of closed G'-conjugacy classes in G' can be identified with the spectrum of

$$F[A']^{W'},$$

that is, with the orbits of W' in A'. This gives the required bijection. We note that the isomorphic algebras $F[A]^{W_H}$ and $F[A']^{W'}$ are polynomial algebras, but the group G' need not be simply connected.

This is the general principle of comparison that guides the particular example to be considered below. Of course, over a global or a local field, it must be modified suitably. Assuming for now its validity over a number field F, it is natural to postulate the existence of a trace formula identity of the form

$$\int K(h_1, h_2)\, dh_1\, dh_2 = \int_{G'(F)\backslash G'(F_{\mathbb{A}})^2} K'(g', g')\, dg'.$$

Here K and K' are the cuspidal kernels, or rather the "discrete parts" of the respective trace formulas. They are associated to functions f and f' on G and G', respectively; the above equality is supposed to be true if f and f' have "matching orbital integrals". The only representations which contribute to the left-hand side are those which are distinguished with respect to H, that is, contain a vector ϕ such that the integral

$$\int_{H(F)\backslash H(F_{\mathbb{A}})} \phi(h)\, dh$$

is nonzero. The equality then should characterize distinguished representations as functorial images from representations of G'.

The purpose of this paper is to explore this idea in a simple case. However, one discovers quickly that such a simple scheme does not work. A more correct formula might take the form

$$\int K(h_1, h_2)\theta(h_1)\, dh_1\, dh_2 = \int K'(g', g')\, dg'$$

or

$$\int K(h_1, h_2)\, dh_1\, dh_2 = \int K'(g', g')\theta'(g')\, dg',$$

where θ and θ' are suitable automorphic forms on H and G' respectively, which serve as "weights" in the formula.

We now describe the situation at hand. We consider a quadratic extension E of F; the group G is the group $GL(2, E)$ regarded as an algebraic group defined over F while G' is the group $GL(2, F)$. We denote by Z the center of G (regarded as an algebraic group over F) and by Z' the center of G'. We also denote by $\eta_{E/F}$ or simply η the quadratic character of F attached to E. If $E = F(\sqrt{\tau})$, we also write $\eta = \eta_\tau$. Let S' be the set of invertible hermitian matrices. For every $\varepsilon \in S'$ we denote by H_ε^0 the corresponding unitary group, by H_ε the similitude group, and by λ_ε the similitude ratio

$$ {}^t\bar{h}\varepsilon h = \lambda_\varepsilon(h)\varepsilon . $$

In particular, for $\varepsilon = w$, where

$$ w = \begin{pmatrix} 0 & 1 \\ 1 & 0 \end{pmatrix}, $$

we will write simply H^0, H, and λ. It will be convenient to introduce the set

$$ S = S'w . $$

Thus the group $GL(2, E)$ operates on S by

$$ s \mapsto w{}^t\bar{g}wsg . $$

It follows that H operates on S as follows:

$$ s \mapsto h^{-1}sh\lambda(h) . $$

It will be convenient to define a modified action of H on S:

$$ s \mapsto h^{-1}sh . $$

We now describe a natural map from the set of conjugacy classes of H on S (i.e. the orbits for the modified action) to the set of the conjugacy classes of $GL(2, F)$. We let S_0 be the intersection of $GL(2, F)$ with S; this is the space of matrices in $GL(2, F)$ which are symmetric with respect to the second diagonal. We first recall that all conjugacy classes in $GL(2, F)$ intersect S_0. More precisely, a semisimple regular element is always conjugate to a matrix in S_0 of the form

$$ \begin{pmatrix} \alpha & \beta_2 \\ \beta_1 & \alpha \end{pmatrix}, $$

with $\beta_1\beta_2 \neq 0$. Moreover, the conjugacy class of such an element is determined by α and the product $\beta_1\beta_2$. A unipotent element is always conjugate to an element of the form

$$\begin{pmatrix} \alpha & \beta \\ 0 & \alpha \end{pmatrix}, \quad \text{or} \begin{pmatrix} \alpha & 0 \\ \beta & \alpha \end{pmatrix},$$

with $\beta \neq 0$. Moreover, the class is determined by α. Finally, a central element is always in S_0. In the case at hand, the comparison principle takes the following form:

LEMMA 1.1. *Every conjugacy class of H in S intersects S_0. Moreover, two elements s_1 and s_2 in S_0 are conjugate under H if and only if they are conjugate under G'.*

Proof. We leave the elementary proof to the reader. □

1.2. Now we outline the two trace formula identities we have in mind. The action of G on S is not transitive. To compensate for this fact, we consider a set of representatives $\{\varepsilon\}$ for the orbits of $G = GL(2, E)$ on S' (or S); the fixator of ε in G is then the unitary group H_ε^0. For each ε we consider the cuspidal kernel K_ε attached to a function f_ε on G and the trivial central character. On the other hand, the kernel K' is the cuspidal kernel attached to a function f' on G' for the central character η, except that the contribution of the representations dihedral with respect to E has been removed. The functions f_ε and f' are products of local functions which have "matching orbital integrals". At almost all places, the local functions are Hecke functions corresponding to one another by the base-change homomorphism. As usual, part of our task will be to prove the fundamental lemma, i.e. that such Hecke functions indeed have matching orbital integrals.

Recall that η is the quadratic character attached to the extension. Recall an automorphic cuspidal representation π of G is H-distinguished if it contains a vector ϕ such that the integral

$$\int_{H(F)Z(F_\mathbb{A})\backslash H(F_\mathbb{A})} \phi(h)\, dh$$

is nonzero. Implicit in this definition is the fact that the central character of the representation π is trivial. Similarly, we say that a representation π is η-distinguished if the integral

$$\int \phi(h)\eta(\lambda(h))\, dh$$

is nonzero for some ϕ in the space of π. Similar definitions apply to the groups H_ε. If a representation is distinguished (resp. η-distinguished) for a group H_ε, it is also distinguished (resp. η-distinguished) with respect to H. A proof is outlined in [JY1].

It is known ([HLR]) that an automorphic representation π of G with trivial central character is distinguished (resp. η-distinguished) with respect to H if and only if it is the base change of a representation of G' with central character η (resp.

trivial). In particular, π cannot be simultaneously distinguished and η-distinguished. Thus the double integral

$$\int\int K_\varepsilon(h_\varepsilon, h)\eta(\lambda(h))\, dh_\varepsilon\, dh$$

is zero. Here the integrals are over the quotients

$$H_\varepsilon(F)Z(F_{\mathbb{A}})\backslash H_\varepsilon(F_{\mathbb{A}}),\ H(F)Z(F_{\mathbb{A}})\backslash H(F_{\mathbb{A}}).$$

In fact, this can be derived from our trace-formula identity. Let $E(h, s)$ be an Eisenstein series on H. We normalize it so that it has a pole at $s = 1$ with constant residue 1. It follows that the sum

$$\sum_\varepsilon \int\int K_\varepsilon(h_\varepsilon, h)E(h, s)\eta(\lambda(h))\, dh_\varepsilon\, dh$$

is actually holomorphic at 1. Its value at $s = 1$ has the form

$$\sum_\varepsilon \int K_\varepsilon(h_\varepsilon, h)\theta(h)\eta(\lambda(h))\, dh_\varepsilon\, dh$$

where θ is defined by

$$E(h, s) = \frac{1}{s-1} + \theta(h) + O(s-1).$$

Our first trace formula is an equality of the form

$$\sum_\varepsilon \int K_\varepsilon(h_\varepsilon, h)\theta(h)\eta(\lambda(h))\, dh_\varepsilon\, dh = \int K'(g', g')\, dg'.$$

We also consider an Eisenstein series $E'(g, s)$ on G' with pole at $s = 1$. Recall (see [J1]) that for an automorphic representation π', and for ϕ' in the space of π', the integral

$$\int \phi'(g)\overline{\phi'}(g)E'(g, s)\eta(\det(g))\, dg$$

is an holomorphic multiple of the L-function

$$L(s, \pi' \otimes \eta \times \tilde{\pi}').$$

In particular, if π' is not a dihedral representation for the extension E, this L-function is holomorphic at $s = 1$. The integral

$$\int K'(g, g)\eta(\det g)E'(g, s)\,dg$$

is thus holomorphic at $s = 1$. Its value has the form

$$\int K'(g, g)\theta'(g)\,dg$$

where θ' is defined in the same way as θ. Our second trace-formula identity has the form

$$\sum_{\varepsilon} \int K_\varepsilon(h_\varepsilon, h)\,dh_\varepsilon\,dh = \int K'(g', g')\theta'(g')\,dg'.$$

We timorously conjecture that similar identities are true for any symmetric space.

1.3. The case at hand has a special added feature. Indeed, we will derive our identities by taking residues in identities valid for all values of a complex parameter s. They read

$$|D_E|^{1/2}|D_F|^{-1}L(s, 1_F)\sum_{\varepsilon} \int K(h_\varepsilon, h)\,dh_\varepsilon E(h, s)\eta(\lambda(h))\,dh$$
$$= L(s, \eta)\int K'(g', g')E'(g', s)\,dg'$$

and

$$L(s, \eta)\sum_{\varepsilon} \int K(h_\varepsilon, h)\,dh_\varepsilon E(h, s)\,dh$$
$$= |D_E|^{1/2}|D_F|^{-1}L(s, 1_F)\int K'(g', g')E'(g', s)\eta(\deg g')\,dg'.$$

Here D_E and D_F are the respective discriminants. The integrals converge absolutely for $\Re s > 1$ and have meromorphic continuation. The kernels K_ε and K' correspond to functions with "matching orbital integrals", the matching being uniform in s. The precise form of our identities will be given in the text.

Let us remark that the right-hand sides of our identities are the kind of integrals considered in [JZ]. We would like to think of them as the Zagier trace formula. The left-hand sides are thus a "relative version" of the Zagier trace formula. The above identity equality involving the Zagier trace formula and the relative version of the Zagier trace formula is directly related to a similar equality involving the Rankin-Selberg L-function for $GL(2)$ and the L-function studied by Asai ([A]). In turn, this relation is the key to the characterization of the distinguished representations given in [HLR]. We do not expect to have a similar identity—valid for all s—for other symmetric spaces.

Another approach to the study of distinguished representations is an identity of the form

$$\sum_{\varepsilon} \int K(h_\varepsilon, n)\, dh_\varepsilon \theta(n)\, dn = \int K'(n_1', n_2')\theta'(n_1' n_2')\, dn_1'\, dn_2'.$$

Here N and N' are the groups of upper triangular matrices with unit diagonals in G and G' respectively; θ and θ' are characters of $N(F_{\mathbb{A}})$ and $N'(F_{\mathbb{A}})$ trivial on the rational points, and the integrals are over

$$N(F)\backslash N(F_{\mathbb{A}})^2,\ N'(F)\backslash N'(F_{\mathbb{A}})^2$$

respectively. See [JY1] for more details. In principle, this formula can be obtained from our second identity by taking the limit as s tends to infinity. This identity was discovered first and should generalize to symmetric spaces where G is quasi-split. It provides information which is different from the information provided by the formula considered here. For more information on this approach, see [F], [I1], [I2], [J2], [J3], [JR], [M], [Y], [JY1], [JY2], and [Z]. The present approach seems to be more geometric and to introduce the weight factors θ which appear as an essential ingredient of the situation.

We also mention another class of problems for symmetric spaces. Let S be the symmetric space of $s \in G$ such that $s\bar{s} = 1$. Now let G_1 be an inner form of $GL(2, E)$ which is defined over F. Consider the symmetric space S_1 of $s_1 \in G_1$ such that $s_1 \bar{s}_1 = 1$, where $\overline{s_1}$ means the conjugate of s_1 for the Galois action which defines G_1 over F. We may think of S_1 as being an inner form of S and compare distinguished representations for G and G_1. This was done in [JL] under restrictive assumptions necessitated by the lack of local information. The purpose of Hakim's thesis was to provide the missing local information ([Ha]). The global application follows easily. It is described in reference [FH] which is largely "based on Hakim's thesis". One expects comparison results between a symmetric space and its inner forms to be valid under much more general assumptions (see [F] for the case of $GL(n)$).

1.4. The material is arranged as follows. We begin with the first trace formula. We prove the fundamental lemma for the unit element of the Hecke algebra in Section 2, and for the general element in Section 3. We establish that we can match arbitrary functions in Section 4; in Section 5 we prove our trace formula under some restrictive assumptions. Finally, in Section 6 we indicate how to modify our arguments to obtain the second trace formula. For simplicity, we prove our formula under restrictive assumptions specified in the text.

In the present work, we consider only representations of G with trivial central character. One should, more generally, consider representations of G with a central character which is the base change of a character of F. Moreover, in the present work, we have singled out a unitary group H, namely the split one. One should also investigate an identity where one unitary group other than the split one is singled out. We intend to come back to these questions in a subsequent work.

2. The fundamental lemma: unit element.

2.1. We consider an unramified quadratic extension E/F of local fields with odd residual characteristic. We let $\eta_{E/F}$ or η be the quadratic character of F attached to E. We denote by R_F (respectively R_E) the ring of integers of F (resp. E), by P_F (respectively P_E) the maximal ideal in that ring. We denote by q_F or simply q the cardinality of R_F/P_F. Thus R_E/P_E has q^2 elements. We denote by α_F the module function on F.

We let P be the group of upper triangular matrices in $GL(2, E)$, regarded as a group defined over F, and let N be its unipotent radical. We set

$$P' = P \cap G', \quad N' = N \cap G', \quad P_H = P \cap H, \quad N_H = N \cap H,$$

$$K = GL(2, R_E), \quad K' = GL(2, R_F),$$

and we let $f(h, s)$ be the function on H defined by

$$f(h, s) = |a|^s, \qquad \text{if } h = \begin{pmatrix} 1 & x \\ 0 & 1 \end{pmatrix} z \begin{pmatrix} a & 0 \\ 0 & 1 \end{pmatrix} k$$

with $k \in K \cap H$, $a \in F^\times$, $x + \bar{x} = 0$ and $z \in Z$. We let Φ be a smooth function of compact support on S. For σ in S_0 but not in P', we set

$$\Omega(\Phi, \sigma, s) = \int \Phi(h^{-1} z \sigma h) \eta(\lambda(h)) f(h, s) \, dh d^\times z. \tag{1}$$

The integral in h is over $Z \backslash H$, the measure to be specified later, the integral in z is over $Z' \simeq F^\times$. It will be proved below that the integral converges absolutely for $\Re s > 0$.

On the other hand, let Φ' be a smooth function of compact support on G'. We set

$$\Omega'(\Phi', \sigma, s) = \int \Phi'(g'^{-1} z \sigma g') f'(g', s) \, dg' \eta(z) \, d^\times z, \tag{2}$$

where $f'(g', s)$ is the function on $GL(2, F)$ defined in the same way as f. The integral is over G'/Z' and $Z' \simeq F^\times$. It converges for $\Re s > 0$.

We shall say Φ and Φ' have *matching orbital integrals* if

$$\Omega(\Phi, \sigma, s) L(s, 1_F) = \eta(\beta_1) \Omega'(\Phi', \sigma, s) L(s, \eta) \tag{3}$$

for all s with $\Re s > 0$ and all σ of the form

$$\sigma = \begin{pmatrix} \alpha & \beta_2 \\ \beta_1 & \alpha \end{pmatrix},$$

with $\beta_1 \beta_2 \neq 0$. The measures are chosen as follows. We choose a τ_0 such that $E = F(\sqrt{\tau_0})$. Then conjugacy by

$$\begin{pmatrix} \sqrt{\tau_0} & 0 \\ 0 & 1 \end{pmatrix}$$

transforms H into the group $GL(2, F)Z$, and we use this map to transport the Haar measure on H/Z to

$$GL(2, F)Z/Z \simeq GL(2, F)/Z'. \tag{4}$$

Thus the measure on H/Z depends only on the measure on G'/Z'.

To prove the convergence of the integrals we proceed as follows. Suppose that $\Re s > 0$. Then we can write

$$L(2s, 1_F)f'(g, s) = \int \Phi_0[(0, x)g]|x|^{2s}\, d^\times x |\deg g|^s,$$

where Φ_0 is the characteristic function of R_F^2. Let σ be an element of S_0 such that $\beta_1 \beta_2 \neq 0$. We denote by T' the torus of G' which centralizes σ. It can be viewed as the multiplicative group of a commutative semisimple algebra L'. Then

$$L(2s, 1)\Omega'(\Phi', \sigma, s) = \int_{T' \backslash G'} \mu(g')\, dg' \int \Phi_0[(0, 1)t'g']|\det t'|^s\, dt'$$

where

$$\mu(g') = \int \Phi'(g'^{-1} z\sigma g')\eta(z)\, d^\times z.$$

The inner integral is actually a Tate integral for the algebra L'. Thus it converges absolutely for $\Re s > 0$. On the other hand, μ is, for fixed σ, a smooth function of compact support on $T' \backslash G'$. This implies that Ω' converges for $\Re s > 0$. Our argument also shows that, for each s, $\Omega'(\Phi', \sigma, s)$ is a smooth function on the set of σ such that $\beta_1 \beta_2 \neq 0$. We can use the isogeny (4) to transport the integral Ω to the group G' and obtain an integral whose convergence is established in a similar way.

A formal computation shows that relation (3) is satisfied for all $(\alpha, \beta_1, \beta_2)$ with $\beta_1 \beta_2 \neq 0$ if it is satisfied for $\alpha \in \{0, 1\}$, $\beta_1 = 1$, and $\beta_2 \neq 0$. We will write $\Omega(\Phi, \beta, s)$

and $\Omega'(\Phi', \beta, s)$ for the integrals when

$$\sigma = \begin{pmatrix} 1 & \beta \\ 1 & 1 \end{pmatrix}.$$

Similarly, we will write $\Omega^\infty(\Phi, \beta, s)$ and $\Omega'^\infty(\Phi', \beta, s)$ for the integrals when

$$\sigma = \begin{pmatrix} 0 & \beta \\ 1 & 0 \end{pmatrix}.$$

Let

$$\sigma = \begin{pmatrix} \alpha & \beta \\ 1 & \alpha \end{pmatrix}.$$

Then

$$\Omega(\Phi, \sigma, s) = |\alpha|^s \eta(\alpha) \Omega(\Phi, \beta\alpha^{-2}, s).$$

If we fix β and let α tend to 0, we obtain, for $|\alpha|$ small enough,

$$\Omega^\infty(\Phi, \beta, s) = |\alpha|^s \eta(\alpha) \Omega(\Phi, \beta\alpha^{-2}, s).$$

A similar relation is true for Ω'. Thus we see that the relation

$$\Omega(\Phi, \beta, s) L(s, 1_F) = \Omega'(\Phi', \beta, s) L(s, \eta) \tag{5}$$

for all β implies that the functions have matching orbital integrals.

PROPOSITION 2.1. *Assume the extension is unramified of odd residual characteristic. Let Φ_0 be the characteristic function of $K \cap S$ and let Φ'_0 be the characteristic function of K'. Then Φ_0 and Φ'_0 have matching orbital integrals.*

2.2. The proof of the proposition is based on the computation of an integral that we now introduce:

$$A(t, s) = \int_{|x| \leqslant 1} \frac{1 - |t - x^2|^{s-1} q^{-s+1}}{1 - q^{-s+1}} \, dx. \tag{6}$$

Here the volume of R_F is one. We let q be the cardinality of the residual field of F and write $A(t, X)$ for $A(t, s)$ if $X = q^{-s}$. We let τ_0 be any unit of F which is not a square in E, and we let ϖ be a uniformizer.

LEMMA 2.2. *Let t be an integer. Then $A(t, s)$ converges absolutely for $\Re s > 0$. It is given by the formulas, where $X = q^{-s}$, $|y| \leqslant 1$,*

$$A(t, s) = \begin{cases} \dfrac{1+X}{1-X^2q} + |y|^{2s-1}\dfrac{X(1+X)(1-Xq)}{(1-X^2q)(1-X)} & \text{if } t = y^2 \\[2ex] \dfrac{1+X}{1-X^2q} - \dfrac{|y|^{2s-1}X(1+Xq)}{1-X^2q} & \text{if } t = y^2\tau_0 \\[2ex] \dfrac{1+X}{1-X^2q} - \dfrac{|y|^{2s-1}X^2q(1+X)}{1-X^2q} & \text{if } t = y^2\varpi. \end{cases}$$

We leave the computation to the reader. □

The following lemma is then easily verified.

LEMMA 2.3. *If τ_0 is any unit which is not a square, then*

$$\frac{A(y^2, -X)}{1-X} = \frac{A(y^2\tau_0, X)}{1+X},$$

$$\frac{A(y^2, X)}{1+X} = \frac{A(y^2\tau_0, -X)}{1-X}.$$

If ϖ and ϖ' are two uniformizers, then

$$\frac{A(y^2\varpi, -X)}{1-X} = \frac{A(y^2\varpi', X)}{1+X}.$$

2.3. We now compute the orbital integrals for

$$\sigma = \begin{pmatrix} 1 & \beta \\ 1 & 1 \end{pmatrix}$$

where β is nonzero. We compute first Ω. We set

$$h = \begin{pmatrix} 1 & x \\ 0 & 1 \end{pmatrix}\begin{pmatrix} t & 0 \\ 0 & 1 \end{pmatrix}k.$$

Then

$$dh = du|t|^{-1}\, d^{\times}t\, dk$$

if $x = \sqrt{\tau_0}u$ and τ_0 is a unit which is not a square. The measure of R_F for du is one, and the measure of $R^\times$ for $d^\times t$ is also one. In passing, we remark that our choice of measures suggests writing $x = u/\sqrt{\tau_0}$ instead. Since τ_0 is a unit, the measure dx is the same. However, when E is ramified, we will take τ_0 to be a uniformizer and will write $x = u/\sqrt{\tau_0}$. Now

$$f(h, s) = |t|^s.$$

We get

$$\Omega(\Phi_0, \beta, s) = \int \Phi_0\left[\begin{pmatrix} z(1-x) & t^{-1}z(\beta - x^2) \\ zt & z(1+x) \end{pmatrix}\right] |t|^{s-1}\eta(t)\, d^\times t\, du\, d^\times z.$$

The integral is zero unless $1 - \beta$ is a norm. Assuming this to be the case, we set

$$|1 - \beta| = |\gamma|^2.$$

The support of the integrand is defined by the conditions

$$|(1-\beta)z^2| = 1, \quad |z| \leqslant 1, \quad |zx| \leqslant 1,$$

$$|z(\beta - x^2)| \leqslant |t| \leqslant |z|^{-1}.$$

Integrating with respect to t first, we get

$$\int \frac{\eta(z)|z|^{1-s} + |z|^{s-1}\eta(z)|\beta - x^2|^{s-1}\eta(\beta - x^2)q^{-s+1}}{1 + q^{-s+1}}\, du\, d^\times z.$$

Integrating over z, we get

$$\eta(\gamma)\int_{|u| \leqslant |\gamma|} \frac{|\gamma|^{s-1} + |\gamma|^{1-s}|\beta - x^2|^{s-1}\eta(\beta - x^2)q^{-s+1}}{1 + q^{-s+1}}\, du.$$

Recall $x = \sqrt{\tau_0}u$. We now change u to $u\gamma$. We arrive at the expression

$$|\gamma|^s\eta(\gamma)\int_{|u| \leqslant 1} \frac{1 + \eta\left(\dfrac{\beta}{\gamma^2} - x^2\right)\left|\dfrac{\beta}{\gamma^2} - x^2\right|^{s-1} q^{-s+1}}{1 + q^{-s+1}}\, du.$$

Thus

$$\Omega(\Phi_0, \sigma, s) = |\gamma|^s\eta(\gamma)\int_{|u| \leqslant 1} \frac{1 + \eta(t - u^2)|t - u^2|^{s-1}q^{-s+1}}{1 + q^{-s+1}}\, du,$$

where

$$t = \frac{\beta}{\gamma^2 \tau_0}.$$

We remark that t is always an integer. Finally,

$$\Omega(\Phi_0, \beta, s) = |\gamma|^s \eta(\gamma) A\left(\frac{\beta}{\gamma^2 \tau_0}, -X\right). \tag{7}$$

2.4. We pass to the computation of the integral Ω'. We set

$$g = \begin{pmatrix} 1 & x \\ 0 & 1 \end{pmatrix} \begin{pmatrix} t & 0 \\ 0 & 1 \end{pmatrix} k.$$

Then

$$dg = dx|t|^{-1}\, d^\times t\, dk, \qquad f'(g, s) = |t|^s.$$

As before, we find

$$\Omega'(\Phi'_0, \beta, s) = \int \Phi'_0\left[\begin{pmatrix} z(1-x) & t^{-1}z(\beta - x^2) \\ zt & z(1+x) \end{pmatrix}\right] |t|^{s-1}\, d^\times t\, dx \eta(z)\, d^\times z.$$

However, this time x is in F. As before, the integral is zero unless

$$|1 - \beta| = |\gamma|^2,$$

and the support of the integrand is defined by

$$|(1-\beta)z^2| = 1, \quad |z| \leqslant 1, \quad |zx| \leqslant 1,$$

$$|z(\beta - x^2)| \leqslant |t| \leqslant |z|^{-1}.$$

Integrating with respect to t first, we get

$$\int \frac{|z|^{-s+1} - |z|^{s-1}|\beta - x^2|^{s-1} q^{-s+1}}{1 - q^{-s+1}}\, dx \eta(z)\, d^\times z.$$

Integrating over z, we get

$$\Omega'(\Phi'_0, \sigma, s) = \eta(\gamma) \int \frac{|\gamma|^{s-1} - |\gamma|^{1-s}|\beta - x^2|^{s-1} q^{-s+1}}{1 - q^{-s+1}}\, dx,$$

the integral over $|x| \leqslant |\gamma|$. If we change x to $x\gamma$, we arrive at

$$|\gamma|^s\eta(\gamma) \int_{|x|\leqslant 1} \frac{1 - |t - x^2|^{s-1} q^{-s+1}}{1 - q^{-s+1}}\, dx,$$

with $t = \beta/\gamma^2$, that is,

$$\Omega'(\Phi_0', \beta, s) = |\gamma|^s\eta(\gamma) A\left(\frac{\beta}{\gamma^2}, X\right). \tag{8}$$

Comparing with Lemma (2.3), we get equation (5) and the proposition. □

3. The fundamental lemma: general element.

3.1. We keep the notations of the previous section. Let f be a bi-K-invariant on G of compact support and f' be a bi-K-invariant function on G' with compact support. We shall say that f' *corresponds* to f if the following condition is satisfied. For $x = q^{-s}$ set $X = q^{-2s}$; let Π be the unramified component of the representation of G induced by $(\alpha_E^s, \alpha_E^{-s})$ and let π be the unramified component of the representation of G' induced by $(\alpha_F^s, \eta\alpha_F^{-s})$. Thus Π is the base change of π, and for any x,

$$\operatorname{Tr} \Pi(f) = \operatorname{Tr} \pi(f'). \tag{9}$$

We note that this relation makes sense if f is invariant under the center of G and f' transforms under the character η of the center of G'. If it is satisfied, we shall also say that f' corresponds to f.

Let Φ_0 be the characteristic function of $K \cap S'$. We define a function Φ_f on S' by

$$\Phi_f(s) = \int_{H^0} f(hg)\, dh$$

if s is of the form $s = {}^t\bar{g}wg$, and

$$\Phi_f(s) = 0$$

if s is not in the orbit of w. We assume that the measure of

$$H^0 \cap K$$

is one. If f_0 is the characteristic function of K, then $\Phi_{f_0} = \Phi_0$. In general, since $f = f_0 * f$, we can write

$$\Phi_f(s) = \int \Phi_0({}^t\bar{g}sg) f(g^{-1})\, dg.$$

The bijection $s \mapsto sw$ allows us to identify S' to S and to regard a function Φ on S' as a function on S that we denote also by Φ. We want to prove the following statement.

PROPOSITION 3.1. *If f' corresponds to f, then the functions Φ_f and f' have matching orbital integrals.*

3.2. We need some preliminaries. For almost all X, the representation Π introduced above is H-distinguished in the sense that there is a nonzero linear μ form on the space of Π such that

$$\mu(\Pi(h)v) = \mu(v),$$

for every vector v in the space of Π and $h \in H$. Let v_0 be a K-fixed vector. Then the function ω defined by

$$\omega(g) = \mu(\Pi(g)v_0)$$

is H-invariant on the left and an eigenfunction of the Hecke operators on the right. In particular, let f_1 be the characteristic function of

$$K\begin{pmatrix} \varpi & 0 \\ 0 & 1 \end{pmatrix}K,$$

where ϖ is a uniformizer. Then

$$\int \omega(gy)f_1(y)\,dy = (X + X^{-1})q\omega(g).$$

Since ω is invariant under the center of G, this relation implies conversely that ω is an eigenfunction of the Hecke operators. Next, let Ω be the function on S' defined by

$$\Omega(s) = \omega(g)$$

if $s = {}^t\bar{g}wg$ for some g, and

$$\Omega(s) = 0 \tag{10}$$

otherwise. Then

$$\Omega(sz) = \Omega(s) \tag{11}$$

for $z \in F^\times$ because the image of H under the similitude ratio map is $F^\times$. Moreover, for all $k \in K$,

$$\Omega({}^t\bar{k}sk) = \Omega(s) \tag{12}$$

and

$$\int \Omega({}^t\bar{g}sg)f_1(g)\,dg = (X + X^{-1})q\Omega(s). \tag{13}$$

We will show that for all X there is a function Ω satisfying conditions (10)–(13), unique within a constant factor; moreover $\Omega(e) \neq 0$.

The theory of elementary divisors shows that a set of representatives for the orbits of K on S' is the set of matrices of the form

$$\begin{pmatrix} \varpi^i & 0 \\ 0 & \varpi^j \end{pmatrix}$$

with $i \geqslant j$. Moreover, such a matrix is in the orbit of w if and only if its determinant is minus a norm, that is, $i + j$ is even. In particular, the function Ω is determined by its values on the above matrices with $i = 2n$, $j = 0$. We denote these values by Ω_n. We denote by P_E the maximal ideal in R_E. Then the above equation is equivalent to the difference equation

$$\Omega_1 + (q^2 - q - 1)\Omega_1 + (q + 1)\Omega_0 = (X + X^{-1})q\Omega_0,$$

$$\Omega_{n-1} + q^2\Omega_{n+1} = (X + X^{-1})q\Omega_n, \qquad \text{if } n > 0.$$

Clearly, these conditions determine the scalars Ω_n uniquely in terms of Ω_0. In particular, we can take $\Omega_0 = 1$. This shows the existence and uniqueness of Ω. We will denote by Ω_X the function we just defined. Solving the difference equation, we find

$$\Omega_n = \frac{qX - 1}{(X + 1)(q - 1)}(Xq^{-1})^n + \frac{qX^{-1} - 1}{(X^{-1} + 1)(q - 1)}(X^{-1}q^{-1})^n. \tag{14}$$

To continue, there is an invariant measure on the orbit of w under G. This measure can be computed as follows

$$\int \Phi(s)\,ds = \sum_{i \geqslant j} \int \Phi\left({}^t\bar{k}\begin{pmatrix} \varpi^i & 0 \\ 0 & \varpi^j \end{pmatrix}k\right) dk\,\gamma(\varpi^{i-j}).$$

To compute γ we write that

$$\int \Phi({}^t\bar{g}sg)f_1(g)\,dg\,ds = \int f_1(g)\,dg \int \Phi(s)\,ds.$$

We take the measure of $K \cap S$ to be 1 so that $\gamma(1) = 1$. Applying the previous relation to the characteristic function of an orbit of K, we find a recurrence relation for $\gamma(\varpi^{2m})$, and then

$$\gamma(1) = 1, \quad \gamma(\varpi^{2m}) = q^{2m}(1 - q^{-1}) \qquad \text{for } m > 0. \tag{15}$$

We also denote by ω_x the spherical function attached to the representation π and by ω_n its value on the matrix

$$\begin{pmatrix} \varpi^n & 0 \\ 0 & 1 \end{pmatrix}.$$

Recall this is

$$\omega_n = \frac{qx^2 + 1}{(q+1)(x^2+1)}(xq^{-1/2})^n + \frac{qx^{-2} + 1}{(q+1)(x^{-2}+1)}(-x^{-1}q^{-1/2})^n. \tag{16}$$

Of course, this can be proved as before by solving a difference equation. We also recall that the Haar measure on G' can be computed like so:

$$\int f'(g')\,dg' = \sum_{i \geqslant j} \iint f'\left(k_1 \begin{pmatrix} \varpi^i & 0 \\ 0 & \varpi^j \end{pmatrix} k_2\right) dk_1\, dk_2 \gamma'(\varpi^{i-j}),$$

where γ' is defined by

$$\gamma'(1) = 1, \quad \gamma'(\varpi^m) = q^m(1 + q^{-1}) \qquad \text{for } m > 0.$$

Recall that $X = x^2$. It is now a simple matter to verify the following relations:

LEMMA 3.2. *Notation being as above, for any $n \geqslant 1$,*

$$\Omega_n \gamma(\varpi^{2n}) + \Omega_{n-1}\gamma(\varpi^{2(n-1)}) = \omega_{2n}\gamma'(\varpi^{2n}) - \omega_{2(n-1)}\gamma'(\varpi^{2(n-1)}),$$

$$\sum_{0 \leqslant i \leqslant n} \Omega_{n-i}\gamma(\varpi^{2(n-i)}) = \sum_{0 \leqslant i \leqslant n} \omega_{2(n-i)}\gamma'(\varpi^{2(n-i)})(-1)^i.$$

3.3. Now suppose that f and f' correspond to one another. Then equality (9) gives

$$\int \Omega_X({}^t\bar{g}wg) f(g)\,dg = \int \omega_x(g') f'(g')\,dg'.$$

In terms of the function $\Phi = \Phi_f$ on S' associated to f, this reads

$$\int \Omega_X(s)\Phi(s)\,ds = \int \omega_x(g') f'(g')\,dg'.$$

Conversely, if this relation is satisfied, then f' is associated to f. If Φ is a K-invariant function of compact support on S' with support in the orbit of w and f' is a Hecke function on G', we shall say that f' *corresponds* to Φ if this relation is satisfied for all x. To prove the proposition it will suffice to show that, given Φ and a corresponding f', then Φ and f' have matching orbital integrals.

Now the previous lemma and the integration formulas imply the following lemma:

LEMMA 3.3. *Let Φ_n be the characteristic function of the set of matrices s in S' with integral entries and $|\det s| = q^{-2n}$. Let Φ'_n be similarly the characteristic function of the set of matrices g' in G' with integral entries and $|\deg g'| = q^{-n}$. Then Φ_n corresponds to Φ'_{2n}.*

Proof. Indeed, the equality to be proved reads

$$\int \Phi_n(s)\Omega_X(s)\,ds = \int \Phi'_{2n}(g')\omega_x(g')\,dg'$$

and follows at once from the previous lemma and the formulas for γ and γ'.

Now the translates of Φ_n under the center form a basis of the space of K-invariant functions on S' (or S) with support in the set of matrices with determinant of even valuation. We see that the proposition will be completely proved if we prove the following lemma:

LEMMA 3.4. *For $n \geqslant 0$ the functions Φ_n and Φ'_{2n} have matching orbital integrals.*

Proof. We regard Φ_n as a function on S. We first compute the orbital integral of Φ_n with $n \geqslant 0$. We proceed as before: the integral is equal to

$$\Omega(\Phi_n, \beta, s) = \int |t|^{s-1}\eta(t)\,d^\times t\,d^\times z\,du$$

where $x = u\sqrt{\tau_0}$ and the integral is over the domain

$$|(1-\beta)z^2| = q^{-2n},$$
$$|z| \leqslant 1, \quad |zx| \leqslant 1,$$
$$|z(\beta - x^2)| \leqslant |t| \leqslant |z|^{-1}.$$

The integral is zero unless

$$|1-\beta| = |\gamma|^2, \quad |\gamma|q^n \geqslant 1.$$

Assuming this condition is satisfied, we find

$$\Omega(\Phi_n, \beta, s) = |\gamma|^s q^{ns}\eta(\gamma)(-1)^n A\left(\frac{\beta\varpi^{2n}}{\gamma^2\tau_0}, -X\right),$$

where A was defined in the previous section.

We pass to the computation of $\Omega'(\Phi'_{2n}, \beta, s)$. We proceed as before: the integral is equal to

$$\Omega'(\Phi'_{2n}, \beta, s) = \int |t|^{s-1}\eta(z)\, d^\times t\, d^\times z\, dx$$

taken over

$$|(1-\beta)z^2| = q^{-2n},$$

$$|z| \leqslant 1, \quad |zx| \leqslant 1,$$

$$|z(\beta - x^2)| \leqslant |t| \leqslant |z|^{-1}.$$

We see the integral is zero unless

$$|1-\beta| = |\gamma|^2, \quad |\gamma| q^n \geqslant 1.$$

Assuming this condition is satisfied, we find

$$\Omega'(\Phi'_{2n}, \beta, s) = |\gamma|^s q^{ns}\eta(\gamma)(-1)^n A\left(\frac{\beta\varpi^{2n}}{\gamma^2}, X\right).$$

Comparing the above explicit formulas and the functional equation of the function A, we see that our functions have indeed matching orbital integrals. This concludes the proof of the lemma and the proposition. □

4. Matching arbitrary functions.

4.1. Now we let E be a quadratic extension of F, a nonarchimedean field of odd residual characteristic. Thus E is an unramified or tamely ramified extension of F. We define the notion of "matching" for this more general situation. Recall that Z denotes the center of G, a group isomorphic to $E^\times$, and that Z' denotes the center of G', a group isomorphic to $F^\times$. As before, we have an isomorphism

$$H/Z \simeq G'/Z',$$

unique within an inner automorphism, which allows us to transport a measure on G'/Z' to H/Z. We define the function f' on G' as before. To complete our definition, we need to specify the function f in terms of the function f'. To that end we choose τ_0 such that $E = F(\sqrt{\tau_0})$; if E is unramified, we take for τ_0 a unit which is not a square, and if E is (tamely) ramified, we take for τ_0 a uniformizer of F. We let ξ be the isomorphism

$$\xi\colon H \to G'Z$$

defined by

$$\xi(g) = \begin{pmatrix} \sqrt{\tau_0} & 0 \\ 0 & 1 \end{pmatrix} g \begin{pmatrix} \sqrt{\tau_0}^{-1} & 0 \\ 0 & 1 \end{pmatrix}.$$

We may view f' as a function on $G'Z$ invariant under Z. Thus we may set

$$f(g) = f'(\xi(g)).$$

Let U be the intersection $Z \cap GL(2, R_E)$. Thus U is isomorphic to the group of units of E and $K'U$ is a maximal compact subgroup of $G'Z$. We set $K_H = \xi^{-1}(K'U)$. Hence K_H is a maximal compact subgroup of H. Note that f and K_H do not depend on the choice of τ_0, provided τ_0 is as above. As before, we consider the integrals

$$\Omega(\Phi, \sigma, s) = \int f(h, s)\Phi[h^{-1}z\sigma h]\eta(\lambda(h))\, dh\, d^\times z,$$

$$\Omega'(\Phi', \sigma, s) = \int f'(g', s)\Phi'[g'^{-1}z\sigma g']\, dg'\eta(z)\, d^\times z.$$

As before, the integrals converge absolutely for $\Re s > 0$ and $\sigma \in S_0$

$$\sigma = \begin{pmatrix} \alpha & \beta_1 \\ \beta_2 & \alpha \end{pmatrix}, \qquad \beta_1\beta_2 \neq 0.$$

Moreover, their interpretation as "Tate integrals" show that they continue analytically as meromorphic functions to the whole complex plane.

We shall say that a smooth function of compact support Φ on S and a smooth function of compact support Φ' on G' have *matching orbital integrals* if, for all σ in S_0 of the above form, relation

$$|\tau_0|^{-1/2}\Omega(\Phi, \sigma, s)L(s, 1_F) = \eta(\beta_1)\Omega'(\Phi', \sigma, s)L(s, \eta)$$

holds for all s with $\Re s > 0$. This definition is independent of the choice of τ_0 and coincides with the definition of the previous section when E is unramified.

As before, we will introduce the notations $\Omega(\Phi, \beta, s)$ and $\Omega^\infty(\Phi, \beta, s)$ and the corresponding notations for Ω'.

In this section, our goal is the following proposition:

PROPOSITION 4.1. *Given a smooth function of compact support Φ on S, there is a smooth function of compact support Φ' on G' with matching orbital integrals. Conversely, given Φ' there is Φ with matching orbital integrals.*

To be definite, we prove the existence of Φ' for a given Φ.

4.2. We first study the asympotic behavior of the integrals $\Omega(\Phi, \beta, s)$ when β tends to 0. Explicitly,

$$\Omega(\Phi, \beta, s) = \int \Phi\left[k^{-1}\begin{pmatrix} z(1-x) & zt^{-1}(\beta - x^2) \\ zt & z(1+x) \end{pmatrix} k\right] \eta(\lambda(k))\, dk |t|^{s-1} \eta(t)\, d^\times t\, dx\, d^\times z.$$

Here $x = u/\sqrt{\tau_0}$ and $dx = du$.

We will denote by $I(\Phi, s)$ the integral corresponding to $\beta = 0$. Explicitly,

$$\begin{aligned} I(\Phi, s) &= \int \Phi\left[h^{-1} z \begin{pmatrix} 1 & 0 \\ 1 & 1 \end{pmatrix} h\right] f(h, s) \eta(\lambda(h))\, dh \\ &= \int \Phi\left[k^{-1}\begin{pmatrix} z(1-x) & -zt^{-1}x^2 \\ zt & z(1+x) \end{pmatrix} k\right] \eta(\lambda(k))\, dk |t|^{s-1} \eta(t)\, d^\times t\, dx\, d^\times z. \end{aligned}$$

This integral is thus a unipotent orbital integral. It will be convenient to introduce the scalar part of the intertwining operator:

$$m(s) = \int_{N_H} f(wnk, s)\, dn = \int_{N_H} f(wn, s)\, dn.$$

Since $\lambda(w) = 1$, we can change h to wh in the integral and then use the Iwasawa decomposition to obtain

$$I(\Phi, s) = m(s) \int \Phi\left[k^{-1} z \begin{pmatrix} 1 & t \\ 0 & 1 \end{pmatrix} k\right] \eta(\lambda(k))\, dk |t|^{s} \eta(t)\, d^\times t\, d^\times z.$$

In particular, this shows that the integral converges absolutely for $\Re s > 0$.

We consider also the integral

$$J(\Phi, s) = \int \Phi\left[h^{-1} z \begin{pmatrix} 1 & 1 \\ 0 & 1 \end{pmatrix} h\right] \eta(\lambda(h)) f(h, s)\, dh\, d^\times z,$$

the integral over the quotient $ZN_H \backslash H$. Explicitly,

$$J(\Phi, s) = \int \Phi\left[k^{-1} z \begin{pmatrix} 1 & t \\ 0 & 1 \end{pmatrix} k\right] \eta(\lambda(k))\, dk\, d^\times z |t|^{1-s} \eta(t)\, d^\times t.$$

This integral converges for $\Re s < 1$.

It will be convenient to introduce the following function ϕ on F:

$$\phi(x) = \int \Phi\left[k^{-1} z \begin{pmatrix} 1 & x \\ 0 & 1 \end{pmatrix} k\right] \eta(\lambda(k))\, dk\, d^\times z. \tag{17}$$

Thus ϕ is a Schwartz-Bruhat function on F such that

$$\phi(ux) = \phi(x)\eta(u)$$

for all $u \in F$ with $|u| = 1$. It is easily seen that the function ϕ is an arbitrary Schwartz-Bruhat function with that property.

In terms of ϕ,

$$I(\Phi, s) = m(s) \int \phi(t)|t|^s \eta(t)\, d^\times t, \tag{18}$$

$$J(\Phi, s) = \int \phi(t)|t|^{1-s}\eta(t)\, d^\times t. \tag{19}$$

The first integral converges for $\Re s > 0$ and the second for $\Re s < 1$. These distributions determine the "Shalika germs" of the orbital integrals. Namely, let $A > 0$ a constant and set for β an integer

$$C(\beta, s) = \int_{|x| \leqslant A} (\eta(\beta - x^2)|\beta - x^2|^{s-1} - \eta(-x^2)|x^2|^{s-1})\, dx \tag{20}$$

where

$$x = \frac{u}{\sqrt{\tau_0}}, \qquad u \in F, dx = du.$$

If we change the value of A, the integral does not change provided $|\beta|$ is small enough.

PROPOSITION 4.2. *Given Φ, for $|\beta|$ sufficiently small and all s with* $0 < \Re s < 1$,

$$\Omega(\Phi, \beta, s) = I(\Phi, s) + C(\beta, s)J(\Phi, s).$$

To prove the proposition, we use the following lemma:

LEMMA 4.3. *Let χ be a fixed character of $F^\times$ of module 1. Given a Schwartz-Bruhat function Ψ on $F \times F$, there are two Schwartz-Bruhat functions $\Psi_1(x, u)$ and $\Psi_2(x, u)$ on F, which are Laurent polynomial in q^{-u}, such that*

$$\int \Psi(t, xt^{-1})\chi(t)|t|^u\, d^\times t = L(u, \chi)\Psi_1(x, u) + L(-u, \chi^{-1})\Psi_2(x, u)\chi(x)|x|^u.$$

Moreover,

$$\Psi_1(0, u) = \frac{\int \Psi(t, 0)\chi(t)|t|^u\, d^\times t}{L(u, \chi)},$$

$$\Psi_2(0, u) = \frac{\int \Psi(0, t)\chi(t)^{-1}|t|^{-u}\, d^\times t}{L(-u, \chi^{-1})}.$$

To be more precise, each function $\Psi_i(x, u)$ is a finite sum of the form

$$\sum_j \phi_j(x) q^{-uj}$$

where the functions ϕ_j are Schwartz-Bruhat functions. The original integral converges for all u provided $x \neq 0$, and the other integrals are meromorphic functions of u. We leave the proof of the lemma to the reader.

Next, we remark that, if $|\beta| < 1$, then the corresponding matrix σ has a determinant of absolute value one. It follows that, for a given Φ, there is a compact set of $F^\times$ with characteristic function μ such that

$$\Omega(\Phi, \beta, s) = \int \Phi[h^{-1} z\sigma h] \eta(\lambda(h)) f(h, s) \, dh \mu(z) \, d^\times z.$$

We apply our lemma (or rather a variant of the lemma with parameters) to the function Ψ defined by

$$\Psi(v, x_1, x_2) = \int \Phi\left[k^{-1}\begin{pmatrix} z(1 + \bar{v}) & zx_2 \\ zx_1 & z(1 + v)\end{pmatrix} k\right] \eta(\lambda(k)) \, dk \mu(z) \, d^\times z.$$

It is a Schwartz-Bruhat function on $E \times F \times F$. In terms of Ψ (recall $x = u/\sqrt{\tau_0}$, $dx = du$),

$$\Omega(\Phi, \beta, s) = \int \Psi(x, t, (\beta - x^2)t^{-1}) |t|^{s-1} \eta(t) \, d^\times t \, dx.$$

Thus we find

$$\Omega(\Phi, \beta, s) = L(s - 1, \eta) \int \Psi_1(x, \beta - x^2, s) \, dx$$
$$+ L(1 - s, \eta) \int \Psi_2(x, \beta - x^2, s) |\beta - x^2|^{s-1} \eta(\beta - x^2) \, dx$$

where the functions Ψ_i are as in the lemma; here, they are Schwartz-Bruhat functions on $E \times F$ depending on s. For $|\beta|$ small enough we get,

$$\Omega(\Phi, \beta, s) = L(s - 1, \eta) \int \Psi_1(x, -x^2, s) \, dx$$
$$+ L(1 - s, \eta) \int \Psi_2(x, -x^2, s) |\beta - x^2|^{s-1} \eta(\beta - x^2) \, dx.$$

Similarly, we get

$$I(\Phi, s) = \int \Psi(x, t, -x^2t^{-1})|t|^{s-1}\eta(t)\, d^\times t\, dx$$

$$= L(s-1, \eta) \int \Psi_1(x, -x^2, s)\, dx$$

$$+ L(1-s, \eta) \int \Psi_2(x, -x^2, s)|x^2|^{s-1}\eta(-x^2)\, dx.$$

On the other hand,

$$J(\Phi, s) = \int \Psi(0, 0, t)|t|^{1-s}\eta(t)\, d^\times t.$$

Thus, by the lemma

$$J(\Phi, s) = L(1-s, \eta)\Psi_2(0, 0, s).$$

To obtain our assertion we write our orbital integral in the form

$$L(s-1, \eta) \int \Psi_1(x, -x^2, s)\, dx + L(1-s, \eta) \int \Psi_2(x, -x^2, s)|x^2|^{s-1}\eta(-x^2)\, dx$$

$$+ L(1-s, \eta) \int \Psi_2(x, -x^2, s)(|\beta - x^2|^{s-1}\eta(\beta - x^2) - |x^2|^{s-1}\eta(-x^2))\, dx.$$

The sum of the two first terms is $I(\Phi, s)$. In the third term the bracket expression vanishes if $|x^2|$ is large in comparison with β. Thus, for β small enough, we may replace $\Psi_2(x, -x^2, s)$ by $\Psi_2(0, 0, s)$ in the third term. We get then

$$\Omega(\Phi, \beta, s) = I(\Phi, s) + C(\beta, s)J(\Phi, s).$$

This proves the proposition. □

4.3. We consider similarly the integrals Ω'. Recall the integral

$$\Omega'(\Phi', \beta, s) = \int \Phi'[g^{-1}z\sigma g]f(g, s)\, dg\eta(z)\, d^\times z$$

where

$$\sigma = \begin{pmatrix} 1 & \beta \\ 1 & 1 \end{pmatrix}.$$

Explicitly,

$$\Omega'(\Phi', \beta, s) = \int \Phi'\left[k^{-1}\begin{pmatrix} z(1-x) & zt^{-1}(\beta - x^2) \\ zt & z(1+x) \end{pmatrix} k\right] dk|t|^{s-1}\, d^\times t\, dx\eta(z)\, d^\times z.$$

As before, we consider the orbital integral corresponding to $\beta = 0$:

$$I'(\Phi', s) = \int \Phi'\left[k^{-1}\begin{pmatrix} z(1-x) & -zt^{-1}x^2 \\ zt & z(1+x) \end{pmatrix} k\right] dk|t|^{s-1}\, d^\times t\, dx\eta(z)\, d^\times z.$$

We denote by N' the intersection of G' and N and denote by $m'(s)$ the scalar part of the intertwining operator

$$m'(s) = \int_{N'} f'(wnk, s)\, dn = \int_{N'} f'(wn, s)\, dn.$$

We have then

$$I'(\Phi', s) = m'(s) \int \Phi'\left[k^{-1}z\begin{pmatrix} 1 & t \\ 0 & 1 \end{pmatrix} k\right] dk|t|^{s}\, d^\times t\eta(z)\, d^\times z.$$

Similarly, we define

$$J'(\Phi', s) = \int \Phi'\left[g'^{-1}z\begin{pmatrix} 1 & 1 \\ 0 & 1 \end{pmatrix} g'\right] f'(g', s)\, dg'\eta(z)\, d^\times z,$$

the integral over $Z'N'\backslash G'$. Explicitly,

$$J'(\Phi', s) = \int \Phi'\left[k^{-1}z\begin{pmatrix} 1 & t \\ 0 & 1 \end{pmatrix} k\right] dk\eta(z)\, d^\times z|t|^{1-s}\, d^\times t.$$

As before, it will be convenient to introduce the following function ϕ' on F:

$$\phi'(x) = \int \Phi'\left[k^{-1}z\begin{pmatrix} 1 & x \\ 0 & 1 \end{pmatrix} k\right] dk\eta(z)\, d^\times z.$$

Thus ϕ' is a Schwartz-Bruhat function on F such that

$$\phi'(ux) = \phi'(x)$$

for all $u \in F$ with $|u| = 1$. As before, the function ϕ' is an arbitrary Schwartz-Bruhat function with that property. In terms of ϕ',

$$I'(\Phi', s) = m'(s) \int \phi'(t)|t|^s \, d^\times t, \tag{21}$$

$$J'(\Phi', s) = \int \phi'(t)|t|^{1-s} \, d^\times t. \tag{22}$$

We define

$$C'(\beta, s) = \int_{|x| \leqslant A} (|\beta - x^2|^{s-1} - |x^2|^{s-1}) \, dx.$$

As before, this does not depend on the value on A, provided $|\beta|$ is small enough and we have the following proposition:

PROPOSITION 4.4. *Given Φ', for $|\beta|$ sufficiently small and all s with $\Re s > 0$,*

$$\Omega'(\Phi', \beta, s) = I'(\Phi', s) + C'(\beta, s)J'(\Phi', s).$$

The proof is the same as before.

4.4. The first step in matching the orbital integrals is the following lemma:

LEMMA 4.5. *Given Φ, there is Φ' such that for all s and all β sufficiently small,*

$$|\tau_0|^{-1/2}\Omega(\Phi, \beta, s)L(s, 1_F) = \Omega'(\Phi', \beta, s)L(s, \eta).$$

Proof. It suffices to show that given Φ there is Φ' with

$$|\tau_0|^{-1/2}I(\Phi, s)L(s, 1_F) = I'(\Phi', s)L(s, \eta), \tag{23}$$

$$|\tau_0|^{-1/2}C(\beta, s)J(\Phi, s)L(s, 1_F) = C'(\beta, s)J'(\Phi', s)L(s, \eta). \tag{24}$$

Now it is easily verified that

$$m(s) = m'(s)|\tau_0|^s.$$

The relation (23) reads

$$|\tau_0|^{s-1/2} \int \phi(x)|x|^s\eta(x) \, d^\times x/L(s, \eta) = \int \phi'(x)|x|^s \, d^\times x/L(s, 1_F).$$

The left-hand side is a Laurent polynomial in q^{-s}. Since ϕ' is an arbitrary Schwartz-Bruhat function such that

$$\phi'(xu) = \phi'(x)$$

for $|u| = 1$, we see that the right-hand side is an arbitrary Laurent polynomial and so we can choose Φ' in such a way that this relation and thus (23) is satisfied. Then the relation (24) will be a consequence of the following identity that we state as a lemma:

LEMMA 4.6. *With the above notations,*

$$|\tau_0|^{s-1} \frac{L(s, 1_F)C(\beta, s)}{L(1-s, 1_F)} = \frac{L(s, \eta)C'(\beta, s)}{L(1-s, \eta)},$$

provided $|\beta|$ *is small enough.*

As before, we set $X = q^{-s}$ and we write $C(\beta, X)$ and $C'(\beta, X)$ for C and C'. The function C' is easily computed: we let τ be a unit of F which is not a square, ϖ a uniformizer of F, and y be an integer of F; then

$$C'(\beta, X) = \begin{cases} -q^{-1}|y|^{2s-1} \dfrac{(Xq-1)^2(1+X)}{(1-X^2q)(1-X)} & \text{if } \beta = y^2 \\[2ex] -q^{-1}|y|^{2s-1} \dfrac{(X^2q^2-1)}{1-X^2q} & \text{if } \beta = y^2\tau \\[2ex] |y|^{2s-1} \dfrac{X(1-Xq)(1+X)}{1-X^2q} & \text{if } \beta = y^2\varpi. \end{cases} \tag{25}$$

In particular, we have the functional equation

$$\frac{1-X^{-1}q^{-1}}{1-X} C'\left(\frac{\beta}{\tau}, -X\right) = \frac{1+X^{-1}q^{-1}}{1+X} C'(\beta, X). \tag{26}$$

Now suppose the extension unramified. Then τ_0 is a unit and

$$\eta(\tau_0) = \eta(-\tau_0) = 1.$$

We have then

$$\begin{aligned} C(\beta, s) &= \int \left[\eta\left(\beta - \frac{u^2}{\tau_0}\right)\left|\beta - \frac{u^2}{\tau_0}\right|^{s-1} - \left|\frac{u^2}{\tau_0}\right|^{s-1}\right] du \\ &= C'(\beta\tau_0, -X). \end{aligned}$$

Thus

$$\frac{L(s,1)}{L(1-s,1)}C(\beta,s)=\frac{1-X^{-1}q^{-1}}{1-X}C'(\beta\tau_0,-X)$$

$$=\frac{1+X^{-1}q^{-1}}{1+X}C'(\beta,X)$$

$$=\frac{L(s,\eta)}{L(1-s,\eta)}C'(\beta,s)$$

by the functional equation (26). This proves the lemma in this case.

Now suppose the extension is ramified. Then τ_0 is a uniformizer of F, and the relation to be proved reads

$$qX\frac{1-X^{-1}q^{-1}}{1-X}C(\beta,s)=C'(\beta,s). \tag{27}$$

We first prove this relation when $\beta=y^2$ for some integer y. Then in the integral for C, both $\beta-x^2$ and $-x^2$ are norms. Thus we get

$$C(y^2,s)=\int\left(\left|y^2-\frac{u^2}{\tau_0}\right|^{s-1}-\left|\frac{u^2}{\tau_0}\right|^{s-1}\right)du$$

$$=X^{-1}q^{-1}\int(|\tau_0y^2-u^2|^{s-1}-|u^2|^{s-1})\,du$$

$$=X^{-1}q^{-1}C'(y^2\tau_0,X)$$

$$=X^{-1}q^{-1}|y|^{2s-1}\frac{X(1-Xq)(1+X)}{1-X^2q}.$$

On the other hand,

$$C'(y^2,s)=-q^{-1}|y|^{2s-1}\frac{(Xq-1)^2(1+X)}{(1-X^2q)(1-X)}$$

so that the relation (27) is easily verified.

For the remaining cases fresh computations are needed:

LEMMA 4.7. *Suppose the extension is ramified so that τ_0 is a uniformizer. Suppose also $\beta=y^2\tau_1$ where τ_1 is not a square and y is an integer. If τ_1 is a unit, then*

$$C(\beta,s)=|y|^{2s-1}\frac{q^{-1}(1+Xq)(X-1)}{1-X^2q}.$$

If τ_1 is a uniformizer, then

$$C(\beta, s) = |y|^{2s-1} \frac{-X(1-X^2)}{1-X^2 q}.$$

The relation (27) follows easily by comparing with the corresponding formula for C'. It remains to establish the lemma.

First suppose τ_1 is a unit. Then $\eta(\tau_1) = -1$. After changing u to $u\tau_0$, we get

$$C(y^2\tau_1, s) = q^{-1} \int_{|u|\leqslant 1} (\eta(y^2\tau_1 - u^2\tau_0)|y^2\tau_1 - u^2\tau_0|^{s-1} - |u^2\tau_0|^{s-1}\eta(-u^2\tau_0))\, du.$$

For $|u| > |y|$ the integrand is zero. Thus the integral simplifies to

$$q^{-1} \int_{|u|\leqslant |y|} (\eta(\tau_1)|y^2\tau_1 - u^2\tau_0|^{s-1} - |u^2\tau_0|^{s-1})\, du.$$

Changing u to yu, we get

$$q^{-1}|y|^{2s-1} \int_{|u|\leqslant 1} (-|\tau_1 - u^2\tau_0|^{s-1} - |u^2\tau_0|^{s-1})\, du,$$

which simplifies further to

$$q^{-1}|y|^{2s-1}\left[-1 - \int_{|u|\leqslant 1} Xq|u^2|^{s-1}\, du\right].$$

This is easily evaluated to give the answer.

Now assume τ_1 is a uniformizer. Recall τ_0 is also assumed to be a uniformizer. We have again

$$C(\beta, s) = q^{-1} \int_{|u|\leqslant 1} (\eta(y^2\tau_1 - \tau_0 u^2)|\tau_1 y^2 - \tau_0 u^2|^{s-1} - |\tau_0 u^2|^{s-1})\, du.$$

As before, the integrand is zero for $|u| > |y|$. After changing u to uy, we find for the integral

$$q^{-1}|y|^{2s-1} \int_{|u|\leqslant 1} (\eta(\tau_1 - \tau_0 u^2)|\tau_1 - \tau_0 u^2|^{s-1} - |\tau_0 u^2|^{s-1})\, du.$$

We write the integral of this difference as a difference of two integrals, the first of which we decompose as the sum of the contributions of the sets

$$|u| = 1, \qquad |u| < 1.$$

In this way we obtain

$$q^{-1}|y|^{2s-1}\left[\int_{|u|=1}\eta(\tau_1-\tau_0u^2)|\tau_1-\tau_0u^2|^{s-1}\,du+X\eta(\tau_1)-\frac{Xq(1-q^{-1})}{1-X^2q}\right].$$

The rest of the computation depends on whether or not τ_1/τ_0 is a square. First assume it is not. Then

$$|\tau_1-\tau_0u^2|^{s-1}=Xq$$

and

$$\eta(\tau_1)=\eta(\tau_1/\tau_0)\eta(-\tau_0)\eta(-1)=-\eta(-1).$$

Thus the above expression becomes

$$q^{-1}|y|^{2s-1}\left[-\eta(-1)Xq\int_{|u|=1}\eta(1-\varepsilon u^2)\,du-\eta(-1)X-\frac{Xq(1-q^{-1})}{1-X^2q}\right],$$

where $\varepsilon=\tau_0/\tau_1$. Thus ε is a unit and not a square. Let k be the residual field of F; thus k has q elements, and we may view η as the unique nontrivial quadratic character of $k^\times$. The integral which remains to be computed can be evaluated as a sum over $k^\times$:

$$q\int_{|u|=1}\eta(1-u^2\varepsilon)\,du=\sum_{x\in k^\times}\eta(1-x^2\varepsilon).$$

Thus the total expression to be evaluated is

$$q^{-1}|y|^{2s-1}\left[-\eta(-1)X\left(1+\sum\eta(1-\varepsilon x^2)\right)-\frac{Xq(1-q^{-1})}{1-X^2q}\right].$$

Since ε is not a square, $1-x^2\varepsilon$ is not zero and the sum over x is also equal to

$$2\#\{x\neq 0\,|\,1-x^2\varepsilon\text{ is a square}\}-(q-1)$$

or, what amounts to the same,

$$\#\{(x,y)\,|\,x^2\varepsilon+y^2=1\}-2-(q-1).$$

If -1 is a square, then $\eta(-1)=1$ and the quadratic form $x^2\varepsilon+y^2$ is equivalent to the norm form for the quadratic extension of k. Thus the above number of pairs is $q+1$, and the sum over x is zero. The first term in the total expression is then $-X$.

If -1 is not a square, then $\eta(-1) = -1$ and the quadratic form $x^2\varepsilon + y^2$ is equivalent to the form xy so that the number of pairs is now $q - 1$. The sum over x is then equal to -2, and the first term in the total expression is still equal to $-X$. Thus the total expression is

$$q^{-1}|y|^{2s-1}\left[-X - \frac{Xq(1-q^{-1})}{1-X^2q}\right].$$

This simplifies to the required expression for Lemma 4.7.

Now assume τ_0/τ_1 is a square. We may assume $\tau_1 = \tau_0$ as well. Then $\eta(\tau_0) = \eta(-1)$. We find

$$q^{-1}|y|^{2s-1}\left[\eta(-1)Xq\int_{|u|=1}\eta(1-u^2)|1-u^2|^{s-1}\,du + X\eta(-1) - \frac{Xq(1-q^{-1})}{1-X^2q}\right].$$

Because the character η is ramified, the contribution of the sets $|u+1| < 1$ and $|u-1| < 1$ vanishes. Thus the integral can be replaced by an integral over the set

$$|u| = |u+1| = |u-1| = 1$$

or by a sum over $k^\times$

$$q\int\eta(1-u^2)|1-u^2|^{s-1}\,du = q\int\eta(1-u^2)\,du = \sum_{x\neq 0,\,1-x^2\neq 0}\eta(1-x^2).$$

In this way the total expression is equal to

$$|y|^{2s-1}\left[\eta(-1)X\left(1 + \sum_{x\neq 0,\,1-x^2\neq 0}\eta(1-x^2)\right) - \frac{Xq(1-q^{-1})}{1-X^2q}\right].$$

Now

$$\begin{aligned}\sum_{x\neq 0,\,1-x^2\neq 0}\eta(1-x^2) &= 2\#\{x\neq 0 \mid 1-x^2\neq 0,\ 1-x^2 \text{ is a square}\}\\ &\quad - \#\{x\neq 0 \mid 1-x^2\neq 0\}\\ &= \#\{(x,y) \mid x^2+y^2=1\} - q - 1.\end{aligned}$$

If -1 is a square, we have $\eta(-1) = 1$; the form $x^2 + y^2$ is then equivalent to xy and the number of pairs is $q - 1$. The sum over x is thus equal to -2, and the first term in the total expression is $-X$. If -1 is not a square, then $\eta(-1) = -1$; the form is equivalent to the norm form and the number of pairs is $q + 1$. Thus the sum is 0, and the first term in the total expression is again $-X$. Thus we find the same result as before and we are done. This completes the proof of Lemmas 4.7 and 4.5.

4.5. To complete the proof of Proposition 4.1, it will suffice to show the following statement. We choose a set of representatives $\{\tau\}$ for the cosets $F^\times/F^{\times 2}$.

LEMMA 4.8. *Given τ and*

$$\sigma_0 = \begin{pmatrix} \alpha_0 & \beta_0\tau \\ \beta_0 & \alpha_0 \end{pmatrix}$$

with $\beta_0 \neq 0$, let T_0 be the torus which centralizes σ_0 in $GL(2, E)$. Its intersection with H is a torus T in H, and its intersection with G' a torus T' in G'. Let U be a compact open set of T' containing σ_0. Assume furthermore that U does not contain a pair of the form

$$\begin{pmatrix} \alpha & \beta\tau \\ \beta & \alpha \end{pmatrix}, \begin{pmatrix} \alpha & -\beta\tau \\ -\beta & \alpha \end{pmatrix},$$

or an element with $\beta \neq 0$. Then there is Φ' such that

$$\eta(\beta)\Omega(\Phi, \sigma, s)L(s, 1) = \Omega'(\Phi', \sigma, s)L(s, \eta)$$

for σ in U, and

$$\Omega'(\Phi', \sigma, s) = 0$$

if σ is not conjugate to an element of $F^\times U$.

Proof. We can write Ω in the form

$$\Omega(\Phi, \sigma, s) = \int_{T\backslash H} \Phi(h^{-1}z\sigma h)\eta(\lambda(h))\, dh \int_{Z\backslash T} f(th, s)\, dt\, d^\times z.$$

The relations

$$\sigma \in U,\ h^{-1}\sigma zh \in \text{Support } \Phi$$

imply that z is in a compact set of $F^\times$ and h in a fixed compact set of $T\backslash H$. Thus, for $\sigma \in U$ we can write

$$\Omega(\Phi, \sigma, s) = \int \phi(h, \sigma)\eta(\lambda(h))f(h, s)\, dh,$$

where ϕ is a smooth function on $T\backslash H \times F^\times U$, such that

$$\phi(h, zu) = \phi(h, u)$$

and whose support has a compact projection on the first factor. Similarly, we can choose a Φ' whose support is compact and contained in the set of conjugates of U in G'. Then Ω' vanishes for σ not conjugate into $UF^\times$. For σ in U, it can be written

$$\Omega'(\Phi', \sigma, s) = \int \phi'(g', \sigma) f(g', s)\, dg',$$

where ϕ' is an arbitrary smooth function on $T'\backslash G' \times UF^\times$ such that

$$\phi'(h, zu) = \eta(z)\phi'(h, u)$$

and whose support has a compact projection on the first factor. Thus it will suffice to show that, given ϕ, we can find ϕ' such that

$$L(s, 1)\eta(\beta) \int \phi(h, \sigma)\eta(\lambda(h)) f(h, s)\, dh = L(s, \eta) \int \phi'(g', \sigma) f'(g', s)\, dh$$

for $\sigma \in U$. In other words, it will suffice to show that, given $h \in H$, there are $g_i \in G'$ and $c_i \in \mathbb{C}$ such that

$$L(s, 1) \int f(th, s)\eta(\lambda(th))\, dt = L(s, \eta) \sum c_i \int f'(t'g_i, s)\, dt'.$$

We can use the isomorphism between H/Z and G'/Z' to reduce this assertion to a question on G': T' is as before but T is now a torus of G'. We can take it to be the torus which centralizes the element

$$\begin{pmatrix} 1 & \tau\tau_0 \\ 1 & 1 \end{pmatrix}$$

in G'. We can take f to be f' and replace λ by det. Furthermore, if Φ_0 is the characteristic function of $R_F \times R_F$ in F^2, we can take

$$f'(g, s) = \int \Phi_0[(0, u)g]|u|^{2s}\, d^\times u |\det g|^s.$$

The above identity reads

$$\int_{T/Z'} f'(th, s)\eta(\det(th))\, dt / L(s, \eta) = \sum c_i \int_{T'/Z'} f(t'g_i, s)\, dt' / L(s, 1).$$

We can view T' as the multiplicative group of an algebra L' isomorphic to $F \times F$ if τ is a square and to $F(\sqrt{\tau})$ if not. In particular (see [JZ]),

$$\int_{T'/Z'} f'(t'g, s)\, dt' = \int_{T'} \Phi_0[(0, 1)t'g]|\det tg|^s\, d^\times t'.$$

is a Tate integral for L' and the trivial character. The corresponding L-function is $L(s, 1)L(s, \eta_\tau)$.

A similar interpretation is available for T as multiplicative group of an algebra L isomorphic to $F \times F$ if $\tau\tau_0$ is a square and to $F(\sqrt{\tau\tau_0})$ if not. Similarly,

$$\int_{T/Z'} f'(th, s)\eta(\det(th))\, dt = \int_T \Phi_0[(0, 1)th]|\det th|^s \eta(\det th)\, d^\times t$$

is a Tate integral for the character $\eta \circ \det$ of L. The corresponding L-function is $L(s, \eta)L(s, \eta_\tau)$.

The proposition will then follow from the next lemma:

LEMMA 4.9. *Let Φ_0 be the characteristic function of R_F^2 in F^2. Assume τ is a unit or a uniformizer. Set $X = q^{-s}$ and*

$$P(X) = \int_{T'} \Phi_0[(0, 1)t'g]|\det t'g|^s\, dt'/L(s, 1)L(s, \eta_\tau).$$

Then

$$P(X) = P(X^{-1}q^{-1}), \tag{28}$$

and any Laurent polynomial satisfying this identity can be written as as sum of expressions of the above type. The same assertion is true for

$$Q(X) = \int_T \Phi_0[(0, 1)tg]\eta(\det tg)|\det tg|^s\, dt/L(s, \eta)L(s, \eta_\tau).$$

4.6. We now prove the lemma. This requires some preparation. It will be convenient to say that a Laurent polynomial is *symmetric* if it satisfies (28). Let ψ_0 be an additive character of F whose conductor is R_F. We define the Fourier transform of a Schwartz-Bruhat function ϕ on F to be

$$\hat{\phi}(x) = \int \phi(y)\psi_0(xy)\, dy,$$

the measure being self-dual. We define the *symplectic* Fourier transform of a Schwartz-Bruhat function Φ on F^2 by

$$\hat{\Phi}(x, y) = \int\int \Phi(u, v)\psi_0(uy - vx)\, du\, dv.$$

It follows that for $g \in SL(2, F)$ (or for $|\det g| = 1$) the function

$$(x, y) \mapsto \Phi_0[(x, y)g]$$

is its own symplectic Fourier transform. Conversely, we have the following statement.

LEMMA 4.10. *Let V be the space of Schwartz-Bruhat functions Φ on F^2 such that*

$$\hat{\Phi} = \Phi, \quad \Phi[u(x, y)] = \Phi[(x, y)] \qquad \textit{for } |u| = 1.$$

Then V is spanned by the functions of the form

$$(x, y) \mapsto \Phi_0[(x, y)g]$$

with $\det g = 1$.

This is a standard result ([Ho]). Similarly, we have the following lemma.

LEMMA 4.11. *Let V_1 be the space of Schwartz-Bruhat functions Φ on F^2 such that*

$$\hat{\Phi}(x, y) = q\Phi(\varpi^{-1}x, \varpi^{-1}y), \quad \Phi[u(x, y)] = \Phi[(x, y)] \qquad \textit{for } |u| = 1.$$

Then V_1 is spanned by the functions of the form

$$(x, y) \mapsto \Phi_0[(x\varpi, y)g]$$

with $\det g = 1$.

4.7. Now we prove the assertions relative to the polynomial P in Lemma 4.9. We can assume that τ is either a unit or a uniformizer.

First suppose τ is a uniformizer. Then the image of T' under det is the norm group for a ramified extension. Its product by the group of units is the whole multiplicative group of F. Thus we can take g in $SL(2, F)$ without restricting the generality. By Lemma 4.10 we can write

$$P(X) = \int_{T'} \Phi[(0, 1)t']|\det t'|^s \, dt'/L(s, 1)L(s, \eta_\tau),$$

where Φ is an arbitrary element of V. We have to show that P is an arbitrary symmetric polynomial. Recall the algebra L' of matrices of the form

$$\begin{pmatrix} a & b\tau \\ b & a \end{pmatrix}.$$

Thus $L' \simeq F(\sqrt{\tau})$ and

$$L(s, 1_{L'}) = L(s, 1)L(s, \eta_\tau).$$

We identify Φ to the function Ψ on L' defined by

$$\Psi(a + b\sqrt{\tau}) = \Phi(b, a).$$

We define an additive character ψ on L' by

$$\psi(z) = \psi_0\left(\mathrm{Tr}\left(\frac{z}{2\sqrt{\tau}}\right)\right).$$

The corresponding Fourier transform is

$$\hat{\Psi}(z) = \int \Psi(u)\psi(uz)\, du,$$

the measure being self-dual. Note that $\hat{\Phi}$ is identified to $\hat{\Psi}(\bar{z})$. We now have

$$P(X) = \int_{L'} \Psi(z)|z|^s\, d^\times z/L(s, 1_{L'}),$$

where Ψ is in the space $V_{L'}$ of functions Ψ such that

$$\Psi(z) = \hat{\Psi}(\bar{z}), \quad \Psi(uz) = \Psi(z), \qquad u \in R_F^\times.$$

Now Φ_0 is its own Fourier transform. The corresponding function on L' is the characteristic function of the ring $R_{L'}$ of integers in L'. It follows that the Tate functional equation takes the form

$$\int_{L'} \Psi(z)|z|^s\, d^\times z/L(s, 1_{L'}) = \int_{L'} \hat{\Psi}(z)|z|^{1-s}\, d^\times z/L(1 - s, 1_{L'}). \tag{29}$$

Since L' is a ramified quadratic extension, the cardinality of its residual field is q. For $\Psi \in V_{L'}$, the left-hand side is thus a symmetric polynomial in $X = q^{-s}$. Moreover, it is easily checked that it is in fact an arbitrary symmetric polynomial. This proves our assertion in this case.

Now assume that τ is a unit but not a square. Then the image of T' under det is the set of elements with even valuation. This time we may assume $|\det g|$ is 1 or q^{-1}. This leads us to consider the polynomials

$$P(X) = \int_{T'} \Phi[(0, 1)t']|\det t'|^s\, dt'/L(s, 1)L(s, \eta_\tau),$$

where Φ is an arbitrary element of V, and

$$P_1(X) = X \int_{T'} \Phi[(0, 1)t']|\det t'|^s\, dt'/L(s, 1)L(s, \eta_\tau),$$

where Φ_1 is an arbitrary element of V_1. We have to show that they are symmetric and that they span, when the functions Φ vary, the space of symmetric polynomials. As before, we introduce the algebra L' and the character ψ, and the Tate functional equation takes the form (29). We can write

$$P(X) = \int_{L'} \Psi(z)|z|^s \, d^\times z/L(s, 1_{L'}),$$

where Ψ is in the space $V_{L'}$ corresponding to V. Since the residual field of L' has q^2 elements, $P(X)$ is an arbitrary symmetric polynomial in X^2. Similarly, the space $V_{1L'}$ of the functions Ψ corresponding to the functions $\Phi \in V_1$ consists of all functions Ψ such that

$$\hat{\Psi}(\bar{z}) = q\Psi(\varpi^{-1}z), \quad \Psi(uz) = \Psi(z) \qquad \text{for } u \in R_F^\times .$$

We then have

$$P_1(X) = X \int_{L'} \Psi(z)|z|^s \, d^\times z/L(s, 1_{L'}),$$

with $\Psi \in V_{1L'}$. Applying the Tate functional equation, we find that P_1 is a symmetric polynomial containing odd-degree monomials only. It is easily seen that P_1 is an arbitrary polynomial with these properties. Our assertion is proved in the case at hand.

We pass to the case where τ is a unit and a square. After a change of variables, we can take $\tau = 1$. Moreover, without a loss of generality, we can take $g \in SL(2, F)$. Thus we can take

$$P(X) = \int_{T'} \Phi[(0, 1)t'] |\det t'|^s \, dt'/L(s, 1_F)L(s, 1_F),$$

where Ψ is in V. We have to show this is an arbitrary symmetric polynomial. The algebra L' is isomorphic to the sum of two copies of F. Correspondingly, we identify Φ to the function Ψ on L' defined by

$$\Psi(u, v) = \Phi(b, a)$$

where

$$u = a + b, \quad v = a - b.$$

We define a Fourier transform on L' by

$$\hat{\Psi}[u, v] = \iint \Psi[u', v'] \psi_0\left(\frac{u'v - v'u}{2}\right) du' \, dv'.$$

The space $V_{L'}$ corresponding to V is actually identical to V. We can take

$$P(X) = \int \Psi[u, v]|uv|^s \, du \, dv/L(s, 1_F)^2$$

where Ψ is in V_L. The Tate functional equation reads

$$\int \Psi[u, v]|uv|^s \, du \, dv/L(s, 1_F)^2 = \int \hat{\Psi}[u, v]|uv|^{1-s} \, du \, dv/L(1 - s, 1_F)^2.$$

This shows that P is indeed symmetric for $\Psi \in V_{L'}$. Now $V_{L'}$ is spanned by functions of the form

$$\Psi(x, y) = \phi(x)\hat{\phi}(y),$$

where ϕ is a Schwartz-Bruhat function such that

$$\phi(ux) = \phi(x)$$

for $|u| = 1$. For such a Ψ, the polynomial P has the form

$$P(X) = S(X)S(X^{-1}q^{-1})$$

where S is an arbitrary Laurent polynomial. Any symmetric polynomial is a linear combination of polynomials of this form, and our assertion follows in this case.

4.8. We pass to the study of the polynomials Q in Lemma 4.9. Here T is the multiplicative group of L, the algebra of matrices of the form

$$\begin{pmatrix} a & b\tau\tau_0 \\ b & a \end{pmatrix}.$$

The character $\eta \circ \det$ is a multiplicative character η_L of L and

$$L(s, \eta)L(s, \eta_\tau) = L(s, \eta_L).$$

If $\tau = 1$, then L is the extension E, $\eta_L = 1$, and we can apply the previous results.

If $\tau = \tau_0\varepsilon^2$ where ε and τ_0 are units, then $\eta = \eta_\tau$ is an unramified character and the algebra L is split; so we can again apply the previous results (with X replaced by $-X$).

If $|\tau| = 1, |\tau_0| = q^{-1}$ or $|\tau| = q^{-1}, |\tau_0| = 1$, then L is a ramified quadratic extension of F but $\eta \circ \det$ is an unramified character of L, so that we can apply again the previous results with X replaced by $-X$.

In the two remaining cases we have $|\tau| = |\tau_0| = q^{-1}$. Suppose first that $\tau = \tau_0 \varepsilon^2$ where ε is a unit. We may as well assume $\varepsilon = 1$. Then L is isomorphic to the sum of two copies of F. A function $\Phi(b, a)$ is identified to the function $\Psi(u, v)$ where

$$u = a + b\tau_0, \quad v = a - b\tau_0.$$

The symplectic Fourier transform becomes the following Fourier transform on L:

$$\hat{\Psi}(u, v) = \int \Psi(u', v')\psi_0\left(\frac{u'v - v'u}{2\tau_0}\right) du'\, dv'.$$

The space V_L corresponding to V is the space of functions Ψ such that

$$\hat{\Psi} = \Psi, \quad \Psi(ux, uy) = \Psi(x, y) \qquad \text{for } u \in R_F^\times.$$

Since η is ramified, the polynomial P has the form

$$P(X) = \int \Psi(u, v)|uv|^s \eta(uv)\, du\, dv,$$

for $\Psi \in V_L$. Now the Tate functional equation reads

$$\int \hat{\Psi}(u, v)|uv|^{1-s}\eta(uv)\, d^\times u\, d^\times v = \eta(-1)\varepsilon(\eta, s, \psi)^2 \int \Psi(u, v)|uv|^s \eta(uv)\, d^\times u\, d^\times v$$

where ψ is the character of F defined by

$$\psi(x) = \psi_0\left(\frac{x}{2\tau_0}\right).$$

If ϕ is the function of support $R^\times$ defined by

$$\phi(u) = \eta(u)$$

for $|u| = 1$, its Fourier transform with respect to ψ has the form $c\phi$ for a suitable c. The function

$$\Psi(x, y) = c\phi(x)\phi(y)$$

is thus in V_L, and the Tate functional equation in fact reads

$$\int \hat{\Psi}(u, v)|uv|^{1-s}\eta(uv)\, d^\times u\, d^\times v = \int \Psi(u, v)|uv|^s \eta(uv)\, d^\times u\, d^\times v.$$

For $\Psi \in V_L$ this is a symmetric polynomial in X. Moreover, just as before, this polynomial is arbitrary.

Finally, we assume $\tau = \varepsilon\tau_0$ where ε is a unit which is not a square. Thus L is another ramified extension of F. A function Φ on F^2 is identified to the function Ψ on L defined by

$$\Psi(a + b\tau_0\sqrt{\varepsilon}) = \Phi(b, a).$$

If we define

$$\psi(z) = \psi_0\left(\mathrm{Tr}\left(\frac{z}{2\tau_0\varepsilon}\right)\right)$$

and

$$\hat{\Psi}(z) = \int \Psi(u)\psi(uz)\,du,$$

then the image V_L of V is the space of functions Ψ such that

$$\hat{\Psi}(z) = \Psi(\bar{z}), \quad \Psi(uz) = \Psi(z)$$

for $u \in R_F^\times$. The character η_L is defined by $\eta_L(z) = \eta(z\bar{z})$. The Tate functional equation reads

$$\int \hat{\Psi}(z)\eta_L(z)|z|^{1-s}\,d^\times z = \varepsilon(s, \eta_L, \psi)\int \Psi(z)\eta_L(z)|z|^s\,d^\times z.$$

We first show that the ε factor is actually one. It suffices to find $\Psi \in V_L$ such that the integral

$$\int \Psi(z)\eta_L(z)|z|\,d^\times z$$

is a positive constant. We first choose the corresponding function Φ. Let ϕ^0 be the characteristic function of $R_F^\times$ and ϕ_0 the characteristic function of R. The Fourier transform of ϕ^0 with respect to ψ_0 is the function ϕ_1 defined by

$$\phi_1(x) = \phi_0(x) - q^{-1}\phi_0(x\varpi).$$

It follows that the function

$$\Phi(x, y) = \phi_1(x)\phi^0(y)$$

is in V. We consider the corresponding function Ψ in V_L. Its support is contained in the set of units of L. Up to a positive factor, the corresponding Tate integral is found to be equal to

$$1 - q^{-1} - \sum_{x \in k^\times} \eta(1 - x^2\varepsilon).$$

As we have seen in the previous section, the sum is actually 0 or -2 and so the total integral is > 0, and our assertion is proved. We finish the proof as before: the polynomial Q is the sum of two polynomials

$$P(X) = \int \Psi(z)\eta_L(z)|z|^s\, d^\times z$$

with $\Psi \in V_L$, and

$$P_1(X) = X \int \Psi(z)\eta'(z)|z|^s\, d^\times z$$

with $\Psi \in V_{1L}$, the space corresponding to V_1. Then P (resp. P_1) is an arbitrary symmetric polynomial containing monomials of even (resp. odd) degree. Our assertion follows. This conclude the proof of the lemma and the proposition. □

4.9. We conclude this section by the following corollary to Proposition 4.1:

LEMMA 4.12. *If Φ and Φ' have matching orbital integrals, then*

$$|\tau_0|^{s-1/2} \int \phi(x)|x|^s\eta(x)\, d^\times L(s, 1) = \int \phi'(x)|x|^s\, d^\times x L(s, \eta). \tag{30}$$

This follows from the proof of the proposition. □

5. Global theory.

5.1. We let E/F be a quadratic extension of number fields. Then S is an algebraic variety defined over F. We set

$$g^* = w^t\bar{g}w.$$

Then $G = GL(2, E)$ operates on S on the right:

$$s \mapsto g^{*}sg.$$

For ε in S we denote by H_ε^0, H_ε, and λ_ε the corresponding unitary group, similitude group, and similitude ratio. Thus for $h \in H_\varepsilon$,

$$h^*\varepsilon h = \lambda_\varepsilon(h)\varepsilon.$$

We say that two matrices s_1 and s_2 are rationally equivalent over F if there is $\gamma \in GL(2, E)$ such that

$$s_1 = \gamma^* s_2 \gamma.$$

We say they are projectively equivalent if there is a $\gamma \in GL(2, E)$ and a $\mu \in F^\times$ such that

$$s_1 = \mu\gamma^* s_2 \gamma.$$

In particular, μ is in the image F^ε of λ_ε if and only if $\mu\varepsilon$ and ε are rationally equivalent. Two matrices are rationally (resp. projectively) equivalent if and only if they are rationally (resp. projectively) equivalent at each place of F which is inert in E.

At a nonarchimedean place, the two definitions of equivalence coincide. Moreover, s_1 is equivalent to s_2 if and only if $\det s_1$ and $\det s_2$ are equal up to the product by a norm. In particular, s belongs to the orbit of e, that is, the form associated to sw is split, if and only if $\det s$ is a norm.

At a real place v of F inert in E, s_1 and s_2 are projectively equivalent if and only if $\det s_1$ and $\det s_2$ are equal up to the product by a norm, that is, have the same sign. Matrices s with positive determinant are those for which sw is split; they form a single rational equivalence class. The set of matrices with negative determinant is the union of two rational classes.

For a place v of F inert in E, set $F_v^\varepsilon = \lambda_\varepsilon(H_\varepsilon(F_v))$. Then F_v^ε is actually equal to $F_v^\times$, except when v is real and inert, and $\det \varepsilon_v < 0$, in which case it consists of the positive elements of $F_v^\times$.

Let $F_{\mathbb{A}}^\varepsilon$ be the product of the F_v^ε. Thus

$$F_{\mathbb{A}}^\varepsilon = \lambda_\varepsilon(H_\varepsilon(F_{\mathbb{A}})).$$

Clearly, $F_{\mathbb{A}}^\varepsilon$ is an open subgroup of finite index in $F_{\mathbb{A}}^\times$. Since $F^\times$ is dense in the product of the $F_v^\times$ with v real, we have

$$F_{\mathbb{A}}^\times = F^\times F_{\mathbb{A}}^\varepsilon.$$

On the other hand,

$$F^\varepsilon = F^\times \cap F_{\mathbb{A}}^\varepsilon.$$

It follows that F^ε is a subgroup of finite index in $F^\times$, and we have a disjoint union

$$F_{\mathbb{A}}^\times = \bigcup_{\xi \in F^\times/F^\varepsilon} \xi F_{\mathbb{A}}^\varepsilon.$$

We have exact sequences

$$1 \to H_\varepsilon^0(F) \to H_\varepsilon(F) \xrightarrow{\lambda_\varepsilon} F^\varepsilon \to 1,$$

$$1 \to H_\varepsilon^0(F_v) \to H_\varepsilon(F_v) \xrightarrow{\lambda_\varepsilon} F_v^\varepsilon \to 1,$$

$$1 \to H_\varepsilon^0(F_{\mathbf{A}}) \to H_\varepsilon(F_{\mathbf{A}}) \xrightarrow{\lambda_\varepsilon} F_{\mathbf{A}}^\varepsilon \to 1.$$

We choose a measure on $H_\varepsilon(F_{\mathbf{A}})$ that we write as a product of local measures. It is assumed that for almost all places the measure of the maximal compact defined below is 1. We choose a measure on $F_v^\times$ in the usual way. It induces a measure on F_v^ε, and we use the local sequence to define a measure on $H_\varepsilon^0(F_v)$.

We choose a system of representatives $\{\varepsilon\}$ for the rational classes. We can take it to be of the form $\varepsilon = \mu\varepsilon'$ where $\{\varepsilon'\}$ is a set of representatives for the projective classes and, for each ε', $\{\mu\}$ is a set of representatives for the cosets of $F^{\varepsilon'}$ in $F^\times$. We choose a function of compact support Φ on $S(F_{\mathbf{A}})$. We assume that Φ is a product of local functions Φ_v. Thus Φ_v is a smooth function of compact support on $S(F_v)$. Furthermore, for almost all v, it is in fact the characteristic function of the set of integral elements of S_v with unit determinant. For each ε, we choose a function f_ε on $GL(2, E_{\mathbf{A}})$, smooth of compact support, such that

$$\int_{H_0^\varepsilon(F_{\mathbf{A}})} f_\varepsilon(h_\varepsilon g)\, dh_\varepsilon = \Phi(g^*\varepsilon g).$$

In order for this integral to be nonzero, there must be an idèle z which is a norm at each place such that $z \det \varepsilon$ is in the image of the support of Φ under det. Let $X(\Phi)$ be the set of ε with this property. This image is a compact set; since $N_{E/F}(E_{\mathbf{A}}^\times)$ is open in $F_{\mathbf{A}}^\times$, this image is contained in a finite union of cosets of $N_{E/F}(E_{\mathbf{A}}^\times)$. A fortiori, det ε belongs to a finite union of cosets of $N_{E/F}(E^\times)$. Writing $\varepsilon = \mu\varepsilon'$, we see that ε' belongs to a finite set. Thus $X(\Phi)$ is a finite set itself.

It will be convenient to choose a finite set X of ε and to choose the function Φ_v for each place v of F inert in E with support in the union of the orbits of the $\varepsilon \in X$ under $GL(2, E_v)$. Then

$$\Phi(g^*\varepsilon g) = 0$$

if ε is not in X. For $\varepsilon \notin X$ we will take $f_\varepsilon = 0$.

Next, to each ε we associate a kernel function

$$K_\varepsilon(g_1, g_2) = \sum_{\xi \in GL(2,E)} f_\varepsilon(g_1^{-1}\xi g_2).$$

We will set

$$K_\varepsilon(g) = \int_{H_\varepsilon(F)\backslash H_\varepsilon(F_{\mathbf{A}})} K_\varepsilon(h, g)\, dh.$$

We claim that

$$\sum_{\varepsilon} K_\varepsilon(g) = \int_{F_{\mathbf{A}}^\times/F^\times} d^\times z \sum_{\xi \in S(F)} \Phi(g^* \xi z g).$$

Indeed, it suffices to prove this for $g = e$. We can write

$$K_\varepsilon(e) = \sum_{\xi \in H_\varepsilon(F)\backslash G(F)} \int \left(\int f_\varepsilon(h_0 h_1 \xi)\, dh_0 \right) dh_1$$

with

$$h_0 \in H_\varepsilon^0(F_{\mathbf{A}}), \quad h_1 \in H_\varepsilon^0(F_{\mathbf{A}})\backslash H_\varepsilon(F_{\mathbf{A}}).$$

Thus we find

$$\begin{aligned} K_\varepsilon(e) &= \sum_{\xi} \int \Phi(\xi^* \varepsilon \xi \lambda(h_1))\, dh_1 \\ &= \sum_{\xi} \int_{F_{\mathbf{A}}^\varepsilon} \Phi(\xi^* \varepsilon \xi z)\, d^\times z. \end{aligned}$$

Replacing ε by $\mu\varepsilon$ does not change the similitude group; thus summing over $\mu \in F^\times/F^\varepsilon$, we find

$$\begin{aligned} \sum_{\mu} K_{\mu\varepsilon}(e) &= \sum_{\xi} \sum_{\mu} \int_{F_{\mathbf{A}}^\varepsilon} \Phi(\xi^* \varepsilon \mu z \xi)\, d^\times z \\ &= \sum_{\xi} \int_{F_{\mathbf{A}}^\times} \Phi(\xi^* \varepsilon z \xi)\, d^\times z \\ &= \int_{F_{\mathbf{A}}^\times/F^\times} \sum_{\xi} \sum_{\mu \in F^\times} \Phi(\xi^* \mu \varepsilon z \xi)\, d^\times z \\ &= \int_{F_{\mathbf{A}}^\times/F^\times} \sum_{\sigma} \Phi(\sigma z)\, d^\times z \end{aligned}$$

where the sum is for σ in the projective class of ε. Summing over the projective classes, we find our result.

5.2. We set

$$K_\Phi(g) = \sum_{\varepsilon} K_\varepsilon(g). \tag{31}$$

We have thus

$$K_\Phi(g) = \int_{F_{\mathbb{A}}^\times/F^\times} \sum_{\xi \in S(F)} \Phi(g^*\xi z g)\, d^\times z. \tag{32}$$

Associated to the function f_ε and the trivial central character, we have in the usual way a cuspidal kernel, a special kernel, and an Eisenstein kernel, K_ε^{cusp}, K_ε^{spe}, and K_ε^{eis}, respectively. We define

$$K_\Phi^{cusp}(g) = \sum_\varepsilon \int K_\varepsilon^{cusp}(h_\varepsilon, g)\, dh_\varepsilon,$$

the integral over $Z(F_{\mathbb{A}})H_\varepsilon(F)\backslash H_\varepsilon(F_{\mathbb{A}})$. We define similarly K_Φ^{spe}, K_Φ^{eis}. We remark that for any cusp form ϕ on $GL(2, E_{\mathbb{A}})$, invariant under the center,

$$\int K_\Phi(g)\phi(g)\, dg = \int K_\Phi^{cusp}(g)\phi(g)\, dg.$$

Thus we may view K_Φ^{cusp} as the projection of K_Φ on the space of cusp forms.

In this section, our goal is to compute the integral

$$I_\Phi^{cusp}(s) = \int K_\Phi^{cusp}(h)E(h, s)\eta(\lambda(h))\, dh, \tag{33}$$

where $E(g, s)$ is an Eisenstein series on the group $H = H_w$, suitably normalized. The integral is over $Z(F_{\mathbb{A}})H(F)\backslash H(F_{\mathbb{A}})$. We first compute formally. We will justify our steps completely under the restrictions mentioned in the introduction.

At this point, we assume that every real place of F splits in E and that every finite place of even residual characteristic splits in E. We need to specify the measures we are using on H and G', the maximal compacts, and the functions f, f' which are used to define the Eisenstein series. For each place v of F inert in E, we choose an element t_v such that

$$E_v = F_v(\sqrt{t_v}). \tag{34}$$

If v is unramified in E, we take for t_v a unit; if v is ramified, we take for t_v a uniformizer of F_v. For v split, we take $t_v = 1$. We set

$$t_0 = \prod t_v. \tag{35}$$

Then for v inert, conjugacy by

$$g_v = \begin{pmatrix} \sqrt{t_v} & 0 \\ 0 & 1 \end{pmatrix} \tag{36}$$

defines an isomorphism

$$\xi_v\colon H_v \to G'_v Z_v .$$

If v is split, then H_v is the set of triples (g_1, g_2, λ) with $g_i \in G'_v$, $\lambda \in F^\times$ such that

$$g_1 = w^t g_2^{-1} w \lambda .$$

On the other hand, $G'_v Z_v$ is isomorphic to the set of pairs

$$(g\lambda_1, g\lambda_2)$$

with $g \in G_v$, $\lambda_i \in F_v^\times$. Thus we have again an isomorphism

$$\xi_v\colon H_v \to G'_v Z_v .$$

The product of the ξ_v defines then an isomorphism

$$\xi\colon H(F_{\mathbb{A}}) \to G'(F_{\mathbb{A}}) Z(F_{\mathbb{A}}).$$

Recall P is the group of upper triangular matrices, P_H its intersection with H, and P' its intersection with G'. Here it will be more convenient to think of P as a group defined over E. For each finite place v of F we set

$$K'_v = GL(2, R_v).$$

For v real we set

$$K'_v = O(2, \mathbb{R}).$$

For v complex we set

$$K'_v = U(2, \mathbb{C}).$$

We regard $Z \simeq E^\times$ as an algebraic group over F, and for each place v of F we denote by U_v the maximal compact subgroup of Z_v. We let K' be the product of the K'_v and U the product of the U_v. We define a function f' on $G'(F_{\mathbb{A}})Z(F_{\mathbb{A}})$ by

$$f'(g'z) = |t|^s$$

for

$$g' = \begin{pmatrix} t & u \\ 0 & 1 \end{pmatrix} k'.$$

Thus we may write f' as the product of local functions f'_v which, for v inert, are the functions introduced in the previous sections. We use the isomorphism ξ to transport $K'U$ and f' to H. We denote by K_H and f the corresponding maximal subgroup and function. Thus

$$f(h, s) = |t|^s \qquad \text{if } h = \begin{pmatrix} tz & x \\ 0 & z \end{pmatrix} k.$$

Then

$$E(h, s) = \sum_{\gamma \in P_H(F)\backslash H(F)} f(\gamma h, s).$$

On $G'(F_{\mathbb{A}})/Z'(F_{\mathbb{A}})$, we choose a measure compatible with the Iwasawa decomposition

$$dg' = dk'|t|^{-1}\, d^\times t\, du$$

if

$$g' = \begin{pmatrix} t & u \\ 0 & 1 \end{pmatrix} k'.$$

Here $d^\times t$ and du are the Tamagawa measures. We transport this measure to $H(F_{\mathbb{A}})/Z(F_{\mathbb{A}})$. The measure needs not be the same as the one we have used to define the kernels in terms of Φ. The integral (33) is defined for $\Re s > 1$. For simplicity, we will drop the index Φ from the kernels. Unfolding the Eisenstein series, we find also for $\Re s > 1$

$$I^{cusp}(s) = \int_{Z(F_{\mathbb{A}})P_H(F)\backslash H(F_{\mathbb{A}})} K^{cusp}(h) f(h, s) \eta(\lambda(h))\, dh.$$

We have

$$K^{cusp} = K - K^{eis} - K^{spe}, \qquad K = K^{reg} + K^{sing},$$

where

$$K^{reg}(g) = \int \sum_{\xi \notin P(E) \cap S(F)} \Phi(g^* \xi z g)\, d^\times z,$$

$$K^{sing}(g) = \int \sum_{\xi \in P(E) \cap S(F)} \Phi(g^* \xi z g)\, d^\times z.$$

Each function of g is invariant under $P(E)$ on the left. It will be convenient to introduce the function $\tilde{K}$ defined by

$$\tilde{K} = K^{sing} - K^{eis} - K^{spe}.$$

Then

$$K^{cusp} = K^{reg} + \tilde{K}.$$

We will denote by $\tilde{I}$ the integral

$$\tilde{I}(s) = \int_{Z(F_{\mathbb{A}})P_H(F)\backslash H(F_{\mathbb{A}})} \tilde{K}(h)f(h, s)\eta(\lambda(h))\, dh.$$

We will define similarly I^{eis}, I^{reg}, and so on. Thus

$$I^{cusp}(s) = I^{reg}(s) + \tilde{I}(s). \tag{37}$$

5.3. We first consider I^{reg}. We need the following lemma which relates conjugation under $P_H(F)$ and $H(F)$ in $S(F)$:

LEMMA 5.1. *Suppose $\xi \notin P(E) \cap S(F)$. Then ξ is conjugate to an element of $S_0(F)$ by an element of $P_H(F)$. Moreover, two elements of $S_0(F)$ which are not in $P(F)$ are conjugate under $H(F)$ if and only if they are conjugate under $P_H(F)$. Finally, if ξ is in $S_0(F) - S_0(F) \cap P(E)$, then its centralizer in $P_H(F)$ is equal to $Z(F)$.* □

According to the lemma, a semisimple element of $S(F) - S(F) \cap P(E)$ can be written in one way in the form

$$\gamma^{-1} z \begin{pmatrix} 1 & \beta \\ 1 & 1 \end{pmatrix} \gamma, \qquad \gamma \in P_H(F)/Z(F), \qquad z \in F^\times, \qquad \beta \in F - \{0, 1\},$$

or in the form

$$\gamma^{-1} z \begin{pmatrix} 0 & \beta \\ 1 & 0 \end{pmatrix} \gamma, \qquad \gamma \in P_H(F)/Z(F), \qquad z \in F^\times, \qquad \beta \in F^\times/F^{\times 2}.$$

A unipotent element in $S(F) - S(F) \cap P(E)$ can be written in one way in the form

$$\gamma^{-1} z \begin{pmatrix} 1 & 0 \\ 1 & 1 \end{pmatrix} \gamma, \qquad \gamma \in P_H(F)/Z(F), \qquad z \in F^\times.$$

Thus the integral $I^{reg}(s)$ can be written as a sum

$$I^{reg}(s) = \sum_{\beta \in F - \{0,1\}} \Omega(\Phi, \beta, s) + \sum_{\beta \in F^\times/F^{\times 2}} \Omega^\infty(\Phi, \beta, s) + I^{uni}(\Phi, s)$$

where we have set

$$\Omega(\Phi, \beta, s) = \int \Phi\left[h^{-1}z\begin{pmatrix}1 & \beta\\ 1 & 1\end{pmatrix}h\right]\eta(\lambda(h))f(h, s)\,dh\,d^{\times}z,$$

$$\Omega^{\infty}(\Phi, \beta, s) = \int \Phi\left[h^{-1}z\begin{pmatrix}0 & \beta\\ 1 & 0\end{pmatrix}h\right]\eta(\lambda(h))f(h, s)\,dh\,d^{\times}z,$$

and

$$I^{uni}(\Phi, s) = \int \Phi\left[h^{-1}z\begin{pmatrix}1 & 0\\ 1 & 1\end{pmatrix}h\right]\eta(\lambda(h))f(h, s)\,dh\,d^{\times}z,$$

where the integral is for $h \in Z(F_{\mathbb{A}})\backslash H(F_{\mathbb{A}})$ and $z \in F_{\mathbb{A}}^{\times}$.

We give another formula for the last integral: we can change h to wh to get

$$I^{uni}(s) = \int \Phi\left[h^{-1}z\begin{pmatrix}1 & 1\\ 0 & 1\end{pmatrix}h\right]\eta(\lambda(h))f(wh, s)\,dh\,d^{\times}z.$$

Integrating via the Iwasawa decomposition, we get

$$I^{uni}(s) = m(s)\int_{F_{\mathbb{A}}^{\times}\times K\times F_{\mathbb{A}}^{\times}} \Phi\left[k^{-1}z\begin{pmatrix}1 & t\\ 0 & 1\end{pmatrix}k\right]\eta(t)|t|^{s}\eta(\lambda(k))\,d^{\times}t\,dk\,d^{\times}z,$$

where we have set

$$m(s) = \int_{\sqrt{\tau_0}F_{\mathbb{A}}} f\left[w\begin{pmatrix}1 & x\\ 0 & 1\end{pmatrix}zk\right]du;$$

here $E = F(\sqrt{\tau_0})$ and

$$x = \frac{u}{\sqrt{t_0}}.$$

For future needs, we compare $m(s)$ with the scalar $m'(s)$ defined by

$$m'(s) = \int_{F_{\mathbb{A}}} f'\left[w\begin{pmatrix}1 & u\\ 0 & 1\end{pmatrix}\right]du.$$

We find

$$m(s) = |t_0|^{s}m'(s). \tag{38}$$

As in the local case, we introduce a Schwartz-Bruhat function ϕ on $F_{\mathbb{A}}$:

$$\phi(x) = \int \Phi\left[k^{-1}z\begin{pmatrix}1 & x\\ 0 & 1\end{pmatrix}k\right]\eta(\lambda(k))\, d^{\times}z\, dk. \tag{39}$$

Then

$$I^{uni}(s) = m(s)\int \phi(t)\eta(t)|t|^{s}\, d^{\times}t. \tag{40}$$

5.4. We now consider the integral $\tilde{I}$. Recall

$$\tilde{I}(s) = \int_{Z(F_{\mathbb{A}})P_H(F)\backslash H(F_{\mathbb{A}})} \tilde{K}(h)f(h, s)\eta(\lambda(h))\, dh.$$

The integral $\tilde{I}$ factors through an integral on $N_H(F)\backslash N_H(F_{\mathbb{A}})$, namely,

$$\int_{u\in F_{\mathbb{A}}/F} \tilde{K}\left[\begin{pmatrix}1 & x\\ 0 & 1\end{pmatrix}g\right] du. \tag{41}$$

Let us agree to denote by dx the measure on $\sqrt{\tau_0}F_{\mathbb{A}}$ such that

$$\int_{\sqrt{\tau_0}F_{\mathbb{A}}/\sqrt{\tau_0}F} dx = 1.$$

Thus we can write $x = \sqrt{\tau_0}v$ and $dx = dv$ where dv is the Tamagawa measure on $F_{\mathbb{A}}$. In terms of the variable u defined by

$$x = \frac{u}{\sqrt{t_0}},$$

we have

$$v = \frac{u}{\sqrt{\tau_0 t_0}}$$

so that

$$dx = dv = |t_0^{-1/2}|\, du.$$

Let ψ_0 be a nontrivial additive character of $E_{\mathbb{A}}$ trivial on $E + F_{\mathbb{A}}$. If μ is a smooth function on $E_{\mathbb{A}}/E$, then

$$\int_{\sqrt{\tau_0}F_{\mathbb{A}}/\sqrt{\tau_0}F} \mu(x)\, dx = \int_{E_{\mathbb{A}}/E} \mu(y)\, dy + \sum_{\alpha\in F^{\times}} \int_{E_{\mathbb{A}}/E} \mu(y)\psi_0(\alpha y)\, dy,$$

where dy is the Tamagawa measure on E, that is, the one for which the measure of $E_{\mathbb{A}}/E$ is 1. Recall

$$\tilde{K}(g) = K^{sing}(g) - K^{spec}(g) - K^{eis}(g).$$

The sum

$$-\sum_{\alpha \in F^\times} \int_{E_{\mathbb{A}}/E} K^{spe}\left[\begin{pmatrix}1 & y\\ 0 & 1\end{pmatrix} g\right] \psi_0(\alpha y)\, dy$$

vanishes. Thus we can evaluate (41) as follows:

$$|t_0|^{1/2} \int_{\sqrt{\tau_0}F_{\mathbb{A}}/\sqrt{\tau_0}F} K^{sing}\left[\begin{pmatrix}1 & x\\ 0 & 1\end{pmatrix} g\right] dx, \tag{42}$$

$$-|t_0|^{1/2} \int_{E_{\mathbb{A}}/E} (K^{spec} + K^{eis})\left[\begin{pmatrix}1 & y\\ 0 & 1\end{pmatrix} g\right] dy, \tag{43}$$

$$-|t_0|^{1/2} \sum_{\alpha \in F^\times} \int_{E_{\mathbb{A}}/E} K^{eis}\left[\begin{pmatrix}1 & y\\ 0 & 1\end{pmatrix} g\right] \psi_0(\alpha y)\, dy. \tag{44}$$

Because of the cuspidality, we can replace $K^{spe} + K^{eis}$ by K in the middle integral. Thus the integral (41) can be written as the sum of three terms:

$$|t_0|^{1/2} \int_{\sqrt{\tau_0}F_{\mathbb{A}}/\sqrt{\tau_0}F} K^{sing}\left[\begin{pmatrix}1 & x\\ 0 & 1\end{pmatrix} g\right] dx, \tag{45}$$

$$-|t_0|^{1/2} \int_{E_{\mathbb{A}}/E} K\left[\begin{pmatrix}1 & y\\ 0 & 1\end{pmatrix} g\right] dy, \tag{46}$$

$$-|t_0|^{1/2} \sum_{\alpha \in F^\times} \int_{E_{\mathbb{A}}/F} K^{eis}\left[\begin{pmatrix}1 & y\\ 0 & 1\end{pmatrix} g\right] \psi_0(\alpha y)\, dy. \tag{47}$$

To evaluate (45) we write

$$K^{sing}(g) = \sum_{\beta \in F} \int \Phi\left[g^* \begin{pmatrix}1 & \beta\\ 0 & 1\end{pmatrix} zg\right] d^\times z + \sum_{\beta \in F,\, \{\alpha \neq \bar{\alpha}\}/F^\times} \int \Phi\left[g^* \begin{pmatrix}\alpha & \beta\\ 0 & \bar{\alpha}\end{pmatrix} zg\right] d^\times z.$$

Integrating over $N_H(F)\backslash N_H(F_{\mathbb{A}})$, we get for (45)

$$|t_0|^{1/2} \sum_{\beta \in F} \int \Phi\left[g^* \begin{pmatrix}1 & \beta\\ 0 & 1\end{pmatrix} zg\right] d^\times z$$

$$+\, |t_0|^{1/2} \sum_{\beta \in F,\, \{\alpha \neq \bar{\alpha}\}/F^\times} \int \Phi\left[g^* \begin{pmatrix}\alpha & \beta + (\alpha - \bar{\alpha})\sqrt{\tau_0}v\\ 0 & \bar{\alpha}\end{pmatrix} zg\right] d^\times z\, dv$$

where dv is the Tamagawa measure on $F_{\mathbb{A}}/F$. The second term simplifies, and we obtain for (45) the expression

$$|t_0|^{1/2} \sum_{\beta\in F} \int_{F_{\mathbb{A}}^\times} \Phi\left[g^*\begin{pmatrix}1 & \beta\\ 0 & 1\end{pmatrix}zg\right] d^\times z$$

$$+ |t_0|^{1/2} \sum_{\{\alpha\neq\bar{\alpha}\}/F^\times} \int_{F_{\mathbb{A}}^\times \times F_{\mathbb{A}}} \Phi\left[g^*\begin{pmatrix}\alpha & v\\ 0 & \bar{\alpha}\end{pmatrix}zg\right] d^\times z\, dv.$$

Now we write again $K = K^{sing} + K^{reg}$ to compute (46). The integral of $-K^{sing}$ gives

$$-|t_0|^{1/2} \sum_{\beta} \int \Phi\left[g^*\begin{pmatrix}1 & \beta + y + \bar{y}\\ 0 & 1\end{pmatrix}zg\right] d^\times z\, dy$$

$$- |t_0|^{1/2} \sum_{\beta\in F, \{\alpha\neq\bar{\alpha}\}/F^\times} \int \Phi\left[g^*\begin{pmatrix}\alpha & \beta + \alpha y + \overline{\alpha y}\\ 0 & \bar{\alpha}\end{pmatrix}zg\right] d^\times z\, dy.$$

This simplifies to

$$-|t_0|^{1/2} \int_{F_{\mathbb{A}}} \Phi\left[g^*\begin{pmatrix}1 & x\\ 0 & 1\end{pmatrix}zg\right] d^\times z\, dx$$

$$- |t_0|^{1/2} \sum_{\{\alpha\neq\bar{\alpha}\}/F^\times} \int_{F_{\mathbb{A}}} \Phi\left[g^*\begin{pmatrix}\alpha & x\\ 0 & \bar{\alpha}\end{pmatrix}zg\right] d^\times z\, dx$$

where dx is the Tamagawa measure on $F_{\mathbb{A}}$. If we combine it with the expression we obtained for (45), we obtain

$$|t_0|^{1/2} \sum_{\beta\in F} \int_{F_{\mathbb{A}}} \Phi\left[g^*\begin{pmatrix}1 & \beta\\ 0 & 1\end{pmatrix}zg\right] d^\times z - |t_0|^{1/2} \int_{F_{\mathbb{A}}} \Phi\left[g^*\begin{pmatrix}1 & x\\ 0 & 1\end{pmatrix}zg\right] d^\times z\, dx.$$

It remains to compute the integral of $-K^{reg}$ on $N(E_{\mathbb{A}})/N(E)$. The matrices of the form

$$\begin{pmatrix}0 & \beta\\ \gamma & 0\end{pmatrix}$$

with $\beta, \gamma \in F^\times$ form a set of representatives for the orbits of $N(E)$ on $S(F) - S(F) \cap P(E)$. It follows that the integral of $-K^{reg}$ is

$$-|t_0|^{1/2} \sum_{\gamma\in F^\times} \int_{N(E_{\mathbb{A}})} \Phi\left[g^*n^*z\begin{pmatrix}0 & 1\\ \gamma & 0\end{pmatrix}ng\right] dn\, d^\times z.$$

To compute (47) we set

$$K^{whi}(g) = \int_{F_{\mathbb{A}}/F} K^{eis}\left[\begin{pmatrix} 1 & x \\ 0 & 1 \end{pmatrix} g\right] \psi_0(x)\, dx\, d^\times z,$$

and we see that (41) is equal to the sum of three terms

$$|t_0|^{1/2}\left\{\sum_{\beta \in F} \int \Phi\left[g^* \begin{pmatrix} 1 & \beta \\ 0 & 1 \end{pmatrix} zg\right] d^\times z \right. \tag{48}$$

$$\left. - \int_{F_{\mathbb{A}}} \Phi\left[g^* \begin{pmatrix} 1 & y \\ 0 & 1 \end{pmatrix} zg\right] d^\times z\, dy\right\} \tag{49}$$

$$- |t_0|^{1/2} \sum_{\gamma \in F^\times} \int_{N(E_{\mathbb{A}})} \Phi\left[g^* n^* z \begin{pmatrix} 0 & 1 \\ \gamma & 0 \end{pmatrix} ng\right] dn\, d^\times z \tag{50}$$

$$- |t_0|^{1/2} \sum_{\alpha \in F^\times} K^{whi}\left[\begin{pmatrix} \alpha & 0 \\ 0 & 1 \end{pmatrix} g\right]. \tag{51}$$

For each integral, z is integrated over $F_{\mathbb{A}}^\times$. To compute $\tilde{I}(s)$ we need to integrate each one of the above functions on $Z(F_{\mathbb{A}})(N_H)(F_{\mathbb{A}})P_H(F)\backslash H(F_{\mathbb{A}})$ against $f(h, s)\eta(\lambda(h))$. To do so, we use the Iwasawa decomposition: we write

$$h = \begin{pmatrix} t & 0 \\ 0 & 1 \end{pmatrix} k.$$

Then

$$dh = |t|^{-1}\, d^\times t\, dk, \qquad \eta(\lambda(h)) f(h, s) = \eta(t)|t|^s.$$

We obtain $\tilde{I}$ as the sum of three new integrals I_i, $1 \leqslant i \leqslant 3$. To compute $I_1(s)$ we recall the Schwartz-Bruhat function introduced earlier:

$$\phi(x) = \int \Phi\left[k^{-1} z \begin{pmatrix} 1 & z \\ 0 & 1 \end{pmatrix} k\right] \eta(\lambda(k))\, d^\times z\, dk.$$

We find

$$I_1(s) = |t_0|^{1/2} \int_{F_{\mathbb{A}}^\times/F^\times} \left[\sum_{\beta \in F} \phi(t^{-1}\beta) - \int \phi(t^{-1}x)\, dx\right] \eta(t)|t|^{s-1}\, d^\times t$$

or, by the Poisson summation formula,

$$|t_0|^{1/2} \int_{F_{\mathbb{A}}^\times/F^\times} \sum_{\beta \neq 0} \hat{\phi}(t\beta)|t|^s \eta(t)\, d^\times t.$$

Finally,

$$I_1(s) = |t_0|^{1/2} \int_{F_{\mathbb{A}}^\times} \hat{\phi}(t)|t|^s \eta(t)\, d^\times t. \tag{52}$$

For $I_2(s)$ we find

$$-|t_0|^{1/2} \int_{F_{\mathbb{A}}^\times/F^\times} \sum_\gamma \Phi\left[k^{-1} n^* \begin{pmatrix} 0 & z \\ \gamma t^2 z & 0 \end{pmatrix} nk\right] \times \eta(\lambda(k))\eta(t)|t|^{s+1}\, d^\times t\, dn\, dk\, d^\times z.$$

This can be further transformed into (see [JZ])

$$I_2(s) = -\frac{1}{2}|t_0|^{1/2} \sum_{\omega^2=\eta} \int \Phi\left[k^{-1} n^* \begin{pmatrix} 0 & z \\ zt & 0 \end{pmatrix} nk\right] \omega(t)|t|^{s+1}\eta(\lambda(k))\, d^\times t\, dn\, dk\, d^\times z. \tag{53}$$

The integral is for $z, t \in F_{\mathbb{A}}^\times$, $n \in N(E_{\mathbb{A}})$, $k \in K$, and the sum over all idèle-class characters ω of F whose square is η. Finally, we have

$$I_3(s) = -|t_0|^{1/2} \int \hat{F}_{\mathbb{A}}^\times K^{whi}\left[\begin{pmatrix} t & 0 \\ 0 & 1 \end{pmatrix} k\right] \eta(t)\eta(\lambda(k))|t|^{s-1}\, d^\times t\, dk. \tag{54}$$

Thus we have

$$I^{cusp}(s) = \sum_{\beta \in F - \{0,1\}} \Omega(\Phi, \beta, s) + \sum_{\beta \in F^\times/F^{\times 2}} \Omega^\infty(\Phi, \beta, s) + I^{uni}(\Phi, s) + \sum_i I_i(s). \tag{55}$$

5.5. We consider similarly a smooth function of compact support Φ' on $G'(F_{\mathbb{A}})$. We associate to Φ' a kernel

$$K'(x, y) = \int_{F_{\mathbb{A}}^\times/F^\times} \sum_{\xi \in GL(2,F)} \Phi'(x^{-1}\xi z y)\eta(z)\, d^\times z$$

and corresponding kernels K'^{cusp}, K'^{spe}, and K'^{eis}. We consider the integral

$$I'^{cusp}(s) = \int K'^{cusp}(g', g')E'(g, s)\, dg' \tag{56}$$

where E' is an Eisenstein series defined as

$$E'(g', s) = \sum_{P'(F)\backslash G'(F)} f'(\gamma g, s).$$

Just as before (see [JZ]), we have an expression

$$I'^{cusp}(s) = \sum_{\beta \in F - \{0,1\}} \Omega'(\Phi', \beta, s) \tag{57}$$

$$+ \sum_{\beta \in F^\times / F^{\times 2}} \Omega'^{\infty}(\Phi', \beta, s) \tag{58}$$

$$+ I'^{uni}(\Phi', s) \tag{59}$$

$$+ \sum_i I'_i(s) \tag{60}$$

where

$$\Omega'(\Phi', \beta, s) = \int \Phi'\left[g'^{-1}\begin{pmatrix} 1 & \beta \\ 1 & 1 \end{pmatrix} zg'\right]\eta(z) f'(g', s)\, dg',$$

$$\Omega'^{\infty}(\Phi', \beta, s) = \int \Phi'\left[g'^{-1}\begin{pmatrix} 0 & \beta \\ 1 & 0 \end{pmatrix} zg'\right]\eta(z) f'(g', s)\, dg',$$

$$I'^{uni}(\Phi', s) = \int \Phi'\left[g'^{-1}\begin{pmatrix} 1 & 0 \\ 1 & 1 \end{pmatrix} zg'\right]\eta(z) f'(g', s)\, dg',$$

and I'_i, $1 \leqslant i \leqslant 3$, is defined below.

We introduce as before a Schwartz-Bruhat function ϕ' on $F_{\mathbb{A}}$,

$$\phi'(x) = \int \Phi'\left[k^{-1} z \begin{pmatrix} 1 & x \\ 0 & 1 \end{pmatrix} k\right]\eta(z)\, d^\times z\, dk,$$

and the scalar part of the intertwining operator

$$\int f'(wn, s)\, dn = m'(s).$$

Then

$$I'^{uni}(\Phi', s) = m'(s) \int \phi'(t) |t|^s\, d^\times t. \tag{61}$$

The remaining integrals are

$$I'_1(s) = \int_{F_{\mathbb{A}}^\times} \hat{\phi}'(t) |t|^s\, d^\times t, \tag{62}$$

$$I'_2(s) = -\frac{1}{2} \sum_{\omega^2 = \eta} \int \Phi'\left[k^{-1} n_1 \begin{pmatrix} 0 & 1 \\ t & 0 \end{pmatrix} z n_2 k\right] \omega(t) |t|^{(s+1)/2}\, d^\times t\, dn_1\, dn_2\, dk\, \eta(z)\, d^\times z. \tag{63}$$

We also define

$$K'^{whi}(g_1, g_2) = \int_{F_{\mathbb{A}}/F} K'^{eis}(n_1 g_1, n_2 g_2)\psi(x_1)\psi(x_2)\, dn_1\, dn_2.$$

Then

$$I'_3(s) = \int K'^{whi}\left[\begin{pmatrix} t & 0 \\ 0 & 1 \end{pmatrix} k, \begin{pmatrix} t & 0 \\ 0 & 1 \end{pmatrix} k\right] |t|^{s-1}\, d^\times t\, dk. \tag{64}$$

5.6. We now compare the two formulas we have obtained. We assume that Φ and Φ' are products of local functions. At a place v of F which is inert in E, we assume that Φ_v and Φ'_v have matching orbital integrals. At a place v of f which splits in E, we can identify the variety S_v to G'_v and then we take $\Phi_v = \Phi'_v$. The functions f_ε are also taken to be product of local functions. At a place v which is inert in E, we take

$$\int_{H^0_{\varepsilon v}} f_{\varepsilon, v}(hg)\, dh = \Phi_v(g^* \varepsilon g).$$

At a place v which splits into two places v_1, v_2 of E, we may identify F_{v_i} to F_v. Then S_v is the set of pairs (s_1, s_2) with $s_1 \in GL(2, F_v)$ and

$$s_2 = w^t s_1 w.$$

Then we take

$$\Phi_v(s_1, s_2) = \Phi'_v(s_1).$$

The image of ε in S_v is $(\varepsilon_1, \varepsilon_2)$ and the corresponding group $H^0_{\varepsilon v}$ is the set of pairs (h_1, h_2) with

$$h_2 = w^t \varepsilon_1^{-1t} h_1^{-1t} \varepsilon_1 w.$$

We demand that

$$\int f_{\varepsilon v_1}(h_1 g_1) f_{\varepsilon v_2}(h_2 g_2)\, dh_1 = \Phi'_v(w^t g_2 w \varepsilon_1 g_1),$$

that is,

$$\int f_{\varepsilon v_1}(h_1 g_1) f_{\varepsilon v_2}(h_2)\, dh_1 = \Phi'_v(\varepsilon_1 g_1).$$

If Φ'_v is a Hecke function and ε_1 is in K'_v, and $f_{\varepsilon v_i}$ are Hecke functions, this relation simplifies. Indeed, since h and ${}^t h$ have the same double coset for the maximal compact of $GL(2, F_v)$, we get

$$\int f_{\varepsilon v_1}(hg) f_{\varepsilon v_2}(h^{-1})\, dh = \Phi'_v(g).$$

In what follows, we will assume that, for a finite place u of F which splits in E, the function Φ'_u and the functions $f_{\varepsilon u_1}$, $f_{\varepsilon u_2}$ are supercuspidal. Thus in the above discussion the kernels K^{eis}, K^{spe}, K'^{eis}, K'^{spe} are zero. Moreover, the integrals $I_2(s)$, $I_3(s)$ and $I'_2(s)$, $I'_3(s)$ are 0.

The formal computations simplify. For $I^{cusp}(s)$ for instance, we find directly that

$$I^{cusp}_{\Phi}(s) = \sum_{\beta \in F - \{0,1\}} \Omega(\Phi, \beta, s) + \sum_{\beta \in F^\times/F^{\times 2}} \Omega^{r\infty}(\Phi, \beta, s) + I^{uni}(\Phi, s) + I_1(\Phi, s) \tag{65}$$

where

$$I_1(\Phi, s) = |t_0|^{1/2} \int_{F^\times_{\mathbb{A}}/F^\times} \left[\sum_{\beta \in F} \phi(t^{-1}\beta)\right] |t|^{s-1}\, d^\times t.$$

However, because of the cuspidality condition, we find

$$\int_{F_{\mathbb{A}}} \Phi\left[g^* \begin{pmatrix} 1 & x \\ 0 & 1 \end{pmatrix} g\right] dx = 0.$$

Thus the integral of ϕ vanishes, and we can write also

$$I_1(s) = |t_0|^{1/2} \int_{F^\times_{\mathbb{A}}/F^\times} \left[\sum_{\beta \in F} \phi(t^{-1}\beta) - \int \phi(t^{-1}x)\, dx\right] \eta(t)|t|^{s-1}\, d^\times t.$$

Thus $I_1(s)$ has the same expression as before.

To justify our formal computation, we need only show that in the above expression for I^{cusp} all the integrals and sums converge absolutely. The above expression for $I_1(s)$ converges absolutely for $\Re s > 1$. Similarly, the expression for I^{uni} converges absolutely for $\Re s > 1$. To prove the convergence of $\Omega(\Phi, \sigma, s)$ we proceed as in the local case, that is, we write it as a Tate integral over the multiplicative group of an algebra L. The trace of the matrix in the integral for $\Omega(\Phi, \beta, s)$ is $2z$ while its determinant is $z^2(1-\beta)$. Thus $z^2(1-\beta)$ remains in a compact set of the idèles while z remains in a compact set of the adèles. The first condition implies in particular that $|z|$ is in a compact set of $\mathbb{R}^\times$. The second condition implies then that z is in a compact set of the idèles. In turn, this implies that $1-\beta$ is in a compact set of the idèles and hence in a finite set. Thus β takes only finitely many values. The determinant of the matrix in $\Omega^\infty(\Phi, \beta, s)$ is $-z^2\beta$. It remains in a fixed compact set of the idèles. This implies that β takes only finitely many values (modulo $F^{\times 2}$). Hence our formal computation is justified, for $\Re s > 1$.

Similarly, we find

$$I'^{cusp}(s) = \sum_{\beta \in F - \{0,1\}} \Omega'(\Phi', \beta, s) \tag{66}$$

$$+ \sum_{\beta \in F^\times/F^{\times 2}} \Omega'^{\infty}(\Phi', \beta, s) \tag{67}$$

$$+ I'^{uni}(\Phi', s) \tag{68}$$

$$+ I'_1(s) \tag{69}$$

for $\Re s > 1$.

5.7.

THEOREM 5.2. *Under the previous assumptions (in particular, the condition of cuspidality at a place u), we have for all s*

$$|t_0|^{-1/2} L(s, 1_F) I^{cusp}(s) = L(s, \eta) I'^{cusp}(s).$$

Proof. We claim that for any place v of F

$$|t_{v0}|^{-1/2} L(s, 1_v) \Omega(\Phi_v, \beta, s) = L(s, \eta_v) \Omega'(\Phi'_v, \beta, s).$$

Indeed, this is the condition of matching orbital integrals if v is inert. Suppose v splits; then $t_{v0} = 1$ by choice, and the character η_v is trivial and the functions Φ_v, Φ'_v identical. The equality is thus trivial. Multiplying these equalities together, we obtain

$$|t_0|^{-1/2} L(s, 1_F) \Omega(\Phi, \beta, s) = L(s, \eta_v) \Omega'(\Phi'_v, \beta, s).$$

Similarly,

$$|t_0|^{-1/2} L(s, 1_F) \Omega^{\infty}(\Phi, \beta, s) = L(s, \eta_v) \Omega'^{\infty}(\Phi', \beta, s).$$

To complete the proof we need to show that

$$|t_0|^{-1/2} L(s, 1_F) I^{uni}(\Phi, s) = L(s, \eta) I'^{uni}(\Phi', s),$$

$$|t_0|^{-1/2} L(s, 1_F) I_1(\Phi, s) = L(s, \eta) I'_1(\Phi', s).$$

Since $m(s) = |t_0|^s m'(s)$, the first relation is equivalent to

$$|t_0|^{s-1/2} L(s, 1_F) \int \phi(t) |t|^s \eta(s)\, d^\times t = L(s, \eta) \int \phi'(t) |t|^s\, d^\times t.$$

The functions ϕ and ϕ' are products of local functions. We claim that, for any place v of F,

$$|t_{0v}|^{s-1/2}L(s, 1_v) \int \phi_v(t)|t|^s\eta(s)\, d^\times t = L(s, \eta_v) \int \phi_v'(t)|t|^s\, d^\times t.$$

Indeed, if v is inert, this is a consequence of the local theory. If v is split, this equality is trivial. The global equality follows.

Now we have

$$|t_0| = \frac{|D_F|^2}{|D_E|}. \tag{70}$$

Using the functional equation of the L-functions, we see the global identity just obtained reads

$$L(1 - s, 1_F) \int \phi(t)|t|^s\eta(t)\, d^\times t = L(1 - s, \eta) \int \phi'(t)|t|^s\, d^\times t,$$

or changing s to $1 - s$,

$$L(s, 1_F) \int \phi(t)|t|^{1-s}\eta(t)\, d^\times t = L(s, \eta) \int \phi'(t)|t|^{1-s}\, d^\times t.$$

Using the global Tate functional equation, we find

$$L(s, 1_F) \int \hat{\phi}(t)|t|^s\eta(s)\, d^\times t = L(s, \eta) \int \hat{\phi}'(t)|t|^s\, d^\times t.$$

Now

$$\begin{aligned} |t_0|^{-1/2}L(s, 1_F)I_1(s) &= |t_0|^{-1/2}L(s, 1_F)|t_0|^{1/2} \int \hat{\phi}(t)|t|^s\eta(t)\, d^\times t \\ &= L(s, 1_F) \int \hat{\phi}(t)|t|^s\eta(t)\, d^\times t \\ &= L(s, \eta) \int \hat{\phi}'(t)|t|^s\, d^\times t \\ &= L(s, \eta)I_1'(s). \end{aligned}$$

This concludes the proof of the theorem. □

5.8. We remark that the identity of the theorem is true for any normalization of the Eisenstein series $E'(g, s)$ provided the Eisenstein series $E(g, s)$ verifies

$$E(g, s) = E'(\xi(g), s).$$

We choose E' in such a way that its residue at 1 is the constant function equal to 1. We write

$$E(g, s) = \frac{1}{s-1} + \theta(g) + (s-1)O(1).$$

We evaluate the residues of the two sides of our identity at $s = 1$. The right-hand side has a pole of order at most 1. Its residue is simply

$$L(1, \eta) \int K'^{cusp}(g', g')\, dg';$$

that is, apart from the factor $L(1, \eta)$, it is the standard trace formula. The left-hand side must have a simple pole at $s = 1$. This shows that I^{cusp} is holomorphic at $s = 1$, or what amounts to the same,

$$\int K^{cusp}(h)\, dh = 0.$$

The residue of the left-hand side is thus

$$|t_0|^{-1/2} \operatorname{Res}_{s=1} L(s, 1_F) \int K^{cusp}(h)\theta(h)\, dh.$$

More explicitly,

$$|t_0|^{-1/2} \operatorname{Res}_{s=1} L(s, 1_F) \sum_{\varepsilon} \int K^{cusp}_{\varepsilon}(h_\varepsilon, h)\theta(h) d_\varepsilon\, dh.$$

We have thus proved the weighted trace formula announced in the introduction.

5.9. We now state some consequences of our formula. Let π be a cuspidal automorphic representation of G', with a supercuspidal component at u, and central character η. The representation $\pi \otimes \eta$ gives the same contribution to our residual trace formula. The sum of these contributions is equal to the contribution of the base change Π of π. It follows that this contribution is nonzero. Thus there must be an ε such that Π is distinguished with respect to H_ε. Moreover, there must be a ϕ in the space of Π such that the integral

$$\int \phi(h)\theta(h)\eta(h)\,dh$$

is nonzero. Of course this result could be obtained directly in terms of the relevant L-functions. It is likely that similar results can be obtained for any symmetric space.

5.10. We comment on what should happen when the assumption of cuspidality at one place is removed. The kernel K_{ε}^{eis} is integrable in the first variable on H_{ε}; on the other hand, K^{eis} is itself an integral over a certain set of quasi characters. The double integral does not converge when $\varepsilon = 1$; when one interchanges the order of integration, one gets a continuous sum, i.e. a convergent integral over a set of quasi characters and a discrete sum. If $\varepsilon \neq 1$, then one gets only a continuous sum. Thus the term $I_3(s)$ is the sum of a continuous part and a discrete part. The discrete part should match the contribution of the dihedral representations on G'. The continuous sum should be equal to the integral $I'_3(s)$. The terms I_2 and I'_2 should be equal.

6. The second trace formula. We now explain how to modify the previous constructions to obtain the second trace formula mentioned in the introduction.

We consider first the local situation: E is an extension of a nonarchimedean field F of odd residual characteristic. The notations f' and τ_0 have the same meaning as before. However, we need to replace τ_0 by its inverse. Thus we set

$$\tau_1 = \tau_0^{-1}.$$

We define an isomorphism

$$\xi_1(g) = \begin{pmatrix} \sqrt{\tau_1} & 0 \\ 0 & 1 \end{pmatrix} g \begin{pmatrix} \sqrt{\tau_1^{-1}} & 0 \\ 0 & 1 \end{pmatrix}$$

of H/Z onto $G'Z/Z$. We set

$$f_1(g) = f'(\xi_1(g)).$$

For Φ smooth of compact support on S, we now set

$$\Omega(\Phi, \sigma, s) = \int \Phi(h^{-1}\sigma z h) f_1(h, s)\,dh\,d^{\times}z.$$

For Φ' smooth of compact support on G', we set

$$\Omega'(\Phi', \sigma, s) = \int \phi'(g^{-1}\sigma z g) f'(g, s)\eta(\det g)\,dg\,\eta(z)\,d^{\times}z.$$

We say that Φ and Φ' have "matching orbital integrals" if

$$|\tau_1|^{-1/2}L(s,\eta)\Omega(\Phi,\sigma,s)=\eta(\beta_1)L(s,1_F)\Omega'(\Phi',\sigma,s) \tag{71}$$

for all

$$\sigma=\begin{pmatrix}\alpha & \beta_2\\ \beta_1 & \alpha\end{pmatrix}$$

with $\beta_1\beta_2\neq 0$. As before, we introduce the notations $\Omega(\Phi,\beta,s)$, $\Omega^\infty(\Phi,\beta,s)$ and the similar notations for Ω'. Just as before, we have the following fundamental lemma:

PROPOSITION 6.1. *If the extension E is unramified, then the functions Φ_0 and Φ'_0 have matching orbital integrals. More generally, if Φ is a function of compact support on S invariant under K and Φ' is a function of compact support on G' bi-invariant under K' which corresponds to Φ, then Φ and Φ' have matching orbital integrals.*

Proof. Recall that τ_1 is then any unit which is not a square. With the notation of Section 3, it will suffice to show that Φ_n and Φ'_{2n} have matching orbital integrals.

We get as before that $\Omega(\Phi_n,\beta,s)$ and $\Omega'(\Phi'_{2n},\beta,s)$ are zero unless

$$|1-\beta|=|\gamma|^2,$$

and then

$$\Omega(\Phi_n,\beta,s)=|\gamma|^s q^{ns}A\left(\frac{\beta\varpi^{2n}}{\gamma^2},X\right),$$

$$\Omega'(\Phi_{2n},\beta,s)=|\gamma|^s q^{ns}A\left(\frac{\beta\varpi^{2n}}{\gamma^2\tau_1},-X\right).$$

Then our assertion follows from the functional equation satisfied by the function A.

Next, we have the following result:

PROPOSITION 6.2. *Given Φ, there is a Φ' with matching orbital integrals and conversely.*

Proof. We only discuss the asymptotic behavior of our integrals when β tends to 0 and check they are compatible.

We first study the asymptotic behavior of the integrals $\Omega(\Phi,\beta,s)$ when β tends to 0. As before, we will denote by $I(\Phi,s)$ thé integral corresponding to $\beta=0$.

$$I(\Phi,s)=\int\Phi\left[h^{-1}z\begin{pmatrix}1 & 0\\ 1 & 1\end{pmatrix}h\right]f(h,s)\,dh.$$

We consider also the integral

$$J(\Phi, s) = \int \Phi\left[h^{-1}z\begin{pmatrix}1 & 1\\ 0 & 1\end{pmatrix}h\right]f(h, s)\, dh\, d^\times z,$$

the integral over the quotient $ZN_H\backslash H$.

It will be convenient to introduce the following function ϕ on F:

$$\phi(x) = \int \Phi\left[k^{-1}z\begin{pmatrix}1 & x\\ 0 & 1\end{pmatrix}k\right] dk\, d^\times z. \tag{72}$$

Thus ϕ is an arbitrary Schwartz-Bruhat function on F such that

$$\phi(ux) = \phi(x)$$

for all $u \in F$ with $|u| = 1$. In terms of ϕ,

$$I(\Phi, s) = m_1(s) \int \phi(t)|t|^s\eta(t)\, d^\times t, \tag{73}$$

$$J(\Phi, s) = \int \phi(t)|t|^{1-s}\, d^\times t. \tag{74}$$

The first integral converges for $\Re s > 0$ and the second for $\Re s < 1$. These distributions determine the "Shalika germs" of the orbital integrals. Namely, let $A > 0$ be a constant and set for β an integer

$$C_1(\beta, s) = \int_{|x| \leqslant A} (|\beta - x^2|^{s-1} - |x^2|^{s-1})\, dx \tag{75}$$

where

$$x = \frac{u}{\sqrt{\tau_1}}, \qquad u \in F, dx = du.$$

PROPOSITION 6.3. *Given Φ, for $|\beta|$ sufficiently small and all s with $0 < \Re s < 1$,*

$$\Omega(\Phi, \beta, s) = I(\Phi, s) + C_1(\beta, s)J(\Phi, s).$$

The proof is the same as before. □

6.1. We consider similarly the integrals Ω'. Recall the integral

$$\Omega'(\Phi', \beta, s) = \int \Phi'[g^{-1}z\sigma g]f(g, s)\eta(\det g)\, dg\, \eta(z)\, d^\times z$$

where

$$\sigma = \begin{pmatrix} 1 & \beta \\ 1 & 1 \end{pmatrix}.$$

As before, we consider the orbital integral corresponding to $\beta = 0$:

$$I'(\Phi', s).$$

Similarly, we define

$$J'(\Phi', s) = \int \Phi'\left[g'^{-1}z\begin{pmatrix} 1 & 1 \\ 0 & 1 \end{pmatrix}g'\right] f'(g', s)\eta(\det g')\, dg'\, \eta(z)\, d^{\times}z.$$

As before, we introduce the following function ϕ' on F:

$$\phi'(x) = \int \Phi'\left[k^{-1}z\begin{pmatrix} 1 & x \\ 0 & 1 \end{pmatrix}k\right]\eta(\det k)\, dk\, \eta(z)\, d^{\times}z.$$

Thus ϕ' is an arbitrary Schwartz-Bruhat function on F such that

$$\phi'(ux) = \eta(u)\phi'(x)$$

for all $u \in F$ with $|u| = 1$. In terms of ϕ', we have

$$I'(\Phi', s) = m'(s)\int \phi'(t)\eta(t)|t|^{s}\, d^{\times}t, \tag{76}$$

$$J'(\Phi', s) = \int \phi'(t)\eta(t)|t|^{1-s}\, d^{\times}t. \tag{77}$$

We define

$$C_1'(\beta, s) = \int_{|x|\leqslant A} (\eta(\beta - x^2)|\beta - x^2|^{s-1} - \eta(-x^2)|x^2|^{s-1})\, dx.$$

As before, this does not depend on the value of A, provided $|\beta|$ is small enough and we have the following proposition:

PROPOSITION 6.4. *Given Φ', for $|\beta|$ sufficiently small and all s with $\Re s > 0$,*

$$\Omega'(\Phi', \beta, s) = I'(\Phi', s) + C_1'(\beta, s)J'(\Phi', s).$$

The proof is the same as before.

6.2. The first step and the only one we discuss in matching the orbital integrals is the following lemma:

LEMMA 6.5. *Given Φ, there is Φ' such that for all s and all β sufficiently small,*

$$|\tau_1|^{-1/2}\Omega(\Phi, \beta, s)L(s, \eta) = \Omega'(\Phi', \beta, s)L(s, 1_F).$$

Proof. It suffices to show that, given Φ, there is Φ' with

$$|\tau_1|^{-1/2}I(\Phi, s)L(s, \eta) = I'(\Phi', s)L(s, 1_F), \tag{78}$$

$$|\tau_1|^{-1/2}C_1(\beta, s)J(\Phi, s)L(s, \eta) = C_1'(\beta, s)J'(\Phi', s)L(s, 1_F). \tag{79}$$

Since

$$m_1(s) = m'(s)|\tau_1|^s,$$

the relation (78) reads

$$\eta(-1)|\tau_1|^{s-1/2}\int \phi(x)|x|^s\, d^\times x\, L(s, \eta) = \int \phi'(x)\eta(x)|x|^s\, d^\times x\, L(s, 1_F).$$

As before, given ϕ, there is ϕ' such that this relation is satisfied. Then the relation (79) will be a consequence of the following identity that we state as a lemma:

LEMMA 6.6. *With the above notation,*

$$\frac{L(s, 1_F)C_1'(\beta, s)}{L(1-s, 1_F)} = \eta(-1)|\tau_1|^{s-1}\frac{L(s, \eta)C_1(\beta, s)}{L(1-s, \eta)},$$

provided $|\beta|$ is small enough.

Indeed, we have

$$C_1(\beta, s) = |\tau_0|^{s-1}C'\left(\frac{\beta}{\tau_0}, s\right),$$

$$C_1'(\beta, s) = \eta(-1)|\tau_0|^{s-1}C\left(\frac{\beta}{\tau_0}, s\right).$$

Thus the above equation is equivalent to the corresponding equation for the pair (C, C').

6.3. Now we let E/F be a quadratic extension of number fields. As before, we assume that every real place of F splits in E and that every finite place of even residual characteristic splits in E. For each place v of F inert in E we choose an element $t_{1,v}$ such that

$$E_v = F_v(\sqrt{t_{1,v}}). \tag{80}$$

If v is unramified in E, we take $t_{1,v}$ of module 1; if v is ramified, we take for $t_{1,v}$ the inverse of a uniformizer of F_v. If v splits in E, we take $t_v = 1$. We set

$$t_1 = \prod t_{1,v}. \tag{81}$$

Conjugacy by

$$\begin{pmatrix} \sqrt{t_{1,v}} & 0 \\ 0 & 1 \end{pmatrix} \tag{82}$$

defines for each inert place v an isomorphism

$$\xi_{1,v}\colon H(F_v) \to G'(F_v)Z(F_v).$$

At a split place we have a natural isomorphism. As before, we define an isomorphism

$$\xi_1\colon H(F_{\mathbb{A}}) \to G'(F_{\mathbb{A}})Z(F_{\mathbb{A}}).$$

We use it to transport $K'U$ and f' to H. We denote by K_H and f_1 the corresponding maximal subgroup and function. Thus

$$f_1(h, s) = |t|^s, \qquad \text{if } h = \begin{pmatrix} tz & x \\ 0 & z \end{pmatrix} k.$$

Then

$$E_1(h, s) = \sum_{\gamma \in P_H(F)\backslash H(F)} f_1(\gamma h, s).$$

On $G'(F_{\mathbb{A}})/Z'(F_{\mathbb{A}})$, we use the same measure dg' as before. We transport this measure to $H(F_{\mathbb{A}})/Z(F_{\mathbb{A}})$. The measure need not be the same as the one we have used to define the kernels in terms of Φ. Given a smooth function of compact support Φ on $S(F_{\mathbb{A}})$, we define as before a cuspidal kernel K^{cusp} and the integral

$$I^{cusp}(s) = \int_{Z(F_{\mathbb{A}})H(F)\backslash H(F_{\mathbb{A}})} K^{cusp}(h) E_1(h, s)\, dh.$$

Similarly, given a smooth function of compact support Φ' on $G'(F_{\mathbb{A}})$, we define the cuspidal kernel K'^{cusp} and the integral

$$I'^{cusp}(s) = \int K'^{cusp}(g', g')\eta(\det g')E'(g, s)\, dg' \tag{83}$$

where E' is the Eisenstein series defined as

$$E'(g', s) = \sum_{P'(F)\backslash G'(F)} f'(\gamma g, s).$$

We assume that Φ and Φ' are product of local functions. At a place v of F inert in E, we assume that Φ_v and Φ'_v have matching orbital integrals in the sense of the present section. At a place v of f which splits in E, we can identify the variety S_v to G'_v and then take $\Phi_v = \Phi'_v$. The functions f_ε are chosen as before. Then we have the following result.

THEOREM 6.7. *Under the previous assumptions (in particular, the condition of cuspidality at a place u), we have for all s*

$$|t_1|^{-1/2} L(s, \eta) I^{cusp}(s) = L(s, 1_F) I'^{cusp}(s).$$

The proof is the same as before. □

6.4. If we normalize the Eisentein series as before and take the residue at $s = 1$ of both sides of our identity, we obtain

$$|t_1|^{-1/2} L(1, \eta) \sum_\varepsilon \int K_\varepsilon(h_\varepsilon, k)\, dh_\varepsilon\, dh$$

$$= \operatorname{Res} L(s, 1)|_{s=1} \int K'_{cusp}(g', g')\theta(g')\eta(\det g')\, dg'.$$

This is the other weighted trace formula mentioned in the introduction.

REFERENCES

[A] T. ASAI, *On certain Dirichlet series associated with Hilbert modular forms and Rankin's method*, Math. Ann. **226** (1977), 81–94.

[F] Y. FLICKER, *On distinguished representations*, J. Reine Angew. Math. **418** (1991), 139–172.

[FH] Y. FLICKER AND J. HAKIM, *Quaternionic distinguished representations*, to appear in Amer. J. Math.

[Ha] J. HAKIM, *Distinguished p-adic representations*, Duke Math. J. **62** (1991), 1–22.

[HLR] G. HARDER, R. LANGLANDS, AND M. RAPOPORT, *Algebraische Zyklen auf Hilbert-Blumenthal Flächen*, J. Reine Angew. Math. **366** (1986), 53–120.

[Ho] R. HOWE, *Theta Series and Invariant Theory*, Proc. Sympos. Pure Math. **33**, Amer. Math. Soc., Providence, 1979, 275–306.

[I1] H. IWANIEC, *Fourier coefficients of modular forms of half integral weight*, Invent. Math. **87** (1987), 385–401.

[I2] ———, *On Waldspurger's theorem*, Acta Arith. **49** (1987), 205–212.

[J1] H. JACQUET, *Automorphic forms on GL(2) II*, Lecture Notes in Math. **278**, Springer-Verlag, New York, 1972.

[J2] ———, *On the nonvanishing of some L-functions*, Proc. Indian Acad. Sci. Math. Sci. **97** (1987), 117–155.

[J3] ———, *Kloosterman integrals for GL(3)*, to appear in Canad. J. Math.
[JL] H. Jacquet and K. F. Lai, *A relative trace formula*, Compositio Math. **54** (1985), 243–310.
[JR] H. Jacquet and S. Rallis, *Kloosterman integrals for skew symmetric matrices*, Pacific J. Math. **154** (1992), 265–283.
[JY1] ———, *Une remarque sur le changement de base quadratique*, C.R. Acad. Sci. Paris Sér. I Math. **311** (1990), 671–676.
[JY2] H. Jacquet and Y. Yangbo, *Relative Kloosterman integrals for GL(3)*, Bull Soc. Math. France **120** (1992), 263–295.
[JZ] H. Jacquet and D. Zagier, *Eisenstein series and the Selberg trace formula II*, Trans. Amer. Math. Soc. **300** (1987), 1–47.
[kL] K. F. Lai, *Algebraic cycles on compact Shimura surfaces*, Math. Z. **189** (1985), 593–602.
[KR] B. Kostant and S. Rallis, *Orbits and representations associated with symmetric spaces*, Amer. J. Math. **93** (1971), 275–306.
[rL] R. Langlands, *Base Change for GL(2)*, Ann. of Math. Stud. **96**, Princeton Univ. Press, Princeton, 1980.
[M] Z. Mao, *Relative Kloosterman integrals for the unitary group*, C.R. Acad. Sci. Paris Sér. I Math. **315** (1992), 381–386.
[R] R. Richardson, *Orbits, invariants, and representations associated to involutions of reductive groups*, Invent. Math. **66** (1982), 287–312.
[Y] Y. Ye, *Kloosterman integrals and base change for GL(2)*, J. Reine Angew. Math. **400** (1989), 57–121.
[Z] D. Zagier, *Modular forms associated to real quadratic fields*, Invent. Math. **30** (1975), 1–46.

Jacquet: Department of Mathematics, Columbia University, New York, New York, 10027, USA
Lai: School of Mathematics and Statistics, The University of Sydney, Sydney NSW 2006, Australia
Rallis: Department of Mathematics, The Ohio State University, Columbus, Ohio 43210, USA

Journal of Number Theory **87**, 109–143 (2001)
doi:10.1006/jnth.2000.2584, available online at http://www.idealibrary.com on IDEAL®

Factorization of Period Integrals

Hervé Jacquet[1]

Columbia University, New York, New York 10027
E-mail: hj@math.columbia.edu

Communicated by S. Rallis

Received January 10, 2000; published online January 30, 2001

TO THE MEMORY OF YASUKO JACQUET

We show that for certain quadratic extensions E/F of number fields the period integral of a cusp form of $GL(3, E)$ over the unitary group H_0 in three variables is a product of local linear forms.

Contents.

1. GLOBAL RESULTS

Let E/F be a quadratic extension of number fields. We will denote by σ the non-trivial element of the Galois group of E/F and will often write $\sigma(z) = \bar{z}$. We will denote by $\mathbb{U}_1$ the unitary group in 1 variable. We assume that every Archimedean place of F *splits* in E. We let H_0 be the unitary group for the 3×3 identity matrix. Recall that a cuspidal automorphic representation Π of $GL(3, E_{\mathbb{A}})$ is said to be *distinguished* by H_0 if the linear form:

$$\mathscr{P}(\phi) := \int_{H_0(F)\backslash H_0(F_{\mathbb{A}})} \phi(h)\, dh,$$

[1] Partially supported by NSF Grant DMS 9619766 and a grant from Banker Trust to the IAS.

0022-314X/01 $35.00

is non identically zero on the space $\mathscr{V}(\Pi)$ of smooth vectors of Π. Our main result is that if this linear form is non-zero, then it can be written as a tensor product of local linear forms. This is a rather startling result, since there is no local property of uniqueness which guarantees in advance the existence of such a decomposition.

In more detail, if Π is distinguished, then the representation Π^σ defined by $\Pi^\sigma(g)=\Pi(g^\sigma)$ is equivalent to Π. Moreover, the following condition is satisfied for every place v_0 of F inert in E; let v be the corresponding place of E and $\mathscr{H}_{v_0}=\mathscr{H}(\Pi_v, H_{0,v})$ be the space of linear forms on the space $\mathscr{V}(\Pi_v)$ of smooth vectors of Π_v which are invariant under H_{0,v_0}. Then $\mathscr{H}_{v_0}\neq 0$. If v_0 is a place of F which split into v_1 and v_2 in E let similarly $\mathscr{H}_{v_0}=\mathscr{H}(\Pi_{v_1}\otimes\Pi_{v_2}, H_{0,v_0})$ be the space of linear forms on the space $\mathscr{V}(\Pi_{v_1}\otimes\Pi_{v_2})$ of smooth vectors for the tensor product $\Pi_{v_1}\otimes\Pi_{v_2}$ which are invariant under H_{0,v_0}. Now H_{0,v_0} is isomorphic to the group of pairs (g_{v_1}, g_{v_2}) such that $g_{v_1}={}^t g_{v_2}^{-1}$, $g_{v_1}\in GL(3, F_{v_0})$. Furthermore $\Pi_{v_1}=\Pi_{v_2}$. Thus the dimension of $\mathscr{H}_{v_0}$ is actually one.

Let S_0 be a finite set of places of F containing all the places at infinity, the even places, and the places which ramify in E. Let S_i be the set of places in S_0 which are inert in E and let S_s be the set of split places. Let S be the set of places of E above a place of S_0. Let Π be a distinguished representation. Suppose that Π is unramified outside S. We let $\mathscr{V}^S(\Pi)$ or simply $\mathscr{V}^S$ be the subspace of vectors in $\mathscr{V}(\Pi)$ which are invariant under $K^S:=\prod_{v\notin S}K_v$, $K_v=GL(3, \mathcal{O}_v)$. We consider the elements of $\mathscr{V}^S(\Pi)$ which are *pure tensors*. They can be described in terms of the Whittaker models as follows. Let ψ be a non-trivial character of $F_{\mathbf{A}}/F$. Set $\psi_E(z)=\psi(z+\bar z)$. Denote by N the group of upper triangular matrices with unit diagonal, and by θ the character of $N(F_{\mathbf{A}})$ defined by:

$$\theta(n)=\psi(n_{1,2}+n_{2,3}).$$

Similarly, define a character $n\mapsto\theta(n\bar n)$ on $N(E_{\mathbf{A}})$ by:

$$\theta(n\bar n)=\psi_E(n_{1,2}+n_{2,3}).$$

For $\phi\in\mathscr{V}(\Pi)$, set

$$\mathscr{W}(\phi)=\int_{N(E)\backslash N(E_{\mathbf{A}})}\phi(n)\,\theta^{-1}(n\bar n)\,dn,\qquad W(g)=\mathscr{W}(\Pi(g)\,\phi).$$

Then

$$\phi(g)=\sum_{\gamma\in N(2,E)\backslash GL(2,E)} W\left[\begin{pmatrix}\gamma & 0\\ 0 & 1\end{pmatrix} g\right],$$

where $N(2)$ denotes the group of upper triangular matrices with unit diagonal in $GL(2)$. If ϕ is a pure tensor in $\mathscr{V}^S$, then the corresponding function W has the form

$$W(g) = \prod W_v(g_v),$$

where W_v is in the Whittaker model $\mathscr{W}(\Pi_v)$ of Π_v for every place v, and, for $v \notin S$, $W_v = W^{K_v}$ is the unique element of $\mathscr{W}(\Pi_v)$ which is invariant under K_v and equal to 1 on K_v. Whenever convenient, we identify the space $\mathscr{V}(\Pi_v)$ with $\mathscr{W}(\Pi_v)$. Then our precise result is the following theorem:

THEOREM 1. *There exist a constant $c \neq 0$ and, for each $v_0 \in S_0$, a non-zero element $\mathscr{P}_{v_0} \in \mathscr{H}_{v_0}$ such that, for any pure tensor ϕ in $\mathscr{V}^S$,*

$$\mathscr{P}(\phi) = c \prod_{v_0 \in S_i} \mathscr{P}_{v_0}(W_v) \prod_{v_0 \in S_s} \mathscr{P}_{v_0}(W_{v_1} \otimes W_{v_2}).$$

We remark that the proof of the theorem will provide us with a specific choice of the local linear forms and also a specific value for the constant c in terms of L-functions.

Since the spaces $\mathscr{H}_{v_0}$ with $v_0 \in S_i$ are not one-dimensional in general, the existence of such a decomposition is *not formal.* Moreover, the theorem implies the non-trivial result that the linear form $\mathscr{P}$ is non-zero on $\mathscr{V}^S$, for any S satisfying the above conditions. A priori, one can only say that if it is non-zero, then it is non-zero on a space $\mathscr{V}^S$, with a large enough S.

The paper is arranged as follows. We review the relative trace formula of [JY] in Section 3. We prove the main theorem in Section 4. In Section 5 and 7 we prove local results. Let E/F be a local quadratic extension of non-Archimedean fields. If Π is a supercuspidal representation of $GL(3, E)$ we prove in Section 5 that the dimension of $\mathscr{H}(\Pi, H_0(F))$ is at most one. In Section 6, we review the local theory for $GL(2)$. In Section 7 we show, that, at least for certain irreducible representations Π of $GL(3, E)$, the dimension of $\mathscr{H}(\Pi, H_0(F))$ is equal to the number of irreducible representations of $GL(3, F)$ (with a given central character) which base change to Π. Conjecturally, this should always be the case. Finally, in Section 8, we state general conjectures.

We note that a recent work of E. Lapid and J. Rogawski treat a related question ([LR]). Roughly speaking, they investigate the notion of distinguished representation and period integrals for non-cuspidal automorphic representations. Finally, we refer to [yF2] for the discussion of representations distinguished by $GL(n, F)$.

We now explain our notations. In general G denotes the group $GL(3)$ regarded as an algebraic group. The context indicates whether we regard G as an algebraic group over F or over E. For instance, if v_0 is a place of F then G_{v_0} denotes the group $GL(3, F_{v_0})$. Likewise if v is a place of E then G_v denotes the group $GL(3, E_v)$. Moreover K_{v_0} and K_v denote the corresponding standard maximal compact subgroups. Throughout the paper S_0 and S are finite sets of places of F and E respectively satisfying the previous conditions. Then G_{S_0} is the product of the groups G_{v_0} with $v_0 \in S_0$ and G^{S_0} the (restricted) product of the groups G_{v_0} with $v_0 \notin S_0$. The notations F_{S_0}, $F^\times_{S_0}$, K_{S_0}, K^{S_0}, G_S, G^S, K_S, K^S have a similar meaning.

2. L^2-NORM OF A PURE TENSOR

We keep to the notations of the introduction. We recall how to compute the L^2-norm of a pure tensor in a cuspidal representation π of $GL(3, F_{\mathbf{A}})$. We assume that π is unramified outside S_0. An invariant scalar product on each space $\mathscr{W}_{v_0}(\Pi_{v_0})$ is given by:

$$(W_1, W_2) = \int_{N(2, F_{v_0})\backslash GL(2, F_{v_0})} W_1 \bar{W}_2 \left[\begin{pmatrix} g & 0 \\ 0 & 1 \end{pmatrix} \right] dg.$$

This is a result of Bernstein ([B]) in the non-Archimedean case and of Baruch in the Archimedean case ([Ba]). More precisely, Baruch shows that the restriction of the unitary representation Π_{v_0} to the group P_{v_0} of triangular matrices with last row $(0, 0, 1)$ is irreducible (as a unitary representation). On the other hand in [JS] it is shown that the irreducible representation τ of P_{v_0} induced by the character θ_{v_0} of N_{v_0} occurs in this restriction. The conclusion follows. In fact this assertion is also a consequence of the global theory as we are going to see (once one knows that the above Hermitian form is defined by a *convergent integral*).

This being so, there is a constant $c(F, S_0)$, which depends on F, S_0, and the choice of the Haar measures, but not on π, such that:

$$\|\phi\|^2 = c(F, S_0)\, L^{S_0}(1, \pi, \mathrm{Ad}) \prod_{v_0 \in S_0} \|W_{v_0}\|^2. \tag{1}$$

Here $L^{S_0}(s, \pi, \mathrm{Ad})$ denotes the partial adjoint L-function attached to π.

Indeed, let $\Phi = \prod \Phi_{v_0}$ be a Schwartz–Bruhat function in three variables where Φ_{v_0} is the characteristic function of $\mathcal{O}^3_{v_0}$ for $v_0 \notin S_0$. Consider the Epstein–Eisenstein series:

$$E(g, \Phi, s) = \int_{F^\times\backslash F^\times_{\mathbf{A}}} \sum_{\xi \in F^3 - \{0\}} \Phi(t\xi g)\, |t|^{3s}\, d^\times t\, |\det g|^s.$$

Then, for $\Re s \gg 0$, if ϕ_1, ϕ_2 are pure tensors,

$$\int \phi_1 \bar{\phi}(g)\, E(g, \Phi, s)\, dg$$

$$= L^{S_0}(s, 1_F)\, L^{S_0}(s, \pi, \mathrm{Ad}) \prod_{v_0 \in S_0} \int W_{1, v_0} \overline{W_{2, v_0}}(g)\, \Phi_{v_0}[(0, 0, 1)\, g]\, |\det g|^s\, dg.$$

Taking the residue at $s = 1$ we obtain

$$\int \phi_1 \bar{\phi}(g)\, dg \prod_{v_0 \in S_0} \int_{F^3_{v_0}} \Phi_{v_0}(x)\, dx$$

$$= \mathrm{Res}_{s=1} L^{S_0}(s, 1_F)\, L^{S_0}(1, \pi, \mathrm{Ad})$$

$$\times \prod_{v_0 \in S_0} \int W_{1, v_0} \overline{W_{2, v_0}} \left[\begin{pmatrix} g & 0 \\ 0 & 1 \end{pmatrix} k \right] \Phi_{v_0}[tk]\, |\det t|^3\, d^\times t\, dk\, dg.$$

In the last integral k is in K_{S_0}, $g \in GL(2, F_{S_0})$, $t \in F^\times_{S_0}$. Now for $v_0 \in S_0$ the integrals

$$\int_{F^3_{v_0}} \Phi_{v_0}(x)\, dx, \qquad \int \Phi_{v_0}[tk]\, |\det t|^3\, d^\times t\, dk$$

are equal (up to a scalar factor). Moreover, the left hand side of the previous formula is an invariant Hermitian form on the space $\mathscr{V}(\Pi)$. If follows that for $v_0 \in S_0$ there is an Hermitian form β_{v_0} on $\mathscr{W}(\Pi_{v_0})$ such that

$$\int W_1 \bar{W}_2 \left[\begin{pmatrix} g & 0 \\ 0 & 1 \end{pmatrix} k \right] \Phi_{v_0}[tk]\, |\det t|^3\, d^\times t\, dk\, dg$$

$$= \beta_{v_0}(W_1, W_2) \int \Phi_{v_0}[tk]\, |\det t|^3\, d^\times t\, dk.$$

It follows in turn that for every smooth function f_{v_0} on K_{v_0} invariant on the left under $P_{v_0} Z_{v_0} \cap K_{v_0}$, Z denoting the center,

$$\int W_1 \bar{W}_2 \left[\begin{pmatrix} g & 0 \\ 0 & 1 \end{pmatrix} k \right] f_{v_0}(k)\, dk\, dg = \beta_{v_0}(W_1, W_2) \int f_{v_0}(k)\, dk.$$

The same relation is then true for every smooth function on K_{v_0}. In turn this implies that the Hermitian form

$$(W_1, W_2)$$

is invariant under K_{v_0} and thus under G_{v_0} a claimed. Then

$$\int W_1 \bar{W}_2\left[\begin{pmatrix} g & 0 \\ 0 & 1 \end{pmatrix} k\right] \Phi_{v_0}[tk]\, |\det t|^3\, d^\times t\, dk\, dg$$

$$= (W_1, W_2) \int \Phi_{v_0}[tk]\, |\det t|^3\, d^\times t\, dk,$$

and relation (1) follows.

Thus we can construct an orthonormal basis of $\mathscr{V}^{S_0}$ as follows: for each $v_0 \in S_0$, we choose an orthonormal basis $(W_{\alpha_{v_0}})$, $\alpha_{v_0} \in A_{v_0}$, of $\mathscr{W}(\Pi_{v_0}, \psi_{v_0})$. We then set $A = \prod_{v_0 \in S_0} A_{v_0}$. For each $\alpha \in A$ we define

$$W_\alpha(g) = \prod_{v_0 \in S_0} W_{\alpha_{v_0}}(g_{v_0}) \prod_{v_0 \notin S_0} W^{K_{v_0}}(g_{v_0})$$

and then we set

$$\phi_\alpha(g) = \frac{1}{\sqrt{c(F, S_0)\, L^{S_0}(1, \pi, \mathrm{Ad})}} \sum_{\gamma \in GL(2, F)} W_\alpha\left[\begin{pmatrix} \gamma & 0 \\ 0 & 1 \end{pmatrix} g\right].$$

Then (ϕ_α) is indeed an orthonormal basis of $\mathscr{V}^{S_0}$.

We introduce the matrix

$$w = \begin{pmatrix} 0 & 0 & -1 \\ 0 & -1 & 0 \\ -1 & 0 & 0 \end{pmatrix}$$

and a Bessel distribution on $GL(3, F_{S_0})$. We let $\mathscr{W}'$ be the linear form on $\mathscr{V}(\pi)$ defined by:

$$\mathscr{W}'(\phi') = \int_{N(F)\backslash N(F_{\mathbb{A}})} \phi'(n)\, \theta^{-1}(n)\, dn.$$

We define the *global Bessel distribution* attached to π as follows. If f' is a smooth function of compact support on $GL(3, F_{S_0})$ we set

$$\mathscr{B}_\pi(f') := \sum_\alpha \mathscr{W}'(\pi(f')\, \phi_\alpha)\, \overline{\mathscr{W}'(\pi(w)\, \phi_\alpha)}. \tag{2}$$

Remark. To make the above sum finite, we have to assume that $f = \prod_{v_0 \in S_0} f_{v_0}$ where f_{v_0} is K_{v_0}-finite if v_0 is Archimedean. However with a little more effort, one can show that there is distribution whose value on a function of this type is given by the above expression.

On the other hand, for every place $v_0 \in S_0$ we denote by $\mathscr{W}_{v_0}$ the local Whittaker linear form (evaluation at e) and introduce the *local Bessel distribution* $\mathscr{B}_{\pi_{v0}}$ or simply $\mathscr{B}_{v_0}$:

$$\mathscr{B}_{v_0}(f'_{v_0}) := \sum_{\alpha_{v0}} \mathscr{W}_{v_0}(\pi_{v_0}(f') W_{\alpha_{v0}}) \overline{\mathscr{W}_{v_0}(\pi_{v_0}(w) W_{\alpha_{v0}})}. \tag{3}$$

Then if $f' = \prod_{v_0 \in S_0} f'_{v_0}$ we have

$$\mathscr{B}_\pi(f') = c(\pi) \prod_{v_0 \in S_0} \mathscr{B}_{v_0}(f'_{v_0}),$$

where we have set

$$c(\pi) := \frac{1}{c(F, S_0)\, L^{S_0}(1, \pi, \mathrm{Ad})}. \tag{4}$$

Similar results are true for a cuspidal automorphic representation Π of $GL(3, E_{\mathbf{A}})$ and a pure tensor $\phi \in \mathscr{V}^S$:

$$\|\phi\|^2 = c(E, S)\, L^{S_0}(1, \pi, \mathrm{Ad}) \prod_{v \in S} \|W_v\|^2 \tag{5}$$

with

$$c(\Pi) := \frac{1}{c(E, S)\, L^{S_0}(1, \Pi, \mathrm{Ad})}. \tag{6}$$

3. MATCHING OF ORBITAL INTEGRALS

The main theorem will a consequence of the relative trace formula of [JY]. We recall the geometric side of the trace formula in question. We fix an idele class character ω of F.

Let F^+ be the set of elements of $F^\times$ which are norm of an element of $E^\times$ and, for each inert place v_0 of F, let $F^+_{v_0}$ be the set of $x \in F^\times_{v_0}$ which are a norm of an element of $E^\times_v$. We set $F^\times_{S_0} = \prod_{v_0 \in S_i} F^+_{v_0} \prod_{v_0 \in S_s} F^\times_{v_0}$. Similarly, let $F^+_{\mathbf{A}}$ be the set of ideles x such that $x_{v_0} \in F^+_{v_0}$ for every inert place v_0. Then $F^+ = F^\times \cap F^+_{\mathbf{A}}$.

Let $\mathfrak{S}$ be the variety of Hermitian matrices in $GL(3, E)$ and $\mathfrak{S}^+(F)$ the set of elements of $\mathfrak{S}(F)$ whose determinant is in F^+. For each inert place v_0, let $\mathfrak{S}^+_{v_0}$ be the set of $s \in \mathfrak{S}_{v_0}$ whose determinant is in $F^+_{v_0}$. Note that $GL(3, \mathcal{O}_v) \cap \mathfrak{S}_{v_0}$ is contained in $\mathfrak{S}^+_{v_0}$ if v_0 is odd, inert, and unramified. Let $\mathfrak{S}^+_{S_0}$ be the set of $s \in \mathfrak{S}_{S_0}$ such that $\det s \in F^+_{S_0}$. Let also $\mathfrak{S}^+_{\mathbf{A}}$ be the set of elements of $\mathfrak{S}(F_{\mathbf{A}})$ whose determinant is in $F^+_{\mathbf{A}}$. We define a distribution

$J(\bullet)$ on $\mathfrak{S}^+_{\mathbf{A}}$ as follows. If Φ is a smooth function of compact support on $\mathfrak{S}^+_{\mathbf{A}}$ we set:

$$J(\Phi) := \int_{N(E_{\mathbf{A}})/N(E)} \int_{F^+_{\mathbf{A}}/F^+} \left(\sum_{\xi \in \mathfrak{S}^+(F)} \Phi(n\xi^t\bar{n}z) \right) \omega(z)\, dz \theta(n\bar{n})\, dn. \tag{7}$$

On the other hand, let $G^+(F)$ be the group of $g \in GL(3, F)$ such that $\det g \in F^+$. Define similarly, for any inert place v_0, the group $G^+_{v_0}$ of $g_{v_0} \in GL(3, F_{v_0})$ with $\det g_{v_0} \in F^+_{v_0}$. Finally, let $G^+(F_{\mathbf{A}})$ be the set of $g \in GL(3, F_{\mathbf{A}})$ such that $\det g$ is in $F^+_{\mathbf{A}}$. We define a distribution $J'(\bullet)$ on $G^+(F_{\mathbf{A}})$:

$$J'(f') := \int_{(N(F_{\mathbf{A}})/N(F))^2} \int_{F^+_{\mathbf{A}}/F^+} \left(\sum_{\xi \in \mathfrak{S}^+(F)} f'(n_1 \xi z^t n_2) \right) \times \omega(z)\, dz \theta(n_1 n_2)\, dn_1\, dn_2. \tag{8}$$

We have a notion of *matching orbital integrals*; if Φ and f' have matching orbital integrals, then:

$$J(\Phi) = J'(f'). \tag{9}$$

In a precise way, we assume that Φ and f' are products of local functions which themselves have *matching orbital integrals*. This means the following. If v_0 is an inert place, we say that Φ_{v_0} and f'_{v_0} have matching orbital integrals, if, for any diagonal matrix a whose determinant is a norm:

$$\int_{N(E_v)} \Phi_{v_0}(na^t\bar{n})\, \theta_{v_0}(n\bar{n})\, dn = \gamma(a, \psi_{v_0}) \int_{N(F_{v_0}) \times N(F_{v_0})} f'_{v_0}(n_1 a^t n_2)\, \theta_{v_0}(n_1 n_2)\, dn_1\, dn_2. \tag{10}$$

Here the *transfer factor* $\gamma(a, \psi_{v_0})$ is defined by:

$$\gamma(\operatorname{diag}(a_1, a_2, a_3), \psi) = \omega_{E/F}(a_2).$$

One can show that, this relation implies, in turn, that there are similar relations between the other relevant orbital integrals [J3]. In an earlier paper [JY4], we have shown that given Φ_{v_0} there is a function f'_{v_0} satisfying the above conditions and conversely. For instance, if v_0 is odd, E_v/F_{v_0} is unramified and the character ψ_{v_0} has for conductor the ring of integers, then, if Φ_{v_0} is the characteristic function of $K_v \cap \mathfrak{S}_{v_0}$ we may take for f'_{v_0} the characteristic function of K_{v_0}.

If f_v is a smooth function of compact support on $GL(3, E_v)$ and Φ_{v_0} is the function on $\mathfrak{S}^+_{v_0}$ defined by

$$\Phi_{v_0}(g^t_v \bar{g}_v) = \int f_v(g_v h_{v_0})\, dh_{v_0},$$

we say that f_v and f'_{v_0} have matching orbital integrals provided Φ_{v_0} and f'_{v_0} do. If f_v is a Hecke function and f'_{v_0} is its image by the base change homomorphism, then f'_{v_0} is supported on $G^+_{v_0}$ and f_v and f'_{v_0} have matching orbital integrals.

If v_0 splits into v_1, v_2 then we may identify $\mathfrak{S}_{v_0}$ to the set of pairs $(s, {}^t s)$ with $s \in GL(3, F_{v_0})$ and H_{v_0} to the set of pairs $(h, {}^t h^{-1})$ with $h \in GL(3, F_{v_0})$. Thus we may identify $GL(3, F_{v_0})$, $\mathfrak{S}_{v_0}$ and H_{v_0}. Then we take the condition of matching orbital integrals to be:

$$f'_{v_0}(g) = \Phi_{v_0}(g). \tag{11}$$

We say that f_{v_0} and $f_{v_1} \times f_{v_2}$ have matching orbital integrals if:

$$f'_{v_0}(g) = \Phi_{v_0}(g) = \int f_{v_0}(gh)\, f_{v_2}({}^t h^{-1})\, dh.$$

For Hecke functions this means that f_{v_0} is the convolution of f_{v_1} and f_{v_2}.

Identity (9) follows readily from the condition of matching.

To the functions f and f' we attach kernels in the usual way:

$$K_f(x, y) := \int_{E^\times_{\mathbb{A}}/E^\times} \sum_{\xi \in GL(3, E)} f(x^{-1}\xi z y)\, \Omega(z)\, dz,$$

$$K_{f'}(x, y) := \int_{F^+_{\mathbb{A}}/F^+} \sum_{\xi \in G^+(F)} f'(x^{-1}\xi z y)\, \Omega(z)\, dz.$$

Then

$$J(\Phi) = \int_{N(E)\backslash N(E_{\mathbb{A}}) \times H(F)\backslash H(F_{\mathbb{A}})} K_f(n, h)\, \theta^{-1}(n\bar{n})\, dn\, dh,$$

$$J'(f') = \int_{N(F)\backslash N(F_{\mathbb{A}}) \times N(F)\backslash N(F_{\mathbb{A}})} K_{f'}(n_1, {}^t n_2)\, \theta^{-1}(n_1)\, \theta(n_2)\, dn_2.$$

From identity (9) follows the equality of the integral of the two kernels. In turn, this implies the equality of the integrals of the corresponding spectral kernels and, finally, the equality of the integrals of the kernels attached to a cuspidal representation π and its base change.

Now let Ω be the base change of ω, that is, $\Omega(z)=\omega(z\sigma(z))$. Let π a cuspidal representation of $GL(3, F_{\mathbb{A}})$ with central character ω and let Π its base change. Let K_f^Π and $K_{f'}^\pi$ be the kernels attached to the representations Π and π respectively:

$$K_f^\Pi(x, y)=\sum \Pi(f)\,\phi_i(x)\,\overline{\phi_i(y)},$$

$$K_{f'}^\pi(x, y)=\sum \pi(f')\,\phi_i'(x)\,\overline{\phi_i'(y)},$$

where the sums are over orthonormal bases of $\mathscr{V}(\Pi)$ and $\mathscr{V}(\pi)$ respectively.

Then, if f and f' have matching orbital integrals,

$$\iint K_f^\Pi(nh)\,\theta^{-1}(n\bar{n})\,dn\,dh=\iint K_{f'}^\Pi(n_1, {}^t n_2)\,\theta^{-1}(n_1)\,\theta(n_2)\,dn_1\,dn_2. \tag{12}$$

Recall that on the space of $\mathscr{V}(\Pi)$ we have introduced the following linear forms:

$$\mathscr{W}(\phi)=\int \phi(n)\,\theta^{-1}(n\bar{n})\,dn, \qquad \mathscr{P}(\phi)=\int \phi(h)\,dh.$$

Thus the integral of K_f^Π can be written as

$$\sum_{\phi_i} \mathscr{W}(\Pi(f)\,\phi_i)\,\overline{\mathscr{P}(\phi_i)}.$$

On the other hand:

$$\begin{aligned}\int_{N(F_{\mathbb{A}})/N(F)} \overline{\phi'}({}^t n')\,\theta(n')\,dn' &= \overline{\int_{N(F_{\mathbb{A}})/N(F)} \phi'({}^t n')\,\theta^{-1}(n')\,dn'}\\ &= \overline{\int_{N(F_{\mathbb{A}})/N(F)} \phi'(w\,{}^t n' w w)\,\theta^{-1}(n')\,dn'}\\ &= \overline{\int_{N(F)\backslash N(F_{\mathbb{A}})} \phi'(n' w)\,\theta^{-1}(n')\,dn'}\\ &= \overline{\mathscr{W}'(\pi(w')\,\phi')}.\end{aligned}$$

Thus the integral of $K_{f'}^\pi$ can also be written:

$$\sum_{\phi_i'} \mathscr{W}'(\pi(f')\,\phi_i')\,\overline{\mathscr{W}'(\pi(w)\,\phi_i')}.$$

We arrive at the identity:

$$\sum_{\phi_i} \mathscr{W}(\Pi(f)\,\phi_i)\,\overline{\mathscr{P}(\phi_i)} = \sum_{\phi'_i} \mathscr{W}'(\pi(f')\,\phi'_i)\,\overline{\mathscr{W}'(\pi(w)\,\phi'_i)}, \tag{13}$$

whenever f and f' have matching orbital integrals.

Remark. If v_0 is an Archimedean place of F, by hypothesis, the place v_0 splits into v_1 and v_2. We assume that the functions f_{v_1} and f_{v_2} are in fact $K_{v_1} \simeq K_{v_2} \simeq K_{v_0}$ finite so as to have only finite sums in the above identity. In fact, both sides may be viewed as distributions and then the identity is true without restriction at the infinite places.

4. PROOF OF THE MAIN THEOREM

We now prove the theorem stated in the first section. Thus we let Π be a distinguished cuspidal representation of $GL(3, E_{\mathbb{A}})$ with central character Ω. It is thus the base change of a *unique* cuspidal representation π of $GL(3, F_{\mathbb{A}})$ with central character ω. We let S_0 and S be as before. If f is a smooth function of compact support on G_S and K_S-finite, we set:

$$\mathscr{R}_{\Pi}(f) = \sum_{\phi_i} \mathscr{W}(\Pi(f)\,\phi_i)\,\overline{\mathscr{P}(\phi_i)},$$

the sum over an orthonormal basis (ϕ_i) of $\mathscr{V}^S$. The sum does not depend on the choice of the orthonormal basis. We think of this linear form as being the *relative Bessel distribution* attached to Π.

Recall

$$G^+_{S_0} := \prod_{v_0 \in S_s} G_{v_0} \prod_{v_0 \in S_i} G^+_{v_0}.$$

We have defined the global Bessel distribution attached to π. We can compute its value on a function f' smooth and of compact support, on the group $G^+_{S_0}$:

$$\mathscr{B}_{\pi}(f') = \sum_{\phi'_i} \mathscr{W}'(\pi(f')\,\phi'_i)\,\overline{\mathscr{W}'(\pi(w)\,\phi'_i)},$$

where the sum is over an orthonormal basis of $\mathscr{V}^{S_0}(\pi)$. Actually, as before, to make the sum finite we have to assume that $f' = \prod_{v_0 \in S_0} f'_{v_0}$ where f_{v_0} is K_{v_0}-finite for v_0 infinite. We think of $\mathscr{B}_{\pi}$ as the *Bessel distribution* attached to π.

If f and f' have matching orbital integrals then it follows from the previous section that:

$$\mathcal{R}_\Pi(f) = \mathcal{B}_\pi(f'). \tag{14}$$

For $v_0 \in S_i$ we define a distribution $\mathcal{R}_v$ on $GL(3, E_v)$ as follows: given a smooth function of compact support f_v, we choose a function f'_{v_0} with matching orbital integrals and we set:

$$\mathcal{R}_v(f_v) = \mathcal{B}_{\pi_{v_0}}(f'_{v_0}).$$

We must check that the right hand side is independent of the choice of f'_{v_0}. But if f''_{v_0} is another choice then f'_{v_0} and f''_{v_0} have the same orbital integrals, or, what amounts to the same, all the orbital integrals of the difference vanish. However, the orbital integrals are weakly dense in the space of distributions on $GL(3, F_{\mathbb{A}})$ which transform on the left and on the right under the character θ of $N(F_{v_0})$ ([GK], principle of localization). Thus the distribution $\mathcal{B}_{\pi_{v_0}}$ takes the same value on both functions, and the distribution $\mathcal{R}_v$ is well defined.

At a place $v_0 \in S_s$ we set

$$\mathcal{R}_{v_0}(f_{v_1} \otimes f_{v_2}) = \mathcal{B}_{v_0}(f'_{v_0}),$$

where f'_{v_0} have matching orbital integrals with $f_{v_1} \otimes f_{v_2}$, that is, $f'_{v_0} = f_{v_1} * {}^t f_{v_2}$.

Recall the decomposition

$$\mathcal{B}_\pi(f') = c(\pi) \prod_{v \in S_0} \mathcal{B}_{v_0}(f'_{v_0}).$$

It follows that we can write:

$$\mathcal{R}_\Pi(f) = c(\pi) \prod_{v_0 \in S_i} \mathcal{R}_{v_0}(f_v) \prod_{v_0 \in S_s} \mathcal{R}_{v_0}(f_{v_1} \otimes f_{v_2}). \tag{15}$$

The theorem will follow from (15) and a careful analysis of the distributions $\mathcal{R}_{v_0}$ (See the next two lemmas.).

LEMMA 1. *For every $v_0 \in S_i$ there is a unique element $\mathcal{P}_{v_0}$ of $\mathcal{H}_{v_0}$ such that*

$$\mathcal{R}_{v_0}(f_v) = \sum_{u_i} \mathcal{W}_v(\Pi_v(f_v)\, u_i)\, \overline{\mathcal{P}_{v_0}(u_i)},$$

where the sum is over an orthonormal basis of $\mathcal{V}(\Pi_v)$.

Proof of the lemma. We first prove the uniqueness. Suppose $\mathscr{P}_0$ is any linear form such that

$$\sum_{u_i} \mathscr{W}_v(\Pi_v(f_v)\, u_i)\, \overline{\mathscr{P}_0(u_i)} = 0,$$

for any function f_v. Our task is to show that, for any vector u_0, $\mathscr{P}_0(u_0) = 0$. We may as well assume that u_0 is a unit vector and even a member of the orthonormal basis (u_i). We then choose a vector u' such that $\mathscr{W}(u') \neq 0$ and a function f_v such that $\Pi_v(f_v)\, u_0 = u'$ and $\Pi_v(f_v)\, u_i = 0$ for $i \neq 0$. We then obtain our conclusion by applying the hypothesis to f_v.

To prove the existence, we fix a place $w_0 \in S_i$ and let w be the corresponding place of E. Let Π^w be the restricted tensor product of the unitary representations Π_v with $v \neq w$. We fix a unitary intertwining operator A: $\Pi^w \otimes \Pi_w \to \Pi$. In a precise way, the space of smooth vectors of Π^w can be identified with the space $\mathscr{V}^w$ spanned by the functions of the form $W^w(g^w) = \prod_{v \neq w} W_v(g_v)$ on the group G^w, the restricted product of the groups G_v with $v \neq w$, where W_v is in $\mathscr{W}(\Pi_v)$ and $W_v = W^{K_v}$ for $v \notin S$. Then A has the form:

$$A(W^w \otimes W_w) = d\phi(g),$$

$$\phi(g) = \sum_{\gamma \in N(2, E)\backslash GL(2, E)} W\left[\begin{pmatrix} \gamma & 0 \\ 0 & 1 \end{pmatrix} g\right],$$

$$W(g) = W^w(g^w)\, W_w(g_w).$$

The constant d is chosen to make the map a unitary operator. Let (u_α) be an orthonormal basis of $\mathscr{W}(\Pi_w)$ and (m^β) be an orthonormal basis of $\mathscr{V}^w$. Then $A(m^\beta \otimes u_\alpha)$ is an orthonormal basis of Π. If we set $\mathscr{W}^w(W^w(g)) = dW^w(e)$ then

$$\mathscr{W}(A(m \otimes u)) = \mathscr{W}^w(m)\, \mathscr{W}_w(u).$$

For every vector m in $\mathscr{V}(\Pi^w)$, the linear form

$$u \mapsto \mathscr{P}(A(m \otimes u))$$

is invariant under $H_0(F_{w_0})$ thus belongs to $\mathscr{H}_{w_0}$. Thus there is a linear map A_H: $\mathscr{V}^w \to \mathscr{H}_{w_0}$ such that

$$\mathscr{P}(A(m \otimes u)) = A_H(m)(u).$$

The global distribution $\mathscr{R}_\Pi$ can be written

$$\mathscr{R}_\Pi(f) = \sum_\beta \sum_\alpha \mathscr{W}^w(\Pi^w(f^w)\, m^\beta)\, \mathscr{W}_w(\Pi_w(f_w)\, u_\alpha)\, \overline{A_H(m^\beta)(u_\alpha)}.$$

Let $\mathscr{P}_\kappa$ be a basis for the space $\mathscr{H}_{w_0}$. Note that, at this point, we do not know that the space is finite dimensional in general. We can write

$$A_H(m) = \sum_\kappa A_\kappa(m)\, \mathscr{P}_\kappa,$$

where the A_κ are suitable linear forms on $\mathscr{V}^w$. Note that for a given m, $A_\kappa(m) = 0$ for all but finitely many κ's. We get:

$$\mathscr{R}_\Pi(f) = \sum_\kappa \left(\sum_\beta \mathscr{W}^w(\Pi^w(f^w)\, m^\beta)\, A_\kappa(m^\beta) \right) \mathscr{R}_\kappa(f_v)$$

where we have set:

$$\mathscr{R}_\kappa(f_v) = \sum_\alpha \mathscr{W}_v(\Pi_v(f_v)\, u_\alpha)\, \overline{\mathscr{P}_\kappa(u_\alpha)}.$$

Now for each $v_0 \in S_i$, $v_0 \neq w_0$ we can choose a function f_v such that $\mathscr{R}_{v_0}(f_v) \neq 0$ and, for $v_0 \in S_s$, functions f_{v_1}, f_{v_2} such that $\mathscr{R}_{v_0}(f_{v_1} \otimes f_{v_2}) \neq 0$. Thus the distribution $\mathscr{R}_{w_0}$ is a linear combination of the distributions $\mathscr{R}_\kappa$:

$$\mathscr{R}_{w_0} = \sum_\kappa c_\kappa \mathscr{R}_\kappa.$$

Now we set

$$\mathscr{P}_{w_0} = \sum_\kappa c_\kappa \mathscr{R}_\kappa,$$

and then $\mathscr{R}_{w_0}$ has the required form. ∎

We need an analog of the previous lemma for a place $v_0 \in S_s$. It is in fact formal. We define an element $\mathscr{P}_{v_0} \in \mathscr{H}_{v_0}$ as follows: we may identify the representations π_{v_1}, π_{v_2}, π_{v_0}; their common space is the space $\mathscr{W}(\pi_{v_0})$. Define an antilinear map A from that space to itself by:

$$A W(g) = \overline{W(w\,{}^t g^{-1})}$$

Then $A\pi_{v_0}(g) = \pi_{v_0}({}^t g^{-1})\, A$ and $A^2 = 1$. Now consider the unitary representation attached to π_{v_0} and let $\mathscr{H}$ be its Hilbert space. Of course, we use the same notation for the unitary representation attached to π_{v_0}. We regard the representation $\pi_{v_0}({}^t g^{-1})$ has a representation on the conjugate Hilbert space. This new representation and the representation π_{v_0} have the same character thus are equivalent. It follows that there is a unitary antilinear operator U such that $U\pi_{v_0}(g) = \pi_{v_0}({}^t g^{-1})\, U$. Then U^2 is a unitary operator which commutes to π_{v_0}. It is therefore a scalar μ with

$|\mu|=1$. Dividing U by the square root of μ we may as well assume $U^2=1$. We have then (on the space of smooth vectors) $A=\lambda U$ with $\lambda \in \mathbb{C}^\times$. Since $A^2=1$ we get $\lambda=\pm 1$ and so A preserves the norm. This being so, for u_1, $u_2 \in \mathscr{W}(\pi_{v_0})$, we set

$$\mathscr{P}_{v_0}(u_1 \otimes u_2)=(u_1, Au_2).$$

Then

$$\mathscr{P}_{v_0}(\pi_{v_0}(g)\,u_1 \otimes u_2)=\mathscr{P}_{v_0}(u_1 \otimes \pi_{v_0}({}^t g)\,u_2).$$

We then define a distribution

$$\mathscr{R}'_{v_0}(f_{v_1} \otimes f_{v_2}) := \sum_{\alpha, \beta} \mathscr{W}_{v_0}(\pi_{v_0}(f_{v_1})\,u_\alpha)\,\mathscr{W}_{v_0}(\pi_{v_0}(f_{v_2})\,m^\beta)\,\overline{\mathscr{P}_{v_0}(u_\alpha \otimes m^\beta)}.$$

The sum is over orthonormal bases (u_α) and (m^β) of π_{v_0}.

LEMMA 2. *In fact:*

$$\mathscr{R}'_{v_0}=\mathscr{R}_{v_0}$$

Proof of the Lemma. Indeed, let us take $m^\beta = A(u_\beta)$. Then

$$\mathscr{P}_{v_0}(u_\alpha \otimes m^\beta)=\langle u_\alpha, A^2 u_\beta\rangle = \langle u_\alpha, u_\beta\rangle = \delta_{\alpha,\beta}.$$

Then:

$$\begin{aligned}\mathscr{R}'_{v_0}(f_{v_1} \otimes f_{v_2}) &= \sum_\alpha \mathscr{W}_{v_0}(\pi_{v_0}(f_{v_1})\,u_\alpha)\,\mathscr{W}_{v_0}(\pi_{v_0}(f_{v_2})\,Au_\alpha)\\ &= \sum_\alpha \mathscr{W}_{v_0}(\pi_{v_0}(f_{v_1})\,u_\alpha)\,\mathscr{W}_{v_0}(A\pi_{v_0}({}^t f^*_{v_2})\,u_\alpha)\\ &= \sum_\alpha \mathscr{W}_{v_0}(\pi_{v_0}(f_{v_1})\,\pi_{v_0}({}^t f_{v_2})\,u_\alpha)\,\mathscr{W}_{v_0}(Au_\alpha).\end{aligned}$$

But $\mathscr{W}_{w_0}(Au)=\overline{\mathscr{W}}_{w_0}(\pi_{v_0}(w)\,u)$. Thus we get at last:

$$\mathscr{R}'_{v_0}(f_{v_1} \otimes f_{v_2}) = \sum_\alpha \mathscr{W}_{v_0}(\pi_{v_0}(f_{v_1} * {}^t f_{v_2})\,u_\alpha)\,\overline{\mathscr{W}_{v_0}}(\pi_{v_0}(w)\,u_\alpha),$$

which is indeed $\mathscr{R}_{v_0}(f_{v_1} \otimes f_{v_2})$. ∎

Now we prove the theorem. Let $\tilde{\mathscr{P}}$ be the linear form on $\mathscr{V}^S(\Pi)$ defined by

$$\tilde{\mathscr{P}}(\phi)=\prod_{v_0 \in S_i} \mathscr{P}_{v_0}(W_v) \prod_{v_0 \in S_s} \mathscr{P}_{v_0}(W_{v_1} \otimes W_{v_2})$$

when ϕ is a pure tensor. Taking into account the computation of the norm of a pure tensor, we get:

$$\sum_{\phi} \mathscr{W}(\Pi(f)\,\phi_i)\,\overline{\tilde{\mathscr{P}}(\phi_i)} = c(\Pi) \prod_{v_0 \in S_i} \mathscr{R}_{v_0}(f_v) \prod_{v_0 \in S_s} \mathscr{R}_{v_0}(f_{v_1} \otimes f_{v_2})$$
$$= \frac{c(\Pi)}{c(\pi)} \mathscr{R}_{\Pi}(f)$$
$$= \frac{c(\Pi)}{c(\pi)} \sum_{\phi} \mathscr{W}(\Pi(f)\,\phi_i)\,\overline{\mathscr{P}(\phi_i)}.$$

Since Π is irreducible (See the proof of uniqueness in Lemma (1)) we get

$$\mathscr{P} = \frac{c(\pi)}{c(\Pi)} \tilde{\mathscr{P}} \tag{16}$$

and the theorem follows.

5. LOCAL RESULTS: SUPERCUSPIDAL CASE

Now we consider a local quadratic extension E/F of non-Archimedean local fields. We only consider irreducible unitary generic representations of $GL(3, E)$. We say that such a representation Π is *distinguished* if the space $\mathscr{H}(\Pi, H)$ of H-invariant linear forms is non-zero. Then the central character Ω of Π is itself distinguished, that is, trivial on $\mathbb{U}_1$. We then fix a character ω of $F^{\times}$ such that $\Omega(z) = \omega(z\bar{z})$.

THEOREM 2. *Suppose Π is supercuspidal. Then Π is distinguished if and only $\Pi^{\sigma} = \Pi$. The dimension of $\mathscr{H}(\Pi, H)$ is then one. Let Ω and ω as above. Then Π is the base change of a unique cuspidal representation π of $GL(3, F)$ with central character ω and there exists a unique element $\mathscr{P}_{\pi}$ of $\mathscr{H}(\Pi, H)$ such that*

$$\sum_{\phi_i} \mathscr{W}(\Pi(f)\,\phi_i)\,\overline{\mathscr{P}_{\pi}(\phi_i)} = \mathscr{B}_{\pi}(f'), \tag{17}$$

each time f and f' have matching orbital integrals.

We first prove a result of density.

LEMMA 3. *If Π is supercuspidal and distinguished, let $\mathscr{H}^c_{\Pi}$ be the space spanned by the linear forms $\mathscr{P}$ defined by:*

$$\mathscr{P}(u) = \int_{H(F)} (\Pi(h)\,u, \tilde{u})\,dh,$$

where $\tilde{u}$ is a (smooth) vector. Then for any $\mathcal{P}_1 \in \mathcal{H}(\Pi, H)$, the space $\operatorname{Ker}(\mathcal{P}_1)$ *contains the intersection* $\bigcap_{\mathcal{P} \in \mathcal{H}_\Pi^c} \operatorname{Ker}(\mathcal{P})$.

Proof. We use Bernstein theory to write the space $\mathcal{C}$ of smooth functions on $GL(3, E_v)$ transforming under the character Ω and compactly supported modulo the center, as a direct sum of bi-invariant subspaces

$$\mathcal{C} = \mathcal{C}(\Pi) \oplus \mathcal{C}^{\Pi},$$

where $\mathcal{C}(\Pi)$ denotes the space spanned by the matrix coefficients of Π_v, that is, the functions of the form

$$g \mapsto (\Pi(g)\, u, \tilde{u}).$$

If we choose a linear basis (u_i) of the space $\mathcal{V}(\Pi)$ we can decompose further $\mathcal{C}(\Pi)$ as direct sum of the invariant spaces $\mathcal{C}_i$,

$$\mathcal{C}(\Pi) = \bigoplus \mathcal{C}_i,$$

where $\mathcal{C}_i$ is the space spanned by the functions of the form: $g \mapsto (\Pi(g)\, u, u_i)$. Choosing an index i_0, we may identify the space $\mathcal{C}_{i_0}$ to the space $\mathcal{V}$ and view $\mathcal{P}_1$ as a linear form on $\mathcal{C}_{i_0}$. We may extend $\mathcal{P}_1$ to $\mathcal{C}$ by demanding that it be zero on $\mathcal{C}_i$ with $i \neq i_0$, and then zero on $\mathcal{C}^{\Pi}$. Thus we now view the given linear form $\mathcal{P}_1$ as a distribution invariant on the right under $H(F)$. Its value on the function $f(g) = \langle \Pi(g)\, u, u_{i_0} \rangle$ is equal to $\mathcal{P}_1(u)$. Since u_{i_0} is invariant under a compact open subgroup K' this distribution is invariant on the left under K'. It follows that there exists a distribution μ on $G(E)/H(F)$ such that, for any $f \in \mathcal{C}$,

$$\mathcal{P}_1(f) = \int \left(\int f(xh)\, dh \right) d\mu(x).$$

Moreover, this distribution is invariant on the left under K'. Thus, if (x_j) is a set of representatives for the double cosets $Z(E)\, K' \backslash G(E)/H(F)$ we have, for suitable constants λ_j,

$$\mathcal{P}_1(f) = \sum_j \lambda_j \int \int f(k' x_i h)\, dh\, dk'.$$

For a given function f, there are only finitely many non-zero terms on the right. Coming back to the original linear form $\mathcal{P}_1$, we see that

$$\mathcal{P}_1(u) = \sum_j \lambda_j \int (\Pi_v(x_j h)\, u, u_{i_0})\, dh,$$

the sum on the right having only finitely many non-zero terms. If we set $u_j = \Pi(x_j^{-1})(u_{i_0})$ we get finally:

$$\mathscr{P}_1(u) = \sum_j \lambda_j \int (\Pi(h)\, u, u_j)\, dh.$$

The lemma follows. ∎

To finish the proof we write our local extension in the form E_v/F_{v_0} where E/F is a quadratic extension of number fields and v_0 a place of F inert in E, v the corresponding place of E. We assume that all the infinite places of F split in E. We write the given supercuspidal representation as Π_v.

Suppose first that $\Pi_v^\sigma = \Pi_v$. Then Π_v is the base change of a supercuspidal representation π_{v_0}. In turn we may write π_{v_0} as the local component of a cuspidal automorphic representation π. Let Π be the base change of π. Then Π_v is the local component of Π at the place v. Since Π is globally distinguished, it follows that Π_v is (locally) distinguished.

Now we suppose that Π_v is distinguished. Then its central character Ω_v is distinguished and we write, as before, $\Omega_v(z) = \omega_{v_0}(z\bar{z})$. We may further choose an idele class character ω of F whose component at v_0 is ω_{v_0}. We then set $\Omega(z) = \omega(z\bar{z})$. We first show that Π_v is the local component at v of a distinguished cuspidal automorphic representation Π with central character Ω. Since $\Pi^\sigma = \Pi$ then, it will follow that, as claimed, $\Pi_v^\sigma = \Pi_v$. From the previous density lemma, it follows there is a smooth vector u_1 in the space of Π_v such that the linear form $\mathscr{P}_1$ defined by

$$\mathscr{P}_1(u) = \int_{H_{v_0}} (\Pi_v(h)\, u, u_1)\, dh$$

is non-zero. For any smooth vector u in the space of Π_v we set:

$$f_v^u(g_v) = (\Pi_v(g_v)\, u, u_1).$$

On the other hand, we let f^v be a smooth function on the group G^v, transforming under the character Ω^v of Z^v, and compactly supported, modulo the center. We set $f(g) = f^v(g^v)\, f_v^u(g_v)$. We define a function ϕ^{u, f^v} on $GL(3, E_{\mathbb{A}})$ as follows:

$$\phi^{u, f^v}(g) = \sum_{\gamma \in Z(E)\backslash G(E)} f(\gamma g).$$

The resulting function is invariant under $G(E)$ on the left, compactly supported modulo $Z(E_{\mathbb{A}})\, G(E)$ and cuspidal. Let (Π_α) be the family of cuspidal representations with central character Ω, and for each α, let $\mathscr{V}_\alpha$ be the space of smooth vectors of Π_α. Let ϕ_α^{u,f^v} be the orthogonal projection of ϕ^{u,f^v} on $\mathscr{V}_\alpha$. Then

$$\phi^{u,f^v}(g) = \sum_\alpha \phi_\alpha^{u,f^v}(g).$$

The series converges in the space of rapidly decreasing functions on the quotient $G(E)\,Z(E_{\mathbb{A}})\backslash G(E_{\mathbb{A}})$. Thus we may write:

$$\int \phi^{u,f^v}(h)\,dh = \sum_\alpha \int \phi_\alpha^{u,f^v}(h)\,dh.$$

Moreover, Schur orthogonality relations imply the existence of a constant $d > 0$ such that

$$\int_{Z_v\backslash G_v} (\Pi_v(g_0 g_v)\,u, u_0)(u_1, \Pi_v(g_v)\,u)\,dg_v = d(u, u)(\Pi_v(g_0)\,u_1, u_0).$$

This implies that

$$\int_{Z_v\backslash G_v} \phi^{u,f^v}(g g_v)(u_1, \Pi_v(g_v)\,u)\,dg_v = d(u, u)\,\phi^{u_1,f^v}(g).$$

Thus, for each α

$$\int_{Z_v\backslash G_v} \phi_\alpha^{u,f^v}(g g_v)(u_1, \Pi_v(g_v)\,u)\,dg_v = d(u, u)\,\phi_\alpha^{u_1,f^v}(g).$$

It follows that if the projection ϕ_α^{u,f^v} is not zero (for some choice of u and f^v) then the representation Π_α has the form $\Pi_v \otimes \Pi_\alpha^v$. We claim further that at least one of the representations Π_α is distinguished. Indeed, suppose not. Thus, for all α,

$$\int \phi_\alpha^{u,f^v}(h)\,dh = 0.$$

It follows that for any function f^v

$$\int \phi^{u,f^v}(h)\,dh = 0.$$

Explicitly:

$$\sum_{\gamma \in G(E)/H(F)\,Z(E)} \int_{H_{v_0}} (\Pi_v(\gamma h_{v_0})\,u, u_0)\,dh_{v_0} \int_{H^{v_0}} f^v(\gamma h^{v_0})\,dh^{v_0} = 0.$$

Now let Φ_{v_0} be the function on $\mathfrak{S}^+_{v_0}$ defined by

$$\Phi_{v_0}(g^t\bar{g}) = \int_{H_{v_0}} (\Pi_v(gh_{v_0})\, u, u_0)\, dh_{v_0}.$$

Similarly, let Φ^{v_0} be the function on $\mathfrak{S}^{v_0,+}$ defined by:

$$\Phi^{v_0}(g^t\bar{g}) = \int_{H^{v_0}} f^v(gh^{v_0})\, dh^{v_0}.$$

Finally let Φ be the product of Φ_{v_0} and Φ^{v_0}. Thus $\Phi(sz) = \Phi(s)\, \omega(z)$ for $z \in F^+_{\mathbf{A}}$. Moreover the support of Φ is contained in a set of the form $MF^+_{\mathbf{A}}$ where M is a compact set. The above relation reads:

$$\sum_{\xi \in S^+(F)/F^+} \Phi(\xi) = 0.$$

We can choose Φ^{v_0} in such a way that the above relation reduces to $\Phi_{v_0}(1) = 0$ or

$$\int_{H_{v_0}} (\Pi_v(h_{v_0})\, u, u_0)\, dh_{v_0} = 0,$$

that is, $\mathscr{P}_1(u) = 0$, which is a contradiction. Thus one of the representations Π_α is distinguished. As we have remarked before, this implies that $\Pi^\sigma_\alpha = \Pi_\alpha$ and thus as claimed, $\Pi^\sigma_v = \Pi_v$.

Finally, it remains to prove that if Π_v is distinguished then the dimension of $\mathscr{H}(\Pi_v, H_v)$ is one. We have just seen that Π_v is the local component of a distinguished cuspidal representation Π, which is itself the base change of a cuspidal representation π. In particular, Π_v is the local base change of the local component π_{v_0}; in fact π_{v_0} is the unique irreducible representation with central character ω_{v_0} whose base change is Π_v. By lemma there is a unique element $\mathscr{P}_0$ of $\mathscr{H}_{\Pi_v}$ such that (17) is satisfied. By the density result, it will suffice to prove that if u is a vector in the kernel of $\mathscr{P}_0$ then $\mathscr{P}_1(u) = 0$ for every element $\mathscr{P}_1$ of $\mathscr{H}^c_{\Pi_v}$. We may assume that $\mathscr{P}_1$ has the form:

$$\mathscr{P}_1(u) = \int_{H_{v_0}} (\Pi_v(h)\, u, u_1)\, dh.$$

Now if Π is *any* cuspidal automorphic representation of $GL(3, E_{\mathbf{A}})$ with local factor Π_v and ϕ is any smooth vector of Π which is a pure tensor of the form $u \otimes u^v$ then

$$\int \phi(h)\, dh = 0.$$

Indeed, this is clear if Π is not distinguished and follows from the factorization of the global period and the uniqueness of π_{v_0} otherwise. We now apply the previous construction. Each function ϕ_α^{u, f^v} corresponds to a pure tensor vector of the form $u \otimes u^v$. Then

$$\int \phi_\alpha^{u, f^v}(h)\, dh = 0.$$

As we have seen this implies that in fact $\mathscr{P}_1(u) = 0$, as claimed. Thus the theorem is completely established.

6. SUPERCUSPIDAL REPRESENTATIONS FOR $GL(2, E)$

We briefly review the case of the group $GL(2, E)$ (Cf. [HLR], [jH], [jHyF], [yF2], [P]). Let H_1 be a split-unitary group in two variables. Denote by $\tilde{H}_1$ the corresponding similitude group and by λ the similitude ratio. Thus $H_1 Z(E)$ has index two in $\tilde{H}_1$. Let h_1 be an element of $\tilde{H}_1 - H_1 Z(E)$. We say that an irreducible admissible (unitary) representation Π of $GL(2, E)$ is distinguished by H_1 if the space $\mathscr{H}(\Pi, H_1)$ of linear forms invariant under H_1 is non-zero.

Proposition 1. *Suppose that Π is supercuspidal. Then Π is distinguished by H_1 if and only if $\Pi^\sigma = \Pi$. Then* $\dim(\mathscr{H}(\Pi, H_1)) = 1$. *Moreover, let π be a supercuspidal representation whose base change is Π and let ω be the central character of π. Then in fact, for $\mathscr{P} \in \mathscr{H}(\Pi, H_1)$ and $h \in \tilde{H}_1$:*

$$\mathscr{P}(\Pi(h)\, u) = \omega\omega_{E/F}((\lambda(h))\, \mathscr{P}(u).$$

Proof of the Proposition. If Π is distinguished by H_1 then its central character Ω is distinguished by $\mathbb{U}_1$ and so has the form $\Omega(z) = \omega(z\bar{z})$ for a suitable ω. For $\mathscr{P} \in \mathscr{H}(\Pi, H_1)$ set

$$\mathscr{P}_\omega(u) = \tfrac{1}{2}(\mathscr{P}(\Pi(h_1)\, u) + \omega(\lambda(h_1))\, \mathscr{P}(u)), \tag{18}$$

$$\mathscr{P}_{\omega_{E/F}\omega}(u) = \tfrac{1}{2}(\mathscr{P}(\Pi(h_1)\, u) + \omega_{E/F}\omega(\lambda(h_1))\, \mathscr{P}(u)). \tag{19}$$

Then, for any $h \in \tilde{H}_1$ and any vector u:

$$\mathscr{P}_\omega(\Pi(h)\, u) = \omega(\lambda(h))\, \mathscr{P}(u), \tag{20}$$

$$\mathscr{P}_{\omega_{E/F}\omega}(\Pi(h)\, u) = \omega_{E/F}\omega(\lambda(h))\, \mathscr{P}(u). \tag{21}$$

We denote by $\mathscr{H}(\Pi, H_1, \omega)$ (resp. $\mathscr{H}(\Pi, H_1, \omega_{E/F}\omega)$ the space of linear forms satisfying (20) (resp. (21)).

LEMMA 4. *The dimension of the space $\mathscr{H}(\Pi, H_1, \omega)$ is at most one.*

Indeed, $\tilde{H}_1$ is conjugate to $GL(2, F)\, Z(E)$ by an element of $GL(2, E)$ and the conjugation takes the similitude ratio to $gz \mapsto \det gz\bar{z}$. Thus it suffices to prove that the space $\mathscr{H}(\Pi, GL(2, F), \omega)$ of linear forms $\mathscr{Q}$ such that

$$\mathscr{Q}(\Pi(g)\, u) = \omega(\det g)\, \mathscr{Q}(u)$$

for $g \in GL(2, F)$ has dimension at most one. We may extend ω to a character ω_1 of $E^\times$ and replace Π by $\Pi \otimes \omega_1^{-1}$. We are then reduced to proving that the space $\mathscr{H}(\Pi \otimes \omega_1^{-1}, GL(2, F))$ of linear forms invariant under $GL(2, F)$ by the representation $\Pi \otimes \omega_1^{-1}$ has dimension at most one; this is known (see [jH] and [yF2]). Moreover if that space is non-zero then (loc. cit.)

$$(\Pi \otimes \omega_1^{-1})^\sigma = \widetilde{\Pi \otimes \omega_1^{-1}} = \Pi \otimes \Omega^{-1} \omega_1^2.$$

This relation is in fact equivalent to $\Pi^\sigma = \Pi$. Thus if Π is distinguished by H_1 then it is invariant by σ.

Now we recall some global results. Let again E/F be a global quadratic extension of number fields. Let Π be a cuspidal representation. Let ω be an idele class character of F. The two following conditions are equivalent ([HLR]): (i) the representation Π is the base change of a cuspidal representation π with central character $\omega\omega_{E/F}$; (ii) the restriction of the central character Ω of Π to $F_{\mathbb{A}}^\times$ is ω^2 and there is ϕ in the space of Π such that

$$\int_{Z(F_{\mathbb{A}})\, G(F) \backslash G(F_{\mathbb{A}})} \phi(g)\, \omega^{-1}(\det g)\, dg \neq 0.$$

Note that the central character Ω verifies then $\Omega(z) = \omega(z\bar{z})$. Thus, the second condition amounts to: (iii) the central character Ω has the form $\Omega = \omega \circ \mathrm{Norm}$ and there is ϕ in the space of Π such that

$$\int_{Z(E_{\mathbb{A}})\, \tilde{H}(F) \backslash H(F_{\mathbb{A}})} \phi(g)\, \omega^{-1}(\lambda(h))\, dh \neq 0.$$

Now we go back to the local problem and again write our local extension in the form E_v/F_{v_0} and write Π_v instead of Π. If Π_v is given and $\mathscr{P}_{\omega_v} \neq 0$, then we can argue as before and find a cuspidal representation Π of which Π_v is the local component at the place v and Π satisfies (ii). Then Π is the base change of a representation π with central character $\omega\omega_{E/F}$. Thus Π_v is the base change of a representation π_{v_0} with central character

$\omega_{v_0}\omega_{E_v/F_{v0}}$. The only other representation of which Π_v is the base change is the representation $\pi_{v_0}\otimes\omega_{E_v/F_{v0}}$ and it has the same central character. Likewise, if $\mathscr{P}_{\omega_{v0}\omega_{E_v}/F_{v0}}\neq 0$ then Π_v is the base change of a representation π_{v_0} with central character ω_{v_0}.

We conclude that $\mathscr{P}_{\omega_v}$ and $\mathscr{P}_{\omega_{v0}\omega_{E_v}/F_{v0}}$ cannot be both non-zero. This already prove that if $\mathscr{H}(\Pi_v, H_{1,v_0})\neq 0$ then it has dimension one. Moreover Π_v is then a base change of a representation π_{v_0}, $\Pi_v=\Pi_v^\sigma$ and the central character has the required properties.

If $\Pi_v^\sigma=\Pi_v$ then, as before, Π_v is the base change of a supercuspidal representation π_{v_0}. We can find a cuspidal representation of which π_{v_0} is a component. We base change π to Π and apply the previously recalled result to conclude that Π is distinguished by H_1 and Π_v by H_{1,v_0}.

Remark. We could argue as in the previous section using the trace formula described in [JY] (Cf. [yF2]).

Now let H_2 be a unitary group which is not split. Let also $\tilde{H}_2$ be the corresponding similitude group and λ the similitude ratio. We define again $\mathscr{H}(\Pi, H_2)$ as the space of linear forms on $\mathscr{V}(\Pi)$ which are invariant under H_2. Then:

PROPOSITION 2. *Suppose that Π is supercuspidal. Then Π is distinguished by H_2 if and only if $\Pi^\sigma=\Pi$. Then* $\dim(\mathscr{H}(\Pi, H_2))=1$. *Moreover, let π be a supercuspidal representation whose base change is Π and let ω be the central character of π. Then in fact, for $\mathscr{P}\in\mathscr{H}(\Pi, H_2)$ and $h\in\tilde{H}_2$:*

$$\mathscr{P}(\Pi(h)\,u)=\omega\omega_{E/F}((\lambda(h))\,\mathscr{P}(u).$$

Let $G'(F)\subset GL(2, E)$ be the multiplicative group of a quaternion algebra. It is known that the dimension of $\mathscr{H}(\Pi, G'(F))$ is at most one. Moreover, $\mathscr{H}(\Pi, G'(F))\neq 0$ if and only if $\mathscr{H}(\Pi, GL(2, F))\neq 0$ (see [jH] and [jHyF]). Arguing as before we can reduce this proposition to the previous one.

7. REPRESENTATIONS INDUCED FROM A CUSPIDAL REPRESENTATION

For the other unitary generic representations of $GL(3, E)$ we propose the following conjecture:

Conjecture 1. Suppose Π is a unitary irreducible generic representation of $GL(3, E)$. Then Π is distinguished by H if and only $\Pi^\sigma=\Pi$. Let Ω and ω as above. For each irreducible admissible representation π of $GL(3, F)$

with central character ω whose base change is Π, there exists a unique element $\mathscr{P}_\pi$ of $\mathscr{H}(\Pi, H)$ such that

$$\sum_{\phi_i} \mathscr{W}(\Pi(f)\,\phi_i)\,\overline{\mathscr{P}_\pi(\phi_i)} = \mathscr{B}_\pi(f'), \tag{22}$$

each time f and f' have matching orbital integrals. Moreover, the linear forms $\mathscr{P}_\pi$ form a basis of $\mathscr{H}(\Pi, H)$

We have established this conjecture when Π is supercuspidal. We prove it in another case. Suppose that Π is induced by a supercuspidal representation. In a precise way, let $P = MU$ be the Levi decomposition of the parabolic subgroup P of type (2, 1) (upper generalized triangular matrices). Let Π_1 is a supercuspidal representation of $GL(2, E)$ and Π_2 a character of $E^\times$. Thus we may regard $\Pi_1 \times \Pi_2$ as a representation of $M(E) \simeq GL(2, E) \times E^\times$. We assume that Π is the corresponding normalized induced representation:

$$\Pi = \mathrm{Ind}(\Pi_1, \Pi_2).$$

Thus Π operates by right shifts on the space of smooth maps ϕ: $GL(3, E) \to \mathscr{V}(\Pi_1)$ such that

$$\phi(ph) = \delta_P(p)^{1/2}\,\Pi_1 \times \Pi_2(p)\,\phi(g)$$

for every $p \in P(E)$; here δ_P is the module of $P(E)$. We will content ourselves with proving the conjecture in this case. We begin with a lemma:

LEMMA 5. *The dimension of $\mathscr{H}(\Pi, H)$ is at most two. Moreover, if Π is distinguished by H then Π_1 is distinguished by a split group in two variables, Π_2 is distinguished by $\mathbb{U}_1$ and $\Pi^\sigma = \Pi$.*

Proof of Lemma. Let P_0 be the group of upper triangular matrices. We first study the orbits of $P_0(E)$ on $\mathfrak{S}(F)$; a system of representatives is given by the following matrices:

$$\begin{pmatrix} 0 & 1 & 0 \\ 1 & 0 & 0 \\ 0 & 0 & b \end{pmatrix}, \begin{pmatrix} b & 0 & 0 \\ 0 & 0 & 1 \\ 0 & 1 & 0 \end{pmatrix}, \begin{pmatrix} 0 & 0 & 1 \\ 0 & b & 0 \\ 1 & 0 & 0 \end{pmatrix}, \begin{pmatrix} b_1 & 0 & 0 \\ 0 & b_2 & 0 \\ 0 & 0 & b_3 \end{pmatrix},$$

where the elements b and b_i take their values in $F^\times/F^+$. Next, a set of representatives for the orbits of $P(E)$ on $\mathfrak{S}(F)$ is given by the matrices:

$$\begin{pmatrix} 0 & 0 & 1 \\ 0 & b & 0 \\ 1 & 0 & 0 \end{pmatrix}, \begin{pmatrix} b_1 & 0 & 0 \\ 0 & 1 & 0 \\ 0 & 0 & b_3 \end{pmatrix},$$

with b, b_1, b_3 as before. A system of representatives for the orbits of $P(E)$ on $\mathfrak{S}^+(F)$ is then given by the following matrices:

$$\sigma_0 = \begin{pmatrix} 0 & 0 & 1 \\ 0 & -1 & 0 \\ 1 & 0 & 0 \end{pmatrix}, \sigma_1 = \begin{pmatrix} -1 & 0 & 0 \\ 0 & 1 & 0 \\ 0 & 0 & -1 \end{pmatrix}, \sigma_2 = \begin{pmatrix} -b & 0 & 0 \\ 0 & 1 & 0 \\ 0 & 0 & -b^{-1} \end{pmatrix},$$

with $b \in F^\times - F^+$. For each element σ_i of the above type, let ξ_i be such that $\xi_i {}^t\overline{\xi}_i = \sigma_i$. Let δ_{ξ_i} be the module of the group $P_{\xi_i} := P(E) \cap \xi_i H(F)\, \xi_i^{-1}$ and $V(\xi_i)$ be the space of linear forms μ on the space $\mathscr{V}(\Pi_1)$ such that

$$\delta_P^{1/2}(p)\, \mu(\Pi_1 \times \Pi_2(p)\, v) = \delta_{\xi_i}(p)\, v,$$

for every vector u and every $p \in P_{\xi_i}$. Then

$$\dim(\mathscr{H}(\Pi, H)) \leqslant \sum_i \dim(V(\xi_i)).$$

Now P_{ξ_0} contains the subgroup

$$\left\{ \begin{pmatrix} 1 & x & \frac{x\bar{x}}{2} \\ 0 & 1 & \bar{x} \\ 0 & 0 & 1 \end{pmatrix} \right\}$$

whose intersection with (or, more correctly, projection on) $M(E)$ contains the unipotent radical U_1 of the parabolic subgroup of type $(1, 1)$ of M. Thus for any $\mu \in V(\xi_0)$, any $u \in U_1$ and any vector v:

$$\mu(\Pi_1(u)\, v) = \mu(v)$$

Because Π_1 is supercuspidal, this implies $\mu = 0$. Thus $V(\xi_0) = 0$.

Next, consider the case of the element σ_1. Then P_{ξ_1} is the set of matrices

$$\begin{pmatrix} h_1 & 0 \\ 0 & h_2 \end{pmatrix}$$

with $h_1 \in H_1(F)$ and $h_2 \in \mathbb{U}_1(F)$, where H_1 is the unitary group in $GL(2, E)$ for the Hermitian matrix

$$\begin{pmatrix} -1 & 0 \\ 0 & 1 \end{pmatrix}.$$

Thus $V(\xi_1)=0$ unless Π_2 is distinguished and then $V(\xi_1)$ may be identified with the space of linear forms μ on $\mathscr{V}(\Pi_1)$ such that

$$\mu(\Pi_1(h_1)\, v)=\mu(v)$$

for any $h_1 \in H_1$ and any vector v. Thus $\dim(V(\xi_1)) \leqslant 1$ by the previous section. Moreover if $V(\xi_1) \neq 0$ then Π_1 is distinguished by H_1 and Π_2 by $\mathbb{U}_1$. In particular, it follows then that $\Pi_1^\sigma = \Pi_1$ and $\Pi_2^\sigma = \Pi_2$, and thus $\Pi^\sigma = \Pi$ as well.

Now consider the case of the element σ_2. We let H_2 be the unitary group for the matrix

$$\begin{pmatrix} -b & 0 \\ 0 & 1 \end{pmatrix}.$$

Then $V(\xi_2)=0$ unless Π_2 is distinguished by $\mathbb{U}_1$ and then $V(\xi_2)$ is isomorphic to the space of linear forms μ on $\mathscr{V}(\Pi_1)$ which are invariant under H_2. Thus $\dim(V(\xi_2)) \leqslant 1$ again. Furthermore, if $V(\xi_2) \neq 0$ then again Π_1 and Π_2 are invariant under σ and so is Π.

Thus we do get $\dim(\mathscr{H}(\Pi, H)) \leqslant 2$. Moreover if Π is distinguished by H then $\Pi^\sigma = \Pi$.

Now suppose that $\Pi^\sigma = \Pi$. Then $\Pi_1^\sigma = \Pi_1$ and $\Pi_2^\sigma = \Pi_2$. The representation Π_1 is the base change of two supercuspidal representations, say π_1^1 and $\pi_1^2 := \pi_{1,v}^1 \otimes \omega_{E/F}$. Let ω_1 be their common central character and set $\omega_2 = \omega\omega_1^{-1}$. Thus Π is the base change of exactly two irreducible representations with central character ω, the central character of Π being $\Omega = \omega \circ \mathrm{Norm}$. The two representations in question are the induced representations

$$\pi^i = \mathrm{Ind}(\pi_1^i, \pi_2), \qquad i=1, 2.$$

It remains to show that there are two elements $\mathscr{P}_i$, $i=1, 2$, *of* $\mathscr{H}(\Pi, H)$ with the following property. Define as before distributions $\mathscr{R}_i$, $i=1, 2$, by

$$\mathscr{R}_i(f)=\mathscr{B}_{\pi^i}(f'),$$

for (f, f') with matching orbital integrals. Then

$$\mathscr{R}_i(f)=\sum \mathscr{W}_v(\Pi_v(f)\, u_\alpha)\, \overline{\mathscr{P}_i(u_\alpha)}.$$

Note that f' is an arbitrary smooth function of compact support on G^+ and the restrictions of the representations π^i to G^+ inequivalent. If follows that the distributions $\mathscr{B}_{\pi^i}$ are linearly independent. The same is true of the distributions $\mathscr{R}_i$. It follows that the linear forms $\mathscr{P}_i$ are linearly independent (see Lemma (1)).

To apply the global theory, we again write our local extension in the form E_v/F_{v_0}, write Π_v for Π and so on. Our assertion follows from the global theory if there exist two cuspidal representations π^i, $i=1,2$, with components $\pi^i_{v_0}$ at v_0. Indeed, the corresponding base change representations Π_i are then cuspidal and distinguished and we can argue as before. Of course, this will not be the case in general.

To remedy the situation, for every $z=q_{v_0}^{-s}$ with $s\in\mathbb{C}$, we set:

$$\pi^{i,z}_{v_0}=\mathrm{Ind}(\pi^i_{1,v_0}\otimes\alpha^s_{v_0},\alpha^{-2s}_{v_0}\pi_{2,v_0}),$$

where α_{v_0} denotes the module of F_{v_0}. We recall a standard lemma:

LEMMA 6. *Fix an index i. Let X be the set of complex numbers z of module* 1 *such that there is a cuspidal automorphic representation π of $GL(3,F_{\mathbb{A}})$ whose component at v_0 is $\pi^{i,z}_{v_0}$. The set X is infinite.*

Proof of the Lemma. For the convenience of the reader we provide a proof. For the proof of the lemma we fix the index i and drop it from the notations and consider the representation

$$\pi^z_{v_0}:=\mathrm{Ind}(\pi_{1,v_0}\otimes\alpha^s_{v_0},\alpha^{-2s}_{v_0}\pi_{2,v_0}).$$

Let $\mathscr{C}(G_{v_0},\omega^{-1}_{v_0})$ be the space of smooth functions transforming under the character $\omega^{-1}_{v_0}$ and compactly supported modulo the center. We recall that as a consequence of Bernstein's theory, there is a bi-invariant subspace $\mathscr{C}_0$ of the space $\mathscr{C}(G_{v_0},\omega^{-1}_{v_0})$ with the following properties. There is a direct sum decomposition

$$\mathscr{C}(G_{v_0},\omega^{-1}_{v_0})=\mathscr{C}_0\oplus\mathscr{C}^0$$

where $\mathscr{C}^0$ is also bi-invariant. Let $f\in\mathscr{C}_0$. For any irreducible admissible representation π_1 of $GL(3,F_{v_0})$ with central character ω_{v_0} we have $\pi_1(f)=0$ unless $\pi=\pi^z_{v_0}$ for some z. If f is given and f_0 is its projection on $\mathscr{C}_0$ in the above decomposition then $\pi^z_{v_0}(f)=\pi^z_{v_0}(f_0)$ for any z. Given $m+1$ complex numbers $z_0,z_1,z_2,\dots,z_m$ one can find $f\in\mathscr{C}(G_{v_0},\omega^{-1}_{v_0})$ such that $\pi^{z_0}_{v_0}(f)\neq 0$ but $\pi^{z_i}_{v_0}(f)=0$, $1\leqslant i\leqslant m$. This follows from the fact the representations are irreducible and inequivalent. Thus there is an element of $\mathscr{C}_0$ with the same property, namely f_0. In particular, suppose Y is a set of complex numbers of module 1 with the following property of density: for $f\in\mathscr{C}_0$ the relations $\pi^y_{v_0}(f)=0$ for all $y \in Y$ imply $f=0$. Then the set Y must be infinite. Thus it suffices to show X has precisely this property of density.

To show this is the case, consider an element $f_0 \in \mathscr{C}_0$. Define a function

$$f_{v_0}(g)=\int_{G_{v_0}} \overline{f_0(gx)}\, f_0(x)\, dx$$

on G_{v_0}. Choose another place v_1 and a cuspidal element f_{v_1} of $\mathscr{C}(G_{v_1}, \omega_{v_1})$ with $f_{v_1}(e) \neq 0$. Choose also an element f^{v_1, v_2} of $\mathscr{C}(G^{v_0, v_1}, \omega^{v_0, v_1})$ and set

$$f=f_{v_0} f_{v_1} f^{v_0, v_1}, \quad f'=\overline{f_0}\, f_{v_1} f^{v_0, v_1}.$$

Consider the sums

$$\phi(g)=\sum_{\gamma \in Z(F)\backslash G(F)} f(\gamma g), \quad \phi'(g)=\sum_{\gamma \in Z(F)\backslash G(F)} f'(\gamma g).$$

Both functions are cuspidal. Let again π_α denote the set of cuspidal representations with central character ω and let ϕ_α, ϕ'_α denote the corresponding orthogonal projections. Then

$$\phi_\alpha=\pi_{\alpha, v_0}(f_0)\, \phi'_\alpha.$$

Thus if $\phi_\alpha \neq 0$ then $\pi_{\alpha, v_0}=\pi^z_{v_0}$ for some $z \in X$. Now suppose that f_0 is such that $\pi^z_{v_0}(f_0)=0$ for all $z \in X$. Then $\phi_\alpha=0$ for all α and so $\phi=0$. In particular

$$\sum f(\gamma)=0,$$

for all choices of f^{v_0, v_1}. This implies $f_{v_0}(e)=0$ and so $f_0=0$. Thus X is infinite, as claimed. ∎

For any z of module 1, let Π^z_v be the base change of the representation $\pi^{i,z}_{v_0}$. Let $\mathscr{R}^{i,z}$ be the distribution corresponding to $\mathscr{B}_{\pi^{i,z}_{v_0}}$. For $z \in X$, there is thus a linear form in $\mathscr{H}(\Pi^z_v, H)$ with the required property. More precisely, let $\mathscr{W}^z$ be any non-zero Whittaker linear form on $\mathscr{V}(\Pi^z_v)$, then there is a unique $\mathscr{P}^i_z \in \mathscr{H}(\Pi^z_v, H)$ such that

$$\mathscr{R}^{i,z}(f)=\sum_\alpha \mathscr{W}^z(\Pi^z_v(f)\, W_\alpha)\, \overline{\mathscr{P}^i_z(W_\alpha)}.$$

Our task is to show that for every z (of module 1) there is such a linear form.

At this point, we may as well revert to a local notation, writing our extension as E/F and writing simply Π^z rather than Π^z_v and so on. We may regard the representations $\pi^{i,z}$ as a fiber bundle of representations. In a precise way, we set $K' = GL(3, \mathcal{O}_F)$ and let $\mathcal{V}'$ be the space of smooth functions $\phi: K' \to \mathcal{V}(\pi_1)$ such that $\phi(mk) = \pi_1^i \times \pi_2(m)$ for $m \in M(F) \cap K'$. Then we may regard all the representations $\pi^{i,z}$ as operating on $\mathcal{V}'$. For every $u \in \mathcal{V}'$ the map $z \mapsto \pi^{i,z}(f')\, u$ takes its values in a fixed finite dimensional vector space and is a polynomial function of z. Similarly, there is a holomorphic family of Whittaker linear forms $\mathcal{W}'_z$ on $\mathcal{V}'$. More precisely, for every $u \in \mathcal{V}'$ the map $z \mapsto \mathcal{W}'_z(u)$ is a polynomial in z. Now suppose that f' is bi-invariant under the compact open subgroup K'_1. Then the Bessel distribution corresponding to $\mathcal{W}'_z$ is given by:

$$\mathcal{B}_{\pi_i^z}(f') = \sum_{\alpha} \mathcal{W}'_z(\pi_i^z(f')\, u_\alpha)\, \overline{\mathcal{W}'_z(u_\alpha)}$$

where u_α is a fixed orthonormal basis of the space of K'_1-invariant vectors in $\mathcal{V}$. It follows that

$$z \mapsto \mathcal{B}_{\pi_i^z}(f')$$

is a polynomial in z. The same is therefore true of the map $z \mapsto \mathcal{R}^{i,z}(f)$.

We introduce the notation of generalized vector and write

$$\mathcal{W}^z(W) = (W, \mathcal{W}^z),\ \mathcal{P}^i_z(W) = (W, \mathcal{P}^i_z).$$

In terms of generalized vectors, we see that for $z \in X$:

$$(\Pi_v^z(f)\, \mathcal{P}^i_z, \mathcal{W}^z) = \mathcal{R}^{i,z}(f).$$

Fix a compact open subgroup K_1 and let dk_1 denote the Haar measure of mass one on K_1. Set

$$f^{K_1}(g) = \int f(gk_1)\, dk_1.$$

Let also $P^z_{K_1}$ denote the corresponding projection operator:

$$P^z_{K_1} = \int \Pi^z(k_1)\, dk_1.$$

Then for *any* z, we can write:

$$\mathcal{R}^{i,z}(f^{K_1}) = \int W_z(g)\, f(g)\, dg$$

where W_z is a function such that

$$W_z(ngk_1)=\theta(n\bar{n})\, W_z(g),$$

for $n\in N(E)$ and $k_1\in K_1$. For every g the map $z\mapsto W_z(g)$ is a polynomial in z.

We claim that W_z belongs to the Whittaker model of Π^z. This is true for $z\in X$. Indeed, for $z\in X$, we have:

$$W_z(g)=(\Pi^z(g)\, P^z_{K_1}\mathcal{P}^i_z,\, \mathcal{W}^z).$$

Again the representations Π^z form a fiber bundle of representations all operating on the same space $\mathcal{V}$ smooth of functions on $K=GL(3,\mathcal{O}_E)$ with values in $\mathcal{V}(\Pi_1)$. There is also an analytic family of Whittaker linear forms $\mathcal{W}^z$ on $\mathcal{V}$. Let e_μ, $1\leqslant\mu\leqslant M$, be a basis of the space of vectors in $\mathcal{V}$ invariant under K_1. The functions W^z_μ defined by:

$$W^z_\mu(g)=\mathcal{W}^z(\Pi^z(g)\, e_\mu)$$

form a basis of the space of K_1 invariant elements in the Whittaker model of Π^z_v. Thus for $z\in X$ we have a unique decomposition:

$$W_z(g)=\sum_{1\leqslant\mu\leqslant M}\lambda_\mu(z)\, W^z_\mu(g).$$

Let z_0 be a point not in X. Now choose M elements (g^ν) such that

$$D(z):=\det(W^z_\mu(g^\nu))$$

is non-zero at z_0. Then $D(z)\neq 0$ on a subset X_0 of X which is also infinite. For $z\in X_0$ the scalars $(\lambda_\mu(z))$ are solutions of the Cramer system

$$W_z(g^\nu)=\sum_{1\leqslant\mu\leqslant M}\lambda_\mu(z)\, W^z_\mu(g^\nu).$$

Thus there exist polynomials (P_μ) such that

$$\lambda_\mu(z)=\frac{P_\mu(z)}{D(z)}$$

for $z\in X_0$. Then

$$W_z(g)=\sum_\mu\frac{P_\mu(z)}{D(z)}\, W^z_\mu(g)$$

for $z\in X_0$. Thus the same relation is true at z_0. Our assertion follows.

This being so the above result for $z=0$ amounts to saying that for every K_1 there is a unique vector u_{K_1} invariant under K_1 such that, for any function f,

$$\mathscr{R}^i(f^{K_1}) = (\Pi(f)\, u_{K_1}, \mathscr{W}).$$

Now if $K_2 \supseteq K_1$ then $(f^{K_2})^{K_1} = f^{K_2}$. Thus

$$\mathscr{R}^i(f^{K_2}) = (\Pi(f^{K_2})\, u_{K_1}, \mathscr{W}) = (\Pi(f)\, P_{K_2} u_{K_1}, \mathscr{W}).$$

It follows that $P_{K_2} u_{K_1} = u_{K_2}$. It follows there is a generalized vector $\mathscr{P}_i$ such that

$$\mathscr{R}^i(f) = (\Pi(f)\, \mathscr{P}_i, \mathscr{W}).$$

By definition the distribution $\mathscr{R}^i$ is invariant under H on the right. Thus the generalized vector $\mathscr{P}_i$ is also invariant and we are done.

8. CONCLUDING REMARKS

The same technics can be used to prove the conjecture for other representations. As a matter of fact, this is done in [LR]. However, it is difficult to prove the conjecture for all representations thus we prefer to limit ourself to the above cases.

One expects the above results to generalize in a straightforward way to the groups $GL(n)$ with n odd. For n even, the situation is more complicated. Even if we assume that the infinite places of F split in E, we have to deal with more than one unitary group. Thus it is reasonable to conjecture that a cuspidal representation which is a base change is distinguished by some unitary group; it is then a separate issue to show that it is in fact distinguished with respect to the quasi-split group H, that is the unitary group for the Hermitian matrix w with entries $w_{i,j} = \delta_{i+j,n+1}$. Assuming that it is the case, it is best to introduce the similitude group $\tilde{H}$ and global period integrals of the form:

$$\int_{Z(E_{\mathbb{A}})\backslash \tilde{H}(E_{\mathbb{A}})} \phi(h)\, \omega^{-1}(\lambda(h))\, dh.$$

Such an integral should be non-identically zero if and only if Π is the base change of a cuspidal representation π with central character $\omega\omega_{E/F}^{n/2}$. It should factor as product of local invariant linear forms.

Now we discuss the local situation for n even. So let E/F be a local non-Archimedean quadratic extension. For every generic irreducible representation Π, we should introduce, for a character ω of $F^\times$, the space $\mathcal{H}(\Pi, \tilde{H}, \omega)$ of linear forms $\mathcal{P}$ such that

$$\mathcal{P}(\Pi(\tilde{h})\,u) = \omega(\lambda(\tilde{h}))\,\mathcal{P}(u).$$

Then it should be that $\mathcal{H}(\Pi, \tilde{H}, \omega) \neq 0$ if and only if Π is the base change of at least one representation π with central character $\omega\omega_{E/F}^{n/2}$. This conjecture is motivated by the following property of the local transfer factor $\gamma(a, \psi_{v_0})$, at an inert place v_0: the transfer factor is a function defined on the group $A(F_{v_0})$ of diagonal matrices with entries in F_{v_0}. If n is even, for any scalar matrix $z \in F_{v_0}^\times$,

$$\gamma(az, \psi) = \gamma(a, \psi)\,\omega_{E/F}(z)^{n/2}.$$

Going back to a local situation, for global purposes, we deal with functions Φ supported on the set of Hermitian matrices in the orbit of w, that is, whose determinant is in $\det wF^+$. The matching functions f' are supported on the set wG^+. It is more convenient to consider the symmetric space $\mathfrak{S}_w$ of matrices s such that $s = s^*$ where we have set $g^* := w{}^tg^\sigma w$. The group $GL(n, E)$ operates on $\mathfrak{S}_w$ by $s \mapsto gsg^*$. We consider the orbit of w, that is, the set $\mathfrak{S}_w^+$ of matrices with $\det s \in F^+$. Then the condition of matching reads

$$\Phi(gg^*) = \int f(gh)\,dh,$$

$$\int \Phi(nawn^*)\,\theta(n\bar{n})\,dn = \gamma(a, \psi) \int f'(n_1awn_2)\,\theta(n_1)\,\theta(n_2)\,dn_1\,dn_2.$$

The relative Bessel distribution is defined in the same way as before but the Bessel distribution is now defined by:

$$\mathcal{B}_\pi(f') := \sum_\alpha \mathcal{W}(\pi(f')\,W_\alpha)\,\overline{\mathcal{W}(W_\alpha)}.$$

The distributions $\mathcal{B}_\pi$ and $\mathcal{B}_{\pi\otimes\omega_{E/F}}$ have the same restriction to G^+. We have the following lemma:

LEMMA 7. *The restriction of $\mathcal{B}_\pi$ to G^+ is non-zero. Let π_j, $1 \leqslant i \leqslant m$ be a family of irreducible generic representations of $GL(n, F)$ such that for any pair (i, j), $i \neq j$, the representations π_i and π_j (resp. π_i and $\pi_j \otimes \omega_{E/F}$) are* *inequivalent. Then the restrictions of the Bessel distributions* $\mathcal{B}_{\pi_i}$ *to* G^+ *are linearly independent.*

Proof of the Lemma. Indeed, the first assertion is clear if the restriction of π to G^+ is irreducible (see Lemma (1)). Suppose it is not. Then π is induced by an irreducible representation π_0 of G^+ and $\pi = \pi \otimes \omega_{E/F}$. Suppose that $W \in \mathcal{W}(\pi)$ is such that $\mathcal{W}(W) \neq 0$. Then the function W_1 defined by:

$$W_1(g) = \tfrac{1}{2}(W(g) + W(g)\, \omega_{E/F}(\det g))$$

is a non-zero element of $\mathcal{W}$ supported on G^+. It follows that

$$\mathcal{W}(\pi) = \mathcal{W}^+ \oplus \pi(r)\, \mathcal{W}^+,$$

where $\mathcal{W}^+$ is the space of elements of $\mathcal{W}(\pi)$ supported on G^+ and $\det r \notin F^+$. It follows that the representation π^+ on $\mathcal{W}^+$ by right shifts is equivalent to π_0 or π_0^r and is irreducible. Since $\mathcal{W}$ vanishes on $\pi(r)\,\mathcal{W}^+$, for f' supported on G^+, we may take the sum defining the Bessel distribution over a basis of $\mathcal{W}^+$. The first assertion follows.

To prove the second assertion, for each i, denote by $\pi_{i,0}$ the restriction of π_i to G^+, if this restriction is irreducible, or the irreducible representation of G^+ which induces π_i if not. Then the representations $\pi_{i,0}$ are irreducible and pairwise inequivalent. By the first part of the proof, the restriction of the Bessel distribution $\mathcal{B}_{\pi_i}$ to G^+ may be viewed as a (generalized) matrix coefficient of $\pi_{i,0}$. The second assertion follows. ∎

The lemma implies that in the (still conjectural) relative trace formula, for a global extension E/F of number fields, where the infinite places of F split in E, a cuspidal automorphic representation of $GL(n, F_{\mathbb{A}})$ cannot give a zero contribution, even if we consider only functions f' supported on $G_{\mathbb{A}}^+$. This will prove that any base change representation of $GL(n, E\mathbb{A})$ is distinguished by a quasi-split unitary group. Note that this argument is insufficient if some real place of F is inert in E.

The lemma also suggest the following conjecture: suppose that $\mathcal{H}(\Pi, \tilde{H}, \omega)$ is not zero. Let $\pi_1, \pi_2, \pi_3, \ldots, \pi_m$ be representations of $GL(n, F)$ with central character $\omega\omega_{E/F}^{n/2}$ which base change to Π. We assume that for any other representation π which base change to Π there is exactly one index i such that either $\pi = \pi_i$ or $\pi = \pi_i \otimes \omega_{E/F}$. Then, for each index i, there is a unique element $\mathcal{P}_i$ of $\mathcal{H}(\Pi, \tilde{H}, \omega)$ such that

$$\mathcal{R}_\Pi(f) = \mathcal{B}_{\pi_i}(f')$$

if f and f' have matching orbital integrals. Moreover, the vectors $\mathcal{P}_i$, $1 \leqslant i \leqslant m$, form a basis of $\mathcal{H}(\Pi, \tilde{H}, \omega)$.

REFERENCES

[AC] J. Arthur and L. Clozel, "Simple Algebras, Base Change, and the Advanced Theory of the Trace Formula," Annals of Math. Studies, Princeton University Press, Princeton, 1989.

[B] J. Bernstein, P-invariant distributions of $GL(n)$ and the classification of unitary representations of $GL(n)$ (non Archimedean case), *in* "Lie Group Representations II, College Park, MD., 1982/1983," Lecture Notes in Mathematics, Vol. 1041, Springer-Verlag, Berlin/New York, 1984.

[Ba] E. M. Baruch, Proof of Kirillov's conjecture, preprint.

[BZ] J. Bernstein and A. V. Zelevinski, Induced representations of reductive p-adic groups I & II, *Ann. Sc. École Norm. Sup.* **10** (1980), 441–472; **13** (1980). 165–210.

[DB] J. Bernstein (rédigé par P. Deligne), "Le "centre" de Bernstein in Représentations des groupes réductifs sur un corps local," Collection Travaux en cours, Hermann, Paris, 1984.

[CHM] W. Casselman, H. Hecht, and Miličié, Bruhat filtrations and Whittaker vectors for real groups, preprint.

[DM] J. Dixmier and P. Malliavin, Factorizations de fonctions et vecteurs indéfiniment différentiables, *Bull. Sci. Math. France* **102** (1978), 305–330.

[yF1] Y. Flicker, Twisted tensors and Euler products, *Bull. Soc. Math. France* **116** (1988), 295–313.

[yF2] Y. Flicker, On distinguished representations, *J. Reine Angew. Math.* **418** (1991), 139–172.

[yH] Y. Hironaka, Spherical functions of Hermitian and symmetric forms I, *Japan Math. Soc.* **14** (1988), 203–223.

[GK] I. M. Gelfand and Kazhdan, Representations of $GL(n, K)$ where K is a local field, *in* "Lie Groups and Their Representations, Proc. Summer School, Bolyai Janos Math. Soc., Budapest, 1971," pp. 95–118, Halsted, New York, 1975.

[jH] J. Hakim, Distinguished p-adic representations, *Duke Math. J.* **62** (1991), 1–22.

[jHyF] J. Hakim and Y. Flicker, Quaternionic distinguished representations, *Am. J. Math.* **116** (1994), 255–278.

[HLR] G. Harder, R. P. Langlands, and M. Rapoport, Algebraische Zyklen auf Hilbert–Blumenthal–Flächen [Algebraic cycles on Hilbert–Blumenthal surfaces] *J. Reine Angew. Math.* **366** (1986), 53–120.

[J1] H. Jacquet, Generic representations, *in* "Non Commutative Harmonic Analysis, Actes Colloq., Marseille-Luminy," Lecture Notes in Mathematics, Vol. 587, Springer-Verlag, Berlin, 1977.

[J2] H. Jacquet, The continuous spectrum of the relative trace formula for $GL(3)$ over a quadratic extension, *Israel J. Math.* **89** (1995), 1–59.

[J3] H. Jacquet, A theorem of density for Kloosterman integrals, *Asian J. Math.* **2** (1988), 759–778.

[JLR] H. Jacquet, E. M. Lapid, and J. Rogawski, Periods of automorphic forms, *J. Amer. Math. Soc.* **12** (1999), 173–240.

[JS] H. Jacquet and J. Shalika, On Euler products and the classification of automorphic representations I, II, *Amer. J. Math.* **103** (1981), 449–558, 777–815.

[JY1] H. Jacquet and Y. Ye, Une remarque sur le changement de base quadratique, *C.R. Acad. Sci. Paris Sér. I, Math.* **311** (1990), 449–558.

[JY2] H. Jacquet and Y. Ye, Relative Kloosterman integrals for $Gl(3)$, *Bull. Soc. Math. France* **120** (1992), 449–558.

[JY3] H. Jacquet and Y. Ye, Distinguished representations and quadratic base change for $GL(3)$, *Trans. Amer. Math. Soc.* **348** (1996), 913–939.

[JY4] H. Jacquet and Y. Ye, Germs of Kloosterman integrals for $GL(3)$, *Trans. Amer. Math. Soc.* **351** (1996), 127–1255–939.

[LR] E. Lapid and J. Rogawski, Stabilization of periods of Eisenstein series on $GL(3)$ relative to $U(3)$, preprint.

[MW] C. Moeglin and J. L. Waldspurger, Le spectre résiduel de $GL(n)$, *Ann. Sci. École Norm. Sup.* **22** (1989), 605–704.

[N1] B. C. Ngo, Lemme fondamental de Jacquet et Ye en caractéristiques égales, *C.R. Acad. Sci. Paris Sér. I* **325** (1997), 307–312.

[N2] B. C. Ngo, Lemme fondamental de Jacquet et Ye en caractéristique positive, *Duke Math. J.* **325** (1997), 307–312.

[N3] B. C. Ngo, Faisceaux pervers, homomorphisme de changement de base et lemme fondamental de Jacquet et Ye, *Ann. Sci. École Norm. Sup.* **32** (1999), 619–679.

[P] D. Prasad, Invariant forms for representations of GL_2 over a local field, *Amer. J. Math.* **114** (1992), 1317–1363.

[Ro1] J. Rogawski, "Automorphic Representations of Unitary Groups in Three Variables," Ann. of Math. Studies, Vol. 123, Princeton University Press, Princeton, 1990.

[Ro2] J. Rogawski, The multiplicity Theorem of A-packets, *in* "The Zeta Functions of Picard Modular Surfaces," Les Publications CRM, Montreal, 1992.

[jS] J. Shalika, The multiplicity one theorem of GL_n, *Ann. of Math.* **100** (1974), 171–193.

[W] N. Wallach, "Real Reductive Groups II," Pure and Applied Mathematics, Vol. 132-II, Academic Press, San Diego, 1992.

[Y1] Y. Ye, Kloosterman integrals and base change, *J. Reine Angew. Math.* **400** (1989), 57–121.

[Y2] Y. Ye, The fundamental lemma of a relative trace formula for $GL(3)$, *Compositio Math.* **89** (1993), 121–162.

[Y3] Y. Ye, An integral transform and its applications, *Math. Ann.* **300** (1994), 405–417.

Printed by Catherine Press, Ltd., Tempelhof 41, B-8000 Brugge, Belgium

Bull. Soc. math. France
129 (1), 2001, p. 33–90

POSITIVITY OF QUADRATIC BASE CHANGE L-FUNCTIONS

BY HERVÉ JACQUET & CHEN NAN

ABSTRACT. — We show that certain quadratic base change L-functions for Gl(2) are non-negative at their center of symmetry.

RÉSUMÉ (*Positivité des fonctions L du changement de base quadratique*)
On montre que certaines des fonctions L de Gl(2) obtenues par changement de base quadratique sont positives en leur centre de symétrie.

Contents

Texte reçu le 23 septembre 1999, révisé le 1er mars 2000

HERVÉ JACQUET, Department of Mathematics, Columbia University, 2990 Boradway, MC 4408, New York, NY 10027 (USA) • *E-mail :* hj@math.columbia.edu
CHEN NAN, Department of Mathematics, Columbia University, 2990 Broadway, New York, NY 10027 (USA) • *E-mail :* nanchen@math.columbia.edu

2000 Mathematics Subject Classification. — 11F67, 11F70, 11F72, 22E55.
Key words and phrases. — L-functions, positivity, quadratic base change.
H.J.: Partially supported by NSF Grant DMS 9619766.

BULLETIN DE LA SOCIÉTÉ MATHÉMATIQUE DE FRANCE 0037-9484/2001/33/$ 5.00

1. The main theorem

Let E/F be a quadratic extension of number fields and $\eta_{E/F}$ or simply η the quadratic character of F attached to E, $\{1,\tau\}$ the Galois group of E/F. We will often write

$$\tau(z) = \bar{z}.$$

We will denote by U_1 the unitary group in one variable, that is, the group of elements of norm 1 in $E^\times$. Suppose that π is an automorphic cuspidal representation of $\mathrm{Gl}(2, F_{\mathbb{A}})$ whose central character ω is trivial on the group of norms. In other words $\omega = 1$ or $\omega = \eta$. We assume that π is not dihedral with respect to E so that the base change representation Π of π to $\mathrm{Gl}(2, E)$ is still automorphic and cuspidal. Let Ω be an idele class character of E whose restriction to $F_{\mathbb{A}}^\times$ is equal to ω. Our main result is the following theorem:

THEOREM 1. — *With the previous notations:* $L\big(\frac{1}{2}, \Pi \otimes \Omega^{-1}\big) \geq 0$.

If $\omega = 1$ and $\Omega = 1$ then $L(s, \Pi) = L(s,\pi)L(s, \pi \otimes \eta)$ and the result has been established by Guo (see [G1], under some restrictions on E/F). As a matter of fact, by using results on averages of L-functions (see [FH]), Guo is able to prove that $L(\frac{1}{2}, \pi) \geq 0$, which then implies our result for $\omega = 1$, $\Omega = 1$, without restriction on E/F. At any rate, Baruch and Mao [BM] have independently established that $L(\frac{1}{2}, \pi) \geq 0$ if $\omega = 1$. However, the present result–where Ω needs not be trivial–is more general, even in the case $\omega = 1$.

Results on the positivity of $\mathrm{Gl}(2)$ L-functions have been considered by many mathematicians (see, for instance, [BFH], [Gr], [K], [Kk], [KS], [KZ], [S], [R], [S], [W3], [Ya]). Specially, the positivity of the twisted L-function at hand has been investigated (for holomorphic forms) in [GZ].

We note that $\Omega^\tau = \Omega^{-1}$ and Π is self-contragredient: $\widetilde{\Pi} = \Pi$. Thus

$$L(s, \Pi \otimes \Omega^{-1}) = L\big(s, \Pi^\tau \otimes (\Omega^{-1})^\tau\big) = L(s, \Pi \otimes \Omega) = L(s, \widetilde{\Pi} \otimes \Omega).$$

Likewise,

$$\epsilon(s, \Pi \otimes \Omega^{-1})\epsilon(1-s, \Pi \otimes \Omega^{-1}) = \epsilon(s, \Pi \otimes \Omega^{-1})\epsilon(1-s, \widetilde{\Pi} \otimes \Omega) = 1.$$

In particular $\epsilon(\frac{1}{2}, \Pi \otimes \Omega^{-1}) = \pm 1$. Thus, despite the fact that $\Pi \otimes \Omega^{-1}$ is not necessarily self-contragredient, the L-function $L(s, \Pi \otimes \Omega^{-1})$ is symmetric:

$$L(s, \Pi \otimes \Omega^{-1}) = \epsilon(s, \Pi \otimes \Omega^{-1})L(1-s, \Pi \otimes \Omega^{-1}).$$

The following lemma is easily verified:

LEMMA 1. — *Let v_0 be a place of F. If v_0 is inert and v is the corresponding place of E then:*

$$L\big(\tfrac{1}{2}, \Pi_v \otimes \Omega_v^{-1}\big) > 0.$$

If v_0 splits into v_1 and v_2 then:

$$L\big(\tfrac{1}{2}, \Pi_{v_1} \otimes \Omega_{v_1}^{-1}\big)L\big(\tfrac{1}{2}, \Pi_{v_2} \otimes \Omega_{v_2}^{-1}\big) > 0.$$

Let S_0 be a finite set of places of F and S the corresponding set of places of E. Set

$$L^S(s,\Pi\otimes\Omega^{-1}) = \prod_{v\not\in S} L(s,\Pi_v\otimes\Omega_v^{-1}).$$

In view of the lemma, the statement of the theorem is equivalent to the positivity of $L^S(\frac{1}{2},\Pi\otimes\Omega^{-1})$.

If Π is dihedral with respect to E, then Π is associated with an idele class character Ξ of E whose restriction to $F_{\mathbb{A}}^\times$ is $\omega\eta$. Thus $\Xi^\tau = \Xi^{-1}$ and

$$L(s,\Pi\otimes\Omega^{-1}) = L(s,\Xi\,\Omega^{-1})L(s,\Xi^\tau\Omega^{-1}) = L(s,\Xi\,\Omega^{-1})L(s,\Xi^{-1}\Omega^{-1}).$$

If Ω is trivial or even quadratic this is ≥ 0. At any rate, in general, $\Xi\,\Omega^{-1}$ and $\Xi^{-1}\Omega^{-1}$ have η for restriction to $F_{\mathbb{A}}^\times$. Thus there are cuspidal representations π_1 and π_2 of $\mathrm{Gl}(2,F_{\mathbb{A}}^\times)$ with trivial central character such that:

$$L(s,\Xi\,\Omega^{-1}) = L(s,\pi_1), L(s,\Xi^{-1}\Omega^{-1}) = L(s,\pi_2)$$

and by the results already quoted each factor is ≥ 0 at $s=\frac{1}{2}$. We will not discuss this case but remark that, by considering the discrete but non-cuspidal terms in our trace formula, we could probably handle this case as well.

The proof of the theorem is based on a careful analysis of the relative trace formula of [J2] (In the case $\Omega = 1$ we could, like Guo, use the simpler trace formula of [J1].) Namely, we consider an inner form G of $\mathrm{Gl}(2,F)$ which contains a torus T isomorphic to $E^\times$. There is then an $\epsilon\in F^\times$, uniquely determined modulo $\mathrm{Norm}(E^\times)$, such that the pair (G,T) is isomorphic to the pair (G_ϵ,T) defined as follows. We denote by $\mathbb{H}_\epsilon$ the semi-simple algebra of matrices $g\in M(2,E)$ of the form

$$g = \begin{pmatrix} a & \epsilon b \\ \bar{b} & \bar{a}\end{pmatrix} \tag{1}$$

and by G_ϵ its multiplicative group. Then

$$T = \left\{ t = \begin{pmatrix} a & 0 \\ 0 & \bar{a}\end{pmatrix}\right\}.$$

We let Z be the center of G_ϵ. We regard Ω as a character of $T(F_{\mathbb{A}})$: $t\mapsto\Omega(a)$. Suppose that f is a smooth function of compact support on $G_\epsilon(F_{\mathbb{A}})$. We form as usual a kernel

$$K_f(x,y) := \int_{Z(F_{\mathbb{A}})/F^\times} \sum_{\xi\in G_\epsilon(F)} f(x^{-1}z\,\xi y)\,\omega(z)\mathrm{d}z$$

and a distribution

$$J_\epsilon(f) := \int_{(Z(F_{\mathbb{A}})T(F)\backslash T(F_{\mathbb{A}}))^2} K_f(t_1,t_2)\,\Omega(t_1)^{-1}\mathrm{d}t_1\,\Omega(t_2)\mathrm{d}t_2.$$

We have a spectral decomposition of the kernel:

$$K_f = \sum_\sigma K_{f,\sigma} + K_{f,\mathrm{cont}}\,;$$

the sum on the right is over all irreducible (cuspidal) automorphic representations σ of G_ϵ: if G_ϵ is not split, by cuspidal we mean an irreducible automorphic representation which is not one-dimensional. The term $K_{f,\mathrm{cont}}$ represents the contribution of the one dimensional representations and the continuous spectrum which is present only if G_ϵ is split, that is, ϵ is a norm. For every σ the kernel $K_{f,\sigma}$ is defined by

$$K_{f,\sigma}(x,y) = \sum_\phi \rho(f)\phi(x)\overline{\phi(y)},$$

the sum over an orthonormal basis of the space of σ. We define then:

$$J_\sigma(f) := \int_{(Z(F_\mathbb{A})T(F)\backslash T(F_\mathbb{A}))^2} K_{f,\sigma}(t_1,t_2)\,\Omega(t_1)^{-1}\mathrm{d}t_1\,\Omega(t_2)\mathrm{d}t_2.$$

This is a distribution of positive type: if $f = f_1 * f_1^*$ where $f_1^*(g) := \overline{f}_1(g^{-1})$ then

$$J_\sigma(f) = \sum \nu\big(\rho(f_1)\phi\big)\,\overline{\nu\big(\rho(f_1)\phi\big)},$$

where we have set

$$\nu(\phi) := \int_{Z(F_\mathbb{A})T(F)\backslash T(F_\mathbb{A})} \phi(t)\Omega(t)^{-1}\mathrm{d}t; \tag{2}$$

thus $J_\sigma(f) \geq 0$. Moreover, if ν is not identically zero on the space of σ, or as we shall say, if σ is *distinguished* by (T,Ω), then every local component σ_{v_0} is *distinguished* by (T_{v_0},Ω_{v_0}), that is, admits a non-zero continuous linear form ν_{v_0} such that $\nu_{v_0}(\pi_{v_0}(t)u) = \Omega_{v_0}(t)\nu_{v_0}(u)$ for all $t \in T_{v_0}$ and all smooth vectors u. The dimension of the space of such linear forms is one. One can then define a local distribution

$$J_{\sigma_{v_0}}(f_{v_0}) = \sum \nu_{v_0}\big(\rho(f_{v_0})u\big)\,\overline{\nu_{v_0}(u)},$$

the sum over an orthonormal basis. The distribution $J_{\sigma_{v_0}}$ is defined within a positive factor. It is of positive type. Normalizing in an appropriate way we get

$$J_\sigma(f) = C(\sigma)\prod_{v_0} J_{v_0}(f_{v_0}), \tag{3}$$

where the constant $C(\sigma)$ is positive. Assuming that $L(\frac{1}{2},\Pi\otimes\Omega^{-1}) \neq 0$ we can find an ϵ such that there is an automorphic representation σ of G_ϵ corresponding to π and distinguished by (T,Ω) (see [J2], [W4]). Another goal of the paper is to obtain an *explicit* decomposition of the above form, with a specific normalization (Theorem 2). The crux of the matter is then to show that $C(\sigma)$ is essentially equal to $L(\frac{1}{2},\Pi\otimes\Omega^{-1})$ which gives the positivity result. Possibly, this can be used to provide lower bounds for $L(\frac{1}{2},\Pi\otimes\Omega^{-1})$.

We note that if ϵ exists then it is unique. Indeed, this follows at once from the following local fact: if v_0 is inert, ϵ not a norm at v_0, π is a square integrable representation of $\mathrm{Gl}(2, F_{v_0})$, and σ the representation of $G_\epsilon(F_{v_0})$ corresponding to π then π and σ cannot be both distinguished by (T_{v_0}, Ω_{v_0}) (see [W4]).

We stress that there is no direct way to compute the constant $C(\sigma)$ because there is no direct relation between the global linear form ν and the local linear forms ν_{v_0}. The situation at hand (a globally defined distribution of positive type decomposed as a product over all places of F of local distributions of positive type, times the appropriate special values of L-functions) is, conjecturally, quite general. In this situation, the positivity of the special value of the L-function follows. One can view this question as a generalization of the problem of computing the Tamagawa number. This is our motivation for investigating in detail the present situation.

We proceed as follows. We introduce the matrices

$$w = \begin{pmatrix} 0 & 1 \\ 1 & 0 \end{pmatrix}, \quad w_\epsilon = \begin{pmatrix} 0 & \epsilon \\ 1 & 0 \end{pmatrix}. \tag{4}$$

It will be more convenient to consider instead the distributions

$$\theta_\epsilon(f) := \int_{(Z(F_\mathbb{A})T(F)\backslash T(F_\mathbb{A}))^2} K_f(t_1, t_2)\, \Omega(t_1 t_2)^{-1} \mathrm{d}t_1 \mathrm{d}t_2,$$

and, for σ an automorphic representation of G_ϵ,

$$\theta_\sigma(f) := \int_{(Z(F_\mathbb{A})T(F)\backslash T(F_\mathbb{A})^2} K_{f,\sigma}(t_1, t_2)\, \Omega(t_1 t_2)^{-1} \mathrm{d}t_1 \mathrm{d}t_2.$$

Thus

$$J_\sigma(f) = \theta_\sigma\big(\rho(w_\epsilon)f\big)$$

and likewise for J_ϵ. We will decompose explicitly θ_σ into a product over all places v_0 of F of local distributions $\theta_{\sigma_{v_0}}$.

To that end, we compute the geometric expression for $\theta_\epsilon(f)$. A set of representatives for the double cosets of $T(F)$ in $G_\epsilon(F)$ is given by the matrices:

$$\mathbf{1}_2, \quad \begin{pmatrix} 0 & \epsilon \\ 1 & 0 \end{pmatrix}, \quad \begin{pmatrix} 1 & \epsilon\beta \\ \overline{\beta} & 1 \end{pmatrix}, \quad \beta \in E^\times / U_1(F).$$

We define orbital integrals. For $\xi \neq 1$ in $\mathrm{Norm}(E^\times)\epsilon$ we write $\xi = \beta\overline{\beta}\epsilon$ and set:

$$H(\xi; f) = \iint f\Big[t_1 \begin{pmatrix} \beta^{-1} & \epsilon \\ 1 & \overline{\beta}^{-1} \end{pmatrix} t_2\Big] \Omega(t_1 t_2^{-1}) \mathrm{d}t_1 \mathrm{d}t_2. \tag{5}$$

Note that the right hand side of the integral depends only on $\beta\overline{\beta}\epsilon$, which justifies the notations. In addition, we define

$$H(\infty; f) := \int f\Big[t_1 \begin{pmatrix} 0 & \epsilon \\ 1 & 0 \end{pmatrix}\Big] \Omega(t_1) \mathrm{d}t_1, \tag{6}$$

and, if Ω^2 is trivial,

$$H(0;f) := \int f(t_1)\Omega(t_1)\,\mathrm{d}t_1. \tag{7}$$

It will be useful to observe that the condition $\Omega^2 = 1$ is equivalent to $\Omega^\tau = \Omega$ or $\Omega_{|U_1}(F_{\mathbb{A}}) = 1$. The integrals are for $t_1 \in T(F_{\mathbb{A}})$ and $t_2 \in T(F_{\mathbb{A}})/Z(F_{\mathbb{A}})$. An idele class character χ of E is *normalized* if it is trivial on the subgroup E_∞^+ of ideles with all finite components trivial and all infinite components equal to some common positive number. For such a character we set $\delta(\chi) = 1$ if $\chi = 1$ and $\delta(\chi) = 0$ otherwise. All idele class characters will be assumed to be normalized. We have then

$$\begin{aligned}\theta_\epsilon(f) = & \sum_{\xi \in \mathrm{Norm}(E^\times)\epsilon - \{1\}} H(\xi; f) \\ & + \big(\delta(\Omega^2)H(0;f) + H(\infty;f)\big)\,\mathrm{vol}\big(T(F_{\mathbb{A}})/T(F)Z(F_{\mathbb{A}})\big).\end{aligned} \tag{8}$$

In what follows we denote by G the group $\mathrm{Gl}(2)$, by A the group of diagonal matrices, by Z the group of scalar matrices, by P the group of upper triangular matrices, by P_1 the subgroup of matrices with $(0,1)$ for second row and by N the subgroup of triangular matrices with unit diagonal. Depending on the context these groups are regarded as algebraic groups over F or E. We now view the group G as an algebraic group defined over E and we let H_0 be the unitary group for the matrix w. We denote by H the corresponding similitude group and by κ the similitude ratio. It is a result of [HLR] that Π is *distinguished* by $(H, \eta\omega)$, in the sense that there is a vector ϕ in the space of Π such that

$$\mathcal{P}(\phi) := \int_{H(F)Z(E_{\mathbb{A}})\backslash H(F_{\mathbb{A}})} \phi(h)\eta\,\omega\big(\kappa(h)\big)\,\mathrm{d}h \neq 0.$$

This condition characterizes representations which are base change of representations of $\mathrm{Gl}(2, F_{\mathbb{A}})$ with central character ω (*loc. cit.*).

We may regard Ω as a character of $A(E_{\mathbb{A}})$ trivial on $Z(E_{\mathbb{A}})$:

$$\Omega\big[(\mathrm{diag}(a_1, a_2)\big] = \Omega(a_1 a_2^{-1}).$$

Then $L(\frac{1}{2}, \Pi \otimes \Omega^{-1}) \neq 0$ if and only if there is ϕ in the space of Π such that

$$\lambda(\phi) := \int_{Z(E_{\mathbb{A}})A(E)\backslash A(E_{\mathbb{A}})} \phi(a)\Omega^{-1}(a)\,\mathrm{d}a \neq 0. \tag{9}$$

These facts suggest the following construction. Let f be a smooth function of compact support on $G(E_{\mathbb{A}})$. Define a kernel

$$K_f(x,y) = \int_{Z(E_{\mathbb{A}})/Z(E)} \sum_{\xi \in G(E)} f(x^{-1}z\xi y)\,\mathrm{d}^\times z,$$

and a distribution:

$$\Theta(f) := \int_{Z(E_{\mathbb{A}})A(E)\backslash A(E_{\mathbb{A}})} \int_{Z(E_{\mathbb{A}})H(F)\backslash H(F_{\mathbb{A}})} K_f(a,h)\Omega^{-1}(a)\eta\omega\big(\kappa(h)\big)\,\mathrm{d}h\mathrm{d}^{\times}a. \tag{10}$$

The outer integral is not convergent and must be regularized. If Π is a cuspidal automorphic representation of $G(E_{\mathbb{A}})$ with trivial central character then we define similarly the distribution Θ_Π. This distribution is non-zero if and only if Π is distinguished by $(H,\omega\eta)$ and $L(\frac{1}{2},\Pi\otimes\Omega^{-1}) \neq 0$. It can decomposed *explicitly* into a product of local distributions.

The distribution Θ can be computed in terms of orbital integrals in the following way. We let $\mathcal{S}$ be the space of invertible Hermitian matrices and $\mathcal{S}_s$ the space of split Hermitian matrices. The group $\mathrm{Gl}(2,E)$ operates on $\mathcal{S}$ by $s \mapsto gs\,{}^t\overline{g}$. We let Φ_f or simply Φ be the function on the space $\mathcal{S}(F_{\mathbb{A}})$ such that

$$\Phi_f(qw\,{}^t\overline{q}) = \int_{H_0(F_{\mathbb{A}})} f(gh_0)\mathrm{d}h_0$$

and Φ_f vanishes outside $\mathcal{S}_s(F_{\mathbb{A}})$. We have

$$\int K_f\Big[\begin{pmatrix} a^{-1} & 0 \\ 0 & 1\end{pmatrix}, h\Big]\eta\omega\big(\kappa(h)\big)\mathrm{d}h$$
$$= \int \sum_{\xi\in\mathcal{S}(F)} \Phi\Big[\begin{pmatrix} a & 0 \\ 0 & 1\end{pmatrix}\xi\begin{pmatrix} \overline{a} & 0 \\ 0 & 1\end{pmatrix}z\Big]\omega\eta(z)\mathrm{d}^{\times}z.$$

This leads us to introduce the action of $E^{\times}\times F^{\times}$ defined by:

$$s \xrightarrow{(\alpha,\zeta)} \begin{pmatrix} \alpha & 0 \\ 0 & 1\end{pmatrix} s \begin{pmatrix} \overline{\alpha} & 0 \\ 0 & 1\end{pmatrix}\begin{pmatrix} \zeta & 0 \\ 0 & \zeta\end{pmatrix}.$$

A system of representatives for the orbits of $E^{\times}\times F^{\times}$ on $\mathcal{S}(F)$ is constituted of the following matrices:

$$\begin{pmatrix} \xi^{-1} & 1 \\ 1 & 1\end{pmatrix},\ \xi\in F^{\times}-1,\quad \begin{pmatrix} \epsilon^{-1} & 0 \\ 0 & 1\end{pmatrix},\ \epsilon\in F^{\times}/\operatorname{Norm}(E^{\times}), \tag{11}$$

$$\begin{pmatrix} 0 & 1 \\ 1 & 0\end{pmatrix}, \tag{12}$$

$$\begin{pmatrix} 0 & 1 \\ 1 & 1\end{pmatrix},\ \begin{pmatrix} 1 & 1 \\ 1 & 0\end{pmatrix}. \tag{13}$$

For $\xi\in F^{\times}-\{1\}$, we set

$$U(\xi;\Phi) := \int_{E_{\mathbb{A}}^{\times}}\int_{F_{\mathbb{A}}^{\times}} \Phi\Big[\begin{pmatrix} z & 0 \\ 0 & z\end{pmatrix}\begin{pmatrix} y & 0 \\ 0 & 1\end{pmatrix}\begin{pmatrix} \xi^{-1} & 1 \\ 1 & 1\end{pmatrix}\begin{pmatrix} \overline{y} & 0 \\ 0 & 1\end{pmatrix}\Big]$$
$$\times\ \eta\,\omega(z)\mathrm{d}^{\times}z\Omega(y)\mathrm{d}^{\times}y.$$

If Ω is trivial on U_1 we define

$$U\Big[\begin{pmatrix}\epsilon^{-1} & 0\\ 0 & 1\end{pmatrix};\Phi\Big] := \int_{F_{\mathbb{A}}^\times}\int_{E_{\mathbb{A}}^\times/U_1(F_{\mathbb{A}})} \Phi\Big[\begin{pmatrix}\epsilon^{-1}y\bar{y}z & 0\\ 0 & z\end{pmatrix}\Big]\omega\eta(z)\Omega(y)\,\mathrm{d}^\times y\,\mathrm{d}^\times z.$$

The stabilizer of (12) in $E_{\mathbb{A}}^\times \times F_{\mathbb{A}}^\times$ is the set of pairs (y,z) with $yz=1$. However, because of our assumptions, the character $(y,z)\mapsto \Omega(y)\omega\eta(z)$ is non-trivial on that subgroup and so the element (12) does not contribute to the sum below. We also introduce two *unipotent* orbital integrals

$$U\Big[\begin{pmatrix}0 & 1\\ 1 & 1\end{pmatrix};\Phi\Big] := \iint \Phi\Big[\begin{pmatrix}0 & y\\ \bar{y} & z\end{pmatrix}\Big]\Omega(y)\,\mathrm{d}^\times y\,\eta(z)\,\mathrm{d}^\times z,$$

$$U\Big[\begin{pmatrix}1 & 1\\ 1 & 0\end{pmatrix};\Phi\Big] := \iint \Phi\Big[\begin{pmatrix}z & y\\ \bar{y} & 0\end{pmatrix}\Big]\Omega(y)\,\mathrm{d}^\times y\,\eta(z)\,\mathrm{d}^\times z.$$

These integrals are improper integrals. For the first one, for instance, we remark that the function ϕ defined by:

$$\phi(z) := \int \Phi\Big[\begin{pmatrix}0 & y\\ \bar{y} & z\end{pmatrix}\Big]\Omega(y)\,\mathrm{d}^\times y$$

is a Schwartz-Bruhat function on $F_{\mathbb{A}}$. The unipotent integral is then the analytic continuation to the point $s=0$ of the Tate integral:

$$\int_{F_{\mathbb{A}}^\times}\phi(z)\eta(z)|z|^s\,\mathrm{d}^\times z.$$

We have then

$$\text{(14)}\quad \Theta(f) = \sum_{\xi\in F^\times-\{1\}} U(\xi;\Phi) + \delta(\Omega^2)\operatorname{vol}\big(U_1(F_{\mathbb{A}})/U_1(F)\big)$$
$$\times \sum_{\epsilon\in F^\times/\operatorname{Norm}(E^\times)} U\Big[\begin{pmatrix}\epsilon^{-1} & 0\\ 0 & 1\end{pmatrix};\Phi\Big] + U\Big[\begin{pmatrix}0 & 1\\ 1 & 1\end{pmatrix};\Phi\Big] + U\Big[\begin{pmatrix}1 & 1\\ 1 & 0\end{pmatrix};\Phi\Big].$$

We say that f and a family of functions (f_ϵ), $\epsilon\in F^\times/\operatorname{Norm}(E^\times)$, have *matching orbital integrals* if

$$U(\xi;\Phi) = H(\xi;f_\epsilon),$$

for $\xi\in\epsilon\operatorname{Norm}(E^\times)-\{1\}$. Implicit in this definition is the fact that $f_\epsilon=0$ for almost all ϵ. We show that for any f there is a family (f_ϵ) with matching orbital integrals. Note that the converse is not true: to have a converse one would need to consider all unitary groups. When f and the family (f_ϵ) have matching orbital integrals, we have then

$$\Theta(f) = \sum_\epsilon \theta_\epsilon(f_\epsilon),$$

that is, the terms corresponding to all orbitals integrals match (see [J2]).

Now we apply standard arguments: by identifying the contribution of the continuous (and residual) spectrum on both sides we arrive at the identity

$$\Theta_\Pi(f) = \sum \theta_{\sigma_\epsilon}(f_\epsilon).$$

In a precise way, the representation Π is the base change of π and $\pi \otimes \eta$. The sum on the right is over all representations σ_ϵ which correspond to π or $\pi \otimes \eta$. As noted above, the sum on the right reduces to two terms, that is, there is a unique ϵ and a representation σ_ϵ of $G_\epsilon(F_{\mathbb{A}})$ which corresponds to π such that

$$\Theta_\Pi(f) = \theta_{\sigma_\epsilon}(f_\epsilon) + \theta_{\sigma_\epsilon \otimes \eta}(f_\epsilon).$$

The distribution Θ_Π can be decomposed explicitly into a product; the above identity allows us to decompose θ_{σ_ϵ} and, finally, prove the theorem.

The method is quite general and should apply to many situations. For instance, the work of [G2] suggests a possible direct generalization of the present set-up to $\mathrm{Gl}(2n)$, the group T being replaced by the group $\mathrm{Gl}(n, E)$ embedded in $\mathrm{Gl}(2n, F)$ and some inner forms of it. [G2] is concerned with the generalization of the simpler trace formula in [J1].

We remark that it would be interesting to compare our explicit result with the results of [W4].

The paper is arranged as follows. In sections 2 and 3 we review the results of [J2] on the matching of orbital integrals, reformulating the results in terms of symmetric spaces; we carefully normalize the various Haar measures. In section 4 we decompose the distribution Θ_Π explicitly into a product of local distributions. This is mainly a review of the material in [HLR]. The heart of the paper is section 5 where we compare the local distributions at hand.

The proof of the main theorem is given in section 6, with some additional comments in section 7. The paper concludes with sections 8 and 9, an appendix where we discuss the absolute convergence of the term coming from the continuous spectrum in the trace formulas at hand. Unfortunately, reference [J2] (also [J1]) is somewhat deficient on this point: the main point is that [J2] omits infinite sums on idele class characters unramified at all places. To make the argument rigorous we introduce, in the case at hand, a new form of truncation which may be more appropriate than the standard truncation operator for the investigation of period integrals (*cf.* [JLRo]) and a new device (special to $\mathrm{Gl}(2)$ or closely related groups) to estimate some period integrals. Note that [G1] which is based on [J1] is also deficient and so is [BM]. Thus we take care at once of a gap in several papers.

Finally, we would like to thank the referee for a careful and thorough reading of the manuscript.

2. Choice of measures and local matching

We review the results of [J2] on local matching. It will be essential to keep track of the choice of the various Haar measures.

2.1. Inert case. — We consider a local quadratic extension E/F. We choose an additive character ψ_F of F and set

$$\psi_E(z) = \psi_F(z + \bar{z}).$$

On the additive groups F and E we consider the self-dual Haar measures. In particular, if we write $E = F[\sqrt{\delta}\,]$ and $z = a + b\sqrt{\delta}$ then $\mathrm{d}z = \mathrm{d}a\mathrm{d}b\,|2|_F\,|\delta|_F^{1/2}$. The multiplicative Tamagawa measures on $F^\times$ and $E^\times$ are respectively:

$$\mathrm{d}^\times x = L(1, 1_F)\frac{\mathrm{d}x}{|x|_F}, \quad \mathrm{d}^\times z = L(1, 1_E)\frac{\mathrm{d}z}{|z|_E}.$$

The following integration formula will be used below:

LEMMA 2. — *One has*

$$\int_{E^\times} \phi(z)\frac{\mathrm{d}z}{|z\bar{z}|_F} = \int_{\mathrm{Norm}(E^\times)} \widetilde{\phi}(x)\frac{\mathrm{d}x}{|x|_F},$$

where $\widetilde{\phi}(x)$ is the function on $\mathrm{Norm}(E^\times)$ *defined by*

$$\widetilde{\phi}(x) := \int_{E^\times/F^\times} \phi\Big(z\frac{u}{\bar{u}}\Big)\mathrm{d}^0u, \quad x = z\,\bar{z},$$

and the measure d^0u is the quotient of $\mathrm{d}z/|z|_E$ by $\mathrm{d}x/|x|_F$.

Since $T(F)$ is isomorphic to $E^\times$ via the map $\mathrm{diag}(a, \bar{a}) \mapsto a$, we obtain the Tamagawa measure on $T(F)$. Likewise the center Z of $\mathrm{Gl}(2)$ is isomorphic to $\mathrm{Gl}(1)$ and so we have the Tamagawa measure on $Z(F)$ and $Z(E)$.

On the group $\mathrm{Gl}(2, F)$ we have the Tamagawa measure

$$\mathrm{d}^\times g = L(1, 1_F)\frac{\mathrm{d}p\mathrm{d}q\mathrm{d}r\mathrm{d}s}{|\det g|_F^2}, \quad g = \begin{pmatrix} p & q \\ r & s \end{pmatrix}.$$

Using the Iwasawa decomposition we can write

$$\mathrm{d}^\times g = \mathrm{d}^\times a\mathrm{d}^\times b\mathrm{d}x\mathrm{d}k,$$

if

$$g = \begin{pmatrix} a & 0 \\ 0 & b \end{pmatrix}\begin{pmatrix} 1 & x \\ 0 & 1 \end{pmatrix}k,$$

where $\mathrm{d}k$ is a Haar measure on the standard maximal compact subgroup K_F of $\mathrm{Gl}(2, F)$. Let Φ be a Schwartz-Bruhat function on F^2. The following lemma is easily verified:

LEMMA 3. — *One has*

$$\iint \Phi(x,y)\mathrm{d}x\mathrm{d}y = \iint \Phi\big[(0,t)k\big]\cdot|t|^2\mathrm{d}^\times t\mathrm{d}k.$$

In particular, in the unramified situation, we can take for Φ the characteristic function of the set $\mathcal{O}_F^2$ and obtain $\mathrm{vol}(K_F) = L(2,1_F)^{-1}$. On the group $\mathrm{Gl}(2,E)$ we also have the Tamagawa measure

$$\mathrm{d}^\times g = L(1,1_E)\frac{\mathrm{d}g}{|\det g|_E^2},\quad \mathrm{d}g = \mathrm{d}p\mathrm{d}q\mathrm{d}r\mathrm{d}s$$

if

$$g = \begin{pmatrix} p & q \\ r & s \end{pmatrix}.$$

On the group G_ϵ we take for the Tamagawa measure the measure:

$$\mathrm{d}^\times g = L(1,1_F)|\epsilon|_F\frac{\mathrm{d}a\mathrm{d}b}{|a\bar{a}-\epsilon b\bar{b}|_F^2},$$

if

$$g = \begin{pmatrix} a & \epsilon b \\ \bar{b} & \bar{a} \end{pmatrix}.$$

If ϵ is a norm the group $G_\epsilon(F)$ is isomorphic to $\mathrm{Gl}(2,F)$ and the isomorphism takes the above measure to the Tamagawa measure on $\mathrm{Gl}(2,F)$. From Lemma 2 we get the following integration formula:

$$(15)\quad \int_{G_\epsilon} F(g)\mathrm{d}^\times g = \frac{1}{L(1,\eta)^2}\int_{x=\epsilon a\bar{a}}\left\{\iint F\left[t_1\begin{pmatrix} a^{-1} & \epsilon \\ 1 & \bar{a}^{-1}\end{pmatrix}t_2\right]\mathrm{d}t_1\mathrm{d}t_2\right\}\frac{\mathrm{d}x}{|1-x|_F^2}$$

with $t_1 \in T(F)$, $t_2 \in T(F)/Z(F)$.

In view of the isomorphism

$$\mathrm{Gl}(2,F)Z(E)/Z(E) \simeq \mathrm{Gl}(2,F)/Z(F)$$

we give to $\mathrm{Gl}(2,F)Z(E)$ the measure defined by

$$\mathrm{d}g = L(1,1_E)\frac{\mathrm{d}a}{|a|_F}\frac{\mathrm{d}b}{|b|_E}\mathrm{d}x\mathrm{d}y$$

if

$$g = \begin{pmatrix} ab & 0 \\ 0 & b \end{pmatrix}\begin{pmatrix} 1 & 0 \\ y & 1 \end{pmatrix}\begin{pmatrix} 1 & x \\ 0 & 1 \end{pmatrix}.$$

On the group H isomorphic to $\mathrm{Gl}(2,F)Z(E)$ the corresponding measure is given by

$$L(1,1_E)\frac{\mathrm{d}a}{|a|_F}\frac{\mathrm{d}b}{|b|_E}\mathrm{d}x_1\mathrm{d}y_1$$

for

$$g = \begin{pmatrix} ab & 0 \\ 0 & b \end{pmatrix} \begin{pmatrix} 1 & 0 \\ y & 1 \end{pmatrix} \begin{pmatrix} 1 & x \\ 0 & 1 \end{pmatrix},$$

with $a \in F^\times$, $b \in E^\times$, $x = x_1\sqrt{\delta}$, $y = y_1\sqrt{\delta}^{-1}$, $x_1, y_1 \in F$. Of course, this measure does not depend on the choice of δ. Using the exact sequence

$$1 \to H_0 \longrightarrow H \xrightarrow{\kappa} F^\times \to 1,$$

we get the Tamagawa measure on H_0:

$$\mathrm{d}h_0 = L(1,\eta)\frac{\mathrm{d}b}{|b|_E}\mathrm{d}x_1\mathrm{d}y_1$$

if

$$h_0 = \begin{pmatrix} b & 0 \\ 0 & \bar{b}^{-1} \end{pmatrix} \begin{pmatrix} 1 & 0 \\ y & 1 \end{pmatrix} \begin{pmatrix} 1 & x \\ 0 & 1 \end{pmatrix}.$$

We need the following integration formula:

LEMMA 4. — *The quotient of the Tamagawa measure on* Gl$(2,E)$ *by the Tamagawa measure on* H_0 *is the following measure on the symmetric space* $\mathcal{S}_s$*:*

$$\mathrm{d}s = L(1,1_F)\frac{\mathrm{d}x\mathrm{d}y\mathrm{d}z}{|xy - z\bar{z}|_F^2}, \quad s = \begin{pmatrix} x & z \\ \bar{z} & y \end{pmatrix}.$$

For f a smooth function of compact support on $G_\epsilon(F)$ we define

$$(16)\qquad H(x;f) := \int_{T(F)/Z(F)} \int_{T(F)} f\Big[t_1\begin{pmatrix} b^{-1} & \epsilon \\ 1 & \bar{b}^{-1} \end{pmatrix}t_2\Big]\Omega(t_1t_2^{-1})\mathrm{d}t_1\mathrm{d}t_2$$

if $x = \epsilon b\bar{b}$, $x \neq 1$. The integrand is a smooth function of b depending only on $b\bar{b}\epsilon$ which justifies the notation. Next we define the orbital integrals for the space $\mathcal{S}(F)$ of invertible Hermitian matrices. As in the global case, the group $E^\times \times F^\times$ operates. Relative to this action we have the local orbital integrals of a function $\Phi \in \mathcal{C}_0^\infty(\mathcal{S}(F))$:

$$U(x;\Phi) := \int_{E^\times}\int_{F^\times} \Phi\Big[\begin{pmatrix} z & 0 \\ 0 & z \end{pmatrix}\begin{pmatrix} y & 0 \\ 0 & 1 \end{pmatrix}\begin{pmatrix} x^{-1} & 1 \\ 1 & 1 \end{pmatrix}\begin{pmatrix} \bar{y} & 0 \\ 0 & 1 \end{pmatrix}\Big] \times \eta\,\omega(z)\mathrm{d}^\times z\Omega(y)\mathrm{d}^\times y.$$

Now let $\{\epsilon_1, \epsilon_2\}$ be a system of representatives for the classes of Norm$(E^\times)$ in $F^\times$, where ϵ_1 is a norm. Let $\Phi \in \mathcal{C}_0^\infty(\mathcal{S}(F))$ and f_{ϵ_i}, $i = 1,2$, be functions in $\mathcal{C}_0^\infty(G_{\epsilon_i}(F))$. We say that Φ and the pair $(f_{\epsilon_1}, f_{\epsilon_2})$ have *matching orbital integrals* if

$$U(x;\Phi) = H(x;f_{\epsilon_i}) \quad \text{for} \quad x \in \epsilon_i\,\mathrm{Norm}(E^\times).$$

PROPOSITION 1. — *Given* Φ *there is a pair* $(f_{\epsilon_1}, f_{\epsilon_2})$ *with matching orbital integrals.*

To prove the result one needs to consider the behavior at infinity of the orbital integrals defined above as well as the other orbital integrals. This is done in [J2]. We will use this result in the following way: f will be a smooth function of compact support on $\mathrm{Gl}(2,E)$ and we will set

$$\Phi_f(s) = \int_{H_0(F)} f(gh_0)\mathrm{d}h_0, \quad s = g\,w\,{}^t\overline{g},$$
$$\Phi_f(s) = 0, \quad s \notin \mathcal{S}_s(F).$$

If the pair $(f_{\epsilon_1}, f_{\epsilon_2})$ has matching orbital integrals with Φ_f we shall say that it has *matching orbital integrals* with f.

Consider now the unramified situation: the residual characteristic is odd, the extension E/F is unramified, the conductor of ψ_F is $\mathcal{O}_F$, ϵ_1 is a unit, Ω is unramified. Then $K_E := \mathrm{Gl}(2,\mathcal{O}_E)$ has volume $L(2,1_E)^{-1}$ and $K_F := \mathrm{Gl}(2,\mathcal{O}_F)$ has volume $L(2,1_F)^{-1}$. Likewise, the group

$$K_0 := \mathrm{Gl}(2,\mathcal{O}_E) \cap H_0(F)$$

is a maximal compact subgroup of $H_0(F)$ with volume $L(2,1_F)^{-1}$ and

$$K_\epsilon := \mathrm{Gl}(2,\mathcal{O}_E) \cap G_\epsilon(F)$$

is a maximal compact subgroup of G_ϵ with volume $L(2,1_F)^{-1}$. The *unramified Tamagawa measures* are obtained by multiplying the Tamagawa measures by the inverse of those volumes. Using the Cartan decomposition and the methods of [J1, pp. 199–204], one can prove the following proposition:

PROPOSITION 2. — *Suppose the situation is unramified. Let q be the cardinality of the residual field of F. For $n \geq 0$ let Φ_{2n} be the characteristic function of the set of matrices $s \in \mathcal{S}(F)$, with integral entries such that $|\det s|_F = q^{-2n}$. Let f_{2n} be the characteristic function of the set of matrices $g \in G_{\epsilon_1}(F)$ with integral entries such that $|\det g|_F = q^{-2n}$. Then Φ_{2n} and the pair $(f_{2n}, 0)$ have matching orbital integrals.*

Finally, we recall the fundamental lemma for the unramified situation. Since ϵ_1 is a norm, there is an isomorphism of $G_{\epsilon_1}(F)$ onto $G(F)$ taking K_{ϵ_1} to K_F and the Hecke algebra $\mathcal{H}(G_{\epsilon_1}(F), K_{\epsilon_1})$ to the Hecke algebra $\mathcal{H}(\mathrm{Gl}(2,F), K_F)$. We have thus a base change homomorphism b from $\mathcal{H}(\mathrm{Gl}(2,E), K_E)$ to $\mathcal{H}(G_{\epsilon_1}(F), K_{\epsilon_1})$.

PROPOSITION 3. — *For $f \in \mathcal{H}(K)$ the function Φ_f and the pair $(b(f), 0)$ have matching orbital integrals.*

Proof. — This is really a reformulation of the corresponding result of [J2]. It can derived more directly from the previous proposition by using the methods of [JLR, pp. 318–322]. □

2.2. Split case. — *Mutatis mutandis*, the above discussion applies to the split case where the quadratic extension is replaced by the algebra $E = F \oplus F$ with F embedded diagonally in E and Galois action $\tau(x,y) = (y,x)$. If we write $z = (x,y)$ then $\psi_E(z) = \psi(x+y)$ and so the self dual Haar measure on E is $\mathrm{d}z = \mathrm{d}x\mathrm{d}y$. To obtain the Tamagawa measure on $\mathrm{Gl}(2,E)$ we must divide the self-dual additive measure $\mathrm{d}g = \mathrm{d}g_1\mathrm{d}g_2$ by $|\det g_1g_2|_F$ and then multiply by $L(1,1_F)^2$. So the isomorphism $\mathrm{Gl}(2,E) \simeq \mathrm{Gl}(2,F) \times \mathrm{Gl}(2,F)$ preserves the Tamagawa measures.

Likewise, the group H_0 becomes the group of pairs (h_1,h_2) with $h_2 = w\,{}^th_1^{-1}w$. Choosing the first factor in the decomposition $E = F\oplus F$, we have an isomorphism $H_0(F) \simeq \mathrm{Gl}(2,F)$ which is also compatible with the Tamagawa measures.

The group $G_\epsilon(F)$ is the group of invertible matrices of the form

$$\begin{pmatrix} a & \epsilon b \\ \bar{b} & \bar{a} \end{pmatrix},$$

with a,b in E. We consider the measure

$$|\epsilon|_F\, L(1,1_F)\frac{\mathrm{d}a\mathrm{d}b}{|a\bar{a} - \epsilon b\bar{b}|_F^2}.$$

Thus $G_\epsilon(F)$ is the group of pairs

$$(g_1,g_2)$$

with $g_2 = w_\epsilon g w_\epsilon^{-1}$. Again $(g_1,g_2) \mapsto g_1$ defines an isomorphism $G_\epsilon(F) \simeq \mathrm{Gl}(2,F)$ which takes the above measure to the Tamagawa measure.

The manifold $S(F)$ is the submanifold of pairs

$$S(F) = \{(s_1,s_2) \mid s_2 = {}^ts_1\}$$

and is again isomorphic to $G(F)$ via $(s_1,s_2) \mapsto s_1$.

The torus $T(F)$ is the subgroup of pairs:

$$t = \left(\begin{pmatrix} a_1 & 0 \\ 0 & a_2 \end{pmatrix}, \begin{pmatrix} a_2 & 0 \\ o & a_1 \end{pmatrix}\right).$$

A character Ω of $T(F)$ trivial on $Z(F)$ has the form

$$\Omega(t) = \Omega_1(a_1)\Omega_2(a_2)$$

with $\Omega_1\Omega_2 = 1$.

In particular, if $f_1 \otimes f_2$ is a product function on $G(E)$ then

$$\Phi_{f_1\otimes f_2}(s_1,s_2) = \int f_1(g_1h_1)f_2(g_2h_2)\mathrm{d}h_0$$

becomes

$$\Phi_{f_1\otimes f_2}(g) = \int f_1(gwh)f_2(w\,{}^th^{-1}w)\mathrm{d}h = f_1 * \tilde{f}_2(gw)$$

where we have set

$$\tilde{f}_2(g) = f_2(w\,{}^t g w).$$

With this identification, the orbital integrals for $\Phi \in \mathcal{C}_0^\infty(\mathcal{S}(F))$ become:

$$U(x;\Phi) = \iiint \Phi\left[\begin{pmatrix} z & 0 \\ 0 & z \end{pmatrix}\begin{pmatrix} a & 0 \\ 0 & 1 \end{pmatrix}\begin{pmatrix} x^{-1} & 1 \\ 1 & 1 \end{pmatrix}\begin{pmatrix} b & 0 \\ 0 & 1 \end{pmatrix}\right] \times \Omega_1(a)\Omega_2(b)\mathrm{d}^\times a\mathrm{d}^\times b\mathrm{d}^\times z.$$

Similarly, for $f_\epsilon \in \mathcal{C}_0^\infty(G_\epsilon(F))$, the orbital integrals take the form:

$$H(x;f_\epsilon) = \iiint f_\epsilon\left[\begin{pmatrix} a & 0 \\ 0 & b \end{pmatrix}\begin{pmatrix} u_1^{-1} & \epsilon \\ 1 & u_2^{-1} \end{pmatrix}\begin{pmatrix} c & 0 \\ 0 & 1 \end{pmatrix}\right] \times \Omega_1(a)\Omega_2(b)\mathrm{d}^\times a\mathrm{d}^\times b\Omega_1(c)^{-1}\mathrm{d}^\times c,$$

with $x = u_1u_2\epsilon$. The condition of matching $U(x;\Phi) = H(x;f_\epsilon)$ is trivially verified with

$$f_\epsilon(g) = \Phi\left[g\begin{pmatrix} \epsilon & 0 \\ 0 & 1 \end{pmatrix}\right].$$

If $\Phi = \Phi_{f_1\otimes f_2}$ this becomes the condition of matching for the split case:

$$f_\epsilon(g) = f_1 * \tilde{f}_2(gw_\epsilon). \tag{17}$$

If F is non-Archimedean, Ω_1 unramified, the functions f_i bi-invariant under $K_F = \mathrm{Gl}(2,\mathcal{O}_F)$, and the Haar measure of K_F is 1, then $\tilde{f}_2 = f_2$, $w_\epsilon \in K_F$ and

$$f = \Phi = f_1 * f_2.$$

3. Global matching

We now consider an extension E/F of number fields.

3.1. Global Haar measures. — On the group $F_{\mathbb{A}}^\times$ we consider the Tamagawa measure $d^\times x$ which is the (convergent) product of the local Tamagawa measures. We let F^1 be the group of ideles of norm 1 and use the exact sequence

$$1 \to F^1 \longrightarrow F_{\mathbb{A}}^\times \to \mathbb{R}_+^\times$$

to define a measure on F^1 for which

$$\mathrm{vol}(F^1/F^\times) = \mathrm{res}_{s=1} L(s, 1_F).$$

Likewise for E.

We also consider a finite set places S_0 of F and the let S be the corresponding set of places of E. We assume that S_0 contains all the places at infinity, the places of even residual characteristic, the finite places which ramify in E, and all the places where the character ψ is ramified (that is, the conductor is not the ring of integers). We also assume that Ω is unramified outside S. We enlarge S_0 as the need arises.

On the group $\mathrm{Gl}(2, F_{\mathbb{A}})$ we consider the Tamagawa measure $\mathrm{d}^{\times}g$ which is the convergent product of the local Tamagawa measures and multiply it by the factor $L^{S_0}(2, 1_F) := \prod_{v_0 \notin S_0} L(2, 1_{F_{v_0}})$. In other words, the measure we are considering is the product of the Tamagawa measures $\mathrm{d}^{\times}g_{v_0}$ for $v_0 \in S_0$ times the measures $\mathrm{d}g_{v_0}$, $v_0 \notin S_0$, for which the measure of $\mathrm{Gl}(2, \mathcal{O}_{v_0})$ is one. Using the Iwasawa decomposition

$$g = \mathrm{diag}(a, b)\begin{pmatrix} 1 & x \\ 0 & 1 \end{pmatrix} k$$

we obtain a measure on the standard maximal compact K:

$$\mathrm{d}g = \mathrm{d}^{\times}a\,\mathrm{d}^{\times}b\,\mathrm{d}x\,\mathrm{d}k.$$

We have then the following identity for a Schwartz-Bruhat function Φ in two variables:

$$\iint \Phi(x, y)\mathrm{d}x\mathrm{d}y = \iint \Phi\big[(0, t)k\big] \cdot |t|^2 \mathrm{d}^{\times}t\,\mathrm{d}k \frac{1}{L^{S_0}(2, 1_F)}.$$

If $\Phi = \prod \Phi_{v_0}$ with Φ_{v_0} the characteristic function of $\mathcal{O}_{v_0}^2$ outside S_0 we get then:

$$\iint \Phi(x, y)\mathrm{d}x\mathrm{d}y = \int_{K_{S_0}} \int_{F_S^{\times}} \Phi_{S_0}\big[(0, t)k\big] \cdot |t|^2 \mathrm{d}^{\times}t\,\mathrm{d}k.$$

We use the similar measure on $\mathrm{Gl}(2, E_{\mathbb{A}})$. Likewise for $H_0(F_{\mathbb{A}})$ and $H(F_{\mathbb{A}})$ we multiply the Tamagawa measure by $L^{S_0}(2, 1_F)$.

3.2. Matching. — As explained in the introduction, we consider a smooth function of compact support f on $G(E_{\mathbb{A}})$ and the corresponding function $\Phi = \Phi_f$ on $\mathcal{S}(F_{\mathbb{A}})$. The function f is a product of local functions f_v. For $v \notin S$ the function f_v is bi-invariant under $K_v := \mathrm{Gl}(2, \mathcal{O}_v)$. We choose a set of representative $\{\epsilon\}$ for $F^{\times}/\operatorname{Norm}(E^{\times})$. For each ϵ we choose a smooth function of compact support f_ϵ on $G_\epsilon(F_{\mathbb{A}})$. It is a product of local functions that we choose as follows. For a place v_0 inert in F and the corresponding place v of E we demand that f_{ϵ,v_0} and Φ_{v_0} have matching orbital integrals, that is, $H(x; f_{\epsilon,v_0}) = U(x; \Phi_{v_0})$ for $x \in \epsilon \operatorname{Norm}(E_v^{\times})$. If $v_0 \notin S_0$ and ϵ is not a norm at the place v_0, we have seen that $H(x; \Phi_{v_0}) = 0$ for $x \in \epsilon \operatorname{Norm}(E_v^{\times})$ and so we take $f_{\epsilon,v_0} = 0$. Thus we have $f_\epsilon = 0$ unless ϵ is a norm at all places not in S_0. Thus there is only a finite set Ξ of ϵ such that $f_\epsilon \neq 0$. There is a finite set $S_0' \supseteq S_0$ of places of F such that the $\epsilon \in \Xi$ are unit at all places not in S_0'. We let S' be the set of places of E above a place in S_0'.

For v_0 inert not in S_0' we may and do assume that $f_{\epsilon,v_0} = b(f_v)$ where v is the corresponding place of E and b is the base change homomorphism. For v_0 split into v_1 and v_2 in E we assume that

$$f_{\epsilon,v_0}(g) = \Phi_{v_0}\Big[g\begin{pmatrix} \epsilon & 0 \\ 0 & 1 \end{pmatrix}\Big].$$

In particular, for $\epsilon \in \Xi$ and $v_0 \notin S_0'$ this means that (with the usual identifications) $f_{\epsilon,v_0} = f_1 * f_2$. This being so, we change our notations and now take the set S_0 to be the set S_0'.

We compare the geometric terms in the two relative trace formulas. As usual we set

$$\mathrm{e}^{H(g)} = |a_1 a_2^{-1}|$$

if $g = nak$, $a = \operatorname{diag}(a_1, a_2)$.

As in [J2], we consider a *compact* truncation along the diagonal. Namely, if ϕ is a function on $A(E_{\mathbb{A}})$ invariant under $A(E)$ and $Z(E_\infty^+)$, we set:

$$\Lambda_c^T \phi(a) = \phi(a) \qquad \text{if } -T < H(a) < T, \tag{18}$$

$$\Lambda_c^T \phi(a) = 0 \qquad \text{otherwise.} \tag{19}$$

We define then

$$\Theta_T(f) := \iint_{Z(E_{\mathbb{A}})H(F)\backslash H(F_{\mathbb{A}})} \Lambda_c^T K_f(a,h)\Omega^{-1}(a)\,\mathrm{d}a\,\omega\eta\bigl(\kappa(h)\bigr)\,\mathrm{d}h.$$

The outer integral is over $A(E_{\mathbb{A}})/A(E)Z(E_{\mathbb{A}})$. This integral is absolutely convergent and equal to:

$$\int_{\mathrm{e}^{-T}\le|a|\le \mathrm{e}^{T}} \int_{F_{\mathbb{A}}^\times/F^\times} \sum_{\xi\in\mathcal{S}(F)} \Phi\Bigl[\begin{pmatrix} a & 0\\ 0 & 1\end{pmatrix}\xi\begin{pmatrix} \bar a & 0\\ 0 & 1\end{pmatrix}z\Bigr]\Omega^{-1}(a)\omega\eta(z)\,\mathrm{d}^\times z.$$

In [J2] it is shown that as $T \to +\infty$ this tends to the right hand side of (14). Thus we define

$$\Theta(f) =: \lim_{T\longrightarrow+\infty} \Theta_T(f),$$

and we arrive at (14). Comparing with (8) we get:

$$\Theta(f) = \sum_{\epsilon\in\Xi} \theta(f_\epsilon).$$

Indeed given $\xi \in F^\times - \{1\}$ we have $\xi \in \epsilon\,\mathrm{Norm}(E^\times)$ for a unique ϵ and

$$U(\xi;\Phi) = H(\xi; f_\epsilon).$$

The matching of the other terms is explained in [J2].

4. Factorization of Θ_Π over E

Most of the material of this section is a review of [HLR] (see also [A], [F2], [FZ]).

We now consider the Epstein Eisenstein series associated to a Schwartz-Bruhat function Φ on $F_{\mathbb{A}}^2$; we assume that Φ is a product, the local component

Φ_{v_0} being the characteristic function of $\mathcal{O}^2_{v_0}$ for all $v_0 \notin S_0$:

$$E(g,\Phi,s) = \int_{F_{\mathbb{A}}^\times/F^\times} \sum_{\xi\in F^2-\{(0,0)\}} \Phi[\xi t g]\cdot |\det t|^{2s}\mathrm{d}^\times t\,|\det g|^s.$$

It is meromorphic, with simple pole at $s=1$ and residue:

$$\frac{1}{2}\operatorname{vol}\Big(\frac{F^1}{F^\times}\Big)\iint \Phi(x,y)\mathrm{d}x\mathrm{d}y = \frac{1}{2}\Big(\frac{F^1}{F^\times}\Big)\int_{K_{S_0}}\int_{F^\times_{S_0}} \Phi_{S_0}\big[(0,t)k\big]\cdot |t|^2\mathrm{d}^\times t\mathrm{d}k.$$

Let Π be the base change of a representation π with central character η and ϕ a cusp form in the space of Π which is invariant under translations by the compact subgroup $K^S := \prod_{v\notin S} K_{E_v}$. Let us compute the integral:

$$\Psi(s,\phi,\Phi) := \int_{G(F)Z(F_{\mathbb{A}})\backslash G(F_{\mathbb{A}})} E(g,\Phi,s)\phi(g)\mathrm{d}g.$$

It is a meromorphic function of s, with simple pole at $s=1$ and residue:

$$\frac{1}{2}\operatorname{vol}\Big(\frac{F^1}{F^\times}\Big)\int_{K_{S_0}}\int_{F^\times_{S_0}} \Phi_{S_0}\big[(0,t)k\big]\cdot |t|^2\mathrm{d}^\times t\mathrm{d}k \int \phi(g)\mathrm{d}g. \tag{20}$$

On the other hand, for $\operatorname{Re} s \gg 0$,

$$\Psi(s,\phi,\Phi) = \Psi(s,W,\Phi)$$

where we have set:

$$W(g) := \int_{E_{\mathbb{A}}} \phi\Big[\begin{pmatrix}1 & x\\ 0 & 1\end{pmatrix} g\Big]\psi_E(-x)\mathrm{d}x$$

and, writing $E = F[\sqrt{\delta}\,]$,

$$\Psi(s,W,\Phi) := \int_{N(F_{\mathbb{A}})\backslash \mathrm{Gl}(2,F_{\mathbb{A}})} W\Big[\begin{pmatrix}\sqrt{\delta}^{\,-1} & 0\\ 0 & 1\end{pmatrix} g\Big]\Phi\big[(0,1)g\big]\cdot |\det g|^s\mathrm{d}^\times g.$$

We assume that W is a product

$$W(g) = \prod_v W_v(g_v)$$

over all places of E, with $W_{v|K_{E_v}} = 1$ outside S. The local component is an element of the local Whittaker model $\mathcal{W}(\Pi_v,\psi_v)$. If v_0 is a place of F we write $E_{v_0} = E\otimes F_{v_0}$. If v_0 is inert in E we also write W_{v_0} for W_v. If v_0 splits into v_1, v_2 we write W_{v_0} for $W_{v_1}\otimes W_{v_2}$, a function on $\mathrm{Gl}(2,E_{v_0})$. We can then set in all cases:

$$\Psi(s,W_{v_0},\Phi_{v_0}) = \int_{N(F_{v_0})\backslash \mathrm{Gl}(2,F_{v_0})} W_{v_0}\Big[\begin{pmatrix}\sqrt{\delta}^{\,-1} & 0\\ 0 & 1\end{pmatrix} g\Big]\Phi_{v_0}\big[(0,1)g\big]\cdot |\det g|^s\mathrm{d}^\times g.$$

Explicitly, if v_0 splits, that is, δ is the square of two elements ϵ_1 and ϵ_2 of F_{v_0}, the above integral is in fact equal to:

$$\int W_{v_1}\left[\begin{pmatrix}\epsilon_1^{-1} & 0\\ 0 & 1\end{pmatrix}g\right]W_{v_2}\left[\begin{pmatrix}\epsilon_2^{-1} & 0\\ 0 & 1\end{pmatrix}g\right]\Phi_{v_0}\left[(0,1)g\right]\cdot|\det g|^s \mathrm{d}^\times g.$$

We recall that S_0 is so large that all the data is unramified outside S_0 (or S).

We introduce the partial Asai L-function attached to Π noted $L^{S_0}(s,\Pi,\mathrm{Asai})$ (see [HLR]). For every place $v_0 \notin S_0$ the integral $\Psi(s, W_{v_0}, \Phi_{v_0})$ is equal to the corresponding local L-factor of the Asai L-function. Thus

$$\Psi(s,\phi,\Phi) = L^{S_0}(s,\Pi,\mathrm{Asai}) \prod_{v_0\in S_0} \Psi(s, W_{v_0}, \Phi_{v_0}).$$

Taking the residue at $s=1$ we obtain that (20) is equal to:

$$\mathrm{res}_{s=1} L^{S_0}(s,\Pi,\mathrm{Asai}) \prod_{v_0\in S_0} \Psi(1, W_{v_0}, \Phi_{v_0}).$$

This equality implies that, as a linear form in W_{v_0}, $\Psi(1,W_{v_0},\Phi_{v_0})$ is invariant under right shifts by $\mathrm{Gl}(2,F_{v_0})$. As a linear form in Φ_{v_0}, it is proportional to

$$\int_{K_{v_0}}\int_{F_{v_0}^\times} \Phi_{v_0}[(0,t)k]\cdot|t|^2\mathrm{d}^\times t\mathrm{d}k.$$

Now let us set:

$$\mathcal{P}_{v_0}(W_{v_0}) := \int_{F_{v_0}^\times} W_{v_0}\begin{pmatrix}\sqrt{\delta}^{-1}a & 0\\ 0 & 1\end{pmatrix}\mathrm{d}^\times a.$$

This is a linear form invariant under $\mathrm{Gl}(2,F_{v_0})$. Moreover, in the Archimedean case, it is continuous for the topology of the smooth vectors. The invariance is well known for v_0 split (see also [B]) and is established in [H] for v_0 inert. At any rate, it follows from the above considerations. Then

$$\Psi(1,W_{v_0},\Phi_{v_0}) = \mathcal{P}_{v_0}(W_{v_0})\int_{K_{v_0}}\int_{F_{v_0}^\times} \Phi_{v_0}\left[(0,t)k\right]\cdot|t|^2\mathrm{d}^\times t\mathrm{d}k.$$

The above residue becomes:

$$\mathrm{res}_{s=1} L^{S_0}(s,\Pi,\mathrm{Asai})\prod_{S_0}\mathcal{P}_{v_0}(W_{v_0})\int_{K_{S_0}}\int_{F_{S_0}^\times} \Phi_{S_0}\left[(0,t)k\right]\cdot|t|^2\mathrm{d}^\times t\mathrm{d}k.$$

Comparing with the previous residue computation, we get:

$$\int \phi(g)\mathrm{d}g = \prod_{v_0\in S_0}\mathcal{P}_{v_0}(W_{v_0})\frac{2\,\mathrm{res}_{s=1} L^{S_0}(s,\Pi,\mathrm{Asai})}{\mathrm{vol}(F^1/F^\times)}. \tag{21}$$

The identity is to be interpreted as the statement that the period integral of ϕ is non-zero if and only the Asai L-function has a pole at $s=1$ and each of the local integral $\mathcal{P}_{v_0}$ does not vanish on W_{v_0}.

REMARK. — The convergence of the integral defining $\mathcal{P}_{v_0}$ follows from the local analog of the integral representation for Whittaker functions given below in (44). Formula (21) implies that the linear form $\mathcal{P}_{v_0}$ is continuous for the topology of smooth vectors, since the left hand side is.

Recall that we write $E = F[\sqrt{\delta}\,]$. Then

$$\begin{pmatrix} \sqrt{\delta} & 0 \\ 0 & 1 \end{pmatrix} H(F) \begin{pmatrix} \sqrt{\delta}^{-1} & 0 \\ 0 & 1 \end{pmatrix} = Z(E)\,\mathrm{Gl}(2,F). \tag{22}$$

Accordingly:

$$\int_{H(F)Z(E_{\mathbb{A}})\backslash H(F_{\mathbb{A}})} \phi(h)\,\mathrm{d}h = \int_{Z(F_{\mathbb{A}})\,\mathrm{Gl}(2,F)\backslash\,\mathrm{Gl}(2,F_{\mathbb{A}})} \phi\Big[g\begin{pmatrix} \sqrt{\delta} & 0 \\ 0 & 1 \end{pmatrix}\Big]\,\mathrm{d}g$$

or

$$\int_{H(F)Z(E_{\mathbb{A}})\backslash H(F_{\mathbb{A}})} \phi(h)\,\mathrm{d}h = \prod_{S_0} \mathcal{P}_{v_0}(W_{v_0})\,\frac{2\,\mathrm{res}_{s=1} L^{S_0}(s,\Pi,\mathrm{Asai})}{\mathrm{vol}(F^1/F^\times)} \tag{23}$$

where here we have set:

$$\mathcal{P}_{v_0}(W_{v_0}) = \int_{F_{v_0}^\times} W_{v_0}\Big[\begin{pmatrix} a & 0 \\ 0 & 1 \end{pmatrix}\Big]\,\mathrm{d}^\times a.$$

We also denote by $L^{S_0}(s,\Pi,\mathrm{Asai};1)$ the previous Asai L-function and by $L(s,\Pi,\mathrm{Asai};\eta)$ the function $L(s,\Pi\otimes\mu,\mathrm{Asai};1)$ where μ is an idele class character of E whose restriction to F is η. Suppose that now Π is the base change of π with trivial central character. Then:

$$\begin{aligned} &\int_{H(F)Z(E_{\mathbb{A}})\backslash H(F_{\mathbb{A}})} \phi(h)\eta(\kappa(h))\,\mathrm{d}h \\ &\qquad = \prod_{S_0} \mathcal{P}_{v_0}(W_{v_0})\,\frac{2\,\mathrm{res}_{s=1} L^{S_0}(s,\Pi,\mathrm{Asai};\eta)}{\mathrm{vol}(F^1/F^\times)} \end{aligned} \tag{24}$$

where here we have set:

$$\mathcal{P}_{v_0}(W_{v_0}) = \int_{F_{v_0}^\times} W_{v_0}\Big[\begin{pmatrix} a & 0 \\ 0 & 1 \end{pmatrix}\Big]\eta(a)\,\mathrm{d}^\times a.$$

From now on we consider both cases at once, that is, Π is the base change of π with central character $\omega = 1$ or $\omega = \eta$.

A similar discussion applies to the computation of the scalar product of two forms ϕ_1, ϕ_2. As usual we assume that the corresponding functions W_i are product and the data is unramified outside S. We introduce a scalar product $\mathcal{B}_v$ on $\mathcal{W}(\Pi_v,\psi_v)$ for any place $v \in S$:

$$\mathcal{B}_v(W_1,W_2) := \int_{E_v^\times} W_1\begin{pmatrix} a_1 & 0 \\ 0 & 1 \end{pmatrix}\overline{W_2\begin{pmatrix} a_1 & 0 \\ 0 & 1 \end{pmatrix}}\,\mathrm{d}^\times a_1.$$

Then:

$$\int \phi_1(g)\overline{\phi_2}(g)\mathrm{d}g = \prod_{v\in S} \mathcal{B}_v\big((W_1)_v,(W_2)_v\big)\frac{2\operatorname{res}_{s=1} L^S(s,\Pi\times\widetilde{\Pi})}{\operatorname{vol}(E^1/E^\times)}.$$

We set:

$$\lambda(\phi) := \int_{E_{\mathbb{A}}^\times/E^\times} \phi\Big[\begin{pmatrix} a & 0\\ 0 & 1\end{pmatrix}\Big]\Omega^{-1}(a)\mathrm{d}^\times a.$$

At a place $v\in S$, we define

$$\lambda_v(W_v) := \int_{E_v^\times} W_v\Big[\begin{pmatrix} a & 0\\ 0 & 1\end{pmatrix}\Big]\Omega^{-1}(a)\mathrm{d}^\times a.$$

Then

$$\lambda(\phi) = \prod_{v\in S}\lambda_v(W_v)\, L^S\big(\tfrac12,\Pi\otimes\Omega^{-1}\big).$$

We consider the global distribution

$$\Theta_\Pi(f) := \sum_\phi \lambda(\Pi(f)\phi)\overline{\int \phi(h)\omega\eta(\kappa(h))\,\mathrm{d}h},$$

where the sum is over an orthonormal basis. Likewise, for a place v_0 inert in E with corresponding place v of E we define

$$\Theta_{\Pi,v_0}(f_v) := \sum_W \lambda_v\big(\Pi_v(f_v)W\big)\overline{\mathcal{P}_v(W)},$$

the sum over an orthonormal basis for the scalar product $\mathcal{B}_v$. Whenever convenient, we write f_{v_0} for f_v

Similarly, at a place v_0 which splits into v_1, v_2 we define

$$\Theta_{\Pi,v_0}(f_{v_1}\otimes f_{v_2}) := \sum_{W_1,W_2}\lambda_{v_1}\big(\Pi_{v_1}(f_{v_1})W_1)\big)\lambda_{v_2}\big(\Pi_{v_2}(f_{v_2})W_2)\big)\overline{\mathcal{P}_{v_0}(W_{v_1}\otimes W_{v_2})}.$$

We again write f_{v_0} for $f_{v_1}\otimes f_{v_2}$.

Now we assume that the function f is a product with f^S the characteristic function of K^S. Then:

$$\Theta_\Pi(f) = \prod_{v_0\in S_0}\Theta_{\Pi,v_0}(f_{v_0})\frac{\operatorname{vol}(E^1/E^\times)\operatorname{res}_{s=1}L^{S_0}(s,\Pi,\mathrm{Asai};\omega\eta)L^S(\frac12,\Pi\otimes\Omega^{-1})}{\operatorname{vol}(F^1/F^\times)\operatorname{res}_{s=1}L^S(s,\Pi\times\widetilde{\Pi})}.$$

Recall that here $\widetilde{\Pi}=\Pi=\Pi^\tau$ and Π is the base change of π with central character ω. We have the relation

$$L^S(s,\Pi\times\Pi^\tau) = L^{S_0}(s,\pi\times\pi)L(s,\pi\times\pi\otimes\eta).$$

If $\omega=1$ we have the relation

$$L^{S_0}(s,\Pi,\mathrm{Asai};\eta) = \frac{L^{S_0}(s,\pi\times\pi\otimes\eta)L^{S_0}(s,1_F)}{L^{S_0}(s,\eta)}.$$

Moreover since π is self contragredient then

$$L^{S_0}(s,\pi\times\pi)=L^{S_0}(s,1_F)L^{S_0}(s,\pi;\mathrm{Ad}).$$

where the last factor on the right is the partial adjoint L-function. Thus

$$\frac{L^{S_0}(s,\Pi,\mathrm{Asai};\eta)}{L^S(s,\Pi\times\widetilde{\Pi})}=\frac{1}{L^{S_0}(s,\eta)L^{S_0}(s,\pi,\mathrm{Ad})}.$$

If $\omega=\eta$ then

$$L^{S_0}(s,\Pi,\mathrm{Asai};1)=\frac{L^{S_0}(s,\pi\times\pi)L^{S_0}(s,1_F)}{L^{S_0}(s,\eta)}.$$

On the other hand

$$L^{S_0}(s,\pi\times\pi\otimes\eta)=L^{S_0}(s,\pi\times\widetilde{\pi})=L^{S_0}(s,1_F)L^{S_0}(s,\pi,\mathrm{Ad}).$$

Thus

$$\frac{L^{S_0}(s,\Pi,\mathrm{Asai};1)}{L^S(s,\Pi\times\widetilde{\Pi})}=\frac{1}{L^{S_0}(s,\eta)L^{S_0}(s,\pi,\mathrm{Ad})}.$$

Also

$$\mathrm{vol}(E^1/E^\times)=L(1,\eta)\,\mathrm{vol}(F^1/F^\times).$$

We can thus simplify our ratio of L-functions to obtain the following result:

PROPOSITION 4. — *One has*

$$\Theta_\Pi(f)=\prod_{v_0\in S_0}\Theta_{\Pi,v_0}(f_{v_0})\frac{L_{S_0}(1,\eta)L^S(\frac{1}{2},\Pi\otimes\Omega^{-1})}{L^{S_0}(1,\pi,\mathrm{Ad})}.$$

5. Local comparisons

As explained in the introduction, when f and the family (f_ϵ) match we have a spectral identity, with a single ϵ on the right, which is uniquely determined by Π:

$$\Theta_\Pi(f)=\theta_{\sigma_\epsilon}(f_\epsilon)+\theta_{\sigma_\epsilon\otimes\eta}(f_\epsilon).$$

We recall the expression for θ_{σ_ϵ}:

$$\theta_{\sigma_\epsilon}(f_\epsilon)=\sum_\phi\int\sigma_\epsilon(f_\epsilon)\phi(t)\Omega(t)^{-1}\,\mathrm{d}t\overline{\int\phi(t)\Omega(t)\mathrm{d}t}.$$

Let us write σ for σ_ϵ.

5.1. Local comparison: inert case. — Now consider a place v_0 of F, inert in E. Let v be the corresponding place of E. If the distribution θ_σ is not identically 0, then σ_{v_0} admits non-zero linear forms transforming under the characters Ω_v and Ω_v^{-1} of T_{v_0}, a group compact modulo the center. Thus there are smooth unit vectors e_T and e'_T such that

$$\sigma_{v_0}(t)e_T = \Omega(t)^{-1}e_T, \quad \sigma_{v_0}(t)e'_T = \Omega(t)e'_T.$$

We may assume that

$$e'_T = \sigma_{v_0}(w_\epsilon)(e_T).$$

We define a local distribution:

$$\theta_{\sigma_{v_0}}(f_{v_0}) = \sum_{u_i} \langle \sigma_{v_0}(f_{v_0})u_i, e'_T\rangle \overline{\langle u_i, e_T\rangle},$$

the sum over an orthonormal basis. This can also be written as

$$\int f(g)\omega_{\sigma_{v_0}}(g)\mathrm{d}g,$$

where we have set

$$\omega_{\sigma_{v_0}}(g) = \langle \sigma_{v_0}(g)e_T, e'_T\rangle. \tag{25}$$

Note that $\omega_{\sigma_{v_0}}(w_\epsilon) = 1$. There is a similar distribution $\theta_{\sigma_{v_0}\otimes\eta_{v_0}}$ defined by:

$$\theta_{\sigma_{v_0}\otimes\eta_{v_0}}(f_{v_0}) = \sum_{u_i} \langle \sigma_{v_0}\otimes\eta_{v_0}(f_{v_0})u_i, e''_T\rangle \overline{\langle u_i, e_T\rangle},$$

as well as a function

$$\omega_{\sigma_{v_0}\otimes\eta_{v_0}}(g) = \langle \sigma_{v_0}\otimes\eta_{v_0}(g)e_T, e''_T\rangle, \tag{26}$$

where we have set:

$$e''_T = \sigma_{v_0}\otimes\eta_{v_0}(w_\epsilon)(e_T) = \eta_{v_0}(-\epsilon)e'_T.$$

Thus

$$\omega_{\sigma_{v_0}\otimes\eta_{v_0}}(g) = \eta_{v_0}(\det g)\eta_{v_0}(-\epsilon)\omega_{v_0}(g). \tag{27}$$

In particular, if $\det g \in -\epsilon\,\mathrm{Norm}(E^\times)$ then $\omega_{\sigma_{v_0}}(g) = \omega_{\sigma_{v_0}\otimes\eta_{v_0}}(g)$.

The global distribution θ_σ decomposes into a tensor product

$$\theta_\sigma(f_\epsilon^{v_0}\otimes f_{\epsilon,v_0}) = \theta^{v_0}(f_\epsilon^{v_0})\theta_{\sigma_{v_0}}(f_{\epsilon,v_0}).$$

Likewise, we have obtained a decomposition of the global distribution Θ_Π into a tensor product $\Theta^v\otimes\Theta_{\Pi_{\sigma_v}}$. The global spectral identity gives then a linear relation of the form:

$$\Theta_{\Pi_{\sigma_v}}(f_v) = C_1\theta_{\sigma_{v_0}}(f_{\epsilon,v_0}) + C_2\theta_{\sigma_{v_0}\otimes\eta_{v_0}}(f_{\epsilon,v_0}).$$

To analyze the situation conveniently, let us go back to a local situation. That is, let E/F be a local quadratic extension and Π an irreducible unitary representation of $G(E)$ with trivial central character. We assume that Π is the base change of a representation π of $G(F)$ with central character ω, and $\omega = 1$ or $\omega = \eta_{E/F}$. We assume that ϵ is such that the representation σ of $G_\epsilon(F)$ corresponding to π has a vector transforming under Ω and Ω^{-1} with $\Omega_{|F^\times} = \omega$. We assume that we have a relation of the form:

$$\Theta_\Pi(f) = C_1\theta_\sigma(f_\epsilon) + C_2\theta_{\sigma\otimes\eta}(f_\epsilon)$$

for any pair of functions (f, f_ϵ) with matching orbital integrals; that is, f correspond to a function Φ on $\mathcal{S}_s$ and $U(x;\Phi) = H(x; f_\epsilon)$ for $x \in \epsilon\,\mathrm{Norm}(E^\times)$. In particular, if f_ϵ is supported on the set of $g \in G_\epsilon$ such that $\det(g) \in -\epsilon\,\mathrm{Norm}(E^\times)$ then this simplifies to

$$\Theta_\Pi(f) = C\int_{G_\epsilon(F)} \omega_\sigma(g) f_\epsilon(g)\,\mathrm{d}g, \quad C = C_1 + C_2. \tag{28}$$

Our goal in this section is to compute the constant C.

PROPOSITION 5. — *Suppose that f and f_ϵ have matching orbital integrals and f_ϵ is supported on the set $\{g \in G_\epsilon(F) \mid \det(g) \in -\epsilon\,\mathrm{Norm}(E^\times)\}$. Then*

$$\Theta_\Pi(f) = \epsilon(1,\eta,\psi_F)2\eta(-\epsilon)L(0,\eta)\theta_\sigma(f_\epsilon).$$

Moreover, we can choose the pair in such a way that both sides are non-zero.

5.2. Computation of Θ_Π. — To prove the proposition, we first consider a sequence of functions (f_j) on $G(E)$ which tends to δ_e. That is, $f_j \geq 0$, $\int f_j \mathrm{d}g = 1$ and $\mathrm{supp} f_j \to e$. We let Φ_j be the associated function on $\mathcal{S}$. It tends to δ_w on $\mathcal{S}(F)$. Note that we can start with an approximation Φ_j of δ_w on $\mathcal{S}$ and then choose for the functions f_j an approximation of δ_e. We also choose a function $\phi \in C_0^\infty(E)$ of small support. We set

$$f^j(g) = \int_E \phi(x) f_j\Big[\begin{pmatrix}1 & x\\ 0 & 1\end{pmatrix} g\Big]\mathrm{d}x \tag{29}$$

and we denote by Φ^j the corresponding function on $\mathcal{S}(F)$.

We first compute $\Theta_\Pi(f^j)$. Recall that the scalar product is given by:

$$\langle W_1, W_2\rangle = \int W_1\overline{W}_2\big[\mathrm{diag}(a,1)\big]\,\mathrm{d}^\times a.$$

and

$$\Theta_\Pi(f) = \sum \lambda\big(\Pi(f)W_i\big)\overline{\mathcal{P}(W_i)},$$

where W_i is an orthonormal basis. We have:

$$\begin{aligned}\lambda(\Pi(f^j)W_i) &= \iint W_i\Big[\begin{pmatrix} a & 0 \\ 0 & 1\end{pmatrix}g\Big]\Omega^{-1}(a)f^j(g)\mathrm{d}^\times a\mathrm{d}g \\ &= \iiint W_i\Big[\begin{pmatrix} a & 0 \\ 0 & 1\end{pmatrix}\begin{pmatrix} 1 & -x \\ 0 & 1\end{pmatrix}g\Big]\Omega^{-1}(a)f_j(g)\phi(x)\mathrm{d}^\times a\mathrm{d}x\mathrm{d}g \\ &= \iint \widehat{\phi}(a)W_i\Big[\begin{pmatrix} a & 0 \\ 0 & 1\end{pmatrix}g\Big]\Omega^{-1}(a)f_j(g)\mathrm{d}^\times a\mathrm{d}g\end{aligned}$$

where $\widehat{\phi}$ denotes the Fourier transform of ϕ.

For now we assume that E is non-Archimedean. We choose ϕ in such a way that $\widehat{\phi}(0) = 0$. Then there is an element $W^0 \in \mathcal{W}(\Pi, \psi)$ such that

$$W^0\Big[\begin{pmatrix} a & 0 \\ 0 & 1\end{pmatrix}\Big] = \overline{\widehat{\phi}(a)\Omega^{-1}(a)}.$$

Then the above expression is

$$\langle \Pi(f_j)W_i, W^0\rangle = \langle W_i, \Pi(f_j^*)W^0\rangle.$$

Since

$$\Pi(f_j^*)W^0 = \sum_i \langle \Pi(f_j^*)W^0, W_i\rangle W_i$$

we see that:

$$\Theta_\Pi(f^j) = \overline{\mathcal{P}\big(\Pi(f_j^*)W^0\big)}.$$

Since f_j is an approximation of δ_e, for j large enough, $\Pi(f_j^*)W^0 = W^0$ and

$$\Theta_\Pi(f^j) = \overline{\mathcal{P}(W^0)} = \overline{\int W^0\Big[\begin{pmatrix} b & 0 \\ 0 & 1\end{pmatrix}\Big]\omega\eta(b)\mathrm{d}^\times b} = \int_{F^\times} \widehat{\phi}(b)\eta(b)\mathrm{d}^\times b.$$

Clearly, we can choose ϕ with $\widehat{\phi}(0) = 0$ in such a way that this is non-zero.

We pass to the Archimedean case. We first explain how to choose the function ϕ. We assume, as we may, that $\psi_F(x) = \exp(2i\pi x)$. Let ϕ_0 be a smooth function of compact support on F with $\int \phi_0(x)\mathrm{d}x \neq 0$. Then the function $\phi_1 = \phi_0 * \phi_0^*$ is smooth of compact support. Its Fourier transform is $\widehat{\phi}_1 = |\widehat{\phi}_0|^2$ and is thus ≥ 0. Moreover, $\widehat{\phi}_1(0) = |\widehat{\phi}_0(0)|^2 > 0$. Now consider the function on E defined by:

$$\phi_2(x+iy) = \phi_1(x)\phi_1(y).$$

The Fourier transform on E of this function is given by

$$\widehat{\phi}_2(x+iy) = 2\widehat{\phi}_1(2x)\widehat{\phi}_1(-2y).$$

Let $Q > 0$ be an integer. There is a smooth function of compact support ϕ on E, whose Fourier transform is given by:

$$\widehat{\phi}(z) = 2^{2Q+1}(z\overline{z})^Q z\widehat{\phi}_1(2x)\widehat{\phi}_1(-2y), \quad z = x+iy.$$

Then

$$\int_{F^\times} \widehat{\phi}(x)\eta(x)\,\mathrm{d}^\times x = \widehat{\phi}_1(0)\int |x|_F^{2Q+1}\widehat{\phi}_1(x)\,\mathrm{d}^\times x > 0.$$

This function, which will be also denoted by ϕ_Q to stress the dependence on Q, will have the required properties, provided the integer Q is sufficiently large.

To continue we recall that the unitary representation Π can be realized in the Hilbert space $\mathcal{H}$ of square integrable functions on $E^\times$. Let P_1 be the group of matrices with last row $(0,1)$. The representation of $P_1(E)$ on $\mathcal{H}$ is the representation induced by the generic character of $N(E)$ determined by ψ_E. It is (topologically) irreducible. The subspace $\mathcal{V}$ of smooth vectors may be viewed as a certain space of smooth functions on $E^\times$ (Kirilov model). If W is in the Whittaker model of Π then the function Φ defined by

$$\Phi(a) = W\Big[\begin{pmatrix} a & 0 \\ 0 & 1 \end{pmatrix}\Big]$$

is in $\mathcal{V}$ and determines W. Let $\mathcal{V}'$ be the topological dual of $\mathcal{V}$ (space of *generalized vectors*). We have the inclusions: $\mathcal{V} \subset \mathcal{H} \subset \mathcal{V}'$. We extend the scalar product $\langle \bullet\,,\bullet \rangle$ on $\mathcal{H}$ to $\mathcal{V}\times\mathcal{V}'$ and $\mathcal{V}'\times\mathcal{V}$. The group and its enveloping algebra operate on $\mathcal{V}$ and $\mathcal{V}'$. For instance, the Whittaker linear from $\mathcal{W}$ is the element of $\mathcal{V}'$ such that, for $\Phi \in \mathcal{V}$:

$$\langle \Phi, \mathcal{W}\rangle = \Phi(e).$$

Likewise, λ and $\mathcal{P}$ are elements of $\mathcal{V}'$ such that

$$\langle \Pi, \lambda\rangle = \int_{E^\times} \Phi(a)\Omega^{-1}(a)\,\mathrm{d}^\times a, \quad \langle \Phi, \mathcal{P}\rangle = \int_{F^\times} \Phi(b)\omega\eta(b)\,\mathrm{d}^\times b,$$

and

$$\Theta_\Pi(f) = \big\langle \Pi(f)\mathcal{P}, \lambda\big\rangle = \big\langle \mathcal{P}, \Pi(f^*)\lambda\big\rangle.$$

We have already remarked that the convergence of these integrals follows at any rate from the local analog of the integral representation for Whittaker functions given below (see (44)). Moreover, the continuity follows for instance from the global theory (since we assume anyway that Π is a component of a cuspidal representation). The continuity of $\mathcal{P}$ means that there is M such that, if (X_α), $1 \le \alpha \le N$, is a basis of the space of elements of the enveloping algebra with (filtration) degree $\le M$, then, for suitable constants $c_\alpha > 0$,

$$\big|\langle \Phi, \mathcal{P}\rangle\big| \le \sum_\alpha c_\alpha \big\|\Pi(X_\alpha)\Phi\big\|.$$

In this sum $X_1 = 1$. Let $\mathcal{H}^N$ be the Hilbert sum of N copies of $\mathcal{H}$ and $\mathcal{V}_0$ the (non-closed) subspace of N−tuples of the form $v = (\Pi(X_\alpha)\phi)$, with $\phi \in \mathcal{V}$. Let $\mathcal{P}_0$ be the linear form on $\mathcal{V}_0$ defined by $\mathcal{P}_0(v) = \langle \phi, \mathcal{P}\rangle$. This linear form is continuous for the topology induced by the topology of $\mathcal{H}^N$, thus extends to the closure of $\mathcal{V}_0$ in $\mathcal{H}^N$ and is therefore given by the scalar product with a vector $(\Phi_\alpha) \in \mathcal{H}^N$. In other words,

$$\langle \Phi, \mathcal{P} \rangle = \sum_\alpha \langle \Pi(X_\alpha)\Phi, \Phi_\alpha \rangle \tag{30}$$

with $\Phi_\alpha \in \mathcal{H}$.

Now the scalar product $\langle \Phi, \mathcal{P} \rangle$ is defined if Φ is a vector of class C^M. We will prove the following assertion:

LEMMA 5. — *If Q is sufficiently large, then the function Φ^0 defined*

$$\Phi^0(z) = \overline{\widehat{\phi}_Q(z)\Omega(z)^{-1}}$$

is square integrable and is, in fact, a vector of class C^M in the representation. Moreover:

$$\mathcal{P}(\Phi^0) = \int_{F^\times} \Phi^0(x)\omega\eta(x)d^\times x = \overline{\int_{F^\times} \widehat{\phi}_Q(x)\eta(x)\mathrm{d}^\times x}.$$

We prove the lemma in the case where $\omega = \eta$. Then the restriction of Ω to $F^\times$ is η. Thus Ω has the form:

$$\Omega(z) = \frac{z^m}{(z\bar{z})^{m/2}},$$

where m is an odd integer. Taking a large enough Q, we see that the function Φ^0 of the above lemma has in turn the form

$$\Phi^0(z) = (z\bar{z})^{\frac{1}{2}}(z\bar{z})^P\Psi(z)$$

where P is another integer which tends to infinity with Q. The function Ψ (which depends on Q) is in $\mathcal{S}(E)$, the space of Schwartz functions on E. Let us denote by $\mathcal{V}(P)$ the space of functions of the above form, with $\Psi \in \mathcal{S}(E)$. It is contained in $\mathcal{H}$. It will suffice to prove the following general lemma:

LEMMA 6. — *If P is large enough the space $\mathcal{V}(P)$ is contained in the space of vectors of class C^M and, for $\Phi \in \mathcal{V}(P)$,*

$$\mathcal{P}(\Phi) = \int_{F^\times} \Phi(x)\omega\eta(x)\mathrm{d}^\times x.$$

Proof. — To prove the lemma we study the action of the enveloping algebra of $\mathrm{Gl}(2,E)$ (viewed as a real Lie group) on $\mathcal{V}$. We regard the space $M(2\times 2,\mathbb{C})$ of 2×2 matrices with complex entries as a real vector space and identify it to the Lie algebra of $\mathrm{Gl}(2,E)$ as a real Lie group. Entries of such a matrix are written in the form $a+b\sqrt{-1}$ with a,b real. We define the following elements of $M(2\times 2,\mathbb{C})\otimes\mathbb{C}$:

$$X_+ = \frac{1}{2}\left\{\begin{pmatrix}0&1\\0&0\end{pmatrix} - i\begin{pmatrix}0&\sqrt{-1}\\0&0\end{pmatrix}\right\},\quad X_- = \frac{1}{2}\left\{\begin{pmatrix}0&0\\1&0\end{pmatrix} - i\begin{pmatrix}0&0\\\sqrt{-1}&0\end{pmatrix}\right\},$$

$$H = \frac{1}{2}\left\{\begin{pmatrix} 1 & 0 \\ 0 & -1 \end{pmatrix} - i\begin{pmatrix} \sqrt{-1} & 0 \\ 0 & -\sqrt{-1} \end{pmatrix}\right\}, \quad T = \frac{1}{2}\left\{\begin{pmatrix} 1 & 0 \\ 0 & 0 \end{pmatrix} - i\begin{pmatrix} \sqrt{-1} & 0 \\ 0 & 0 \end{pmatrix}\right\}.$$

We define similarly conjugate elements $\overline{X}_+$ and so on, by replacing i by $-i$ in the above formulas. They commute to the previous ones. The Casimir element

$$\Omega = X_+X_- + X_-X_+\frac{1}{2}H^2$$

is in the center of the enveloping algebra and $T - \frac{1}{2}H$ in the center of the Lie algebra. Thus X_+X_- can be expressed in terms of T and elements of the center of the enveloping algebra. As a consequence,

$$\Pi(X_+)\Pi(X_-) = R\big(\Pi(T)\big),$$

where R is a polynomial (of degree 2). We have then, for $\Phi \in \mathcal{V}$,

$$\Pi(X_+)\Phi(z) = 2i\pi z\Phi(z), \quad \Pi(T)\Phi(z) = z\frac{\partial\Phi(z)}{\partial z},$$

$$2i\pi z\Pi(X_-)\Phi(z) = \Pi(X_+)\Pi(X_-)\Phi(z) = R\big(\Pi(T)\big)\Phi(z).$$

In other words (after a change of notations) we see that there is a polynomial R such that

$$\Pi(X_-)\Phi(z) = z^{-1}R\Big(z\frac{\partial}{\partial z}\Big)\Phi(z).$$

Similarly, there is a polynomial R' such that

$$\Pi(\overline{X}_-)\Phi(z) = \overline{z}^{-1}R'\Big(\overline{z}\frac{\partial}{\partial \overline{z}}\Big)\Phi(z).$$

Now the space $\mathcal{V}(P)$ is invariant under the operators of multiplication by $z, \overline{z}$ and the differential operators $z\partial/\partial z$, $\overline{z}\partial/\partial\overline{z}$. This implies that in fact $\mathcal{V}(P)$ is contained in the space of smooth vectors for the group $P_1(E)$. Now multiplication by z^{-1} sends $\mathcal{V}(P)$ to $\mathcal{V}(P-1)$. It follows that if P is sufficiently large and $\Phi \in \mathcal{V}(P)$ the vector $\Pi(X_-)\Phi$ which, *a priori*, is only a vector in $\mathcal{V}'$ is in fact in $\mathcal{H}$, more precisely, in $\mathcal{V}(P-1)$. Moreover, it is given by the same formula as above. Likewise for $\Pi(\overline{X}_-)\Phi$. The first assertion of the lemma follows. To prove the second assertion, we let $\mathcal{V}_0$ be the subspace of Φ in $\mathcal{V}(P)$ for which the function Ψ is flat at $0 \in E$, that is, all its derivatives vanish at 0. From the above analysis, it follows that such a vector is in $\mathcal{V}$. Then $\mathcal{P}(\Phi)$ is given by the integral over $F^\times$. Now recall the decomposition (30). From the above computation, there exist (non commutative) polynomials P_α in $(z^{-1}, \overline{z}^{-1}, z\partial/\partial z, \overline{z}\partial/\partial\overline{z})$ such that

$$\mathcal{P}(\Phi) = \sum_\alpha \int (P_\alpha\Phi)(z)\,\overline{\Phi_\alpha(z)}\mathrm{d}^\times z.$$

This applies in particular to a $\Phi \in \mathcal{V}(P)$ for P large enough. Thus we have to show that the difference

$$\int_{F^\times} |x|^{2P+1}\Psi(x)\mathrm{d}^\times x - \sum_\alpha \int_{E^\times} \big(P_\alpha\big((z\bar{z})^{\frac{1}{2}+P}\Psi(z)\big)\big)\overline{\Phi}_\alpha(z)\mathrm{d}^\times z$$

vanishes for all $\Psi \in \mathcal{S}(E)$, if P is large enough. At any rate, this is true for Ψ flat at 0. Since the difference is a distribution (linear form in Ψ), the difference is at most a linear combination of derivatives of Ψ at 0. Thus it will vanish if Ψ vanishes at sufficiently high order at 0. Thus it might not vanish on $\mathcal{V}(P)$ but it will vanish on the smaller space $\mathcal{V}(P_1)$ for $P_1 \gg P$. We are done. □

In summary then, we have found a smooth function of small support ϕ on E and an element $\Phi^0 \in \mathcal{H}$ of class C^M such that

$$\lambda\Big(\int_{N(E)} \phi(z)\Pi(n^{-1})\mathrm{d}n\Phi\Big) = \langle \Phi, \Phi_0\rangle, \quad n = \begin{pmatrix} 1 & z \\ 0 & 1\end{pmatrix}, \ \Phi \in \mathcal{V},$$

$$\mathcal{P}(\Phi^0) = \overline{\int_{F^\times} \widehat{\phi}(b)\eta(b)\mathrm{d}^\times b} \neq 0.$$

Suppose that f_j is a smooth approximation of δ_e. From (30) we get:

$$\mathcal{P}\big(\Pi(f_j)\Phi^0\big) = \sum_\alpha \int \langle \Pi(X_\alpha)\Pi(g)\Phi^0, \Phi_\alpha\rangle f_j(g)\mathrm{d}g$$

If we write

$$\mathrm{Ad}\, g^{-1}X_\alpha = \sum_\beta \lambda_{\alpha,\beta}(g)X_\beta$$

this becomes

$$\sum_{\alpha,\beta} \int \lambda_{\alpha,\beta}(g)\big\langle \Pi(X_\beta)\Phi^0, \Pi(g)^{-1}\Phi_\alpha\big\rangle f_j(g)\mathrm{d}g.$$

As $j \to +\infty$ this tends to

$$\sum_{\alpha,\beta} \lambda_{\alpha,\beta}(e)\big\langle \Pi(X_\beta)\Phi^0, \Phi_\alpha\big\rangle = \sum_\alpha \big\langle \Pi(X_\alpha)\Phi_0, \Phi_\alpha\big\rangle = \mathcal{P}(\Phi^0).$$

We thus consider an approximation Φ_j of δ_w on $\mathcal{S}$. We can choose an approximation f_j of δ_e on $G(E)$ such that Φ_j is the function associated to f_j. Just as in the $p-$adic case, we define f^j in terms of ϕ and Φ^j. Then

$$\lambda\big(\Pi(f^j)\Phi\big) = \lambda\Big(\int_{N(E)} \phi(z)\Pi(n^{-1})\mathrm{d}n\Pi(f_j)\Phi\Big) = \big\langle \Phi(f_j)\Phi, \Phi_0\big\rangle,$$

or

$$\Pi(f^{j,*})\lambda = \Pi(f_j^*)\Phi_0.$$

Then

$$\Theta_\Pi(f) = \big\langle \mathcal{P}, \Pi(f^{j,*})\lambda\big\rangle = \big\langle \mathcal{P}, \Pi(f_j^*)\Phi_0\big\rangle$$

which tends to $\langle\mathcal{P},\Phi_0\rangle = \overline{\mathcal{P}(\Phi_0)}$. Thus we find again

$$\lim_{j\to+\infty}\Theta_\Pi(f^j) = \int_{F^\times}\widehat{\phi}(b)\eta(b)\mathrm{d}^\times b.$$

We introduce the following function on F:

$$\phi_1(x) = \int\phi\big(x+y\sqrt{\delta}\big)|2|_F\cdot|\delta|_F^{\frac{1}{2}}\mathrm{d}y. \tag{31}$$

Then, denoting by $\mathcal{F}$ the Fourier transform for a function on F, we get for $x\in F$:

$$\widehat{\phi}(x) = \mathcal{F}(\phi_1)(2x).$$

Thus, in all cases,

$$\lim_{j\to+\infty}\Theta_\Pi(f^j) = \eta(2)\int_{F^\times}\mathcal{F}(\phi_1)(x)\eta(x)\mathrm{d}^\times x. \tag{32}$$

5.3. Computation on G_ϵ. — We now compute the orbital integral of Φ^j. After a change of variables, we get

$$U(x;\Phi^j) = \iint\Phi^j\Big[\begin{pmatrix} y\overline{y}z^{-1}x^{-1} & y\\ \overline{y} & z\end{pmatrix}\Big]\Omega(y)\eta(z)\mathrm{d}^\times y\mathrm{d}^\times z.$$

From (29) we get

$$\Phi^j(s) = \int\Phi_j\Big[\begin{pmatrix}1 & t\\ 0 & 1\end{pmatrix}s\begin{pmatrix}1 & 0\\ \overline{t} & 1\end{pmatrix}\Big]\phi(t)\mathrm{d}t.$$

After changing t to $-t$ we find that the orbital integral of Φ^j is equal to

$$\iiint\Phi_j\Big[\begin{pmatrix} y\overline{y}z^{-1}x^{-1}-t\overline{y}-\overline{t}y+t\overline{t}z & y-tz\\ \overline{y}-\overline{t}z & z\end{pmatrix}\Big]\phi(-t)\Omega(y)\eta(z)\mathrm{d}t\mathrm{d}^\times y\mathrm{d}^\times z.$$

Recall ϕ has compact support and the support of Φ_j tends to w. Thus there is a sequence $\{c_j\}$ of positive real numbers such that $\lim_{j\to+\infty}c_j = +\infty$ and the non-vanishing of $U(x;\Phi^j)$ implies $|x|>c_j$. Now we choose a function f^j_ϵ which matches f^j (or Φ^j). For $x\in\epsilon\,\mathrm{Norm}(E^\times)$ we write $x=\epsilon a\overline{a}$ and then

$$U(x;\Phi^j) = H(x;f^j_\epsilon) = \iint f^j_\epsilon\Big[t_1\begin{pmatrix}a^{-1} & \epsilon\\ 1 & \overline{a}^{-1}\end{pmatrix}t_2\Big]\Omega(t_1t_2^{-1})\mathrm{d}t_1\mathrm{d}t_2.$$

Suppose the integral is non-zero. Then $|x|>c_j$. If j is large enough, then in the integrand, the determinant of the matrix $g=t_1(*)t_2$ belongs to $-\epsilon\,\mathrm{Norm}(E^\times)$. Thus we may and will assume that f^j_ϵ is supported on the set $\{g\mid\det g\in -\epsilon\,\mathrm{Norm}(E^\times)\}$. From (15) we get

$$\int\omega_\sigma(g)f^j_\epsilon(g)\mathrm{d}^\times g = \frac{1}{L(1,\eta)^2}\int_{x=\epsilon a\overline{a}}\omega_\sigma\Big[\begin{pmatrix}a^{-1} & \epsilon\\ 1 & \overline{a}^{-1}\end{pmatrix}\Big]U(x;\Phi^j)\frac{\mathrm{d}x}{|1-x|^2}.$$

Now we define a function ω on $\epsilon\,\mathrm{Norm}(E^\times)-\{1\}$ by

$$\omega(x):=\omega_\sigma\Big[\begin{pmatrix} a^{-1} & \epsilon \\ 1 & \bar a^{-1}\end{pmatrix}\Big],\quad \text{for } x=\epsilon a\bar a.$$

One can check that indeed the right hand side depends only on x. The function $\omega(x^{-1})$ extends to a smooth function near 0 on $\epsilon\,\mathrm{Norm}(E)$, whose value at 0 is

$$\omega_\sigma\Big[\begin{pmatrix} 0 & \epsilon \\ 1 & 0\end{pmatrix}\Big]=1.$$

We denote by $\widetilde\omega$ a smooth function on F, supported on a neighborhood of 0, and equal to $\omega(x^{-1})|1-x|^{-2}$ for x near 0 in $\epsilon\,\mathrm{Norm}(E)$. In particular $\widetilde\omega(0)=1$. Thus, for large enough j,

$$\begin{aligned}\int\omega_\sigma(g)f_\epsilon^j(g)\mathrm{d}g&=\frac{1}{L(1,\eta)^2}\int_{\epsilon\,\mathrm{Norm}(E^\times)}U(x;\Phi^j)\frac{\omega(x)\mathrm{d}x}{|1-x|^2}\\&=\frac{1}{L(1,\eta)^2}\int_{\epsilon\,\mathrm{Norm}(E^\times)}U(x^{-1};\Phi^j)\widetilde\omega(x)\mathrm{d}x.\end{aligned}$$

Equivalently,

$$\int\omega_\sigma(g)f_\epsilon^j(g)\mathrm{d}g=\frac{1}{2L(1,\eta)^2}\big[I_1+\eta(\epsilon)I_\eta\big]\tag{33}$$

where we have set

$$\begin{aligned}I_{\left\{\substack{1\\ \eta}\right\}}&:=\int_E U(x^{-1};\Phi^j)\Big\{\begin{matrix}1\\ \eta(x)\end{matrix}\Big\}\widetilde\omega(x)\mathrm{d}x\\&=\int\Phi_j\Big[\begin{pmatrix} y\bar y z^{-1}x-t\bar y-\bar t y+t\bar t z & y-tz\\ \bar y-\bar t z & z\end{pmatrix}\Big]\\&\qquad\times\phi(-t)\widetilde\omega(x)\Omega(y)\eta(z)\Big\{\begin{matrix}1\\ \eta(x)\end{matrix}\Big\}\mathrm{d}t\mathrm{d}^\times y\mathrm{d}^\times z\mathrm{d}x.\end{aligned}$$

To compute this, we first change x to $xz/y\bar y$:

$$\begin{aligned}&=\int\Phi_j\Big[\begin{pmatrix} x-t\bar y-\bar t y+t\bar t z & y-tz\\ \bar y-\bar t z & z\end{pmatrix}\Big]\phi(-t)\widetilde\omega\Big(\frac{xz}{y\bar y}\Big)\\&\qquad\times\Omega(y)\Big\{\begin{matrix}\eta(z)\\ \eta(x)\end{matrix}\Big\}\mathrm{d}t|y|_E^{-1}\mathrm{d}^\times y|z|_F\mathrm{d}^\times z\mathrm{d}x.\end{aligned}$$

Next we change y to $y+tz$ to get:

$$\begin{aligned}&\int\Phi_j\Big[\begin{pmatrix} x-t\bar y-\bar t y-t\bar t z & y\\ \bar y & z\end{pmatrix}\Big]\phi(-t)\widetilde\omega\Big(\frac{xz}{(y+tz)(\bar y+\bar t z)}\Big)\\&\qquad\times\Omega(y+tz)\Big\{\begin{matrix}\eta(z)\\ \eta(x)\end{matrix}\Big\}\mathrm{d}t|y+tz|_E^{-2}|y|_E\mathrm{d}^\times y|z|_F\mathrm{d}^\times z\mathrm{d}x.\end{aligned}$$

Now we change t to $t/\overline{y}$ to get:

$$\int \Phi_j\Big[\begin{pmatrix} x-(t+\overline{t}+t\overline{t}z/y\overline{y}) & y \\ \overline{y} & z\end{pmatrix}\Big]\phi\Big(\frac{-t}{\overline{y}}\Big)\widetilde{\omega}\Big(\frac{xz}{(y+tz/\overline{y})(\overline{y}+\overline{t}z/y)}\Big)$$
$$\times\,\Omega\Big(y+\frac{tz}{\overline{y}}\Big)\begin{Bmatrix}\eta(z)\\ \eta(x)\end{Bmatrix}\mathrm{d}t\big|y+tz/\overline{y}\big|_E^{-2}\mathrm{d}^\times y|z|_F\mathrm{d}^\times z\mathrm{d}x.$$

We finally change z to $zy\overline{y}$:

$$(34)\quad I_{\{{1\atop \eta}\}}=\int \Phi_j\Big[\begin{pmatrix} x-(t+\overline{t}+t\overline{t}z) & y \\ \overline{y} & zy\overline{y}\end{pmatrix}\Big]\phi\Big(\frac{-t}{\overline{y}}\Big)\widetilde{\omega}\Big(\frac{xz}{(1+tz)(1+\overline{t}z)}\Big)$$
$$\times\Omega(y)\frac{\Omega(1+tz)}{|1+tz|_E^2}\begin{Bmatrix}\eta(z)\\ \eta(x)\end{Bmatrix}\mathrm{d}t|y|_E^{-1}\mathrm{d}^\times y|z|_F\mathrm{d}^\times z\mathrm{d}x.$$

To evaluate the limit of the integrals we remind ourselves that Φ_j is an approximation of δ_w on $\mathcal{S}_s$. In particular (see Lemma 4)

$$L(1,1_F)\int \Phi_j\Big[\begin{pmatrix} x & y\\ \overline{y} & z\end{pmatrix}\Big]\frac{\mathrm{d}x\mathrm{d}z\mathrm{d}y}{|xz-y\overline{y}|_F^2}=1,$$

with $x,z\in F$ and $y\in E$. If we change z to $zy\overline{y}$ and take into account the relations $L(1,1_E)=L(1,1_F)L(1,\eta)$ and $\mathrm{d}^\times y=L(1,1_E)\mathrm{d}y\,|y|_E^{-1}$ we get:

$$(35)\qquad \int \Phi_j\Big[\begin{pmatrix} x & y\\ \overline{y} & zy\overline{y}\end{pmatrix}\Big]\frac{\mathrm{d}x\mathrm{d}z\mathrm{d}^\times y}{|1-xz|_F^2}=L(1,\eta).$$

We are going to see that the limit of the integral I_1 corresponding to the factor $\eta(z)$ in (34) is 0. We prove this in the Archimedean case, leaving the easier non-Archimedean case to the reader. We change x to $x+(t+\overline{t}+t\overline{t}z)$ to arrive (up to a multiplicative constant) at

$$\int \Psi_j(x,z,y)A(x,z,y)\eta(z)\mathrm{d}x\mathrm{d}z\mathrm{d}^\times y$$

where we have set

$$A(x,z,y)=\big|1-xz\big|_F^2\cdot\Omega(y)|y|_E^{-1}\int\phi\Big(\frac{-t}{\overline{y}}\Big)$$
$$\times\,\widetilde{\omega}\Big(\frac{xz+(1+tz)(1+\overline{t}z)-1}{(1+tz)(1+\overline{t}z)}\Big)\frac{\Omega(1+tz)}{|1+tz|_F^2}\mathrm{d}t,$$

and

$$\Psi_j(x,z,y)=\Phi_j\Big[\begin{pmatrix} x & y\\ \overline{y} & zy\overline{y}\end{pmatrix}\Big]\frac{1}{|1-xz|_F^2}\cdot$$

Recall that Φ_j is an arbitrary smooth approximation of δ_w. Equivalently, Ψ_j is an arbitrary approximation of $L(1,\eta)\delta_{(0,0,1)}$ on the space $F\times F\times E$ for the measure $\mathrm{d}x\mathrm{d}z\mathrm{d}^\times y$. Thus, there is no harm in assuming that our approximation of δ satisfies:

$$\Psi_j(x,z,y)=\Psi_j(x,-z,y).$$

For ϕ with small support, A is continuous near $(0,0,1)$ on $F\times F\times E$. We can write $A=A_{+}+A_{-}$ where $A_{\pm}(x,-z,y)=\pm A_{\pm}(x,z,y)$. In view of our assumption on Ψ_j the integral of $A_{+}(x,z,y)\eta(z)$ against Ψ_j is 0. Now $A_{-}(x,z,y)\eta(z)$ is continuous and vanishes at $(0,0,1)$. We are integrating on $F\times F\times E$ against an approximation of a multiple of $\delta_{(0,0,1)}$. The limit of the integral is thus 0, as claimed.

To compute the integral I_η corresponding to the factor $\eta(x)$ in (34), we first replace t by a new variable s such that:

$$s+\overline{s}=t+\overline{t}+t\overline{t}z,\quad s-\overline{s}=t-\overline{t}.$$

Explicitly, for ϕ of small support:

$$t=p+q\sqrt{\delta},\quad s=p_1+q_1\sqrt{\delta},\quad q=q_1,$$

$$p=\frac{-1+\sqrt{1+2p_1z+q_1^2\delta z^2}}{z}=\frac{2p_1+q_1^2\delta z}{1+\sqrt{1+2p_1z+q_1^2\delta z^2}}\mathrm{d}t$$
$$=\mathrm{d}s|1+2p_1z+q_1^2\delta z^2|_F^{-\frac{1}{2}}.$$

Then the integral becomes:

$$I_\eta=\int\Phi_j\Big[\begin{pmatrix}x-2p_1 & y\\ \overline{y} & zy\overline{y}\end{pmatrix}\Big]\phi\Big(\frac{-t}{\overline{y}}\Big)$$
$$\times\widetilde{\omega}\Big(\frac{xz}{(1+tz)(1+\overline{t}z)}\Big)\Omega(y)|y|_E^{-1}\frac{\Omega(1+tz)}{|1+tz|_E^2}$$
$$\times\eta(x)|1+2p_1z+q_1^2\delta z^2|_F^{-\frac{1}{2}}\mathrm{d}s\mathrm{d}^\times y|z|_F\mathrm{d}^\times z\mathrm{d}x,$$

where now

$$t=\frac{-1+\sqrt{1+2p_1z+q_1^2\delta z^2}}{z}+q_1\sqrt{\delta}.$$

We change p_1 to $p_1+\frac{1}{2}x$ and then p_1 to $-\frac{1}{2}p_1$:

$$\int\Phi_j\Big[\begin{pmatrix}p_1 & y\\ \overline{y} & zy\overline{y}\end{pmatrix}\Big]\phi\Big(\frac{-t}{\overline{y}}\Big)\widetilde{\omega}\Big(\frac{xz}{(1+tz)(1+\overline{t}z)}\Big)$$
$$\times\Omega(y)|y|_E^{-1}\frac{\Omega(1+tz)}{|1+tz|_E^2}\eta(x)\cdot|1-p_1z+xz+q_1^2\delta z^2|_F^{-\frac{1}{2}}$$
$$\times\mathrm{d}p_1\mathrm{d}q_1|\delta|_F^{\frac{1}{2}}\mathrm{d}^\times y|z|_F\mathrm{d}^\times z\mathrm{d}x,$$

where

$$t=\frac{-1+\sqrt{1-p_1z+zx+q_1^2\delta z^2}}{z}+q_1\sqrt{\delta}$$
$$=\frac{-p_1+x+q_1^2\delta z}{1+\sqrt{1-p_1z+zx+q_1^2\delta z^2}}+q_1\sqrt{\delta}.$$

Since $|z|_F \mathrm{d}^\times z \mathrm{d}x = \mathrm{d}z |x|_F \mathrm{d}^\times x$, this can be written as

$$\int \Phi_j\Big[\begin{pmatrix} p_1 & y \\ \overline{y} & zy\overline{y}\end{pmatrix}\Big] A(p_1,z,y) \frac{\mathrm{d}p_1 \mathrm{d}z \mathrm{d}^\times y}{|1-p_1 z|_F^2}$$

with

$$A(p_1,z,y) := |1-p_1 z|_F^2 \int \phi\Big(\frac{-t}{\overline{y}}\Big)\widetilde{\omega}\Big(\frac{xz}{(1+tz)(1+\overline{t}z)}\Big)$$
$$\times |1-p_1 z + xz + q_1^2\delta z^2|_F^{-\frac{1}{2}} \Omega(y)|y|_E^{-1}$$
$$\times \frac{\Omega(1+tz)}{|1+tz|_E^2}\eta(x)|x|_F \mathrm{d}^\times x |\delta|_F^{\frac{1}{2}} \mathrm{d}q_1.$$

For ϕ of small support, A is continuous on $F\times F\times E$ near $(0,0,1)$ with

$$A(0,0,1) = \iint \phi\big(-\tfrac{1}{2}x - q_1\sqrt{\delta}\,\big)\eta(x)|x|\mathrm{d}^\times x |\delta|_F^{\frac{1}{2}} \mathrm{d}q_1$$
$$= \eta(-2)\int \phi_1(x)\eta(x)|x|_F \mathrm{d}^\times x,$$

the last equation by definition of ϕ_1 (see (31)). Thus (see (35))

$$\lim_{j\to+\infty} I_\eta = L(1,\eta)\eta(-2)\int \phi_1(x)\eta(x)|x|_F \mathrm{d}^\times x$$

and we find from (33)

$$\lim_{j\to+\infty}\int \omega_\sigma(g) f_\epsilon^j(g)\mathrm{d}g = \frac{\eta(2\epsilon)}{2L(1,\eta)}\int \phi_1(x)\eta(x)|x|_F \mathrm{d}^\times x. \tag{36}$$

We compare this with (32):

$$\lim_{j\longrightarrow+\infty} \Theta_\Pi(f^j) = \eta(2)\int \mathcal{F}(\phi_1)(x)\eta(x)\mathrm{d}^\times x,$$

which, by the Tate functional equation, is equal to:

$$= \eta(2)\epsilon(1,\eta,\psi_F)\frac{L(0,\eta)}{L(1,\eta)}\int \phi_1(x)|x|_F\eta(x)\mathrm{d}^\times x$$
$$= \epsilon(1,\eta,\psi_F)2\eta(-\epsilon)L(0,\eta)\lim_{j\to+\infty}\int \omega_\sigma(g) f_\epsilon^j(g)\mathrm{d}g.$$

Proposition 5 follows.

5.4. Local comparison: split case. — We now consider the similar decomposition at a split place v_0. The isomorphism of (2.2) takes $G_\epsilon(F_{v_0})$ to $\mathrm{Gl}(2,F_{v_0})$, T_{v_0} to the group of A_{v_0} diagonal matrices and the character Ω to the character $\mathrm{diag}(a_1,a_2)\mapsto \Omega_{v_1}(a_1 a_2^{-1})$. We may realize the representation σ_{v_0} in its Whittaker model. Then we have again a decomposition

$$\theta_\sigma(f^{v_0}\otimes f_{v_0}) = \theta^{v_0}(f^{v_0})\theta_{\sigma_{v_0}}(f_{v_0}),$$

where the local distribution at the place v_0 is now given by

$$\theta_{\sigma_{v_0}}(f_{v_0}) = \sum_W \int \sigma_{v_0}(f_{\epsilon,v_0})W(a)\Omega_{v_1}(a)^{-1}\mathrm{d}^\times a \overline{\int W(aw_\epsilon^{-1})\Omega_{v_1}(a)^{-1}\mathrm{d}^\times a}.$$

On the other hand, recall the distribution

$$\Theta_{\Pi,v_0}(f_{v_1}\otimes f_{v_2}) = \sum_{W_1,W_2} \lambda_{v_1}\big(\Pi_{v_1}(f_{v_1})W_1\big)\big)\lambda_{v_2}\big(\Pi_{v_2}(f_{v_2})W_2\big)\big)\overline{\mathcal{P}_{v_0}(W_1\otimes W_2)}.$$

PROPOSITION 6. — *If $f_{v_1}\otimes f_{v_2}$ and f_{ϵ,v_0} match then*

$$\Theta_{\Pi_{v_0}}(f_{v_1}\otimes f_{v_2}) = \theta_{\sigma_{v_0}}(f_{\epsilon v_0}).$$

The distribution $\Theta_{\Pi_{v_0}}$ can be simplified as follows. Let us identify E_{v_1}, E_{v_2} and F_{v_0}. The representations Π_{v_1} and Π_{v_2} are in fact identical. The Hermitian form $\mathcal{P}_{v_0}(W_1\otimes W_2)$ is invariant under the unitary group at v_0. It follows that

$$\Theta_{\Pi,v_0}(f_{v_1}\otimes f_{v_2}) = \sum_{W_1,W_2} \lambda_{v_1}(\Pi_{v_1}(f_{v_1}*\tilde{f}_{v_2})W_1))\lambda_{v_2}(W_2))\overline{\mathcal{P}_{v_0}(W_1\otimes W_2)}$$

where we have set

$$\tilde{f}(g) = f(w\,{}^t g w).$$

Since Π_{v_1} is unitary and self contragredient, for $W\in\mathcal{W}(\Pi_{v_1},\psi_{v_0})$ the function $g\mapsto \overline{W}(\mathrm{diag}(-1,1)g)$ is still in the same space. Thus we can choose the functions W_1 in some orthonormal basis and choose then for the functions W_2 the functions of the form

$$g \longmapsto \overline{W}_1\big(\mathrm{diag}(-1,1)g\,\mathrm{diag}(-1,1)\big).$$

Since $\Omega_{v_2}\Omega_{v_1}=1$ our distribution reads then

$$\sum_{W_1,W_2}\int \Pi_{v_1}(f_{v_1}*\tilde{f}_{v_2})W_1(a_1)\Omega_{v_1}^{-1}(a_1)\mathrm{d}^\times a_1 \overline{\int W_2(a_2)\Omega_{v_1}^{-1}(a_2)\mathrm{d}^\times a_2}$$
$$\times\int_{F_{v_0}^\times} W_1\overline{W}_2\begin{pmatrix} b & 0\\ 0 & 1\end{pmatrix}\mathrm{d}^\times b,$$

where W_1, W_2 vary independently in the *same* orthonormal basis. The orthogonality relations give:

$$\sum_W \int \Pi_{v_1}(f_{v_1}*\tilde{f}_{v_2})W(a_1)\Omega_{v_1}^{-1}(a_1)\mathrm{d}^\times a_1 \overline{\int W(a_2)\Omega_{v_1}^{-1}(a_2)\mathrm{d}^\times a_2}.$$

Recall that the matching condition (17) amounts to:

$$f_{v_1}*\tilde{f}_{v_2}(g) = f_{\epsilon,v_0}(gw_\epsilon^{-1}).$$

Using the fact that we can change the orthonormal basis, we see that the above expression becomes, after a change of variables,

$$\sum_W \int \Pi_{v_1}(f_{\epsilon,v_0}) W(a_1)\Omega_{v_1}^{-1}(a_1)\mathrm{d}^\times a_1 \overline{\int W(a_2 w_\epsilon^{-1})\Omega_{v_1}^{-1}(a_2)\mathrm{d}^\times a_2}$$

which is equal to $\theta_{\sigma_\epsilon,v_0}(f_{v_0})$.

6. Global comparison

We go back to the global relation:

$$\Theta_\Pi(f) = \theta_\sigma(f_\epsilon) + \theta_{\sigma\otimes\eta}(f_\epsilon).$$

We assume f is a product with f^S the characteristic function of K^S. For v_0 inert not in S_0 the function f_{ϵ,v_0} is the characteristic function of $\mathrm{Gl}(2,\mathcal{O}_E) \cap G_\epsilon(F)$ and in particular supported on the set of g with $\det g \in -\epsilon\,\mathrm{Norm}(E_v^\times)$. Let us make the assumption that for any v_0 inert in S_0 the function f_{ϵ,v_0} is also supported on the set of g with $\det g \in -\epsilon\,\mathrm{Norm}(E_v^\times)$. Then the function $\rho(w_\epsilon)^{-1} f_\epsilon$ is supported on the set of g with $\det g \in \mathrm{Norm}(E_{\mathbb{A}}^\times)$ so that the distributions J_σ and $J_{\sigma\otimes\eta}$ take the same value on that function. Equivalently, $\theta_\sigma(f_\epsilon) = \theta_{\sigma\otimes\eta}(f_\epsilon)$ so that

$$\theta_\epsilon(f_\epsilon) = \frac{1}{2}\Theta_\Pi(f).$$

At a split place $v_0 \in S_0$, we have

$$\Theta_{\Pi_{v_0}}(f_{v_0}) = \theta_{\sigma_{v_0}}(f_{v_0}).$$

At a place $v_0 \in S_0$ inert in E, we have (Proposition 5)

$$\Theta_{\Pi_v}(f_v) = \epsilon(1,\eta_{v_0},\psi_{v_0})\, 2\eta_{v_0}(-\epsilon) L(0,\eta_{v_0})\, \theta_{\sigma_{v_0}}(f_{\epsilon,v_0}).$$

Then

$$\theta_\sigma(f_\epsilon) = \frac{1}{2}\Theta_\Pi(f) = \frac{1}{2} \prod_{\substack{v_0\in S_0,\\ \text{inert}}} \epsilon(1,\eta_{v_0},\psi_{v_0}) 2\eta_{v_0}(-\epsilon) L(0,\eta_{v_0})$$
$$\times \frac{L_{S_0}(1,\eta) L^S(\frac{1}{2},\Pi\otimes\Omega^{-1})}{L^{S_0}(1,\pi,\mathrm{Ad})} \prod_{v_0\in S_0} \theta_{\sigma_{v_0}}(f_{v_0}).$$

At this point we go back to distributions of positive type:

$$J_\sigma(f) := \theta_\sigma\big(\rho(w_\epsilon)f\big), \quad \widetilde{J}_{\sigma_{v_0}}(f) := \theta_{\sigma_{v_0}}\big(\rho(w_\epsilon)f\big).$$

Explicitly:

$$J_\sigma(f) = \sum \int \sigma(f)\phi(t)\Omega(t)^{-1}\mathrm{d}t \overline{\int \phi(t)\Omega^{-1}(t)\mathrm{d}t}.$$

For v_0 inert:

$$\widetilde{J}_{\sigma_{v_0}}(f_{v_0}) = \int_{G_{\epsilon v_0}} f_{v_0}(g)\langle\sigma_{v_0}(g)e'_T, e'_T\rangle \mathrm{d}g.$$

For v_0 split:

$$\widetilde{J}_{\sigma_{v_0}}(f) = \sum_W \int \rho(f)W(a)\Omega_{v_1}^{-1}(a)\mathrm{d}a \overline{\int W(a)\Omega_{v_1}^{-1}(a)\mathrm{d}a}.$$

Because S_0 is large enough the product of the factors $\eta_{v_0}(-\epsilon)$ is one.

THEOREM 2. — *For any smooth function $f = \prod_{v_0 \in S_0} f_{v_0} f^{S_0}$ of compact support:*

$$J_\sigma(f) = \prod_{v_0 \in S_0} \widetilde{J}_{\sigma_{v_0}}(f_{v_0}) \times \frac{1}{2}\Big(\prod_{\substack{v_0 \in S_0,\\ \text{inert}}} \epsilon(1, \eta_{v_0}, \psi_{v_0}) 2L(0, \eta_{v_0})\Big) \times \frac{L_{S_0}(1,\eta)L^S(\frac{1}{2}, \Pi \otimes \Omega^{-1})}{L^{S_0}(1, \pi, \mathrm{Ad})}.$$

Indeed, we know in advance that the left-hand side decomposes into a product $C\prod_{v_0 \in S_0} \widetilde{J}_{\sigma_{v_0}}(f_{v_0})$ for a suitable constant C. To evaluate C we choose the data as above. The result follows. Moreover, since the factors of C other than the one we are interested in are > 0, we conclude that $L^S(\frac{1}{2}, \Pi \otimes \Omega^{-1}) > 0$, as was claimed.

7. Concluding remarks

The distributions $\widetilde{J}_{\sigma_{v_0}}$ have the following property. We let f_j be an approximation of δ on G_{ϵ, v_0}.

If v_0 is split, we choose a Schwartz-Bruhat function ϕ on F_{v_0} such that $\widehat{\phi}(0) = 0$. We also choose also an isomorphism of $G_\epsilon(F_{v_0})$ with $\mathrm{Gl}(2, F_{v_0})$ and set:

$$f^j(g) = \int f_j\Big[\begin{pmatrix} 1 & -x \\ 0 & 1 \end{pmatrix} g\Big]\phi(x)\mathrm{d}x$$

We have then

$$\lim_{j \to +\infty} \widetilde{J}_{\sigma_{v_0}}(f^j) = \int \widehat{\phi}(x)\mathrm{d}^\times x.$$

Now the Tate functional equation gives:

$$\int \widehat{\phi}(x)\mathrm{d}^\times x = \epsilon(1, 1_{v_0}, \psi_{v_0})\, L(1-s, 1_{v_0}) \int \phi(x)|x|^{s-1}\mathrm{d}x\big|_{s=1}.$$

To have our local distribution independent of the choice of the additive character it is reasonable to set

$$\widetilde{J}_{\sigma_{v_0}}(f) = \epsilon(1, 1_{v_0}, \psi_{v_0}) J_{\sigma_{v_0}}(f).$$

This new distribution is still of positive type and now:

$$\lim_{j\to+\infty} J_{\sigma_{v_0}}(f^j) = L(s, 1_{v_0}) \int \phi(x)|x|^s \mathrm{d}x\Big|_{s=0}.$$

Note that the last expression is a multiple of $\int \phi(x)\log(|x|)\mathrm{d}x$. For instance in the non-Archimedean case, the expression is $-\int \phi(x)\nu(x)\mathrm{d}x$ where ν is the valuation. In the p-adic case there is an asymptotic theory of spherical characters (see [RR]). This is worked in [F4] for the case of the space $\mathrm{Gl}(2,F)/A(F)$. Let us assume for simplicity that $\Omega = 1$. The result is then that a spherical character like $J_{\sigma_{v_0}}$ can be represented, near the origin, as a unique linear combination

$$c_1 \int f\Big[\begin{pmatrix} 1 & u \\ v & 1 \end{pmatrix} a\Big] \mathrm{d}a\mathrm{d}u\mathrm{d}v - c_2 \int f\Big[\begin{pmatrix} 1 & u \\ v & 1 \end{pmatrix} a\Big] \mathrm{d}a\nu(uv)\mathrm{d}u\mathrm{d}v.$$

For the distribution $J_{\sigma_{v_0}}$ we have just defined, the above assertion amounts to saying that the constant c_2 is 1.

On the other hand, if v_0 is inert, then $\widetilde{J}_{\sigma_{v_0}}$ is a smooth function and we have simply: $\lim_{j\to+\infty} \widetilde{J}_{\sigma_{v_0}}(f_j) = 1$.

Finally, the reader can check that the left hand side of the formula in the theorem does not depend on the choice of the Haar measure on the group $G_\epsilon(F_{\mathbb{A}})$, and the formulas for a finite set of places S_0 and a larger set S_0' are compatible.

8. Appendix: Continuous spectrum over E

8.1. A diagonal truncation operator. — To prepare for the computation of the integral of the spectral kernel, we introduce another *diagonal truncation* operator for functions ϕ on $G(E_{\mathbb{A}})$ invariant under $Z(E_{\mathbb{A}})G(E)$:

$$\Lambda_d^T \phi(a) := \phi(a) - \sum_{\gamma\in A(E)\backslash \mathrm{Norm}(A)} \phi_N(\gamma a)\widehat{\tau}_P\big(H(\gamma a) - T\big).$$

Here $\widehat{\tau}_P(x) = 1$ if $x > 0$ and equal 0 otherwise. The sum is over the normalizer of $A(E)$ in $G(E)$ modulo $A(E)$, thus, has only two elements, that we can take to be e and w. The result of the truncation is a function on $A(E_{\mathbb{A}})$ invariant under $A(E)Z(E_{\mathbb{A}})$. It is useful to remember the following facts:

(i) $\widehat{\tau}_P(H(a) - T) = 1 \iff |a_1/a_2| > \mathrm{e}^T$;

(ii) for any function f invariant under $Z(E_{\mathbb{A}})$, $f(a) = f(wa^{-1}w)$;

(iii) $H(wa) = H(a^{-1}) = -H(a)$. Thus we can write more explicitly

$$\Lambda_d^T\phi(a) = \phi(a) - \phi_N(a)\widehat{\tau}_P(H(a) - T) \\ - \phi_N(a^{-1}w)\widehat{\tau}_P(-H(a) - T),$$

$$\begin{aligned} \Lambda_d^T\phi(a) &= \phi(a) && \text{if } -T < H(a) < T, \\ \Lambda_d^T\phi(a) &= \phi(a) - \phi_N(a) && \text{if } H(a) \geq T, \\ \Lambda_d^T\phi(a) &= \phi(a^{-1}w) - \phi_N(a^{-1}w) && \text{if } H(a) \leq -T. \end{aligned}$$

If ϕ is of slow increase (as well as all its derivatives) then the function $a \mapsto \Lambda_d^T\phi(\mathrm{diag}(a,1))$ is rapidly decreasing for $|a|$ large and $|a|$ small. Furthermore, if Λ_c^T denotes the compact truncation operator introduced in (18), then

$$\lim_{T\to+\infty} \int (\Lambda_d^T - \Lambda_c^T)\phi(a)\Omega^{-1}(a)\,\mathrm{d}^\times a = 0.$$

Consider again a smooth function of compact support f on $\mathrm{Gl}(2, E_{\mathbb{A}})$. We have a standard majorization [Ar, Lemma 4.3]:

$$\big|\rho_1(X_1)\rho_2(X_2)K_f(x,y)\big| \leq C(f)\|x\|. \tag{37}$$

Here X_1 and X_2 are elements of the enveloping algebra of G_∞ acting as left invariant differential operators on the first and second variables respectively. For given X_1 and X_2 the scalar $C(f)$ is bounded for f in a bounded set. Thus the function

$$g \longmapsto \int K_f(g,h)\omega\eta\big(\kappa(h)\big)\,\mathrm{d}h$$

is of slow increase as well as all its derivatives. Thus we arrive at the following lemma:

LEMMA 7. — *One has*

$$\Theta(f) = \lim_{T\to+\infty} \iint \Lambda_{1,d}^T K_f(a,h)\Omega^{-1}(a)\,\mathrm{d}a\,\omega\kappa(h)\,\mathrm{d}h.$$

In [JL] and [JLRo] we have introduced a *mixed truncation operator* Λ_m^T for the group $\mathrm{Gl}(2,F)$. It satisfies the following easy lemma:

LEMMA 8. — *Suppose that ϕ is a smooth function on $\mathrm{Gl}(2, E_{\mathbb{A}})$ invariant under the center, uniformly bounded with derivatives of slow increase. Then*

$$\int_{Z(F_{\mathbb{A}})\,\mathrm{Gl}(2,F)\backslash\,\mathrm{Gl}(2,F_{\mathbb{A}})} \phi(h)\,\mathrm{d}h = \lim_{T\to+\infty} \int \Lambda_m^T\phi(h)\,\mathrm{d}h.$$

Proof. — Since ϕ is integrable over the quotient we may write

$$\int \Lambda_m^T\phi(h)\,\mathrm{d}h = \int \phi(h)\,\mathrm{d}h - \int \sum_{\gamma\in P(F)\backslash\,\mathrm{Gl}(2,F)} \phi_N(\gamma h)\widehat{\tau}_P(\gamma h - T)\,\mathrm{d}h.$$

In this expression the second term can be computed as

$$\int_{P(F)Z(F_{\mathbb{A}})\backslash \mathrm{Gl}(2,F_{\mathbb{A}})} \phi_N(h)\widehat{\tau}_P(h-T)\mathrm{d}h.$$

Since ϕ_N is also uniformly bounded this is majorized by a constant multiple of

$$\int_{P(F)Z(F_{\mathbb{A}})\backslash \mathrm{Gl}(2,F_{\mathbb{A}})} \widehat{\tau}_P(h-T)\mathrm{d}h$$

itself a multiple of e^{-T}. The lemma follows. □

Using (22) we may define a mixed truncation operator Λ_m^T for the group $H(F)$ which also satisfies the above lemma. It is easy to conclude from (37) that

$$K_f(g,h) = \lim_{T_2\to+\infty} \Lambda_{2,m}^{T_2} K_f(g,h),$$

uniformly for g and h in compact sets. Since the computation of $\Lambda_d^{T_2}\phi(a)$ depends only on the values of ϕ on a compact set (depending on a) we get also

$$\Lambda_{1,d}^{T_1} K_f(a,h) = \lim_{T_2\to+\infty} \Lambda_{1,d}^{T_1}\Lambda_{2,m}^{T_2} K_f(a,h),$$

uniformly for a and h in compact sets. Our next task is the following lemma:

LEMMA 9. — *Given T_1:*

$$\iint \Lambda_{1,d}^{T_1} K_f(a,h)\Omega^{-1}(a)\mathrm{d}a\,\omega\eta\big(\kappa(h)\big)\mathrm{d}h$$
$$= \lim_{T_2\to+\infty} \iint \Lambda_{1,d}^{T_1}\Lambda_{2,m}^{T_2} K_f(a,h)\Omega^{-1}(a)\mathrm{d}a\,\omega\eta\big(\kappa(h)\big)\mathrm{d}h.$$

For the proof of the lemma we use again (37) to show that $|\Lambda_{1,d}^{T_1}K_f(a,g)|$ is bounded by $C|a|^{-N}$ for $H(a) \geq T_1$ and $H(a) \leq -T_1$ and by a constant for $-T_1 < H(a) < T_1$. The derivatives with respect to the second variable satisfy similar majorizations. It follows that the function defined by

$$m(g) = \int \Lambda_{1,d}^{T_1} K_f(a,g)\Omega^{-1}(a)\mathrm{d}a$$

satisfy the conditions of the previous lemma. Moreover:

$$\Lambda_m^{T_2} m(g) = \int \Lambda_{1,d}^{T_1}\Lambda_{2,m}^{T_2} K_f(a,g)\Omega^{-1}(a)\mathrm{d}a.$$

Thus the lemma follows from the previous one. In conclusion:

LEMMA 10. — *Set*

$$\Theta_{T_1,T_2}(f) = \iint \Lambda_{1,d}^{T_1}\Lambda_{2,m}^{T_2} K_f(a,h)\Omega^{-1}(a)\mathrm{d}a\,\omega\eta\big(\kappa(h)\big)\mathrm{d}h.$$

Then:

$$\Theta(f) = \lim_{T_1\to+\infty}\lim_{T_2\to+\infty} \Theta_{T_1,T_2}(f).$$

8.2. The continuous kernel over E. — We let f be a smooth function of compact support on $G(E_{\mathbb{A}})$ which, for now, is K-finite, where K is the standard maximal compact subgroup. Then K_f has a spectral expression

$$K_f = \sum_{\Pi} K_{f,\Pi},$$

where the sum is over all classes Π of cuspidal data for the parabolic subgroups P of G. We set

$$\Theta_{\Pi,T_1,T_2}(f) = \iint \Lambda_{1,d}^{T_1}\Lambda_{2,m}^{T_2} K_{f,\Pi}(a,h)\Omega^{-1}(a)\,\mathrm{d}a\,\omega\eta\big(\kappa(h)\big)\,\mathrm{d}h.$$

Standard estimates [Ar, Lemma 4.4] allow us to write

$$\Theta_{T_1,T_2}(f) = \sum_{\Pi} \Theta_{\Pi,T_1,T_2}(f).$$

If $P = G$, then Π is a cuspidal automorphic representation with trivial central character and we have simply $\Theta_{\Pi,T_1,T_2}(f) = \Theta_{\Pi}(f)$.

If P is the minimal parabolic subgroup, then Π is an equivalence class of pairs of (normalized) characters (Π_1,Π_2) of $E_{\mathbb{A}}^{\times}/E^{\times}$; the equivalence relation is $(\Pi_1,\Pi_2) \simeq (\Pi_2,\Pi_1)$. Here we need only consider pairs such that $\Pi_1\Pi_2 = 1$. The expression for $K_{f,\Pi}$ is [Ar, p. 935]

$$K_{f,\Pi}(x,y) = \frac{1}{2}\sum_{(\Pi_1,\Pi_2)\in\Pi}\int\sum_{\phi} E(x, I_{\lambda,\Pi_1,\Pi_2}(f)\phi;\lambda,\Pi_1,\Pi_2) \overline{E(y,\phi;\lambda,\Pi_1,\Pi_2)}\,\mathrm{d}\lambda + \cdots$$

where the dots represent residual terms. Thus the outer sum has either two terms for $(\Pi_1,\Pi_2),(\Pi_2,\Pi_1)$ (the terms being in fact equal) or just one term (Π_1,Π_1) (where $\Pi_1^2 = 1$).

It is a little simpler to sum over all pairs $\Pi = (\Pi_1,\Pi_2)$ with $\Pi_2 = \Pi_1^{-1}$ and for such a pair define $K_{f,\Pi}$ by

$$K_{f,\Pi}(x,y) = \frac{1}{2}\int\sum_{\phi} E(x, I_{\lambda,\Pi}(f)\phi;\lambda,\Pi)\overline{E(y,\phi;\lambda,\Pi)}\,\mathrm{d}\lambda + \frac{\delta(\Pi_1\Pi_2^{-1})}{\mathrm{vol}(G(E)\backslash G^1)}\int f(g)\Pi_1(\det g)\,\mathrm{d}g\,\Pi_1(xy^{-1}).$$

We are using standard notations. We use the same notation Π for the representation induced by the pair Π. In particular

$$E(g,\phi;\lambda,\Pi) = \sum_{P(E)\backslash G(E)} \phi(\gamma g)\mathrm{e}^{\langle\lambda+\rho,H(\gamma g)\rangle}.$$

The variable λ is identified to a complex number:

$$\mathrm{e}^{\langle\lambda+\rho,H(\gamma g)\rangle} = |a_1a_2^{-1}|_E^{\lambda+\frac{1}{2}}.$$

We often drop Π from the notation.

Then, apart from a residual term, $\Theta_{\Pi,T_1,T_2}(f)$ is given by:

$$\frac{1}{2}\sum_{\Pi}\int\sum_{\phi}\Big(\int\Lambda_d^{T_1}E(a,I_\lambda(f)\phi;\lambda)\Omega^{-1}(a)\mathrm{d}a\Big)$$
$$\times\overline{\Big(\int\Lambda_m^{T_2}E(h,\phi;\lambda)\omega\eta\big(\kappa(h)\big)\mathrm{d}h\Big)}\mathrm{d}\lambda.$$

From now on, we will assume that f is a convolution product of three K-finite functions: $f=f_1*f_2^*$, $f_1=f_1^1*f_1^2$. Later, we will assume that f_2 is itself a convolution product of sufficiently many K-finite functions. Then this is:

$$\frac{1}{2}\int\sum_{\phi,\phi'}\langle I_\lambda(f_1^2)\phi,\phi'\rangle\Big(\int\Lambda_d^{T_1}E(a,I_\lambda(f_1^1)\phi';\lambda)\Omega^{-1}(a)\mathrm{d}^\times a\Big)$$
$$\times\overline{\Big(\int\Lambda_m^{T_2}E(h,I_\lambda(f_2)\phi;\lambda)\omega\eta\big(\kappa(h)\big)\mathrm{d}h\Big)}\mathrm{d}\lambda.$$

Our goal is now to obtain an *absolutely convergent* expression for the spectral terms.

We will establish convergence in the case $\omega=\eta$. Then, by hypothesis, the restriction of Ω to $F_{\mathbb{A}}^\times$ is η. The case $\omega=1$ is similar. Also, to establish the convergence, it will be more convenient to replace the unitary similitude group by the group $\mathrm{Gl}(2,F)$. Then the above expression is (with $h\in Z(F_{\mathbb{A}})G(F)\backslash G(F_{\mathbb{A}})$):

$$\frac{1}{2}\sum_{\Pi}\int\mathrm{d}\lambda\sum_{\phi,\phi'}\langle I_\lambda(f_1^2)\phi,\phi'\rangle \tag{38}$$
$$\times\Big(\int\Lambda_d^{T_1}E(a,I_\lambda(f_1^1)\phi';\lambda)\Omega^{-1}(a)\mathrm{d}^\times a\Big) \tag{39}$$
$$\times\overline{\Big(\int\Lambda_m^{T_2}E(h,I_\lambda(f_2)\phi;\lambda)\mathrm{d}h\Big)}\mathrm{d}\lambda. \tag{40}$$

In this part of the paper, we do not pay much attention to the normalization of the Haar measures.

8.3. Period integral over *A*. — Let (Π_1,Π_2) be a pairs of idele class characters with $\Pi_2=\Pi_1^{-1}$. We set

$$\nu=\begin{pmatrix}1&1\\0&1\end{pmatrix}.$$

Then, in the domain of convergence we get:

$$E(g,\phi;\lambda)=\sum_{\alpha\in Z(E)\backslash A(E)}\phi(w\nu\alpha g)\mathrm{e}^{\langle\lambda+\rho,H(w\nu\alpha g)\rangle}$$
$$+\phi(g)\mathrm{e}^{\langle\lambda+\rho,H(g)\rangle}+\phi(wg)\mathrm{e}^{\langle\lambda+\rho,H(wg)\rangle}.$$

On the other hand:

$$E_N(g,\phi;\lambda)=\phi(g)\mathrm{e}^{\langle\lambda+\rho,H(g)\rangle}+M(w,\lambda)\phi(g)\mathrm{e}^{\langle-\lambda+\rho,H(g)\rangle}.$$

Then:

$$\begin{aligned}\Lambda_d^T E(a,\phi;\lambda) = \sum_\alpha \phi(w\nu\alpha a)\mathrm{e}^{\langle\lambda+\rho,H(w\nu\alpha a)\rangle} \\ + \phi(a)\mathrm{e}^{\langle\lambda+\rho,H(a)\rangle}\big(1-\widehat{\tau}_P(H(a)-T)\big) \\ + \phi(wa)\mathrm{e}^{\langle\lambda+\rho,\mathrm{H(wa)}\rangle}\big(1-\widehat{\tau}_\mathrm{P}(\mathrm{H(wa)}-\mathrm{T})\big) \\ - M(w,\lambda)\phi(a)\mathrm{e}^{\langle\lambda+\rho,H(a)\rangle}\big((\widehat{\tau}_P(H(a)-T)\big) \\ - M(w,\lambda)\phi(wa)\mathrm{e}^{\langle-\lambda+\rho,H(wa)\rangle}\big(\widehat{\tau}_P(H(wa)-T)\big).\end{aligned}$$

A simple computation gives

$$(41)\quad \int \Lambda_d^T E(a,\phi,\lambda,\Pi)\Omega^{-1}(a)\mathrm{d}^\times a = \mu(\phi;\lambda,\Pi_1,\Pi_2) \\ + \big(\delta(\Pi_1\Omega^{-1})\phi(\mathrm{e}) + \delta(\Pi_2\Omega^{-1})\phi(w)\big)\frac{\mathrm{e}^{T(\frac12+\lambda)}}{\frac12+\lambda} \\ + \big(\delta(\Pi_2\Omega^{-1})M(w,\lambda)\phi(\mathrm{e}) + \delta(\Pi_1\Omega^{-1})M(w,\lambda)\phi(w)\big)\frac{\mathrm{e}^{T(\frac12-\lambda)}}{\frac12-\lambda}$$

where we have set $a = \mathrm{diag}(a_1, 1)$ and

$$\mu(\phi;\lambda,\Pi_1,\Pi_2) = \int_{E_\mathbb{A}^\times} \phi(w\nu a)\Omega^{-1}(a_1)\mathrm{e}^{\langle\lambda+\rho,H(w\nu a)\rangle}\mathrm{d}^\times a_1.$$

This formula gives the analytic continuation of μ. However, we will need to have more information on $\mu(I_{\lambda,\Pi}(f)\phi;\lambda,\Pi)$ where f is a smooth function of compact support on $G(E_\mathbb{A})$. We write $f = f_S f^S$ where f_S is a smooth function of compact support on $G(E_S)$ and f^S is the characteristic function of $K^S := \prod_{v\notin S} K_v$ and $K_v = \mathrm{Gl}(2,\mathcal{O}_v)$. We may then assume that ϕ is invariant under K^S. We have, for $g \in G_S$,

$$I_{\lambda,\Pi}(f)\phi(g)\mathrm{e}^{\langle\lambda+\rho,H(g)\rangle} = \int_{G_S} \phi(gx)\mathrm{e}^{\langle\lambda+\rho,H(gx)\rangle} f_0(x^{-1})\mathrm{d}x$$

where we have set $f_0 = \check{f}$. This can be written as

$$\int_{G_S} \phi(x)\mathrm{e}^{\langle\lambda+\rho,H(x)\rangle} f_0(x^{-1}g)\mathrm{d}x$$

or

$$\int f_1\Big[n\begin{pmatrix} a_1 & 0 \\ 0 & a_2\end{pmatrix} g\Big]\mathrm{d}n\, |a_2|^{\lambda+\frac12}\Pi_2(a_2)^{-1}|a_1|^{-\lambda-\frac12}\Pi_1(a_1)^{-1}\mathrm{d}^\times a_1,$$

where we have set

$$f_1(g) = \int_K \phi(k) f_0(k^{-1}g)\mathrm{d}k.$$

At this point we introduce new functions on G_S:

$$f_1(g;\lambda,\Pi_1) := f_1(g)\Pi_1(\det g)^{-1}|\det g|^{-\lambda-\frac12},$$

$$\phi_S[g;\lambda,\Pi_1] := \int_{E_S^\times}\int_{N_S} f_1\Big[n\begin{pmatrix} a_1 & 0\\ 0 & 1\end{pmatrix} g;\lambda,\Pi_1\Big]\,\mathrm{d}n\,\mathrm{d}^\times a_1.$$

We let P_1 be the group of matrices with second row $(0,1)$. Clearly ϕ_S is invariant on the left under $P_1(E_S)$. Thus there is a Schwartz-Bruhat function $\Phi_S[(x,y);\lambda,\Pi_1]$ on $E_S\times E_S$ such that, for $g\in G_S$,

$$\Phi_S[(0,1)g;\lambda,\Pi_1] = \phi_S[g;\lambda,\Pi_1].$$

The support of Φ_S is contained in a compact set of $E_S^2-\{(0,0)\}$ which is independent of λ. We have then for $g\in G_S$:

$$I_{\lambda,\Pi}(f)\phi(g)\mathrm{e}^{\langle\lambda+\rho,H(g)\rangle} = \int_{E_S^\times}\Phi_S[(0,t)g;\lambda,\Pi_1]|t|^{2\lambda+1}\Pi_1\Pi_2^{-1}(t)\,\mathrm{d}^\times t\,\Pi_1(g)\,|\det g|^{\lambda+\frac12}.$$

To obtain a formula valid for all g we introduce

$$\Phi\big[(x,y);\lambda,\Pi_1\big] := \Phi_S\big[(x_S,y_S);\lambda,\Pi_1\big]\prod_{v\notin S}\Phi_v(x_v,y_v),$$

where Φ_v, $v\notin S$, is the characteristic function of $\mathcal{O}_v^2$. Then, for $g\in G(E_{\mathbb{A}})$,

$$\text{(42)}\quad I_{\lambda,\Pi}(f)\phi(g)\mathrm{e}^{\langle\lambda+\rho,H(g)\rangle} = \frac{\Pi_1(g)\,|\det g|^{\lambda+\frac12}}{L^S(2\lambda+1,\Pi_1\Pi_2^{-1})}\int_{E_{\mathbb{A}}^\times}\Phi\big[(0,t)g;\lambda,\Pi_1\big]\cdot|t|^{2\lambda+1}\Pi_1\Pi_2^{-1}(t)\,\mathrm{d}^\times t.$$

After a change of variables we find:

$$\text{(43)}\quad \mu(I_{\lambda,\Pi}(f)\phi;\lambda,\Pi) = \frac{1}{L^S(2\lambda+1,\Pi_1\Pi_2^{-1})} \times\iint\Phi\big[(a_1,a_2);\lambda,\Pi_1\big]\Omega^{-1}\Pi_1(a_1)|a_1|^{\lambda+\frac12} \times\mathrm{d}^\times a_1\,\Omega\Pi_2^{-1}(a_2)|a_2|^{\lambda+\frac12}\,\mathrm{d}^\times a_2.$$

The computations are justified for $\operatorname{Re}\lambda\gg 0$. For a given Π, a given ϕ and a given f, this expression gives the analytic continuation of μ as a meromorphic function of λ, which has no singularity on the line $\operatorname{Re}\lambda=0$. It has a zero at $\lambda=0$ if $\Pi_1=\Pi_2$, that is, if $\Pi_1^2=1$. Moreover, this function of λ is at most of polynomial increase on the line $\operatorname{Re}\lambda=0$. All its derivatives have the same property.

However, for our purposes, we will need to have uniform estimates. We proceed as follows. First we take f in a fixed bounded set $\mathcal{B}$ of the space of smooth functions of compact support. This means that $f=f_Sf^S$ as before, with S fixed; the function f_S has support in a fixed set Ξ, open and relatively compact; we may as well assume $K_S\Xi=\Xi K_S=\Xi$; the function is invariant on the right and the left under a compact open normal subgroup K' of $\prod_{v\in S,\text{finite}}K_v$ (and thus under the subgroup $K'' = K'K^S$ of K^∞). For each element X of the enveloping algebra of G_∞ there is a constant C_X such that

$$\sup\bigl|\rho(X)f\bigr| \leq C_X.$$

Consider the function:

$$f_1(g) = \int \phi(k) f_0(k^{-1}g)\phi(k)\mathrm{d}k.$$

The integral does not change if we replace ϕ by ϕ_1 defined by

$$\phi_1(k) = \mathrm{vol}(K'')^{-1} \int \phi(kk'')\mathrm{d}k''.$$

Then ϕ_1 is invariant on the right and therefore on the left under K''. Thus the same is true of f_1. Moreover $f_1 = f_{1,S} f^S$. For any X,

$$\sup\bigl|\rho(X)f_1\bigr| \leq \|\phi\| C_X.$$

Finally, the support of $f_{1,S}$ is still contained in Ξ. If $I_{\lambda,\Pi}(f) \neq 0$ for a $f \in \mathcal{B}$, then Π_1 is unramified outside S and, for each finite $v \in S$, $\Pi_{1,v|\mathcal{O}_v^\times}$ belongs to a finite set. We now consider the function $f_1(g;\lambda,\Pi_1)$. It has a support contained in Ξ. It is invariant under a compact subgroup K_0' of $\prod_{v\in S,\text{finite}} K_v$ (depending only on K'). Suppose λ is in a strip $a \leq \operatorname{Re}\lambda \leq b$. Then, for any X, there is a polynomial $P_X(\Pi_1,\lambda)$ such that

$$\sup\bigl|\rho(X)f_1(\bullet;\lambda,\Pi_1)\bigr| \leq \|\phi\| P_X(\Pi_1,\lambda).$$

By a *polynomial* we means a function $P(\Pi_1,\lambda)$ of the following form. We may identify the enveloping algebra of $E_\infty^\times$ to the space of distributions with support $\{e\}$. Thus if f is a smooth function on $E_\infty^\times$ or $E_\mathbb{A}^\times$ and X in the enveloping algebra we can define $\langle X, f\rangle$. Then a polynomial is a function of the form $P(\Pi_1,\lambda) = \langle X, f\rangle$ where $f(a) = \Pi_1(a)|a|^\lambda$ and X is in the enveloping algebra. Note that such a function depends only on λ and the infinite components of Π_1. In fact, if we write, for v real,

$$\Pi_{1,v}(x) = |x|_v^{iu_v}\Bigl(\frac{x}{|x|}\Bigr)^{\epsilon_v}, \quad \epsilon_v = 0,1, \quad d_v = 1,$$

and, for v complex,

$$\Pi_{1,v}(z) = (z\bar{z})^{iu_v}\frac{z^{n_v}}{(z\bar{z})^{n_v/2}}, \quad n_v \in \mathbb{Z}, \quad d_v = 2,$$

with $\sum_{v\in\infty} d_v u_v = 0$, then a polynomial $P(\Pi_1,\lambda)$ is just a polynomial in the variables (n_v, u_v) and λ. To allow for the case where λ is complex, we will say that any linear combination of products $P(\Pi_1,\lambda)\,\overline{Q(\Pi_1,\lambda)}$, where P and Q are polynomials, is also a polynomial. For instance, for any integer N, $(1+\lambda\,\bar{\lambda})^N$ is a polynomial in that sense. It will be understood that all polynomials considered are positive.

This being so, all derivatives of $f_1(\bullet;\lambda,\Pi_1)$ have the same properties (with different polynomials). A similar assertion is valid for the function $\phi_S(\bullet;\lambda,\Pi_1)$ except that its support is contained in a fixed set, compact modulo $P_1(E_S)$.

The Schwartz-Bruhat function $\Phi_S(\bullet;\lambda,\Pi_1)$ is compactly supported: its support is contained in a fixed open set of E_S^2-0, relatively compact in E_S^2-0. It is invariant on the right under K_0'. Thus its support is a compact set of E_S^2 and it is invariant under translation by a compact open subgroup of $E_{S_{\text{finite}}}$. Take λ in a strip. Then $\Phi_S[\bullet;\lambda,\Pi_1]$ is bounded by a bound for $\phi_S(\bullet;\lambda,\Pi_1)$. Since G_∞ operates on E_∞^2 every element X of the enveloping algebra defines a differential operator X_a on E_∞^2 (with non constant coefficients). We have

$$(X_a\Phi_S)\big[(0,1)g;\lambda,\Pi_1\big]=\big(\rho(X)\phi_S\big)(g;\lambda,\Pi_1).$$

Consider a differential operator ξ on E_∞^2 with constant coefficients. On the complement of $(0,0)$ we can write $\xi=\sum_i c_iX_{i,a}$ where the $X_\imath$ are elements of the enveloping algebra and the functions c_i are smooth functions on $E_\infty^2-(0,0)$. Using the majorization for the functions $\rho(X_i)\phi_S$ we conclude that

$$\big|\xi\Phi_S[(x,y);\lambda,\Pi_1]\big|\le\|\phi\|P_\xi(\Pi_1,\lambda).$$

Thus the functions $\Phi_S[(x,y);\lambda,\Pi_1]$ remain in a bounded set of $\mathcal{C}_c^\infty(E_S^2)$ and *a fortiori* of the Schwartz-Bruhat space $\mathcal{S}(E_S^2)$. Note that the functions

$$\xi\Phi_S\big[(x,y);\lambda,\Pi_1\big]$$

are continuous in (x,y,λ) and holomorphic in λ. In (43) the analytic continuation of the Tate integral is obtained by using the Poisson summation formula. Thus in a strip, say $-\frac14\le\operatorname{Re}\lambda\le\frac14$, we have then a uniform estimate:

$$\Big|\iint\Phi\big[(a_1,a_2);\lambda,\Pi_1\big]\Omega^{-1}\Pi_1(a_1)|a_1|^{\lambda+\frac12}\,\mathrm{d}^\times a_1\Omega\Pi_2^{-1}(a_1)|a_1|^{\lambda+\frac12}\,\mathrm{d}^\times a_1\Big| \le\|\phi\|P(\Pi_1,\lambda).$$

We have similar estimates for the derivatives by Cauchy integral formula.

We stress that we need only consider characters Π_1 such that $I_{\lambda,\Pi}(f)\ne0$ for $f\in\mathcal{B}$. Thus for $v\notin S$ the character $\Pi_{1,v}$ is unramified and for v finite in S the restriction of $\Pi_{1,v}$ to $\mathcal{O}_v^\times$ belongs to a finite set. If furthermore $\Pi_1^2=1$ this forces Π_1 to belong to a finite set. There is then a polynomial $P(\Pi_1,\lambda)$ such that, on the line $\operatorname{Re}\lambda=0$, for characters Π_1 of the above type,

$$\Big|\frac{1}{L^S(2\lambda+1,\Pi_1^2)}\Big|$$

is bounded by $P(\Pi_1,\lambda)$. Combining with the previous inequality, we get:

LEMMA 11. — *Suppose f is in a bounded set $\mathcal{B}$. Then there is a polynomial $P(\Pi_1,\lambda)$ such that, for* $\operatorname{Re}\lambda=0$, *and all characters* Π_1:

$$\big|\mu(I_{\lambda,(\Pi_1,\Pi_1^{-1})}(f)\phi;\lambda,\,\Pi_1,\Pi_1^{-1})\big|\le P(\Pi_1,\lambda)\|\phi\|.$$

On the other hand, we can obtain similar estimates for the functions

$$L^S(2\lambda+1,\Pi_1^2)$$

in a strip (with modification for a pole at $\lambda=0$ if $\Pi_1^2=1$), hence also for their derivatives. Thus we have also similar estimates for the derivative of the expression in the proposition on the line $\operatorname{Re}\lambda=0$. We can use the notations of generalized vectors (dual to the smooth vectors). There is a generalized vector $\mu_{\lambda,\Pi}$ such that

$$\mu\big(I_{\lambda,\Pi}(f)\phi;\lambda,\Pi\big)=\big\langle\phi,I_{\lambda,\Pi}(f^*)\mu_{\lambda,\Pi}\big\rangle$$

and our estimate amounts to

$$\big\|I_{\lambda,\Pi}(f^*)\mu_{\lambda,\Pi}\big\|\leq P(\Pi_1,\lambda).$$

Remark. — We define the Whittaker linear form $\mathcal{W}$ on the space of $I_{\lambda,\Pi}$ by analytic continuation of the integral

$$\mathcal{W}(\phi;\lambda,\Pi)=\int_{E_{\mathbb{A}}}\phi(wn)e^{\langle\lambda+\rho,H(wn)\rangle}\psi_E(-x)\mathrm{d}x,\quad n=\begin{pmatrix}1&x\\0&1\end{pmatrix}.$$

After a change of variables, we easily find

$$(44)\quad \mathcal{W}\Big[I_{\lambda,\Pi}\begin{pmatrix}a&0\\0&1\end{pmatrix}I_{\lambda,\Pi}(f)\phi\Big]$$
$$=\frac{\Pi_1(a)\,|a|^{\lambda+\frac12}}{L^S(2\lambda+1,\Pi_1\Pi_2^{-1})}\int_{E_{\mathbb{A}}^\times}\widetilde{\Phi}\big[(at,t^{-1};\lambda,\Pi_1\big]\cdot|t|^{2\lambda}\Pi_1\Pi_2^{-1}(t)\mathrm{d}^\times t$$

where $\widetilde{\Phi}$ denotes the Fourier transform with respect to the second variable. This integral is absolutely convergent and gives the analytic continuation of $\mathcal{W}$.

8.4. Intertwining period and intertwining operator. — We need to consider only pairs $\Pi=(\Pi_1,\Pi_2)$ (with $\Pi_2=\Pi_1^{-1}$) such that the integral of a truncated Eisenstein series over $\mathrm{Gl}(2,F_{\mathbb{A}})$ is non-zero. Then (see [JL], [HLRo]) either Π_1 (and Π_2) have a trivial restriction to $F_{\mathbb{A}}^\times$ or $\Pi_2^\tau=\Pi_1^{-1}$ (or, what amounts to the same, $\Pi_1=\Pi_1^\tau$). Of course the two conditions can be both satisfied.

In this section, we assume that $\Pi_1^\tau=\Pi_1$. Then we can define an *intertwining period* as follows (see [JL] and [JLRo]):

$$\mathcal{P}(\phi;\lambda,\Pi):=\int_{T(F_{\mathbb{A}})\backslash G(F_{\mathbb{A}})}\phi(\eta h)e^{\langle\lambda+\rho,H(\eta h)\rangle}\mathrm{d}h.$$

Here $\eta\in G(E)$ is an element such that $\eta\overline{\eta}^{-1}=w$ and T is the torus such that

$$T(F)=\eta^{-1}A(E)\eta\cap\mathrm{Gl}(2,F).$$

The integral converges for $\operatorname{Re}\lambda \gg 0$ and can be analytically continued. Recall the formula (42). We let $\Phi_\eta[m;\lambda,\Pi_1]$ be the Schwartz-Bruhat function on $M(2\times 2, F_{\mathbb{A}})$ defined by

$$\Phi_\eta[m;\lambda,\Pi_1] := \Phi[(0,1)\eta m;\lambda,\Pi_1].$$

Then we get:

$$\mathcal{P}(I_{\lambda,\Pi}(f)\phi;\lambda,\Pi) = \frac{1}{L^S(2\lambda+1,\Pi_1^2)} \times \int_{\mathrm{Gl}(2,F_{\mathbb{A}})} \Phi_\eta[m;\lambda,\Pi_1]\,\Pi_1(\det m)|\det m|_F^{2\lambda+1}\,\mathrm{d}m.$$

The integral is a Tate integral for the group $\mathrm{Gl}(2,F)$. Using the Iwasawa decomposition we see that the Tate integral is an entire multiple of

$$L\big(2\lambda+1,\Pi_{1|F_{\mathbb{A}}^\times}\big)L\big(2\lambda,\Pi_{1|F_{\mathbb{A}}^\times}\big).$$

In particular, it has no pole on the line $\operatorname{Re}\lambda = 0$ except at $\lambda = 0$ when $\Pi_{1|F_{\mathbb{A}}^\times} = 1$. Because we assume that $\Pi_1^\tau = \Pi_2^{-1} = \Pi_1$ this implies that $\Pi_1^2 = 1$. Then the above ratio is an entire multiple of

$$\frac{L_{S_0}(2\lambda+1,1_F)L(2\lambda,1_F)}{L^{S_0}(2\lambda+1,\eta)}.$$

These observations combined with standard estimates for L-functions can be used to prove the following lemma.

LEMMA 12. — *Let $\mathcal{B}$ be a bounded set. Then there is a polynomial $P(\Pi_1,\lambda)$ such that, for all Π_1 with $\Pi_1^\tau = \Pi_1$ and $\Pi_{1|F_{\mathbb{A}}^\times} \neq 1$, all λ with $\operatorname{Re}\lambda = 0$, and all $f\in\mathcal{B}$:*

$$\big|\mathcal{P}(I_{\lambda,\Pi}(f)\phi;\lambda,\Pi)\big| \le P(\Pi_1,\lambda)\|\phi\|.$$

This can expressed again in terms of generalized vectors. There is a generalized vector $\mathcal{P}_{\lambda,\Pi}$ such that

$$\mathcal{P}\big(I_{\lambda,\Pi}(f)\phi;\lambda,\Pi\big) = \big\langle\phi, I_{\lambda,\Pi}(f^*)\mathcal{P}_{\lambda,\Pi}\big\rangle.$$

Then:

$$\big\|I_{\lambda,\Pi}(f^*)\mathcal{P}_{\lambda,\Pi}\big\| \le P(\Pi_1,\lambda).$$

Combining with the results of the previous subsection, we arrive at the following result.

PROPOSITION 7. — *Let $\mathcal{B}$ be a bounded set of the space of smooth functions of compact support. Then there is a polynomial $P(\Pi_1,\lambda) > 0$ such that, for any function f of the form $f = f_1 * f_2 * f_3$ with $f_1, f_3\in\mathcal{B}$, any Π_1 with $\Pi_1^\tau = \Pi_1$, any λ with $\operatorname{Re}\lambda = 0$:*

$$\big|\langle I_{\lambda,\Pi}(f)\mathcal{P}_{\lambda,\Pi},\mu_{\lambda,\Pi}\rangle\big| \le P(\Pi_1,\lambda)\big\|I_{\lambda,\Pi}(f_2)\big\|.$$

For the proof we remark that the expression in the proposition has the form

$$\langle I_{\lambda,\Pi}(f_2)I_{\lambda,\Pi}(f_3)\mathcal{P}_{\lambda,\Pi}, I_{\lambda,\Pi}(f_1^*)\mu_{\lambda,\Pi}\rangle$$

so that our assertion follows from the previous lemmas, for the characters Π_1 whose restriction to $F_{\mathbb{A}}^\times$ is non-trivial. On the other hand, if this restriction is trivial then $\Pi_1^2 = 1$. We need only consider those Π_1 such that $I_{\lambda,\Pi_1}(f_1) \neq 0$ for $f_1 \in \mathcal{B}$. There are only finitely many such Π_1. So we do not need to have estimates uniform in this case. We have written

$$\mu_{\lambda,\Pi}(I_{\lambda,\Pi}(f)\phi) = L^S(2\lambda+1, 1_E)^{-1}\widetilde{\mu}_{\lambda,\Pi}(I_{\lambda,\Pi}(f)\phi)$$

where $\widetilde{\mu}(\bullet)$ has no singularity on $\operatorname{Re}\lambda = 0$. Since $\mathcal{P}_{\Pi_1,\lambda}(I_{\lambda,\Pi}(f)\phi))$ has a pole at $\lambda = 0$ in this case, we need to modify the previous lemma and replace $\overline{\mathcal{P}_{\Pi_1,\lambda}(I_{\lambda,\Pi}(f)\phi)}$ by its product with $L^S(2\lambda+1, 1_E)^{-1}$ which has a zero at $\lambda = 0$. The product has then no singularity at $\lambda = 0$. It is then easy to obtain the required estimates in this case.

Using similar arguments, we can obtain estimates for the intertwining operator, more precisely, for expressions of the form

$$\langle M(w,\lambda)I_{\lambda,\Pi}(f)(\phi), \phi'\rangle.$$

8.5. Continuous contribution. — We study the terms corresponding to pairs Π with $\Pi_2^\tau = \Pi_1^{-1}$, that is, $\Pi_1^\tau = \Pi_1$. Each such term will ultimately contribute an integral. We first consider the terms with $\Pi_1 \neq \Omega$, (a condition equivalent to $\Pi_2 \neq \Omega$ since $\Omega^\tau = \Omega^{-1}$) and $\Pi_{1|F_{\mathbb{A}}^\times} \neq 1$. Then in fact

$$\int \Lambda_d^{T_1}E(a,\phi,\lambda)\Omega^{-1}(a)\mathrm{d}a, \qquad \int \Lambda_m^{T_2}E(h_1,\phi,\lambda)\mathrm{d}h_1$$

are independent of T_1 and T_2, being equal respectively, to $\mu(\phi;\lambda,\Pi)$ and $\mathcal{P}(\phi;\lambda,\Pi)$. Thus:

$$\Theta_{\Pi,T_1,T_2}(f) = \frac{1}{2}\int \sum_{\phi,\phi'} \langle I_\lambda(f_2^2)\phi,\phi'\rangle\, \mu(I_\lambda(f_1^1)\phi',\lambda)\,\overline{\mathcal{P}(I_\lambda(f_2)\phi,\lambda)}\mathrm{d}\lambda.$$

Note that if $\Pi_1 = \Pi_2$ then the expression for $K_{f,\Pi}(x,y)$ contains a residual term: $\int \Pi_1(\det g)f(g)\mathrm{d}g\,\Pi_1(\det x)\overline{\Pi_1(\det y)}$. However, since $\Omega \neq \Pi_1$, the integral of the truncation of this term over A against Ω^{-1} contains as a factor the integral of $\Pi_1\Omega^{-1}$ over the quotient of the ideles of absolute value 1 and is thus 0. With the notation of generalized vectors, the above integral becomes:

$$\int \langle I_\lambda(f)\mathcal{P}_{\lambda,\Pi}, \mu_{\lambda,\Pi}\rangle\mathrm{d}\lambda.$$

For $f = f_1^1 * f_1^2 * f_2^*$:

$$\langle I_\lambda(f)P_{\lambda,\Pi}, \mu_{\lambda,\Pi}\rangle = \langle I_\lambda(f_1^2)I_\lambda(f_2^*)P_{\lambda,\Pi}, I_\lambda(f_1^{1*})\mu_{\lambda,\Pi}\rangle.$$

By Proposition 7, there is then a polynomial $P(\Pi_1,\lambda)$ such that

$$\big|\langle I_\lambda(f)P_{\lambda,\Pi},\mu_{\lambda,\Pi}\rangle\big|\le P(\Pi_1,\lambda)\big\|I_{\lambda,\Pi}(f_1^2)\big\|.$$

In fact, this is true for f_1^1, f_2^* smooth in a bounded set and any smooth function f_1^2.

We will make repeated use of the following lemma:

LEMMA 13. — *Suppose that $\mathcal{B}$ is a bounded set of the space of smooth functions of compact support and $P(\Pi_1,\lambda)$ a polynomial. Then there is a constant $C>0$ such that, for any $f\in\mathcal{B}$,*

$$\sum_{\Pi_1}\int P(\Pi_1,\lambda)\big\|I_{\lambda,\Pi_1,\Pi_1^{-1}}(f)\big\|\mathrm{d}\lambda<C,$$

the sum over all *(normalized) idele class characters of E.*

Proof of the Lemma. — The operator $I_{\lambda,\Pi_1,\Pi_1^{-1}}(f)$ is represented by the following kernel function on $K\times K$:

$$H_{f,\lambda,\Pi}(k_1,k_2):=\int f\Big[k_1^{-1}\begin{pmatrix}a&0\\0&1\end{pmatrix}\begin{pmatrix}b&0\\0&b\end{pmatrix}\begin{pmatrix}1&x\\0&1\end{pmatrix}k_2\Big]\Pi_1(a)|a|_E^{\lambda+\frac12}\mathrm{d}^\times a\,\mathrm{d}^\times b\,\mathrm{d}x.$$

Thus its operator norm is bounded by the supremum over $K\times K$ of

$$\Big|\int\phi(a:k_1:k_2)\Pi_1(a)|a|_E^\lambda\mathrm{d}^\times a\Big|$$

where

$$\phi(a:k_1:k_2):=\int f\Big[k_1^{-1}\begin{pmatrix}a&0\\0&1\end{pmatrix}\begin{pmatrix}b&0\\0&b\end{pmatrix}\begin{pmatrix}1&x\\0&1\end{pmatrix}k_2\Big]|a|_E^{\frac12}\mathrm{d}^\times b\,\mathrm{d}x.$$

For $f\in\mathcal{B}$, $k_i\in K$, the function $a\mapsto\phi(a:k_1:k_2)$ remains in a bounded set $\mathcal{B}_0$ of the space of smooth functions of compact support on $E_{\mathbb{A}}^\times$. Thus it will suffice to show there is C such that, for $\phi\in\mathcal{B}_0$,

$$\sum_{\Pi_1}\int P(\Pi_1,\lambda)\Big|\int\phi(a)\Pi_1(a)|a|_E^\lambda\mathrm{d}^\times a\Big|\mathrm{d}\lambda<C.$$

This is clear if $P=1$. In general

$$\begin{aligned}P(\Pi_1,\lambda)\Big|\int\phi(a)\Pi_1(a)|a|_E^\lambda\mathrm{d}^\times a\Big|&=\Big|P(\Pi_1,\lambda)\int\phi(a)\Pi_1(a)|a|_E^\lambda\mathrm{d}^\times a\Big|\\&=\Big|\int\phi_1(a)\Pi_1(a)|a|_E^\lambda\mathrm{d}^\times a\Big|\end{aligned}$$

where $\phi_1=\rho(X)\phi$ for a suitable element X of the enveloping algebra of $E_\infty^\times$. Since ϕ_1 remains in another bounded set, our assertion follows. □

From our observations on the independence of the integrals on the truncation parameters, we see that we can sum over all Π_1 of the above type and get

$$\lim_{T_1\to+\infty}\lim_{T_2\to+\infty}\sum_{\Pi}\Theta_{\Pi,T_1,T_2}(f)=\sum_{\Pi_1}\frac{1}{2}\int\langle I_{\lambda,\Pi}(f)\mathcal{P}_{\lambda,\Pi},\mu_{\lambda,\Pi}\rangle\mathrm{d}\lambda,$$

with

$$\sum_{\Pi_1}\int\big|\langle I_{\lambda,\Pi}(f)\mathcal{P}_{\lambda,\Pi},\mu_{\lambda,\Pi}\rangle\big|\mathrm{d}\lambda<+\infty.$$

Moreover, if $\mathcal{B}$ is a bounded set, there is $C>0$ such that, for f of the form $f=f_1*f_2*f_3$, with each f_i in $\mathcal{B}$, then

$$\sum_{\Pi_1}\int\big|\langle I_{\lambda,\Pi}(f)\mathcal{P}_{\lambda,\Pi},\mu_{\lambda,\Pi}\rangle\big|\mathrm{d}\lambda<C.$$

Next, we discuss the contribution of the term $\Pi_1=\Omega$, if any. Since we are assuming that $\Omega^\tau=\Omega^{-1}$ in any case and for now we are assuming that $\Pi_1^\tau=\Pi_2^{-1}$ this implies that $\Pi_2=\Pi_1=\Omega$ so that Ω is a quadratic character satisfying $\Omega_{|F_{\mathbb{A}}^\times}=\eta$. Moreover $\Pi_{1|F_{\mathbb{A}}^\times}$ is then non-trivial. The expression for K_Π contains again a residual term; however its truncated integral over $\mathrm{Gl}(2,F_{\mathbb{A}})$ is zero. Indeed, it contains as a factor:

$$\int_{Z(F_{\mathbb{A}})G(F)\backslash G(F_{\mathbb{A}})}\eta(\det h)\mathrm{d}h-\int_{Z(F_{\mathbb{A}})A(F)N(F_{\mathbb{A}})\backslash G(F_{\mathbb{A}})}\eta(\det h)\widehat{\tau}_P\big(H(h)-T\big)\mathrm{d}h$$

which is zero. Then $\int\Lambda_m^{T_2}E(h_1,\phi,\lambda)\mathrm{d}h_1$ is again independent of T_2 but

$$\int\Lambda_d^{T_1}E(a,\phi,\lambda)\Omega^{-1}(a)\mathrm{d}a$$

does depend on T_1. We have then for $f=f_1^1*f_1^2*f_2^*$, all K-finite functions:

$$\begin{aligned}\Theta_{\Pi,T_1,T_2}(f)&=\frac{1}{2}\int\langle I_\lambda(f)\mathcal{P}_{\lambda,\Pi},\mu_{\lambda,\Pi}\rangle\mathrm{d}\lambda\\&+\tfrac{1}{2}\int\sum_{\phi}\big(I_\lambda(f_1^1*f_1^2)\phi(\mathrm{e})+I_\lambda(f_1^1*f_1^2)\phi(w)\big)\\&\qquad\times\frac{\mathrm{e}^{T_1(\frac12+\lambda)}}{\frac12+\lambda}\overline{\mathcal{P}(I_\lambda(f_2)\phi,\lambda)}\mathrm{d}\lambda\\&+\tfrac{1}{2}\int\sum_{\phi}\big((I_{-\lambda}(f_1^1*f_1^2)M(w,\lambda)\phi(\mathrm{e})+I_{-\lambda}(f_1^1*f_1^2)M(w,\lambda)\phi(w)\big)\\&\qquad\times\frac{\mathrm{e}^{T_1(\frac12-\lambda)}}{\frac12-\lambda}\overline{\mathcal{P}(I_{-\lambda}(f_2)M(w,\lambda)\phi,-\lambda)}\mathrm{d}\lambda\end{aligned}$$

where for the last expression we have used the functional equation of the period $\mathcal{P}$ (inherited from the functional equation of the Eisenstein series, see [JLRo]) to insert an intertwining operator. In the last expression, the sum over

ϕ does not depend on the choice of the basis. Thus we can ignore the intertwining operator. In the second expression, we can also change λ to $-\lambda$; we see that the second expression is equal to the third. That is:

$$\Theta_{\Pi,T_1,T_2}(f) = \frac{1}{2}\int \langle I_\lambda(f)P_{\lambda,\Pi}, \mu_{\lambda,\Pi}\rangle \mathrm{d}\lambda$$
$$+\int \sum_\phi \Big\{\big(I_{-\lambda}(f_1^1 * f_1^2)\phi(\mathrm{e}) + I_{-\lambda}(f_1^1 * f_1^2)\phi(w)\big)\frac{\mathrm{e}^{T_1(\frac{1}{2}-\lambda)}}{\frac{1}{2}-\lambda}\overline{\mathcal{P}(I_{\bar\lambda}(f_2)\phi,\bar\lambda)}\Big\}\mathrm{d}\lambda.$$

We will move the integral to the line $\operatorname{Re}\lambda = 1$ rather than to the line $\operatorname{Re}\lambda = \frac{1}{2}$ as we did in [J2] (a similar contour integration for Gl(2) is done in [F3]). The term $\overline{\mathcal{P}(*,\bar\lambda)}$ is the product of a Tate integral which is entire, hence bounded in vertical strips and the factor $(L^S(2\lambda+1, 1_E))^{-1}$. Thus this moving of the contour of integration is legitimate. In doing so we pick up a residue at $\lambda = \frac{1}{2}$. The integral on the line $\operatorname{Re}\lambda = 1$ tends to 0 as T_1 tends to infinity. So we are left with

$$\lim_{T_1\to+\infty}\lim_{T_2\to+\infty}\Theta_{\Pi,T_1,T_2}(f) = \frac{1}{2}\int \langle I_\lambda(f)P_{\lambda,\Pi}, \mu_{\lambda,\Pi}\rangle \mathrm{d}\lambda$$
$$+\sum_\phi \big(I_{-\frac{1}{2}}(f_1^1 * f_1^2)\phi(\mathrm{e}) + I_{-\frac{1}{2}}(f_1^1 * f_1^2)\phi(w)\big)\overline{\mathcal{P}(I_{\frac{1}{2}}(f_2)\phi,\tfrac{1}{2})}.$$

We view again $\mathcal{P}$ as a generalized vector in the (non-unitary) representation $I_{-\frac{1}{2},\Omega,\Omega}$. Then $I_{-\frac{1}{2},\Omega,\Omega}(f)\mathcal{P}$ is a smooth vector in this representation and the second term is the sum $I_{-\frac{1}{2},\Omega,\Omega}(f)\mathcal{P}(e) + I_{-\frac{1}{2},\Omega,\Omega}(f)\mathcal{P}(w)$.

Finally, we study the contribution of the terms where $\Pi_1^\tau = \Pi_1$ and $\Pi_{1|F_{\mathbb{A}}^\times}=1$. Then $\Pi_1 \neq \Omega$. Also $\Pi_1^2 = 1$. Thus, for a given f, or even f in a given bounded set, there are only *finitely* many such Π_1 with $I_{\lambda,\Pi_1,\Pi_2}(f) \neq 0$. Thus the sum $\sum \Theta_{\Pi,T_1,T_2}(f)$ is finite, and in the sum, each term is itself written as a finite sum of terms. The kernel $K_{f,\Pi}$ has once more a residual component whose truncated integral over the diagonal is 0. This time $\int \Lambda_d^{T_1} E(a,\phi,\lambda)\Omega^{-1}(a)\mathrm{d}a$ is independent of T_1 but

$$(45)\quad \int \Lambda_m^{T_2} E(h_1,\phi,\lambda)\mathrm{d}h_1 = \frac{\mathrm{e}^{\lambda T_2}}{\lambda}\int_{K\cap H_1}\phi(k)\mathrm{d}k$$
$$+\frac{\mathrm{e}^{-\lambda T_2}}{-\lambda}\int_{K\cap H_1} M(w,\lambda)\phi(k)\mathrm{d}k + \mathcal{P}(\phi,\lambda,\Pi).$$

Since $\Pi_1 = \Pi_2$ we have $M(w,0) = -1$ and the two first terms have the same residue at $\lambda = 0$. The last term has also a simple pole at $\lambda = 0$. However the factor (39) reduces to $\mu(I_\lambda(f_1^1)\phi,\lambda,\Pi)$ which has a zero at $\lambda = 0$ because in (43) the reciprocal of the L-factor does. Thus the poles of the terms in (40) are compensated by the zero of (39). As T_2 tends to infinity the terms containing

$e^{\pm\lambda T_2}$ tend to 0 and we find the contribution of those finitely many terms to be the same as before (see Proposition 7).

8.6. Discrete terms. — Finally, we discuss the terms whose contributions will be purely discrete. We consider those Π such that $\Pi_{1|F_{\mathbb{A}}^\times} = 1$ but $\Pi_1^\tau \neq \Pi_1$. This condition amounts to $\Pi_1^2 \neq 1$. We also have then $\Pi_1 \neq \Omega$ and the integral $\int \Lambda_d^{T_1} E(a,\phi,\lambda)\Omega^{-1}(a)\mathrm{d}a$ does not depend on T_1. However, we have then, in the sense of analytic continuation,

$$(46)\qquad \begin{cases} \displaystyle\int \Lambda_m^{T_2} E(h_1,\phi,\lambda)\mathrm{d}h_1 = \frac{e^{\lambda T_2}}{\lambda}\mathcal{P}_c(\phi) + \frac{e^{-\lambda T_2}}{-\lambda}\mathcal{P}_c\big(M(\lambda,w)\phi\big), \\ \displaystyle\mathcal{P}_c(\phi) = \int_{K_F}\phi(k)\mathrm{d}k, \end{cases}$$

where K_F denotes the standard maximal compact subgroup of $\mathrm{Gl}(2,F_{\mathbb{A}})$. Since the left hand side is holomorphic at $\lambda = 0$ we get (See also [JL, Lemma 7], for a computational proof.)

$$(47)\qquad \mathcal{P}_c(\phi) = \mathcal{P}_c\big(M(0,w)\phi\big).$$

It follows that

$$\Theta_{\Pi,T_1,T_2}(f) = \frac{1}{2}\sum_\phi \int \mu(I_\lambda(f)\phi)\overline{\Big(\frac{e^{\lambda T_2}}{\lambda}\mathcal{P}_c(\phi) + \frac{e^{-\lambda T_2}}{-\lambda}\mathcal{P}_c\big(M(\lambda,w)\phi\big)\Big)}\mathrm{d}\lambda.$$

Again, in terms of generalized vectors, this can be written in the form:

$$\frac{1}{2}\int\Big(\big\langle I_{\lambda,\Pi}(f)\mathcal{P}_c,\mu_{\lambda,\Pi}\big\rangle\frac{e^{-\lambda T_2}}{-\lambda} + \big\langle I_{\lambda,\Pi}(f)M(w,\lambda)^*\mathcal{P}_c,\mu_{\lambda,\Pi}\big\rangle\frac{e^{\lambda T_2}}{\lambda}\Big)\mathrm{d}\lambda.$$

Consider then the functions:

$$\Phi_1(\lambda) := \sum_\Pi\big\langle I_{\lambda,\Pi}(f)\mathcal{P}_c,\mu_{\lambda,\Pi}\big\rangle,\quad \Phi_2(\lambda) := \sum_\Pi\big\langle I_{\lambda,\Pi}(f)M(w,\lambda)^*\mathcal{P}_c,\mu_{\lambda,\Pi}\big\rangle,$$

the sums over all the Π of the above type.

LEMMA 14. — *The above series converge absolutely, uniformly on compact sets, and define continuous functions on the line* $\operatorname{Re}\lambda = 0$. *Moreover* $\Phi_1(0) = \Phi_2(0)$. *If f_2 is the convolution of sufficiently many K-finite functions, the functions are differentiable on the line* $\operatorname{Re}\lambda = 0$.

Indeed, for the first assertion, we write as before $f_1^1 * f_1^2 * f_2^*$. Then

$$\big\langle I_{\lambda,\Pi}(f)\mathcal{P}_c,\mu_{\lambda,\Pi}\big\rangle = \big\langle I_{\lambda,\Pi}(f_1^2)I_{\lambda,\Pi}(f_2^*)\mathcal{P}_c, I_{\lambda,\Pi}(f_1^{1*})\mu_{\lambda,\Pi}\big\rangle.$$

As before, there is a polynomial $P(\Pi_1,\lambda)$ such that

$$\big\|I_{\lambda,\Pi}(f_1^{1*})\mu_{\lambda,\Pi}\big\| \le P(\Pi_1,\lambda).$$

It is easy to see that

$$\big\|I_{\lambda,\Pi}(f_2^*)\mathcal{P}_c\big\| \le C$$

where C is a constant. Hence

$$\big\|\langle I_{\lambda,\Pi}(f)\mathcal{P}_c, \mu_{\lambda,\Pi}\rangle\big\| \le CP(\Pi_1,\lambda)\big\|I_{\lambda,\Pi}(f_1^2)\big\|.$$

Since

$$\sum_{\Pi_1} P(\Pi_1,\lambda)\big\|I_{\lambda,\Pi}(f_1^2)\big\| < +\infty$$

the first assertion follows for Φ_1 (with uniform convergence on compact sets, even for f_1^2, f_1^2, f_2 smooth in a bounded set). Since the intertwining operator is unitary, the argument for Φ_2 is similar. For the differentiability, we proceed as follows. We do assume that the functions are K-finite. Then we need only consider those Π such that the corresponding operators are non-zero. This means that the restriction of Π_1 to the maximal compact subgroup of $E_{\mathbb{A}}^\times$ takes only finitely values. For the purpose of convergence, we may as well fix this restriction. Then all the representations $I_{\lambda,\Pi}$ operate on the same Hilbert of functions on the maximal compact subgroup K. Thus it suffices to show that

$$\sum_{\phi,\phi'}\sum_{\Pi}\langle I_{\lambda,\Pi}(f_1^2)\phi,\phi'\rangle\mu\big(I_{\lambda,\Pi}(f_1^1)\phi',\lambda,\Pi\big)\,\overline{\mathcal{P}_c(I_{\lambda,\Pi}(f_2)\phi)}$$

is differentiable, the sum over ϕ,ϕ' being *finite*. We may as well fix ϕ and ϕ'. At this point, we do assume that $f_2 = f_2^1 * f_2^2$ where the functions f_2^i, $i = 1,2$, are K-finite. Then the terms $\langle I_{\lambda,\Pi}(f_1^2)\phi,\phi'\rangle$ and $\overline{\mathcal{P}_c(I_{\lambda,\Pi}(f_2))\phi)}$ are bounded by constant multiples of $\|I_{\lambda,\Pi}(f_1^2)\|$ and $\|I_{\lambda,\Pi}(f_2^1)\|$, respectively. Their respective derivatives are bounded, uniformly in (Π_1,λ). As for $\mu(I_{\lambda,\Pi}(f_1^1)\phi',\lambda,\Pi)$ it is bounded as well as its derivatives by a polynomial in (Π_1,λ). Our assertion follows for Φ_1. For Φ_2, we observe that

$$\begin{aligned}\sum_{\Pi_1}&\langle I_{\lambda,\Pi}(f_1^2)\phi,\phi'\rangle\mu(I_{\lambda,\Pi}(f_1^1)\phi',\lambda,\Pi)\,\overline{\mathcal{P}_c(I_{\lambda,\Pi}(f_2)M(w,\lambda)\phi)}\\ &= \sum_{\phi''}\sum_{\Pi_1}\langle I_{\lambda,\Pi}(f_1^1)\phi,\phi'\rangle\mu\big(I_{\lambda,\Pi}(f_1^2)\phi',\lambda,\Pi\big)\\ &\qquad\times\overline{\mathcal{P}_c(I_{\lambda,\Pi}(f_2^1))\phi'')}\cdot\overline{\big\langle I_{\lambda,\Pi}(f_2^2)M(w,\lambda)\phi,\phi''\big\rangle}\end{aligned}$$

with a finite sum over ϕ''. We proceed as before, this time assuming that, in turn, f_2^1 is the convolution of two K-finite functions. □

We have then, at least if f_2 is the convolution of sufficiently many K-finite functions,

$$\sum_{\Pi}\Theta_{\Pi,T_1,T_2}(f) = \int\Big(\Phi_1(\lambda)\frac{\mathrm{e}^{-\lambda T_2}}{-\lambda} + \Phi_2(\lambda)\frac{\mathrm{e}^{\lambda T_2}}{\lambda}\Big)\,\mathrm{d}\lambda.$$

This expression is independent of T_1. So we only need to take the limit as $T_2 \to +\infty$. It is of the form $c\Phi_1(0)$.

8.7. Conclusion. — We are now ready to state the main result of this section:

THEOREM 3. — *For any smooth function of compact support f:*

$$\Theta(f) = \sum_{\Pi\,\mathrm{cuspidal}} \Theta_\Pi(f) + \sum_{\Pi_1^\tau=\Pi_1} \frac{1}{2}\sum_{\Pi}\int \langle I_{\lambda,\Pi}(f)\mathcal{P}_{\lambda,\Pi}, \mu_{\lambda,\Pi}\rangle d\lambda$$
$$+ c \sum_{\substack{\Pi_1|F_{\mathbb{A}}^\times=1\\ \Pi_1^\tau\neq\Pi_1}} \langle I_{0,\Pi}(f)\mathcal{P}_c, \mu_{0,\Pi}\rangle$$
$$+ c_1\big\{\big(I_{-\frac{1}{2},\Omega,\Omega}(f)\mathcal{P}\big)(\mathrm{e}) + \big(I_{-\frac{1}{2},\Omega,\Omega}(f)\big)\mathcal{P}(w)\big\}.$$

The expression is absolutely convergent:

$$\sum_{\Pi_1^\tau=\Pi_1}\int \Big|\langle I_{\lambda,\Pi}(f)\mathcal{P}_{\lambda,\Pi}, \mu_{\lambda,\Pi}\rangle\Big| \mathrm{d}\lambda < +\infty,$$
$$\sum_{\substack{\Pi_1|F_{\mathbb{A}}^\times=1\\ \Pi_1^\tau\neq\Pi_1}} \big|\langle I_{0,\Pi}(f)P_c, \mu_{0,\Pi}\rangle\big| < +\infty.$$

By [DM], every smooth function of compact support can be written as a finite sum of triple convolution products. Thus it suffices to prove our assertion when $f = f_1 * f_2 * f_3$. If each f_i remains in a bounded set both sides are well defined and bounded. Since the space of smooth functions of compact support is a bornological space, each side may be regarded as a distribution on the product of three copies of the group. We have shown both sides agree when each f_i is K-finite and f_3 is itself the convolution of sufficiently many K-finite functions. Thus they agree for arbitrary f_i and our assertion follows. Note that the right hand side is in fact a distribution, as follows for instance from the kernel theorem.

9. Appendix: Continuous spectrum over F

We pass to the much easier discussion of the convergence of the spectral expression for G_ϵ. For simplicity we only consider the case $\omega = 1$, $\Omega = 1$. The only difficult case is the case where ϵ is a norm. Of course, no truncation is needed. We then identify G_ϵ with $\mathrm{Gl}(2, F)$. Consider a pair of characters $\pi = (\pi_1, \pi_2)$ with $\pi_1\pi_2 = 1$. Since $\mathrm{Gl}(2, F) = P(F)T(F)$ we have:

$$\int E(t, \phi; \pi, \lambda)\mathrm{d}t = \nu(\phi; \lambda, \pi)$$

where

$$\nu(\phi;\pi,\lambda) := \int_{T(F)Z(F_{\mathbb{A}}))\backslash T(F_{\mathbb{A}})} \phi(t)\mathrm{e}^{\langle\lambda+\rho,H(t)\rangle}\,\mathrm{d}t.$$

Using again the notation of generalized vectors, what we have to show is that

$$\sum \int \langle I_{\lambda,\pi}(f)\nu_{\lambda,\pi},\nu_{\lambda,\pi}\rangle \mathrm{d}\lambda$$

is absolutely convergent. This can be seen directly when f is K-finite. For our purposes this is not quite enough since we want to use smooth functions of compact support. We use again an integral representation of the form (42):

$$(48)\quad I_{\lambda,\Pi}(f)\phi(g) = \frac{\pi_1(g)\,|\det g|^{\lambda+\frac{1}{2}}}{L^{S_0}(2\lambda+1,\pi_1\pi_2^{-1})} \times \int_{E_{\mathbb{A}}^{\times}} \Phi\big[(0,t)g;\lambda,\pi_1\big]\cdot |t|^{2\lambda+1}\pi_1\pi_2^{-1}(t)\mathrm{d}^{\times}t.$$

Let us introduce the Schwartz-Bruhat function

$$\phi[t;\lambda,\pi_1] := \Phi\big[(0,1)t;\lambda,\pi_1\big]$$

on the simple algebra of which T is the multiplicative group. Then

$$\nu(I_{\lambda,\pi}(f)\phi) = \frac{1}{L^{S_0}(2\lambda+1,\pi_1\pi_2^{-1})}\int \phi(t)\pi_1(\det t)\,|\det t|_F\,\mathrm{d}t.$$

The integral is then a Tate integral for E. In particular, it is a holomorphic multiple of $L(\lambda+\frac{1}{2},\pi_1\circ\mathrm{Norm})$. We can then finish the proof as before.

BIBLIOGRAPHY

[A] ASAI (T.) – *On certain Dirichlet series associated with Hilbert modular forms*, Math. Ann., t. **226** (1977), pp. 81–94.

[Ar] ARTHUR (J.) – *A trace formula for reductive groups I: Terms associated to classes in G(ℚ)*, Duke Math. J., t. **45** (1978), 911-952.

[B] BERNSTEIN (J.) – *P-invariant distributions of* Gl(n) *and the classification of unitary representations of* Gl(n) *(non Archimedean case)*, in *Lie Group Representations II, College Park, Md., 1982/1983*, Lecture Notes in Math. **1041**, Springer-Verlag, 1984.

[BFH] BUMP (D.), FRIEDBERG (S.), HOFFSTEIN (J.) – *Eisenstein series on the metaplectic group and non-vanishing theorems for automorphic L-functions*, Ann. of Math., t. **131** (1990), pp. 53–127.

[BM] BARUCH (E.M.), MAO (Z.) – *Central critical value of automorphic L-functions*, preprint

[CHM] CASSELMAN (W.), HECHT (H.), MILIČIC (D.) – *Bruhat filtrations and Whittaker vectors for real groups*, preprint.

[DM] DIXMIER (J.), MALLIAVIN (P.) – *Factorizations de fonctions et vecteurs indéfiniment différentiables*, Bull. Soc. Math. France, t. **102** (1978), pp. 305–330.

[F1] FLICKER (Y.) – *Twisted tensors and Euler products*, Bull. Soc. Math. France, t. **116** (1988), pp. 295–313.

[F2] FLICKER (Y.) – *On distinguished representations*, J. reine angew Math., t. **418** (1991), pp. 139–172.

[F3] FLICKER (Y.) – *A Fourier summation formula for the symmetric space* $\mathrm{Gl}(n)/\mathrm{Gl}(n-1)$, Comp. Math., t. **88** (1992), pp. 39–117.

[F4] FLICKER (Y.) – *Orbital Integrals on symmetric spaces and spherical characters*, J. Algebra, t. **184** (1996), pp. 705–754.

[FZ] FLICKER (Y.), ZINONIEV (D.) – *On the poles of twisted L-functions*, Proc. Japan Acad. Sci., t. **71** (1995), pp. 114–116.

[FH] FRIEDBERG (S.), HOFFSTEIN (J.) – *Non vanishing theorem for automorphic L-functions for* Gl(2), Ann. of Math., t. **142** (1995), pp. 385–423.

[G1] GUO (J.) – *On the positivity of the central critical value of automorphic L-functions for* Gl(2), Duke Math. J., t. **83** (1996), pp. 157–189.

[G2] GUO (J.) – *On a generalization of a result of Waldspurger*, Canad. J. Math., t. **48** (1996), pp. 105–142.

[Gr] GROSS (B.) – *Heights and the special values of L series*, in *Number Theory (Montreal, Que., 1985)*, 115–187, CMS Conf. Proc. **7**, Americ. Math. Soc., Providence, RI, 1987.

[GZ] GROSS (B.), ZAGIER (D.) – *Heegner points and derivatives of automorphic L-functions for* Gl(2), Invent. Math., t. **84** (1986), pp. 225–320.

[H] HAKIM (J.) – *Distinguished p-adic representations*, Duke Math. J., t. **62** (1991), pp. 1–22.

[HLR] HARDER (G.), LANGLANDS (R.P.), RAPOPORT (M.) – *Algebraische Zyklen auf Hilbert-Blumenthal-Flächen*, J. reine angew. Math., t. **366** (1986), pp. 53–120.

[J1] JACQUET (H.) – *Sur un résultat de Waldspurger*, Ann. Scient. École Norm. Sup., t. **19** (1986), pp. 185–229.

[J2] JACQUET (H.) – *Sur un résultat de Waldspurger II*, Comp. Math., t. **63** (1987), pp. 315–389.

[JL] JACQUET (H.), LAI (K.) – *A relative trace formula*, Comp. Math., t. **54** (1995), pp. 243–310.

[JLRa] JACQUET (H.), LAI (K.), RALLIS (S.) *A trace formula for symmetric spaces*, Duke Math. J., t. **70** (1993), pp. 305–371.

[JLRo] JACQUET (H.), LAPID (E.), ROGAWSKI (J.) – *Periods of automorphic forms*, J. Amer. Math. Soc., t. **12** (1999), pp. 173–240.

[JY] JACQUET (H.), YE (Y.) – *Une remarque sur le changement de base quadratique*, C. R. Acad. Sci. Paris Sér. I Math., t. **311** (1990), pp. 671–676.

[K] KHONEN (W.) – *Fourier coefficients of modular forms of half integral weight*, Math. Ann., t. **271** (1985), pp. 237–268.

[Kk] KAMAL KHURI-MAKDSI – *On the Fourier coefficients of non holomorphic Hilbert modular forms of half integral weight*, Duke Math. J., t. **84** (1996), pp. 399–452.

[KS] KATOK (S.), SARNAK (P.) – *Heegner points, cycles, and Maass forms*, Israel J. Math., t. **84** (1993), pp. 193–227.

[KZ] KOHNEN (W.), ZAGIER (D.) – *Values of L-series of modular forms at the center of the critical strip*, Invent. Math., t. **64** (1981), pp. 175–198.

[R] ROHRLICH (D.) – *Non vanishing of L-functions for* Gl(2), Invent. Math., t. **97** (1989), pp. 381–403.

[RR] RADER (C.), RALLIS (S.) – *Spherical characters on p-adic symmetric spaces*, Amer. J. Math., t. **118** (1996), pp. 91–178.

[S] SHIMURA (S.) – *On Fourier coefficients of Hilbert modular forms of half-integral weight*, Duke Math. J., t. **71** (1993), pp. 501–557.

[W1] WALDSPURGER (J.-L.) – *Corespondances de Shimura*, J. Math. Pures Appl., t. **59** (1980), pp. 1–132.

[W2] WALDSPURGER (J.-L.) – *Sur les coefficients de Fourier des formes modulaires de poids demi-entier*, J. Math. Pures Appl., t. **60** (1981), pp. 375–484.

[W3] WALDSPURGER (J.-L.) – *Quelques propriétés arithmétiques de certaines formes automorphes sur* Gl(2), Comp. Math., t. **54** (1985), pp. 121–171.

[W4] WALDSPURGER (J.-L.) – *Sur les valeurs de fonctions L-automorphes en leur centre de symétrie*, Comp. Math., t. **54** (1985), pp. 173–242.

[W5] WALDSPURGER (J.-L.) – *Correspondances de Shimura et quaternions*, Forum Math., t. **3** (1991), pp. 219–307.

[Wa] WALLACH (N.) – *Real Reductive Groups II*, Pure and Applied Mathematics, Vol **132**-II (1192), Academic Press.

[Ya] YANG (T.) – *Theta liftings and Hecke L-functions*, J. reine angew. Math., t. **485** (1997), pp. 25–53.

Available online at www.sciencedirect.com
SCIENCE DIRECT
ELSEVIER
COMPTES RENDUS MATHEMATIQUE
C. R. Acad. Sci. Paris, Ser. I 336 (2003) 121–124

Théorie des groupes/Théorie des nombres

Facteurs de transfert pour les intégrales de Kloosterman

Transfer factors for Kloosterman integrals

Hervé Jacquet

Math. Department, Columbia University, MC 4408, New York, NY 10027, USA

Reçu le 9 décembre 2002 ; accepté le 16 décembre 2002

Présenté par Hervé Jacquet

Résumé

On calcule les facteurs de transfert qui relient les intégrales de Kloosterman absolues et les intégrales de Kloosterman relatives à une extension quadratique. ***Pour citer cet article : H. Jacquet, C. R. Acad. Sci. Paris, Ser. I 336 (2003).***

Abstract

We compute the transfer factors which relate absolute Klosterman integrals and Kloosterman integrals relative to a quadratic extension. ***To cite this article: H. Jacquet, C. R. Acad. Sci. Paris, Ser. I 336 (2003).***

1. Enoncé du résultat

On complète les résultats de [1] dont on reprend les notations. Soient E/F une extension de corps locaux non-archimédiens et $\eta : F^\times \to \{\pm 1\}$ le caractère quadratique correspondant. On note $x \mapsto \bar{x}$ l'élément non-trivial du groupe de Galois. Soit ψ un caractère additif non trivial de F et $r \geqslant 1$ un entier. Soit N_r le groupe des matrices trigonales strictes dans $\mathrm{GL}(r, F)$ et $A_r(F)$ l'ensemble des matrices diagonales dans $M(r \times r, F)$. Soit $\theta : N_r(F) \to \mathbb{C}^\times$ le caractère défini par $\theta(n) = \psi(\sum n_{i,i+1})$. Ainsi $(n_1, n_2) \mapsto \theta(n_1 n_2)$ est un caractère du groupe produit $N_r(F) \times N_r(F)$. On fait opérer le groupe $N_r(F) \times N_r(F)$ sur $M(r \times r, F)$ par : $g \mapsto {}^t n_1 g n_2$. Si g est une matrice $r \times r$ on désigne par $\Delta_i(g)$ le déterminant de la sous-matrice formée des i premières lignes et i premières colonnes de g. Les fonctions Δ_i sont des invariants de l'action. Si Φ est une fonction lisse à support compact sur $M(r \times r, F)$ et $g \in M(r \times r, F)$ on définit les intégrales de Kloosterman :

$$\Omega(\Phi, \psi : g) = \int \Phi({}^t n_1 g n_2) \theta(n_1 n_2) \, \mathrm{d}n_1 \, \mathrm{d}n_2;$$

Adresse e-mail : hj@math.columbia.edu (H. Jacquet).

1631-073X/03/$ – see front matter
doi:10.1016/S1631-073X(03)00012-8

l'intégrale est prise sur le quotient du produit $N_r(F) \times N_r(F)$ par le stabilisateur de g dans le produit ; l'élément g est *pertinent*, c'est-à-dire que $\theta(n_1 n_2)$ est trivial sur le stabilisateur de g. Si g est pertinent et $\Delta_r(g) = 0$ alors $\Delta_{r-1}(g) \neq 0$. En particulier, pour $a \in A_r(F)$, $a = \mathrm{diag}(a_1, a_2, \ldots, a_r)$, a est pertinent lorsque $a_1 a_2 \cdots a_{r-1} \neq 0$; alors on écrit aussi :

$$\Omega(\Phi, \psi : a_1, a_2, \ldots, a_r) = \int_{N_r(F)\times N_r(F)} \Phi({}^t n_1 a n_2)\theta(n_1 n_2)\, \mathrm{d}n_1\, \mathrm{d}n_2.$$

On définit $\sigma_r : A_r(F) \to F$ par

$$\sigma_r\big(\mathrm{diag}(a_1, a_2, \ldots, a_r)\big) = a_1(a_1 a_2) \cdots (a_1 a_2 \cdots a_{r-1})$$

et on pose

$$\widetilde{\Omega}(\Phi, \psi : a) = \left|\sigma_r(a)\right|_F \Omega(\Phi, \psi : a).$$

De même, soit $H(r \times r, E/F)$ l'espace des matrices hermitiennes dans $M(r \times r, E)$. Soit $n \mapsto \theta(n\bar{n})$ le caractère du groupe $N_r(E)$ défini par $\theta(n\bar{n}) = \psi(\sum n_{i,i+1} + \bar{n}_{i,i+1})$. Le groupe $N_r(E)$ opère par $g \mapsto {}^t\bar{n} g n$. Soit Ψ une fonction lisse à support compact sur $H(r \times r, E/F)$. On définit les integrales de Kloosterman relatives :

$$\Omega(\Psi, E/F, \psi : g) = \int \Psi({}^t\bar{n} g n)\theta(n\bar{n})\, \mathrm{d}n;$$

l'intégrale est prise sur le quotient de $N_r(E)$ par le stabilisateur de g ; l'élement g est pertinent, c'est-à-dire que $\theta(n\bar{n})$ est trivial sur le stabilisateur de g.

Pour m entier, on désigne par w_m la matrice de permutation $m \times m$ dont les éléments anti-diagonaux sont 1. Les éléments de la forme

$$g = \begin{pmatrix} w_{m_1} a_1 & 0 & \cdots & 0 \\ 0 & w_{m_2} a_2 & \cdots & 0 \\ & & \ddots & \\ 0 & 0 & \cdots & w_{m_j} a_j \end{pmatrix} \tag{1}$$

avec $m_1 + m_2 + \cdots + m_j = r$, $a_1, a_2, \ldots, a_j \in F$, $\Delta_r(g) \neq 0$ ou $\Delta_{r-1}(g) \neq 0$, forment un sytème commun de représentants pour les deux ensembles d'orbites pertinentes.

On pose aussi

$$\widetilde{\Omega}(\Psi, E/F, \psi : a) = \eta\big(\sigma_r(a)\big)\left|\sigma_r(a)\right|_F \Omega(\Psi, E/F, \psi : a).$$

On écrit $\Phi \overset{\psi}{\leftrightarrow} \Psi$ et on dit que Φ correspond à Ψ par ψ si $\widetilde{\Omega}(\Phi, \psi : a) = \widetilde{\Omega}(\Psi, E/F, \psi : a)$ pour tout a pertinent dans $A_r(F)$. On a prouvé que tout Φ correspond à un Ψ et réciproquement [1,2].

Théorème 1.1. *Pour g de la forme* (1) *il existe un facteur de transfert $\gamma(g, \psi)$ tel que si $\Phi \overset{\psi}{\leftrightarrow} \Psi$ alors*

$$\Omega(\Phi, \psi : g) = \gamma(g, \psi)\Omega(\Psi, E/F, \psi : g).$$

Ceci est trivial pour $r = 1$. On peut donc supposer cette assertion établie pour $r' < r$. Si g est diagonal, on a déja $\gamma(g, \psi) = \eta(\sigma_{r-1}(g))$; en particulier $\gamma(g, \psi) = 1$ si $r = 1$. Supposons que g soit de la forme

$$g = \begin{pmatrix} wa & 0 \\ 0 & w'a' \end{pmatrix} \tag{2}$$

avec wa relevant dans $\mathrm{GL}(n, F) \cap H(n \times n, E/F)$, $w'a'$ relevant dans $M(m \times m, F) \cap H(m \times m, E/F)$ et $r = n + m$. Si g est diagonal, c'est-à-dire si $w = 1$ et $w' = 1$, alors

$$\gamma(g, \psi) = \gamma(a, \psi)\gamma(a', \psi)\eta(\det a)^m. \tag{3}$$

On introduit des intégrales orbitales partielles. Soit $U_{n,m}$ le radical unipotent du sous groupe parabolique standard de type (n, m). On pose :

$$\Omega\left[\Phi, \psi : \begin{pmatrix} A_n & 0 \\ 0 & B_m \end{pmatrix}\right] = \int_{U_{n,m}(F)\times U_{n,m}(F)} \Phi\left[{}^t u_2 \begin{pmatrix} A_n & 0 \\ 0 & B_m \end{pmatrix} u_1\right] \theta(u_1 u_2)\, du_1\, du_2.$$

On peut définir les intégrales orbitales de cette fonction pour l'action de $(N_n(F) \times N_n(F)) \times N_m(F) \times N_m(F)$. En particulier, pour g de la forme (2), l'intégrale orbitale évaluée sur le couple $(wa, w'a')$ n'est autre que $\Omega(\Phi, \psi : g)$. Des considérations analogues s'appliquent à

$$\Omega\left[\Psi, E/F, \psi : \begin{pmatrix} A_n & 0 \\ 0 & B_m \end{pmatrix}\right] = \int_{U_{n,m}(E)} \Psi\left[{}^t \bar{u} \begin{pmatrix} A_n & 0 \\ 0 & B_m \end{pmatrix} u\right] \theta(u\bar{u})\, du.$$

Il résulte de (3) que la fonction

$$(A_n, B_m) \mapsto \Omega\left[\Phi, \psi : \begin{pmatrix} A_n & 0 \\ 0 & B_m \end{pmatrix}\right]$$

sur $\mathrm{GL}(n, F) \times M(m \times m, F)$ correspond par ψ à la fonction

$$(A_n, B_m) \mapsto \eta\big(\det(A_n)\big)^m \Omega\left[\Psi, E/F, \psi : \begin{pmatrix} A_n & 0 \\ 0 & B_m \end{pmatrix}\right]$$

sur $(\mathrm{GL}(n, E) \cap H(n \times n, E/F)) \times H(m \times m, E/F)$. Par l'hypothèse de récurrence, cela implique l'existence du facteur $\gamma(g, \psi)$ avec

$$\gamma(g, \psi) = \gamma(wa, \psi)\gamma(w'a', \psi)\eta\big(\det(wa)\big)^m.$$

Pour le cas de $g = w_r a$ on utilise une formule élémentaire. On définit des transformées de Fourier :

$$\check{\Phi}(x) = \int_{M(r\times r, F)} \Phi(y)\psi\big(-\mathrm{Tr}(y w_r x_r w_r)\big)\, dy,$$

$$\check{\Psi}(x) = \int_{H(r\times r, E/F)} \Psi(y)\psi\big(-\mathrm{Tr}(y w_r x_r w_r)\big)\, dy.$$

Alors [2] :

$$|a|^{m^2-1}\Omega(\Phi, \psi : w_r a) = \int \Omega\left[\check{\Phi}, \bar{\psi} : \begin{pmatrix} -w_{m-1}a^{-1} & 0 \\ 0 & b \end{pmatrix}\right] db \tag{4}$$

et on a une formule correspondante pour Ψ. Comme $\check{\Phi} \overset{\psi}{\leftrightarrow} c(E/F, \psi)^{r(r-1)/2}\check{\Psi}$, ou $c(E/F, \psi)$ est la constante de Weil, on en déduit l'existence du facteur $\gamma(w_r a, \psi)$ avec :

$$\gamma(w_r a, \psi) = \gamma(-w_{r-1}a^{-1}, \bar{\psi})c(E/F, \psi)^{r(r-1)/2}\eta\big(\det(-w_{r-1}a^{-1})\big).$$

On a $\gamma(w_1 a, \psi) = 1$. Rappelons que

$$c(E/F, \psi)^2 = \eta(-1), \qquad c(E/F, \psi)c(E/F, \bar{\psi}) = 1.$$

On peut calculer $\gamma(w_r a, \psi)$ par récurrence. Ses 8 premières valeures sont

$$1, \quad c(E/F, \psi)\eta(-a), \quad \eta(a), \quad 1, \quad \eta(-1), \quad c(E/F, \psi)\eta(-a), \quad \eta(-a), \quad 1.$$

De plus $\gamma(w_9 a, \psi) = 1$. Comme les facteurs $c(E/F, \psi)^{r(r-1)/2}$ et $\eta(\det(-w_{r-1}a^{-1}))$ sont des fonctions périodiques de r de période 8, on en conclut que $\gamma(w_r a, \psi)$, vu comme fonction de r, est périodique de période 8.

2. Densité des intégrales diagonales

Théorème 2.1. *Si les intégrales orbitales diagonales de Φ s'annulent, alors toutes les intégrales orbitales de Φ s'annulent.*

Comme plus haut on procède par récurrence sur r. Il suffit donc de prouver que $\Omega(\Phi, \psi : w_r a) = 0$. Rappelons la formule d'inversion [2]

$$\widetilde{\Omega}(\check{\Phi}, \psi : a_1, a_2, \ldots, a_r) = \int \widetilde{\Omega}(\Phi, \psi : p_1, p_2, \ldots, p_r)\psi\left(-\sum_{i=1}^{i=r} p_i a_{r+1-i} + \sum_{i=1}^{i=r-1} \frac{1}{p_i a_{r-i}}\right) \mathrm{d}p_r\, \mathrm{d}p_{r-1} \cdots \mathrm{d}p_1. \tag{5}$$

Elle montre que les intégrales orbitales diagonales de $\check{\Phi}$ sont nulles. Il en est donc de même de l'intégrale

$$\Omega\left[\check{\Phi}, \overline{\psi} : \begin{pmatrix} -w_{m-1}a^{-1} & 0 \\ 0 & b \end{pmatrix}\right].$$

La conclusion suit de la formule (4). Un résultat analogue s'applique aux intégrales orbitales relatives.

Remarque 1. Les résultats précédents sont déjà dans [3]. Toutefois, pour simplifier, dans [3], on avait traité seulement le cas d'une extension non-ramifiée de caractéristique 0. En tout cas, les démonstrations présentées ici sont plus simples.

Remerciement

Recherche financée en partie par NSF DMS 96 19766.

Références

[1] H. Jacquet, Transfert lisse d'intégrales de Kloosterman, C. R. Acad. Sci. Paris Sér. I 335 (2002) 229–232.
[2] J. H. Jacquet, Smooth Transfer of Kloosterman integrals, Duke Math. J., to appear.
[3] H. Jacquet, A theorem of density for Kloosterman integrals, Asian J. Math. 2 (1998) 759–778.

SMOOTH TRANSFER OF KLOOSTERMAN INTEGRALS

HERVÉ JACQUET

Abstract

We establish the existence of smooth transfer between absolute Kloosterman integrals and Kloosterman integrals relative to a quadratic extension.

Contents

1. Introduction

Let E/F be a quadratic extension of local non-Archimedean fields, let $\eta : F^\times \to \{\pm 1\}$ be the corresponding quadratic character, and let $\psi : F \to \mathbb{C}^\times$ be a nontrivial character. Let N_m be the subgroup of upper triangular matrices in $\mathrm{GL}(m)$ with unit diagonal, and let A_m be the group of diagonal matrices. We define a character $\theta : N_m(F) \to \mathbb{C}^\times$ by $\theta(n) = \psi\bigl(\sum n_{i,i+1}\bigr)$. The group $N_m(F) \times N_m(F)$ operates on $\mathrm{GL}(m, F)$ by $g \mapsto {}^t n_1 g n_2$, and the orbits that intersect $A_m(F)$ form a dense open subset. We define the diagonal orbital integrals of a smooth function of compact support Φ on $\mathrm{GL}(m, F)$:

$$\Omega(\Phi, \psi : a) := \int_{N_m(F)\times N_m(F)} \Phi({}^t n_1 a n_2)\theta(n_1 n_2)\, dn_1\, dn_2. \tag{1}$$

DUKE MATHEMATICAL JOURNAL
Vol. 120, No. 1, © 2003
Received 28 May 2002. Revision received 5 November 2002.
2000 *Mathematics Subject Classification.* Primary 11F70; Secondary 11F72, 11L05.
Author's work partially supported by National Science Foundation grant number DMS-9619766.

Likewise, we let $H(m \times m, E/F)$ be the space of Hermitian matrices $m \times m$ and set $S_m(F) = H(m \times m, E/F) \cap \mathrm{GL}(m, E)$. We define a character $n \mapsto \theta(n\overline{n})$ of $N_m(E)$ by $\theta(n\overline{n}) = \psi\big(\sum(n_{i,i+1} + \overline{n_{i,i+1}})\big)$. The group $N_m(E)$ operates on $S_m(F)$ by $s \mapsto {}^t\overline{n}sn$, and the orbits that intersect $A_m(F)$ form a dense open subset of $S_m(F)$. We define the diagonal orbital integrals of a smooth function of compact support Ψ on $S_m(F)$:

$$\Omega(\Psi, E/F, \psi : a) := \int_{N_m(E)} \Phi({}^t\overline{n}an)\theta(n\overline{n})\, dn. \tag{2}$$

We say that Φ *matches* Ψ for ψ if for every $a \in A_m(F)$,

$$\Omega(\Phi, \psi : a) = \gamma(a, \psi)\Omega(\Psi, E/F, \psi : a),$$

where

$$\gamma(a, \psi) := \eta(a_1)\eta(a_1a_2)\cdots\eta(a_1a_2, \dots, a_{m-1}) \quad \text{if } a = \mathrm{diag}(a_1, a_2, \dots, a_m).$$

In this paper we prove that for every Φ there is a matching Ψ, and conversely.

Suppose that E/F is unramified, and suppose that the conductor of ψ is $\mathcal{O}_F$. Let Φ_0 be the characteristic function of the set of matrices with integral entries in $M(m \times m, F)$. Similarly, let Ψ_0 be the characteristic function of the set of matrices with integral entries in $H(m \times m, E/F)$. The *fundamental lemma* asserts that the restriction of Φ_0 to $\mathrm{GL}(m, F)$ matches the restriction of Ψ_0 to $S_m(F)$. This has been established by Ngô [10] in the case of positive characteristic.

In general, if Φ matches Ψ, then there are similar relations between the other orbital integrals (see [3], [4]).

Within the context of the relative trace formula, the problem at hand is the analogue of the transfer problem for the ordinary trace formula. Indeed, when E/F is an extension of function fields, the result presented in this paper, together with the fundamental lemma of Ngô [9], [10], implies the existence of global identities of the form

$$\int_{N_m(F)\times N_m(F)\backslash N_m(F_{\mathbb{A}})\times N_m(F_{\mathbb{A}})} \theta(n_1n_2)\, dn_1\, dn_2\Big(\sum_{\xi\in Gl(n,F)} \Phi({}^tn_1\xi n_2)\Big)$$
$$= \int_{N_m(E)\backslash N_m(E_{\mathbb{A}})} \theta(n\overline{n})\, dn\Big(\sum_{\xi\in S_m(F)} \Psi({}^t\overline{n}\xi n)\theta(n\overline{n})\, dn\Big).$$

In turn, the identities are a crucial step in extending the results of [5] to $\mathrm{GL}(m)$.

Our method is first to linearize the problem by replacing $\mathrm{GL}(m, F)$ and $S_m(F)$ by the vector spaces $M(m \times m, F)$ and $H(m \times m, E/F)$, respectively. We then establish a simple relation between the orbital integrals of a function on either space and its Fourier transform. It follows that if Φ matches Ψ, then the Fourier transform of Φ matches the Fourier transform of Ψ. This proves the existence of many pairs of matching functions and, in turn, is used to prove the existence of the transfer in a very simple way.

Our method is suggested by the work of J. Waldspurger [13]. Waldspurger considers orbital integrals in the context of the Lie algebras and derives the existence of the transfer from the fundamental lemma (assumed to hold) and global identities. He uses the Fourier transform on the Lie algebras in an essential way. It is reasonable to ask whether, conversely, the results of the paper can be used to prove the fundamental lemma in the context of the relative trace formula. We answer this question in the affirmative for the case of $m = 2, 3$ in Section 9. In a forthcoming paper (see [7]), we use the previous results to prove the fundamental lemma for all m.

2. Orbital integrals in $M(m \times m, F)$

We now consider the action of $N_m(F) \times N_m(F)$ on $M(m \times m, F)$ given by $r \mapsto {}^t n_1 r n_2$. We denote by θ_0 the algebraic character $\theta_0 : N(F) \to F$ defined by

$$\theta_0(n) = \sum_{i=1}^{i=m-1} n_{i,i+1}.$$

We say that an element $x \in M(m \times m, F)$ (or its orbit) is *relevant* if the character $(n_1, n_2) \mapsto \theta_0(n_1 n_2)$ is trivial on the stabilizer of x in $N(F) \times N(F)$.

We let $\Delta_i(x)$, $1 \le i \le m$, be the determinant of the matrix obtained by removing the last $m - i$ rows and the last $m - i$ columns of a matrix x. In particular, $\Delta_m(x) = \det x$. The functions Δ_i are invariants of the action of $N_m(F) \times N_m(F)$. For $1 \le i \le m - 1$, we denote by O_i the set defined by $\Delta_i \neq 0$. We denote by B_m the group of upper triangular matrices in $\mathrm{GL}(m)$, by A_m the group of diagonal matrices, and by $\alpha_1, \alpha_2, \ldots, \alpha_{m-1}$ the simple roots of A_m corresponding to B_m. We denote by N_α the root group corresponding to a root α. We identify the Weyl group of A_m with the group W_m of permutation matrices, and we let $\mathscr{A}_m$ be the set of diagonal matrices in $M(m \times m, F)$. We often drop the index m from the notation.

PROPOSITION 1

In $M(m \times m, F)$ the set defined by $\Delta_m = 0$, $\Delta_{m-1} = 0$ contains no relevant orbit.

Proof

We begin with an elementary lemma.

LEMMA 1

Any $r \in M(m \times m, F)$ can be written (in possibly several ways) in the form

$$r = {}^t n w b$$

with $w \in W_m$, $n \in N_m(F)$, and b an upper triangular matrix.

Proof

We let

$$\mathscr{F}_0 = \{V_0 \subset V_1 \subset V_2 \subset \cdots \subset V_n\}$$

be the canonical flag. Thus V_i is the space spanned by the first i vectors of the canonical basis. Let $r \in M(m \times m, F)$ be given. Consider the sequence of subspaces

$$rV_0 \subseteq rV_1 \subseteq rV_2 \subseteq \cdots \subseteq rV_m.$$

It need not be a flag, but, inductively, it is easy to show that there is a flag

$$\mathscr{F} = \{U_0 \subset U_1 \subset U_2 \subset \cdots \subset U_m\}$$

such that

$$rV_i \subseteq U_i, \quad 0 \leq i \leq n.$$

By the standard Bruhat decomposition,

$$\mathscr{F} = {}^t nw\mathscr{F}_0$$

for suitable w and $n \in N_m(F)$. Thus

$$({}^t nw)^{-1} rV_i \subseteq V_i, \quad 0 \leq i \leq n.$$

Since $({}^t nw)^{-1} r$ stabilizes $\mathscr{F}_0$, it is an upper triangular matrix. This proves the lemma. □

To prove our assertion, it suffices then to consider the case of a relevant element of the form

$$r = wb$$

with $w \in W_m$ and b an upper triangular matrix such that $\det b = 0$. Our task is then to show that $\Delta_{m-1}(wb) \neq 0$. We remark that if the column of index i, $i < m$, of the matrix r is zero, then r is irrelevant because then $rn_{\alpha_i} = r$ for all $n_{\alpha_i} \in N_{\alpha_i}(F)$. In particular, the first diagonal entry $b_{1,1}$ of b is nonzero. Then at the cost of multiplying b by a suitable element of N_{α_1} on the right, we may assume that $b_{1,2} = 0$. Again, since the second column of b cannot be zero, we must have $b_{2,2} \neq 0$. Inductively, we see that we may assume b to have the form

$$b = \begin{pmatrix} b' & 0 \\ 0 & 0 \end{pmatrix},$$

where b' is an invertible diagonal matrix of size $(m-1)\times(m-1)$. Thus the last row of b is zero. Since wb is relevant, the first $m-1$ rows of wb cannot be zero. This forces w to have the form

$$w = \begin{pmatrix} w' & 0 \\ 0 & 1 \end{pmatrix},$$

where w' is a permutation matrix in $\mathrm{GL}(m-1)$. Now

$$\Delta_{m-1}(wb) = \det b' \det w'.$$

Thus $\Delta_{m-1}(wb) \neq 0$, as claimed. □

Now it is elementary that every element x such that $\det x = 0$ but $\Delta_{n-1}(x) \neq 0$ is in the orbit of an element of the form

$$\begin{pmatrix} x' & 0 \\ 0 & 0 \end{pmatrix}.$$

Moreover, x is relevant if and only if x' is a relevant element in $\mathrm{GL}(m-1, F)$. Also,

$$a = \mathrm{diag}(a_1, a_2, \ldots, a_m) \in \mathscr{A}_m$$

is relevant if and only if $a_1 a_2 \cdots a_{m-1} \neq 0$.

Now we recall the classification of the relevant orbits in $\mathrm{GL}(m, F)$ (cf. [1], [12]). Every orbit has a unique representative of the form wa with $w \in W$ and $a \in A_m(F)$. If this element is relevant, then for every pair of positive roots (α_1, α_2) such that $w\alpha_2 = -\alpha_1$, for $n_{\alpha_i} \in N_{\alpha_i}(F)$ we have

$$^t n_{\alpha_1} w a n_{\alpha_2} = wa \Rightarrow \theta_0(n_{\alpha_1} n_{\alpha_2}) = 0. \tag{3}$$

This condition implies that if α_1 is simple, then α_2 is simple, and conversely. Thus w and its inverse have the property that if they change a simple root to a negative one, then they change it to the opposite of a simple root. Let S be the set of simple roots α such that $w\alpha$ is negative. Then S is also the set of simple roots α such that $w^{-1}\alpha$ is negative and $wS = -S$. Let M be the standard Levi subgroup determined by S. Thus S is the set of simple roots of M for the torus A, w is the longest element of $W \cap M$, and $w^2 = 1$. This being so, if α_2 is simple, then condition (3) implies $\alpha_2(a) = 1$. Thus a is in the center of M.

Let w_m be the $m \times m$ permutation matrix with antidiagonal entries equal to 1. We see that elements of the form

$$g = \begin{pmatrix} w_{m_1} a_1 & 0 & \cdots & 0 \\ 0 & w_{m_2} a_2 & \cdots & 0 \\ \cdots & & & \\ 0 & 0 & \cdots & w_{m_r} a_r \end{pmatrix},$$

with $m_1 + m_2 + \cdots + m_r = m$, $a_1, a_2, \ldots, a_r \in F^\times$, form a set of representatives for the relevant orbits in $\mathrm{GL}(m, F)$. A set of representatives for the relevant orbits in $M(n \times n, F)$ is formed of the same elements, except that $a_r = 0$ is allowed when $m_r = 1$ (and $w_{m_r} = 1$).

We set $\theta = \psi \circ \theta_0$. We define the orbital integral of a relevant element x:

$$\Omega[\Phi, \psi : x] = \int \Phi({}^t n_1 x n_2)\theta(n_1 n_2)\, dn_1\, dn_2;$$

the integral is over the quotient of $N_m(F) \times N_m(F)$ by the stabilizer of x. The normalization of the measure is described below.

It is convenient to introduce the *intermediate orbital integrals*. Let m, n be two integers greater than zero. For $\Phi \in \mathcal{S}(M((m+n) \times (m+n), F))$, $A_n \in \mathrm{GL}(n, F)$, $B_m \in M(m \times m, F)$, we set

$$\Omega_m^n\left[\Phi, \psi : \begin{pmatrix} A_n & 0 \\ 0 & B_m \end{pmatrix}\right] := \int \Phi\left[\begin{pmatrix} 1_n & 0 \\ Y & 1_m \end{pmatrix}\begin{pmatrix} A_n & 0 \\ 0 & B_m \end{pmatrix}\begin{pmatrix} 1_n & X \\ 0 & 1_m \end{pmatrix}\right] \times \psi\left[\operatorname{Tr} \epsilon_n^m X + \operatorname{Tr} Y \tilde{\epsilon}_n^m\right] dX\, dY. \tag{4}$$

Here $\epsilon = \epsilon_n^m$ is the matrix with m rows and n columns whose first row is the row matrix of size n,

$$\epsilon_n = (0, 0, \ldots, 0, 1),$$

and all other rows are zero. Likewise, $\tilde{\epsilon} = \tilde{\epsilon}_n^m$ is the matrix with n rows and m columns whose first column is the column matrix

$$\tilde{\epsilon}_n = \begin{pmatrix} 0 \\ 0 \\ \cdots \\ 0 \\ 1 \end{pmatrix},$$

and all other columns are zero. If $X = (x_{i,j})$, then the measure dX is the tensor product of the self-dual Haar measure $dx_{i,j}$.

The above integral can we written more explicitly as

$$\int \Phi\left[\begin{pmatrix} A_n & A_n X \\ Y A_n & B_m + Y A_n X \end{pmatrix}\right] \psi\left[\operatorname{Tr}(\epsilon X) + \operatorname{Tr}(Y \tilde{\epsilon})\right] dX\, dY,$$

or, after a change of variables,

$$\Omega_m^n\left[\Phi, \psi : \begin{pmatrix} A_n & 0 \\ 0 & B_m \end{pmatrix}\right] = |\det A_n|^{-2m} \int \Phi\left[\begin{pmatrix} A_n & X \\ Y & B_m + Y A_n^{-1} X \end{pmatrix}\right] \times \psi\left[\operatorname{Tr}(\epsilon A_n^{-1} X) + \operatorname{Tr}(Y A_n^{-1} \tilde{\epsilon})\right] dX\, dY.$$

The previous form shows that the integral converges and defines a smooth function on $\mathrm{GL}(n, F) \times M(m \times m, F)$. More precisely, the following result is easily obtained.

LEMMA 2

If Φ is supported on O_n, then $\Omega_m^n[\Phi, \psi : \bullet]$ is a smooth function of compact support on $\mathrm{GL}(n, F) \times M(m \times m, F)$. Conversely, any smooth function of compact support on $\mathrm{GL}(n, F) \times M(m \times m, F)$ is equal to $\Omega_m^n[\Phi, \psi : \bullet]$ for a suitable function Φ supported on O_n.

We also introduce the *normalized* intermediate orbital integrals

$$\tilde{\Omega}_m^n\left[\Phi, \psi : \begin{pmatrix} A_n & 0 \\ 0 & B_m \end{pmatrix}\right] := |\det A_n|^m \times \Omega_m^n\left[\Phi, \psi : \begin{pmatrix} A_n & 0 \\ 0 & B_m \end{pmatrix}\right]. \tag{5}$$

If Ψ is a smooth function of compact support on $\mathrm{GL}(n, F) \times M(m \times m, F)$, we can define its orbital integrals relative to the action $(N_n \times N_n) \times (N_m \times N_m)$ on $\mathrm{GL}(n, F) \times M(m \times m, F)$. They are denoted by $\Omega(\Psi, \psi : x_n, x_m)$, where x_n is relevant in $\mathrm{GL}(n, F)$ and x_m relevant in $M(m \times m, F)$. In particular, consider the case of the function

$$\Psi(A_n, B_m) = \Omega_m^n\left[\Phi, \psi : \begin{pmatrix} A_n & 0 \\ 0 & B_m \end{pmatrix}\right]$$

and a relevant element x of $M((n+m) \times (n+m), F)$ of the form

$$x = \begin{pmatrix} x_n & 0 \\ 0 & x_m \end{pmatrix},$$

where x_n is relevant in $\mathrm{GL}(n, F)$ and x_m is relevant in $M(m \times m, F)$. We then have the following reduction formula:

$$\Omega[\Phi, \psi : x] = \Omega[\Psi, \psi : x_n, x_m]. \tag{6}$$

This formula reduces the computation of the orbital integrals (and the normalization of the measures) to the case of an element of the form $w_n a$, where a is a scalar matrix. Then

$$\Omega[\Phi, \psi : w_n a] = \int \Phi\left[\begin{pmatrix} 0 & 0 & 0 & \cdots & 0 & a \\ 0 & 0 & 0 & \cdots & a & ax_{2,n} \\ \cdots & \cdots & \cdots & \cdots & \cdots & \cdots \\ 0 & 0 & a & \cdots & ax_{n-2,n-1} & ax_{n-2,n} \\ 0 & a & ax_{n-1,3} & \cdots & ax_{n-1,n-1} & ax_{n-1,n} \\ a & ax_{n,2} & ax_{n,3} & \cdots & ax_{n,n-1} & ax_{n,n} \end{pmatrix}\right]$$

$$\times \psi\Big(\sum_{i=2}^{i=n} x_{i,n-i+2}\Big) \otimes dx_{i,j},$$

where the measures $dx_{i,j}$ are self-dual.

We also denote by $\Omega(\Phi, \psi : a_1, a_2, \dots, a_n)$ the orbital integral $\Omega(\Phi, \psi : a)$, where $a = \mathrm{diag}(a_1, a_2, \dots, a_n)$. We introduce normalized diagonal orbital integrals as follows. For $a \in \mathscr{A}_m(F)$ we set

$$\sigma_n(a) := \Delta_1(a)\Delta_2(a)\cdots\Delta_{n-1}(a).$$

Thus

$$\sigma_n(a) = a_1^{n-1}a_2^{n-2}\cdots a_{n-1}$$

and a is relevant if $\sigma_n(a) \neq 0$. For a relevant, we set

$$\tilde{\Omega}(\Phi, \psi : a) := |\sigma_n(a)|\Omega(\Phi, \psi : a). \tag{7}$$

We have a reduction formula for these normalized orbital integrals. In a precise way, for $\Phi \in \mathscr{S}(M(n+m)\times(n+m), F)$, set

$$\Psi(A_n, B_m) = \tilde{\Omega}_m^n\left[\Phi, \psi : \begin{pmatrix} A_n & 0 \\ 0 & B_m \end{pmatrix}\right].$$

Consider a diagonal matrix

$$a = \begin{pmatrix} a_n & 0 \\ 0 & a_m \end{pmatrix}$$

with $a_n \in A_n(F)$ and a_m relevant in $\mathscr{A}_m(F)$. The normalized orbital integral of Ψ on the pair (a_n, a_m) is defined to be

$$\tilde{\Omega}[\Psi, \psi : a_n, a_m] = |\sigma_n(a_n)|\,|\sigma_m(a_m)|\Omega[\Psi, \psi : a_n, a_m].$$

It follows from the relation

$$(\det a_n)^m\sigma_n(a_n)\sigma(a_m) = \sigma_{n+m}(a)$$

that it is equal to $\tilde{\Omega}(\Phi, \psi : a)$.

3. Weil formula: The split case

Our goal is to obtain a simple relation between the orbital integrals of a function and the orbital integrals of its Fourier transform. Our main tool is the Weil formula for the Fourier transform of a character of second order. We state the form of the Weil formula that we are using. We consider the direct sum

$$M(n\times m, F) \oplus M(m\times n, F),$$

the first index being the number of rows. A function Φ on that space is written as

$$\Phi\begin{pmatrix} 0_{n,n} & X \\ Y & 0_{m,m} \end{pmatrix}.$$

Thus the matrix written is a square matrix of size $(n+m)\times(n+m)$ with zero diagonal blocks, $X \in M(n \times m, F)$, and $Y \in M(m \times n, F)$. The Fourier transform of such a function is the function defined by

$$\hat{\Phi}\begin{pmatrix} 0_{n,n} & X \\ Y & 0_{m,m} \end{pmatrix} := \int \Phi\begin{pmatrix} 0 & U \\ V & 0 \end{pmatrix} \psi\left[-\operatorname{Tr}\left(\begin{pmatrix} 0 & X \\ Y & 0 \end{pmatrix}\begin{pmatrix} cc0 & U \\ V & 0 \end{pmatrix}\right)\right] dU\, dV$$

$$= \int \Phi\begin{pmatrix} 0 & U \\ V & 0 \end{pmatrix} \psi\big[-\operatorname{Tr}(YU) - \operatorname{Tr}(XV)\big]\, dU\, dV. \tag{8}$$

PROPOSITION 2

For $A \in \mathrm{GL}(n, F)$, $B \in \mathrm{GL}(m, F)$, and any Schwartz function Φ on $M(n \times m, F) \oplus M(m \times n, F)$, we have

$$\int \Phi\begin{pmatrix} 0 & X \\ Y & 0 \end{pmatrix} \psi\,(\operatorname{Tr} BYAX)\; dX\, dY$$

$$= |\det A|^{-m} |\det B|^{-n} \int \hat{\Phi}\begin{pmatrix} 0 & X \\ Y & 0 \end{pmatrix} \psi(-\operatorname{Tr} B^{-1}YA^{-1}X)\, dX\, dY.$$

Proof

Assume first that A and B are unit matrices. Set

$$\Phi_1(U, V) = \int \Phi\begin{pmatrix} 0 & V \\ Y & 0 \end{pmatrix} \psi\big(-\operatorname{Tr}(YU)\big)\, dU.$$

Then the left-hand side of the identity reduces to

$$\int \Phi_1(-X, X)\, dX.$$

On the other hand, using the (partial) Fourier inversion formula, we see that the right-hand side is equal to

$$\int \Phi_1(Y, -Y)\, dY.$$

Our assertion then follows. In general, we remark that $\operatorname{Tr}(BYAX) = \operatorname{Tr} YAXB$. We change X to XB^{-1} and Y to YA^{-1} in the left-hand side, and we change Y to BY and X to AX in the right-hand side. We see that we are reduced to the case of $A = 1$, $B = 1$ applied to the function

$$(X, Y) \mapsto |\det A|^{-m} |B|^{-n} \Phi\begin{pmatrix} 0 & XB^{-1} \\ YA^{-1} & 0 \end{pmatrix},$$

the Fourier transform of which is the function

$$(X, Y) \mapsto \hat{\Phi}\begin{pmatrix} 0 & AX \\ BY & 0 \end{pmatrix}.$$

Our assertion follows. □

We record a simple consequence of this formula. For $P \in M(m \times n, F)$ and $Q \in M(n \times m, F)$,

$$\int \Phi \begin{pmatrix} 0 & X \\ Y & 0 \end{pmatrix} \psi\big(\mathrm{Tr}(PX) + \mathrm{Tr}(YQ) + \mathrm{Tr}(BYAX)\big)\, dX\, dY$$

$$= |\det A|^{-m} |\det B|^{-n}$$

$$\times \int \hat{\Phi} \begin{pmatrix} 0 & X \\ Y & 0 \end{pmatrix} \psi\big(- \mathrm{Tr}\, B^{-1}(Y+P)A^{-1}(X+Q)\big)\, dX\, dY. \tag{9}$$

4. Inversion formula for the orbital integrals

For $\Phi \in \mathscr{S}(M(n \times n, F))$, we define the Fourier transform of Φ to be

$$\hat{\Phi}(X) = \int \Phi(Y)\psi[-\mathrm{Tr}(XY)]\, dY.$$

The measure is self-dual. Recall that $w_n \in W_n$ is the permutation matrix with unit antidiagonal. We set

$$\check{\Phi}(X) = \hat{\Phi}(w_n X w_n).$$

Our goal in this section is to obtain a relation between the orbital integrals of Φ and the orbital integrals of $\check{\Phi}$. We first establish such a relation for the intermediate orbital integrals.

PROPOSITION 3

Let $n \geq 1$, $m \geq 1$ be two integers. Let $\Phi \in \mathscr{S}(M((m+n) \times (m+n), F))$. Then

$$\int \left(\int \tilde{\Omega}^n_m \left[\Phi, \psi : \begin{pmatrix} A_n & 0 \\ 0 & B_m \end{pmatrix} \right] \psi\, [-\mathrm{Tr}\, B_m w_m C_m w_m]\, dB_m \right)$$

$$\times\, \psi\big[\mathrm{Tr}(w_m C_m^{-1} w_m \epsilon^m_n A_n^{-1} \tilde{\epsilon}^m_n)\big]\psi\big[-\mathrm{Tr}(A_n w_n D_n w_n)\big]\, dA_n$$

$$= \tilde{\Omega}^m_n \left[\check{\Phi}, \overline{\psi} : \begin{pmatrix} C_m & 0 \\ 0 & D_n \end{pmatrix} \right].$$

The integrals are for $B_m \in M(m \times m, F)$ and $A_n \in M(n \times n, F)$.

Proof

We consider the partial Fourier transform Θ (with respect to B_m) of the normalized intermediate orbital integral of Φ:

$$\Theta\,(A_n, C_m) := \int \tilde{\Omega}^n_m \left[\Phi, \psi : \begin{pmatrix} A_n & 0 \\ 0 & B_m \end{pmatrix} \right] \psi\, [-\mathrm{Tr}\, B_m C_m]\, dB_m.$$

If we change C_m to $w_m C_m w_m$ and D_n to $w_n D_n w_n$, we see that the identity of the proposition amounts to

$$\begin{aligned}&\int \Theta(A_n, C_m)\psi[\operatorname{Tr} C_m^{-1}\epsilon_n^m A_n^{-1}\tilde{\epsilon}_n^m]\psi[-\operatorname{Tr} A_n D_n]\,dA_n\\ &\quad = |\det C_m|^n \int \hat{\Phi}\begin{pmatrix} D_n + w_n Y w_m C_m w_m X w_n & w_n Y w_m C_m \\ C_m w_m X w_n & C_m \end{pmatrix}\\ &\qquad \times \psi[-\operatorname{Tr}\epsilon_m^n X - \operatorname{Tr} Y\tilde{\epsilon}_m^n]\,dX\,dY. \end{aligned} \tag{10}$$

Changing X to $w_m Y w_n$ and Y to $w_n X w_m$ and noting the relations

$$w_n \epsilon_m^n w_m = \tilde{\epsilon}_n^m, \qquad w_m \tilde{\epsilon}_m^n w_n = \epsilon_n^m,$$

we see that the right-hand side of the last formula can also be written

$$\int \hat{\Phi}\begin{pmatrix} D_n + X C_m Y & X C_m \\ C_m Y & C_m \end{pmatrix}\psi[-\operatorname{Tr} Y\tilde{\epsilon}_n^m - \operatorname{Tr}\epsilon_n^m X]\,dX\,dY. \tag{11}$$

From now on, let us write ϵ for ϵ_n^m and $\tilde{\epsilon}$ for $\tilde{\epsilon}_n^m$. After a change of variables, we find that Θ is also equal to

$$\begin{aligned}&|\det A_n|^m \int \Phi\left[\begin{pmatrix} A_n & A_n X \\ Y A_n & B_m \end{pmatrix}\right]\\ &\qquad \times \psi[\operatorname{Tr}(\epsilon X) + \operatorname{Tr}(Y\tilde{\epsilon}) + \operatorname{Tr}(C_m Y A_n X)]\,dX\,dY\\ &\qquad \times \psi[-\operatorname{Tr}(C_m B_m)]\,dB_m. \end{aligned}$$

After a new change of variables, this becomes

$$\begin{aligned}&|\det A_n|^{-m} \int \Phi\left[\begin{pmatrix} A_n & X \\ Y & B_m \end{pmatrix}\right]\\ &\qquad \times \psi\big[\operatorname{Tr}(\epsilon A_n^{-1} X) + \operatorname{Tr}(Y A_n^{-1}\tilde{\epsilon}) + \operatorname{Tr}(C_m Y A_n^{-1} X)\big]\,dX\,dY\\ &\qquad \times \psi\big[-\operatorname{Tr}(C_m B_m)\big]\,dB_m. \end{aligned}$$

We now introduce a partial Fourier transform of Φ:

$$\begin{aligned}&\Phi_1\begin{pmatrix} A_n & U \\ V & C_m \end{pmatrix}\\ &\quad := \int \Phi\begin{pmatrix} A_n & X \\ Y & B_m \end{pmatrix}\psi\left[-\operatorname{Tr}\begin{pmatrix} 0 & X \\ Y & B_m \end{pmatrix}\begin{pmatrix} cc0 & U \\ V & C_m \end{pmatrix}\right]dX\,dY\,dB_m. \end{aligned}$$

By the Weil formula, the previous expression is then

$$|\det C_m|^{-n} \int \Phi_1\begin{pmatrix} A_n & X \\ Y & C_m \end{pmatrix}\psi\big[-\operatorname{Tr} C_m^{-1}(Y + \epsilon A_n^{-1})A_n(X + A_n^{-1}\tilde{\epsilon})\big]\,dX\,dY.$$

Expanding ψ, we find

$$\begin{aligned}&|\det C_m|^{-n}\psi\big[-\operatorname{Tr}(C_m^{-1}\epsilon A_n^{-1}\tilde{\epsilon})\big]\\&\qquad\times\int\Phi_1\begin{pmatrix}A_n & X\\ Y & C_m\end{pmatrix}\\&\qquad\times\psi\big[-\operatorname{Tr}(C_m^{-1}YA_nX)-\operatorname{Tr}(C_m^{-1}Y\tilde{\epsilon})-\operatorname{Tr}(C_m^{-1}\epsilon X)\big]\,dX\,dY.\end{aligned}$$

Changing variables once more, we obtain, at last,

$$\begin{aligned}\Theta(A_n,C_m)&=|\det C_m|^{n}\psi\big[-\operatorname{Tr}(C_m^{-1}\epsilon A_n^{-1}\tilde{\epsilon})\big]\\&\quad\times\int\Phi_1\begin{pmatrix}A_n & XC_m\\ C_mY & C_m\end{pmatrix}\\&\quad\times\psi\big[-\operatorname{Tr}C_mYA_nX-\operatorname{Tr}(Y\tilde{\epsilon})-\operatorname{Tr}(\epsilon X)\big]\,dX\,dY\end{aligned}$$

To obtain (10), we must multiply by

$$\psi\big[\operatorname{Tr}(C_n^{-1}\epsilon A_n^{-1}\tilde{\epsilon})\big]$$

and take the Fourier transform of the result with respect to A_n. The Fourier transform is evaluated at D_n. We indeed obtain (10). The proof shows that for $C_m\in \mathrm{GL}(m,F)$, the integral converges as an iterated integral. □

Our main result in this section is the following inversion formula for diagonal orbital integrals.

THEOREM 1
For $\Phi\in\mathscr{S}(M(n\times n,F))$,

$$\begin{aligned}&\tilde{\Omega}(\check{\Phi},\overline{\psi}:a_1,a_2,\dots,a_n)\\&\quad=\int\tilde{\Omega}(\Phi,\psi:p_1,p_2,\cdots,p_n)\\&\qquad\times\psi\Big(-\sum_{i=1}^{i=n}p_ia_{n+1-i}+\sum_{i=1}^{i=n-1}\frac{1}{p_ia_{n-i}}\Big)\,dp_n\,dp_{n-1}\cdots dp_1,\end{aligned}$$

where the multiple integral is only an iterated integral.

Proof
For $n=1$ the formula is just the definition of the Fourier transform. Thus we may assume that $n>1$ and that our assertion is true for $n-1$. We prove it for n. We apply the formula of Proposition 3 to the pair of integers $(n-1,1)$. We obtain, for fixed

$a_1 \in F^\times$,

$$\int \left(\int \tilde{\Omega}_1^{n-1} \left[\Phi, \psi : \begin{pmatrix} A_{n-1} & 0 \\ 0 & p_n \end{pmatrix} \right] \psi\,[-p_n a_1]\, dp_n \right) \\ \times \psi\left[\mathrm{Tr}(a_1^{-1} \epsilon_{n-1} A_{n-1}^{-1} \tilde{\epsilon}_{n-1}) \right] \psi\left[-\mathrm{Tr}(A_{n-1} w_{n-1} D_{n-1} w_{n-1}) \right] dA_{n-1} \\ = \tilde{\Omega}_{n-1}^1 \left[\check{\Phi}, \overline{\psi} : \begin{pmatrix} a_1 & 0 \\ 0 & D_{n-1} \end{pmatrix} \right].$$

This formula shows that the function $\Phi_1(A_{n-1})$ defined by

$$\Phi_1(A_{n-1}) := \psi\left[\mathrm{Tr}(a_1^{-1} \epsilon_{n-1} A_{n-1}^{-1} \tilde{\epsilon}_{n-1}) \right] \\ \times \int \tilde{\Omega}_1^{n-1} \left[\Phi, \psi : \begin{pmatrix} A_{n-1} & 0 \\ 0 & p_n \end{pmatrix} \right] \psi\,[-p_n a_1]\, dp_n,$$

which a priori is defined only on $\mathrm{GL}(n-1, F)$, extends in fact to a Schwartz-Bruhat function on $M((n-1) \times (n-1), F)$, still denoted by Φ_1 such that $\check{\Phi}_1 = \Theta$, where Θ is the Schwartz function on $M((n-1) \times (n-1), F)$ defined by

$$\Theta(D_{n-1}) = \tilde{\Omega}_{n-1}^1 \left[\check{\Phi}, \overline{\psi} : \begin{pmatrix} a_1 & 0 \\ 0 & D_{n-1} \end{pmatrix} \right].$$

The induction hypothesis applied to Φ_1 gives

$$\tilde{\Omega}(\Theta, \overline{\psi} : a_2, a_3, \ldots, a_n) \\ = \int \tilde{\Omega}(\Phi_1, \psi : p_1, p_2, \ldots, p_{n-1}) \\ \times \psi\Big(-\sum_{i=1}^{i=n-1} p_i a_{n+1-i} + \sum_{i=1}^{i=n-2} \frac{1}{p_i a_{n-i}} \Big)\, dp_{n-1} \cdots dp_1. \tag{12}$$

On the other hand, with

$$a = \mathrm{diag}(a_1, a_2, \ldots, a_m), \qquad a' = \mathrm{diag}(a_2, a_3, \ldots, a_n),$$

we have

$$\tilde{\Omega}(\Theta, \overline{\psi} : a_2, a_3, \ldots, a_n) \\ = |\sigma_{n-1}(a')| \Omega(\Theta, \overline{\psi} : a') \\ = |a_1|^{n-1} |\sigma_{n-1}(a')| \\ \times \int_{N_{n-1}(F) \times N_{n-1}(F)} \Omega_{n-1}^1 \left[\check{\Phi}, \overline{\psi} : \begin{pmatrix} a_1 & 0 \\ 0 & {}^t u_2 a' u_1 \end{pmatrix} \right] \overline{\theta}(u_1 u_2)\, du_1\, du_2 \\ = |\sigma_n(a)| \Omega(\check{\Phi}, \overline{\psi} : a) \\ = \tilde{\Omega}(\check{\Phi}, \overline{\psi} : a_1, a_2, \ldots, a_n).$$

Hence we get

$$\begin{aligned}
&\tilde{\Omega}(\check{\Phi},\overline{\psi} : a_1, a_2, \dots, a_n)\\
&\quad = \int \tilde{\Omega}(\Phi_1, \psi : p_1, p_2, \dots, p_{n-1})\\
&\quad\quad \times \psi\Big(- \sum_{i=1}^{i=n-1} p_i a_{n+1-i} + \sum_{i=1}^{i=n-2} \frac{1}{p_i a_{n-i}}\Big) dp_{n-1} \cdots dp_1. \qquad (13)
\end{aligned}$$

To complete the proof, we have to compute the orbital integral of the function $A_{n-1} \mapsto \Phi_1(A_{n-1})$ on the matrix $p' = \operatorname{diag}(p_1, p_2, \dots, p_{n-1})$. Note that the factor

$$\psi\big[\operatorname{Tr}(a_1^{-1} \epsilon_{n-1} A_{n-1}^{-1} \tilde{\epsilon}_{n-1})\big],$$

which appears in the definition of Φ_1, is constant on the orbits of $N_{n-1}(F) \times N_{n-1}(F)$ and, in particular, takes the value $\psi[1/(a_1 p_{n-1})]$ on the orbit of p'. Observe, furthermore, that if ϕ is the characteristic function of a compact open set in $F^\times$, then the function

$$(A_{n-1}, p_n) \mapsto \phi(\det A_{n-1}) \tilde{\Omega}_1^{n-1}\left[\Phi, \psi : \begin{pmatrix} A_{n-1} & 0 \\ 0 & p_n \end{pmatrix}\right]$$

is a smooth function of compact support on $\mathrm{GL}(n-1, F) \times F$. In particular,

$$\int\int \tilde{\Omega}_1^{n-1}\left[\Phi, \psi : \begin{pmatrix} {}^t u_1 x u_2 & 0 \\ 0 & p_n \end{pmatrix}\right] \theta(u_1 u_2) \psi[-p_n a_1]\, dp_n\, du_1\, du_2$$

is an absolutely convergent integral where the order of integration can be reversed. Moreover, for a given p_n,

$$\begin{aligned}
&|\sigma_{n-1}(p')| \int \tilde{\Omega}_1^{n-1}\left[\Phi, \psi : \begin{pmatrix} {}^t u_1 p' u_2 & 0 \\ 0 & p_n \end{pmatrix}\right] \theta(u_1 u_2)\, du_1\, du_2\\
&\qquad\qquad\qquad\qquad = \tilde{\Omega}\,[\Phi, \psi : p_1, p_2, \dots, p_n].
\end{aligned}$$

It follows that

$$
\begin{aligned}
&\tilde{\Omega}[\Phi_1, \psi : p'] \\
&\quad = \psi\left[\frac{1}{a_1 p_{n-1}}\right] |\sigma_{n-1}(p')| \\
&\qquad \times \int\int \tilde{\Omega}_1^{n-1}\left[\Phi, \psi : \begin{pmatrix} {}^t u_1 p' u_2 & 0 \\ 0 & p_n \end{pmatrix}\right] \theta(u_1 u_2)\psi[-p_n a_1]\, du_1\, du_2\, dp_n \\
&\quad = \psi\left[\frac{1}{a_1 p_{n-1}}\right] |\sigma_{n-1}(p')| |\det p'| \\
&\qquad \times \int \left(\Omega_1^{n-1}\left[\Phi, \psi : \begin{pmatrix} {}^t u_1 p' u_2 & 0 \\ 0 & p_n \end{pmatrix}\right] \theta(u_1 u_2)\psi[-p_n a_1]\, du_1\, du_2 \right) dp_n \\
&\quad = \psi\left[\frac{1}{a_1 p_{n-1}}\right] |\sigma_n(p)| \int \Omega[\Phi, \psi : p_1, p_2, \ldots, p_n]\psi[-p_n a_1]\, dp_n \\
&\quad = \psi\left[\frac{1}{a_1 p_{n-1}}\right] \int \tilde{\Omega}[\Phi, \psi : p_1, p_2, \ldots, p_n]\psi[-p_n a_1]\, dp_n.
\end{aligned}
$$

Upon inserting this result in the right-hand side of (13), we find the identity of the theorem. □

There is another more elementary formula that comes directly from the Fourier inversion formula.

PROPOSITION 4
For $\Phi \in \mathscr{S}(M(n \times n, F))$, $n > 1$, *the function*

$$\phi(a) = \Omega\,[\Phi, \psi : w_n a]$$

is a smooth function of compact support on $F^\times$. *Furthermore,*

$$\phi(a) = |a|^{-n^2+1} \int \Omega\left[\check{\Phi}, \overline{\psi} : \begin{pmatrix} -w_{n-1} a^{-1} & 0 \\ 0 & b \end{pmatrix}\right] db.$$

Proof

After a change of variables, we find

$$\phi(a) = |a|^{(n-n^2)/2} \int \Phi\left[\begin{pmatrix} 0 & 0 & 0 & \cdots & 0 & a \\ 0 & 0 & 0 & \cdots & a & x_{2,n} \\ \cdots & \cdots & \cdots & \cdots & \cdots & \cdots \\ 0 & 0 & a & \cdots & x_{n-2,n-1} & x_{n-2,n} \\ 0 & a & x_{n-1,3} & \cdots & x_{n-1,n-1} & x_{n-1,n} \\ a & x_{n,2} & x_{n,3} & \cdots & x_{n,n-1} & x_{n,n} \end{pmatrix}\right]$$
$$\times \psi\left(a^{-1}\sum_{i=2}^{i=n} x_{i,n-i+2}\right) \otimes dx_{i,j}.$$

Since $|\Delta_n|$ is bounded above on the support of Φ, the above integral vanishes for $|a|$ sufficiently large. On the other hand, since the integrand is a smooth function of the $x_{i,j}$, the integral vanishes for $|a|$ small enough.

We pass to the second assertion. In terms of $\check{\Phi}$ we find that ϕ is equal to

$$\int \check{\Phi}\left[\begin{pmatrix} 0 & 0 & 0 & \cdots & -a^{-1} & x_{1,n} \\ \cdots & \cdots & \cdots & \cdots & \cdots & \cdots \\ 0 & 0 & -a^{-1} & \cdots & x_{n-3,n-1} & x_{n-3,n} \\ 0 & -a^{-1} & x_{n-2,3} & \cdots & x_{n-2,n-1} & x_{n-2,n} \\ -a^{-1} & x_{n-1,2} & x_{n-1,3} & \cdots & x_{n-1,n-1} & x_{n-1,n} \\ x_{n,1} & x_{n,2} & x_{n,3} & \cdots & x_{n,n-1} & x_{n,n} \end{pmatrix}\right]$$
$$\times |a|^{(n-n^2)/2}\psi\left(a\sum_{i=1}^{i=n} x_{i,n-i+1}\right) \otimes dx_{i,j}.$$

We define a function Φ_1 on $\mathrm{GL}(n-1, F)$ by

$$\Phi_1(A_{n-1}) = \int \Omega_1^{n-1}\left[\check{\Phi}, \overline{\psi} : \begin{pmatrix} A_{n-1} & 0 \\ 0 & b \end{pmatrix}\right] db.$$

We remark that if f is a smooth function of compact support on $F^\times$, then the product $f(\det A_{n-1})\Phi_1(A_{n-1})$ is a smooth function of compact support on $\mathrm{GL}(n-1, F)$. It easily follows that

$$\int \Omega\left[\check{\Phi}, \overline{\psi} : \begin{pmatrix} -w_{n-1}a^{-1} & 0 \\ 0 & b \end{pmatrix}\right] db = \Omega(\Phi_1; \overline{\psi} : -w_{n-1}a^{-1}).$$

Now

$$\begin{aligned}\Phi_1(A_{n-1}) &= \int \check{\Phi}\left[\begin{pmatrix} A_{n-1} & A_{n-1}X \\ YA_{n-1} & b+YA_{n-1}X \end{pmatrix}\right] \psi(-\epsilon_{n-1}X - Y\tilde{\epsilon}_{n-1})\, dX\, dY\, db \\ &= |\det A_{n-1}|^{-2} \\ &\quad \times \int \check{\Phi}\left[\begin{pmatrix} A_{n-1} & X \\ Y & b \end{pmatrix}\right] \psi(-\epsilon_{n-1}A_{n-1}^{-1}X - YA_{n-1}^{-1}\tilde{\epsilon}_{n-1})\, dX\, dY\, db.\end{aligned}$$

Now

$$\Omega[\Phi_1, \overline{\psi} : -w_{n-1}a^{-1}] = \int \Phi_1(A_{n-1})\psi\Big(-\sum_{i=2}^{i=n-1} x_{i,n-i+1}\Big)\, dx_{i,j},$$

where A_{n-1} is the matrix

$$\begin{pmatrix} 0 & 0 & 0 & \cdots & -a^{-1} \\ \cdots & \cdots & \cdots & \cdots & \cdots \\ 0 & 0 & -a^{-1} & \cdots & -a^{-1}x_{n-3,n-1} \\ 0 & -a^{-1} & -a^{-1}x_{n-2,3} & \cdots & -a^{-1}x_{n-2,n-1} \\ -a^{-1} & -a^{-1}x_{n-1,2} & -a^{-1}x_{n-1,3} & \cdots & -a^{-1}x_{n-1,n-1} \end{pmatrix}.$$

After a change of variables, we find

$$|a|^{((n-1)^2-(n-1))/2} \int \Phi_1(A_{n-1})\psi\Big(a\sum_{i=2}^{i=n-1} x_{i,n-i+1}\Big)\, dx_{i,j},$$

where now A_{n-1} is the matrix

$$\begin{pmatrix} 0 & 0 & 0 & \cdots & -a^{-1} \\ \cdots & \cdots & \cdots & \cdots & \cdots \\ 0 & 0 & -a^{-1} & \cdots & x_{n-3,n-1} \\ 0 & -a^{-1} & x_{n-2,3} & \cdots & x_{n-2,n-1} \\ -a^{-1} & x_{n-1,2} & x_{n-1,3} & \cdots & x_{n-1,n-1} \end{pmatrix}.$$

Combining with the previous formula for Φ_1, we find that its orbital integral on $-w_{n-1}a^{-1}$ is

$$|a|^{((n-1)^2-(n-1))/2+2(n-1)} \int \check{\Phi}(A_n)\psi\Big(a\sum_{i=1}^{i=n-1} x_{i,n-i+1}\Big)\, dx_{i,j},$$

where A_n is now the matrix

$$\begin{pmatrix} 0 & 0 & 0 & \cdots & -a^{-1} & x_{1,n} \\ \cdots & \cdots & \cdots & \cdots & \cdots & \cdots \\ 0 & 0 & -a^{-1} & \cdots & x_{n-3,n-1} & x_{n-3,n} \\ 0 & -a^{-1} & x_{n-2,3} & \cdots & x_{n-2,n-1} & x_{n-2,n} \\ -a^{-1} & x_{n-1,2} & x_{n-1,3} & \cdots & x_{n-1,n-1} & x_{n-1,n} \\ x_{n,1} & x_{n,2} & x_{n,3} & \cdots & x_{n,n-1} & x_{n,n} \end{pmatrix}.$$

Comparing with the last expression for $\phi(a)$, we obtain our assertion. □

5. Orbital integrals in $H(n \times n, E/F)$

We let E/F be a quadratic extension of F and denote by $H(n \times n, E/F)$ the space of $n \times n$ Hermitian matrices, that is, the matrices m such that ${}^t m = \overline{m}$. The group $N_n(E)$ operates on $H(n \times n, E/F)$ by $m \mapsto {}^t\overline{u}mu$. Viewed as functions on $H(n \times n, E/F)$, the functions Δ_i are invariants of this action. For $1 \leq i < n$, we let O'_i be the subset of $H(n \times n, E/F)$ defined by $\Delta_i \neq 0$. We define an algebraic character $\theta_1 : N(E) \to F$ by $\theta_1(u) = \sum_{i=1}^{i=n-1} u_{i,i+1}$. Often we write $\theta_1(u) = \theta_0(u\overline{u})$. We say that an element x (or its orbit) is relevant if θ_1 is trivial on the stabilizer of x in $N_n(E)$.

PROPOSITION 5

In $H(n \times n, E/F)$ the set defined by $\Delta_{n-1} = 0$, $\Delta_n = 0$ contains no relevant orbit.

Proof

Every $s \in H(n \times n, E/F)$ has the form $s = {}^t nwb$ with $n \in N_n(E)$, $w \in W_n$, and b upper triangular. Suppose that $\det s = 0$, and suppose that s is relevant. We have to show that $\Delta_{n-1}(s) \neq 0$. Now, if a column of s with index $i < n$ is zero, then the row with index i is also zero because s is Hermitian, and then s is irrelevant since it is fixed by the group $N_{\alpha_i}(E)$. In particular, $b_{1,1} \neq 0$. As before, at the cost of replacing s by ${}^t\overline{n}_1 s n_1$, we may assume that $s = {}^t nwb$, where b has the form

$$b = \begin{pmatrix} b' & 0 \\ 0 & 0 \end{pmatrix},$$

where b' is an invertible diagonal matrix in $\mathrm{GL}(n-1, E)$. In particular, the last row of b is zero. We may replace s by the element $wb\overline{n}^{-1}$, and the last row of $b\overline{n}^{-1}$ is again zero. Since the rows of $wb\overline{n}^{-1}$ with index less than n cannot be zero, w must have the form

$$w = \begin{pmatrix} w' & 0 \\ 0 & 0 \end{pmatrix}$$

with $w' \in W_{n-1}$. Then $\Delta_{n-1}(wb\overline{n}^{-1}) = \det b \det w' \neq 0$. as claimed. □

As before, an element x such that $\det x = 0$ but $\Delta_{n-1}(x) \neq 0$ is in the same orbit as an element of the form

$$\begin{pmatrix} x' & 0 \\ 0 & 0 \end{pmatrix}.$$

Moreover, x is relevant if and only if x' is a relevant element in $\mathrm{GL}(m-1, E) \cap H((m-1) \times (m-1), E/F)$.

Now we recall the classification of the relevant orbits in $\mathrm{GL}(n-1, E) \cap H((n-1) \times (n-1), E/F)$. Every orbit has a unique representative of the form wa with $w \in W$, $w^2 = 1$, $a \in A_m(E)$, and $waw = \overline{a}$ (cf. [11]). Suppose that α is a simple root such that $w\alpha = -\beta$ where β is positive. For $n_\alpha \in N_\alpha$, define $n_\beta \in N_\beta$ by

$$^t\overline{n}_\beta wan_\alpha = wa.$$

Then

$$^t\overline{n}_\alpha wan_\beta - wa.$$

There exists an element $n_{\alpha+\beta} \in N_{\alpha+\beta}$ (i.e., $n_{\alpha+\beta} = 1$ if $\alpha + \beta$ is not a root) such that $n = n_\alpha n_\beta n_{\alpha+\beta}$ satisfies

$$^t\overline{n}wan = wa.$$

If wa is relevant, this relation implies

$$\theta_0(n_\alpha \overline{n}_\alpha n_\beta \overline{n}_\beta) = 0.$$

If β is not simple, this leads to a contradiction. Thus β is simple. Since $w^2 = 1$, we see that, as before, there is a standard Levi subgroup M such that w is the longest element in $W_m \cap M$. The above relation also implies that a is in the center of M and in $A(F)$.

We conclude that the set of representatives for the relevant orbits of $N_m(F) \times N_m(F)$ in $M(m \times m, F)$ given in Section 2 is also a set of representatives for the relevant orbits of $N_m(E)$ in $H(m \times m, E/F)$.

We set $\theta(u\overline{u}) = \psi(\theta_0(u\overline{u}))$. The orbital integrals

$$\Omega[\Phi, \psi, E/F : x] = \int \Phi(^t\overline{u}xu)\theta(u\overline{u})\, du$$

of a relevant element x are defined as before: the integral is over the quotient of $N_m(E) \cap M$ by the stabilizer of x. Our goal is to study these orbital integrals in complete analogy with the previous discussion.

It is convenient to introduce the *intermediate orbital integrals*. Let m, n be two integers greater than zero. For $\Phi \in \mathscr{S}(H((m+n) \times (m+n), E/F))$, $A_n \in S_n(F)$,

$B_m \in H(m \times m, E/F)$, we set

$$\Omega^n_m\left[\Phi, E/F, \psi : \begin{pmatrix} A_n & 0 \\ 0 & B_m \end{pmatrix}\right]$$
$$:= \int \Phi\left[\begin{pmatrix} 1_n & 0 \\ {}^t\overline{X} & 1_m \end{pmatrix}\begin{pmatrix} A_n & 0 \\ 0 & B_m \end{pmatrix}\begin{pmatrix} 1_n & X \\ 0 & 1_m \end{pmatrix}\right]$$
$$\times \psi\left[\operatorname{Tr}(\epsilon X) + \operatorname{Tr}({}^t\overline{X}\tilde{\epsilon})\right] dX. \tag{14}$$

Here ϵ and $\tilde{\epsilon}$ are as above. Note that $\tilde{\epsilon} = {}^t\epsilon$. If $X = (x_{i,j})$, then the measure dX is the tensor product of the self-dual Haar measures $dx_{i,j}$. The integral converges and defines a smooth function on $S_n(F) \times M(m \times m, F)$. In particular, we have the following easy result.

LEMMA 3
If Φ is supported on O'_n, then the function $\Omega^n_m(\Phi, \psi : \bullet)$ is a smooth function of compact support on $S_n(F) \times M(m \times m, F)$. Conversely, every smooth function of compact support on that space is of this form for a suitable Φ supported on O'_n.

After a change of variables, we find, keeping in mind that $\det A_n \in F^\times$,

$$\Omega^n_m\left[\Phi, E/F, \psi : \begin{pmatrix} A_n & 0 \\ 0 & B_m \end{pmatrix}\right]$$
$$= |\det A_n|_F^{-2m} \int \Phi\left[\begin{pmatrix} A_n & X \\ {}^t\overline{X} & B_m + {}^t\overline{X}A_n^{-1}X \end{pmatrix}\right]$$
$$\times \psi\left[\operatorname{Tr}(\epsilon A_n^{-1} X) + \operatorname{Tr}({}^t\overline{X}A_n^{-1}\tilde{\epsilon})\right] dX.$$

We also introduce the *normalized* intermediate orbital integrals

$$\tilde{\Omega}^n_m\left[\Phi, E/F, \psi : \begin{pmatrix} A_n & 0 \\ 0 & B_m \end{pmatrix}\right]$$
$$:= \eta(\det A_n)^m |\det A_n|_F^m \times \Phi^n_m\left[\Phi, E/F, \psi : \begin{pmatrix} A_n & 0 \\ 0 & B_m \end{pmatrix}\right]. \tag{15}$$

One can define the orbital integrals of a smooth function of compact support Ψ on $S_n(F) \times H(m \times m, E/F)$ relative to the action of $(N_n(E)) \times (N_m(E))$ on $S_n(F) \times H(m \times m, E/F)$. They are denoted by $\Omega(\Psi, \psi : x_n, x_m)$, where x_n is relevant in $S_n(F)$ and x_m is relevant in $H(m \times m, E/F)$. In particular, consider the case of the function

$$\Psi(A_n, B_m) = \Omega^n_m\left[\Phi, \psi, E/F : \begin{pmatrix} A_n & 0 \\ 0 & B_m \end{pmatrix}\right]$$

and a relevant element x of $H(m \times m, E/F)$ of the form

$$x = \begin{pmatrix} x_n & 0 \\ 0 & x_m \end{pmatrix},$$

where x_n is relevant in $S_n(F)$ and x_m is relevant in $H(m \times m, E/F)$. We then have the reduction formula

$$\Omega[\Phi, E/F, \psi : x] = \Omega[\Psi, E/F, \psi : x_n, x_m]. \tag{16}$$

This formula reduces the computation of the orbital integrals (and the normalization of the measures) to the case of an element of the form $w_n a$, where $a \in F^\times$ is a scalar matrix. Then

$$\begin{aligned} &\Omega[\Psi, E/F, \psi : w_n a] \\ &= \int \Phi\left[\begin{pmatrix} 0 & 0 & 0 & \cdots & 0 & a \\ 0 & 0 & 0 & \cdots & a & ax_{2,n} \\ \cdots & \cdots & \cdots & \cdots & \cdots & \cdots \\ 0 & 0 & a & \cdots & ax_{n-2,n-1} & ax_{n-2,n} \\ 0 & a & ax_{n-1,3} & \cdots & ax_{n-1,n-1} & ax_{n-1,n} \\ a & ax_{n,2} & ax_{n,3} & \cdots & ax_{n,n-1} & ax_{n,n} \end{pmatrix}\right] \\ &\quad \times \psi\Big(\sum_{i=2}^{i=n} x_{i,n-i+2}\Big) \otimes dx_{i,j}, \end{aligned}$$

where $x_{i,i} \in F$, $x_{i,j} \in E$, $x_{i,j} = \overline{x}_{j,i}$, where the measures $dx_{i,j}$, $i < j$, on E and the measures $dx_{i,i}$ on F are self-dual.

We also denote by $\Omega(\Phi, E/F, \psi : a_1, a_2, \dots, a_n)$ the orbital integral $\Omega(\Phi, E/F, \psi : a)$, where $a = \operatorname{diag}(a_1, a_2, \dots, a_n)$ is relevant. We introduce normalized diagonal orbital integrals as follows:

$$\begin{aligned} &\tilde{\Omega}(\Phi, E/F, \psi : a_1, a_2, \dots, a_n) \\ &\quad := \eta\big(\sigma_n(a)\big)|\sigma_n(a)|\Omega(\Phi, E/F, \psi : a_1, a_2, \dots, a_n). \end{aligned} \tag{17}$$

As before, there is a reduction formula for the normalized diagonal orbital integrals.

6. Weil formula: The twisted case

As before, our goal is to obtain a simple relation between the orbital integrals of a function and the orbital integrals of its Fourier transform. Again, our main tool is the Weil formula for the Fourier transform of a character of second order. We first recall the one-variable case. Define the Fourier transform of $\Phi \in \mathscr{S}(E)$ by

$$\hat{\Phi}(z) = \int_E \Phi(u)\psi(-uz - \overline{uz})\, du.$$

Then, for $a \in F^\times$,

$$\int \hat{\Phi}(z)\psi(az\bar{z})\,dz = |a|_F^{-1}\eta(a)c(E/F,\psi)\int \Phi(z)\psi\Big(-\frac{z\bar{z}}{a}\Big)\,dz.$$

We may take this formula as a definition of the constant $c(E/F,\psi)$. We also have

$$\int \Phi(z)\psi(az\bar{z})\,dz = |a|_F^{-1}\eta(a)c(E/F,\psi)\int \hat{\Phi}(z)\psi\Big(-\frac{z\bar{z}}{a}\Big)\,dz.$$

Applying the formula twice, we get the relation

$$c(E/F,\psi)c(E/F,\overline{\psi}) = 1.$$

More generally, if Φ is a Schwartz function on the space of column vectors of size n with entries in E, we define its Fourier transform, a function on the space of row vectors of size n, by

$$\hat{\Phi}(X) = \int \Phi(U)\psi[-ZU - \overline{ZU}]\,dU.$$

PROPOSITION 6

For every matrix $A \in S_n(F)$, we have

$$\int_{E^n} \hat{\Phi}(Z)\psi[ZA\,{}^t\overline{Z}]\,dZ = |\det A|_F^{-1}\eta(\det A)c(E/F,\psi)^n \int_{E^n} \Phi(X)\psi[-{}^t\overline{X}A^{-1}X]\,dX.$$

Proof

If A is a diagonal matrix (with diagonal entries in $F^\times$), the formula follows at once from the case of $n = 1$. In general, we may write

$$A = Ma\,{}^t\overline{M},$$

with $M \in \mathrm{GL}(n, E)$, a diagonal. Then the left-hand side is

$$|\det M|_E^{-1}\int \hat{\Phi}(ZM^{-1})\psi[Za\,{}^t\overline{Z}]\,dZ.$$

Since $Z \mapsto |\det M|_E^{-1}\hat{\Phi}(ZM^{-1})$ is the Fourier transform of $X \mapsto \Phi(MX)$, this is equal to

$$\eta(\det a)|\det a|^{-1}c(E/F,\psi)^n\int \Phi(MX)\psi(-{}^t\overline{X}a^{-1}X)\,dX.$$

After a change of variables, this becomes

$$\eta(\det a)|\det M|_E^{-1}|\det a|^{-1}c(E/F,\psi)^n\int\Phi(X)\psi(-{}^t\overline{X}A^{-1}X)\,dX.$$

Since $|\det A|_F = |\det a|_F|\det M|_E$, our assertion follows. □

On the space $H(n\times n, E/F)$, we define the Fourier transform by

$$\hat{\Psi}(X)=\int\Psi(Y)\psi\big[-\mathrm{Tr}(XY)\big]\,dY,$$

where the measure is self-dual. Thus if $Y=(y_{i,j})$, then $y_{i,i}\in F$, $y_{j,i}=\overline{y}_{i,j}$ for $i<j$, and the measure dY is the tensor product of the self-dual Haar measures $dy_{i,i}$ and $dy_{i,j}$, $i<j$. Note that $\mathrm{Tr}(XY)$ is indeed in F. In $H((n+m)\times(n+m), E/F)$ we consider the subspace of matrices with zero diagonal $n\times n$ and $m\times m$ blocks. The Fourier transform of a function in that space is then defined by

$$\hat{\Phi}\begin{pmatrix}0_{n,n} & {}^t\overline{X}\\ X & 0_{m,m}\end{pmatrix}=\int\Phi\begin{pmatrix}0_{n,n} & V\\ {}^t\overline{V} & 0_{m,m}\end{pmatrix}\psi\big[-\mathrm{Tr}(XV+\overline{XV})\big]\,dV.$$

This being so, the form of Weil formula that we are using is as given in the next proposition.

PROPOSITION 7

Let A, B be Hermitian matrices in $\mathrm{GL}(n,E)$ *and* $\mathrm{GL}(m,E)$, *respectively. Then*

$$\int\hat{\Phi}\begin{pmatrix}0_{n,n} & {}^t\overline{Z}\\ Z & 0_{m,m}\end{pmatrix}\psi\big[\mathrm{Tr}(BZA\,{}^t\overline{Z})\big]\,dZ$$
$$=\eta(\det A)^m|\det A|_F^{-m}\eta(\det B)^n|\det B|_F^{-n}c(E/F,\psi)^{mn}$$
$$\times\int\Phi\begin{pmatrix}0_{n,n} & X\\ {}^t\overline{X} & 0_{m,m}\end{pmatrix}\psi\big[-\mathrm{Tr}(B^{-1}\,{}^t\overline{X}A^{-1}X)\big]\,dX.$$

Proof

Let us observe that $\mathrm{Tr}(BZA\,{}^t\overline{Z})$ is indeed in F. As before, we may reduce the computation to the case where B is the diagonal matrix $\mathrm{diag}(b_1,b_2,\dots,b_m)$. We may regard Z as a matrix

$$Z=\begin{pmatrix}Z_1\\ Z_2\\ \dots\\ Z_m\end{pmatrix},$$

where each Z_i is a row of size n. Then the formula follows from the previous formula applied to the matrices $b_1A, b_2A,\dots,b_mA$. □

7. Inversion formula in the Hermitian case

As before, we set

$$\check{\Psi}(X) = \hat{\Psi}(w_m X w_m).$$

We then have the following result.

PROPOSITION 8

Let $n \geq 1$, $m \geq 1$ be two integers. Let $\Psi \in \mathscr{S}\big(H((m+n)\times(m+n), E/F)\big)$. Then we have

$$\int \left(\int \tilde{\Omega}_m^n \left[\Psi, E/F, \psi : \begin{pmatrix} A_n & 0 \\ 0 & B_m \end{pmatrix} \right] \psi[-\operatorname{Tr} B_m w_m C_m w_m]\, dB_m \right) \\ \times \psi\big[\operatorname{Tr}(w_m C_m^{-1} w_m \epsilon A_n^{-1} \tilde{\epsilon})\big] \psi\big[-\operatorname{Tr}(A_n w_n D_n w_n)\big]\, dA_n \\ = c(E/F, \psi)^{mn} \tilde{\Omega}_n^m \left[\check{\Psi}, E/F, \overline{\psi} : \begin{pmatrix} C_m & 0 \\ 0 & D_n \end{pmatrix} \right].$$

Proof

The proof follows step by step the proof of the corresponding result for $M((n+m)\times(n+m), F)$. We consider the partial Fourier transform Θ (with respect to B_n) of the normalized intermediate orbital integral of Ψ:

$$\Theta(A_n, C_m) := \int \tilde{\Omega}_m^n \left[\Psi, E/F, \psi : \begin{pmatrix} A_n & 0 \\ 0 & B_m \end{pmatrix} \right] \psi[-\operatorname{Tr} B_m C_m]\, dB_m.$$

The critical step is as follows. We get for Θ:

$$\eta(\det A_n)^m |\det A_n|^{-m} \int \Psi \left[\begin{pmatrix} A_n & X \\ {}^t\overline{X} & B_m \end{pmatrix} \right] \\ \times \psi\big[\operatorname{Tr}(\epsilon A_n^{-1} X) + \operatorname{Tr}({}^t\overline{X} A_n^{-1} \tilde{\epsilon}) + \operatorname{Tr}(C_m {}^t\overline{X} A_n^{-1} X)\big]\, dX \\ \times \psi\big[-\operatorname{Tr}(C_m B_m)\big]\, dB_m.$$

We now introduce a partial Fourier transform of Ψ:

$$\Psi_1 \begin{pmatrix} A_n & U \\ {}^t\overline{U} & C_m \end{pmatrix} := \int \Psi \begin{pmatrix} A_n & X \\ {}^t\overline{X} & B_m \end{pmatrix} \psi \left[-\operatorname{Tr} \begin{pmatrix} 0 & X \\ {}^t\overline{X} & B_m \end{pmatrix} \begin{pmatrix} 0 & U \\ {}^t\overline{U} & C_m \end{pmatrix} \right] dX\, dB_m.$$

By the Weil formula, the previous expression is then

$$|\det C_m|^{-n} \eta(\det C_m)^n c(E/F, \psi)^{mn} \\ \times \int \Psi_1 \begin{pmatrix} A_n & X \\ {}^t\overline{X} & C_m \end{pmatrix} \psi\big[-\operatorname{Tr} C_m^{-1}({}^t\overline{X} + \epsilon A_n^{-1}) A_n (X + A_n^{-1}\tilde{\epsilon})\big]\, dX.$$

The rest of the proof is identical to the proof of Proposition 3. □

Our main result in this section is the following inversion formula.

THEOREM 2

For $\Phi \in \mathscr{S}(H(n \times n, E/F))$,

$$c(E/F,\psi)^{(n(n-1))/2}\tilde{\Omega}(\check{\Phi}, \overline{\psi}, E/F : a_1, a_2, \dots, a_n)$$
$$= \int \tilde{\Omega}(\Phi, \psi, E/F : p_1, p_2, \dots, p_n)$$
$$\times \psi\Big(-\sum_{i=1}^{i=n} p_i a_{n+1-i} + \sum_{i=1}^{i=n-1} \frac{1}{p_i a_{n-i}}\Big)\, dp_n\, dp_{n-1} \cdots dp_1,$$

where the integral is only an iterated integral.

Proof

Again the formula is trivial for $n = 1$, so we may assume that $n > 1$ and that our assertion is true for $n-1$. We prove it for n. We apply the formula of Proposition 8 to the pair of integers $(n-1, 1)$. We obtain

$$\int \Big(\int \tilde{\Omega}_1^{n-1}\Big[\Psi, E/F, \psi : \begin{pmatrix} A_{n-1} & 0 \\ 0 & p_n \end{pmatrix}\Big] \psi[-p_n a_1]\, dp_n\Big)$$
$$\times \psi\big[\operatorname{Tr}(a_1^{-1}\epsilon_{n-1}A_{n-1}^{-1}\tilde{\epsilon}_{n-1})\big]\psi\big[-\operatorname{Tr}(A_{n-1}w_{n-1}D_{n-1}w_{n-1})\big]\, dA_{n-1}$$
$$= c(E/F, \psi)^{n-1}\tilde{\Omega}_{n-1}^1\Big[\check{\Psi}, E/F, \overline{\psi} : \begin{pmatrix} a_1 & 0 \\ 0 & D_{n-1} \end{pmatrix}\Big].$$

We set

$$\Psi_1(A_{n-1}) := \psi\big[\operatorname{Tr}(a_1^{-1}\epsilon_{n-1}A_{n-1}^{-1}\tilde{\epsilon}_{n-1})\big]$$
$$\times \int \tilde{\Omega}_1^{n-1}\Big[\Psi, E/F, \psi : \begin{pmatrix} A_{n-1} & 0 \\ 0 & p_n \end{pmatrix}\Big][-p_n a_1]\, dp_n.$$

Then $\check{\Psi}_1(D_{n-1}) = \Theta(D_{n-1})$, where

$$\Theta(D_{n-1}) = c(E/F, \psi)^{n-1}\tilde{\Omega}_{n-1}^1\Big[\check{\Psi}, E/F, \overline{\psi} : \begin{pmatrix} a_1 & 0 \\ 0 & D_{n-1} \end{pmatrix}\Big].$$

The induction hypothesis applied to Ψ_1 gives

$$c(E/F,\psi)^{(n-1)(n-2)/2}\tilde{\Omega}(\Theta, E/F, \overline{\psi} : a_2, a_3, \dots, a_n)$$
$$= \int \tilde{\Omega}(\Psi_1, E/F, \psi : p_1, p_2, \dots, p_{n-1})$$
$$\times \psi\Big(-\sum_{i=1}^{i=n-1} p_i a_{n+1-i} + \sum_{i=1}^{i=n-2} \frac{1}{p_i a_{n-i}}\Big)\, dp_{n-1} \cdots dp_1. \qquad (18)$$

The rest of the proof is the same as before once we observe that

$$c(E/F, \psi)^{n-1} c(E/F, \psi)^{(n-1)(n-2)/2} = c(E/F, \psi)^{n(n-1)/2}. \qquad \square$$

We leave it to the reader to formulate and prove the analogue of Proposition 4 in the present situation.

8. Smooth matching

Definition 1

We say that a function $\Phi \in \mathscr{S}(M(n \times n, F))$ and a function $\Psi \in \mathscr{S}(H(n \times n, E/F))$ have matching orbital integrals for ψ, and we write $\Phi \overset{\psi}{\longleftrightarrow} \Psi$ if for every diagonal matrix $a \in \mathscr{A}_n$ with $\Delta_{n-1}(a) \neq 0$,

$$\tilde{\Omega}(\Phi, \psi; a) = \tilde{\Omega}(\Psi, E/F, \psi : a).$$

From the inversion formula for the orbital integrals, we have at once the following result.

PROPOSITION 9

If $\Phi \overset{\psi}{\longleftrightarrow} \Psi$, *then* $\check{\Phi} \overset{\overline{\psi}}{\longleftrightarrow} c(E/F, \psi)^{n(n-1)/2} \check{\Psi}$, *and conversely.*

Now we can formulate our main result.

THEOREM 3

Given $\Phi \in \mathscr{S}(M(n \times n, F))$, *there is* $\Psi \in \mathscr{S}(H(n \times n, E/F))$ *with matching orbital integrals for* ψ, *and conversely.*

Proof

We treat the case of Φ. The case of Ψ is similar. The case of $n = 1$ being trivial, we may assume that $n > 1$ and that our assertion is proved for $n' < n$. For $i < n$ we can consider smooth functions of compact support on $\mathrm{GL}(i, F) \times M((n-i) \times (n-i), F)$ and $S_i(F) \times H((n-i) \times (n-i), E/F)$, respectively, and their orbital integrals for the action of the groups $N_i(F) \times N_i(F) \times N_{n-i}(F) \times N_{n-i}(F)$ and $N_i(E) \times N_{n-i}(E)$, respectively. It follows from the induction hypothesis that any function on the first space matches a function on the second space for ψ. Now let O_i (resp., O'_i) be the open set of $M(n \times n, F)$ (resp., $H(n \times n, E/F)$) defined by $\Delta_i \neq 0$. If Φ is supported on O_i, then its normalized intermediate orbital integral

$$\tilde{\Omega}^i_{n-i}[\Phi, \psi : g_i, m_{n-i}]$$

is a smooth function of compact support $\Phi_{i,n-i}$ on $\mathrm{GL}(i, F) \times M((n-i) \times (n-i), F)$; it matches a function of compact support $\Psi_{i,n-i}$ on $S_i(F) \times H((n-i) \times (n-i), E/F)$, which in turn is the normalized intermediate orbital integral of a function Ψ supported on O'_i. Now consider diagonal matrices $a_i \in A_i$, $a_{n-i} \in M((n-i) \times (n-i), F)$, with $\Delta_{n-i-1}(a_{n-i}) \neq 0$. Then we have

$$\tilde{\Omega}\left[\Phi, \psi : \begin{pmatrix} a_i & 0 \\ 0 & a_{n-i} \end{pmatrix}\right] = \tilde{\Omega}\left[\Phi_{i,n-i}, \psi : a_i, a_{n-i}\right].$$

There is a similar formula for Ψ. Thus, in fact, $\Phi \overset{\psi}{\longleftrightarrow} \Psi$. Hence our assertion is true for Φ supported on O_i, $i < n$.

Next, suppose that Φ is such that

$$\Omega(\Phi, \psi : w_n a) = 0$$

for all $a \in F^\times$. Our classification of relevant orbits shows then that all the integrals of Φ over relevant orbits contained in the closed set Q defined by

$$\Delta_1 = \Delta_2 = \cdots = \Delta_{n-1} = 0$$

vanish. Hence Φ has the same orbital integrals as a function supported on the complement of Q. We may assume that the support of Φ is contained in the complement of Q. Using a partition of unity, we see that we can write

$$\Phi = \sum_{i=1}^{i=n-1} \Phi_i,$$

where Φ is supported on O_i. Then $\Phi_i \overset{\psi}{\longleftrightarrow} \Psi_i$, where the function Ψ_i is supported on O'_i and

$$\Phi \overset{\psi}{\longleftrightarrow} \sum_{i=1}^{i=n-1} \Psi_i.$$

Now let Φ be an arbitrary function. Then

$$\phi(a) := |a|^{(n+1)(n-1)} \Omega[\Phi, \psi : w_n a]$$

is a smooth function of compact support on $F^\times$. Let Φ_1 be a function of compact support contained in O_{n-1}. Then

$$\Omega^1_{n-1}\left[\Phi_1, \overline{\psi} : \begin{pmatrix} g_{n-1} & 0 \\ 0 & b \end{pmatrix}\right]$$

is an arbitrary smooth function of compact support on $\mathrm{GL}(n-1, F) \times F$. In particular, we can choose Φ_1 so that

$$\phi(a) = \int \Omega\left[\Phi_1, \overline{\psi} : \begin{pmatrix} -a^{-1} w_{n-1} & 0 \\ 0 & p \end{pmatrix}\right] dp.$$

Let Φ_2 be the function such that $\check{\Phi}_2 = \Phi_1$. Since Φ_1 matches a function for $\overline{\psi}$, it follows from Proposition 9 that $\Phi_2 \overset{\psi}{\longleftrightarrow} \Psi_2$ for a suitable function Ψ_2. On the other hand, by Proposition 4 applied to Φ_2,

$$\Omega(\Phi, \psi : w_n a) = \Omega(\Phi_2, \psi : w_n a),$$

so that

$$\Omega(\Phi - \Phi_2, \psi : w_n a) = 0.$$

By the previous case, it follows that $\Phi - \Phi_2 \overset{\psi}{\longleftrightarrow} \Psi_3$ for a suitable function Ψ_3. Then $\Phi \overset{\psi}{\longleftrightarrow} \Psi_2 + \Phi_3$ and our assertion follows. □

9. Application to the fundamental lemma

Suppose that E/F is unramified, and suppose that the conductor of ψ is $\mathcal{O}_F$. Let Φ_0 be the characteristic function of the set of matrices with integral entries in $M(m \times m, F)$. Similarly, let Ψ_0 be the characteristic function of the set of matrices with integral entries in $H(m \times m, E/F)$. The *fundamental lemma* asserts that Φ_0 matches Ψ_0. We prove this for $m = 2$ and $m = 3$. The case where $|\det a| = 1$ is already known, but the proof presented below is much simpler. First $\tilde{\Omega}(\Phi_0, \psi : a) = \tilde{\Omega}(\Phi_0, \overline{\psi} : a)$, and likewise for Ψ_0. It follows that the difference

$$\omega(a) := \tilde{\Omega}(\Phi_0, \psi : a) - \tilde{\Omega}(\Psi_0, E/F, \psi : a)$$

satisfies the functional equation

$$\omega(a_1, a_2, \ldots, a_m) = \int \omega(p_1, p_2, \ldots, p_m) \times \psi\Big(-\sum_{i=1}^{i=m} p_i a_{m+1-i} + \sum_{i=1}^{i=m-1} \frac{1}{p_i a_{m-i}}\Big) dp_m\, dp_{m-1} \cdots dp_1. \tag{19}$$

Moreover, it is supported on the set $|a_1 a_2 \cdots a_m| \le 1$. The fundamental lemma asserts that $\omega = 0$.

For $m = 2$ it is easily checked that

$$\int \Omega(\Phi_0, \psi : a_1, a_2)\, da_2$$

is 1 if $|a_1| = 1$ and 0 otherwise, and likewise for

$$\int \Omega(\Phi_0, E/F, \psi : a_1, a_2)\, da_2.$$

Thus

$$\int \omega(a_1, a_2) da_2 = 0.$$

If we set

$$\mu(a_1, b_1) := \int \omega(a_1, p_2)\psi(-p_2 b_1)\, dp_2,$$

then (19) is equivalent to the relation

$$\mu(a_1, -b_1) = \mu(b_1, a_1)\psi\Big(\frac{1}{a_1 b_1}\Big). \tag{20}$$

Moreover, $\mu(a_1, b_1 + t) = \mu(a_1, b_1)$ for $|t| \leq |a_1|$. Since $\mu(a_1, 0) = 0$, we see that $\mu(a_1, b_1) \neq 0$ implies $|b_1| > |a_1|$. By (20), $\mu(a_1, b_1) \neq 0$ also implies $|b_1| > |a_1|$. Thus $\mu = 0$ or, equivalently, $\omega = 0$, and we are done.

For $m = 3$ we first establish the following lemma.

LEMMA 4
If

$$\int \omega(a_1, a_2, a_3)\, da_3 = 0,$$

then $\omega = 0$.

Proof
Indeed, if we set

$$\mu(a_1, a_2, b_1) := \int \omega(a_1, a_2, a_3)\psi(-a_3 b_1)\, da_3$$

and

$$\sigma(a_1, a_2, b_1) := \mu(a_1, a_2, b_1)\psi\Big(\frac{1}{a_2 b_1}\Big),$$

then (19) is equivalent to

$$\sigma(a_1, a_2, -b_1) = \int \sigma(b_1, p_2, a_1)\psi(p_2 a_2)\, dp_2. \tag{21}$$

The condition on the support of ω implies that $\mu(a_1, a_2, b_1 + t) = \mu(a_1, a_2, b_1)$ for $|t| \leq |a_1 a_2|$. Under the assumption of the lemma, we thus have $\mu(a_1, a_2, b_1) = 0$ for $|b_1| \leq |a_1 a_2|$. Equivalently, $\sigma(a_1, a_2, b_1) \neq 0$ implies $|a_2| \leq |\varpi b_1 a_1^{-1}|$. From (21) we have $\sigma(a_1, a_2 + t, b_1) = \sigma(a_1, a_2, b_1)$ for $|t| \leq |\varpi^{-1} b_1 a_1^{-1}|$. Thus $\sigma = 0$ and $\omega = 0$. □

Thus it suffices to prove the relation

$$\int \Omega(\Phi_0, \psi : a_1, a_2, a_3)\, da_3 = \eta(a_2) \int \Omega(\Psi_0, E/F, \psi : a_1, a_2, a_3)\, da_3.$$

Now $\int \Omega_1^2(\Phi_0, \psi : g, a_3)\, da_3$ is $|\det g|_F^{-2}$ times the characteristic function Φ_1 of the set of $g \in M(2 \times 2, F)$ such that

$$\|g\| \leq 1, \qquad \|(0,1)g^{-1}\| \leq 1, \qquad \|g^{-1\,t}(0,1)\| \leq 1.$$

Likewise, $\int \Omega_1^2(\Psi_0, E/F, \psi : g, a_3)\, da_3$ is $|\det g|_F^{-2}$ times the characteristic function Ψ_1 of the set of $g \in H(2 \times 2, E/F)$ such that

$$\|g\| \leq 1, \qquad \|(0,1)g^{-1}\| \leq 1.$$

Thus it suffices to show that

$$\Omega(\Phi_1, \psi : a_1, a_2) = \eta(a_2)\Omega(\Psi_1, E/F, \psi : a_1 a_2). \tag{22}$$

If $|a_1 a_2| = 1$, then this relation is equivalent to

$$\Omega(\Phi', \psi : a_1, a_2) = \eta(a_1)\Omega(\Psi', E/F, \psi : a_1 a_2),$$

where Φ' is the characteristic function of $\mathrm{GL}(2, \mathcal{O}_F)$ and Ψ' is the characteristic function of $\mathrm{GL}(2, \mathcal{O}_E) \cap H(2 \times 2, E/F)$; in turn, this relation is a special case of the fundamental lemma for $m = 2$.

Now

$$\Omega(\Phi_1, \psi : a_1, a_2) = \int \psi(x_1 + x_2)\, dx_1\, dx_2$$

over the set

$$|a_2| \geq 1, \qquad |a_1 x_i| \leq 1, \qquad |x_i| \leq |a_1|^{-1}, \qquad |a_2 + a_1 x_1 x_2| \leq 1.$$

If $|a_2| = 1$, then this is 1. If $|a_2| > 1$, then this is zero unless $|a_1 a_2| = 1$. A similar remark applies to $\Omega(\Psi_0, E/F, \psi : a_1, a_2)$. The relation (22) follows. We are done.

10. Concluding remarks

More generally, one can use Propositions 3 and 4 to prove inductively the theorem of density established in [3]; in a precise way, if Φ is such that $\Omega(\Phi, \psi : a) = 0$ for all diagonal matrices a, then $\Omega(\check{\Phi}, \overline{\psi} : a) = 0$, and $\Omega(\Phi, \psi : g) = 0$, $\Omega(\check{\Phi}, \overline{\psi} : g) = 0$ for all relevant g. Similarly, one can prove that if $\Phi \overset{\psi}{\longleftrightarrow} \Psi$, then for every g in the common set of representatives for the two sets of orbits,

$$\Omega(\Phi, \psi : g) = \gamma(g, \psi)\Omega(\Psi, E/F, \psi : g),$$

where the constant $\gamma(g, \psi)$ (the *transfer factor*) does not depend on the functions, a result that also follows from [3]. This gives another way to compute the transfer factors.

Since the determinant is an invariant of the two actions, it follows from the theorem that for any smooth function of compact support on $\mathrm{GL}(m, F)$ there is a function of compact support Ψ on $S_m(F)$ such that $\Phi \overset{\psi}{\longleftrightarrow} \Psi$.

Except for the last section, the previous discussion applies directly to the extension $\mathbb{C}/\mathbb{R}$.

Suppose that E/F is unramified, and suppose that the conductor of ψ is $\mathcal{O}_F$. Then $c(E/F, \psi) = 1$. If E/F is a quadratic extension of global fields and ψ is a nontrivial character of $F_{\mathbb{A}}/F$, then the product of the constants $c(E_v/F_v, \psi_v)$ over all places v of F inert in E is 1. Likewise, for g relevant in $\mathrm{GL}(n, F) \cap S_n(F)$ the product of the local transfer factors is 1.

References

[1] D. BUMP, S. FRIEDBERG, and D. GOLDFELD, "Poincaré series and Kloosterman sums" in *The Selberg Trace Formula and Related Topics (Brunswick, Maine, 1984)*, Contemp. Math. **53**, Amer. Math. Soc., Providence, 1986, 39–49. MR 87j:11043

[2] H. JACQUET, *The continuous spectrum of the relative trace formula for* GL(3) *over a quadratic extension*, Israel J. Math. **89** (1995), 1–59. MR 96a:22029

[3] ———, *A theorem of density for Kloosterman integrals*, Asian J. Math. **2** (1998), 759–778. MR 2001e:11054

[4] H. JACQUET and Y. YE, *Relative Kloosterman integrals for* GL(3), Bull. Soc. Math. France **120** (1992), 263–295. MR 94c:11047

[5] ———, *Distinguished representations and quadratic base change for* GL(3), Trans. Amer. Math. Soc. **348** (1996), 913–939. MR 96h:11041

[6] ———, *Germs of Kloosterman integrals for* GL(3), Trans. Amer. Math. Soc. **351** (1999), 1227–1255. MR 99j:11053

[7] ———, *Kloosterman integrals over a quadratic extension*, to appear in Ann. of Math. (2).

[8] E. LAPID and J. ROGAWSKI, *Periods of Eisenstein series: The Galois case*, Duke Math. J. **120** (2003), 153–226.

[9] B. C. NGÔ, *Faisceaux pervers, homomorphisme de changement de base et lemme fondamental de Jacquet et Ye*, Ann. Sci. École Norm. Sup. (4) **32** (1999), 619–679. MR 2001g:11076

[10] ———, *Le lemme fondamental de Jacquet et Ye en caractéristique positive*, Duke Math. J. **96** (1999), 473–520. MR 2000f:11059

[11] T. A. SPRINGER, "Some results on algebraic groups with involutions" in *Algebraic Groups and Related Topics (Kyoto/Nagoya, 1983)*, Adv. Stud. Pure Math. **6**, North-Holland, Amsterdam, 1985, 525–543. MR 86m:20050

[12] G. STEVENS, *Poincaré series on* GL(r) *and Kloostermann sums*, Math. Ann. **277** (1987), 25–51. MR 88m:11031

[13] J. L. WALDSPURGER, *Le lemme fondamental implique le transfert*, Compositio Math. **105** (1997), 153–236. MR 98h:22023

[14] Y. YE, *The fundamental lemma of a relative trace formula for* GL(3), Compositio Math. **89** (1993), 121–162. MR 95b:22023

[15] ———, *An integral transform and its applications*, Math. Ann. **30** (1994), 405–417. MR 95j:11045

Department of Mathematics, Columbia University, Mail Code 4408, New York, New York 10027, USA; hj@math.columbia.edu

Ann. Scient. Éc. Norm. Sup.,
4e série, t. 38, 2005, p. 609 à 669.

KLOOSTERMAN IDENTITIES OVER A QUADRATIC EXTENSION II

BY HERVÉ JACQUET[1]

To the memory of Yasuko Jacquet

ABSTRACT. – We prove certain identities between Kloosterman integrals. They constitute the fundamental lemma of a relative trace formula for Hecke functions. The main application of the trace formula in question is the following result. Let E/F be a quadratic extension of number fields. A cuspidal automorphic representation of $GL(n, E_{\mathbb{A}})$ is distinguished by some unitary group if and only if it is the base change of an automorphic cuspidal representation of $GL(n, F_{\mathbb{A}})$.

RÉSUMÉ. – Nous prouvons certaines identités entre intégrales de Kloosterman. Elles constituent le lemme fondamental d'une formule des traces relatives pour les fonctions de Hecke générales. L'application principale de cette formule des traces est le résultat suivant. Soit E/F une extension quadratique de corps de nombres. Une représentation automorphe cuspidale de $GL(n, E_{\mathbb{A}})$ est distinguée par un groupe unitaire si et seulement si c'est le changement de base d'une représentation automorphe cuspidale de $GL(n, F_{\mathbb{A}})$.

Contents

[1] Partially supported by NSF grant DMS-02-45310.

ANNALES SCIENTIFIQUES DE L'ÉCOLE NORMALE SUPÉRIEURE
0012-9593/04/

1. The main result

Let F be a non-Archimedean local field of odd residual characteristic and characteristic 0. We denote by ϖ a uniformizer, by q the cardinality of the residual field κ and by $|x|_F$ or simply $|x|$ the absolute value of an element x. We consider the quadratic unramified extension E/F. Thus ϖ is also a uniformizer for E. We let $\eta_{E/F}$ or simply η be the corresponding unramified quadratic character of $F^\times$. When necessary we use the notation $|x|_F$ and $|z|_E$ to denote the absolute value on F and E respectively. However we adopt the following convention: when we write an inequality (or an equality) such as $|z| \leqslant |\varpi^m|$ the same absolute value is used on the two sides of the inequality: thus the inequality stands for either $|z|_E \leqslant |\varpi^m|_E$ or $|z|_F \leqslant |\varpi^m|_F$. We denote by σ the Galois conjugation of E/F. We also write $\sigma(z) = \bar{z}$. Thus $|z|_E = |z\bar{z}|_F$. We let ψ_F or simply ψ be an additive character of F whose conductor is the ring of integers $\mathcal{O}_F$ of F. Then $\psi_E(z) := \psi_F(z + \bar{z})$ is an additive character of E whose conductor is $\mathcal{O}_E$.

We denote by G_n or simply G the group $GL(n)$. We let N_n or simply N be the subgroup of upper triangular matrices with unit diagonal and A_n or simply A the subgroup of diagonal matrices. We define a character

$$\theta : N_n(F) \to \mathbb{C}^\times$$

by

$$\theta(u) = \psi\Big(\sum_{1 \leqslant i \leqslant n-1} u_{i,i+1}\Big).$$

Let $\mathcal{S}(M(n \times n, F))$ be the space of locally constant compactly supported functions on $M(n \times n, F)$. We define the (diagonal) orbital integral of a function $\Phi \in \mathcal{S}(M(n \times n, F))$. It is the function on $A_n(F)$ defined by

$$\Omega[\Phi, \psi : a] := \int \Phi({}^t u_1 a u_2)\theta(u_1 u_2)\, du_1\, du_2.$$

The integral is over $N_n(F) \times N_n(F)$ and the Haar measure du is normalized by the condition

$$\int_{N_n(\mathcal{O}_F)} du = 1.$$

In fact, the orbital integral is defined on the set of diagonal matrices $a \in M(n \times n, F)$, with $a_1 a_2 \cdots a_{n-1} \neq 0$. We let Φ_0 be the characteristic function of $M(n \times n, \mathcal{O}_F)$. We remark that if we define Φ^t by $\Phi^t(X) = \Phi({}^tX)$ then

$$\Omega[\Phi, \psi : a] = \Omega[{}^t\Phi, \psi : a].$$

Similarly, we let $M_h(E/F, n)$ be the space of $n \times n$ Hermitian matrices, that is, the matrices $x \in M(n \times n, E)$ such that ${}^tx = \bar{x}$. We let Ψ_0 be the characteristic function of

$$M_h(E/F, n) \cap M(n \times n, \mathcal{O}_E).$$

We define a character

$$u \mapsto \theta(u\bar{u})$$

of $N_n(F)$ by

$$\theta(u\bar{u}) = \psi\Bigg(\sum_{1 \leqslant i \leqslant n-1} u_{i,i+1} + \bar{u}_{i,i+1}\Bigg).$$

The notation is justified by the fact that $u\bar{u}$ is the product of an element of $N(F)$ and an element of the derived group of $N(E)$.

If Ψ is in $\mathcal{S}(M_h(E/F, n))$ we define its (diagonal) orbital integral. It is the function on $A_n(F)$ defined by

$$\Omega[\Psi, E/F, \psi : a] := \int \Psi({}^t\bar{u}au)\theta(u\bar{u})\, du.$$

The integral is over $N_n(E)$ and the Haar measure du is normalized by the condition

$$\int_{N_n(\mathcal{O}_E)} du = 1.$$

Again the integral is in fact defined on the set of diagonal matrices $a \in M(n \times n, F)$ such that $a_1a_2 \cdots a_{n-1} \neq 0$. Again

$$\Omega\big[{}^t\Psi, E/F, \psi : a\big] = \Omega[\Psi, E/F, \psi : a].$$

We say that Ψ and Φ *match* and we write $\Psi \leftrightarrow \Phi$ if

(1) $$\Omega[\Psi, E/F, \psi : a] = \gamma(a)\Omega[\Phi, \psi : a]$$

where γ is the *transfer factor* defined by

(2) $$\gamma(a) =: \eta(a_1)\eta(a_1a_2)\cdots\eta(a_1a_2\cdots a_{n-1}).$$

We set

$$K_F := GL(n, \mathcal{O}_F), \qquad K_E := GL(n, \mathcal{O}_E),$$

we denote by $\mathcal{H}(GL(n, F))$ or simply $\mathcal{H}_F$ the Hecke algebra, that is, the convolution algebra of bi-K_F-invariant functions of compact support on $GL(n, E)$. We define similarly $\mathcal{H}(GL(n, E)) = \mathcal{H}_E$.

We recall the base change homomorphism

$$b:\mathcal{H}(GL(n,E)) \to \mathcal{H}(GL(n,F))$$

of the Hecke algebras. If $\mathcal{S}(f)$ denotes the Satake transform of f then

$$\mathcal{S}(b(f))(x) = \mathcal{S}(f)(x^2).$$

Equivalently, if π is an unramified representation of $G(F)$ corresponding to the character $f \mapsto \mathcal{S}(f)(x)$, we write $\mathcal{S}(f)(\pi)$ for $\mathcal{S}(f)(x)$. Then if Π is the representation base change of π we have

$$\mathcal{S}(f)(\Pi) = \mathcal{S}(b(f))(\pi).$$

Our main result is as follows (*fundamental lemma for Hecke functions*).

THEOREM 1. – *For any $f \in \mathcal{H}(GL(n,E))$ the function*

$$X \mapsto \int_{GL(n,E)} \Psi_0({}^t\bar{g}Xg) f(g)\,dg \tag{3}$$

matches the function

$$X \mapsto \int_{GL(n,F)} \Phi_0(Xg) b(f)(g)\,dg. \tag{4}$$

We remark that the same result is true in the case of positive characteristic. This can be derived from the work of Ngô [15]. The proof we give should extend to the case of characteristic, provided the necessary modifications be made in the global theory.

Of course the Haar measures are normalized by the condition that the measures of K_E and K_F are 1. If f is the unit element of the Hecke algebra this result was proved in [6, (8)].

We remark that the second function has the same orbital integral as the function

$$X \mapsto \int_{GL(n,F)} \Phi_0(gX) b(f)(g)\,dg.$$

For any function f on a group we define the function $\check{f}$ by $\check{f}(g) = f(g^{-1})$. We first note that $b(\check{f}) = b(f)\check{}$. Thus, at the cost of replacing the function f by the function $\check{f}$ we may formulate the result as the matching of the functions

$$X \mapsto \int_{GL(n,E)} \Psi_0({}^t\bar{g}Xg) f(g^{-1})\,dg$$

and

$$X \mapsto \int_{GL(n,F)} \Phi_0(Xg) b(f)(g^{-1})\,dg.$$

Now let us apply the theorem to a function f supported on the set $\{g \in GL(n,E)\colon |\det g|_E = |\varpi^m|_E\}$. This means that the Satake transform $\mathcal{S}(f)$ is a homogeneous Laurent polynomial of degree m. Thus $\mathcal{S}(b(f))$ is homogeneous of degree $2m$, that is, the function $b(f)$ is supported on the set $\{g \in GL(n,F)\colon |\det g|_F = |\varpi^{2m}|_F\}$. Next the functions

$$X \mapsto \mathbf{1}_{\{X\colon |\det X|_E = |\varpi^{-2m}|_E\}} \int\limits_{GL(n,E)} \Psi_0({}^t\bar{g}Xg) f(g)\, dg$$

and

$$X \mapsto \mathbf{1}_{\{X\colon |\det X|_F = |\varpi^{-2m}|_F\}} \int\limits_{GL(n,F)} \Phi_0(Xg) b(f)(g)\, dg$$

match because the determinant is constant on the orbits. In other words, the functions

$$\int\limits_{GL(n,E)} \Psi^1({}^t\bar{g}Xg) f(g)\, dg$$

and

$$\int\limits_{GL(n,F)} \Phi^1(Xg) b(f)(g)\, dg$$

match, where we have denoted by Ψ^1 and Φ^1 the characteristic functions of $K_E \cap H(n \times n, E/F)$ and K_F respectively. We have established this when f is supported on a set $\{g \in GL(n,E)\colon |\det g|_E = |\varpi^m|_E\}$. By linearity, it is true for all f.

Equivalently, for any f, the functions

$$\int\limits_{GL(n,E)} \Psi^1({}^t\bar{g}Xg) f(g^{-1})\, dg$$

and

$$\int\limits_{GL(n,F)} \Phi^1(Xg) b(f)(g^{-1})\, dg$$

match. This is the result established by Ngô in the case of positive characteristic in [15]. With minor modifications in the global theory our proof applies to that case as well.

The above result (for general Hecke functions) was first established by Ye Yangbo in the case $n = 3$. His slightly different approach to the problem influenced me, as well as the notes he sent me. The reduction formulas are directly borrowed from his notes.

I am very grateful to Erez Lapid for useful discussions and a careful reading of the manuscript. His insistence that a global argument could be used to finish the proof was very encouraging. Moreover, his result in [10] is used in an essential way in the present paper. I also thank the referee for a very careful reading of the manuscript.

The material is arranged as follows. In Section 3 we explain the global consequences of our main result. This will serve as an introduction. In Section 4 we reformulate the result as a matching of linear combination of characteristic functions of lattices, namely

$$\Psi_\mu \leftrightarrow \sum_\lambda \theta_\mu^\lambda \delta_F^{1/2}(\varpi^{2\mu-\lambda}) \Phi_\lambda(X),$$

with specific constants θ_μ^λ (see below for notation). In Sections 5–11 we develop the machinery which allows us to conclude that there is a matching of this form, where the constants θ_μ^λ are replaced by unknown constants ξ_μ^λ. This amounts to saying there is a linear map $\beta:\mathcal{H}_E \to \mathcal{H}_F$ such that

$$X \mapsto \int_{GL(n,E)} \Psi_0({}^t\bar{g}Xg)f(g)\,dg$$

and

$$X \mapsto \int_{GL(n,F)} \Phi_0(Xg)\beta(f)(g)\,dg$$

match. The assertion that $\theta_\mu^\lambda = \xi_\mu^\lambda$ is then equivalent to the equality $\beta = b$. We check that β coincides with b on the unit element and on a set of generators of the Hecke algebra in Section 12. To finish the proof we discuss a *simple trace formula* in Section 13, where the base change b is replaced by the linear map β. We conclude in Section 14 that β is an algebra homomorphism. Thus $\beta = b$. In Section 15 we discuss three lemmas about local distributions. In Section 16 we briefly discuss more precise global results. We also indicate how to remove the restrictions on the global quadratic extension of Section 2.

2. General matching

For the convenience of the reader we review the general notion of matching [6, (5) and (6)]. Thus we let E/F be an arbitrary quadratic extension of non-Archimedean local fields. We let ψ_F or simply ψ be a non-trivial additive character of F. Then dx is the self dual Haar measure on F and dz the Haar measure on E self dual with respect to the character $\psi_E(z) = \psi_F(z + \bar{z})$.

We let $S(F)$ be the space of invertible Hermitian matrices. The group $G(E)$ operates on $S(F)$ by

$$s \mapsto {}^t\bar{g}sg.$$

For $\Psi \in C_c^\infty(S(F))$ we define as before the diagonal orbital integral

$$\Omega[\Psi, E/F, \psi : a].$$

The measure du on $N_n(E)$ is the product of the self dual Haar measures $du_{i,j}$. More generally, we say that an element s of $S(F)$ (or its orbit under $N(E)$) is *relevant* if the character

$$u \mapsto \sum_i \psi(u_{i,i+1} + \bar{u}_{i,i+1})$$

from $N(E)$ to F is trivial on the stabilizer of s in $N(E)$. A system of representatives for the relevant orbits consists of the elements $w_M a$. Here M is a standard Levi-subgroup of $GL(n)$ containing A, that is, M is a bloc diagonal product of general linear groups. The group $A_M \subseteq A$ is the center of M, the matrix belongs to A_M and w_M the longest element of the Weyl group of

M, viewed as a group of permutation matrices. In particular w_G also noted w_n is the matrix

(5)
$$w_n := \begin{pmatrix} 0 & 0 & \dots & 0 & 0 & 1 \\ 0 & 0 & \dots & 0 & 1 & 0 \\ 0 & 0 & \dots & 1 & 0 & 0 \\ \dots & & & & & \\ 0 & 1 & \dots & 0 & 0 & 0 \\ 1 & 0 & \dots & 0 & 0 & 0 \end{pmatrix}$$

If M is of type $(n_1, n_2, \dots, n_r)$ with $n_1 + n_2 + \dots + n_r = n$ then M consists of diagonal matrices with block entries of size $(n_1, n_2, \dots, n_r)$. The matrix w_M is the matrix

(6)
$$w_M := \begin{pmatrix} w_{n_1} & 0 & \dots & 0 & 0 & 0 \\ 0 & w_{n_2} & 0 & \dots & 0 & 0 \\ \dots & & & & & \\ 0 & 0 & \dots & 0 & w_{n_{r-1}} & 0 \\ 0 & 0 & \dots & 0 & 0 & w_{n_r} \end{pmatrix}$$

We define orbital integrals

(7)
$$\Omega[\Psi, \psi : w_M a] = \int \Psi\left[{}^t\bar{u} w_M a u\right] \theta(u\bar{u})\, du.$$

The integral is over the quotient of $N(E)$ by the stabilizer of $w_M a$.

Similarly we let the group $G(F) \times G(F)$ operate on $G(F)$ by

$$g \mapsto {}^t g_1 g g_2.$$

We define the relevant elements of $GL(n, F)$ for the action of $N(F) \times N(F)$. The above matrices $w_M a$ form again a system of representatives for the relevant orbits. We define orbital integrals

(8)
$$\Omega[\Phi, \psi : w_M a] = \int \Phi\left[{}^t u_1 w_M a u_2\right] \theta(u_1 u_2)\, du_1\, du_2.$$

The integral is over the quotient of $N(F)^2$ by the stabilizer of $w_M a$.

We say that Ψ matches Φ and we write $\Psi \leftrightarrow \Phi$ if

$$\Omega[\Psi, \psi : a] = \gamma(a)\Omega[\Phi, \psi : a]$$

where $\gamma(a)$ is defined as before. Then there are transfer factors $\gamma(w_M a, \psi)$ such that

$$\Omega[\Psi, \psi : w_M a] = \gamma(w_M a, \psi)\Omega[\Psi, \psi : w_M a].$$

Thus the factor $\gamma(a, \psi) = \gamma(a)$ does not depend on ψ. Every Ψ matches a Φ and conversely.

We also need the same notions for the trivial case where E is replaced by $F \oplus F$ with Galois action $\sigma(x_1, x_2) = (x_2, x_1)$. The space $S(F)$ is now the set of pairs (s_1, s_2) with $s_1 = {}^t s_2$. The action of $GL(n, E) = GL(n, F) \times GL(n, F)$ is by

$$(s_1, s_2) \mapsto \left({}^t g_2 s_1 g_1, {}^t g_1 s_2 g_2\right).$$

We say that $\Psi \in C_c^\infty(S(F))$ and $\Phi \in C_c^\infty(G(F))$ match if

$$\int \Psi\left[{}^t n_2 a n_1, {}^t n_1 a n_2\right]\theta(n_1 n_2)\,dn_1\,dn_2 = \int \Phi\left[{}^t n_2 a n_1\right]\theta(n_1 n_2)\,dn_1\,dn_2.$$

This condition is verified if

$$\Phi(g) = \Psi({}^t g, g)$$

or

$$\Phi(g) = \Psi(g, {}^t g).$$

The other orbital integrals are then equal.

3. Applications

Our main result plus the recent work of Lapid [10] on the fine spectral expansion of the relative trace formula imply the following statement. Let E/F be a quadratic extension of number fields. We let $\eta_{E/F}$ or simply η be the quadratic idele class character of F attached to E. We fix a non-trivial character ψ of $F_{\mathbb{A}}/F$. For technical reasons, we assume for now that every real place of F splits in E. If H is a unitary group in n-variables, and Π an automorphic cuspidal representation of $GL(n, E_{\mathbb{A}})$ we say that Π is *distinguished* by H if there is an automorphic form ϕ in the space of Π such that

$$\int_{H(F)\backslash H(F_{\mathbb{A}})} \phi(h)\,dh \neq 0.$$

We have then the following result.

THEOREM 2. – *Suppose that Π is the base change of an automorphic cuspidal representation π of $GL(n, F_{\mathbb{A}})$. Then there is a unitary group H such that Π is distinguished by H. Conversely, if an automorphic cuspidal representation Π of $GL(n, E_{\mathbb{A}})$ is distinguished by some unitary group then it is the base change of an automorphic cuspidal representation π of $GL(n, F_{\mathbb{A}})$.*

We first prove the second statement. Indeed, suppose that Π is distinguished by the stabilizer H of an element $\xi \in S(F)$. If v_1, v_2 are places of E above the same place v of F then the tensor product $\Pi_{v_1} \otimes \Pi_{v_2}$ admits a non-zero linear form μ invariant under the group of pairs $(\xi_1^{-1}{}^t h^{-1}\xi_1, h)$. Since $g \mapsto \Pi_{v_1}({}^t g^{-1})$ is contragradient to Π_{v_1} we conclude that $\Pi_{v_1} = \Pi_{v_2}$. On the other hand, if v is a place of F inert in F, w the corresponding place of E and Π_w is unramified, then Π_w is invariant under the non-trivial element of $\mathrm{Gal}(E_w/F_v)$. Thus Π and Π^σ have the same components at almost all places. It follows that $\Pi = \Pi^\sigma$. By [1] Π is then the base change of a cuspidal automorphic representation of $GL(n, F_{\mathbb{A}})$. (This is a standard argument first found in [5].)

We describe in more detail the (relative) trace formula which gives this result. The method of proof we follow is perhaps not the simplest. However it can be adapted to the case of the *simple* but delicate relative trace formula that we need to prove completely our main result.

We let F^+ be the subgroup of $F_{\mathbb{A}}^\times$ whose finite components are 1 and whose infinite components are all equal to the same positive number. Thus F^+ is isomorphic to $\mathbb{R}^{\times +}$. For $z \in F^+$ we often write z for $z1_n$. We define E^+ similarly.

3.1. The relative trace formula over E/F

We denote by S the algebraic variety of $n \times n$ invertible Hermitian matrices. Let Ψ be a smooth function of compact support on the space $S(F_{\mathbb{A}})$. We consider the function

$$K_\Psi(g) := \sum_{\epsilon \in S(F)} \Psi\left[{}^t\bar{g}\epsilon g\right]$$

and the integral

$$(9) \qquad \int_{E^+} \int_{N_n(E)\backslash N_n(E_{\mathbb{A}})} K_\Psi(uz)\theta(u\bar{u})\, du\, dz$$
$$= \int_{F^+} \int_{N_n(E)\backslash N_n(E_{\mathbb{A}})} \sum_{\epsilon \in S(F)} \Psi[{}^t\bar{u}z\epsilon u]\theta(u\bar{u})\, dz\, du.$$

Of course the Haar measures have to be normalized suitably. In this article we are only interested in qualitative results and so we do not pay much attention to the normalization of the measures. This expression can be computed in terms of orbital integrals (for the relevant orbits). Thus it is a sum

$$(10) \qquad \sum_M \sum_{\alpha \in A_M(F)} \int \Omega[\Psi, \psi : w_M \alpha z]\, dz.$$

The first sum is over all Levi-subgroups M of $GL(n)$ containing A. The global orbital integral $\Omega[\Psi, \psi : w_M \alpha z]$ is over the quotient of $N(E_{\mathbb{A}})$ by the stabilizer of $w_M \alpha$.

We consider a set of representatives $\{\xi\}$ for the orbits of the right action of $GL(n, E)$ on $S(F)$. For each ξ we denote by H^ξ its stabilizer and we choose a function f^ξ on $GL(n, E_{\mathbb{A}})$, smooth of compact support, such that

$$(11) \qquad \int_{H^\xi(F_{\mathbb{A}})} f^\xi(hg)\, dh = \Psi\left({}^t\bar{g}\xi g\right).$$

We introduce the usual kernel function

$$K_{f^\xi}(x, y) := \sum_{\gamma \in GL(n,E)} f^\xi(x^{-1}\gamma y).$$

Then

$$K_\Psi(g) = \sum_\xi \sum_{\gamma \in H^\xi(F)\backslash GL(n,E)} \int_{H^\xi(F_{\mathbb{A}})} f^\xi(h\gamma g)\, dh$$
$$= \sum_\xi \int_{H^\xi(F)\backslash H^\xi(F_{\mathbb{A}})} K_{f^\xi}(h, g)\, dh.$$

Thus (9) is equal to

$$(12) \qquad \sum_\xi \int_{E^+} \int_{H^\xi(F)\backslash H^\xi(F_{\mathbb{A}})} \int_{N(E)\backslash N(E_{\mathbb{A}})} K_{f^\xi}(zh, u)\theta(u\bar{u})\, dh\, du\, dz.$$

Now we make more precise the relation between the function Ψ and the functions f^ξ. We assume that Ψ is decomposable. Then the functions f^ξ are decomposable.

To that end, we consider a place v of F inert and unramified in E and the corresponding place w of E. We consider a ξ which is in $K_w := GL(n, \mathcal{O}_w)$. Let Ψ_v^1 be the characteristic function of $K_w \cap S(F_v)$ and f_w a Hecke function. Assume that Ψ_v has the form

$$\Psi_v(X) = \int_{G_w} \Psi_v^1({}^t\bar{x}Xx) f_w(x^{-1})\, dx.$$

Since the set $K_w \cap S(F_v)$ is a single orbit of K_w we can write

$$\Psi_v^1({}^t\bar{g}\xi g) = \int_{H_v^\xi} f_w^1(hg)\, dh$$

where f_w^1 is the unit element of the Hecke algebra $\mathcal{H}_w$. On the other hand,

$$f_w(g) = \int f_w^1(x) f_w(x^{-1}g)\, dx.$$

Thus

$$\int_{H_v^\xi} f_w(hg)\, dh = \int \Psi_v^1({}^t\bar{x}\,{}^t\bar{g}\xi gx) f_w(x^{-1})\, dx = \Psi_v[{}^t\bar{g}\xi g].$$

In other words we may and will take $f_w^\xi = f_w$.

If v splits into v_1, v_2 then the space $S(F_v)$ of Hermitian matrices at the place v is the set of pairs

$$(g_1, g_2), \quad g_2 = {}^t g_1, \quad g_i \in GL(n, F_v).$$

In particular ξ viewed as an element of $S(F_v)$ has the form $\xi = (\xi_1, \xi_2)$, ${}^t\xi_2 = \xi_1$. The group H_v^ξ is the set of pairs (h_1, h_2) with $h_1 = \xi_1^{-1}\,{}^t h_2^{-1}\xi_1$, $h_2 \in GL(n, F_v)$. The relation between Ψ_v and the function f_{v_1,v_2}^ξ is

$$\Psi_v[({}^t g_2, {}^t g_1)(\xi_1, \xi_2)(g_1, g_2)] = \int f_{v_1,v_2}^\xi(h_1 g_1, h_2 g_2)\, dh$$

or more simply

$$\Psi_v[{}^t g\xi_1, \xi_2 g] = \int_{GL(n,F_v)} f_{v_1,v_2}^\xi(\xi_1^{-1}\,{}^t h^{-1}\xi_1, hg)\, dh$$

or, after a change of variables,

$$\Psi_v[{}^t g, g] = \int_{GL(n,F_v)} f_{v_1,v_2}^\xi(\xi_1^{-1}\,{}^t h^{-1}, hg)\, dh.$$

Again if the function $g \mapsto \Psi[{}^t g, g]$ is a Hecke function and $\xi_1 \in K_v$ we may and will take for $f^{\xi}_{v_1,v_2}$ a Hecke function f_{v_1,v_2} independent of ξ. Then

$$\Psi_v\left[{}^t g, g\right] = \int f^{\xi}_{v_1,v_2}\left(h^{-1}, hg\right) dh.$$

Now we appeal to the result of [10]. The expression (12) admits a *fine spectral expansion*. We only write down the *discrete* part of this spectral expansion. It has the form

$$\sum_{\xi}\sum_{\Pi} \mathcal{R}^{\xi}_{\Pi}(f^{\xi}). \tag{13}$$

The sum is over all ξ. The representations Π which appear are, on the one hand, the cuspidal automorphic representations and, on the other hand, certain Eisenstein automorphic representations. As usual the representations Π are normalized by the condition that their central character is trivial on E^+. If Π is a cuspidal automorphic representation, then the *relative Bessel* distribution $\mathcal{R}^{\xi}_{\Pi}$ is defined by

$$\mathcal{R}^{\xi}_{\Pi}(f^{\xi}) = \sum_i \int_{H^{\xi}(F)\backslash H^{\xi}(F_{\mathbb{A}})} \Pi(f^{\xi})\phi_i(h)\, dh \overline{\int \phi_i(u)\bar{\theta}(u\bar{u})\, du},$$

where the sum is over an orthonormal basis $\{\phi_i\}$ of Π. If Π is an Eisenstein automorphic representation then n is even and Π of the form $\Pi_1 \boxplus \Pi_1^{\sigma}$ where Π_1 is a cuspidal automorphic representation of $GL(n/2, E_{\mathbb{A}})$ with Π_1^{σ} not equivalent to Π_1. The distribution $\mathcal{R}^{\xi}_{\Pi}$ is defined similarly, except that the integral over the unitary group is now a regularized integral [10].

3.2. The relative trace formula over F

Now we consider a smooth function of compact support Φ on $GL(n, F_{\mathbb{A}})$. We associate to Φ the usual kernel

$$K_{\Phi}(x, y) := \sum_{\gamma \in GL(n,F)} \Phi(x^{-1}\gamma y).$$

We consider the expression

$$\begin{aligned} &\int_{F^+}\int_{(N_n(F)\backslash N(F_{\mathbb{A}}))^2} K_{\Phi}\left({}^t u_1^{-1}, u_2 z\right)\theta(u_1)\theta(u_2)\, du_1\, du_2\, dz \\ &= \int_{F^+}\int_{(N_n(F)\backslash N(F_{\mathbb{A}}))^2} \sum_{\epsilon \in GL(n,F)} \Phi[{}^t u_1 z \epsilon u_2]\theta(u_1 u_2)\, du_1\, du_2\, dz. \end{aligned} \tag{14}$$

As before it can be computed in terms of orbital integrals

$$\sum_M \sum_{\alpha} \int \Omega[\Phi, \psi : w_M \alpha z]\, dz. \tag{15}$$

The expression (14) admits a fine spectral expansion. The discrete part of it is equal to

$$\sum_{\pi} \mathcal{B}_{\pi}(\Phi), \tag{16}$$

where the sum is over all irreducible cuspidal automorphic representations π. As usual the representations π are normalized by the condition that their central character is trivial on F^+. The *Bessel* distribution $\mathcal{B}_\pi$ is defined by

$$\mathcal{B}_\pi(\Phi) := \sum_i \int \big(\pi(\Phi)\phi_i\big)\big({}^t u_1^{-1}\big)\theta(u_1)\,du_1 \overline{\int \phi_i(u)\bar{\theta}(u_2)\,du_2},$$

where the sum is over an orthonormal basis $\{\phi_i\}$ of π. Here again the residual discrete spectrum does not contribute because it is degenerate.

In turn, the global Bessel function is, up to a global constant, a product of local Bessel functions $\mathcal{B}_{\pi_v}$ (suitably normalized).

$$\mathcal{B}_{\pi_v}(\Phi_v) = \sum_{W_i} \big(\pi_v(\Phi_v)W_i\big)(w_n)\overline{W}_i(e).$$

Here W_i is an orthonormal basis of the Whittaker model of π_v and w_n is the matrix (5).

To continue we consider a cuspidal automorphic representation π^0 of $GL(n, F_{\mathbb{A}})$. If n is even, we assume that π^0 is not automorphically induced from a cuspidal automorphic representation of $GL(n/2, E_{\mathbb{A}})$, or equivalently, that $\pi^0 \not\simeq \pi^0 \otimes \eta_{E/F}$. Thus the base change Π^0 of π^0 is cuspidal. We want to show that Π^0 is distinguished by some unitary group. To that end, we let S be a finite set of places of F containing all the infinite places, all the even places, all the inert places which are ramified in E, all the places v where the character ψ_v has a conductor not equal to $\mathcal{O}_v$, and all the finite places where π^0 is ramified.

We will choose the functions Φ_v for v not in S to be Hecke functions. We note that if Φ_v is a Hecke function and Φ_v^1 the characteristic function of $K_v := GL(n, \mathcal{O}_v)$ then

$$\Phi_v(X) = \int \Phi_v^1(Xg)\Phi_v(g^{-1})\,dg.$$

By regrouping the terms for which π_∞ has a given value, our sum of Bessel distributions can be written as an absolutely convergent sum

$$\sum c(\pi_\infty)\mathcal{B}_{\pi_\infty}(\Phi_\infty).$$

It will be convenient to use a principle of *infinite linear independence* for such sums.

LEMMA 1. – *If*

$$\sum_{\pi_\infty} c(\pi_\infty)\mathcal{B}_{\pi_\infty}(\Phi_\infty) = 0$$

for all Φ_∞ *then all the coefficients* $c(\pi_\infty)$ *are* 0.

Proof. – We defer the proof to Section 14. □

Now we consider the sum of the distributions $\mathcal{B}_\pi(\Phi)$ for all π such that $\pi_\infty = \pi_\infty^0$, namely,

$$\sum_{\pi_\infty = \pi_\infty^0} \mathcal{B}_\pi(\Phi).$$

We further fix Φ_v for v finite in S so as to have

$$\mathcal{B}_{\pi_v^0}(\Phi_v) \neq 0.$$

Then the above sum is *finite*. We label the representations which appear in the sum π^α, $0 \leqslant \alpha \leqslant r$. They are unramified at all places $v \notin S$. Next, by the strong multiplicity one, we can choose a finite set of places $T_1 = \{v_\alpha \mid 1 \leqslant \alpha \leqslant r\}$ not in S with the property that

$$\forall \alpha, \quad \pi^0_{v_\alpha} \not\simeq \pi^\alpha_{v_\alpha}.$$

For each $\alpha > 0$, we choose then a Hecke function Φ_{v_α} such that

$$\mathcal{S}(\Phi_{v_\alpha})(\pi^0_{v_\alpha}) \neq 0, \qquad \mathcal{S}(\Phi_{v_\alpha})(\pi^\alpha_{v_\alpha}) = 0.$$

With this choice of functions the above sum reduces to the single term $\mathcal{B}_{\pi^0}(\Phi)$. We set $S_1 = S \cup T_1$. Furthermore the single term is in fact a product

$$\mathcal{B}_{\pi^0}(\Phi) = c\mathcal{B}_{\pi^0_\infty}(\Phi_\infty) \prod_{v \notin S_1} \mathcal{S}(\Phi_v)(\pi^0_v) \tag{17}$$

with $c \neq 0$.

3.3. Comparison

If Ψ and Φ satisfy a *matching condition* we have the equality

$$\begin{aligned} \int_{F^+ \times N(E)\backslash N(E_{\mathbb{A}})} & K_\Psi(zu)\theta(u\bar{u})\, dz\, du \\ = \int_{F^+ \times (N(F)\backslash N(F_{\mathbb{A}}))^2} & K_\Phi({}^t u_1^{-1}, u_2 z)\, dz\, \theta(u_1)\, du_1\, \theta(u_2)\, du_2 \end{aligned} \tag{18}$$

that is, the expression (14) is equal to the expression (9) (geometric identity). We now describe the matching condition in detail. We assume Ψ and Φ are decomposable.

Consider first a place v inert in E and let w be the place above v. Then the condition of matching $\Phi_v \leftrightarrow \Psi_v$ has been recalled in Section 2. The product of the transfer factors $\gamma(w_M\alpha, \psi_v)$ over all inert places v is 1 for $\alpha \in A_M(F)$.

At a place v which splits into v_1, v_2 the condition of matching is

$$\Phi_v(g) = \Psi_v({}^t g, g).$$

Note that this definition depends on an ordering of the set $\{v_1, v_2\}$.

At every place $v \in S_1$ we take a function Ψ_v which matches the function Φ_v that we have chosen previously. We note that for an inert place $v \in T_1$ the function Φ_v is a Hecke function but the function Ψ_v needs not be invariant under K_w and the functions f^ξ_w need not be Hecke functions.

Recall that for $v \notin S_1$ we choose Φ_v to be a Hecke function. We make this choice and the choice of the matching function Ψ_v more precise.

Now suppose v is not in S_1 and inert. Let w be the place of E above v. We will let f_w be a Hecke function. We take

$$\Psi_v[X] = \int \Psi^1_w[{}^t\bar{x}Xx] f_w(x^{-1})\, dx,$$

$$\Phi_v[X] = \int \Phi^1_v[Xx] b(f_w)(x^{-1})\, dx.$$

Indeed our main result asserts that $\Psi_v \leftrightarrow \Phi_v$. In fact we have $\Phi_v = b(f_w)$. In particular, Ψ_v is supported on the set of elements whose determinant is a norm of the extension E_w/F_v.

Now let Ξ be the set of ξ such that $\det(\xi)$ is a norm at each place v inert and not in S_1. The set Ξ is *finite* and only the elements of Ξ appear in our formula. We will let T_2' be the (finite) set of inert places $v \notin S_1$ for which at least one $\xi \in \Xi$ is not in K_w, where w is above v. If v is such a place and w the corresponding place of E we take f_w to be the unit element of the Hecke algebra. Then $\Phi_v = \Phi_v^1$ and $\Psi_v = \Psi_v^1$. We note that the functions f_w^ξ need not be Hecke functions.

If v is not in S_1 and splits into v_1, v_2 then Φ_v is a Hecke function and we take

$$\Psi_v\left[{}^t g, g\right] = \Phi_v(g).$$

Let T_2'' be the finite set of split places $v \notin S_1$ such that ξ_1 is not in $K_{v_1} = K_v$ for at least one $\xi \in \Xi$. Again, the functions f_{v_1,v_2}^ξ need not be Hecke functions. For $v \notin T_2''$ we take $f_{v_1,v_2}^\xi = f_{v_1,v_2}$ to be a Hecke function and we get

$$\Phi_v(g) = \int f_{v_1,v_2}\left(h^{-1}, hg\right) dh.$$

Thus Φ_v is the Hecke function base change of f_{v_1,v_2}.

We now set $T_2 = T_2' \cup T_2''$ and $S_2 = S_1 \cup T_2$. Suppose that v is not in S_2 and inert. Let w be the corresponding place of E. Then all the functions f_w^ξ, $\xi \in \Xi$, are equal to the same Hecke function f_w and Φ_v is the base change of f_w. Suppose that v is not in S_2 and splits into v_1, v_2. Then all the functions f_{v_1,v_2}^ξ, $\xi \in \Xi$, are equal to the same Hecke function f_{v_1,v_2} and f_v is the base change of f_{v_1,v_2}. The functions f_w and f_{v_1,v_2} are arbitrary, subject to the condition that they are almost all equal to the unit element. Let us call S_2^E the set of places of E above a place of S_2. Thus we may set

$$f^{S_2^E} := \bigotimes f_w \otimes \bigotimes f_{v_1.v_2},$$

the tensor product being over all finite places of E not in S_2^E. This is an element of the Hecke algebra $\mathcal{H}^{S_2^E}$, tensor product of the Hecke algebras over all places not in S_2^E. Then the function

$$\Phi^{S_2} := \bigotimes_{v \notin S_2} \Phi_v$$

is in the Hecke algebra $\mathcal{H}^{S_2}$. It is the base change of $f^{S_2^E}$.

The functions Φ and Ψ being chosen in this way, we get the equality of (15) and (10). It follows that the two spectral expansions are equal as well. We can equate the discrete parts of the spectral expansions to get the following identity:

$$\sum_\pi \mathcal{B}_\pi(\Phi) = \sum_\xi \sum_\Pi \mathcal{R}_\Pi^\xi\left(f^\xi\right). \tag{19}$$

Indeed, we recall the standard argument. Fix a place v of F split and not in S_2. Let v_1, v_2 be the two places above v. Take $f_{v_1,v_2} = f_v \otimes f_0$, where f_v is in $\mathcal{H}_v$, the Hecke algebra for $GL(n, F_v)$, and f_0 is the unit element of $\mathcal{H}_v$. Thus f_v is the image of f_{v_1,v_2} under the base change map homomorphism. The Satake transform of f_v may be viewed as a continuous function on the set X_v of unitary, irreducible, unramified representations of $GL(n, F_v) \simeq GL(n, F_{v_1})$. Both spectral expressions can then be viewed as measures on X_v. To obtain the claimed identity we equate the discrete parts of the measures. We have just obtained the identity under some restrictions on f_{v_1,v_2}. Since v is arbitrary, we obtain the identity in general.

The representations Π for which the distribution $\mathcal{R}_\Pi^\xi$ is non-zero are distinguished by H^ξ. As before (see the argument after Theorem 2), if v_1, v_2 are above the same place v then $\Pi_{v_1} = \Pi_{v_2}$ (this is a standard argument).

If $\mathcal{R}_\Pi^\xi \neq 0$ then we can identify $GL(n, E_\infty)$ to the product $GL(n, F_\infty) \times GL(n, F_\infty)$ and write Π_∞ as tensor product $\pi_\infty \otimes \pi_\infty$. In other words Π_∞ is the base change of π_∞. By Lemma 21, viewed as a distribution at infinity, $\mathcal{R}_\Pi^\xi(f^\xi)$ is a multiple of $\mathcal{B}_{\pi_\infty}(\Phi_\infty)$. Using Lemma 1 we get

$$\sum_{\Pi_\infty = \pi_\infty^0 \otimes \pi_\infty^0} \sum_{\xi \in X} \mathcal{R}_\Pi^\xi(f^\xi) = \sum_{\pi_\infty = \pi_\infty^0} \mathcal{B}_\pi(\Phi) = \mathcal{B}_{\pi^0}(\Phi). \tag{20}$$

Recall we have fixed the functions Φ_v at all finite places in S_2. Thus the functions f_w^ξ for w finite in S_2^E are also fixed. Thus the sum on the left is a *finite* sum. Since $\Phi_v = \Phi_v^1$ for every $v \in T_2$ we get from (17)

$$\mathcal{B}_{\pi^0}(\Phi) = c\mathcal{B}_{\pi_\infty^0}(\Phi_\infty) \prod_{v \notin S_2} \mathcal{S}(\Phi_v)(\pi_v^0). \tag{21}$$

On the left-hand side of (20)

$$f^\xi = f_{S_2^E}^\xi \otimes f^{S_2^E}.$$

Thus

$$\mathcal{R}_\Pi^\xi(f^\xi) = \prod_{w \notin S_2^E} \mathcal{S}(f_w)(\Pi_w) C(\Pi, \xi)$$

where we have set

$$C(\Pi, \xi) = \mathcal{R}_\Pi^\xi(f_{S_2^E}^\xi \otimes \mathbf{1}^{S_2^E}),$$

and $\mathbf{1}^{S_2^E}$ is the unit element of $\mathcal{H}^{S_2^E}$. The only terms which appear on the left of (20) are those Π for which

$$\prod_{w \notin S_2^E} \mathcal{S}(f_w)(\Pi_w) = \prod_{v \notin S_2} \mathcal{S}(\Phi_v)(\pi_v^0).$$

By the strong multiplicity result, the only Π satisfying this condition is Π^0, the base change of π^0. Thus $\mathcal{R}_{\Pi^0}^\xi(f^\xi) \neq 0$ for at least one ξ and we are done.

We remark that we could also use the above trace formula to prove that a distinguished representation is a base change.

4. The set up

We go back to the notations of Section 1. In particular E/F is an unramified quadratic extension. We denote by $\mathcal{P}^n$ or simply $\mathcal{P}$ the set of n-tuples $\lambda \in \mathbb{Z}^n$ such that

$$\lambda_1 \geqslant \lambda_2 \geqslant \cdots \geqslant \lambda_n$$

and by $\mathcal{P}_+$ the subset of $\lambda \in \mathcal{P}$ such that $\lambda_n \geqslant 0$. We set

$$|\lambda| = \sum_{1\leqslant i\leqslant n} \lambda_i, \qquad \tilde{\lambda} = (-\lambda_n, -\lambda_{n-1}, \dots, -\lambda_1).$$

We denote by $\mathcal{P}_m$ the subset of $\lambda \in \mathcal{P}$ such that $|\lambda| = m$. We recall the standard partial order on $\mathcal{P}$. We write

$$\lambda \preceq \mu$$

if

$$\begin{aligned}
&\lambda_1 \leqslant \mu_1, \qquad \lambda_1 + \lambda_2 \leqslant \mu_1 + \mu_2, \dots, \\
&\lambda_1 + \lambda_2 + \dots + \lambda_{n-1} \leqslant \mu_1 + \mu_2 + \dots + \mu_{n-1}, \\
&|\lambda| = |\mu|.
\end{aligned}$$

If $\lambda \in \mathbb{Z}^n$ we set

$$\varpi^\lambda = \operatorname{diag}(\varpi^{\lambda_1}, \varpi^{\lambda_2}, \dots, \varpi^{\lambda_n}).$$

If f is a Hecke function, we set, for $\lambda \in \mathbb{Z}^n$,

$$\Phi_f(\lambda) = \Phi_f(\varpi^\lambda) := \int_{N_n(F)} f(u\varpi^\lambda)\, du,$$

$$H_f(\lambda) := \Phi_f(\varpi^\lambda)\delta_F^{-1/2}(\varpi^\lambda).$$

Here δ_F is the module of the group $A(F)N(F)$. Thus if du is the Haar measure of $N_n(F)$ then

$$d(aua^{-1}) = \delta_F(a)\, du.$$

The function $\lambda \mapsto H_f(\lambda)$ is compactly supported and symmetric and arbitrary. It is determined by its values on $\mathcal{P}$.

Let ω_x^F be the zonal spherical function of parameters $x = (x_1, x_2, \dots x_n)$. We recall its definition. Let ϕ_0 be the function such that

$$\begin{aligned}
&\phi_0(uak) = \delta_F^{1/2}(a)|a_1|^{s_1}|a_2|^{s_2}\cdots|a_n|^{s_n}, \\
&u \in N_n(F), \quad a \in A_n(F), \quad k \in K_F.
\end{aligned}$$

Here the $s_i \in \mathbb{C}$ are determined by the condition $|\varpi|^{s_i} = x_i$. Then

$$\omega_x^F(g) = \int_{K_F} \phi_0(kg)\, dg.$$

For any Hecke function f the Satake transform is defined by

$$\mathcal{S}(f)(x) := \int f(g)\omega_x^F(g)\, dg.$$

Let $\mathbb{S}_n$ be the subalgebra of symmetric elements in

$$\mathbb{C}[X_1, X_2, \dots, X_n, (X_1X_2\cdots X_n)^{-1}].$$

The maximal spectrum $\mathrm{Spec}[\mathbb{S}_n]$ of $\mathbb{S}_n$ is thus the set of the orbits of the permutation group in $(\mathbf{C}^\times)^n$. If π is the unramified representation which admits the spherical function ω_x^F as a matrix coefficient we also write $\mathcal{S}(f)(\pi)$ for $\mathcal{S}(f)(x)$.

Recall that

$$\begin{aligned}\mathcal{S}(f)(x) &= \int f(g)\phi_0(g)\,dg\\ &= \sum_{\lambda\in\mathbb{Z}^n} \Phi_f(\varpi^\lambda)\delta_F^{-1/2}(\varpi^\lambda)\left|\varpi^{\lambda_1}\right|^{s_1}\left|\varpi^{\lambda_2}\right|^{s_2}\cdots\left|\varpi^{\lambda_n}\right|^{s_n}\\ &= \sum_{\lambda\in\mathbb{Z}^n} \Phi_f(\varpi^\lambda)\delta_F^{-1/2}(\varpi^\lambda)x_1^{\lambda_1}x_2^{\lambda_2}\cdots x_n^{\lambda_n}\\ &= \sum_{\lambda\in\mathbb{Z}^n} H_f(\lambda)x^\lambda\\ &= \sum_{\lambda\in\mathcal{P}} H_f(\lambda)m_\lambda(x)\end{aligned}$$

where we have set $x^\lambda = x_1^{\lambda_1}x_2^{\lambda_2}\cdots x_n^{\lambda_n}$ and m_λ denotes the monomial symmetric function.

If f is a Hecke function on $GL(n,F)$ we set

$$W_f(g) := \int_{N_n(F)} f(ug)\theta^{-1}(u)\,du. \tag{22}$$

In particular, if $g = a \in A_n$ then this depends only on the absolute values of the entries of a. This allows us to write for any $\lambda \in \mathbb{Z}^n$,

$$W_f(\lambda) := W_f(\varpi^\lambda). \tag{23}$$

Of course $W_f(\lambda) = 0$ unless $\lambda \in \mathcal{P}$. We denote by W_x^F the elementary Whittaker function of parameters $x = (x_1, x_2, \ldots, x_n)$. Thus

$$W_x^F(ug) = \theta(u)^{-1}W_x^F(g), \qquad W_x^F(e) = 1,$$

and, for any Hecke function f,

$$\int W_x^F(hg)f(g)\,dg = W_x^F(h)\int f(g)\omega_x^F(g)\,dg.$$

In particular at $x = e$ we get

$$\int W_x^F(g)f(g)\,dg = \int f(g)\omega_x(g)\,dg. \tag{24}$$

Let $\sigma_\lambda(x)$ be the Schur function attached to a $\lambda \in \mathcal{P}$. Recall the Casselman–Shalika–Shintani formula [2,17]:

$$W_x^F(\omega^\lambda) = \sigma_\lambda(x)\delta_F^{1/2}(\varpi^\lambda).$$

The left-hand side of (24) is

$$\int\limits_{A_n(F)} W_x^F(a)W_f(a)\delta_F^{-1}(a)\,da = \sum_{\lambda\in\mathcal{P}} W_x^F(\varpi^\lambda)W_f(\lambda)\delta_F^{-1}(\varpi^\lambda)$$

$$= \sum_{\lambda\in\mathcal{P}} \sigma_\lambda(x)W_f(\lambda)\delta_F^{-1/2}(\varpi^\lambda)$$

The right-hand side of (24) is $\sum_{\lambda\in\mathcal{P}} H_f(\lambda)m_\lambda(x)$. Thus for any r we have

$$\sum_{\lambda\in\mathcal{P}_r} \sigma_\lambda(x)W_f(\lambda)\delta_F^{-1/2}(\varpi^\lambda) = \sum_{\lambda\in\mathcal{P}_r} m_\lambda(x)H_f(\lambda).$$

Now

$$\sigma_\lambda = \sum_{\mu\preceq\lambda} K_\lambda^\mu m_\mu$$

with $K_\lambda^\lambda = 1$ and

$$m_\lambda = \sum_{\mu\preceq\lambda} A_\lambda^\mu \sigma_\mu$$

with $A_\lambda^\lambda = 1$. Because H_f and W_f are compactly supported, we can write

$$H_f(\mu) = \sum_{\lambda\succeq\mu} K_\lambda^\mu W_f(\lambda)\delta_F^{-1/2}(\varpi^\lambda),$$

$$W_f(\lambda)\delta_F^{-1/2}(\varpi^\lambda) = \sum_{\mu\succeq\lambda} A_\mu^\lambda H_f(\mu).$$

In particular, given $\mu\in\mathcal{P}$, we denote by f_μ^F the unique Hecke function such that

(25) $$W_{f_\mu^F}(\lambda) = \delta_{\lambda,\mu}.$$

The functions f_μ^F, $\mu\in\mathcal{P}$, form a linear basis of the space of Hecke functions.

Then

$$\mathcal{S}(f)(x) = \int f(g)W_x^F(g)\,dg = \sum_{\lambda\in\mathcal{P}} W_f(\lambda)\delta_F^{-1/2}(\varpi^\lambda)\sigma_\lambda(x).$$

In particular, for the function f_λ^F, we see that

$$\mathcal{S}(f_\lambda^F)(x) = \delta_F^{-1/2}(\varpi^\lambda)\sigma_\lambda(x).$$

The same applies to E, with $\psi_E(z) = \psi_F(z+\bar{z})$. In particular

$$W_f(a) = \int\limits_{N(E)} f(ua)\theta^{-1}(u\bar{u})\,du.$$

Let b denote the base change homomorphism $b:\mathcal{H}(GL(E))\to\mathcal{H}(GL(F))$. By definition

(26) $$\mathcal{S}(b(f))(x) = \mathcal{S}(f)(x^2)$$

where on the left the Satake transform is for the group $GL(n,F)$ and on the right for the group $GL(n,E)$. Equivalently,

$$\int_{GL(n,E)} f(g) W^E_{x^2}(g)\, dg = \int_{GL(n,F)} b(f)(g) W^F_x(g)\, dg$$

or

$$\int_{A_n(E)} W_f(a) W^E_{x^2}(a) \delta_E^{-1}(a)\, da = \int_{A_n(F)} W_{b(f)}(a) W^F_x(a) \delta_F^{-1}(a)\, da$$

or

$$\sum_\lambda W_f(\varpi^\lambda) W^E_{x^2}(\varpi^\lambda) \delta_E(\varpi^\lambda)^{-1} = \sum_\lambda W_{b(f)}(\varpi^\lambda) W^F_x(\varpi^\lambda) \delta_F(\varpi^\lambda)^{-1}.$$

Thus

$$\sum_\lambda W_f(\lambda) \delta_E(\varpi^\lambda)^{-1/2} \sigma_\lambda(x^2) = \sum_\lambda W_{b(f)}(\lambda) \delta_F(\varpi^\lambda)^{-1/2} \sigma_\lambda(x).$$

In particular, for $f = f^E_\mu$ we get,

$$\delta_E^{-1/2}(\varpi^\mu) \sigma_\mu(x^2) = \sum_\lambda W_{b(f^E_\mu)}(\lambda) \delta_F^{-1/2}(\varpi^\lambda) \sigma_\lambda(x).$$

To exploit this relation, we appeal to a simple lemma.

LEMMA 2. – *For every $\lambda \in \mathcal{P}$ there are unique constants θ^μ_λ, $\mu \preceq 2\lambda$, such that*

$$\sigma_\lambda(x^2) = \sum_{\mu \preceq 2\lambda} \theta^\mu_\lambda \sigma_\mu(x).$$

In particular $\theta^{2\lambda}_\lambda = 1$.

Proof. – For the sake of completeness we give a proof. Indeed

$$\sigma_\lambda(x) = \sum_{\mu \preceq \lambda} K^\mu_\lambda m_\mu(x).$$

Thus

$$\sigma_\lambda(x^2) = \sum_{\mu \preceq \lambda} K^\mu_\lambda m_\mu(x^2).$$

Now

$$m_\mu(x^2) = m_{2\mu}(x) = \sum_{\nu \preceq 2\mu} A^\nu_{2\mu} \sigma_\nu(x).$$

Our assertion follows with

$$\theta^\nu_\lambda = \sum_{\mu,\lambda} K^\mu_\lambda A^\nu_{2\mu}. \qquad \square$$

The relation

$$\sigma_\mu(x_1^2, x_2^2, \dots, x_n^2) = \sum_\lambda W_{b(f_\mu^E)}(\lambda)\delta_F^{-1/2}(\varpi^{\lambda-2\mu})\sigma_\lambda(x)$$

is thus equivalent to

$$W_{b(f_\mu^E)}(\lambda)\delta_F^{-1/2}(\varpi^{\lambda-2\mu}) = \theta_\mu^\lambda.$$

Alternatively,

$$b(f_\mu^E) = \sum_{\lambda \preceq 2\mu} \theta_\mu^\lambda \delta_F^{1/2}(\varpi^{\lambda-2\mu}) f_\lambda^F. \tag{27}$$

Next consider the orbital integrals of

$$\Psi(X) = \int \Psi_0({}^t\bar{g}Xg) f(g)\, dg$$

where f is a Hecke function. We have

$$\begin{aligned}\Omega(\Psi, E/F, \psi : a) &= \int \Psi[{}^t\bar{u}au]\theta(u\bar{u})\, du \\ &= \int \Psi_0[{}^t\bar{g}ag]\left(\int f_E(ug)\theta^{-1}(u\bar{u})\, du\right) dg \\ &= \int \Psi_0[{}^t\bar{g}ag] W_f(g)\, dg.\end{aligned}$$

Using the Iwasawa decomposition, we get

$$= \sum_{\lambda \in \mathcal{P}} \delta_E(\varpi^\lambda)^{-1} W_f(\lambda) \int \Psi_0[\varpi^{\lambda t}\bar{u}au\varpi^\lambda]\theta(u\bar{u})\, du.$$

This is nothing but the orbital integral of the function

$$X \mapsto \sum_{\lambda \in \mathcal{P}} \delta_E(\varpi^\lambda)^{-1} W_f(\lambda)\Psi_0[\varpi^\lambda X \varpi^\lambda].$$

Likewise, the orbital integral of the function Φ defined by

$$\Phi(X) = \int \Phi_0(Xg) b(f)(g)\, dg$$

can be computed as the orbital integral of the function

$$X \mapsto \sum_{\lambda \in \mathcal{P}} \delta_F(\varpi^\lambda)^{-1} W_{b(f)}(\lambda)\Phi_0[X\varpi^\lambda].$$

To prove the theorem, by linearity, we may assume that $f = f_\mu^E$. Then we will have to prove that the function

$$X \mapsto \delta_E^{-1}(\varpi^\mu)\Psi_0(\varpi^\mu X \varpi^\mu)$$

matches the function

$$X \mapsto \sum_{\lambda \preceq 2\mu} \delta_F^{-1}(\varpi^\lambda)\delta_F^{1/2}(\varpi^{\lambda-2\mu})\theta_\mu^\lambda \Phi_0(X\varpi^\lambda).$$

We will set

$$\Psi_\mu(X) := \Psi_0(\varpi^\mu X \varpi^\mu), \tag{28}$$

$$\Phi_\lambda(X) := \Phi_0(X\varpi^\lambda), \tag{29}$$

$$\Phi'_\mu(X) := \sum_\lambda \theta_\mu^\lambda \delta_F^{1/2}(\varpi^{2\mu-\lambda})\Phi_\lambda(X). \tag{30}$$

Our main result will be the consequence of the following proposition.

PROPOSITION 1. – *For every μ the functions*

$$\Psi_\mu(X), \quad \Phi'_\mu(X)$$

match.

Before embarking on the proof we make some formal remarks. We define the Fourier transforms of Schwartz–Bruhat functions on $M(n \times n, F)$ and $M_n(E/F, n)$ respectively by

$$\widehat{\Phi}(X) = \int \Phi(Y)\psi(\mathrm{Tr}[-w_n X w_n Y])\,dY,$$

$$\widehat{\Psi}(X) = \int \Psi(Y)\psi(\mathrm{Tr}[-w_n X w_n Y])\,dY.$$

Then the Fourier transform of Ψ_μ is $q^{2n|\mu|}\Psi_{\tilde{\mu}}$. Likewise the Fourier transform of Φ_λ is $q^{n|\lambda|}\Phi_{\tilde{\lambda}}$. Also the Fourier transform of Φ'_μ is $q^{2n|\tilde{\mu}|}\Phi'_{\tilde{\mu}}$. For the last relation we need the fact that

$$\theta_\mu^\lambda = \theta_{\tilde{\mu}}^{\tilde{\lambda}}.$$

Now let us make some remarks about the action of the scalar matrices $z \in F^\times$. We have $\gamma(az) = \gamma(a)\gamma(z)$. Thus if $\Psi \leftrightarrow \Phi$ then the functions

$$X \mapsto \Psi(Xz)\gamma(z), \qquad X \mapsto \Phi(Xz)$$

match. In particular the functions

$$X \mapsto \Psi(Xz^2), \qquad X \mapsto \Phi(Xz^2)$$

match. As a consequence if we know the proposition for

$$\mu = (\mu_1, \mu_2, \ldots, \mu_n)$$

we also know the proposition for

$$(\mu_1 + m, \mu_2 + m, \ldots, \mu_n + m)$$

for any $m \in \mathbb{Z}$. In particular, we may assume $\mu_n = 0$.

5. Reduction formula

Consider an element of $\mathcal{P}^n$ of the form (μ,ν) with $\mu \in \mathcal{P}^r$, $\nu \in \mathcal{P}^s$ and $n = r + s$. We first prove the following reduction formula. We let ϵ be the matrix with r columns and s rows such that

$$\theta\begin{pmatrix} 1_r & X \\ 0 & 1_s \end{pmatrix} = \psi(\mathrm{Tr}(\epsilon X)).$$

Thus

$$\epsilon = \begin{pmatrix} 0 & \dots & 0 & 1 \\ 0 & \dots & 0 & 0 \\ \dots & \dots & \dots & \dots \\ 0 & \dots & 0 & 0 \end{pmatrix}.$$

Likewise we let η be the transpose of ϵ. Then

$$\theta\left[\begin{pmatrix} 1_r & 0 \\ Y & 1_s \end{pmatrix}^t\right] = \psi(\mathrm{Tr}(Y\eta)).$$

LEMMA 3. – *For $A \in GL(|\mu|, F)$ and $B \in M(|\nu| \times |\nu|, F)$ set*

$$\Phi(A:B) = \int \Phi_{(\mu,\nu)}\begin{bmatrix} A & AX \\ YA & B + YAX \end{bmatrix} \psi(\mathrm{Tr}\,\epsilon X)\psi(\mathrm{Tr}\,Y\eta)\,dX\,dY.$$

If

$$|\det A| = q^{|\mu|}$$

then

$$\Phi(A:B) = q^{-s|\mu|+r|\nu|}\Phi_\mu(A)\Phi_\nu(B).$$

Proof. – Explicitly

$$\Phi(A:B) = \int \Phi_0\begin{bmatrix} A\varpi^\mu & AX\varpi^\nu \\ YA\varpi^\mu & B\varpi^\nu + YAX\varpi^\nu \end{bmatrix} \psi(\mathrm{Tr}\,\epsilon X)\psi(\mathrm{Tr}\,Y\eta)\,dX\,dY.$$

Under the assumptions of the lemma both sides of the identity to be proved are zero unless $A\varpi^\mu \in GL(r, \mathcal{O}_F)$. Assume this is the case. Then $\|Y\| \leqslant 1$. Moreover $\|AX\varpi^\nu\| \leqslant 1$. Hence $\|YAX\varpi^\nu\| \leqslant 1$. Moreover

$$AX\varpi^\nu = A\varpi^\mu\varpi^{-\mu}X\varpi^\nu.$$

Thus in fact

$$\|\varpi^{-\mu}X\varpi^\nu\| \leqslant 1.$$

Writing $X = (x_{i,j})$ we have

$$|x_{i,j}\varpi^{-\mu_i+\nu_j}| \leqslant 1.$$

Since $\mu_i \geqslant \nu_j$, we have $\|X\| \leqslant 1$ which implies $\psi(\epsilon X) = 1$. After integrating over X and Y we find

$$q^{-s|\mu|+r|\nu|}\Phi_0\begin{bmatrix} A\varpi^\mu & 0 \\ 0 & B\varpi^\nu \end{bmatrix} = q^{-s|\mu|+r|\nu|}\Phi_\mu(A)\Phi_\nu(B). \qquad \square$$

We can compute the orbital integral of $\Phi_{(\mu,\nu)}$ as the orbital integral of $\Phi(A:B)$ on the product group $GL(r,F)\times GL(s,F)$. We arrive at the following reduction formula:

LEMMA 4. – *Suppose*

$$|a_1a_2\cdots a_r| = \left|\det \varpi^{-\mu}\right|.$$

Then

$$\begin{aligned}&\Omega(\Phi_{(\mu,\nu)},\psi : a_1,a_2,\ldots,a_r,a_{r+1},a_{r+2}\ldots,a_n)\\&\quad = q^{-s|\mu|+r|\nu|}\Omega(\Phi_\mu,\psi : a_1,a_2,\ldots,a_r)\Omega(\Phi_\nu,\psi : a_{r+1},a_{r+2},\ldots,a_n).\end{aligned}$$

Similarly let

$$\Psi(A:B) = \int \Psi_{(\mu,\nu)}\begin{bmatrix} A & AX\\ {}^t\overline{X}A & B+{}^t\overline{X}AX\end{bmatrix}\psi(\operatorname{Tr}\epsilon X)\psi(\operatorname{Tr}{}^t\overline{X}\eta)\,dX.$$

LEMMA 5. – *Suppose*

$$|\det A|_F = \left|\varpi^{-2\mu}\right|_F.$$

Then

$$\Psi(A:B) = q^{-2s|\mu|+2r|\nu|}\Psi_\mu(A)\Psi_\nu(B).$$

Proof. – Explicitly

$$\begin{aligned}\Psi(A:B) = \int \Psi_{(\mu,\nu)}&\begin{bmatrix}\varpi^\mu A\varpi^\mu & \varpi^\mu AX\varpi^\nu\\ \varpi^{\nu t}\overline{X}A\varpi^\mu & \varpi^\nu B\varpi^\nu + \varpi^{\nu t}\overline{X}AX\varpi^\nu\end{bmatrix}\\&\times\psi(\operatorname{Tr}\epsilon X)\psi(\operatorname{Tr}{}^t\overline{X}\eta)\,dX.\end{aligned}$$

Under the assumption bot sides of the identity are 0 unless $\varpi^\mu A\varpi^\mu \in GL(r,\mathcal{O}_E)$. Assume this is the case. Then the matrix $\varpi^{-\mu}X\varpi^\nu$ is integral. It follows that the matrix

$$\varpi^{\nu t}\overline{X}AX\varpi^\nu = \varpi^{\nu t}\overline{X}\varpi^{-\mu}\varpi^\mu A\varpi^\mu\varpi^{-\mu}X\varpi^\nu$$

is integral. The matrix $X=(x_{i,j})$ is itself integral with $|x_{i,j}|_E \leqslant |\varpi^{\mu_i-\nu_j}|_E$. Integrating over X we get our result. □

LEMMA 6. – *Suppose*

$$|a_1a_2\cdots a_r|_F = \left|\det\varpi^{-2\mu}\right|_F.$$

Then

$$\begin{aligned}&\Omega(\Psi_{(\mu,\nu)},E/F,\psi : a_1,a_2,\ldots,a_r,a_{r+1},a_{r+2},\ldots,a_n)\\&\quad = q^{-2s|\mu|+2r|\nu|}\Omega(\Psi_\mu,E/F,\psi : a_1,a_2,\ldots,a_r)\Omega(\Psi_\nu,E/F,\psi : a_{r+1},a_{r+2},\ldots,a_n).\end{aligned}$$

6. The Kloosterman transform

As in [6, (5)], we introduce the *normalized* orbital integrals

$$\begin{aligned}\widetilde{\Omega}[\Phi,\psi : a] &= |a_1||a_1a_2|\cdots|a_1a_2\cdots a_{n-1}|\Omega[\Phi,\psi : a],\\ \widetilde{\Omega}[\Psi,E/F,\psi : a] &= \eta(a_1)\eta(a_1a_2)\cdots\eta(a_1a_2\cdots a_{n-1})\\&\qquad\times|a_1||a_1a_2|\cdots|a_1a_2\cdots a_{n-1}|\Omega[\Psi,E/F,\psi : a].\end{aligned}$$

We will denote by $\mathcal{I}_n$ the space of functions ω on $(F^\times)^{n-1} \times F$ which are normalized orbital integrals, that is, of the form

$$\omega(a_1, a_2, \dots, a_n) = \widetilde{\Omega}(\Phi, \psi : a),$$

for a suitable $\Phi \in \mathcal{S}(M(n \times n, F))$. Recall that it is also the set of normalized orbital integrals

$$\widetilde{\Omega}[\Psi, E/F, \psi : a].$$

By conjugating by an appropriate diagonal matrix, we see that the space does not change if we replace ψ by $\bar{\psi}$. If ω is in the space $\mathcal{I}_n$ its *Kloosterman* transform $\mathcal{K}_{n,\psi}(\omega)$ is a function in the same space. To define the Kloosterman transform, we define inductively two sequences of functions. First we set

$$\sigma_0(a_1, a_2, \dots, a_n) := \mu_0(a_1, a_2, \dots, a_n) := \omega(a_1, a_2, \dots, a_n).$$

Then we set

$$\mu_1(a_1, a_2, \dots, a_{n-1}, b_1) := \int \sigma_0(a_1, a_2, \dots, a_{n-1}, a_n)\psi(-a_n b_1)\, da_n,$$

$$\sigma_1(a_1, a_2, \dots, a_{n-1}, b_1) = \mu_1(a_1, a_2, \dots, a_{n-1}, b_1)\psi\left(\frac{1}{a_{n-1}b_1}\right).$$

Inductively, if $1 \leqslant i \leqslant n-1$ and we have defined

$$\sigma_i(a_1, a_2, \dots, a_{n-i}, b_i, b_{i-1}, \dots, b_1),$$

then we define

$$\begin{aligned}&\mu_{i+1}(a_1, a_2, \dots, a_{n-i-1}, b_{i+1}, b_i, \dots, b_1)\\ &\quad := \int \sigma_i(a_1, a_2, \dots, a_{n-i}, b_i, b_{i-1}, \dots, b_1)\psi(-a_{n-i}b_{i+1})\, da_{n-i}\end{aligned}$$

and

$$\begin{aligned}&\sigma_{i+1}(a_1, a_2, \dots, a_{n-i-1}, b_{i+1}, b_i, \dots, b_1)\\ &\quad := \mu_{i+1}(a_1, a_2, \dots, a_{n-i-1}, b_{i+1}, b_i, \dots, b_1)\psi\left(\frac{1}{a_{n-i-1}b_{i+1}}\right).\end{aligned}$$

In particular

$$\sigma_n(b_n, b_{n-1}, \dots, b_1) := \mu_n(b_n, b_{n-1}, \dots, b_1).$$

Note that our definition of σ_n and σ_0 is in accordance with the convention that an empty product has the value 1. We emphasize that the integral defining μ_{i+1} is *absolutely convergent.* Moreover, for fixed i, the functions σ_i and μ_i have the *same support.* We set then

$$\mathcal{K}_{n,\psi}(b_1, b_2, \dots, b_n) = \mu_n(b_n, b_{n-1}, \dots, b_1).$$

For $n = 1$ the Kloosterman transform is just the ordinary Fourier transform.

Just as for the ordinary Fourier transform, there is an inversion formula. More precisely, let us set

$$\check{\mu}_{n-i}(b_1, b_2, \dots, b_i, a_{n-i}, a_{n-i-1}, \dots, a_1) := \sigma_i(a_1, a_2, \dots, a_{n-i}, b_i, b_{i-1}, \dots, b_1),$$
$$\check{\sigma}_{n-i}(b_1, b_2, \dots, b_i, a_{n-i}, a_{n-i-1}, \dots, a_1) := \mu_i(a_1, a_2, \dots, a_{n-i}, b_i, b_{i-1}, \dots, b_1).$$

Then

$$\begin{aligned}&\check{\sigma}_{n-i}(b_1, b_2, \dots, b_i, a_{n-i}, a_{n-i-1}, \dots, a_1)\\ &\quad = \check{\mu}_{n-i}(b_1, b_2, \dots, b_i, a_{n-i}, a_{n-i-1}, \dots, a_1)\bar{\psi}\left(\frac{1}{b_i a_{n-i}}\right)\end{aligned}$$

and, from the Fourier inversion formula,

$$\begin{aligned}&\int \check{\sigma}_{n-i-1}(b_1, b_2, \dots, b_i, b_{i+1}, a_{n-i-1}, \dots a_1)\psi(b_{i+1}a_{n-i})\, db_{i+1}\\ &\quad = \check{\mu}_{n-i}(b_1, b_2, \dots, b_i, a_{n-i}, a_{n-i-1}, \dots, a_1).\end{aligned}$$

In particular,

$$\check{\sigma}_n(a_n, a_{n-1}, \dots, a_1) = \mu_0(a_1, a_2 \dots, a_n).$$

In other words, the composition $\mathcal{K}_{n,\bar{\psi}} \circ \mathcal{K}_{n,\psi}$ is the identity.

There is a principle of symmetry: we can exchange the variables (a_*) and (b_*), the left and the right, and the character ψ and the character $\bar{\psi}$.

Then the main result of [6, (5)] are the following identities:

$$\mathcal{K}_{n,\psi}\widetilde{\Omega}[\Phi, \psi; \bullet] = \widetilde{\Omega}[\widehat{\Phi}, \bar{\psi} : \bullet], \tag{31}$$

$$\mathcal{K}_{n,\psi}\widetilde{\Omega}[\Psi, E/F, \psi; \bullet] = \widetilde{\Omega}[\widehat{\Psi}, E/F, \bar{\psi} : \bullet]. \tag{32}$$

7. Definition of the diagrams

For $n = 1$ the Fourier transform has the following property. If ϕ is supported on the set $|a| \leqslant |\varpi^r|$ and its Fourier transform $\hat{\phi}$ is supported on the set $|b| \leqslant |\varpi^t|$ and $t + r > 0$ then ϕ and $\hat{\phi}$ are 0. In other words the support of ϕ and $\hat{\phi}$ are empty so that the above support conditions are trivially verified. Our goal in this section is to extend this result to the Kloosterman transform.

To present the proof in a convenient way, we introduce the notion of a *Kloosterman diagram*. Each diagram has two rows consisting of indexed boxes such as

$$\overset{k}{\boxed{r}}, \qquad \overset{k}{\boxed{=r}}$$

where r is an integer. In the bottom row, the boxes are a shorthand notation for

$$|a_1 a_2 \cdots a_k| \leqslant |\varpi^r|, \qquad |a_1 a_2 \cdots a_k| = |\varpi^r|$$

respectively. In the top row they are a short hand notation for

$$|b_k b_{k-1} \cdots b_1| \leqslant |\varpi^r|, \qquad |b_k b_{k-1} \cdots b_1| = |\varpi^r|$$

respectively. In each diagram the indices are increasing in the bottom row and decreasing in the top row. Indices in boxes in the same column add up to $n + 1$. For consistency we introduce dummy boxes

$$\overset{0}{\boxed{=0}}$$

Thus, in the bottom row say, a diagram

$$\begin{array}{cc} k & k+1 \\ \boxed{=r} & \boxed{s} \end{array}$$

stands for

$$|a_1a_2\cdots a_k| = |\varpi^r|, \qquad |a_{k+1}| \leqslant |\varpi^{s-r}|.$$

All our diagrams will have the following form:

$$\begin{array}{ccccccccccc} & n-u & n-u-1 & n-u-2 & & v+3 & v+2 & v+1 & v \\ & \boxed{t_{n-u}} & \boxed{t_{n-u-1}} & \boxed{t_{n-u-2}} & \cdots & \boxed{t_{v+3}} & \boxed{t_{v+2}} & \boxed{t_{v+1}} & \boxed{=\theta} \\ u & u+1 & u+2 & u+3 & & n-v-2 & n-v-1 & n-v & \\ \boxed{=\rho} & \boxed{r_{u+1}} & \boxed{r_{u+2}} & \boxed{r_{u+3}} & \cdots & \boxed{r_{n-v-2}} & \boxed{r_{n-v-1}} & \boxed{r_{n-v}} & \end{array} \tag{33}$$

Here $0 \leqslant u \leqslant n$, $0 \leqslant v \leqslant n$, $u + v \leqslant n$. We shall call the integer $n - u - v \geqslant 0$ the *length* of the diagram. We shall say that the diagram is *reduced* if $\rho = 0, \theta = 0$. The reduced diagram attached to the above diagram is the one obtained by subtracting θ from each entry in the first row and ρ from each entry in the second row. The bottom row is a shorthand notation for the conditions

$$|a_1a_2\cdots a_u| = |\varpi^{\rho}|, \tag{34}$$

$$|a_1a_2\cdots a_i| \leqslant |\varpi^{r_i}|, \qquad u+1 \leqslant i \leqslant n-v. \tag{35}$$

Of course if $u = 0$ then $\rho = 0$, in accordance with the convention that an empty product is 1. Likewise the top row is a shorthand notation for the conditions

$$|b_1b_2\cdots b_v| = |\varpi^{\theta}|, \tag{36}$$

$$|b_1b_2\cdots b_i| \leqslant |\varpi^{t_i}|, \qquad v+1 \leqslant i \leqslant n-u. \tag{37}$$

In what follows we let ω be a function in the space $\mathcal{I}_n$. We will consider such a diagram. That is, we will have chosen (sometimes only implicitly) $a_1, a_2, \ldots, a_u$ and $b_1, b_2, \ldots, b_v$ satisfying (34) and (36) respectively. We will say that the diagram *holds* if the conditions indicated by the bottom row hold on the support of the function

$$(a_{u+1}, a_{u+2}, \ldots, a_{n-v}) \mapsto \mu_v(a_1, a_2, \ldots, a_u, a_{u+1}, a_{u+2}, \ldots, a_{n-v}, b_v, b_{v-1}, \ldots, b_1)$$

while the conditions indicated by the top row hold on the support of the function

$$\begin{aligned} &(b_{n-u}, b_{n-u-1}, \ldots, b_{v+1}, b_v, b_{v-1}, \ldots, b_1) \\ &\quad \mapsto \mu_{n-u}(a_1, a_2, \ldots, a_u, b_{n-u}, b_{n-u-1}, \ldots, b_v, b_{v-1}, \ldots, b_1). \end{aligned}$$

Of course we could replace μ_v and μ_{n-u} by σ_v and σ_{n-u} respectively. We shall say that the diagram holds *trivially* if the above functions are in fact identically 0. That is, for the given $(a_1, a_2, \ldots, a_u)$ and $(b_1, b_2, \ldots, b_v)$,

$$\mu_v(a_1, a_2, \ldots, a_u, a_{u+1}, a_{u+2}, \ldots, a_{n-v}, b_v, b_{v-1}, \ldots, b_1) = 0$$

for all $(a_{u+1}, a_{u+2}, \ldots, a_{n-v})$ and similarly for μ_{n-u}. In fact, the vanishing condition for μ_v implies the vanishing condition for μ_{n-u} and conversely.

A simple property of the diagrams is the following one. Suppose that the diagram (33) holds. Fix i, $v < i < n - u$. Choose $b_i, b_{i-1}, \ldots, b_{u+1}$ such that

$$|b_ib_{i-1}\cdots b_{v+1}| = |\varpi^{t_i-\theta}|$$

or equivalently

$$|b_i b_{i-1} \cdots b_1| = |\varpi^{t_i}|. \tag{38}$$

Then the following diagram holds:

$$\begin{array}{ccccccccc} & \scriptstyle n-u & \scriptstyle n-u-1 & \scriptstyle n-u-2 & & \scriptstyle i+2 & \scriptstyle i+1 & \scriptstyle i \\ & \boxed{t_{n-u}} & \boxed{t_{n-u-1}} & \boxed{t_{n-u-2}} & \cdots & \boxed{t_{i+2}} & \boxed{t_{i+1}} & \boxed{=t_i} \\ \scriptstyle u & \scriptstyle u+1 & \scriptstyle u+2 & \scriptstyle u+3 & & \scriptstyle n-i-1 & \scriptstyle n-i & \\ \boxed{=\rho} & \boxed{r_{u+1}} & \boxed{r_{u+2}} & \boxed{r_{u+3}} & \cdots & \boxed{r_{n-i-1}} & \boxed{r_{n-i}} & \end{array} \tag{39}$$

Indeed, for the bottom row this follows from the fact that μ_i is obtained from μ_v by repeatedly multiplying by a non-zero factor and taking a Fourier transform. For the top row this is trivial. In particular, if the diagram (39) holds trivially for all choices of $(b_i, b_{i-1}, \ldots, b_{v+1})$ satisfying the above condition, then the diagram (33) holds with t_i replaced by $t_i + 1$.

By symmetry, a similar assertion is true for a diagram of the form:

$$\begin{array}{ccccccccc} & \scriptstyle n-i & \scriptstyle n-i-1 & \scriptstyle n-i-2 & & \scriptstyle v+3 & \scriptstyle v+2 & \scriptstyle v+1 & \scriptstyle v \\ & \boxed{t_{n-i}} & \boxed{t_{n-i-1}} & \boxed{t_{n-i-2}} & \cdots & \boxed{t_{v+3}} & \boxed{t_{v+2}} & \boxed{t_{v+1}} & \boxed{=\theta} \\ \scriptstyle i & \scriptstyle i+1 & \scriptstyle u+2 & \scriptstyle u+3 & & \scriptstyle n-v-2 & \scriptstyle n-v-1 & \scriptstyle n-v & \\ \boxed{=r_i} & \boxed{r_{i+1}} & \boxed{r_{u+2}} & \boxed{r_{u+3}} & \cdots & \boxed{r_{n-v-2}} & \boxed{r_{n-v-1}} & \boxed{r_{n-v}} & \end{array} \tag{40}$$

We shall say that diagrams of the form (39) or (40) are subdiagrams of the diagram (33).

8. The weight of a diagram of length $\leqslant 2$

We shall associate to every diagram D of length an integer $w(D)$, its *weight*. By definition the weight of D is the weight of the corresponding reduced diagram. If D is reduced then $w(D)$ is the sum of the entries of D. If D has length one, that is, has the form

$$\begin{array}{ccc} & \scriptstyle v+1 & \scriptstyle v \\ & \boxed{t} & \boxed{=\theta} \\ \scriptstyle u & \scriptstyle u+1 & \\ \boxed{=\rho} & \boxed{r} & \end{array} \tag{41}$$

with $u + v + 1 = n$, then

$$w(D) = t - \theta + r - \rho.$$

If D has length 2, that is, has the form:

$$\begin{array}{cccc} & \scriptstyle v+2 & \scriptstyle v+1 & \scriptstyle c \\ & \boxed{t_2} & \boxed{t_1} & \boxed{=\theta} \\ \scriptstyle u & \scriptstyle u+1 & \scriptstyle u+2 & \\ \boxed{=\rho} & \boxed{r_1} & \boxed{r_2} & \end{array} \tag{42}$$

then

$$w(D) = t_1 - \theta + t_2 - \theta + r_1 - \rho + r_2 - \rho.$$

PROPOSITION 2. – *If the weight of the diagram of length* 1 *or* 2 *is strictly positive and the diagram holds, then it holds trivially.*

Before we start with the proof and the proof of a corresponding result for certain diagrams of length n (Proposition 3), we remark that we are only interested in the case of a reduced diagram. We introduce the more general diagrams for the purpose of proving the propositions by induction on the length of the diagram. As a matter of fact, the assertion for a given diagram of length $n-u-v<n$ is equivalent to the similar assertion for the space $\mathcal{I}_{n-u-v}$ and the Fourier–Kloosterman transform in $n-u-v$ variables, the integers t_i being replaced by $t_i-\theta$, the integer r_i by $r_i-\rho$ and the integers θ and ρ by 0.

Proof of the proposition. – We first prove the proposition for a diagram of length 1.

LEMMA 7. – *Suppose that the following diagram holds:*

$$
\begin{array}{cc}
v+1 & v \\
\boxed{t} & \boxed{=\theta}
\end{array}
\tag{43}
$$

$$
\begin{array}{cc}
u & u+1 \\
\boxed{=\rho} & \boxed{r}
\end{array}
$$

with $u+v+1=n$. *If* $r+t-\rho-\theta>0$ *the diagram holds trivially.*

The assumption is that the function

$$a_{u+1}\mapsto\sigma_v(\bullet,a_{u+1},\bullet)$$

is supported on the set $|a_{u+1}|\leqslant|\varpi^{r-\theta}|$ while its Fourier transform, namely the function

$$b_{v+1}\mapsto\mu_{v+1}(\bullet,b_{v+1},\bullet),$$

is supported on the set $|b_{v+1}|\leqslant|\varpi^{t-\theta}|$. The lemma follows from the uncertainty principle.

Next, we prove the proposition for a diagram of length 2. As we explained before, this is equivalent to the proposition with $n=2$ and $\theta=\rho=0$. For clarity, we formulate this as a lemma.

LEMMA 8. – *Suppose* $n=2$ *and the following diagram holds:*

$$
\begin{array}{ccc}
2 & 1 & 0 \\
\boxed{t_2} & \boxed{t_1} & \boxed{=0}
\end{array}
\tag{44}
$$

$$
\begin{array}{ccc}
0 & 1 & 2 \\
\boxed{=0} & \boxed{r_1} & \boxed{r_2}
\end{array}
$$

If $r_1+r_2+t_1+t_2>0$ *then the diagram holds trivially, that is,* $\omega=0$.

Proof. – We recall the situation at hand. We are given

$$\omega(a_1,a_2)=\sigma_0(a_1,a_2)$$

supported on the set

$$|a_1|\leqslant|\varpi^{r_1}|,\qquad |a_1a_2|\leqslant|\varpi^{r_2}|.$$

We define

$$\mu_1(a_1,b_1):=\int\sigma_0(a_1,a_2)\psi(-a_2b_1)\,da_2,$$

$$\sigma_1(a_1,b_1)=\mu_1(a_1,b_1)\psi\left(\frac{1}{a_1b_1}\right),$$

$$\mu_2(b_2,b_1)=\int\sigma_1(a_1,b_1)\psi(-a_1b_2)\,da_1$$

and μ_2 is supported on the set

$$|b_1|\leqslant|\varpi^{t_1}|,\qquad |b_1b_2|\leqslant|\varpi^{t_2}|.$$

Thus, viewed as a function of a_1, the function $\sigma_1(a_1,b_1)$ has a module of continuity equal to $|b_1\varpi^{-t_2}|$ (i.e. is invariant under the translations belonging to $\mathcal{O}_F\, b_1\varpi^{-t_2}$). Similarly, viewed as a function of b_1, the function $\mu_1(a_1,b_1)$ has a module of continuity equal to $|a_1\varpi^{-r_2}|$.

We first observe that, by Lemma 7, for a given b_1 with $|b_1|=|\varpi^{t_1+s}|$, $s\geqslant 0$, the function $\mu_1(a_1,b_1)$ vanishes identically unless $r_1\leqslant t_1+s-t_2$.

Now fix a_1 with $|a_1|=|\varpi^{r_1}|$. Let $s\geqslant 0$ be an integer such that

$$\mu_1(a_1,b_1)\neq 0\Longrightarrow|b_1|\leqslant|\varpi^{t_1+s}|.$$

In other words, the following diagram holds:

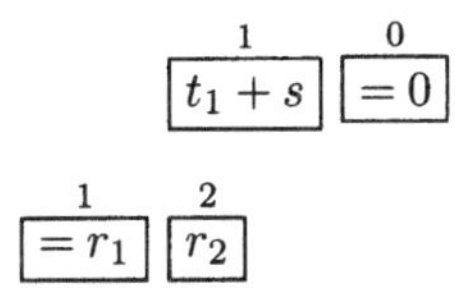

Then by Lemma 7, $t_1+s\leqslant r_1-r_2$. From now on we restrict b_1 to the shell $|b_1|=|\varpi^{t_1+s}|$. We choose ϵ such that

$$|\epsilon|=|\varpi^{t_1+s-t_2}|,\qquad |a_1+\epsilon|=|a_1|=|\varpi^{r_1}|$$

which is always possible since $t_1+s-t_2\geqslant r_1$ and the residual characteristic is not 2. By the continuity of σ_1 we have for all b_1 in the shell

$$\sigma_1(a_1+\epsilon,b_1)=\sigma_1(a_1,b_1),$$

that is,

$$\mu_1(a_1+\epsilon,b_1)\psi\left(\frac{1}{(a_1+\epsilon)b_1}\right)=\mu_1(a_1,b_1)\psi\left(\frac{1}{a_1b_1}\right).$$

We claim that $\mu_1(a_1,b_1)=0$ for all b_1 in the shell. We proceed by contradiction and fix a b_1 such that $\mu_1(a_1,b_1)\neq 0$. Then the above relation reads

$$\frac{\mu_1(a_1+\epsilon,b_1)}{\mu_1(a_1,b_1)}=\psi\left(\frac{\epsilon}{(a_1+\epsilon)a_1b_1}\right).$$

Next, we let η be such that

$$|\eta|\leqslant|\varpi^{r_1-r_2}|,\qquad |b_1+\eta|=|b_1|=|\varpi^{t_1+s}|.$$

Again this is possible because $r_1 - r_2 \geqslant t_1 + s$. Then

$$\mu_1(a_1, b_1 + \eta) = \mu_1(a_1, b_1), \qquad \mu_1(a_1 + \epsilon, b_1 + \eta) = \mu_1(a_1 + \epsilon, b_1)$$

from which we get, for all such η,

$$\psi\left(\frac{\epsilon\eta}{(a_1 + \epsilon)a_1(b_1 + \eta)b_1}\right) = 1.$$

If $r_1 - r_2 > t_1 + s$ then

$$\frac{\eta}{b_1 + \eta}$$

is an arbitrary element of the ideal

$$\varpi^{r_1 - r_2 - t_1 - s}\mathcal{O}_F$$

while

$$\left|\frac{\epsilon}{a_1(a_1 + \epsilon)b_1}\right| = \left|\varpi^{-2r_1 - t_2}\right|.$$

We get then a contradiction since

$$-2r_1 - t_2 + r_1 - r_2 - t_1 - s = -(r_1 + r_2 + t_1 + t_2) - s < 0.$$

If on the contrary $r_1 - r_2 = t_1 + s$ then

$$2r_1 + t_2 = r_1 + r_2 + t_1 + t_2 + s > 0$$

and

$$k := \frac{\epsilon}{a_1(a_1 + \epsilon)b_1}$$

verifies

$$|k| = \left|\varpi^{-t_2 - 2r_1}\right| > 1.$$

On the other hand we take $|\eta| \leqslant |b_1|$, $|b_1 + \eta| = |b_1|$. Thus

$$\frac{\eta}{b_1 + \eta}$$

is an integer of the form

$$\frac{\xi}{1 + \xi}, \quad |\xi| \leqslant 1, \quad |\xi + 1| = 1$$

and we find that

$$\psi\left(k\frac{\xi}{1 + \xi}\right) = 1$$

for all such ξ. Suppose $|k| \geqslant |\varpi^{-2}|$. Any element of $\varpi\mathcal{O}_F$ is of the form $\xi/(1 + \xi)$ with $\xi \in \varpi\mathcal{O}_F$. Thus we get a contradiction. If $|k| = |\varpi^{-1}|$ then in the above formula $\psi(*)$ depends only on ξ modulo $\varpi\mathcal{O}$. Passing to the residual field κ of F we see that, for a non-trivial additive character ψ_0 of κ we have

$$\psi_0\left(\frac{\xi}{1+\xi}\right)=1$$

for all $\xi\in\kappa$, $\xi\neq 0,-1$. Equivalently

$$\psi(\theta)=1$$

for all $\theta\neq 1$. Since

$$\sum_{\theta\in\kappa}\psi(\theta)=0$$

we get $\psi(1)=1-q$. This is a contradiction since $\psi(1)$ is a complex root of 1. □

We conclude that $\mu_1(a_1,b_1)=0$ for $|b_1|=|\varpi^{t_1+s}|$. Thus we may replace s by $s+1$. Inductively, it follows that $\mu_1(a_1,b_1)=0$ for all b_1. Hence in fact $\mu_1(a_1,b_1)=0$ for all $|a_1|=|\varpi^{r_1}|$. Thus in the diagram of the lemma we may replace r_1 by r_1+1. The new diagram holds and has positive weight. Thus we may replace r_1 by an arbitrary large integer. We conclude that the support of ω is empty as claimed. □

9. Slanted diagrams

From now on we suppress the indices in the diagrams if this does not create confusion. We shall say that a diagram is *slanted with slope* m if it has the following form

$$\begin{array}{cccccccc} & \boxed{\rho+m} & \boxed{r_1+m} & \cdots & \boxed{r_{k-2}+m} & \boxed{r_{k-1}+m} & \boxed{=r_k+m} \\ \boxed{=\rho} & \boxed{r_1} & \cdots & \boxed{r_{k-2}} & \boxed{r_{k-1}} & & \boxed{r_k} \end{array} \tag{45}$$

If a diagram is slanted then the corresponding reduced diagram is also slanted (with a different slope).

LEMMA 9. – *Suppose that the above slanted diagram holds. Suppose further that the* (*non-slanted*) *diagram obtained by replacing the entry r_j by r_j+1 holds. Then the slanted diagram obtained by replacing also r_j+m by r_j+m+1 holds.*

Proof. – We may assume $\rho=0$. Our assertion amounts then to saying the following. Suppose that the following diagram holds.

$$\begin{array}{cccccccc} & \boxed{m} & \boxed{r_1+m} & \cdots & \boxed{r_{j-2}+m} & \boxed{r_{j-1}+m} & \boxed{=r_j+m} \\ \boxed{=0} & \boxed{r_1} & \cdots & \boxed{r_{j-2}} & \boxed{r_{j-1}} & & \boxed{r_j+1} \end{array} \tag{46}$$

Then it holds trivially.

Consider first the case $j=1$. The diagram at hand has length one. and weight 1:

$$\begin{array}{ccc} & \boxed{m} & \boxed{=r_1+m} \\ \boxed{=0} & \boxed{r_1+1} & \end{array}$$

Thus it holds trivially. Hence we may assume $j > 1$ and the result true for $j-1$. Assume that the diagram (46) holds. The right subdiagram of length 1 of diagram (46) has weight 1 thus holds trivially. Thus, in fact, the following diagram holds.

$$\begin{array}{ccccccc} & \boxed{m} & \boxed{r_1+m} \cdots & \boxed{r_{j-2}+m} & \boxed{r_{j-1}+m} & \boxed{=r_j+m} \\ \boxed{=0} & \boxed{r_1} \cdots & \boxed{r_{j-2}} & \boxed{r_{j-1}+1} & \boxed{r_j+1} & \end{array}$$

By the induction hypothesis the following diagram holds trivially.

$$\begin{array}{cccccc} & \boxed{m} & \boxed{r_1+m} & \boxed{r_2+m} \cdots & \boxed{r_{j-2}+m} & \boxed{=r_{j-1}+m} \\ \boxed{=0} & \boxed{r_1} & \boxed{r_2} & \cdots & \boxed{r_{j-2}} & \boxed{r_{j-1}+1} \end{array}$$

Thus in fact the following diagram holds.

$$\begin{array}{ccccccc} & \boxed{m} & \boxed{r_1+m} \cdots & \boxed{r_{j-2}+m} & \boxed{r_{j-1}+m+1} & \boxed{=r_j+m} \\ \boxed{=0} & \boxed{r_1} \cdots & \boxed{r_{j-2}} & \boxed{r_{j-1}+1} & \boxed{r_j+1} & \end{array}$$

Thus we have replaced r_{j-1} in the original diagram by $r_{j-1}+1$. Inductively, we can replace r_{j-1} by an arbitrarily large integer. That is, the original diagram holds trivially, as was claimed. □

We will use the following elementary combinatorial lemma.

LEMMA 10. – *If D is slanted of length $l \geqslant 2$ then the sum of the weights of the proper subdiagrams of D is equal to $\frac{l-2}{2} w(D)$.*

Proof. – We may replace D by the corresponding reduced diagram D'. Indeed $w(D) = w(D')$ by definition. Moreover, if D_1 is a subdiagram of D and D'_1 the corresponding subdiagram of D' then $w(D_1) = w(D'_1)$. Thus we may assume D has the form

$$\begin{array}{ccccccccc} & \boxed{m} & \boxed{x_1+m} \cdots & \boxed{x_{j-1}+m} & \boxed{x_j+m} \cdots & \boxed{x_n+m} & \boxed{=0} \\ \boxed{=0} & \boxed{x_1} \cdots & \boxed{x_{j-1}} & \boxed{x_j} & \cdots & \boxed{x_n} & \boxed{-m} \end{array} \tag{47}$$

where $n = l-1$. We may regard the x_i's and m as variables. Thus the sum of the weights of the proper subdiagrams of D is a function $s(x_1, x_2, \ldots, x_n; m)$. Clearly s is linear:

$$s(x_1, x_2, \ldots, x_n; m) = \sum_{1 \leqslant j \leqslant n} A_j x_j + Km.$$

To compute A_j we remark that there are $(n-1)$ subdiagrams containing the diagonal pattern

$$\begin{array}{cc} \cdots & \boxed{x_j+m} \\ \boxed{x_j} & \cdots \end{array}$$

They contribute $(n-1)2x_j$ to s. There is one subdiagram containing the pattern

$$\begin{array}{ccc} & & \boxed{x_j+m}\ \boxed{\cdots} \\ \boxed{=x_j}\ \boxed{\cdots} & & \boxed{\cdots} \end{array}$$

It contributes $-(n+1-j)x_j + x_j$ to s. There is one subdiagram with the pattern

$$\begin{array}{l} \boxed{\cdots}\ \boxed{\cdots}\ \boxed{=x_j+m} \\ \boxed{\cdots}\ \boxed{x_j} \end{array}$$

It contributes $-jx_j + x_j$to s. The total contribution is $(n-1)x_j$. Thus $A_j = n-1$ for all j. To compute K we remark that after changing $x_j \to x_j - m$ and reflecting the diagram we get

$$s(x_1 - m, x_2 - m, \ldots, x_n - m; m) = s(x_n, \ldots, x_2, x_1; -m).$$

Thus

$$Km - n(n-1)m = -Km$$

or

$$K = \frac{n(n-1)}{2}.$$

Thus

$$\begin{aligned} &s(x_1, x_2, \ldots, x_n : K) \\ &= (n-1)\sum_j x_j + \frac{n(n-1)}{2}m = \frac{n-1}{2}\Big(2\sum x_j + nm\Big) = \frac{l-2}{2}w(D). \qquad \square \end{aligned}$$

PROPOSITION 3. – *Suppose that D is a slanted diagram of positive weight. If it holds, then it holds trivially.*

Proof. – We already know the proposition if the length is $\leqslant 2$. Thus we may assume the length k to be $\geqslant 3$ and our assertion true for any slanted diagram of shorter length. Consider a diagram D of the form

$$(48)\qquad \begin{array}{l} \quad\ \boxed{\rho+m}\ \boxed{r_1+m}\ \cdots\ \boxed{r_{k-2}+m}\ \boxed{r_{k-1}+m}\ \boxed{=r_k+m} \\ \boxed{=\rho}\ \boxed{r_1}\ \cdots\ \boxed{r_{k-2}}\ \boxed{r_{k-1}}\ \boxed{r_k} \end{array}$$

Assume that D holds for a function ω. Suppose that $w(D) > 0$. Then, by the previous lemma, at least one of the subdiagrams of D (of length > 1) has positive weight, that is, holds trivially. By symmetry we may assume that it is a right subdiagram. This means that in the diagram we may increase one entry, r_j say, by 1 and the new diagram holds. But then by Lemma 9 equation (46) we may also increase the entry $r_j + m$ by 1 to obtain a new slanted diagram which also holds. The weight of the new diagram has increased by 2. Inductively we construct a sequence D_j of slanted diagrams with increasing weight; one passes from D_j to D_{j+1} by increasing two entries (on the same 45 degree diagonal) by 1. Moreover $\lim_{j\to\infty} w(D_j) = +\infty$. It follows that there are a subsequence j_s of j and an index i independent of s such that the entry in the ith position in the lower line tends to infinity. This implies that $\omega = 0$ and proves the proposition. $\square$

LEMMA 11. – *Suppose that the following slanted diagram holds*:

$$\begin{array}{ccccccccc} & \boxed{r} & \boxed{r} & \cdots & \boxed{r} & \boxed{r} & \boxed{r-k} & \boxed{\cdots} \\ & 1 & 2 & & j-1 & j & j+1 & \\ \boxed{=0} & \boxed{0} & \boxed{0} & \cdots & \boxed{0} & \boxed{-k} & \boxed{\cdots} & \end{array}$$

Here $j \geqslant 2$, $r \geqslant 0$ and $0 \leqslant k \leqslant r$. Then in fact the following diagram holds.

$$\begin{array}{ccccccccc} & \boxed{r} & \boxed{r} & \cdots & \boxed{r} & \boxed{r} & \boxed{r} & \boxed{\cdots} \\ & 1 & 2 & & j-1 & j & j+1 & \\ \boxed{=0} & \boxed{0} & \boxed{0} & \cdots & \boxed{0} & \boxed{0} & \boxed{\cdots} & \end{array}$$

Proof. – Our assertion is vacuous if $k = 0$. Thus we may assume $0 < k \leqslant r$ and our assertion established for $k-1$. Consider the slanted diagram

$$\begin{array}{cccccccc} & \boxed{r} & \boxed{r} & \cdots & \boxed{r} & \boxed{r} & \boxed{=r-k} \\ \boxed{=0} & \boxed{0} & \boxed{0} & \cdots & \boxed{0} & \boxed{-k} & \end{array}$$

We claim it holds trivially. Indeed, the corresponding reduced diagram is

$$\begin{array}{cccccccc} & \boxed{k} & \boxed{k} & \cdots & \boxed{k} & \boxed{k} & \boxed{=0} \\ \boxed{=0} & \boxed{0} & \boxed{0} & \cdots & \boxed{0} & \boxed{-k} & \end{array}$$

It is slanted with a positive weight so holds trivially. Thus in fact the following diagram holds:

$$\begin{array}{cccccccc} & \boxed{r} & \boxed{r} & \cdots & \boxed{r} & \boxed{r} & \boxed{r-k+1} \\ & 1 & 2 & & j-1 & j & j+1 \\ \boxed{=0} & \boxed{0} & \boxed{0} & \cdots & \boxed{0} & \boxed{-k} & \boxed{\cdots} \end{array}$$

and by Lemma 9 and symmetry, the slanted diagram

$$\begin{array}{cccccccc} & \boxed{r} & \boxed{r} & \cdots & \boxed{r} & \boxed{r} & \boxed{r-k+1} \\ & 1 & 2 & & j-1 & j & j+1 \\ \boxed{=0} & \boxed{0} & \boxed{0} & \cdots & \boxed{0} & \boxed{1-k} & \boxed{\cdots} \end{array}$$

holds as well. Now we can apply the induction hypothesis to the integer $k-1$ to obtain the conclusion of the lemma. $\square$

10. Linear independence

In this section we prove that the orbital integrals of the functions Φ_λ, $\lambda \in \mathcal{P}$ are linearly independent. Here and below we use the following simple but crucial fact. Let $\Delta_r(x)$ be the minor of a $n \times n$ matrix x formed with the first r rows and r columns. Then Δ_r is constant on the orbits of $N(F) \times N(F)$ (or $N(E)$). In particular, on the support of $\Omega[\Phi_\lambda, \psi : a]$

$$|a_1 a_2 \cdots a_r| \leqslant \left|\varpi^{-\lambda_1 - \lambda_2 - \cdots - \lambda_r}\right|, \quad 1 \leqslant r \leqslant n.$$

LEMMA 12. – *The functions $a \mapsto \Omega[\Phi_\lambda, \psi : a]$, $\lambda \in \mathcal{P}$ are linearly independent.*

Proof. – Let $\mu > \lambda$ denote the total lexicographic order on $\mathcal{P}$ defined by the components. Thus

$$\mu > \lambda$$

means that there is an integer s, $1 \leqslant s \leqslant n$ such that

$$\mu_1 = \lambda_1, \qquad \mu_2 = \lambda_2, \qquad \ldots, \qquad \mu_{s-1} = \lambda_{s-1}, \qquad \mu_s > \lambda_s.$$

Consider a linear relation

$$\sum_{\lambda \in X} c_\lambda \Omega[\Phi_\lambda, \psi : a] = 0,$$

where the sum is over a finite subset X of $\mathcal{P}$. We have to show that all c_λ are 0. The proof is by induction on the cardinality of X. The statement is vacuous if X is empty. Thus we may assume that X is not empty and our assertion is true for a set with fewer elements. Let μ be the largest element of X for the above order. Let us restrict the relation to the multishell defined by

$$|a_i| = \left|\varpi^{-\mu_i}\right|, \quad 1 \leqslant i \leqslant n.$$

If $\lambda < \mu$ then for at least one $r \leqslant n$

$$\lambda_1 + \lambda_2 + \cdots + \lambda_r < \mu_1 + \mu_2 + \cdots + \mu_r.$$

Since

$$|a_1 a_2 \cdots a_r| \leqslant \left|\varpi^{-\lambda_1 - \lambda_2 - \cdots - \lambda_r}\right|$$

on the support of $\Omega[\Phi_\lambda, \psi : a]$, this function vanishes on the shell. On the other hand the function $\Omega[\Phi_\mu, \psi : a]$ does not vanish on the shell by the reduction formula. In fact, it takes a constant non-zero value. Thus $c_\mu = 0$. Hence we may replace the set X by $X - \{\mu\}$ and apply the induction hypothesis. Our assertion follows. □

LEMMA 13. – *The restrictions of the functions $a \mapsto \Omega[\Phi_\lambda, \psi : a]$, $\lambda \in \mathcal{P}_m$ to the set*

$$\{a\colon |a_1 a_2 \cdots a_n| = \left|\varpi^{-m}\right|\}$$

are linearly independent.

Proof. – This follows from the proof of the previous lemma. □

11. The main lemma

Let $\mathcal{F}(\lambda)$ be the space of (normalized) orbital integrals ω which admit the diagram defined by λ and $\tilde{\lambda}$, that is, the diagram

$$
\begin{array}{ccccccc}
 & \overset{n}{\boxed{|\lambda|}} & \overset{n-1}{\boxed{\lambda_2+\lambda_3+\cdots+\lambda_n}} & \cdots & \overset{2}{\boxed{\lambda_{n-1}+\lambda_n}} & \overset{1}{\boxed{\lambda_n}} & \overset{0}{\boxed{=0}} \\
\overset{0}{\boxed{=0}} & \overset{1}{\boxed{-\lambda_1}} & \overset{2}{\boxed{-\lambda_1-\lambda_2}} & \cdots & \overset{n-1}{\boxed{-\lambda_1-\lambda_2\cdots-\lambda_{n-1}}} & \overset{n}{\boxed{-|\lambda|}} &
\end{array}
\tag{49}
$$

Thus the diagram is slanted with slope $|\lambda|$. Thus $\omega \in \mathcal{F}(\lambda)$ means that the function ω is supported on the set defined by

$$|a_1| \leqslant |\varpi^{-\lambda_1}|, \quad |a_1a_2| \leqslant |\varpi^{-\lambda_1-\lambda_2}|, \quad \ldots, \quad |a_1a_2\cdots a_n| \leqslant |\varpi^{-\lambda_1-\lambda_2\cdots-\lambda_n}|,$$

while the Kloosterman transform $\mathcal{K}\omega$ of ω is supported on the set defined by

$$|b_1| \leqslant |\varpi^{\lambda_n}|, \quad |b_1b_2| \leqslant |\varpi^{\lambda_n+\lambda_{n-1}}|, \quad \ldots, \quad |b_1b_2\cdots b_n| \leqslant |\varpi^{\lambda_n+\lambda_{n-1}+\cdots+\lambda_1}|.$$

If $\mu \preceq \lambda$ then $\tilde{\mu} \preceq \tilde{\lambda}$. It follows that $\mathcal{F}(\mu) \subseteq \mathcal{F}(\lambda)$.

Since the Kloosterman transform of $\widetilde{\Omega}[\Phi_\mu, \psi : a]$ is, up to a scalar factor, $\widetilde{\Omega}[\Phi_{\tilde{\mu}}, \bar{\psi} : b]$ we see that the function $\widetilde{\Omega}[\Phi_\mu, \psi : a]$ belongs to $\mathcal{F}(\mu)$ and to $\mathcal{F}(\lambda)$ for $\mu \preceq \lambda$.

In this section we prove the following key result.

PROPOSITION 4. – *The functions*

$$a \mapsto \widetilde{\Omega}[\Phi_{\lambda'}, \psi : a], \quad \lambda' \preceq \lambda.$$

form a linear basis of the space $\mathcal{F}(\lambda)$.

Our assertion follows from the uncertainty principle for $n = 1$. Thus we may assume $n \geqslant 2$ and our assertion established for $n-1$. We note that if $\lambda = 0$ then our assertion has already been established (*fundamental lemma for the unit element*, [6, (8)]). The space $\mathcal{F}(0)$ is then the one dimensional space spanned by $\widetilde{\Omega}[\Phi_0, \psi : a]$. By homogeneity, our assertion follows when $\lambda_1 = \lambda_n$.

If $\lambda_1 - \lambda_n = 1$ then the relation $\lambda' \preceq \lambda$ implies $\lambda' = \lambda$. Thus the proposition asserts that $\mathcal{F}(\lambda)$ is the one dimensional space spanned by the function $a \mapsto \widetilde{\Omega}[\Phi_\lambda, \psi : a]$. The proof of Proposition 4 will occupy the rest of Section 11.

11.1. The first step

Let ω be in $\mathcal{F}(\lambda)$. Fix a_1 with $|a_1| = |\varpi^{-\lambda_1}|$ and consider the function

$$(a_2, a_3, \ldots, a_n) \mapsto \omega(a_1, a_2, \ldots, a_n).$$

It is supported on the set

$$|a_2| \leqslant |\varpi^{-\lambda_2}|, \quad |a_1a_2| \leqslant |\varpi^{-\lambda_2-\lambda_3}|, \quad \ldots, \quad |a_2a_3\cdots a_n| \leqslant |\varpi^{-\lambda_2-\lambda_3-\cdots-\lambda_n}|.$$

On the other hand, its Kloosterman transform is the function

$$(b_1, b_2, \dots, b_{n-1}) \mapsto \mu_{n-1}(a_1, b_{n-1}, b_{n-2}, \dots, b_1)$$

which is supported on the set

$$|b_1| \leqslant |\varpi^{\lambda_n}|, \quad |b_1 b_2| \leqslant |\varpi^{\lambda_n + \lambda_{n-1}}|, \quad \dots, \quad |b_1 b_2 \cdots b_{n-1}| \leqslant |\varpi^{\lambda_n + \lambda_{n-1} + \dots + \lambda_2}|.$$

In other words, it belongs to the space $\mathcal{F}(\mu)$ where we have set

$$\mu = (\lambda_2, \lambda_3, \dots, \lambda_n).$$

By the induction hypothesis for Proposition 4, we have

$$\omega(a_1, a_2, \dots, a_n) = \sum_{\mu' \preceq \mu} c^{\mu'}(a_1) \widetilde{\Omega}[\Phi_{\mu'} : a_2, a_3, \dots, a_n], \tag{50}$$

where the functions $c^{\mu'}(a_1)$ are uniquely determined functions on the shell $|a_1| = |\varpi^{-\lambda_1}|$.

LEMMA 14. – *With the above notation, the functions $c^{\mu'}(a_1)$ are actually constant on the shell $|a_1| = |\varpi^{-\lambda_1}|$.*

Proof. – To prove Lemma 14, we first prove an auxiliary lemma.

LEMMA 15. – *Let $\omega \in \mathcal{F}(\lambda)$. Fix $b_{n-1}, b_{n-2}, \dots, b_1$ with*

$$|b_1| \leqslant |\varpi^{\lambda_n}|, \quad |b_1 b_2| \leqslant |\varpi^{\lambda_n + \lambda_{n-1}}|, \quad |b_1 b_2 \cdots b_{n-2}| \leqslant |\varpi^{\lambda_n + \lambda_{n-1} + \dots + \lambda_3}|,$$

$$|b_1 b_2 \cdots b_{n-1}| = |\varpi^{|\mu|}|.$$

Then the function

$$a_1 \mapsto \sigma_{n-1}(a_1, b_{n-1}, b_{n-2}, \dots, b_1)$$

is constant on the set

$$|a_1| \leqslant |\varpi^{-\lambda_1}|$$

and the function

$$a_1 \mapsto \mu_{n-1}(a_1, b_{n-1}, b_{n-2}, \dots, b_1)$$

is constant on the shell

$$|a_1| = |\varpi^{-\lambda_1}|.$$

Proof. – Indeed, the function

$$a_1 \mapsto \sigma_{n-1}(a_1, b_{n-1}, b_{n-2}, \dots, b_1)$$

is supported on the set

$$|a_1| \leqslant |\varpi^{-\lambda_1}|.$$

Under the assumption on the b_i's its Fourier transform, that is, the function

$$b_n \mapsto \mathcal{K}(\omega)(b_1, b_2, \dots, b_n)$$

is supported on the set

$$|b_n| \leqslant |\varpi^{\lambda_1}|.$$

The first assertion follows from the uncertainty principle. Since

$$\sigma_{n-1}(a_1, b_{n-1}, b_{n-2}, \ldots, b_1) = \mu_{n-1}(a_1, b_{n-1}, b_{n-2}, \ldots, b_1)\psi\left(\frac{1}{a_1 b_{n-1}}\right)$$

for the second assertion it will suffice to prove that, on the support of the function, the factor ψ is 1 for $|a_1| = |\varpi^{-\lambda_1}|$, or, what amounts to the same,

$$|b_{n-1}| \geqslant |\varpi^{\lambda_1}|.$$

Indeed

$$|b_{n-1}| = |\varpi^{\lambda_n+\lambda_{n-1}+\cdots+\lambda_3+\lambda_2}||b_{n-2}b_{n-3}\cdots b_1|^{-1}.$$

Now

$$|b_{n-2}b_{n-3}\cdots b_1| \leqslant |\varpi^{\lambda_n+\lambda_{n-1}+\cdots+\lambda_3}|.$$

Hence

$$|b_{n-1}| \geqslant |\varpi^{\lambda_2}| \geqslant |\varpi^{\lambda_1}|$$

the last inequality because $\lambda_1 \geqslant \lambda_2$. This concludes the proof of Lemma 15. □

We now go back to the problem of proving that the functions $c^{\mu'}$ we introduced in (50) are constant. To that end, we take the Kloosterman transform of the identity (50) with respect to the variables $(a_2, a_3, \ldots, a_n)$. We get

$$\mu_{n-1}(a_1, b_{n-1}, b_{n-2}, \ldots, b_1) = \sum_{\mu' \preceq \mu} c^{\mu'}(a_1) q^{n|\mu|} \widetilde{\Omega}[\Phi_{\tilde{\mu}'}, \bar{\psi} : b_1, b_2, \ldots, b_{n-1}].$$

Now let us restrict this relation to the set of b_i's in Lemma 15. The left-hand side is independent of a_1 in the shell. On the other hand, the restriction of the functions $\widetilde{\Omega}[\Phi_{\tilde{\mu}'}, \bar{\psi} : b_1, b_2, \ldots, b_{n-1}]$ to the set of the b_i's in Lemma 15 are linearly independent by Lemma 13. It follows that the functions $c^{\mu'}(a_1)$ are indeed constant on the shell. This establishes Lemma 14. □

To continue, we denote by $c^{\mu'}$ their constant values of the functions $c^{\mu'}$. Thus the relation

$$\omega(a_1, a_2, \ldots, a_n) = \sum_{\mu' \preceq \mu} c^{\mu'} \widetilde{\Omega}[\Phi_{\mu'} : a_2, a_3, \ldots, a_n]$$

holds for a_1 in the shell $|a_1| = |\varpi^{-\lambda_1}|$. If $\mu' \preceq \mu$ then $\lambda' := (\lambda_1, \mu')$ satisfies $\lambda' \preceq \lambda$. In addition, by the reduction formula, on the shell,

$$\widetilde{\Omega}[\Phi_{\lambda'} : a_1, a_2, \ldots, a_n] = q^{|\mu|} \Omega[\Phi_{\mu'} : a_2, a_3, \ldots, a_n].$$

After a change of notation we see that we can write, for a_1 in the shell $|a_1| = |\varpi^{-\lambda_1}|$,

$$\omega(a_1, a_2, \ldots, a_n) = \sum_{\lambda' \preceq \lambda, \lambda'_1 = \lambda_1} c^{\lambda'} \widetilde{\Omega}[\Phi_{\lambda'} : a_1, a_2, \ldots, a_n].$$

Let us set

$$\omega_1(a) := \omega(a) - \sum_{\lambda' \preceq \lambda \lambda'_1 = \lambda_1} c^{\lambda'} \widetilde{\Omega}[\Phi_{\lambda'} : a_1, a_2, \ldots, a_n]. \tag{51}$$

This function is still in $\mathcal{F}(\lambda)$. However its support is contained in the set

$$|a_1| \leqslant \left|\varpi^{1-\lambda_1}\right|.$$

That is, the function ω_1 admits the diagram

$$\begin{array}{ccccccc} & \boxed{|\lambda|} & \boxed{\lambda_2 + \cdots + \lambda_n} & \cdots & \boxed{\lambda_{n-1} + \lambda_n} & \boxed{\lambda_n} & \boxed{=0} \\ \boxed{=0} & \boxed{1-\lambda_1} & \boxed{-\lambda_1 - \lambda_2} & \cdots & \boxed{-\lambda_1 - \lambda_2 \cdots - \lambda_{n-1}} & \boxed{-|\lambda|} & \end{array}$$

By Lemma 9 it follows that it admits the diagram

$$\begin{array}{ccccccc} & \boxed{|\lambda|} & \boxed{1 + \lambda_2 + \cdots + \lambda_n} & \cdots & \boxed{\lambda_{n-1} + \lambda_n} & \boxed{\lambda_n} & \boxed{=0} \\ \boxed{=0} & \boxed{1-\lambda_1} & \boxed{-\lambda_1 - \lambda_2} & \cdots & \boxed{-\lambda_1 - \lambda_2 \cdots - \lambda_{n-1}} & \boxed{-|\lambda|} & \end{array} \tag{52}$$

Proposition 4 will now follow from the following lemma.

LEMMA 16. – *The function ω_1 belongs to some space $\mathcal{F}(\lambda')$ with $\lambda' \prec \lambda$.*

If $\lambda_1 - \lambda_2 \geqslant 2$ we can set

$$\lambda' = (\lambda_1 - 1, \lambda_2 + 1, \lambda_3, \ldots, \lambda_n).$$

Clearly λ' is still in $\mathcal{P}$. We have $\lambda' \prec \lambda$ and the function ω_1 is in $\mathcal{F}(\lambda')$. The lemma follows in this case. From now on we assume that $\lambda_1 - \lambda_2 \leqslant 1$.

11.2. The case of a fundamental weight

Suppose that
$\lambda_1 - \lambda_n \leqslant 1$. Then

$$\omega_1(a) = \omega(a) - c\widetilde{\Omega}[\Phi_\lambda : a]$$

with $c \neq 0$. Lemma 16 amounts to saying that $\omega_1 = 0$ or, equivalently, Proposition 4, amounts to saying that $\mathcal{F}(\lambda)$ is the one dimensional space spanned by $\widetilde{\Omega}[\Phi_\lambda : a]$.

Assume first $\lambda_1 - \lambda_n = 1$. Note that we may replace λ by $\tilde{\lambda}$ and use homogeneity. In particular, we may assume

$$\lambda = (\underbrace{1, 1, \ldots, 1}_{r}, \underbrace{0, 0, \ldots, 0}_{s}), \quad r \geqslant s.$$

We have to show that the function ω_1 of (51) is 0.

For instance if

$$\lambda = (1, 1, 1, 1, 0, 0)$$

then the diagram associated with λ is

$$\begin{array}{cccccccc} & \boxed{4} & \boxed{3} & \boxed{2} & \boxed{1} & \boxed{0} & \boxed{0} & \boxed{=0} \\ \boxed{=0} & \boxed{-1} & \boxed{-2} & \boxed{-3} & \boxed{-4} & \boxed{-4} & \boxed{-4} & \end{array}$$

then the diagram for the function ω_1 is

$$\boxed{4}\boxed{4}\quad\boxed{2}\quad\boxed{1}\quad\boxed{0}\quad\boxed{0}\quad\boxed{=0}$$
$$\boxed{=0}\boxed{0}\boxed{-2}\boxed{-3}\boxed{-4}\boxed{-4}\boxed{-4}$$

Applying repeatedly Lemma 11 we see that the following diagram holds.

$$\boxed{4}\boxed{4}\boxed{4}\boxed{4}\boxed{4}\boxed{4}\quad\boxed{=0}$$
$$\boxed{=0}\boxed{0}\boxed{0}\boxed{0}\boxed{0}\boxed{0}\boxed{-4}$$

Since this slanted diagram has positive weight it holds trivially, that is, $\omega_1 = 0$. In general, applying repeatedly Lemma 11 we see that ω_1 admits a slanted diagram whose entries in the first row are all equal to r and all entries in the last row are 0 except one which is $-r$. This diagram has positive weight and the conclusion follows again.

If $\lambda_1 - \lambda_n = 0$, as noted earlier, we have proved our assertion previously in [8, (8)]. For the sake of completeness we reproduce the proof. We may as well assume $\lambda = (0,0,\ldots,0)$. If

$$\lambda = (0,0,0,0,0,0)$$

for instance, then the diagram associated to λ is

$$\boxed{0}\boxed{0}\boxed{0}\boxed{0}\boxed{0}\boxed{0}\boxed{=0}$$
$$\boxed{=0}\boxed{0}\boxed{0}\boxed{0}\boxed{0}\boxed{0}\boxed{0}$$

and ω_1 admits the diagram

$$\boxed{0}\boxed{1}\boxed{0}\boxed{0}\boxed{0}\boxed{0}\boxed{=0}$$
$$\boxed{=0}\boxed{1}\boxed{0}\boxed{0}\boxed{0}\boxed{0}\boxed{0}$$

This is a slanted diagram of slope 0 and weight 2. Thus $\omega_1 = 0$. Thus we have proved Lemma 16 or Proposition 4 when $\lambda_1 - \lambda_n \leqslant 1$.

11.3. The second step

LEMMA 17. – *Suppose that $\lambda \in \mathcal{P}^n$ verifies*

$$\lambda_1 - \lambda_r \leqslant 1$$

for some r with $1 < r \leqslant n-1$. Suppose that ω is an element of $\mathcal{F}(\lambda)$ such that $\omega = 0$ for $|a_1| = |\varpi^{-\lambda_1}|$. Then $\mathcal{K}\omega = 0$ for

$$|b_1 b_2 \cdots b_{n-r}| = \left|\varpi^{\lambda_n + \lambda_{n-1} + \cdots + \lambda_{r+1}}\right|$$

and $\omega = 0$ for

$$|a_1 a_2 \cdots a_r| = \left|\varpi^{-\lambda_1 - \lambda_2 - \cdots - \lambda_r}\right|.$$

Proof. – We set

$$\mu = (\lambda_1, \lambda_2, \dots, \lambda_r), \qquad \nu = (\lambda_{r+1}, \lambda_{r+2}, \dots, \lambda_n).$$

Thus $\lambda = (\mu, \nu)$. Fix $b_1, b_2, \dots, b_{n-r}$ with

$$|b_1, b_2 \cdots b_{n-r}| = \left|\varpi^{\lambda_n + \lambda_{n-1} + \cdots + \lambda_{r+1}}\right|.$$

The function

$$(a_1, a_2, \dots, a_r) \mapsto \sigma_{n-r}(a_1, a_2, \dots, a_r, b_{n-r}, b_{n-r-1}, \dots, b_1)$$

belongs to $\mathcal{F}(\mu)$. By the previous section

$$\sigma_{n-r}(a_1, a_2, \dots, a_r, b_{n-r}, b_{n-r-1}, \dots, b_1) = M(b_1, b_2, \dots, b_{n-r})\widetilde{\Omega}[\Phi_\mu, \psi : a_1, a_2, \dots, a_r].$$

By the assumption, the left-hand side vanishes if $|a_1| = |\varpi^{-\lambda_1}|$. On the other hand $\widetilde{\Omega}[\Phi_\mu, \psi : a_1, a_2, \dots, a_r]$ takes a non-zero constant value on the set

$$|a_1| = \left|\varpi^{-\lambda_1}\right|, \qquad |a_1 a_2| = \left|\varpi^{-\lambda_1 - \lambda_2}\right|, \qquad |a_1 a_2 \cdots a_r| = \left|\varpi^{-\lambda_1 - \lambda_2 - \cdots - \lambda_r}\right|.$$

Thus in fact M is identically 0 and we conclude that

$$\sigma_{n-r}(a_1, a_2, \dots, a_r, b_{n-r}, b_{n-r-1}, \dots, b_1) = 0$$

for $|b_1, b_2 \cdots b_{n-r}| = |\varpi^{\lambda_n + \lambda_{n-1} + \cdots + \lambda_{r+1}}|$ which is equivalent to the first assertion. Moreover, the function μ_{n-r} has the same property.

Now fix $a_1, a_2, \dots, a_r$ with

$$|a_1 a_2 \cdots a_r| = \left|\varpi^{-\lambda_1 - \lambda_2 - \cdots - \lambda_r}\right|.$$

Then the function

$$(a_{r+1}, a_{r+2} \dots, a_n) \mapsto \omega(a_1, a_2, \dots, a_r, a_{r+1}, \dots, a_n)$$

belongs to $\mathcal{F}(\nu)$. By the induction hypothesis for Proposition 4, it can be written as

$$\sum_{\nu' \preceq \nu} C^{\nu'}(a_1, a_2, \dots, a_r)\widetilde{\Omega}[\Phi_{\nu'}, \psi; a_{r+1}, a_{r+2}, \dots, a_n]. \tag{53}$$

Taking the Kloosterman transform of this identity we get

$$\begin{aligned}&\mu_{n-r}(a_1, a_2, \dots, a_r, b_{n-r}, b_{n-r-1}, \dots, b_1)\\ &\quad = \sum_{\nu' \preceq \nu} C^{\nu'}(a_1, a_2, \dots, a_r) q^{2(n-r)|\nu|}\widetilde{\Omega}[\Phi_{\widetilde{\nu'}}, \bar{\psi} : b_1, b_2, \dots b_{n-r}].\end{aligned}$$

Since this vanishes on the set

$$|b_1, b_2 \cdots b_{n-r}| = \left|\varpi^{\lambda_n + \lambda_{n-1} + \cdots + \lambda_{n-r+1}}\right|$$

by the first part of the lemma and the functions

$$\widetilde{\Omega}[\Phi_{\widetilde{\nu'}}, \bar{\psi} : b_1, b_2, \dots b_{n-r}]$$

are linearly independent on that set, we conclude that the functions

$$C^{\nu'}(a_1, a_2, \dots, a_r)$$

all vanish. The lemma follows. □

11.4. Reduction to the case of a fundamental weight

Now we prove Lemma 16 for a λ such that $\lambda_1 - \lambda_2 \leqslant 1$ and $\lambda_1 - \lambda_n > 1$. Next, consider the largest integer $1 < s < n$ such that $\lambda_1 - \lambda_s \leqslant 1$. Thus we can write

$$\lambda = (\mu, \nu)$$

with

$$\mu = (\lambda_1, \lambda_2, \dots, \lambda_s), \qquad \lambda_1 - \lambda_s \leqslant 1, \qquad \lambda_1 - \lambda_{s+1} \geqslant 2.$$

By homogeneity we may assume

$$\mu = (\underbrace{\overbrace{1, 1, 1, \dots, 1}^{s_1}, \overbrace{0, 0 \dots, 0}^{s_2}}_{s})$$

with $s_1 > 0$, $s_2 \geqslant 0$ and

$$\lambda_{s+1} \leqslant -1.$$

We consider an ω in $\mathcal{F}(\lambda)$. As before, we define ω_1 by (51). Thus ω_1 is in $\mathcal{F}(\lambda)$ and $\omega_1 = 0$ for $|a_1| = |\varpi^{-\lambda_1}|$. Then we claim that ω_1 is in $\mathcal{F}(\lambda')$ where λ' is obtained by replacing $\lambda_{s_1} = 1$ by $\lambda_{s_1} = 0$ and λ_{s+1} by $\lambda_{s+1} + 1$. Of course $\lambda' \prec \lambda$. This will prove Lemma 16 in this case. We illustrate the case $s_1 = 2$, $s_2 = 2$, $n = 6$. Thus

$$\lambda = (1, 1, 0, 0, \lambda_5, \lambda_6)$$

with $\lambda_5 < 0$. We set $a := \lambda_5 + \lambda_6$. The diagram associated to λ is

$$\begin{array}{cccccccc} & \boxed{|\lambda|} & \boxed{1+a} & \boxed{a} & \boxed{a} & \boxed{a} & \boxed{\lambda_6} & \boxed{=0} \\ \boxed{=0} & \boxed{-1} & \boxed{-2} & \boxed{-2} & \boxed{-2} & \boxed{-2-\lambda_5} & \boxed{-|\lambda|} & \end{array}$$

Both ω and ω_1 admit this diagram. But by the previous lemma for $r = 2, 3, 4$ the function ω_1 admits in fact the diagram

$$\begin{array}{cccccccc} & \boxed{|\lambda|} & \boxed{1+a} & \boxed{1+a} & \boxed{1+a} & \boxed{1+a} & \boxed{\lambda_6} & \boxed{=0} \\ \boxed{=0} & \boxed{-1} & \boxed{-1} & \boxed{-1} & \boxed{-1} & \boxed{-2-\lambda_5} & \boxed{-|\lambda|} & \end{array}$$

This is the diagram attached to λ' where

$$\lambda' = (1,0,0,0,\lambda_5+1,\lambda_6).$$

Hence $\omega_1 \in \mathcal{F}(\lambda')$. In general we apply the previous lemma with $s_1 \leqslant r \leqslant s$.

This concludes the proof of Lemma 16 and hence the proof of Proposition 4.

12. The map β

Recall that the Kloosterman transform of the function $\widetilde{\Omega}[\Psi_\mu, E/F, \psi : a]$ is, up to a positive constant, the function $\widetilde{\Omega}[\Psi_{\bar\mu}, E/F, \bar\psi : a]$. Thus the function $\widetilde{\Omega}[\Psi_\mu, E/F, \psi : a]$ belongs to the space $\mathcal{F}(2\mu)$. By Proposition 4 there exist unique constants ξ_μ^λ, $\lambda \preceq 2\mu$ such that

$$\widetilde{\Omega}[\Psi_\mu, E/F, \psi : a] = \sum_\lambda \xi_\mu^\lambda \delta_F^{1/2}(\varpi^{2\mu-\lambda})\widetilde{\Omega}[\Phi_\lambda, \psi : a],$$

or, equivalently,

$$\Psi_\mu \leftrightarrow \sum_\lambda \xi_\mu^\lambda \delta_F^{1/2}(\varpi^{2\mu-\lambda})\Phi_\lambda. \tag{54}$$

Motivated by (27), we define a linear map

$$\beta : \mathcal{H}_E \to \mathcal{H}_F$$

by

$$\beta(f_\mu^E) = \sum_{\lambda \preceq 2\mu} \xi_\mu^\lambda \delta_F^{1/2}(\varpi^{\lambda-2\mu}) f_\lambda^F.$$

We denote by $\beta_{\mathcal{S}}$ the corresponding linear transformation of the algebra $\mathbb{S}_n$ of symmetric polynomials to itself, that is,

$$\mathcal{S}(\beta(f)) = \beta_{\mathcal{S}}(\mathcal{S}(f)).$$

Since

$$\mathcal{S}(f_\lambda^F) = \delta_F^{-1/2}(\varpi^\lambda)\sigma_\lambda, \qquad \mathcal{S}(f_\mu^E) = \delta_E^{-1/2}(\varpi^\mu)\sigma_\mu$$

we have

$$\beta_{\mathcal{S}}(\sigma_\mu) = \sum_{\lambda \preceq 2\mu} \xi_\mu^\lambda \sigma_\lambda.$$

If we consider similarly the map $b_{\mathcal{S}}$ then

$$b_{\mathcal{S}}(\sigma)(x) = \sigma_\mu(x^2) = \sum \theta_\mu^\lambda \sigma_\lambda(x).$$

Our definition amounts to saying that the function

$$X \mapsto \int \Psi_0[{}^t\bar{x}Xx] f_\lambda^E(x)\, dx$$

matches the function

$$X \mapsto \int \Phi_0[Xx]\beta(f_\lambda^E)(x)\,dx.$$

By linearity, for any $f \in \mathcal{H}_E$, the function

$$X \mapsto \int \Psi_0\left[{}^t\bar{x}Xx\right]f(x)\,dx$$

matches the function

$$X \mapsto \int \Phi_0[Xx]\beta(f)(x)\,dx.$$

Our main result is then equivalent to the assertion that $\beta = b$ or, what amounts to the same,

$$\forall\mu, \forall\lambda \preceq 2\mu, \quad \xi_\mu^\lambda = \theta_\mu^\lambda. \tag{55}$$

We first list the properties of β which follow from the definitions.

We have proved that $\Psi_0 \leftrightarrow c\Phi_0$ where c is a constant. Using the reduction formula we see that the orbital integrals of Ψ_0 and Φ_0 agree on $A \cap K_F$. Thus $\Psi_0 \leftrightarrow \Phi_0$, that is, the image of the unit element of $\mathcal{H}_E$ under β is the unit element of $\mathcal{H}_F$.

We have $\sigma_{\tilde{\mu}}(x) = \sigma_\mu(x^{-1})$. Also from the properties of the Kloosterman transform we have $\xi_\mu^\lambda = \xi_{\tilde{\mu}}^{\tilde{\lambda}}$. If $P \in \mathbb{S}_n$ is a symmetric polynomial, let us set $\check{P}(x) = P(x^{-1})$. Clearly $\beta_{\mathcal{S}}(\check{P}) = \beta_{\mathcal{S}}(P)\check{}$. In addition $\mathcal{S}(\check{f}) = \mathcal{S}(f)\check{}$. We conclude that

$$\beta(f)\check{} = \beta(\check{f}).$$

Next, let us set

$$\mu_0 = (\underbrace{1, 1, \ldots, 1}_{n}).$$

We have

$$\sigma_{\lambda+\mu_0}(x) = \sigma_\lambda(x)x_1x_2\ldots x_n.$$

On the other hand

$$\begin{aligned}
\Omega[\Psi_{\mu+\mu_0} : a] &= \Omega\left[\Psi_\mu : a\varpi^{2\mu_0}\right] \\
&= \gamma\left(a\varpi^{2\mu_0}\right)\sum_\lambda \xi_\mu^\lambda \delta_F^{1/2}\left(\varpi^{2\mu-\lambda}\right)\Omega\left[\Phi_\lambda : a\varpi^{2\mu_0}\right] \\
&= \gamma(a)\sum_\lambda \xi_\mu^\lambda \delta_F^{1/2}\left(\varpi^{2\mu-\lambda-2\mu_0}\right)\Omega[\Phi_{\lambda+2\mu_0} : a].
\end{aligned}$$

It follows that

$$\xi_{\mu+\mu_0}^{\lambda+2\mu_0} = \xi_\mu^\lambda \tag{56}$$

and

$$\beta_{\mathcal{S}}(\sigma_{\lambda+\mu_0})(x) = \beta_{\mathcal{S}}(\sigma_\lambda)(x)(x_1x_2\cdots x_n)^2. \tag{57}$$

In particular, since $b_{\mathcal{S}}(\sigma_0) = \beta_{\mathcal{S}}(\sigma_0)$ we have also $b_{\mathcal{S}}(\sigma_{\mu_0}) = \beta_{\mathcal{S}}(\sigma_{\mu_0})$. More generally

$$b_{\mathcal{S}}(\sigma_\mu) = \beta_{\mathcal{S}}(\sigma_\mu)$$

if $\mu_1 - \mu_n = 0$.

The function σ_λ is homogeneous of degree $|\lambda|$. Thus it is clear that if P is homogeneous of degree m then $\beta_S(P)$ is homogeneous of degree $2m$. Equivalently, if $f \in \mathcal{H}_E$ is supported on the set

$$\{g \in GL(n,E)\colon |\det g|_E = |\varpi^m|_E\}$$

then $\beta(f)$ is supported on the set

$$\{g \in GL(n,F)\colon |\det g|_F = |\varpi^{2m}|_F\}.$$

By the same argument we used for b, it follows that for any Hecke function $f \in \mathcal{H}_E$ the function

$$X \mapsto \int \Psi^1[{}^t\bar{x}Xx]f(x)\,dx$$

matches the function

$$X \mapsto \int \Phi^1[Xx]\beta(f)(x)\,dx.$$

Now the functions σ_μ with $\mu_1 - \mu_n = 0$ together with the functions σ_λ with $\lambda_1 - \lambda_n$ generate the algebra $\mathbb{S}_n$. The following proposition shows that β_S and b_S agree on a set of generators.

PROPOSITION 5. – *Suppose μ is such that $\mu_1 - \mu_n = 1$. Then, for all λ,*

$$\xi_\mu^\lambda = \theta_\mu^\lambda.$$

Proof. – Again our assertion is vacuous if $n = 1$. Thus we may assume $n > 1$ and our assertion true for $n - 1$. By homogeneity we may as well assume that

$$\mu = (\overbrace{\underbrace{1,1,\ldots,1}_{r},0,0,\ldots,0}^{n}).$$

If μ has the above form then any $\lambda \preceq 2\mu$ has the form

$$\lambda = (\overbrace{\underbrace{2,2,\ldots,2}_{s},\underbrace{1,1,\ldots,1}_{2t},0,0,\ldots 0}^{n})$$

with

$$0 \leqslant t \leqslant r, \qquad s = r - t, \qquad t \leqslant n - r.$$

LEMMA 18. – *With the above notation*

$$\theta_\mu^\lambda = (-1)^t.$$

Proof. – Replacing μ by $\tilde{\mu}$ and using homogeneity we may assume $2r \leqslant n$. Then the inequality $t \leqslant n - r$ is superfluous. We first recall a property of the Schur functions. Given $\mu \in \mathcal{P}_n$ we say that a $\lambda \in \mathcal{P}_{n-1}$ *interlaces* μ and we write $\lambda \lhd \mu$ if

$$\mu_1 \geqslant \lambda_1 \geqslant \mu_2 \geqslant \lambda_2 \geqslant \cdots \geqslant \mu_{n-1} \geqslant \lambda_{n-1} \geqslant \mu_n.$$

Suppose $x = (y,1)$, $y = (y_1, y_2, \dots, y_{n-1})$. Then we have the branching formula.

$$\sigma_\mu(x) = \sum_{\lambda \lhd \mu} \sigma_\lambda(y). \tag{58}$$

This follows for instance from the fact that σ_λ can be viewed as the character of an irreducible representation π_μ of $GL(n,\mathbb{C})$. The formula amounts to saying that the restriction of π_μ to $GL(n-1,\mathbb{C})$ is the direct sum of the representations π_λ with $\lambda \lhd \mu$ (branching rule, see for instance [4, Theorem 8.11]).

The lemma is a tautology for $n = 1$. Thus we may assume that $n > 1$ and our assertion established for $n-1$. Consider the formula defining the constants θ_μ^λ.

$$\sigma_\mu(x^2) = \sum_{\lambda \preceq 2\mu} \theta_\mu^\lambda \sigma_\lambda(x). \tag{59}$$

We may assume that $x = (y,1)$ because the resulting formula still defines the constants uniquely.

Suppose first that $r < n - r$. Then if $\lambda \preceq 2\mu$ we have $s + 2t \leqslant 2s + 2t = 2r < n$. In other words λ has at least one trailing 0. Using the branching formula for our μ, we can write the left-hand side of (59) as

$$\sigma_\mu(x^2) = \sigma_{\mu^1}(y^2) + \sigma_{\mu^2}(y^2)$$

where

$$\mu^1 = (\overbrace{\underbrace{1,1,\dots,1}_{r},0,0,\dots,0}^{n-1})$$

and

$$\mu^2 = (\overbrace{\underbrace{1,1,\dots,1}_{r-1},0,0,\dots,0}^{n-1}).$$

We can apply the induction hypothesis. We have

$$\sigma_{\mu^1}(y^2) = \sum_{\lambda^1 \preceq 2\mu^1} (-1)^t \sigma_{\lambda^1}(y)$$

where

$$\lambda^1 = \lambda_t^1 = (\overbrace{\underbrace{2,2,\dots,2}_{s_1},\underbrace{1,1,\dots,1}_{2t},0,0,\dots 0}^{n-1})$$

with

$$0 \leqslant t \leqslant r, \qquad s = r - t.$$

We stress that $|\lambda^1| = 2r$. On the other hand,

$$\sigma_{\mu^2}(y^2) = \sum_{\lambda^2 \preceq 2\mu^2} \theta_{\mu^2}^{\lambda^2} \sigma_{\lambda^2}(y)$$

with

$$\lambda^2=\lambda^2_{t_2}=(\overbrace{\underbrace{2,2,\ldots,2}_{s_2},\underbrace{1,1,\ldots,1}_{2t_2},0,0,\ldots 0}^{n-1})$$

with

$$0\leqslant t_2\leqslant r-1,\qquad s_2=r-1-t_2.$$

We stress that such a λ^2 verifies $|\lambda^2|=2r-2$. On the right-hand side of (59) we can write

$$\sum_{\lambda\preceq 2\mu}\theta^\lambda_\mu\sum_{\lambda'\lhd\lambda}\sigma_{\lambda'}(y).$$

For each $\lambda\preceq 2\mu$ there is a unique $\lambda^0\lhd\lambda$ with $|\lambda^0|=|\lambda|=2r$. If

$$\lambda=\lambda_t=(\overbrace{\underbrace{2,2,\ldots,2}_{s},\underbrace{1,1,\ldots,1}_{2t},0,0,\ldots 0}^{n})$$

where $0\leqslant t\leqslant r$, $s=r-t$, it is

$$\lambda^0_t=(\overbrace{\underbrace{2,2,\ldots,2}_{s},\underbrace{1,1,\ldots,1}_{2t},0,0,\ldots 0}^{n-1}).$$

In other words λ^0_t is obtained by dropping the last trailing 0 from λ. Comparing terms $\sigma_\nu(y)$ with $|\nu|=2r$ we get

$$\sum_{0\leqslant t\leqslant r}(-1)^t\sigma_{\lambda^1_t}(y)=\sum_{0\leqslant t\leqslant r}\theta^{\lambda_t}_\mu\sigma_{\lambda^0_t}(y).$$

Comparing the coefficients in this formula we conclude that $\theta^{\lambda_t}_\mu=(-1)^t$ as was claimed.

Now suppose $r=n-r$. Thus

$$\mu=(\underbrace{1,1,\ldots,1}_{r},\underbrace{0,0,\ldots,0}_{r}).$$

The λ such that $\lambda\preceq 2\mu$ are of two types, those λ for which there are at least a trailing 0 and one single element

$$\lambda^p=(\underbrace{1,1,\ldots,1}_{n}).$$

The λ with a trailing 0 have the form

$$\lambda=\lambda_t=(\overbrace{\underbrace{2,2,\ldots,2}_{s},\underbrace{1,1,\ldots,1}_{2t},0,0,\ldots 0}^{n})$$

with

$$0\leqslant t<r,\qquad s=r-t.$$

We have

$$\sigma_\mu(x^2) = \sum_{0 \leqslant t \leqslant r-1} \theta_\mu^{\lambda_t} \sigma_{\lambda_t}(x) + \theta_\mu^{\lambda^p} \sigma_{\lambda^p}(x).$$

Again we set $x = (y, 1)$ and we use the branching formula and the induction hypothesis. Here the only element which interlaces λ^p is

$$\lambda^{p0} = (\underbrace{1, 1, \ldots, 1}_{n-1}).$$

Then

$$\sigma_{\lambda^p}(x) = \sigma_{\lambda^{p0}}(y)$$

and $|\lambda^{p0}| = n - 1 = 2r - 1$. When we compare terms $\sigma_\nu(y)$ with $|\nu| = 2r$ this term does not give a contribution. Thus we have again $\theta_\mu^{\lambda_t} = (-1)^t$. Thus we may write

$$\sigma_\mu(x^2) = \sum (-1)^t \sigma_\lambda(x) + \theta_\mu^{\lambda^p} \sigma_{\lambda^p}(x). \tag{60}$$

where

To compute the remaining constant we apply the same approach as before, but this time we look for terms containing $\sigma_{\lambda^{p0}}(y)$. Note that $|\lambda^{p0}| = n - 1 = 2r - 1$. Consider the left-hand side of (60). It can be written as before

$$\sigma_{\mu^1}(y^2) + \sigma_{\mu^2}(y^2).$$

In applying the induction hypothesis to $\sigma_{\mu^1}(y^2)$ we find terms $\sigma_\nu(y)$ with $|\nu| = 2r$. Thus we cannot find the term $\sigma_{\lambda^{p0}}(y)$. In applying the induction hypothesis to $\sigma_{\mu^2}(y^2)$ we find terms containing σ_ν with $|\nu| = 2r - 2$. Thus we cannot find the term $\sigma_{\lambda^{p0}}(y)$ on the left-hand side. We apply formula (58) to the right-hand side of (60). We remark that there are only two λ such that $\lambda^{p0} \lhd \lambda$, namely λ^p and

$$\lambda^{pp} := (2, \underbrace{1, 1, \ldots, 1}_{2r-2}, 0).$$

Thus we find

$$(-1)^{r-1} \sigma_{\lambda^{p0}}(y) + \theta^{\lambda^p} \sigma_{\lambda^{p0}}(y) = 0$$

or

$$\theta^{\lambda^p} = (-1)^r.$$

This concludes the proof of the lemma. □

It remains to prove that $\xi_\mu^\lambda = \theta_\mu^\lambda$. Replacing μ by $\tilde{\mu}$ and using homogeneity we may assume $2r \leqslant n$.

We pass to the computation of the constants ξ_μ^λ. We first compute the coefficient $\xi_\mu^{\lambda_r}$ in the case $2r = n$. We state this as a lemma.

LEMMA 19. – *Assume $2r = n$. Then, with the above notation,*

$$\xi_\mu^{\lambda_r} = (-1)^r.$$

Proof. – The matching (54) implies an equality of the corresponding orbital integrals of the functions on the left and right-hand sides, up to the transfer factor. We consider the orbital integrals for the matrix $w_n z$ where $z \in F^\times$. We find

$$\Omega[\Psi_\mu, E/F, \psi : w_n z] = \gamma(w_n z, \psi) \sum_{\lambda \preceq 2\mu} \xi_\mu^\lambda \delta_F^{1/2}(\varpi^{2\mu-\lambda}) \Omega[\Phi_\lambda, \psi : w_n z].$$

It is immediately verified that for any λ of the form $\lambda = (*, *, \dots, *, 0)$, the integral $\Omega[\Phi_\lambda : w_n z]$ vanishes unless $|z| = 1$. We take $|z| = |\varpi^{-1}|$. These integrals vanish then. On the other hand we have then $\Omega[\Phi_{\lambda^p} : w_n z] = 1$. We claim that

$$\Omega[\Psi_\mu : w_n z] = q^{-r^2}.$$

We illustrate the case $n = 4$ but the computation is general.

$$\Omega[\Psi_\mu : w_4 z] = \int \Psi_0 \begin{bmatrix} 0 & 0 & 0 & z\varpi \\ 0 & 0 & z\varpi & z\varpi x_1 \\ 0 & z\varpi & zx_2 & zu \\ z\varpi & z\varpi \bar{x}_1 & z\bar{u} & zw \end{bmatrix} \psi(x_1 + \bar{x}_1)\psi(x_2)\, dx_1\, dx_2\, du\, dw.$$

Here x_1, u are in E and x_2, w are in F. After integrating we find the stated result. On the other hand, the transfer factor $\gamma(w_n z, \psi)$, viewed as a function of n, has period 8 for an arbitrary quadratic extension (see [6, (8)]). Here the extension is unramified so that the Weil constant $c(E/F, \psi)$ is 1. Thus the factor is in fact periodic of period 4. Its first 4 values (starting with $n = 1$) are

$$1, \quad \eta(z), \quad \eta(z), \quad 1.$$

Thus here $\gamma(w_n z, \psi) = (-1)^r$. It follows that $\xi_\mu^{\lambda^p} = (-1)^r$. □

It remains to compute the remaining coefficients ξ_μ^λ. Now consider the formula which defines the constants ξ_μ^λ.

$$\Omega[\Psi_\mu : a] = \gamma(a) \sum_{\lambda \preceq 2\mu} \xi_\mu^\lambda \delta_F^{1/2}(\varpi^{2\mu-\lambda}) \Omega[\Phi_\lambda : a].$$

Let us set

$$a = (a_1, a'), \qquad |a_1| = |\varpi^{-2}|.$$

and apply the reduction formulas. If $2r = n$ the term corresponding to λ^p on the right is then 0. Thus in all cases the non-zero terms on the right correspond to λ's of the following form. Let us set

$$\mu = (1, \mu').$$

Then

$$\lambda = (2, \lambda'), \qquad \lambda' \preceq 2\mu'.$$

We get

$$q^{-2(n-1)\mu_1 + 2|\mu'|} \Omega[\Psi_{\mu'} : a'] = \gamma(a') \sum q^{-(n-1)\lambda_1 + |\lambda'|} \xi_\mu^\lambda \delta_F^{1/2}(\varpi^{2\mu-\lambda}) \Omega[\Phi_{\lambda'} : a'].$$

After simplification we get

$$\Omega[\Psi_{\mu'}:a'] = \gamma(a') \sum_{\lambda' \preceq 2\mu'} \Omega[\Phi_{\lambda'}:a']\xi_\mu^\lambda \delta_F^{1/2}(\varpi^{2\mu-\lambda}).$$

In addition $\delta_F^{1/2}(\varpi^{2\mu-\lambda}) = \delta_F^{1/2}(\varpi^{2\mu'-\lambda'})$ since $\det(\varpi^{2\mu'-\lambda'}) = 1$. It follows that $\xi_\mu^\lambda = \xi_{\mu'}^{\lambda'}$. On the other hand from Lemma 18 we have $\theta_\mu^\lambda = \theta_{\mu'}^{\lambda'}$. Applying the induction hypothesis we get $\xi_\mu^\lambda = \theta_\mu^\lambda$ as was claimed. This concludes the proof of Proposition 5. □

We have proved that β_S and b_S agree on a set of generators of $\mathrm{Spec}[\mathbb{S}_n]$ as an algebra. To show that $\beta = b$ it will suffice to show that β_S is a homomorphism of algebras. This is equivalent to saying that for any $x \in \mathrm{Spec}[\mathbb{S}_n]$, the map

$$P \mapsto \beta_S(P)(x)$$

is a character of the algebra $\mathbb{S}_n$. Obviously it suffices to prove it for x in a Zariski dense subset. Thus we see that our assertion that $\beta = b$, or equivalently, the main theorem, will be established if we prove that,

$$f \mapsto \mathcal{S}\big(\beta(f)\big)(x)$$

is a character for all x in a Zariski dense subset. This will be taken up in the next section.

13. A simple trace formula

In this section we consider again a quadratic extension of *number fields* E/F. We assume that every real place of F splits in E. We consider a cuspidal automorphic representation π^0 of $GL(n, F_{\mathbb{A}})$. We assume that at a fixed split place v_{cusp} the local component of π^0 is supercuspidal. We let S be a finite set of places of F containing the places at infinity, the even places, the places v where the local component π_v^0 is ramified, all the places of F inert and ramified in E, all the finite places v where the character ψ_v has a conductor different from $\mathcal{O}_v$. In particular S contains v_{cusp}. For $s \in S$ we take Φ_v so that

$$\mathcal{B}_{\pi_v^0}(\Phi_v) \neq 0.$$

We take, as we may, the function $\Phi_{v_{\text{cusp}}}$ to be a supercuspidal function. At each inert place $v \notin S$ we have the linear map β_v:

$$\beta_v : \mathcal{H}_w \to \mathcal{H}_v,$$

where w is the place above v. Recall that β_v takes the unit element to the unit element. Moreover, β_v is so defined that it satisfies the matching assertion of the main theorem.

We fix a place v_{rel} inert and not in S. We denote by w_{rel} the corresponding place of E. Our goal in this section is to prove the following proposition.

PROPOSITION 6. – *The map*

$$f \mapsto \mathcal{S}\big(\beta_{v_{\text{rel}}}(f)\big)\big(\pi_{v_{\text{rel}}}^0\big)$$

is a character of $\mathcal{H}_{w_{\text{rel}}}$.

Proof. – We follow step by step the construction of Section 3. Simply we replace at each inert place $v \notin S$ the base change map b_v by β_v. The set of places T_1 may be chosen not to contain $v_{\rm rel}$. It is convenient to have the set T_2 not to contain $v_{\rm rel}$. For that purpose we appeal to a lemma.

LEMMA 20. – *One can choose the representatives ξ in such a way that $\xi_{v_{\rm rel}}$ is in $K_{w_{\rm rel}}$.*

Proof. – Recall that the only ξ which appear have $\det \xi$ is a norm in the extension $E_{w_{\rm rel}}/F_{v_{\rm re}}$. At any rate there is $g \in GL(n, E_{w_{\rm rel}})$ such that

$$\xi = {}^t\bar{g}g.$$

Thus there is a sequence $\gamma_i \in GL(n, E)$ such that

$$\lim {}^t\bar{\gamma}_i \xi \gamma_i = 1$$

where the limit is in the space $S(F_{v_{\rm rel}})$. Since $K_{w_{\rm rel}} \cap S(F_{v_{\rm rel}})$ is open, we have

$${}^t\bar{\gamma}_i \xi \gamma_i \in K_{w_{\rm rel}} \cap S(F_{v_{\rm rel}})$$

for some i. $\square$

Since $\Phi_{v_{\rm cusp}}$ is a supercuspidal function the function $f^\xi_{v_{\rm cusp,1}, v_{\rm cusp,2}}$ can be taken to be supercuspidal, as function of the second variable.

Because of the supercuspidality condition the expression (12) is equal to the sum

$$\sum_{\xi \in \Xi} \sum_{\Pi} \mathcal{R}^\xi_\Pi(f^\xi)$$

where the sum is over all cuspidal automorphic representations Π with a supercuspidal component at $v_{\rm cusp,2}$ (and hence at $v_{\rm cusp,1}$). Likewise, the expression (14) is equal to

$$\sum_\pi \mathcal{B}_\pi(\Phi)$$

where the sum is over all cuspidal automorphic representations π with a supercuspidal component at $v_{\rm cusp}$.

At this point we have the analogue of (21):

$$\sum_{\Pi_\infty = \pi^0_\infty \otimes \pi^0_\infty} \sum_{\xi \in \Xi} \mathcal{R}^\xi_\Pi(f^\xi) = c\mathcal{B}_{\pi^0_\infty}(\Phi_\infty) \prod_{v \notin S_2} \mathcal{S}(\Phi_v)(\pi^0_v). \tag{61}$$

The sum on the left is finite. We remind ourselves of the relation between the functions. Let U_2 be the set of places of E above a place in S_2. Each function f^ξ is a tensor product. The factors f^ξ_u corresponding to a finite place in $u \in U_2$ are fixed. All the factors f^ξ_u corresponding to a place $u \notin U_2$ are equal to an arbitrary Hecke function f_u. If $v \notin S_2$ splits into v_1, v_2 then Φ_v is the base change of $f_{v_1} \otimes f_{v_2}$, that is, the convolution $f_{v_1} * f_{v_2}$. If $v \notin S_2$ is inert and $w \notin U_2$ is the corresponding place of E then $\Phi_v = \beta_v(f_w)$. We fix a function Φ_∞ such that

$$\mathcal{B}_{\pi^0_\infty}(\Phi_\infty) \neq 0.$$

We fix then correspondingly Ψ_∞ and the functions f^ξ_∞.

We label the representations on the left-hand side Π^γ, $0 \leqslant \gamma \leqslant s$. We set

$$C^\gamma = \sum_{\xi \in \Xi} \mathcal{R}^\xi_{\Pi^\gamma}\left(f^\xi_{U_2} \otimes \mathbf{1}^{U_2}\right).$$

Then the left-hand side takes the form

$$\sum_{0 \leqslant \gamma \leqslant s} C^\gamma \prod_{u \notin U_2} \mathcal{S}(f_u)(\Pi^\gamma_u).$$

If we choose $\Phi_v = \Phi^1_v$ for all $v \notin S_2$ we obtain on the right of Equation (61) a non-zero result namely

$$c\mathcal{B}_{\pi^0_\infty}(\Phi_\infty).$$

Thus there is a least one representation Π^0 such that $C^0 \neq 0$

$$C^0 := \sum_{\xi \in \Xi} \mathcal{R}^\xi_{\Pi^0}(f^\xi), \qquad f_u = f^1_u, \quad u \notin U_2.$$

Then $C^0 \neq 0$. We can choose a finite set of places $T_3 = \{u_\gamma \mid 1 \leqslant \gamma \leqslant s\}$ not containing w_{rel} such that

$$\Pi^0_{u_\gamma} \not\simeq \Pi^\gamma_{u_\gamma}.$$

As before, we use strong multiplicity one. We choose then Hecke functions f_{u_γ} such that

$$\forall \gamma, \mathcal{S}(f_{u_\gamma})(\Pi^0_{u_\gamma}) = 1, \qquad \mathcal{S}(f_{u_\gamma})(\Pi^\gamma_{u_\gamma}) = 0.$$

If u_γ is above an inert place of F we denote this place by v_γ. If u_γ is above a place v_γ which splits into u_γ, u'_γ then $\Pi^\gamma_{u_\gamma} \simeq \Pi^\gamma_{u'_\gamma}$. Thus we may as well assume that u'_γ is not in T_3. Now at all places $u \notin U_2 \cup T_3$ and distinct from w_{rel} we take $f_u = f^1_u$. Then the left-hand side of Equation (61) reduces to

$$C^0 \mathcal{S}(f_{w_{\text{rel}}})(\Pi^0_{w_{\text{rel}}}).$$

The right-hand side of Equation (61) is then the product of the following factors

$$\begin{aligned}
&\mathcal{S}\left(\beta_{v_{\text{rel}}}(f_{w_{\text{rel}}})\right)\left(\pi^0_{v_{\text{rel}}}\right),\\
&c\mathcal{B}_{\pi_\infty}(f_\infty),\\
&\prod_{v_\gamma \text{ split}} \mathcal{S}(f_{u_\gamma})(\pi^0_{v_\gamma}),\\
&\prod_{v_\gamma \text{ inert}} \mathcal{S}\left(\beta_{v_\gamma}(f_{u_\gamma})\right)\left(\pi^0_{v_\gamma}\right).
\end{aligned}$$

Since $C^0 \neq 0$ we see that the three last factors are non-zero and we conclude that

$$\mathcal{S}\left(\beta_{v_{\text{rel}}}(f_{w_{\text{rel}}})\right)\left(\pi^0_{v_{\text{rel}}}\right) = c_1 \mathcal{S}(f_{w_{\text{rel}}})\left(\Pi^0_{w_{\text{rel}}}\right).$$

By taking $f_{w_{\text{rel}}}$ to be the unit element we see that $c_1 = 1$. This concludes the proof of the proposition. □

14. End of proof

Now we let E/F be an unramified quadratic extension of local fields. We can always write it as $E_{w_{\mathrm{rel}}}/F_{v_{\mathrm{rel}}}$ for a suitable quadratic extension of number fields of the type considered in the previous section. We can choose ψ so that the conductor of $\psi_{v_{\mathrm{rel}}}$ is $\mathcal{O}_{v_{\mathrm{rel}}}$. Consider the representations π^0 of the type discussed in the previous section. The Langlands class $x \in \mathbb{C}^{\times n}$ of their components $\pi^0_{v_{\mathrm{rel}}}$ is Zariski dense.

For the convenience of the reader we recall the proof. We have to prove that if $f_{v_{\mathrm{rel}}}$ is a non-zero element of $\mathcal{H}_{v_{\mathrm{rel}}}$ then there exists a π^0 of the above type such that $\pi^0_{v_{\mathrm{rel}}}(f_{v_{\mathrm{rel}}}) \neq 0$. To that end we set

$$f^2_{v_{\mathrm{rel}}}(g) = \overline{f_{v_{\mathrm{rel}}}(g)},$$
$$f^1_{v_{\mathrm{rel}}}(x) = \int_{G_{v_{\mathrm{rel}}}} f^2_{v_{\mathrm{rel}}}(xg) f_{v_{\mathrm{rel}}}(g)\,dg.$$

Thus $f^1_{v_{\mathrm{rel}}}(e) \neq 0$. For each finite place v we choose a smooth function of compact support f_v such that $f_v(e) \neq 0$. At the place v_{cusp} we furthermore assume that $f_{v_{\mathrm{cusp}}}$ is a cuspidal function. Of course, at almost all places, f_v is the characteristic function of K_v. We let f_∞ be a smooth function of compact support on $Z_+\backslash G_\infty$. We consider the following functions, where $i = 1, 2$,

$$f^i(g) = f_\infty(g_\infty) f^i_{v_{\mathrm{rel}}}(g_{v_{\mathrm{rel}}}) \prod_{v \neq v_{\mathrm{rel}}} f_v(g_v),$$
$$\phi^i(g) = \sum_{\gamma \in G(F)} f^i(\gamma g).$$

If we write $f^1(g) = f_\infty(g_\infty) f^\infty(g^\infty)$ we see that $f^\infty(\gamma) \neq 0$ implies that γ is in a discrete subset Γ of G_∞ containing e. We choose f_∞ so that $f_\infty(e) \neq 0$ and $f_\infty(\gamma) \neq 0, \gamma \in \Gamma$, imply $\gamma = e$. Thus $\phi^1(e) \neq 0$. The functions ϕ^1 and ϕ^2 belong to the closure of the space spanned by the cusp forms. Moreover $\phi^1(g) = \int \phi^2(gx) f_{v_{\mathrm{rel}}}(x)\,dx$. Therefore there is at least one automorphic cuspidal representation π^0 such that the image of ϕ^1 under the orthogonal projection P on the space of π^0 is non-zero. We have

$$P\phi^1(g) = \int (P\phi^2)(gx) f_{v_{\mathrm{rel}}}(x)\,dx,$$

that is,

$$\pi^0_{v_{\mathrm{rel}}}(f_{v_{\mathrm{rel}}}) P\phi^2 = P\phi^1.$$

This implies that $\pi^0_{v_{\mathrm{rel}}}(f_{v_{\mathrm{rel}}}) \neq 0$ and we are done.

Thus the map

$$f \mapsto S\big(\beta(f)\big)(x)$$

is a character of $\mathcal{H}_{w_{\mathrm{rel}}}$ for x in a Zariski dense set. As we have seen this implies that $\beta = b$.

15. Three lemmas on Bessel distributions

15.1. Proof of Lemma 1

We prove Lemma 1. For the proof of this lemma we drop the index infinity. We will need the positive type Bessel distribution

$$\mathcal{P}_\pi(\Phi) = \sum \pi(\Phi) W_i(e) \overline{W}_i(e).$$

Note that

$$\mathcal{B}_\pi(\Phi) = \mathcal{P}_\pi(\Phi')$$

where $\Phi'(g) = \Phi(w_n g)$. Thus we can rewrite the assumption as

$$\sum_\pi c(\pi)\mathcal{P}_\pi(\Phi) = 0$$

for all Φ. By taking the functions Φ to be of positive type we see that we have

$$\sum_\pi \Re c(\pi)\mathcal{P}_\pi(f) = 0, \qquad \sum_\pi \Im c(\pi)\mathcal{P}_\pi(f) = 0$$

Thus it suffices to prove the lemma when the coefficients are real. We derive a contradiction if not all terms are 0. We can regroup together the terms with $c(\pi) > 0$ and the terms with $c(\pi) < 0$. After a change of notation, we have now an identity

$$\sum_{\pi \in X} c(\pi)\mathcal{P}_\pi(f) = \sum_{\pi \in Y} c(\pi)\mathcal{P}_\pi(f),$$

where X and Y are disjoint, in general, infinite sets and $c(\pi) > 0$ for all π. We then derive a contradiction. At the cost of changing the Hilbert norm on π we can rewrite this expression in the form

$$\sum_{\pi \in X} \mathcal{P}_\pi(f) = \sum_{\pi \in Y} \mathcal{P}_\pi(f).$$

Recall that to a distribution of positive type θ we can associate a unitary representation in the following way. Indeed, on the space $\mathcal{C}_c^\infty(G_\infty)$

$$(f_1, f_2) \mapsto \theta(f_2^* * f_1)$$

is a positive semi-definite Hermitian form, invariant under right shifts. Passing to the quotient we get a positive definite Hermitian form on a quotient of $\mathcal{C}_c^\infty(G_\infty)$. Then we complete the space for this norm. We obtain a Hilbert space on which G_∞ acts. Applying this construction to $\mathcal{P}_\pi$ we recover π. In a more precise way, we can regard the Whittaker linear form λ_π as a generalized vector (i.e. a vector which has scalar product with smooth vectors). Then

$$\mathcal{P}_\pi(f^* * f) = \big(\pi(f)\lambda_\pi, \pi(f)\lambda_\pi\big).$$

Applying this construction to the two sides of our equality we recover the direct Hilbert sums

$$\bigoplus_{\pi \in X} \pi, \qquad \bigoplus_{\pi \in Y} \pi$$

respectively. To see it is so consider the direct sum

$$\mathcal{H} := \bigoplus_{\pi \in X} \pi.$$

Recall $\mathcal{H}$ can be viewed as the space of t-uples $(v_\pi)_{\pi \in X}$ such that

$$\sum_\pi (v_\pi, v_\pi) < +\infty.$$

Since

$$\sum_{\pi \in X} \mathcal{P}_\pi(f^* * f) = \sum (\pi(f)\lambda_\pi, \pi(f)\lambda_\pi) < +\infty$$

we can consider the closure $\mathcal{V}$ in $\mathcal{H}$ of the space spanned by t-uples

$$(\pi(f)\lambda_\pi)_{\pi \in X}.$$

Since the projection of $\mathcal{V}$ on each π is non-zero, we see that $\mathcal{V}$ is the Hilbert direct sum of the $\pi \in X$. On the other hand calling θ the sum of the $\mathcal{P}_\pi$ we have

$$\theta(f^* * f) = \sum_{\pi \in X} (\pi(f)\lambda_\pi, \pi(f)\lambda_\pi).$$

Thus $\theta(f^* * f) = 0$ if and only if $\pi(f)\lambda_\pi = 0$ for all π. Our assertion follows. Thus we have

$$\bigoplus_{\pi \in X} \pi = \bigoplus_{\pi \in Y} \pi.$$

Since $X \cap Y = \emptyset$ this is a contradiction.

15.2. A formal lemma

We now go back to the situation discussed before (20). Consider one of the distributions $\mathcal{R}^\xi_\Pi$. We write f for the function f^ξ. Let ∞_F be the set of infinite places of F. Each $v \in \infty_F$ splits into v_1, v_2 and $\Pi_{v_1} \simeq \Pi_{v_2}$. Call π_v the equivalence class of this representation. The component f_{v_1,v_2} of f is arbitrary. We set

$$\Phi_v(g) = \int_{GL(n,F_v)} f_{v_1,v_2}(\xi^{-1\,t}h^{-1}, hg).$$

We can define the local Bessel distribution $\mathcal{B}_{\pi_v}$. On the other hand we have chosen a finite set R of finite places of E and for each $v \in R$ we have fixed the local component f_v. At the finite places $v \notin R$ the function f_v is a Hecke function.

LEMMA 21. – *With the above notations, there is a constant c such that*

$$\mathcal{R}^\xi_\Pi(f) = c \prod_{v \in \infty_F} \mathcal{B}_{\pi_v}(\Phi_v) \prod_{v\,\text{finite}\,\notin R} \mathcal{S}(f_v)(\Pi_v).$$

Proof. –

$$\mathcal{R}^{\xi}_{\Pi}(f) = \sum \mu(\Pi(f)\phi_i) \overline{\int \phi_i(u)\bar{\theta}(u\bar{u})\,du}.$$

Here

$$\mu(\phi) = \int_{H^{\xi}(F)\backslash H^{\xi}(F_{\mathbb{A}})} \phi(h)\,dh.$$

In view of our choice of functions, the functions ϕ_i belong to a sub vector space V invariant under $\prod_{v\in\infty_E} G_v$. In turn V is a finite direct sum of subspaces V_i invariant and irreducible under $\prod_{v\in\infty_E} G_v$. The representation of $\prod_{v\in\infty_E} G_v$ on each V_i is equivalent to

$$\bigotimes_{v\in\infty_F} \Pi_{v_1} \otimes \Pi_{v_2}.$$

The restriction μ_i of μ to V_i is a constant c_i times a tensor product

$$\bigotimes_{v\in\infty_F} \mu_v$$

where μ_v is a linear form on $\pi_v \otimes \pi_v$ invariant under H^{ξ}_v. Likewise the restriction of the linear form

$$\phi \mapsto \int \phi(u)\bar{\theta}(u\bar{u})\,du$$

is a constant times a tensor product

$$\bigotimes_{v\infty_F} \mu_{v,1} \otimes \mu_{v,1}$$

where $\mu_{v,1}$ is a linear form on the space of π_v transforming under the character θ_v. It follows that

$$\mathcal{R}^{\xi}_{\Pi}(f) = c \prod_{v\in\infty_F} \mathcal{R}_v(f_{v_1,v_2}) \prod_{v \text{ finite } \notin R} \mathcal{S}(f_v)(\Pi_v)$$

where

$$\mathcal{R}_v(f_{v_1,v_2}) = \sum \mu_v\big(\pi_v \otimes \pi_v(f_{v_1,v_2})\phi_i\big)\overline{\mu_{v,1} \otimes \mu_{v,1}(\phi_i)}$$

and ϕ_i is an orthonormal basis of $\pi_v \otimes \pi_v$.

To conclude we apply another formal lemma, valid for any local field (see [6, (7)]).

LEMMA 22. – *Suppose that F is a local field. Let π be a unitary irreducible generic representation of $GL(n,F)$. Let λ be a linear form on $\pi\otimes\pi$ invariant under the group $H^{\xi}(F)$. Let μ_1 be a linear form on π transforming under the character θ. Consider the distributions*

$$\mathcal{R}(f) = \sum \lambda\big(\pi\otimes\pi(f)\phi_i\big)\overline{\mu_1 \otimes \mu_1(\phi_i)},$$

$$\mathcal{B}(\Phi) = \sum \mu_2\big(\pi(\Phi)\phi_i\big)\overline{\mu_1(\phi_i)}.$$

Then there is a constant c such that

$$\mathcal{R}(f) = c\mathcal{B}(\Phi)$$

where

$$\Phi(g) = \int f(\xi^{-1t}h^{-1}, hg)\, dh.$$

15.3. A lemma of Shalika

LEMMA 23. – *Let π be a unitary irreducible generic representation of $GL(n,\mathbb{R})$. The Bessel distribution $\mathcal{B}_\pi$ has a non-zero restriction to the open set*

$${}^tN(\mathbb{R})A(\mathbb{R})N(\mathbb{R}).$$

Proof. – We may replace $\mathcal{B}_\pi$ by the positive type Bessel distribution

$$\mathcal{P}_\pi(f) = \sum \pi(f)W_i(e)\overline{W_i(e)}.$$

Our assertion is then that the restriction of $\mathcal{P}_\pi$ to the open set

$$N(\mathbb{R})w_n A(\mathbb{R})N(\mathbb{R})$$

is non-zero. Suppose it is. Then $\mathcal{P}_\pi$ is a distribution supported on the complement of that set. Moreover it is an eigenfunction of the Casimir operator and transforms on both sides under non-degenerate characters of $N_n(F)$. By Proposition 2.10 of [16] it follows that $\mathcal{P}_\pi = 0$, a contradiction. □

16. Concluding remarks

We briefly indicate how to obtain more precise global results. Assume n is even. Let $\widetilde{H}^\xi$ be the similitude group corresponding to H^ξ and $\lambda : \widetilde{H}^\xi \to F^\times$ the similitude ratio. Let χ be an idele class character of F. We say that an automorphic representation Π is distinguished by $(\widetilde{H}^\xi, \chi)$ if the central character of Π is the character $z \mapsto \chi(z\bar{z})$ and the following integral is non-zero for at least one ϕ in the space of Π:

$$\int_{Z(E_\mathbb{A})\widetilde{H}^\xi(F)\backslash \widetilde{H}^\xi(F_\mathbb{A})} \phi(\tilde{h})\chi^{-1}(\lambda(\tilde{h}))\, d\tilde{h}.$$

PROPOSITION 7. – *A representation Π is the base change of a representation π of central character ω if and only if it is distinguished by a pair $(\widetilde{H}^\xi, \omega\eta^{n/2})$.*

Proof (Sketch). – We simply indicate which modifications must be made to the proof of Theorem 2. At a place v of F inert in E we have

$$\gamma_v(az) = \gamma_v(a)\eta_v^{n/2}(z).$$

Thus globally if $\Psi \leftrightarrow \Phi$ then for each $z \in F_\mathbb{A}^\times$ the function

$$X \mapsto \Psi(Xz)$$

matches the function

$$X \mapsto \eta(z)^{n/2}\Phi(Xz).$$

For each $z \in F_{\mathbb{A}}^{\times}$ we have

$$\int K_{\Psi}(zu)\theta(u\bar{u})\,du = \eta^{n/2}(z)\int K_{\Phi}(u_1,u_2z)\theta(u_1)^{-1}\theta(u_2)\,du_2.$$

Integrating over z we get instead of (18)

$$\int\limits_{F_{\mathbb{A}}^{\times}\times N(E)\backslash N(E_{\mathbb{A}})} K_{\Psi}(zu)\theta(u\bar{u})\chi(z)\,dz\,du$$

$$= \int\limits_{F_{\mathbb{A}}^{\times}\times(N(F)\backslash N(F_{\mathbb{A}}))^2} K_{\Phi}(u_1,u_2z)\theta(u_1)^{-1}\theta(u_2)\,du_2\chi\eta^{n/2}(z)\,dz.$$

Now we choose a set of representatives $\{\xi\}$ for the orbits of $F^{\times}\times GL(n,E)$ in $S(F)$, where $F^{\times}$ operates by multiplication. For each ξ we replace (11) by

$$\int f^{\xi}(\tilde{h}g)\chi(\lambda(\tilde{h}))\,d\tilde{h} = \int \Psi\left[{}^t\bar{g}\xi zg\right]\chi(z)\,dz.$$

We now define a new kind of relative Bessel distributions

$$\mathcal{MR}_{\Pi}^{\xi}(f^{\xi}) = \sum \int \Pi(f^{\xi})\phi_i(\tilde{h})\chi(\lambda(\tilde{h}))\,d\tilde{h}\,\overline{\int \phi_i(u)\bar{\theta}(u\bar{u})\,du}.$$

We now get instead of (19)

$$\sum_{\xi}\sum_{\Pi}\mathcal{MR}_{\Pi}^{\xi}(f^{\xi}) = \sum_{\pi}\mathcal{B}_{\pi}(\Phi). \tag{62}$$

The end of the proof is the same as before. □

Historical remarks. – For $n=2$, the study of the poles of the Asai L-function in [5] proves that the base change Π of a representation π with central character ω is distinguished by the character $\omega\eta\circ\det$ of the group $GL(2,F_{\mathbb{A}})$ and *not* distinguished by the character $\omega\circ\det$ of $GL(2,F_{\mathbb{A}})$ (this fact is used explicitly in [6, (1), middle of p. 388]). Equivalently, the representation Π is distinguished by the character $\omega\eta$ of the split unitary similitude group but not distinguished by the character ω. We recall the proof. Let ω_1 be an idele class character of F. Let Π be a cuspidal automorphic representation of $GL(2,E_{\mathbb{A}})$ with central character Ω. Assume that the restriction of Ω to F is ω_1^2. Recall the Asai L-function $L(s,\Pi,\omega_1^{-1},\text{Asai})$. In this context, it is essentially defined by the condition that an integral of the form

$$\int\limits_{Z(F_{\mathbb{A}})GL(2,F)\backslash GL(2,F_{\mathbb{A}})} \phi(g)E(g,\Phi,s)\omega_1^{-1}(\det g)\,dg$$

is an entire multiple of the L function. Here ϕ is in the space of Π and $E(g,\Phi,s)$ is the Eisenstein series on $GL(n,F_{\mathbb{A}})$ associated with a Schwartz–Bruhat function Φ on $F_{\mathbb{A}}^n$. In particular the (partial) Asai L–function has a pole at $s=1$ if and only if Π is distinguished by $(GL(2,F_{\mathbb{A}}),\omega_1)$. Now if Π is the base change of π with central character ω then we have the following relations between (partial) L functions:

$$L(s,\Pi,\omega^{-1}\eta,\text{Asai}) = L(s,\pi\times\tilde{\pi}\otimes\eta)\frac{L(s,1_F)}{L(s,\eta)},$$

$$L(s,\Pi,\omega^{-1},\text{Asai}) = L(s,\pi\times\tilde{\pi})\frac{L(s,\eta)}{L(s,1_F)},$$

$$L(s,\Pi\times\widetilde{\Pi}) = L(s,\pi\times\tilde{\pi})L(s,\pi\times\tilde{\pi}\otimes\eta).$$

The function $L(s,\pi\times\tilde{\pi})$ has a pole at $s=1$ while the last relation shows that the function $L(s,\pi\times\tilde{\pi}\otimes\eta)$ does not have a pole at $s=1$ and does not vanish at $s=1$. The assertion follows.

The result of Proposition 7 is conjectured in [8, (1)] in the context of $GL(n)$ and reproved by a relative trace formula as in the original work of Ye [18, (1)] in the context of $n=2$. At the time, the general conjecture with the exponent $n/2$ was motivated by the results of [5] as referenced in [6, (1)] in the case $n=2$ and by preliminary computations of Ye in the case $n=4$.

On the other hand, the author of [3] discusses distinction by the unitary group, first in the context of $GL(2)$. Earlier, he had proved the result of [5] that we recalled in the case $\omega=\eta$ but not in the case $\omega=1$. For the case $\omega=1$ he only discusses an example in [3]. [2]

It is natural to conjecture that one can choose the unitary group which distinguishes Π to be quasi-split.

THEOREM 3. – *Suppose each real place of F splits in E. Suppose n is even. If Π is a base change then Π is distinguished by a quasi-split unitary group.*

Proof. – The group H^ξ is quasi-split if and only if at every inert place v, $\det\xi$ is equal to $\det w_n=(-1)^{n/2}$ times a norm of E_w/F_v. We restrict the function Ψ_v by demanding that it be supported on the set S'_v of elements whose determinant is $(-1)^{n/2}$ times a norm. We take the function Φ_v to be supported on the set G'_v of elements g such that $\det g$ is $(-1)^{n/2}$ times a norm. As before, we introduce the function Φ' defined by $\Phi'[g]=\Phi[w_n g]$. We now get a spectral identity

$$\sum_{\Pi}\mathcal{R}^\xi_\Pi(f_\xi)=\sum_{\pi}\mathcal{P}_\pi(\Phi')$$

where

$$\mathcal{P}_\pi(\Phi')=\sum\int\phi_i(u)\theta(u_1)\,du_1\overline{\int\phi_(u)\bar{\theta}(u_2)\,du_2}.$$

The function Φ'_v is now supported on the group G^+_v of elements whose determinant is a norm. Moreover,

$$\mathcal{P}_\pi(\Phi')=\mathcal{P}_{\pi\otimes\eta}(\Phi').$$

To conclude we have to see that the local Bessel distribution $\mathcal{P}_{\pi_v}$ has a non-zero restriction to G^+_v. This is checked in [6, (7)].

If the quadratic extension E/F is arbitrary we have only a partial result.

THEOREM 4. – *Suppose that Π is an automorphic cuspidal representation of $GL(n,E_{\mathbb{A}})$. Then Π is distinguished by a unitary group if and only if it is a base change.*

We proceed as before except that at any inert real place v we choose the function Φ_v supported on the open set ${}^tN_vA_vN_v$. Let w be the place of E above v. It is elementary that there is a

[2] In connection with this example the author makes a claim of priority. However, he does not acknowledge earlier references [5] and [6, (1)].

function Ψ_v supported on the open set of H_v of matrices of the form

$$^t\bar{u}au, \quad u \in N_w,\ a \in A_v$$

which matches Φ_v. Equating the discrete parts of the spectral expansions in the standard way, we get again:

$$\sum_{\pi} \mathcal{B}_{\pi}(\Phi) = \sum_{\xi} \sum_{\Pi} \mathcal{R}^{\xi}_{\Pi}(f^{\xi}). \tag{63}$$

Fix a representation π_0 of $GL(n, F_{\mathbb{A}})$ such that $\pi_0 \otimes \eta_{E/F}$ is not equivalent to π_0. Thus the base change Π_0 of π_0 is automorphic cuspidal. Say n is even. Then using the principle of linear independence of the characters of the Hecke algebra we get

$$\mathcal{B}_{\pi_0}(\Phi) + \mathcal{B}_{\pi_0 \otimes \eta_{E/F}}(\Phi) = \sum_{\xi} \mathcal{R}^{\xi}_{\Pi_0}(f^{\xi}).$$

As before

$$\mathcal{B}_{\pi_0}(\Phi) = \prod_v \mathcal{B}_{\pi_{0,v}}(\Phi_v), \qquad \mathcal{B}_{\pi_0 \otimes \eta_{E/F}}(\Phi) = \prod_v \mathcal{B}_{\pi_{0,v} \otimes \eta_v}(\Phi_v).$$

At a real place v inert in E, by Lemma 23 we choose Φ_v so that

$$\mathcal{B}_{\pi_{0,v}}(\Phi_v) \neq 0.$$

There is at least one finite place u inert in E such that

$$\pi_{0,u} \not\simeq \pi_{0,u} \otimes \eta_u.$$

We choose Φ_u so that

$$\mathcal{B}_{\pi_{0,u}}(\Phi_u) \neq 0, \qquad \mathcal{B}_{\pi_{0,u} \otimes \eta_u}(\Phi_u) = 0.$$

Then we find

$$\mathcal{B}_{\pi_0}(\Phi) = \sum_{\xi} \mathcal{R}^{\xi}_{\Pi_0}(f^{\xi}).$$

We conclude that $\mathcal{R}^{\xi}_{\Pi_0}(f^{\xi})$ is non-zero for at least one ξ and the theorem follows. If n is odd the proof is similar but simpler. □

REFERENCES

[1] ARTHUR J., CLOZEL L., Simple Algebras, Base Change, and the Advanced Theory of the Trace Formula, Ann. of Math. Stud., vol. **120**, Princeton University Press, Princeton, NJ, 1989.

[2] CASSELMAN W., SHALIKA J., The unramified principal series of p-adic groups. II. The Whittaker function, *Compositio Math.* **41** (1980) 207–231.

[3] FLICKER Y., On distinguished representations, *J. reine angew. Math.* **418** (1991) 139–172.

[4] GOODMAN R., WALLACH N., Representations and Invariants of the Classical Groups, Encyclopedia of Mathematics and its Applications, Cambridge University Press, Cambridge, 1998.

[5] HARDER G., LANGLANDS R., RAPOPORT M., Algebraische Zyklen auf Hilbert–Blumenthal–Flächen, *J. reine angew. Math.* **366** (1986) 53–120.

[6] (1) JACQUET H., Sur un résultat de Waldspurger II, *Compositio Math.* **63** (1987) 315–389;
(2) JACQUET H., The continuous spectrum of the relative trace formula for $GL(3)$ over a quadratic extension, *Israel J. Math.* **89** (1995) 1–59;
(3) JACQUET H., A theorem of density for Kloosterman integrals, *Asian J. Math.* **2** (1998) 759–778;
(4) JACQUET H., Transfert lisse d'intégrales de Kloosterman, *C. R. Acad. Sci. Paris* **335** (2002) 229–302;
(5) JACQUET H., Smooth transfer of Kloosterman integrals, *Duke Math. J.* **120** (2003) 121–152;
(6) JACQUET H., Facteurs de transfert pour les intégrales de Kloosterman, *C. R. Acad. Sci. Paris. Ser. I* **336** (2003) 121–124;
(7) JACQUET H., Factorization of period integrals, *J. Number Theory* **87** (1) (2001) 109–143;
(8) JACQUET H., Kloosterman Identities over a quadratic extension, *Ann. of Math.*, submitted for publication.

[7] JACQUET H., RALLIS S., Kloosterman integrals for skew symmetric matrices, *Pacific J. Math.* **154** (2) (1992) 265–283.

[8] (1) JACQUET H., YE Y., Une remarque sur le changement de base quadratique, *C. R. Acad. Sci. Paris Sér. I Math.* **311** (1990) 671–676;
(2) JACQUET H., YE Y., Relative Kloosterman integrals for $GL(3)$, *Bull. Soc. Math. France* **120** (1992) 449–558;
(3) JACQUET H., YE Y., Distinguished representations and quadratic base change for $GL(3)$, *Trans. Amer. Math. Soc.* **348** (1996) 913–939;
(4) JACQUET H., YE Y., Germs of Kloosterman integrals for $GL(3)$, *Trans. Amer. Math. Soc.* **351** (1999) 1227–1255.

[9] KOTTWITZ R., Base change for unit elements of Hecke algebras, *Compositio Math.* **60** (1986) 237–250.

[10] LAPID E., On the fine spectral expansion of Jacquet's relative trace formula, Preprint.

[11] LAPID E., ROGAWSKI J., Periods of Eisenstein series: the Galois case, *Duke Math. J.* **120** (2003) 153–226.

[12] LAPID E., ROGAWSKI J., Stabilization of periods of integrals of Eisenstein series and Bessel distributions on $GL(3)$ relative to $U(3)$, *Doc. Math.* **5** (2000) 317–350.

[13] MACDONALD I.G., Symmetric Functions and Orthogonal Polynomials, Amer. Math. Soc. Univ. Lecture Ser., vol. **12**, 1991.

[14] MAO Z., RALLIS S., Howe duality and the trace formula, *Pacific J. Math.* **191** (2) (1999) 329–346.

[15] (1) NGÔ B.C., Le lemme fondamental de Jacquet et Ye en caractéristique positive, *Duke Math. J.* **96** (1999) 473–520;
(2) NGÔ B.C., Faisceaux pervers, homomorphisme de changement de base et lemme fondamental de Jacquet et Ye, *Ann. Sci. École Norm. Sup.* **32** (1999) 619–679.

[16] SHALIKA J., The multiplicity one theorem for GL_n, *Ann. of Math.* **100** (2) (1974) 171–193.

[17] SHINTANI T., On an explicit formula for class-1 Whittaker functions on GL_n over P-adic fields, *Proc. Japan Acad.* **52** (1976) 180–182.

[18] (1) YE Y., Kloosterman integrals and base change for $gl(2)$, *J. reine angew. Math.* **400** (1989) 57–121;
(2) YE Y., The fundamental lemma of a relative trace formula for $GL(3)$, *Compositio Math.* **89** (1993) 121–162;
(3) YE Y., An integral transform and its applications, *Math. Ann.* **30** (1994) 405–417.

(Manuscrit reçu le 22 décembre 2004;
accepté, après révision, le 15 avril 2005.)

Hervé JACQUET
Mathematics Department MC 4408,
Columbia University,
New York, NY 10027, USA
E-mail: hj@math.columbia.edu

Acknowledgments

The American Mathematical Society gratefully acknowledges the kindness of these institutions in granting the following permissions:

Academie des Sciences

Hervé Jacquet, *Représentations distingués pour le groupe orthogonal*, C. R. Acad. Sci. Paris Sér. I Math. **312**(13), 957-961.

Hervé Jacquet, *Facteurs de transfert pour les intégrales de Kloosterman*, C. R. Math. Acad. Sci. Paris **336**(2), 121–124, 2003.

Compositio Mathematica

H. Jacquet and K. F. Lai, *A relative trace formula*, Compositio Math. **54**(2), 243-310, 1985.

Duke University Press

Hervé Jacquet, King F. Lai, and Stephen Rallis, *A trace formula for symmetric spaces*, Duke Math. J. **70**(2), 305–372, 1993. Duke University Press. All rights reserved. Reprinted by permission of the publisher.

Hervé Jacquet, *Smooth transfer of Kloosterman integrals*, Duke Math. J. **120**(1), 121–152, 2003. Duke University Press. All rights reserved. Reprinted by permission of the publisher.

Ecole Normale Supérievre

Stephen Gelbart and Hervé Jacquet, *A relation between automorphic representations of GL(2) and GL(3)*, Ann. Sci. Ecole Norm. Sup. (4), **11**(4), 471-542, 1978.

Hervé Jacquet, *Sur un résultat de Waldspurger*, Ann. Sci. École Norm. Sup. (4), **19**(2), 185-229, 1986.

Hervé Jacquet, *Kloosterman identities over a quadratic extension. II*, Ann. Sci. École Norm. Sup. (4), **38**(4), 609–669, 2005.

Elsevier Ltd

Hervé Jacquet, *Factorization of period integrals*, J. Number Theory **87**(1), 109-143, 2001.

Societé Mathématique de France

Hervé Jacquet and Nan Chen, *Positivity of quadratic base change L-functions*, Bull. Soc. Math. France **129**(1), 33-90, 2001.

Springer Science & Business Media

With kind permission from Springer Science + Business Media. Hervé Jacquet and Joseph A. Shalika, *A non-vanishing theorem for zeta functions of* GL_n, Invent. Math. **38**(1), 1-16, 1976.

With kind permission from Springer Science + Business Media. H. Jacquet, I. I. Piatetski-Shapiro, and J. Shalika, *Conducteur des représentations du groupe linéaire*, Math. Ann. **256**(2), 199- 214, 1981.

With kind permission from Springer Science + Business Media. Hervé Jacquet and Joseph Shalika, *A lemma on highly ramified* ϵ*-factors*, Math. Ann. **271**(3), 319-332, 1985.

The Johns Hopkins University Press

Copyright © 1981 The Johns Hopkins Univeristy Press. H. Jacquet and J. A. Shalika's article, *On Euler products and the classification of automorphic representations. I*, was first published in American Journal of Mathematics **103**(3), 499-558, 1981. Reprinted with permission by The Johns Hopkins University Press.

Copyright © 1981 The Johns Hopkins Univeristy Press. H. Jacquet and J. A. Shalika's article, *On Euler products and the classification of automorphic forms. II*, was first published in American Journal of Mathematics **103**(4), 777-815, 1981. Reprinted with permission by The Johns Hopkins University Press.

ISBN 978-0-8218-5356-6

9 780821 853566

CWORKS/23